# ELECTRONIC MEASUREMENTS AND INSTRUMENTATION

**For Engineering Students preparing for B.E./B.Tech.; AMIE-Section B (India) and Competitive Examinations**



**Er. R.K. RAJPUT**

*M.E. (Hons.) Gold Medallist; Grad.-Mech. Engg. & Elect. Engg.; MIE (India); MSESI; MISTE; CE (India)*

***Recipient of :***

*"Best Teacher (Academic) Award"*

***"Distinguished Author Award"***

*"Jawahar Lal Nehru Memorial Gold Medal" for an Outstanding Research Paper (Institution of Engineers—India)*

***Principal*** *(Formerly)*

- *Thapar Polytechnic College;*
- *Punjab College of Information Technology;*

**PATIALA**

S. CHAND
PUBLISHING

**S Chand And Company Limited**

**(ISO 9001 Certified Company)**

**S Chand And Company Limited**
**(ISO 9001 Certified Company)**
***Head Office:*** Block B-1, House No. D-1, Ground Floor, Mohan Co-operative Industrial Estate, New Delhi – 110 044 | Phone: 011-66672000
***Registered Office:*** A-27, 2nd Floor, Mohan Co-operative Industrial Estate, New Delhi – 110 044
Phone: 011-49731800
www.**schandpublishing.com**; e-mail: **info@schandpublishing.com**

**Branches**

Ahmedabad : Ph: 27542369, 27541965; ahmedabad@schandpublishing.com
Bengaluru : Ph: 22354008, 22268048; bangalore@schandpublishing.com
Bhopal : Ph: 4274723, 4209587; bhopal@schandpublishing.com
Bhubaneshwar : Ph: 2951580; bhubaneshwar@schandpublishing.com
Chennai : Ph: 23632120; chennai@schandpublishing.com
Guwahati : Ph: 2738811, 2735640; guwahati@schandpublishing.com
Hyderabad : Ph: 40186018; hyderabad@schandpublishing.com
Jaipur : Ph: 2291317, 2291318; jaipur@schandpublishing.com
Jalandhar : Ph: 4645630; jalandhar@schandpublishing.com
Kochi : Ph: 2576207, 2576208; cochin@schandpublishing.com
Kolkata : Ph: 23357458, 23353914; kolkata@schandpublishing.com
Lucknow : Ph: 4003633; lucknow@schandpublishing.com
Mumbai : Ph: 25000297; mumbai@schandpublishing.com
Nagpur : Ph: 2250230; nagpur@schandpublishing.com
Patna : Ph: 2260011; patna@schandpublishing.com
Ranchi : Ph: 2361178; ranchi@schandpublishing.com
Sahibabad : Ph: 2771238; info@schandpublishing.com



*First Edition 2008*
*Second Revised Edition 2009*
*Reprints 2010, 2011, 2012*
*Third Edition 2015*
*Reprint 2018*
*Reprint with Corrections 2020*

**ISBN:** 978-81-219-2917-2 **Product Code:** H3MEI53ETRN10ENAC15R

PRINTED IN INDIA

By Vikas Publishing House Private Limited, Plot 20/4, Site-IV, Industrial Area Sahibabad, Ghaziabad – 201 010 and Published by S Chand And Company Limited, A-27, 2nd Floor, Mohan Co-operative Industrial Estate, New Delhi – 110 044

## *Other Books by the Same Author*

1. Electrical Measurements and Measuring Instruments
2. Electrical and Electronic Measurements and Instrumentation
3. **Questions and Answers** in "Electrical Engineering"
4. **Questions and Answers** in "Electronics and Communication Engineering"
5. Mechatronics

# PREFACE TO THE THIRD EDITION

I take pleasure in presenting the **"Third Edition"** of this book on "Electronic Measurements and Instrumentation". The warm reception which the previous editions have received is a matter of much satisfaction to me.

In this edition, the book has been *completely updated by adding new topics in various chapters*. Besides this, two new chapters namely : **"Microprocessors and Microcontrollers"** (***Chapter-13***) and **"Universities Questions (Latest) with Solutions"** (***Chapter-14***) have been added to make the book still more useful to the readers.

Any suggestions for the improvement of the book are most welcome.

**Er. R.K. RAJPUT**
*(Author)*

# PREFACE TO THE FIRST EDITION

This treatise on "**Electronic Measurements and Instrumentation**" contains comprehensive treatment of the subject matter in simple, lucid and direct language. It also envelopes a large number of solved problems. The book covers exhaustively the syllabi of various Indian Universities in this subject.

The book comprises *two parts* : ***Part – I*** deals with **"Electronic measurements and Measuring Instruments"** while ***Part – II*** includes **"Electronic Instrumentation"**. The book contains 12 chapters; all the chapters are effectively saturated with much needed text, supported by simple and self explanatory figures and an ample number of solved examples, wherever required. At the end of each chapter, ***Highlights, Objective Type Questions, Theoretical Questions*** and ***Unsolved examples*** have been added to make the book a comprehensive and a complete unit in all respects.

The author's thanks are due to his wife Ramesh Rajput for extending all cooperation during the preparation of the manuscript and its proof reading.

In the end the author wishes to express his gratitude to the Managing Director and all the concerned staff members of S. Chand & Company Ltd. who were involved so dedicately in the publication of this book.

Although every care has taken to make the book free of errors both in text as well as in solved examples, yet the author shall feel obliged if any errors present are brought to his notice. Constructive criticism of the book will be warmly received.

**Er. R.K. RAJPUT**
*(Author)*

# CONTENTS

# Introduction to S.I. Units and Conversion Factors

## A. INTRODUCTION TO S.I. UNITS

S.I., the international system of units is divided into three classes :

1. Base unit
2. Derived units
3. Supplementary units.

From the scientific point of view division of S.I. unit into these classes is to a certain extent arbitrary, because it is not essential to the physics of the subject. Nevertheless the General Conference, considering the advantages of a single, practical, world-wide system for international relations, for teaching and for scientific work, decided to base the international system on choice of six well-defined units given in Table 1 below:

**Table 1. S.I. Base Units**

| *Quantity* | *Name* | *Symbol* |
|---|---|---|
| length | metre | m |
| mass | kilogram | kg |
| time | second | s |
| electric current | ampere | A |
| thermodynamic temperature | kelvin | K |
| luminous intensity | candela | cd |
| amount of substance | mole | mol |

The second class of S.I. units contains derived units, i.e., units which can be formed by combining base units according to the algebraic relations linking the corresponding quantities. Several of these algebraic expressions in terms of base units can be replaced by special names and symbols can themselves be used to form other derived units.

Derived units may, therefore be classified under three headings. Some of them are given in Tables 2, 3 and 4.

**Table 2. Examples of S.I. derived units expressed in terms of base units**

| *Quantity* | *S.I. Unit* | |
|---|---|---|
| | *Name* | *Symbol* |
| area | square metre | $m^2$ |
| volume | cubic metre | $m^3$ |
| speed, velocity | metre per second | m/s |
| acceleration | metre per second squared | $m/s^2$ |
| density, mass density | kilogram per cubic metre | $kg/m^3$ |
| concentration (of amount of substance) | mole per cubic metre | $mol/m^3$ |
| activity (radioactive) | 1 per second | $s^{-1}$ |
| specific volume | cubic metre per kilogram | $m^3/kg$ |
| luminance | candela per square metre | $cd/m^2$ |

## Table 3. Derived Units with Special Names

| *Quantity* | *S.I. Unit* | | | |
|---|---|---|---|---|
| | *Name* | *Symbol* | *Expression in terms of other units* | *Expression in terms of S.I. base units* |
| frequency | hertz | Hz | — | $s^{-1}$ |
| force | newton | N | — | $m.kg.s^{-2}$ |
| pressure | pascal | Pa | $N/m^2$ | $m^{-1}.kg.s^{-2}$ |
| energy, work, quantity of heat power | joule | J | N.m | $m^2.kg.s^{-2}$ |
| radiant flux, quantity of electricity | watt | W | J/s | $m^2.kg.s^{-3}$ |
| electric charge | coloumb | C | A.s. | s.A |
| electric tension, electric potential | volt | V | W/A | $m^2.kg.s^{-3}.A^{-1}$ |
| capacitance | farad | F | C/V | $m^{-2}.kg^{-1}.s^4$ |
| electric resistance | ohm | Ω | V/A | $m^{-2}$, $kgs^{-3}.A^{-2}$ |
| conductance | siemens | S | A/V | $m^{-2}$. $kg^{-1}.s^3.A^{-1}$ |
| magnetic flux | weber | Wb | VS | $m^2.kg.s^{-2}.A^{-1}$ |
| magnetic flux density | tesla | T | $Wb/m^2$ | $kg.s^{-2}.A^{-1}$ |
| inductance | henry | H | Wb/A | $m^2kg.s^{-2}.A^{-2}$ |
| luminous flux | lumen | lm | — | cd.sr |
| illuminance | lux | lx | — | $m^{-2}$.cd.sr |

## Table 4. Examples of S.I. Derived Units expressed by means of Special Names

| *Quantity* | *S.I. Unit* | | |
|---|---|---|---|
| | *Name* | *Symbol* | *Expression in terms of S.I. base units* |
| dynamic viscosity | pascal second | Pa-s | $m^{-1}.kgs^{-1}$ |
| moment of force | metre newton | N.m | $m^{-2}.kgs^{-2}$ |
| surface tension | newton per metre | N/m | $kg.s^{-2}$ |
| heat flux density, irradiance | watt per square metre | $W/m^2$ | $kg.s^{-3}$ |
| heat capacity, entropy | joule per kelvin | J/K | $m^2,kg.s^{-2}.K^{-1}$ |
| specific heat capacity, specific entropy | joule per kilogram\ kelvin | J/(kg.K) | $m^2.s^{-2}.K^{-1}$ |
| specific energy | joule per kilogram | J/kg | $m^2.s^{-2}$ |
| thermal conductivity | watt per metre kelvin | W/(m.K) | $m.kg.s^{-3}$, $K^{-1}$ |
| energy density | joule per cubic metre | $J/m^3$ | $m^{-1}.kg.s^{-2}$ |
| electric field strength | volt per metre | V/m | $m.kg.s^{-3}.A^{-1}$ |
| electric charge density | coloumb per cubic metre | $C/m^3$ | $m^{-3}$.s.A |
| electric flux density | coloumb per square metre | $C/m^2$ | $m^{-2}$.s.A |
| permitivity | farad per metre | F/m | $m^{-3}.kg^{-1}.s^4$ |
| current density | ampere per square metre | $A/m^2$ | — |
| magnetic field strength | ampere per metre | H/m | — |
| permeability | henry per metre | H/m | $m.kg.s^{-2}.A.^{-2}$ |
| molar energy | joule per mole | J/mol | $m^2.kg.s^{-2}mol^{-1}$ |
| molar heat capacity | joule per mole kelvin | J/(mol.K) | $m^2.kg.s^{-2}.K^{-1}$ $mol^{-1}$ |

The S.I. units assigned to third class called "supplementary units" may be regarded either as base units or as derived units. Refer to Table 5 and Table 6.

**Table 5. S.I. Supplementary Units**

| *Quantity* | *S.I.Units* | |
|---|---|---|
| | *Name* | *Symbol* |
| plane angle | radian | rad |
| solid angle | steradian | sr |

**Table 6. Examples of S.I. Derived Units formed by Using Supplementary Units**

| *Quantity* | *S.I.Units* | |
|---|---|---|
| | *Name* | *Symbol* |
| angular velocity | radian per second square | rad/s |
| angular acceleration | radian per second squared | $rad/s^2$ |
| radiant intensity | watt per steradian | W/sr |
| radiance | watt per square metre steradian | $W\text{-}m^{-2}.sr^{-1}$ |

**Table 7. S.I. Prefixes**

| *Factor* | *Prefix* | *Symbol* | *Factor* | *Prefix* | *Symbol* |
|---|---|---|---|---|---|
| $10^{12}$ | tera | T | $10^{-1}$ | deci | d |
| $10^{9}$ | giga | G | $10^{-2}$ | centi | c |
| $10^{6}$ | mega | M | $10^{-3}$ | milli | m |
| $10^{3}$ | kilo | k | $10^{-6}$ | micro | μ |
| $10^{2}$ | hecto | h | $10^{-9}$ | nano | n |
| $10^{1}$ | deca | da | $10^{-12}$ | pico | p |
| | | | $10^{-15}$ | fasnot | f |
| | | | $10^{-18}$ | atto | a |

## B. DEFINITIONS

The S.I. seven base units and two supplementary units are defined below:

*(i)* **Metre.** The metre is the length equal to 1, 650, 763.73 wavelengths in vacuum of the radiation corresponding to the transition between the levels $2p_{10}$ and $5d_5$ of the krypton-86 atom.

*(ii)* **Kilogram.** One kilogram is equal to the mass of the international prototype of the kilogram.

*(iii)* **Second.** The second is the duration of 9, 192, 631, 770 periods of the radiation corresponding to the transition between the two hyperfine levels of the ground state of the Cesium-133 atom.

*(iv)* **Ampere.** One ampere is that constant current which, if maintained in two straight parallel conductors of infinite length, of negligible circular cross-section and placed one meter apart in vacuum, would produce between these conductors a force equal to $2 \times 10^{-7}$ newton per metre of length.

*(v)* **Kelvin.** The kelvin is the fraction $\frac{1}{273.16}$ of thermodynamic temperature of the triple point of water.

(*vi*) **Candela.** The candela is the luminous intensity, in the perpendicular direction, of a surface of a $\frac{1}{600,000}$ square metre of a black body at a temperature of freezing platinum under a pressure of 101, 325 newton per square metre.

(*vii*) **Mole.** The mole is the amount of substance of a system which contains as many elementary entities as there are atoms in 0.012 kg, carbon 12. When the mole is used, the elementary entities must be specified and may be atoms, molecules, ions, electrons, other particles or specified groups of such particles.

(*viii*) **Radian.** The radian is the plane angle between two radii of a circle that cut off on the circle an arc equal in length to the radius.

(*ix*) **Steradian.** The steradian is the solid angle which having its vertex in centre of a sphere, cuts off an area of the surface of the sphere equal to that of a square with sides of length equal to the radius of the sphere.

## Some Other Definitions

**Newton.** The newton (N) is a derived unit of force and is defined as the unit of force which when acting on a mass of 1 kilogram gives it an acceleration of one metre per second per second. Since acceleration due to gravity equals 9.81 $m/s^2$, kilogram force equals 9.81 newtons.

**Joule.** The joule (J) is a derived unit of energy, work or quantity of heat and is defined as the work done when a force of one newton acts so as to cause a displacement of one metre. Energy is defined as the capacity to do work. A unit of energy in nuclear physics is the electron volt (eV) which is defined as the energy gained by an electron in rising through a potential difference of one volt.

$$1 \text{ eV} = 1.6021 \times 10^{-19} \text{ J.}$$

**Watt.** The watt (W) is a unit of power (i.e., rate of doing work)

$$\text{Power in watts} = \frac{\text{work (or energy) in joules}}{\text{time in seconds}}$$

Thus 1 watt equals 1 joule/sec.

$$1 \text{ kilo watt-hour (kWh)} = 1000 \text{ watt-hours}$$
$$= 3600000 \text{ joules.}$$

**Coulomb.** The coulomb (C) is the derived unit of charge. It is defined as *the quantity of electricity passing a given point in a circuit when a current of 1 A is maintained for 1 second.*

$$Q = I \cdot t$$

Where $Q$ = charge in coulombs,

$I$ = current in amperes, and

$t$ = time in second.

1 coulomb represents $6.24 \times 10^{18}$ electrons.

**Ohm.** The ohm (Ω) is the unit of electric resistance and is defined as *the resistance in which a constant current of 1 A generates heat at the rate of 1 watt.*

**Siemen.** The siemen is a unit of electric conductance (i.e., reciprocal of resistance). If a circuit has a resistance of 5 ohms, its conductance is 0.2 siemens. A more commonly used name for siemen is mho (℧).

**Volt.** The volt is a unit of potential difference and electromotive force. It is defined as *the difference of potential across a resistance of 1 ohm carrying a current of 1 ampere.*

**Hertz.** The hertz (Hz) is a unit of frequency. 1 Hz = 1 cycle per second.

**Horse-power.** It is a practical unit of mechanical output. BHP (British horse power or brake horse power) equals 746 watts. The metric horse power equals 735.5 watts. To avoid confusion between BHP and metric horse power, the mechanical output of machines in S.I. units, is expressed in watts or kilowatts.

## C. SALIENT FEATURES OF S.I. UNITS

The salient features of S.I. units are as follows :

1. It is a coherent system of units, i.e., product of quotient of any two base quantities results in a unit resultant quantity. For example, unit length divided by unit time gives unit velocity.
2. It is a rationalised system of units, applicable to both, magnetism and electricity.
3. It is a non-gravitational system of units. It clearly distinguishes between the units of mass and weight (force) which are kilogram and newton respectively.
4. All the units of the system can be derived from the base and supplementary units.
5. The decimal relationship between units of same quantity makes possible to express any small or large quantity as a power of 10.
6. For any quantity, there is one and only one S.I. unit. For example, joule is the unit of energy of all forms such as mechanical, heat, chemical, electrical and nuclear. However, kWh will also continue to be used as unit of electrical energy.

### Advantages of S.I. Units:

1. Units for may different quantities are related through a series of simple and basic relationship.
2. Being an absolute system, it avoids the use of factor 'g' i.e. acceleration due to gravity in several expressions in physics and engineering which had been a nuisance in all numericals in physics and engineering.
3. Being a rationalised system, it ensures all the advantages of rationalised MKSA system in the fields of electricity, magnetism, electrical engineering and electronics.
4. Joule is the only sole unit of energy of all forms and watt is the sole unit of power hence a lot of labour is saved in calculations.
5. It is a coherent system of units and involves only decimal co-efficients. Hence it is very convenient and quick system for calculations.
6. In electricity, all the practical units like volt, ohm, ampere, henry, farad, coulomb, joule and watt accepted in industry and laboratories all over the world for well over a century have become absolute in their own right in the S.I. system, without the need for any more practical units.

### Disadvantages:

1. The non-S.I. time units 'minute' and 'hour' will still continue to be used until the clocks and watches are all changed to kilo seconds and mega seconds etc.
2. The base unit kilogram (kg) includes a prefix, which creates an ambiguity in the use of multipliers with gram.
3. S.I. units for energy, power and pressure (i.e., joule, watt and pascal) are too small to be expressed in science and technology, and, therefore, in such cases the use of larger units, such as MJ, kW, kPa will have to be made.

4. There are difficulties with regard to developing new S.I. units for apparent and reactive energy while joule is the accepted unit for active energy in S.I. systems.

## D. CONVERSION FACTORS

**1. Force:**

$$1 \text{ newton} = \text{kg-m/sec}^2 = 0.012 \text{ kgf}$$
$$1 \text{ kgf} = 9.81 \text{ N}$$

**2. Pressure:**

$$1 \text{ bar} = 750.06 \text{ mm Hg} = 0.9869 \text{ atm} = 10^5 \text{ N/m}^2$$
$$= 103 \text{ kg/m-sec}^2$$
$$1 \text{ N/m}^2 = 1 \text{ pascal} = 10^{-5} \text{ bar} = 10^{-2} \text{ kg/m-sec}^2$$
$$= 1.01325 \times 10^5 \text{ N/m}^2$$

**3. Work, energy or Heat:**

$$1 \text{ joule} = 1 \text{ newton metre} = 1 \text{ watt-sec}$$
$$= 2.7778 \times 10^{-7} \text{ kW-h} = 0.239 \text{ cal}$$
$$= 0.239 \times 10^{-3} \text{ kcal}$$
$$1 \text{ cal} = 4.184 \text{ joule} = 1.1622 \times 10^{-6} \text{ kWh}$$
$$1 \text{ kcal} = 4.184 \times 10^3 \text{ joule} = 427 \text{ kgf m}$$
$$= 1.1622 \times 10^{-3} \text{ kWh}$$
$$1 \text{ kwh} = 8.6042 \times 10^5 \text{ cal} = 860.42 \text{ kcal}$$
$$= 3.6 \times 10^6 \text{ joule}$$
$$1 \text{ kgf.m} = \left(\frac{1}{427}\right) \text{kcal} = 9.81. \text{ joules}$$

**4. Power:**

$$1 \text{ watt} = 1 \text{ Joule/sec} = 0.86 \text{ kcal/h}$$
$$1 \text{ h.p.} = 75 \text{ m kgf/sec} = 0.1757 \text{ kcal/sec}$$
$$= 735.5 \text{ watt}$$
$$1 \text{ kW} = 1000 \text{ watts} = 860 \text{ kcal/h}$$

**5. Specific heat:**

$$1 \text{ kcal/kg-°K} = 0.4184 \text{ joules/kg-K}$$

**6. Thermal conductivity:**

$$1 \text{ watt/m-K} = 0.8598 \text{ kcal/h-m-°C}$$
$$1 \text{ kcal/h-m-°C} = 1.16123 \text{ watt/m-K}$$
$$= 1.16123 \text{ joules/s-m-K}$$

**7. Heat transfer co-efficient:**

$$1 \text{ watt/m}^2\text{-K} = 0.86 \text{ kcal/m}^2\text{-h-°C}$$
$$1 \text{ kcal/m}^2\text{-h-°C} = 1.163 \text{ watt/m}^2\text{-K}$$

The following conversion factors may be used to convert the quantities in non-S.I. units into S.I. units.

| *To convert* | *To* | *Multiply by* |
|---|---|---|
| angstroms | m | $10^{-10}$ |
| atmospheres | $kg/m^2$ | 10332 |
| bars | $kg/m^2$ | $1.02 \times 10^4$ |
| Btu | joules | 1054.8 |
| Btu | kWh | $2.928 \times 10^{-4}$ |
| circular mils | $m^2$ | $5.067 \times 10^{-10}$ |
| cubic feet | $m^3$ | 0.02831 |
| dynes | newtons | $10^{-5}$ |
| ergs | joules | $10^{-7}$ |
| ergs | kWh | $0.2778 \times 10^{-13}$ |
| feet | m | 0.3048 |
| foot-pounds | joules | 1.356 |
| foot-pounds | kg-m | 0.1383 |
| gauss | tesla | $10^{-4}$ |
| grams (force) | newton | $9.807 \times 10^{-3}$ |
| horse power (metric) | watts | 735.5 |
| lines/sq. inch | tesla | $1.55 \times 10^{-5}$ |
| Maxwell | webers | $10^{-8}$ |
| mho | siemens | 1 |
| micron | metre | $10^{-6}$ |
| miles | km | 1.609 |
| mils | cm | $2.54 \times 10^{-3}$ |
| poundals | newton | 0.1383 |
| pounds | kilogram | 0.454 |
| pounds (force) | newtons | 0.448 |
| pounds/sq. ft. | $N/m^2$ | 47.878 |
| pounds/sq. inch. | $N/m^2$ | 6894.43 |

## E. IMPORTANT ENGINEERING CONSTANTS AND EXPRESSIONS IN S.I. UNITS

| *Engineering constants and expressions* | *M.K.S. system* | *S.I. units* |
|---|---|---|
| 1. Value of $g_0$ | 9.81 kg-m/kgf-$sec^2$ | 1 kg-m/N-$sec^2$ |
| 2. Universal gas constant | 848 kgf-m/kg mole-°K | $848 \times 9.81 = 8341$ J/kg-mole-°C ($\because$ 1 kgf-m = 9.81 joules) |
| 3. Gas constant (R) | 29.27 kgf m/kg-°K | $\frac{8314}{29} = 278$ joules/kg-K |
| 4. Specific heat (for air) | for air $c_v = 0.17$ kcal/kg-°K $c_v = 0.24$ kcal/kg-°K | for air $c_v = 0.17 \times 4.184 = 0.71128$ kJ/kg-K $c_p = 0.24 \times 4.184 = 1$ kJ/kg-K |
| 5. Flow through nozzle-Exist velocity ($C_2$) | $91.5\sqrt{U}$ where U is in kcal | $44.7\sqrt{U}$ where U is in kJ |
| 6. Refrigeration 1 ton | = 50 kcal/min | = 210 kJ/min |
| 7. **Heat transfer** Stefan Boltzmann Law is given by : | $Q = \sigma T^4$ kcal/$m^2$-h where $\sigma = 4.9 \times 10^{-8}$ kcal/h-$m^2$-°$K^4$ | $Q = \sigma T^4$ watts/$m^2$-h where $\sigma = 5.67 \times 10^{-8}$ watt/$m^2K^4$ |

## F. DIMENSIONS OF QUANTITIES

Different units can be represented dimensionally in terms of units of length L, mass M, time T and current I. The dimensions can be derived as under:

1. Velocity = length/time = $L/T = LT^{-1}$
2. Acceleration = velocity/time = $LT^{-1}/T = LT^{-2}$
3. Force = mass × acceleration = $MLT^{-2}$
4. Charge (coluomb) = current × time = $IT$
5. Work or energy = force × distance = $ML^2T^{-2}$
6. EMF or potential = work/charge = $ML^2T^{-2}/IT = ML^2I^{-1}T^{-3}$
7. Power = work/time = $ML^2T^{-2}/T = ML^2T^{-3}$
8. Current density = current/area = $I/L^2 = IL^{-2}$
9. Resistance = emf/current = $ML^2I^{-1}T^{-3}/I = ML^2I^{-2}T^{-3}$
10. Electric flux density = electric flux or charge/area = $IT/L^2 = ITL^{-2}$
11. MMF = current × number of turns = $I$
12. Conductance = I/resistance = $1/ML^2I^{-2}T^{-3} = I^2T^3M^{-1}L^{-2}$
13. Electric field intensity = volt/metre = $ML^2I^{-1}T^{-3}/L = MLI^{-1}T^{-3}$
14. Resistivity = $\dfrac{\text{resistant} \times \text{area}}{\text{length}} = (ML^2I^{-2}T^{-3})\,(L^2)/L = ML^3I^{-2}T^{-3}$
15. Magnetic field intensity (H) = MMF/length = $I/L = IL^{-1}$
16. Magnetic flux = emf × time = $(ML^2I^{-1}T^{-3})\,(T) = ML^2I^{-1}T^{-2}$
17. Magnetic flux intensity = magnetic flux/area = $(ML^2I^{-1}T^{-2})/L^2 = MI^{-1}T^{-2}$
18. Impedence = emf/current = $ML^2I^{-2}T^{-3}$
19. Admittance = 1/impedence = $I^2T^3M^{-1}L^{-2}$
20. Inductance = magnetic flux/current = $ML^2T^{-2}I^{-1}/I = ML^2T^{-2}I^{-2}$
21. Capacitance = electric charge/potential = $IT/ML^2T^{-3}I^{-1} = M^{-1}L^{-2}T^4I^2$

# PART I : *Electronic Measurements and Measuring Instruments*

1. Concepts of Measurements and Electromechanical Instruments.
2. Electronic Instruments.
3. Signal Generators.
4. Signal Analysers.
5. Digital Measurements and Measuring Instruments.
6. Instrument Transformers.

CHAPTER

# 1 Concepts of Measurements and Electromechanical Instruments



## 1.1 INTRODUCTION TO MEASUREMENT AND INSTRUMENTATION

### 1.1.1. Measurement

***Measurement** is the act, or the result of a quantitative comparison between a predetermined standard and an unknown magnitude.* The procedure and apparatus employed for obtaining the comparison, however, must be *provable*; the procedure for this is called *calibration.*

Probably it has rightly been said: *"whatever exists, exists in some amount"*. The determination of the amount is what measurement is all about.

**Measurand.** *The physical quantity or the characteristic condition which is the object of measurement in an instrumentation system is variously termed as "measurand", "measurement variable", "instrumentation variable" or "process variable".*

The measurand may be :

— Fundamental quantity, *e.g.,* length, mass and time;

— Derived quantity, *e.g.,* speed, velocity, acceleration, pressure.

The word *measurand* is used to designate the particular physical parameter being observed and quantified; that is, the *input quantity* to the measuring process. The act of measurement produces a *result.* Fig. 1.1 shows a *fundamental measuring process.*

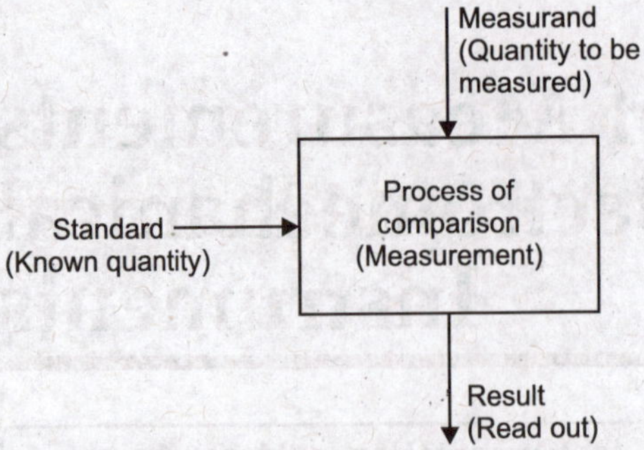


Fig. 1.1. Fundamental measuring process.

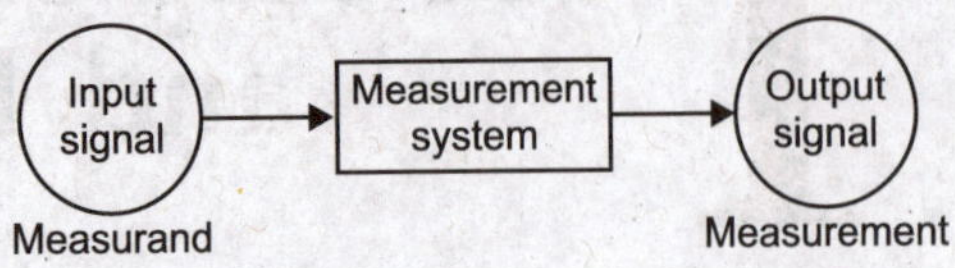


Fig. 1.2. Measurand and Measurement.

— Whereas the input to a measuring system is known as *measurand* the output is called *measurement* (See Fig. 1.2).

The *basic requirements* for getting meaningful result of measurement are:

(*i*) The standard employed for comparison purpose *must be accurately defined* and should be *commonly acceptable*.

(*ii*) The standard must be of the *same character* as the measurand and usually, but not always, is prescribed and defined by a legal or recognized agency or organisation *e.g.*, Indian Bureau of Standards (IBS) or the International Organisation of Standards (ISO).

(*iii*) The apparatus used and method adopted for the comparison purposes must be *provable*.

- Several measurands of interest *vary with time in a continuous manner over a range of magnitude.*

For example, the speed of the automobile, as it starts from rest has some magnitude at every instant during its motion. A sensor that responds to velocity will produce an output signal having a line variation analogous to the time change in automobile's speed. Such a signal is referred to as **analog signal** *because it is analogous to a continuous physical process.*

*The analog signal, which has a value at every instant of time, becomes a* **digital signal.**

— A digital signal is a *set of discrete numbers each corresponding to the value of the analog signal at a single specific instant of time.*

— The digital signal contains *no information* about the value of the analog signal at times *other than sample times.*

The mechanical quantities such as pressure, stress, strain, temperature, fluid-flow rates behave time-wise in an analog manner. However, by converting an analog signal to an equivalent digital signal for the purpose of signal conditioning and/or read out oftenly following distinct *advantages* are obtained:

(*i*) *Noise problems are reduced or sometimes eliminated completely.*

(*ii*) *Data transmission becomes simpler.*

(*iii*) *Computers are designed to process digital information, and direct numerical display or recording is more easily accomplished by manipulating digital quantities.*

### 1.1.2. Instrumentation

*The technology of using instruments to measure and control the physical and chemical properties of materials is called* ***"instrumentation"***.

- When the instruments are used for the *measurement and control of industrial manufacturing, conversion or treatment process,* the term **"process instrumentation"** is used.

- *When the measuring and controlling instruments are combined so that measurements provide impulses for remote automatic action, the result is called a* ***control system.***

## 1.2 SIGNIFICANCE OF MEASUREMENTS

The significance of measurements is discussed below:

- Measurements play a very significant role in every branch of scientific research and engineering processes which include the following:
  — Control systems;
  — Process instrumentation;
  — Data reduction.
- The whole area of automation or automatic control is based on measurements. The very concept of control is based on the comparison of the actual condition and the desired performance. *The exactness of error depends on the precision and accuracy of measurement made.*
- The measurements confirm the validity of a hypothesis and also add to it its understanding. This eventually leads to new discoveries that require new and sophisticated measuring techniques.
- Through measurements a product can be designed or a process be operated with maximum efficiency, minimum cost, and with desired degree of reliability and maintainability.

## 1.3 STANDARDS OF MEASUREMENT

*A known accurate measure of physical quantity is termed as* **standard**.

As the measurement is a process of comparison, therefore, regardless of our measurement method, we must employ a basis of comparison—*standardized units.* The standard must be precisely defined, and because different systems of units exist, the method of conversion from system to system must be mutually agreed upon.

— It is of significant importance that a relationship between the standards and the read out scale of each measuring system must be established through a process known as *calibration.*

The standards of measurement are often *classified* as :

1. International standards
2. Primary standards
3. Secondary standards
4. Working standards.

*The* **measurement standard** *is the physical embodiment of the unit of measurement as well as that of its submultiple value.*

### 1.3.1. International Standards

These standards, defined by *'International agreement'* are *periodically evaluated and checked by absolute measurements in terms of fundamental units of physics.*

These standards are not available to ordinary users for measurement of calibrations.

**1. International ohms :**

It is defined as *the resistance offered by a column of mercury having a mass of 14.4,521 gms, uniform cross-sectional area and length of 106.300 cm, to the flow of constant current at the melting point of ice.*

**2. International ampere :**

It is defined as an *unvarying current, when passed through a solution of silver nitrate in water, prepared in accordance with stipulated specification, deposits silver at the rate of 0.00111800 gms.*

**3. Absolute units:**

In 1948, International units were *replaced* by absolute units. These units are more accurate than International units, and differ slightly from them.

*Examples* : (i) International ohm = 1.00049 Absolute ohm, and (ii) International ampere = 0.99985 Absolute ampere.

### 1.3.2. Primary standards

*The highest standard of either a base unit or a derived unit is called a* **primary standard.**

- These standards essentially are copies of international prototypes and are kept throughout the world in national standard laboratories of similar standing.
- These standardized units are :
  — *Quite stable*
  — *Independent*
  — *Invariant*
  — *Not relative but finite.*

While setting up the primary standards, the following points are given due considerations:

(*i*) Long time stability of the material.
(*ii*) Low temperature coefficient.
(*iii*) No deterioration in its characteristics with changes in humidity, temperature and other environmental conditions.
(*iv*) Rigidity of construction.
(*v*) Accuracy of machining.
(*vi*) Most sophisticated techniques to be used for measuring physical dimensions which determine the accuracy of the standard.

- The primary standards *constitute the ultimate basis of reference* and are used for the purpose of *"verification and calibration of secondary standards"*.

— These standards have the *"highest possible accuracy"* but are *very expensive* to own and maintain.

— The primary standards are *not available for use outside the national laboratories.*

### 1.3.3. Secondary standards

The secondary standards are the *reference calibrated standards designed and calibrated from the primary standards.*

- These are sent periodically to the national standard laboratories for their calibration (the interval of calibration depends upon the accuracy and the type of standard being maintained).
- These standards are kept by the measurement laboratories and the industrial organisations to check and calibrate the general tools for their accuracy and precision.
- These standards have an accuracy of *one order lower than that of the secondary standards.*
- These are the normal standards which are *used by the workers and technicians who actually carry out the measurements.*

### Electrical Standards :

- All electrical measurements are based on fundamental quantities *I*, *R* and *V*. A systematic measurement depends upon the definition of these quantities. These quantities are related to each other through Ohm's law,

$$V = IR.$$

— It is sufficient to define two parameters to obtain the definition of the third. Hence in electrical measurements, it is possible to assign values to the remaining standard, by defining units of the other two standards. Standards of e.m.f. and resistance are therefore usually maintained at the National laboratory. The base value of the other standards are defined from these two standards.

**Time Standards:**

- Time defined in terms of *rotation of earth* is called ***Universal time (UT).*** Time defined in terms of the *earth's orbital motion* is called ***Ephemersis time (ET).*** Both UT and ET are determined by astronomical observations.
- Since frequency is the inverse of the time interval, time can be calibrated in terms of frequency. The first *atomic clock* was based on *Cesium atom* (using periodic atomic vibrations as a standard).
- The International Committee of Weights and Measures *defines the second* in terms of *'Cesium transitions'*, assigning a value of 9, 192, 631, 770 Hz to the hyperfine transitions of the Cesium atom unperturbed by external fields.

## 1.4 MODES OF MEASUREMENT

Following are the *three* modes of measurement :

1. Primary measurements.
2. Secondary measurements.
3. Tertiary measurements.

**1. Primary measurements :**

- In this case the *sought value of a parameter is determined by comparing it directly with "reference standards".*
- There is *no conversion* of measurand in terms of length.

*Examples :*

(*i*) Measurement of time by counting the number of strokes of a clock.

(*ii*) Matching of two lengths when determining the length of an object with a ruler.

(*iii*) Matching of two colours when judging the temperature of a red hot steel.

**2. Secondary measurements :**

*The indirect measurements involving 'one translation' are called* ***secondary measurements.***

*Examples :*

(*i*) The pressure measurement by manometers.

(*ii*) The temperature measurement by mercury-in-glass thermometers.

**3. Tertiary measurements:**

*The indirect measurements involving 'two conversions' are called* ***tertiary measurements.***

*Examples :*

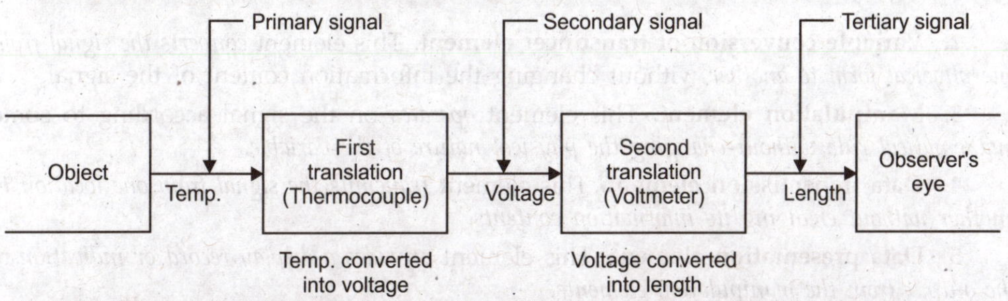


**Fig. 1.3.** Tertiary measurement : Measurement of temperature by a thermocouple.

(*i*) The measurement of temperature of an object by thermocouple (Fig. 1.3). The primary signal (temperature of object) is transmitted to a thermocouple which generates a *voltage* which is a function of the temperature (first translation/conversion). The voltage in turn is applied to a voltmeter through a pair of wires. The second conversion is then voltage into length. The tertiary signal is transmitted to brain of the observer.

(*ii*) The measurement of static pressure by a Bourdon tube pressure gauge.

(*iii*) The measurement of the speed of a rotating shaft by means of an electric tachometer.

The unit of a measuring system where translation of measurand takes place is called the *transducer or translator*.

- Majority of measurement systems are tertiary systems and they include the whole range of mechanical, electrical, pneumatic electro-mechanical and electro-pneumatic instruments.

  Measurements may also be divided into following *two categories* :

  1. **Contact type.** In this case the sensor of the measuring device *contacts* the controlled medium.
  2. **Non-contact type.** Here the sensor *does not contact* the controlled medium.

- Non-contact measurements include *optical, radioactive and others.*

## 1.5 GENERALISED MEASUREMENT SYSTEM AND ITS FUNCTIONAL ELEMENTS

Fig. 1.4, shows a measurement system (generalized) with different components (called *elements*).

The various elements are :

**1. Primary sensing element.** It is an element that is sensitive to the measured variable. The sensing elements *sense the condition, state or value of the process variable by extracting a small part of energy from the measurand, and then produces an output which reflects this condition, state or value of the measurand.*

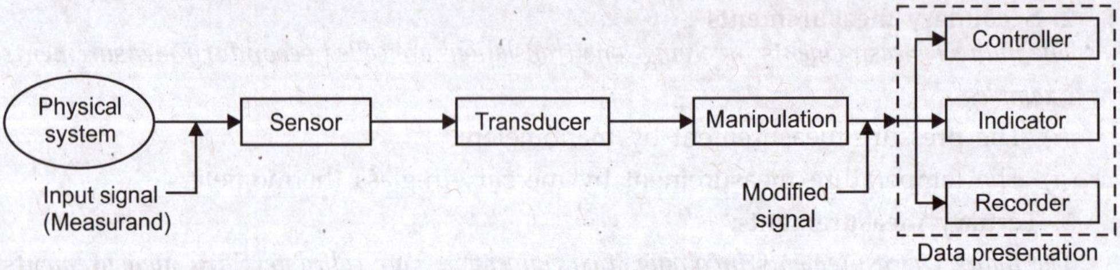


**Fig. 1.4.** Generalised measurement system.

**2. Variable conversion of transducer element.** This element *converts the signal from one physical form to another* without changing the information content of the signal.

**3. Manipulation element.** This element *operates* on the signal according to some *mathematical rule without changing the physical nature of the variable.*

**4. Data transmission element.** This element *transmits the signal from one location to another without changing its information contents.*

**5. Data presentation element.** This element *provides a display record or indication of the output from the manipulation elements.*

## Stages of the general measurement system :

The general measuring system consists of the following *three stages :*

1. *Stage-1 :*

- It senses desired input to exclusion of all others and *provides analogous output.*

*"Types and Examples"* :

*Mechanical :* Contacting spindle, spring-mass, elastic devices (*e.g.,* Bourdon tube for pressure, proving ring for force), gyro.

*Electrical :* Contacts, resistance, capacitance, inductance, piezoelectric crystals and polymers, thermocouple, semiconductor junction, etc.

*Optical :* Photographic film, photoelectric diodes and transistors, photomultiplier tubes, holographic plates.

*Hydro-pneumatic :* Buoyant float, orifice, venturi, vane, propeller.

2. *Stage-2 :*

- *It modifies transduced signal into form usable by final stage.*
- It *usually increases amplitude and/or power* depending on requirement.
- It may also *selectively filter unwanted component or convert signal into pulsed form.*

*"Types and Examples"* :

*Mechanical :* Gearing, cranks, slides, connecting links, cams, etc.

*Electrical :* Amplifying or attenuating systems, bridges, filters, telemetering systems, various special purpose I.C. (Integrated circuit) devices.

*Optical :* Mirrors, lenses, optical filters, optical fibers, spatial filters (pin hole, slit).

*Hydro-pneumatic:* Piping, valving, dashpots, plenum chambers.

3. *Stage-3 :*

- It *provides an indication or recording* in form that can be evaluated by an unaided human sense or by a controller.
- It *records data* digitally on a computer.

*"Types and Examples"* :

*Indicators Displace types :* Moving pointer and scale, moving scale and index, light beam and scale, electron beam and scale (oscilloscope), liquid column.

*Digital types :* Direct alphanumeric readout.

*Recorder :* Digital printing, inked pen and chart, direct photography, magnetic recording (hard disk or type).

*Processor and computers :* Various types of computing systems, either special purpose or general, used to feed read out/recording devices and/or controlling systems.

*Controllers :* All types.

## Illustration of measuring systems :

***I. Bourdon tube pressure gauge:***

Fig. 1.5 shows the simple Bourdon tube pressure gauge – an example of a simple measurement system.

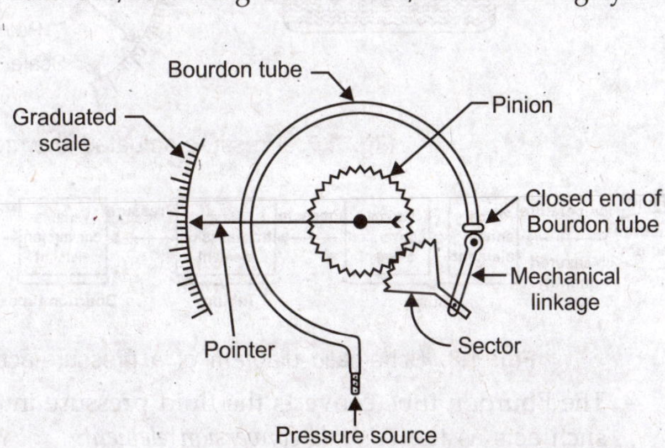


**Fig. 1.5.** Simple Bourdon tube pressure gauge.

— In this case the Bourdon tube acts as the primary sensing element and a variable conversion element. It senses the input (pressure) quantity. On account of the pressure, the closed end of the Bourdon tube is displaced and thus the pressure is converted into a small displacement.

— The closed end of the Bourdon tube is connected to a gearing arrangement through mechanical linkage. The gearing arrangement *amplifies* the small displacement and consequently the pointer rotates through a large angle. Thus, the mechanical linkage acts as a data transmission element while the gearing arrangement acts as a data manipulation element.

— The final data presentation stage consists of a pointer and dial arrangement, which when calibrated with known pressure inputs, gives an indication of the pressure signal applied to the Bourdon tube.

Fig. 1.6 shows the schematic diagram of this measurement system.

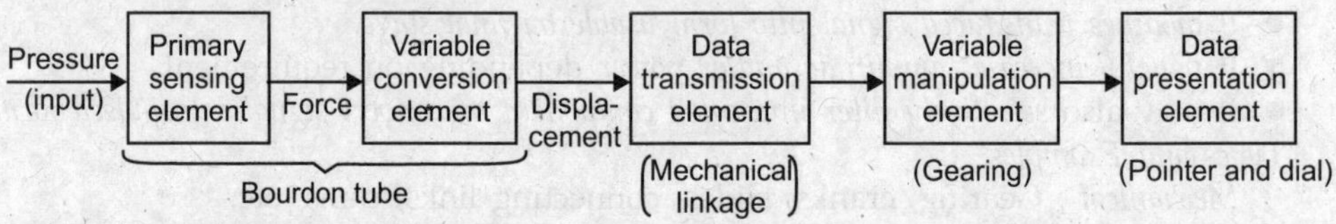


**Fig. 1.6.** Schematic diagram of a Bourdon tube pressure gauge.

***II. Pressure actuated thermometer :***

Fig. 1.7 shows a pressure-actuated thermometer.

— The liquid-filled bulb acts as the *primary sensor* and *variable conversion element* since a temperature change results in pressure build-up within the bulb, due to the constrained thermal expansion of filling liquid.

— The pressure tubing is employed to transmit the pressure to the Bourdon tube and thus functions as the *data transmission element.*

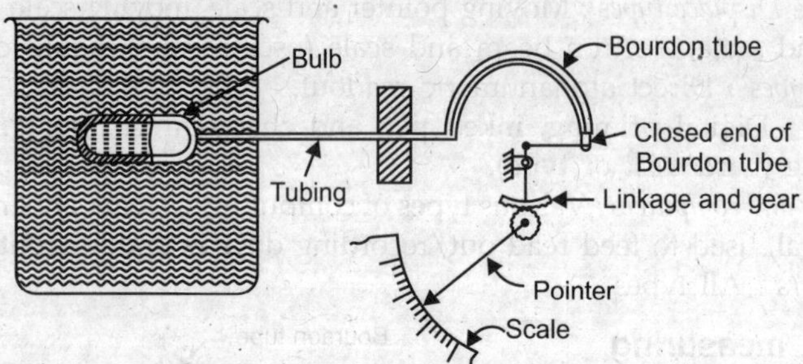


**Fig. 1.7.** Pressure-actuated thermometer.

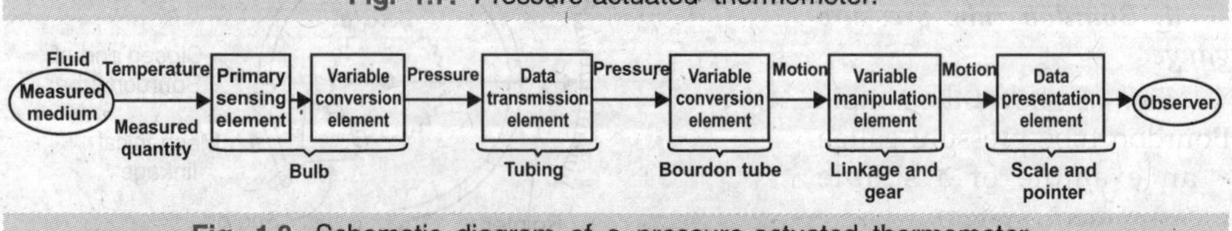


**Fig. 1.8.** Schematic diagram of a pressure-actuated thermometer.

— The Bourdon tube converts the fluid pressure into displacement of its tip, and as such acts as the *variable conversion element.*

— The displacement is manipulated by the linkage and gear (*manipulation elements*) to give a larger pointer motion.

— The scale and the pointer serve as the *data presentation element.*

Fig. 1.8 shows the schematic diagram of a pressure-actuated thermometer.

## 1.6 INSTRUMENTS

### 1.6.1. Introduction

The **instrument** *serves as an extension of human faculties and enables the man to determine the value of unknown quantity or variable which his unaided human faculties cannot measure.* A measuring instrument provides information about the physical value of some variable being measured.

— The instrument would sense a physical parameter (*e.g.*, velocity, pressure, temperature, etc.), process and translate it into a format and range which can be interpreted by the observer.

— The instrument *must also provide the controls* by which the operator can obtain, manipulate and respond to the information.

- An instrument, in *simple cases,* consists of a *single unit* which gives an output reading or signal according to the unknown variable (measurand) applied to it. Under *more complex measurement situations,* a measuring instrument may consist of *transducing elements* which convert the measurand to analogous form. The *analogous signal* is then processed by some intermediate means and then fed to the end devices for presenting the measurement results for the purpose of display, record and control.

- The instruments made by man not only retain their characteristics for extended periods of time but are also accurate and sensitive in their response.

- *Measurements* involve the use of *instruments* as a physical means of determining quantities and variables.

### 1.6.2. Classification of Instruments

The instruments may be *classified* as follows :

1. Absolute and secondary instruments.
2. Analog and digital instruments.
3. Mechanical, electrical and electronic instruments.
4. Manual and automatic instruments.
5. Self-contained and remote indicating instruments.
6. Self-operated and power-operated instruments.
7. Deflection and null output instruments.

**1. Absolute and secondary instruments :**

***(i) Absolute instruments :***

*These instruments give the magnitude of the quantity under measurement in terms of physical constant of the instrument.*

*Examples :* Tangent galvanometer and Rayleigh's current balance.

- These instruments are *seldom used except in standard institutions.*

***(ii) Secondary instruments :***

*These instruments are so constructed that the quantity being measured can only be measured by observing the output indicated by the instrument.*

These instruments are *calibrated* by comparison with an absolute instrument or another secondary instrument which has already been calibrated against an absolute instrument.

*Examples* : Voltmeter, glass thermometer, pressure gauge etc.

- The secondary instruments find *wide use in every sphere of measurement.*

**2. Analog and digital instruments :**

***(i) Analog instruments :***

The signals of an analog unit vary in a continuous fashion and can take on infinite number of values in a given range.

*Examples* : Fuel gauge, ammeters and voltmeters, wrist watch, speedometer of an automobile, etc.

***(ii) Digital instruments :***

Signals varying in *discrete steps* and taking on a finite number of different values in a given range are *digital signals* and the corresponding instruments are of *digital type.*

*Examples* : Odometer of an automobile, calibrated balance of a platform scale, timer on a score board, etc.

- The digital devices have the following *advantages* :
  — High accuracy;
  — High speed;
  — Elimination of human operational errors.
- Owing to the application of digital computers for data handling, reduction and in automatic controls, the importance of *digital instrumentation* is increasing very fast. Thus, it becomes necessary to have both analog-to-digital (A/D) converters at input to the computers and digital-to-analog (D/A) converts at the output of the computers.

**3. Mechanical, electrical and electronic instruments :**

***(i) Mechanical instruments:***

- These instruments are very reliable for static and stable conditions.
- They are unable to respond rapidly to the measurements of dynamic and transient conditions due to the fact that they have moving parts that are rigid, heavy and bulky and consequently have a large mass. Mass presents inertia problems and hence these instruments cannot faithfully follow the rapid changes which are involved in dynamic instruments.
- Most of the mechanical instruments cause noise pollution.

***(ii) Electrical instruments :***

- The electrical methods of indicating the output of detectors *are more rapid than that of mechanical methods.*
- Unfortunately an electrical system normally depends upon a mechanical meter as an indicating device. This mechanical movement has some inertia due to which the *frequency response* of these instruments is *poor.*

***(iii) Electronic instruments :***

- Most of the scientific and industrial instruments require *very fast responses.* Such requirements cannot be met with by mechanical and electrical instruments.
- These instruments use *semiconductor devices. In electronic devices, since the only movement involved is that of electrons, the response time is extremely small owing to very small inertia of electrons.*
- With the use of electronic devices *very weak signals can be detected by using pre-amplifiers and amplifiers.*

**4. Manual and automatic instruments :**

- In case of *manual* instruments services of an operator are required.

  *Example* :Measurement of temperature by a resistance thermometer incorporating a Wheatstone bridge in its circuit (an operator is required to indicate the temperature being measured).

- In an *automatic* type of instrument, an operator is *not* required.

  *Example* : Measurement of temperature by mercury-in-glass thermometer.

**5. Self-contained and remote indicating instruments :**

- A self-contained instrument *has all its different elements in one physical assembly.*
- In a remote indicating instrument, the primary sensing element *may be located at an adequate long distance* from the secondary indicating element. Such types of instruments are *finding wide use in the modern instrumentation technology.*

**6. Self-operated and power-operated instruments :**

- A self-operated instrument does not require any outside power for its operation; the output energy is supplied wholly or almost wholly by the input signal.

  *Examples* : Mercury-in-glass thermometer, Dial indicator etc.

  The power operated instruments are those instruments which require some auxiliary source of power such as electricity, compressed air, hydraulic supply etc. for their operation. In such cases, the input signal supplies only an insignificant portion of the output power.

  *Example* : Electromechanical measurement system shown in Fig. 1.9.

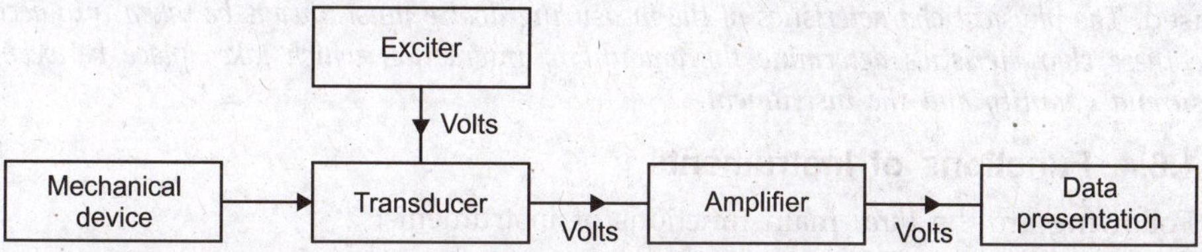


**Fig. 1.9.** Electromechanical measurement system.

**7. Deflection and null output instruments :**

- In a ***"deflection type"*** *instrument, the deflection of the instrument provides a basis for determining the quantity under measurement.* The measured quantity produces some physical effect which deflects or produces a mechanical displacement of the moving system of the instrument. An opposing effect is built in the instrument which tries to oppose the deflection or the mechanical displacement of the moving system. The balance is achieved when opposing effect *equals* to cause producing the deflection or mechanical movement. The deflection or mechanical displacement at the point of balance then gives the value of measured quantity (which depends upon the calibration of the instrument).

  *Examples* : Permanent magnet moving coil (PMMC) ammeter etc.

  — These instruments are more suited for measurements under *dynamic conditions* than null type of instruments whose intrinsic response is slower.

- In a ***"null type"*** *of instrument, a zero or null indication leads to determination of the magnitude of measured quantity.* The null condition is dependent upon some other known conditions.

  *Example* : D.C. potentiometer.

*Advantages :*

(*i*) The "null type" instruments are *more accurate* than deflection type instruments.

(*ii*) These instruments can be *highly sensitive* as compared with deflection type instruments.

### 1.6.3. Factors Relating to Selection of Instruments

The following are some important factors/points which are very essential to be considered for the proper selection of a correct instrument for any application :

1. The *accuracy* expected from the instrument.

   The following most significant and fundamental rule, which is perhaps the most important rule in measurement must always be kept in mind :

   *"Never demand an accuracy of measurement higher than that of which is really needed, and never forget that each degree of accuracy, if demanded, is likely to have a disproportionate effect on the complexity and cost of measuring apparatus".*
2. *When are the final data required* (*i.e.,* at the time of taking the measurement, or later one ?
3. The *cost criterion* (*i.e.,* the cost of the measuring process).
4. *In what form should the data be displayed* (*i.e.,* indicating, recording, integrating etc.)?
5. Whether quantity to be measured has *constant value or is it a time variant* ?

— In the measurement of time variant quantities the manner in which the quantity to be measured changes has a significant effect on the type of measuring instruments to be used. *The physical characteristics of the measuring device must always be taken into account since these characteristics determine the amount of interaction which takes place between the measurand quantity and the instrument.*

### 1.6.4. Functions of Instruments

Following are the *three* main functions of instruments:

1. Indicating function.
2. Recording function.
3. Controlling function.

**1. Indicating function :**

- The value of the quantity can be read by the movement of the needle on a calibrated scale provided on the instrument. Readings can be taken to any fraction within the limitations of the instrument and human eye.

  *Examples :*

  (*i*) The deflection of a pointer of a *'speedometer'* indicates the speed of the automobile at the moment.

  (*ii*) The use of *'pressure gauge'* for indicating pressure.
- These days digital display (*i.e.,* the reading in numericals) is also commonly used. In fact where highly accurate and exact readings are desired, use of digital readouts is recommended.

**2. Recording function :**

- In several cases the instrument continuously record, with pen and ink, the value of measured quantity against some other variable or against time. A graph paper is provided whose speed can be adjusted. A recording instrument can be used to record more than one number of quantities also, in which case, either the recording of different parameters may be done by different pens or same pen may record one parameter after other.

*Example* : A 'potentiometric type of recorder' used for monitoring temperature records the instantaneous values of temperatures on a strip chart recorder.

**3. Controlling function :**

- The controlling function is perhaps one of the most important functions particularly in the field of industrial control processes. Here, the information is used by the instrument/system to control the original measured quantity.

  *Examples* :

  (*i*) *Floats* for liquid level control.

  (*ii*) *Thermostats* for temperature control.

## 1.7 INPUT-OUTPUT CONFIGURATIONS OF MEASURING INSTRUMENTS AND MEASURING SYSTEMS

Fig. 1.10 shows the generalised input-output configuration of measurement systems. The various inputs to the measurement system are *classified* as follows :

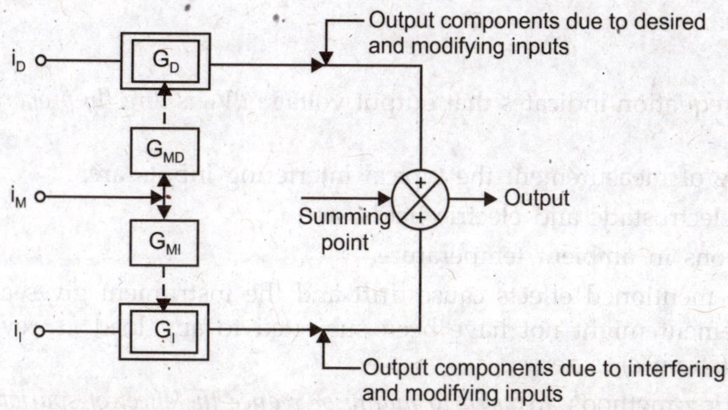


**Fig. 1.10.** Generalised input-output configuration of measurement systems.

1. Desired inputs.
2. Interfering inputs.
3. Modifying inputs.

**1. Desired inputs :**

*The quantities for which the instrument or the measurement system is specifically designed to measure and respond are called* ***desired inputs.***

The desired input $i_D$ produces an output component according to an input-output relation symbolised by $G_D$; here $G_D$ (*Transfer function*) represents the mathematical operation necessary to obtain the necessary output from the input.

- The transfer function $G_D$ is a representative of a wide range of functions from a *constant in the case of linear systems to a statistical function used for statistical measurements.*

**2. Interfering inputs :**

*The quantities to which an instrument or a measurement system becomes unintentionally sensitive are called* ***interfering inputs.***

The interfering inputs $i_I$ would produce an output component according to input-output relation symbolised by $G_I$.

**3. Modifying inputs :**

*The inputs which cause a change in input-output relationships for either desired input or interfering input or for both are called* ***modifying inputs.***

The modifying input $i_M$ would cause a change in $G_D$ and/or $G_I$. The specific manner in which $i_M$ affects $G_D$ and $G_I$ is represented by $G_{MD}$ and $G_{MI}$ respectively.

For illustration, let us consider *strain measurement by a strain gauge.* When the gauge is strained due to mechanical loading, the change in output voltage is given by,

$$dV_0 = \frac{V_s}{4} \cdot \frac{dR}{R}$$

where, $dV_0$ = Output voltage,

$V_s$ = Supply voltage (battery), and

$dR$ = Change in resistance of the strain gauge which constitutes one arm of the wheatstone bridge circuit (assuming, $R_1 = R_2 = R_3 = R_4 = R$)

In terms of strain $e$ and strain gauge factor, $F$,

$$dV_0 = \frac{V_s}{4} \cdot Fe$$

The above equation indicates that output voltage $dV_0$ is *directly proportional to applied strain.*

In this case of measurement the typical interfering inputs are:

- Stray electrostatic and electric fields, and
- Variations in ambient temperature.

The above mentioned effects cause drift and the instrument gives an output even though the element might not have been subjected to any load *i.e.*, even though the desired input (strain $e$) is zero.

The following methods are *used to nullify or reduce the effect of spurious inputs on the indicated output of an instrument:*

1. Signal filtering.
2. Compensation by opposing inputs.
3. Output correction.
4. Use of high amplification and stable/accurate feedback device.
5. Use of instrument elements which are inherently sensitive only to the desired input (*Example :* The strain gauge may be made of a material with a very low temperature coefficient of resistance and temperature independent gauge factor).

## 1.8 APPLICATIONS OF MEASUREMENT SYSTEMS

The instruments and measurement systems are employed for the following *applications :*

1. Monitoring of processes and operations.
2. Control of processes and operations.
3. Experimental engineering analysis.

**1. Monitoring of processes and operations :**

In these types of applications the measuring instruments *simply indicate the value or condition of parameter under study and do not serve any control function.*

*Examples :*

(*i*) Water and electric energy meters installed in homes.

(*ii*) An ammeter or a voltmeter indicates the value of current or voltage being monitored at a particular instant.

2. **Control of processes and operations :**

There is a strong association between measurements and control. The instruments find a very useful application in automatic control systems.

- A common example is the typical refrigeration system which employs a thermostatic control. A temperature measuring device (often a bimetallic element) senses the room temperature, thus providing the information necessary for proper functioning of the control system.

3. **Experimental engineering analysis :**

The engineering problems can be solved by theoretical as well as experimental methods; several applications require the use of both the methods.

Following are the *uses* of experimental engineering analysis :

(*i*) To determine system parameters, variable and performance indices.

(*ii*) To formulate the generalised empirical relationship in cases where there is no proper theoretical backing.

(*iii*) To test the validity of theoretical predictions.

(*iv*) To solve mathematical relationships with the help of analogies.

## 1.9 DEFINITIONS RELATING TO MEASURING INSTRUMENTS

Some important terms used in connection with measuring instruments are discussed below :

1. **True or actual value.** *The actual magnitude of a signal input to a measuring system which can only be approached and never evaluated is termed as "true or actual value".*
2. **Indicated value.** *It is the magnitude of a variable indicated by a measuring instrument.*
3. **Correction.** *The revision applied to the critical value so that the final result obtained improves the worth of the result is called "correction".*
4. **Overall error.** *It is the difference of the scale reading and the true value.*

   — When the instrument is properly designed and correctly adjusted the consistent bias in error is very rare.
5. **Range.** *The region between the limits within which an instrument is designed to operate for measuring, indicating or recording a physical quantity is called the "range of the instrument".*
6. **Sensitivity.** *The ratio of output response to a specified change in the output is called "sensitivity".*
   - The minimum change in the measured variable which produces an effective response of the instrument is called *"Resolution sensitivity"*. It is also called *"discrimination"*.
   - The lowest level of measured variable which produces effective response of the instrument is called *"Threshold sensitivity"*.
7. **Scale sensitivity.** *It is defined as the ratio of a change in scale reading to the corresponding change in pointer deflection.*

8. **Scale readability.** The scale readability (in analog instruments) *indicates the closeness with which the scale can be read.*
9. **Repeatability.** It is defined as the *variation of scale reading*; it is random in nature.
   - *It is a measure of closeness with which a given input can be measured over and over again.*
10. **Accuracy.** It may be defined as *conformity with or closeness to an accepted standard value (true value).*
    - Accuracy of an instrument is influenced by factors like static error, dynamic error, reproducibililty, dead zone.
11. **Uncertainty.** Uncertainty denotes the *range of error, i.e.,* the region in which one guesses the error to be.
12. **Precision.** It refers to the *degree of agreement* within a group measurements.
    - It is usually expressed in *terms of the deviation in measurement.*
13. **Drift.** *An undesired gradual departure of the instrument output over a period of time that is unrelated to changes in input, operating conditions or lead is called "drift".*
14. **Linearity or non-linearity.** *Deviation of transducer output curve from a specified straight line.* The *"non-linearity"* may be: (*i*) *Terminal linearity* (deviation from a straight line through the end points; (*ii*) *Best-fit linearity* (deviation from the straight line which gives minimum errors, both plus and minus.
15. **Dead zone.** *It is the range within which variable can vary without being detected.*
16. **Dead time.** *It is the time before the instrument begins to respond after the measured quantity has been changed.*
17. **Speed of response.** *The quickness of an instrument to read the measured variable is called "speed of response".*
18. **Reproducibility.** *The degree of closeness with which the same value of a variable may be measured at different times is called "reproducibility".*
19. **Tolerance.** *It is the range of inaccuracy which can be tolerated in measurements.*
20. **Backlash.** It is defined as the *maximum distance or angle through which any part of a mechanical system may be moved in one direction without applying appreciable force or motion to the next part in a mechanical system.*
21. **Stiction (static friction).** *It is the force or torque that is necessary just to initiate motion from rest.*
22. **Noise.** It may be defined *extraneous disturbance generated in a measuring system which conveys no meaningful information w.r.t. desired signal.*

## 1.10 STATIC CHARACTERISTICS

*Measurement of applications in which parameter of interest is more or less constant; or varies very slowly with time are called* **static measurements.** *A set of criteria (e.g., "accuracy", "error", "reproducibility", "drift", "sensitivity", "dead zone") that provide meaningful description of measurements under static conditions are called* **static characteristics.**

The main *static characteristics* may be summed up as follows :

(*i*) Accuracy (*ii*) Sensitivity
(*iii*) Reproducibility (*iv*) Drift
(*v*) Static error (*vi*) Dead zone.

### 1.10.1. Accuracy, Errors and Correction

**Accuracy :**

*It is the degree of correctness with which a measuring means yields the "true value" with reference to accepted engineering standards.*

Accuracy is determined as the maximum amount by which the result differs from the true value (the average of an infinite number of measured values when the average deviation due to the various contributing factors tends to zero). It is *almost impossible to determine experimentally the true value. The true value is not indicated by any measurement system due to the loading effects, lags and mechanical problems* (*e.g., wear, hysteresis, noise pick-up etc.*).

*Accuracy of the measured signal* depends upon the following *factors :*

- Variation of the signal being measured;
- Intrinsic accuracy of the instrument itself;
- Accuracy of the observer;
- Whether or not the quantity is being truly impressed upon the instrument.

**Errors :**

The accuracy of any measurement system is measured in terms of its *errors.* Since it is impossible to measure the true value of a quantity, as such, an approximation of the "*true value*" obtained by sufficiently extended series of measurements and also taking into parameters and conditions to which corrections may be applied, we obtain, what is called "*best-measured value*" of the quantity. *The difference between the best measured value and the true value of the quantity is known as* **static error.**

The absolute value of error does not indicate precisely the accuracy of measurements. It is the *relative static error* which provides the quality of measurement. The **relative static error** *is defined as the ratio of absolute static error to the true value of the quantity under measurement.*

Let, $V_m$ = Measured value of quantity,

$V_t$ = True value of quantity,

$E_s$ = Static (or absolute) error of measurement, and

$E_r$ = Relative static error.

Then,
$$E_s = V_m - V_t \quad ...(1.1)$$

and,
$$E_r = \frac{\text{Absolute error}}{\text{True value}} = \frac{E_s}{V_t} \quad ...(1.2)$$

or, %age relative error, $\% E_r = E_r \times 100$ ...(1.3)

We have,
$$V_t = V_m - E_s \quad \text{...[From eqn. (1.1)]}$$
$$= V_m - E_r \cdot V_t \quad \text{...[From eqn. (1.2)]}$$

or,
$$V_t + E_r \cdot V_t = V_m$$

or,
$$V_t = \frac{V_m}{1 + E_r} \quad ...(1.4)$$

However, when $E_r$ is small, $E_r << 1$,

Eqn. (1.4) may be written as :

$$V_t = V_m (1 - E_r) \quad ...(1.5)$$

- The error may be either *+ve* or *–ve*. For *+ve* static errors the instrument reads *high* whereas it reads *low* for *–ve* static errors.

**Correction.**

*The difference between the true value and the measured value of a quantity is called* **static correction** ($C_s$).

*i.e.,*

$$C_s = V_t - V_m \; (= -E_s) \qquad ...(1.6)$$

Eqn. (1.6) indicates that the static correction of the instrument reading is of the same magnitude as the error, but *opposite in sign* (*i.e.,* $C_s = -E_s$).

**"Accuracy"** may be *expressed as :*

(*i*) Point accuracy, or

(*ii*) Percent of scale range, or

(*iii*) Percent of true value.

**(*i*) *Point accuracy.*** In this case, the accuracy of an instrument is stated for one or more points in a range.

*Examples :*

(*a*) A given thermometer may be stated to read ± 0.2°C between 50°C and 100°C;

(*b*) A scale of length may be read within ± 0.2 mm.

**(*ii*) *Percent of scale range.*** Here, the error is calculated on the basis of maximum value of the scale.

$$\% \text{ Error} = \left(\frac{V_m - V_t}{\text{Max. scale value}}\right) \times 100 \qquad ...(1.7)$$

Eqn. (1.7) refers to the intrinsic accuracy of the instrument itself and does not include procedural or personal performance.

**(*iii*) *Percent of true value.*** In this case the absolute error of measurement is expressed as %age of true value of the unknown quantity.

*i.e.,*

$$\%\text{Error} = \left(\frac{V_m - V_t}{V_t}\right) \times 100 \qquad ...(1.8)$$

— Here, with a drop in true value, the size of error decreases.

— In this type of expression of accuracy, the percentage error is the maximum for any point in the instrument range.

The percent accuracies on the basis of actual scale or full-scale reading are normally used to express *equipment accuracies* and do not include procedural or personnel performance which is actually included in the total error.

**Example 1.1.** *A voltmeter reads 112.68 V. If the true value of the voltage is 112.6 V, determine the following :*

(*i*) *The static error.*

(*ii*) *The static correction for the voltmeter.*

**Solution.** *Given :* $V_m$ = 112.68 volts; $V_t$ = 112.6 volts.

**(*i*) Static error, $E_s$ :**

$$E_s = V_m - V_t = 112.68 - 112.6 = +0.08 \text{ V} \quad \textbf{(Ans.)}$$

**(*ii*) Static correction, $C_s$ :**

$$C_s = -E_s = \mathbf{-\ 0.08\ V} \quad \textbf{(Ans.)}$$

**Example 1.2.** *A thermometer reads 92.35°C and the static correction given in the correction curves is – 0.07°C. Determine the true value of the temperature.*

**Solution.** *Given :* $V_m = 92.35°C; \quad C_s = -0.07°C.$

**True value, $V_t$ :**

We know that, $E_s = V_m - V_t$

or, $-C_s = V_m - V_t$ $\quad (\because C_s = -E_s)$

or, $V_t = V_m + C_s$ ...[Eqn (1.6)]

$= 92.35 - 0.07 = \mathbf{92.28°C}$ **(Ans.)**

**Example 1.3.** *An analog indicating instrument with a scale range of 0–5.0V shows a voltage of 2.65 V. The true value of a voltage is 2.70 volts.*

*(i) What are the values of absolute error and correction ?*

*(ii) Express the error as a function of the true value and full scale deflection.*

**Solution.** *Given :* $V_m = 2.65$ V; $V_t = 2.70$ V

**(i) Absolute error, $E_s$ :**

$$E_s = V_m - V_t = 2.65 - 2.70 = \mathbf{-0.05\ V} \quad \textbf{(Ans.)}$$

**Absolute correction, $C_s$ :**

$$C_s = -E_s = -(-0.05) = \mathbf{0.05\ V} \quad \textbf{(Ans.)}$$

**(ii) Relative error, $E_r$ :**

As a function of the *true value* :

$$E_r = \frac{E_s}{V_t} = \frac{-0.05}{2.70} = -0.0185, \text{ or, } \mathbf{-1.85\%} \quad \textbf{(Ans.)}$$

As a function of *full scale deflection* :

$$E_s = \frac{E_s}{(5.0-0)} = \frac{-0.05}{5} = -0.01, \text{ or, } \mathbf{-1\%} \quad \textbf{(Ans.)}$$

**Example 1.4.** *A pressure indicator showed a reading as 42 bar on a scale range of 0–50 bar. If the true value was 41.4 bar, determine.*

*(i) Static error,*

*(ii) Static correction, and*

*(iii) Relative static error.*

**Solution.** *Given :* $V_m = 42$ bar ; $V_t = 41.4$ bar ; Scale range = 0–50 bar

**(i) Static error, $E_s$ :**

$$E_s = V_m - V_t = 42 - 41.4 \text{ bar} = \mathbf{+\ 0.6\ bar} \quad \textbf{(Ans.)}$$

**(ii) Static correction, $C_s$ :**

$$V_s = -E_s = -(+\ 0.6) = \mathbf{-0.6\ bar} \quad \textbf{(Ans.)}$$

**(iii) Relative error, $E_r$ :**

$$E_r = \frac{E_s}{V_t} = \frac{0.6}{41.4} = 0.0145 \text{ or } \mathbf{1.45\%} \quad \textbf{(Ans.)}$$

**Example 1.5.** *A pressure gauge of range 50 bar is stated to have an error of ± 0.15 bar when calibrated by the manufacturer. Determine :*

*(i) The percentage error on the basis of maximum scale value.*

*(ii) The possible error as a percentage of the indicated value when a reading of 10 bar is obtained in a test.*

**Solution.** *Given :* Pressure range = 50 bar ; Error = ± 0.15 bar

*(i) Percentage error on the basis of maximum scale value*

$$= \pm\frac{0.15}{50}\times 100 = \pm 0.3\% \quad \textbf{(Ans.)}$$

(*ii*) *Percentage error on the basis of indicated value of 10 bar pressure*

$$= \pm\frac{0.15}{10}\times 100 = \pm 1.5\% \quad \textbf{(Ans.)}$$

The above results indicate that the pressure gauge is *more unreliable at the lower end of its gauge.*

**Example 1.6.** *A system comprising a transmitter, a relay and a receiver element has been used to measure the pressure at a remote point. If the specified accuracy limits for transmitter, relay and receiver are within ± 0.3%, ±1.4% and ± 0.9% estimate the maximum possible error and root-square accuracy of the measurement system.*

**Solution.** Maximum possible error (*least accuracy*) = ± (0.3 + 1.4 + 0.9) = **± 2.6% (Ans.)**

*Least root-square accuracy or root-sum square error*

$$= \pm\sqrt{(0.3)^2 + (1.4)^2 + (0.9)^2} = \pm 1.69\% \quad \textbf{(Ans.)}$$

The above results indicate that the error is possibly as large as ± 2.6% but probably not larger than ± 1.69%,

- In an instrument system, *root-square accuracy is more justifiable because it is not probable that all the elements of the system will have the greatest static error at the same point and at the same time.*

### 1.10.2. Static Calibration

**Static calibration** *is a process by which all the static performance characteristics are obtained in one form or another.* In general, static calibration refers to a situation in which all inputs, whether desirable, interfering or modifying except one are kept at some constant values. Then the one input under study is varied over some range of constant values, which causes the output (s) to vary over some range of constant values. Thus an *output-input relationship is developed which comprise a static calibration valid under the stated constant conditions of all the other inputs.*

- The procedure of calibration involves a comparison of the particular instrument with either :
  (*i*) a *primary standard.*
  (*ii*) a *secondary standard with a higher accuracy than the instrument to be calibrated, or*
  (*iii*) an *instrument of known accuracy.*

— All working instruments must be calibrated against some reference instruments *which have a higher accuracy.* The reference instruments in turn must be calibrated against instruments of still higher grade of accuracy or against primary standards or against other standards of known accuracy.

- The following **steps** are necessary in *performing a calibration :*
  1. *Examine the construction of the instrument,* and identify and list all the *possible inputs.*
  2. Decide as best as possible which of the *inputs will be significant* in the application for which instrument is to be calibrated.
  3. Procure apparatus that *will allow to vary all the significant inputs over the ranges considered necessary.* Procure standards to measure each input.
  4. By holding some inputs constant, varying others, and recording the output(s), *develop the desired static input-output relations.*

### 1.10.3. Range and Span

**Range.** *The difference between the largest and the smallest reading of the instrument is called the "Range of an instrument".* The range is expressed by stating the lower and upper values.

**Span** *represents the algebraic difference between the upper and lower range values of the instrument.*

If the highest point of calibration is $S_{max}$ units while the lowest is $S_{min}$ units and that the calibration is continuous between the points, then we say that the *instrument range is between* $S_{min}$ *and* $S_{max}$.

The *instrument* ***span*** is given by, $S_{max} - S_{min}$.

The above definitions apply both to analog as well as digital instruments.

**Error calibration curve :** Refer to Fig. 1.11.

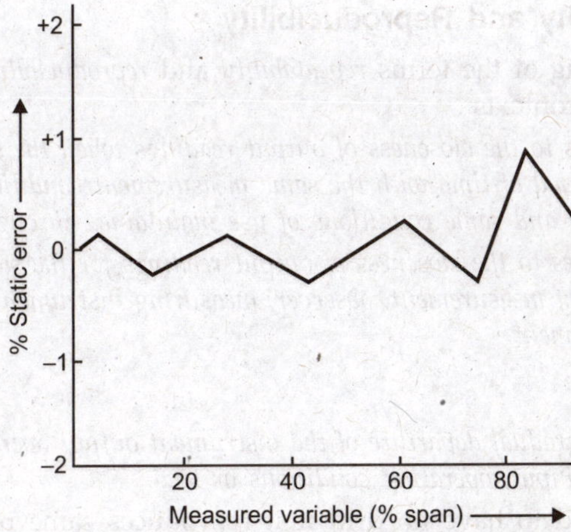


**Fig. 1.11.** Error calibration curve.

- *It is an error curve which can be used for correcting instrument readings.*
- The curve is plotted by calibrating the instrument against a suitable standard at a number of points on the scale.

  *Examples :*

  (*i*) *Range* : 2 kN/m$^2$ to 50 kN/m$^2$;

  *Span* : 50 − 2 = 48 kN/m$^2$

  (*ii*) *Range* : −5°C to 90°C ;

  *Span* : 90 − (−5) = 95°C.
- While determining the range of the instrument the frequency range must be taken into consideration. *"Frequency range"* is defined as *frequency over which measurements can be performed with a specified degree of accuracy.*

  *Example :* A moving voltmeter may have a range of 0–120 V range and 0–125 Hz frequency range with an accuracy of 0.5 or 1 percent of full scale reading.

**Example 1.7.** *A pressure gauge is calibrated 5 bar to 60 bar. The accuracy is specified with ± 0.20 percent of instrument span. What is the maximum static error ?*

**Solution.** Span of pressure gauge = 60 − 5 = 55 bar

$\therefore$ Maximum static error = $\frac{\pm 0.20 \times 55}{100}$ = **±0.11 bar** **(Ans.)**

### 1.10.4. Scale Readability

The term scale readability is frequently used in analog type of instruments. *It indicates the closeness with which the scale of an analog type of instrument can be read.*

The readability depends upon many *factors* such as :

(*i*) Number of graduations;
(*ii*) Spacing of graduations;
(*iii*) Size of pointer;
(*iv*) Parallax effects;
(*v*) Discriminating power of the observer.

The readability, thus, is logically implied by the number of significant figures; the higher the number of significant figures, the better the readability of the instrument scale.

### 1.10.5. Repeatability and Reproducibility

Although the meaning of the terms *repeatability* and *reproducibility* is same but they are applied in different contexts.

**Repeatability** *pertains to the closeness of output readings when the same input is applied repetitively over a short period of time with the same measurement conditions, same instrument and observer, same location and same conditions of use maintained throughout.*

**Reproducibility** *relates to the closeness of output readings for the same input when there are changes in the method of measurement, observer, measuring instrument, location, conditions of use and time of measurement.*

### 1.10.6. Drift

***Drift** is an undesired gradual departure of the instrument output over a period of time that is unrelated to changes in input, operating conditions or load.*

An instrument is said to have no drift if it reproduces same readings at different times for same variation in measured variable.

The drift may be caused by the following *factors* :

(*i*) High mechanical stresses developed in some parts of instruments and systems.
(*ii*) Wear and tear.
(*iii*) Mechanical vibrations.
(*iv*) Temperature changes.
(*v*) Stray electric and magnetic fields.
(*vi*) Thermal e.m.fs.

**Examples of drift :**

- Drift occurs in *"resistance thermometers and thermocouples" due to the contamination of the metal and a change in its atomic or metallurgical structure.*
- Drift may occur in *"flow meters" due to wear and erosion of the orifice plate, nozzle or venturimeter.*

**Classification of drift:**

Drift may be *classified* as follows :

1. Zero drift; 2. Span drift; 3. Zonal drift.

**1. Zero drift :**

Drift is called *zero drift if the whole of instrument calibration gradually shifts over by the same amount.*

— It may be due to *permanent set or slippage* and can be corrected by *shifting pointer position.*

2. **Span drift :**

*If the calibration from zero upwards changes proportionally, it is called "span (or sensitivity) drift".*

— It may be due to change in spring gradient etc.

3. **Zonal drift :**

*When the drift occurs only over a portion of span of an instrument, it is called "zonal drift".*

— In industrial instruments, drift is an undesirable quantity since it is rarely apparent and cannot be easily compensated for.

- *Drift occurs very slowly and can be checked only by periodic inspection and maintenance.*

### 1.10.7. Accuracy and Precision

**Accuracy.** *The closeness with which an instrument reading approaches the true value of the quantity being measured is called* ***accuracy***. Accuracy is determined as the maximum amount by which the result differs from the true value. Accuracy of an instrument is influenced by factors like static error, dynamic error, reproducibility, dead zone.

**Precision.** The term 'precise' means clearly or sharply defined. *"Precision" is a measure of reproducibility of measurement.* It is usually expressed in terms of the deviation in measurement.

— ***Significant figure*** *is an indication of the precision of measurement.* Significant figures convey actual information regarding the magnitude of precision.

Following **examples** will illustrate the difference in meaning of the two terms *accuracy* and *precision :*

(*i*) Let us consider the case of an ammeter having high precision which can take readings to 1/150 of an ampere, but has a wrong zero adjustment. Now we can take readings to 1/150 of an ampere which are consistent and clearly defined since the ammeter is *precise* as ever. However, the readings taken with this ammeter are *not accurate*, since they do not confirm to truth because of faulty adjustment of zero.

(*ii*) Take the case of micrometer which is normal in every respect but its anvil is displaced from its true position. When the readings are taken by this instrument, these would be clearly defined and consistent and we would say that micrometer is as *precise* as ever. However, the readings do not conform to the truth as anvil is displaced from its true position. Thus, the readings taken by this micrometer are *not accurate.*

- It is worth noting that *precision is a necessary prerequisite to accuracy but it does not guarantee accuracy.*

### 1.10.8. Sensitivity

*The ratio of the magnitude of output signal to the input signal or response of measuring system to the quantity being measured is called* **sensitivity.**

It is represented by the slope of the calibration curve if the ordinates are expressed in actual units.

— When the calibration curve is linear (Fig. 1.12*a*), the sensitivity is constant. However, if the calibration curve is non-linear (Fig. 1.12*b*), the sensitivity is different at different points, being the slope of curve in various points.

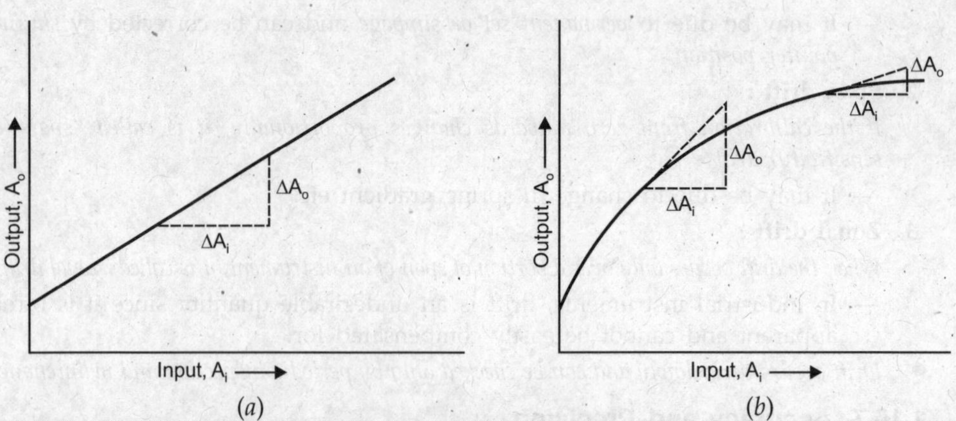


**Fig. 1.12.** Static sensitivity.

Static sensitivity, in general, is defined as :

$$\text{Static sensitivity } (S) = \frac{\text{Infinitesimal change in output}}{\text{Infinitesimal change in input}} = \frac{\Delta A_o}{\Delta A_i} \quad ...(1.9)$$

Similarly, *Inverse sensitivity* or *deflection factor* or *scale factor* $= \dfrac{\Delta A_i}{\Delta A_o}$ ...(1.10)

- *The sensitivity of an instrument should be high and thus the range should not be high in comparison to the value-being measured.*
- Sensitivity has a wide range of units, and these depend upon the instrument or measurement system being investigated.

The *following points are worth noting :*

— In case the form of input to and output from the measurement system used with electrical/electronic equipment is same, the term **gain** is used instead of *sensitivity.*

— The *increase* in displacement with the optical and mechanical instruments is described by the term **amplification.**

— When the input or output signal is *changing with time,* the term **transfer function** or **transfer operator** is used (in place of the terms *sensitivity, gain or amplification*— which apparently mean the same and describe a relationship between output and input).

**Example 1.8.** *A pressure gauge, which has linear calibration curve, has radius of scale line as 90 mm and pressure of zero to 60 pascals is displayed over an arc of 300°. Determine the sensitivity of the gauge as a ratio of scale length to pressure.*

**Solution.** Full scale deflection $= 300° = 300 \times \dfrac{\pi}{180} = \dfrac{5\pi}{3}$ rad.

$$\text{Length of scale} = \frac{5\pi}{3} \times 90 = 150\ \pi \text{ mm}$$

$$\therefore \quad \textbf{Sensitivity} = \frac{\text{Scale length}}{\text{Calibration pressure}}$$

$$= \frac{150\pi}{60} = \mathbf{2.5\pi \text{ mm/Pa}} \quad \textbf{(Ans.)}$$

**Example 1.9.** *In a Wheatstone bridge, a change of 6W in the unknown arm of the bridge is required to produce a change in deflection of 2.4 mm of the galvanometer. Determine :*

*(i) Sensitivity;* *(ii) Deflection factor.*

**Solution.** *Given* : Magnitude of output response = 2.4 mm;

Magnitude of input = 6 Ω

∴ (*i*) **Sensitivity** $= \frac{2.4}{6} =$ **0.4 mm/Ω (Ans.)**

and, (*ii*) **Deflection factor** $= \frac{6}{2.4} =$ **2.5 Ω/mm (Ans.)**

**Example 1.10.** *A pressure measuring system consists of a piezoelectric transducer, a charge amplifier and an ultra violet charge recorder; their sensitivities are 6.8 pC/bar, 0.0032 V/pC and 16 mm/V respectively. Determine the deflection on the chart for a pressure change of 20 bar.*

**Solution.** With the elements arranged in series, the *overall sensitivity* of the measuring system ($S$)

= Sensitivity of piezoelectric transducer ($S_1$) × sensitivity of the charge amplifier ($S_2$) × sensitivity of ultra violet charge recorder ($S_3$).

$$= 6.8\frac{pC}{\text{bar}} \times 0.0032\frac{V}{pC} \times 16\frac{\text{mm}}{V} = 0.348\frac{\text{mm}}{\text{bar}}$$

Also, $\textit{Sensitivity} = \dfrac{\text{Change of output signal}}{\text{Change of input signal}}$

or, $0.348 = \dfrac{\text{Output}}{20 \text{ bar}}$

∴ *Deflection on the chart* = 0.348 × 20 = **6.96 mm (Ans.)**

### 1.10.9. Linearity

*The ability to reproduce the input characteristics symmetrically is called* **linearity**. It can be expressed by the straight line equation.

*The linearity is simply a measure of maximum deviation of any of the calibration points from the straight line* (drawn by using the method of least square from the given calibration data). Fig. 1.13 shows the actual calibration curve and a straight line drawn from the origin using method of least squares.

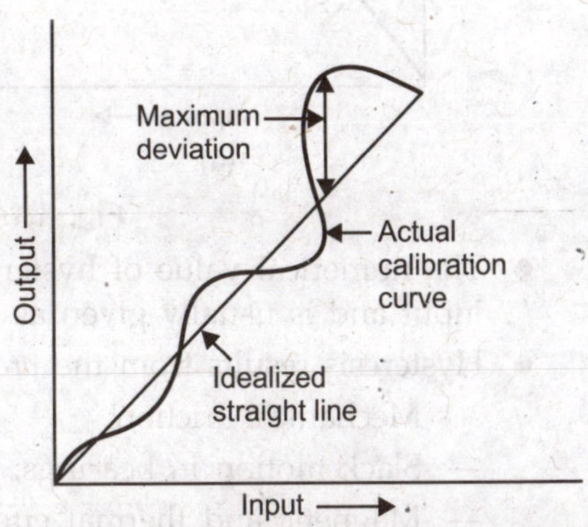


**Fig. 1.13.** Linearity exhibited as deviation between actual calibration curve and idealized straight line relationship.

*Any departure from straight line relationship is non-linearity.* The non-linearity may be due to the following *factors* :

(*i*) Viscous flow or creep;

(*ii*) Non-linear elements in the measurement device;

(*iii*) Mechanical hysteresis;

(*iv*) The elastic after-effects in the mechanical system.

The two common types of non-linearity are :

1. **Terminal linearity.** It is the deviation from a straight line through the end points.
2. **Best fit linearity.** It is the deviation from the straight line which gives minimum errors, both plus and minus.

- **Linearity** (*i.e.*, output is linearly proportional to input) is considered to be one of the best characteristics of an instrument or measurement system. This is so, because the conversion from a scale reading to the corresponding value of input quantity is very convenient. However, it maybe understood that a non-linear behaviour does not essentially lead to inacuracy; an instrument having non-linear calibration curve may be highly accurate as ever.
- Since accuracy and linearity are related closely to each other, it is better to keep the non-linearity as small as possible so as to result in small errors.

### 1.10.10. Hysteresis

*The maximum differences in output at any measured value within the specified range when approaching the point first with increasing and then with decreasing input may be termed as* **hysteresis.**

- It is a phenomenon which shows different output effects when loading and unloading. It is *non-coincidence of loading and unloading curves.*

Figure 1.14(*a*) shows output and input curves (loading and unloading) for an instrument which has no friction due to sliding parts. The non-coincidence of loading and unloading curves is on account of *internal friction or hysteresis damping.*

Figure 1.14(*b*) shows the input-output relations of instruments which do not have internal friction but have external sliding friction.

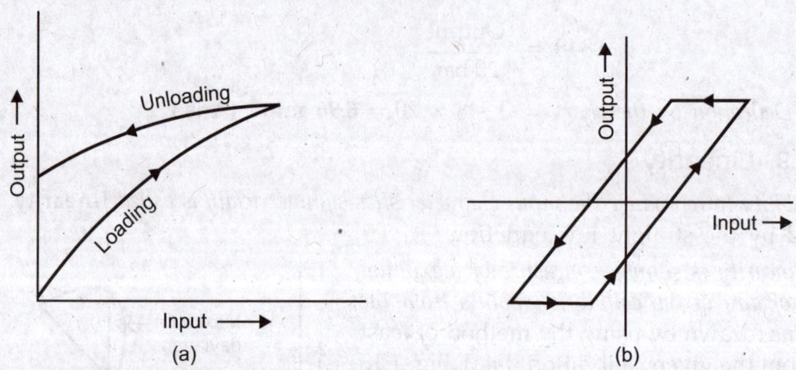


**Fig. 1.14.** Hysteresis effects.

- The numerical value of hysteresis can be specified in terms of either output or input and is usually given as *%age of full scale.*
- Hysteresis results from the *presence of irreversible phenomenon such as :*
  — Mechanical friction;
  — Slack motion in bearings;
  — Magnetic and thermal effects.

### 1.10.11. Threshold and Resolution

**Threshold :**

*The minimum value below which no output change can be detected when the input of an instrument is increased gradually from zero is called the* **threshold** *of the instrument.* Thus threshold defines minimum value of input which is necessary to cause a detectable change from zero output.

— In a *digital system,* threshold is the input signal necessary to cause one least significant digit of the output reading to change.

- Threshold may be caused by *backlash or internal noise.*

**Resolution or Discrimination :**

*When the input is slowly increased from some arbitrary (non-zero) input value, it is observed that the output does not change at all until a certain increment is exceeded; this increment is called* **Resolution or discrimination** *of the instrument.* Thus resolution defines the smallest change of input for which there will be a change of output.

— In case of *analog instruments,* the resolution is determined by the observer's ability to judge the position of a pointer on a scale. Resolution is usually reckoned to be no better than ± 0.2 of the smallest division of the scale.

— In case of digital instruments, resolution is determined by the number of neon tubes taken to show the measured value.

- **Threshold** *defines the smallest measurable input* while the **resolution** *defines the smallest measurable input change.*
- *"Threshold"* and *"resolution"* may be *expressed as an actual value or as a fraction or percentage of full scale value.*

**Example 1.11.** *A transducer measures a range of 0–200 N force with a resolution of 0.15 percent of full scale. What is the smallest change which can be measured by this transducer ?*

**Solution.** *Given :* Range of force : 0 – 200 N

Resolution : 0.15% of full scale

*Smallest change* which can be measured = $\frac{0.15}{100} \times 200$ = **0.3 N** **(Ans.)**

**Example 1.12.** *Determine the resolution of a moving coil voltmeter having a uniform scale with 50 divisions; the full scale reading is 50 V and $\frac{1}{10}$ of a scale division can be estimated with a fair degree of certainty.*

**Solution.** One scale division = $\frac{50}{50}$ = 1 V

∴ *Resolution* = $\frac{1}{10} \times 1$ = **0.1 V** **(Ans.)**

**Example 1.13.** *Determine the resolution of a digital voltmeter which has read-out range 0 to 9999 counts and its full scale reading is 9.999 V.*

**Solution.** *Resolution of the digital voltmeter*

$$= \frac{1}{9999} \times 9.999$$

= 0.001 V or **1 mV** **(Ans.)**

### 1.10.12. Dead Zone and Dead Time

**Dead zone :**

Refer to Fig. 1.15. *The largest change of input quantity for which there is no output of the instrument is termed as* ***dead zone.*** It may occur due to friction in the instrument which does not allow pointer to move till sufficient driving force is developed to overcome the friction loss.

● *"Dead zone" is caused by backlash and hysteresis in the instrument.*

**Dead time :**

Refer to Fig. 1.15. *The time required by a measurement system to begin to respond to a change in the measurand is termed as **dead time.*** It represents the time before the instrument begins to respond after the measured quantity has been changed.

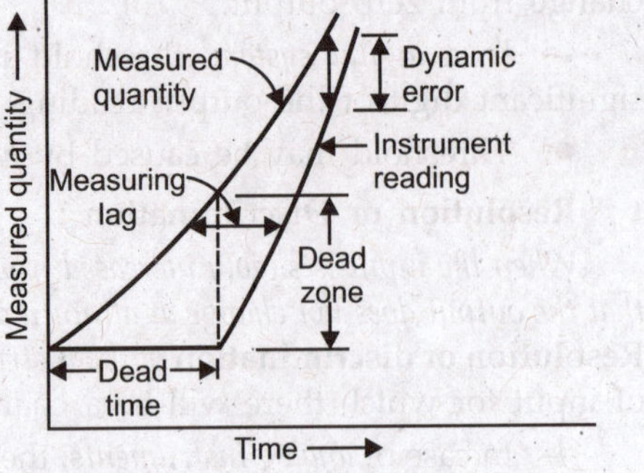


Fig. 1.15. Dead zone and dead time.

**Example 1.14.** *The calibration range of a certain pyrometer is 300°C to 800°C. If the dead zone in it is 0.11 percent of span, determine the temperature change which might occur before it is detected.*

**Solution.** Span of the pyrometer = 800 – 300 = 500°C

$$\therefore \quad \text{Dead zone} = \frac{0.11}{100} \times 500 = 0.55°\text{C}$$

*Hence, a change of 0.55°C must occur before it is detected.* **(Ans.)**

### 1.10.13. Loading Effects

Under ideal conditions an element used for signal sensing, conditioning, transmission and detection should not change/distort the original signal. The sensing element should not use any energy or take least energy from the process so as not to change the parameter being measured. However, under practical condition, it has been observed that the introduction of any element in a system results invariably *in extraction of the energy* from the system thereby *distorting the original signal.* This distortion may take the form of *attenuation, waveform distortion, phase shift etc.* Consequently the ideal measurements become impossible.

*The incapability of the system to faithfully measure the input signal in undistorted form is called the **loading effect.*** This results in *loading errors.*

● The loading effects, in a measurement system, not only occur in *detector-transducer stage* but also occur in *signal conditioning and signal presentation stages* as well. The loading problem is carried right down to the basic elements themselves.

The loading effects may occur on account of both electrical and mechanical elements; these are *due to impedances of the various elements connected in a system.* The mechanical impedances may be treated similar to electrical impedances.

● *Loading effects occur due to extraction of energy from the measurand* (as earlier stated). The transfer of energy requires the specification of two variables. The physical variables which determine the flow of energy in all dynamical systems can be classified as :

(*i*) Through variables; (*ii*) Across variables.

— *"Through variables"*, also called flow variables or pervariables, are those which can be specified and measured at *one point in a space.* These are extensive variables in the sense, that their magnitude depends upon the extent of the system taking part in the energy transfer.

— *"Across variables"*, also called effort variables and transvariables are those which can be specified by *two points in a space;* usually one point is a reference point. These are intensive variables whose magnitude is independent of the medium being considered.

*All objects in a measurement system or any dynamical system involve a measurand or defined relationship between an "across variable" and a "through variable".*

**Static stiffness and static compliance :**

In some instruments when input signal is applied to them under *static conditions*, the output signal is a faithful representation of the input signal. But when these instruments are subjected to *dynamic conditions* (*i.e.*, going from one steady state to another one) they produce loading effects since energy is required to perform these operations. Under such conditions concepts of static stiffness and static compliance (instead of the concepts of impedance and admittance) need to be introduced.

It may be understood that *while impedance and admittance are concerned with* **power drain,** *stiffness and compliance are concerned with* **energy drain.**

**Mechanical Stiffness,** in *general*, is defined as :

$$K_g = \frac{\text{Across variable}}{\int (\text{Through variable})\, dt} \qquad ...(1.11)$$

**Compliance** is defined as *inverse of stiffness* (as admittance is inverse of impedance) is given by

$$C_g = \frac{1}{K_g} = \frac{\text{Through variable}}{\int (\text{Across variable})\, dt} \qquad ...(1.12)$$

**Example 1.15.** *(a) Discuss the concept of impedance loading and impedance matching as applied to a measurement system.*

*(b) An unknown resistance is connected in series with a milliammeter of very small internal resistance. A voltmeter with internal resistance 125 kW is connected across the unknown resistance.*

*Determine the loading error when :*

*(i) The voltmeter and milliammeter read 180 V and 6 mA respectively.*

*(ii) The same voltmeter and milliammeter with another resistance read 60 V and 1.2 A respectively. Comment upon the results.*

**Solution.** (*a*) **Impedance loading and impedance matching :**

In a generalised measurement system, the sensing element senses the condition, state or value of the process variable by *extracting a small part of energy from the measurand*, and then processes an output which reflects this condition, state or value of the measurand. Due to this energy extraction the measurand is always disturbed by the measurement act and consequently a perfect measurement becomes *theoretically impossible.*

*Examples :* (*i*) The reading shown by the hand tachometer would vary with the pressure with which it is pressed against the shaft; (*ii*) A milliammeter would introduce additional resistance in the circuit and thereby change the flow of current by a significant amount.

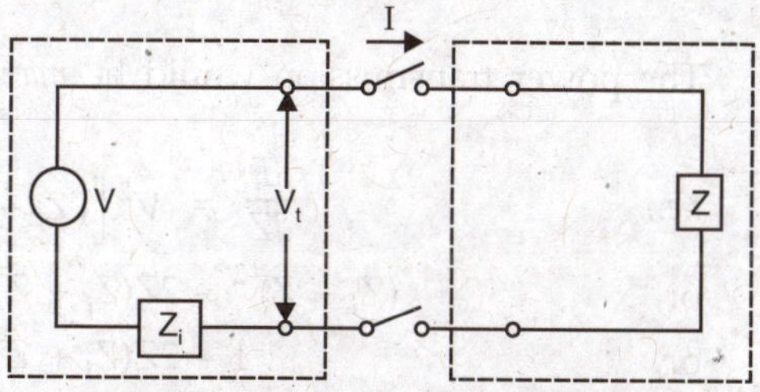


**Fig. 1.16.** Loading effect of recorder on the transducer element.

The *interaction of system* (mentioned above) is referred to as *"loading effect"* and the *term loading when applied to a general electrical system is called* **impedance loading.** While designing an instrument efforts should be made to ensure that the *alteration due to system interaction is minimal.*

Refer to Fig. 1.16. Consider a transducer element with internal impedance, $Z_i$ (= $R_i$ + $jX_i$), $R$ and $X$ being the resistance and reactance respectively.

Let, $V$ = Voltage developed in the transducer, and

$V_t$ = Terminal voltage.

Then, when output terminals are open (*i.e.*, no-load condition),

$$V_t = V \quad ...(i)$$

When the transducer is connected in series with a recorder of input impedance $Z$ (= $R + jX$), a current flows through the circuit, the value of which is given as,

$$I = \frac{V}{Z_i + Z}$$

and,

$$V_t = V - I Z_i = V - \left(\frac{V}{Z_i + Z}\right) Z_i$$

$$= V\left[1 - \frac{Z_i}{Z_i + Z}\right] = \frac{Z}{Z_i + Z} \cdot V$$

*i.e.*,

$$V_t = \frac{Z}{Z_i + Z} \cdot V \quad ...(ii)$$

[In ideal cases, when impedance loading is zero (*i.e.*, $Z_i$ = 0), eqn. (*ii*) transforms to eqn. (*i*)]

From eqn. (*ii*), it may be observed that for the *loading effect to be minimal, the following conditions should be satisfied :*

(*i*) $Z_i$ (transducer impedance) should be *small.*

(*ii*) $Z$ (impedance of the measuring device) must be *high.* (It is due to this reason that vacuum tube voltmeters having an internal resistance to the tune of 10 MΩ are used in the measurement of the voltage).

The *power delivered* from the transducer to the external load,

$$P = \frac{V_t^2}{Z}$$

$$= \left(\frac{Z}{Z_i + Z} \cdot V\right)^2 \cdot \frac{1}{Z} \quad \text{...[Substituting for } V_t \text{ from eqn. } (ii)]$$

$$= V^2 Z (Z_i + Z)^{-2}$$

The power transmission would be *maximum* when, $\frac{dP}{dZ} = 0$

*i.e.*,

$$\frac{dP}{dZ} = V^2\left[(Z_i + Z)^{-2} \times 1 + Z \times (-2)(Z_i + Z)^{-3}\right] = 0$$

or, $$(Z_i + Z)^{-2} - 2Z(Z_i + Z)^{-3} = 0$$

or, $$1 - 2Z(Z_i + Z)^{-1} = 0$$

or, $$(Z_i + Z) - 2Z = 0$$

or, $$Z = Z_i \quad ...(iii)$$

Eqn. (*iii*) indicates that maximum power will be delivered from the transducer when the impedance of the external load just *matches* its internal impedance. This aspect is referred to as ***impedance matching*** *in electrical circuits.*

If $$Z_i = R_i \text{ and } Z = R$$

Then, $$V_t = \frac{R}{R_i + R'}, \text{ and } R = R_i.$$

(*b*) Refer to Fig. 1.17. *Given* : Internal resistance of the voltmeter, $R_v$ = 125 kΩ;

Readings of voltmeter and ammeter in two cases :

(*i*) 180 V, 6 mA; (*ii*) 60 V, 1.2 A

(*i*) Total resistance of the circuit,

$$R_T = \frac{V}{I} = \frac{180}{6 \times 10^{-3}} = 30 \text{ k}\Omega$$

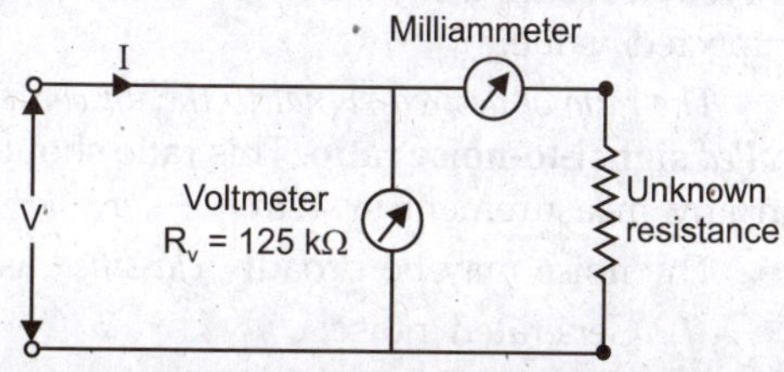


**Fig. 1.17.**

Since the milliammeter is stated to have negligible resistance, the total resistance as calculated above equals the apparent resistance ($R_{appt.}$) of the resistor.

*i.e.,* $$R_{appt.} = 30 \text{ k}\Omega$$

The actual value of resistance of the resistor is obtained from the condition that voltmeter is connected is parallel with the resistor, *i.e.,*

$$\frac{1}{R_T} = \frac{1}{R_{act.}} + \frac{1}{R_v}$$

or, $$\frac{1}{R_{act.}} = \frac{1}{R_T} - \frac{1}{R_v} = \frac{R_v - R_T}{R_T \times R_v}$$

or, $$R_{act.} = \frac{R_T \times R_v}{R_v - R_T} = \frac{30 \times 125}{125 - 30} = 39.474 \text{ k}\Omega$$

∴ $$\textit{Percentage loading error} = \frac{R_{act.} - R_{appt.}}{R_{act.}}$$

$$= \frac{39.474 - 30}{39.474} \times 100 = \mathbf{24\%} \quad \textbf{(Ans.)}$$

(*ii*) $$\text{Total resistance, } R_T = \frac{60}{1.2} = 50 \ \Omega$$

*i.e.,* $$R_{appt.} = 50 \ \Omega$$

$$R_{act.} = \frac{R_T \times R_v}{R_v - R_T} = \frac{50 \times 125}{125 - (50/1000)} = 50.02 \ \Omega$$

∴ $$\textit{Percentage loading error} = \frac{50.02 - 50}{50.02} \times 100 = \mathbf{0.04\%} \quad \textbf{(Ans.)}$$

From the above results it may be concluded that :

- When a voltmeter with high internal resistance is connected across two points in a *high resistance circuit,* the loading effect is appreciable and, therefore, the voltmeter gives *misleading voltage reading.*

- The same voltmeter has a *negligible loading error* when connected across a *low resistance circuit,* and it gives more *reliable voltage reading.*

### 1.10.14. Noise

**Noise** *is a spurious current or voltage extraneous to the current or voltage of interest in an electrical or electronic circuit.* It does not convey any useful information. It is an unwanted signal superimposed upon the signal of interest thereby causing a deviation of the output from its expected value.

*The ratio of desired signal to the unwanted signal is called* **signal-to-noise ratio.** This ratio should be large in any measurement system.

The noise may be broadly *classified* as follows :

(*i*) Generated noise;

(*ii*) Conducted noise;

(*iii*) Radiated noise.

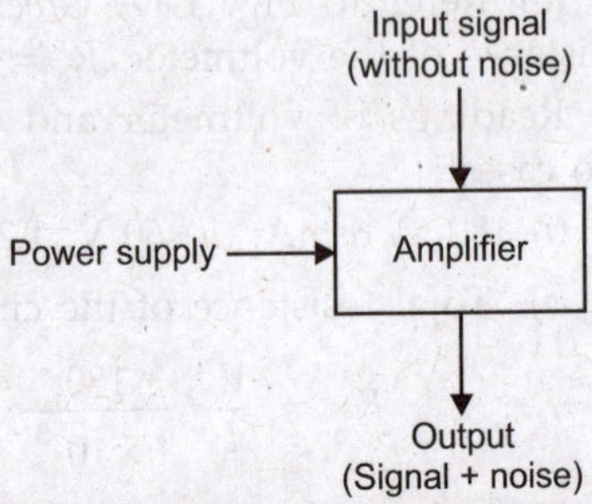


**Fig. 1.18.** Block diagram of an amplifier for illustration of different types of noise signals.

Refer to Fig. 1.18. It shows a block diagram of an amplifier for illustration of different types of noise signals.

**(*i*) Generated noise :**

- The internally generated noise is on account of components like resistors, capacitors, transistors, etc.
- The temperature dependent noise in a resistor which increases with internal heating ($I^2R$ loss) or with an increase in the ambient temperature is called *"Johnson noise"*.
- The wide band noise, produced within a resistor, due to vibrations produced by the thermal effects, covering a wide frequency range is called *"White noise"*.

  The internally generated noise in resistor can be *reduced* by :

  (*a*) lowering the internal temperature;

  (*b*) using special film and glass substrates.

  The internally generated noise in semiconductor and vacuum tube devices on account of random movement of charges is called *"Shot noise"*. It is difficult to reduce this type of noise; however, it can be reduced to some extent by the use of selective filtering.
- The internally generated noise may also be caused by the changing electric fields in the region between plates of a capacitor and changing magnetic fields around in an inductor. In this case noise signal has a fixed frequency (since fields change in orderly manner) and as such its magnitude can be reduced by using a filter tuned to this frequency.

**(*ii*) Conducted noise :**

- This type of noise may be caused by the power supply (to the amplifier) which may have, ripples or random deviations.
- 50 Hz power supply and the harmonics contained in it are one of the common sources for conducted noise.
- It can be reduced by using filters in the leads to trap out the noise.

(*iii*) **Radiated noise :** Radiated noise consists of unwanted signals (relating to electric or magnetic fields or disturbances in the environments around the amplifier) which are radiated into the interior of the amplifier.

- The electromagnetic impulses radiated from ignition wiring of spark plugs are one of the common sources of radiated noise.
- It can be *reduced by proper shielding.*

## 1.11 DYNAMIC CHARACTERISTICS

When an instrument is required to measure a time-varying process variable, one has to be concerned with *"dynamic characteristics", these characteristics quantify the dynamic relation between the input and output.*

### 1.11.1. Dynamic Response

*The evaluation of the ability of a system to faithfully transmit and present all the pertinent information included in the input signal and to exclude all else, is called the* **response.**

*The behaviour of the system when inputs vary with time (i.e., inputs are dynamic in nature) and so does the output, is called* **dynamic response.** Such a response in measurement systems is found to occur in industrial, aerospace and biological applications.

The dynamic inputs are of two types :

(*i*) Steady state periodic; (*ii*) Transient.

The *"steady state periodic"* quantity is one whose magnitude has a definite repeating time, whereas the time variation of a ***"transient"*** magnitude does not repeat.

- The response of a measurement system subjected to a time varying input can be divided into the following two parts :

  (*i*) Steady state response; (*ii*) Transient response.

  ***"Steady state response"*** is simply the response when time reaches infinity.

  ***"Transient response"*** in measurement systems, is defined as the part of response which goes to zero as time becomes large.

- **Periodically varying inputs may be :**

  — A *purely sinusoidal input* like the A.C. voltage with a frequency of 50 Hz.

  or

  — A complex time varying repetitive input which is a *combination of the fundamental* and *harmonics.*

When the systems are subjected to periodically varying inputs, they exhibit in their response a magnitude and phase relationship which is *different* from that of the input signal because of the *energy storage elements. The output is not a faithful representation of the input.* The harmonic contents of the input are not reproduced in the output both in magnitude and phase relationship as in the input. *The output is distorted both in terms of magnitude as well as phase relationships.*

### 1.11.2. Dynamic Characteristics of a Measurement System

The *dynamic characteristics* of a measurement system are :

1. Speed of response ... *Desirable*
2. Measuring lag ... *Undesirable*
3. Fidelity ... *Desirable*
4. Dynamic error ... *Undesirable*

**1. Speed of response.** The *"speed of response"* or *"responsiveness"* is defined as the *rapidity with which a measurement system responds to changes in the measured quantity.*

**2. Measuring lag.** It refers to *retardation or delay in the response of a measurement system to changes in measured quantity.*

The lag *is caused by conditions such as capacitance, inertia, or resistance.*

The measuring lags are of the following two types :

(i) ***Retardation type lag.*** In this type of measuring lag the response of the measurement system *begins immediately after a change in measured quantity has occurred.*

(ii) ***Time delay type lag.*** In this case the response of the measurement system *begins after a dead time after the application of the input.* If the measured quantity varies at a fast rate, the dead time has a severe adverse effect on the performance of the system. The measurement lags of this type are *very small* and are of the order of a fraction of a second and hence can be *ignored.*

**3. Fidelity.** It is defined as the *degree to which a measurement system indicates changes in the measured quantity without any dynamic error.* It refers to the ability of the system to reproduce the output in the same form as the input.

**4. Dynamic error.** The dynamic error, also called **"measurement error"**, is the *difference between the true value of the quantity changing with time and the value indicated by the measurement system if no static error is assumed* (see Fig. 1.15.)

- *The maximum amount by which the pointer moves beyond the steady state is called "overshoot".*

### 1.11.3. 'Dynamic' Analysis of Measurement Systems

The dynamic behaviour of measurement systems is studied in the following two domains :

1. Time domain analysis.
2. Frequency domain analysis.

**1. Time domain analysis :**

In this the input signal is applied to the measurement system and the behaviour of the system is studied as a *function of time.* The dynamic response of the system to different types of inputs, which are a function of time is analysed at different intervals of time after the application of the input signal. In most cases, the actual input signals vary in random fashion with respect to time and therefore cannot be mathematically defined. Consequently the performance of a system can be analysed (in the time domain analysis) by using the following standard *test signals/inputs* :

(i) Step input;
(ii) Ramp input;
(iii) Parabolic input;
(iv) Impulse input.

**2. Frequency domain analysis :**

This type of analysis of a system *pertains to the steady state response of the system to a sinusoidal input.* Here, the system is subjected to a sinusoidal input and the system response is studied with *frequency as the independent variable.*

- *Frequency response.* It is the maximum frequency of the measured variable that an instrument is capable of following without error. The usual requirement is that the frequency of measurand should not exceed 60 per cent of the natural frequency of the measuring instrument.

**Standard test signals/inputs :**

The most common standard inputs used for dynamic analysis are discussed below :

1. ***Step function :*** Refer to Fig. 1.19(*a*). It is a *sudden change from one steady value to another.*

It is mathematically represented by the relationship :

$$x = 0 \quad \text{at} \quad t < 0$$
$$x = x_c \quad \text{at} \quad t \geq 0$$

where $x_c$ is a constant value of the input signal $x_i$.

- The *"transient response" indicates the capacity of the system to cope with change in the input signal.*

2. ***Ramp or linear function :*** Refer to Fig. 1.19(*b*). In this case, the *input varies linearly with time.*

This input is mathematically represented as :

$$x = 0 \quad \text{at} \quad t < 0$$
$$x = \psi \quad \text{at} \quad t \geq 0$$

where, $\psi$ is the slope of the input versus time relationship.

- *The ramp-response becomes indicative of the steady state error in following the changes in the input signal.*

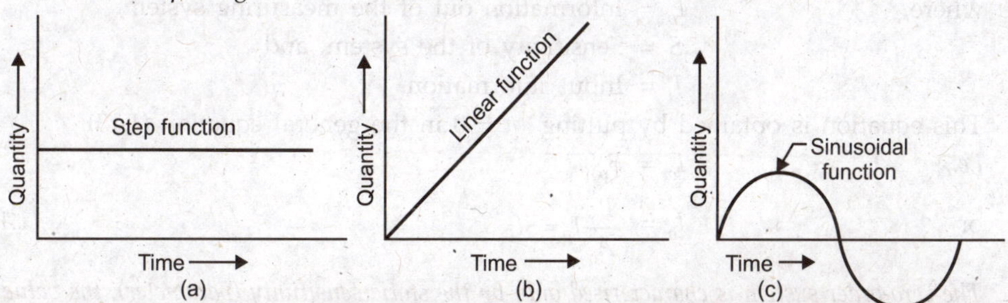


**Fig. 1.19.** Standard input functions.

3. ***Sinusoidal function :*** Refer to Fig. 1.19(*c*). In this case the *input varies sinusoidally with a constant maximum amplitude.*

$$I_i = A \sin \omega t$$

where, $A$ = Amplitude, and

$\omega$ = Frequency in rad/s.

- *The frequency or harmonic response is a measure of the capability of the system to respond to inputs of cyclic nature.*

A general measurement system can be mathematically described by the following differential equation:

$$(A_n D^n + A_{n-1} D^{n-1} + .... + A_1 D + A_0) I_o = (B_m D^m + B_{m-1} D^{m-1} + .... + B_1 D + B_0) I_i \quad ...(1.13)$$

where, $A$'s and $B$'s = Constants, depending upon the physical parameters of the system,

$D^k$ = Operative derivative of the order $k$,

$I_o$ = The information out of the measurement system, and

$I_i$ = The input information.

The order of the measurement system is generally classified by the value of the power of $n$.

- *Zero-order system* : $n = 0$ and $A_1, A_2, A_3 .... A_n = 0$
- *First-order system* : $n = 1$ and $A_2, A_3, A_4 .... A_n = 0$
- *Second-order system* : $n = 2$ and $A_3, A_4, A_5 .... A_n = 0$

The above method of classification is used for most of the instruments and systems.

Although general equation can be solved by various methods, we shall be using methods of $D$-operator for getting its solution.

### 1.11.4. Zero, First and Second Order Systems

#### 1.11.4.1. Zero-order systems

Figure 1.20, shows the block diagram of a 'Zero-order system'. In this case the output of the measuring system (ideal) is directly proportional to input, no matter how the input varies. The *output is faithful reproduction of input without any distortion or time lag.*

Fig. 1.20. Block diagram for Zero-order system.

The behaviour of the zero-order system is represented by the following mathematical solution :

$$I_o = S\, I_i \quad ....(1.14)$$

where, $I_o$ = Information out of the measuring system,

$S$ = Sensitivity of the system, and

$I_i$ = Input information.

This equation is obtained by putting $n = 0$ in the general equation (1.13),

*i.e.*, $$A_o I_o = B_o I_i$$

or $$I_o = \frac{B_o}{A_o} I_i = S I_i \quad ...(1.15)$$

*The zero-order system is characterised only by the static sensitivity (parameter), the value of which is obtained through the process of static calibration.*

***Examples of zero-order system :***

- Mechanical levers;
- Amplifiers;
- Potentiometer (It gives an output voltage which is proportional to wiper's displacement) etc.

#### 1.11.4.2. First-order systems

Figure 1.21, shows the block diagram of a 'First-order system'.

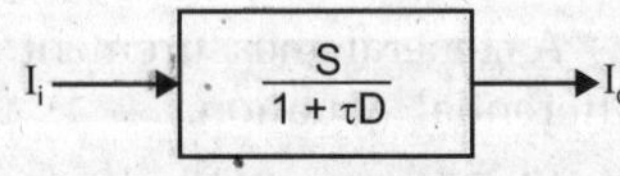


Fig. 1.21. Block diagram for First-order system.

The behaviour of a first-order system is given by following first-order differential equation :

$$A_1 \frac{dI_o}{dt} + A_o I_o = B_o I_i \quad ...(1.16)$$

(This equation is obtained by inserting $n = 1$ in the general equation).

Eqn. (1.16) may be written in standard form as follows:

$$\frac{A_1}{A_o}\frac{dI_o}{dt} + I_o = \frac{B_o}{A_o} I_i \qquad ...(1.17)$$

or,
$$\tau \frac{dI_o}{dt} + I_o = SI_i \qquad ...(1.18)$$

where,
$$\tau = \frac{A_1}{A_0} = \text{Time constant, and}$$

$$S = \frac{B_o}{A_o} = \text{Sensitivity.}$$

Using D-operator, we get

$$\left[\text{where, } D = \frac{d}{dt}, \text{ and } D^2 = \frac{d^2}{dt^2}\right]$$

$$\tau DI_o + I_o = SI_i$$

or,
$$I_o(\tau D + 1) = SI_i$$

or,
$$\frac{I_o}{I_i} = \frac{S}{I + \tau D} \qquad ...(1.19)$$

Equation (1.19) gives the standard from of *transfer operator* for first-order system.

***Examples of first-order system :***

- Velocity of a true falling mass;
- Air pressure build-up in bellow;
- Measurement of temperature by mercury-in-glass thermometers;
- Thermistors and thermocouples;
- Resistance-capacitance network.

**1.11.4.3. Second-order system**

Fig. 1.22 shows the block diagram of 'Second-order system':

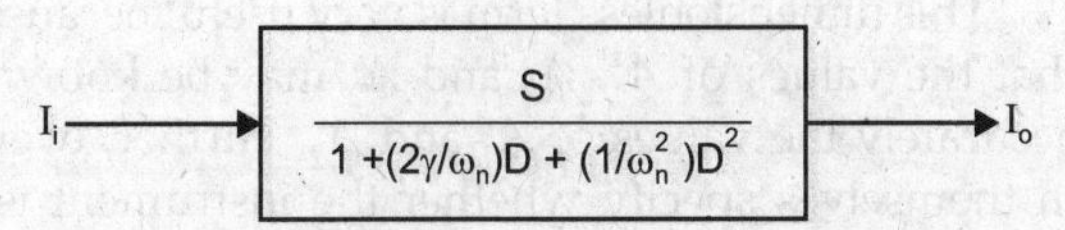


**Fig. 1.22.** Block diagram for Second-order system.

The behaviour of a second-order system is given by the following differential equation (obtained by putting $n = 2$ in the general equation);

$$A_2 \frac{d^2 I_o}{dt^2} + A_1 \frac{dI_o}{dt} + A_o I_o = B_o I_i \qquad ...(1.20)$$

Dividing the above equation by $A_o$, we have

$$\frac{A_2}{A_o}\frac{d^2 I_o}{dt^2} + \frac{A_1}{A_0}\frac{dI_o}{dt} + I_o = \frac{B_o}{A_o} I_i \qquad ...(1.21)$$

Let,
$$\omega_n = \sqrt{\frac{A_o}{A_2}} = \text{Undamped natural frequency, rad/s.}$$

$$\gamma = \frac{A_1}{2\sqrt{A_0 A_2}} = \text{Damping ratio, dimensionless, and}$$

$$S = \frac{B_o}{A_o} = \text{Static sensitivity or steady-state gain.}$$

Then, by substituting these values in eqn. 1.21, we get

$$\frac{I}{\omega_n^2} \cdot \frac{d^2 I_o}{dt^2} + \frac{2\gamma}{\omega_n} \cdot \frac{dI_o}{dt} + I_o = SI_i \quad \text{...(1.22)}$$

or, in terms of D-operator, we have

$$\left( \frac{D^2}{\omega_n^2} + \frac{2\gamma}{\omega_n} D + 1 \right) I_o = SI_i$$

or,

$$\frac{I_o}{I_i} = \frac{S}{\frac{1}{\omega_n^2} D^2 + \frac{2\gamma}{\omega_n} D + 1} \quad \text{...(1.23)}$$

***Examples of second-order system :***

- Piezoelectric pickup;
- Spring-mass system (used for acceleration and force measurements);
- Pen control system on X-Y plotters;
- U.V. galvanometer, etc.

**Damping ratio :**

In the design of instruments a term which is very frequently used is the "*damping ratio*" ($\gamma$) defined as the *ratio of the actual value of coefficient of viscous friction in movement and the value required to produce critical damping.*

*i.e.,*

$$\gamma = \frac{A_1}{2\sqrt{A_o A_2}}$$

This dimensionless term is very useful because to determine its value, it is not necessary that the values of $A_1$, $A_0$ and $A_2$ may be known. In practice it is not easy to determine accurately the values of $A_1$ and $A_2$. Further, even if these values are known, they do not in themselves specify whether the instrument is under, over or critically damped, since a numerical calculation has to be performed with them first. Therefore, designers find "*damping ratio*" as a very convenient measure of the amount of damping present in the movement.

The terms *damping ratio* ($\gamma$) and *underdamped natural frequency* ($\omega_n$) immediately conjure up a physical picture of the response of an instrument and both of the quantities are very easy to measure. Thus $\gamma$ and $\omega_n$ easily do away with quantities $A_2$, $A_1$ and $A_0$.

### 1.11.5. First-order System Responses

The complete solution of an equation which describes the dynamical behaviour of a system consists of the following *two parts :*

(*i*) **Complementary function.** *It corresponds to the short time or transient response.*

(*ii*) **Particular integral.** *It refers to the long time steady state response.*

The transfer operator form of the first-order system is given by :

$$\frac{I_o}{I_i} = \frac{S}{I + \tau D}$$

When $S$ (static sensitivity or steady state gain) equals *unity*, we get

$$(I + \tau D)I_o = I_i \qquad ...(1.24)$$

Now we shall obtain the solution of this equation for different standard inputs (The solutions are not mathematically rigorous, but are practical).

***Transient response (Complementary function) :***

The transient response from the auxiliary equation is obtained by putting input $I_i$ equal to zero :

*i.e.,* $$(1 + \tau D)I_{o,t} = 0 \qquad ...(1.25)$$

(subscript $t$ refers to the transient value)

Let the solution be of the form :

$$I_{o,\,t} = A\,e^{mt}$$

where, $m$ is an algebraic variable.

or, $$(1 + \tau D)A\,e^{mt} = 0$$

or, $$A\,e^{mt} + \tau \cdot \frac{d}{dt}(A\,e^{mt}) = 0$$

or, $$A\,e^{mt} + \tau \cdot A\,m\,e^{mt} = 0$$

$$A\,e^{mt}\,(1 + \tau \cdot m) = 0$$

$$\therefore \quad m = -\frac{1}{\tau}$$

Then, $$I_{o,\,t} = A\,e^{mt} = Ae^{-t/\tau} \qquad ...(1.26)$$

The transient response of a first-order system is same for different standard inputs.

***Steady state response (Particular integral) :***

The steady state response is given by

$$(1 + \tau D)I_{o,s} = I_i \qquad ...(1.27)$$

(Subscript $s$ refers to the steady state value)

or, $$I_{o,s} = (1 + \tau D)^{-1}\,I_i$$

$$= (1 - \tau D + \text{terms in } D^2 \text{ and higher})\,I_i \qquad ...(1.28)$$

**1. Step input :**

Since the input $I_i$ is a step of constant magnitude; its differential equals zero, and subsequently, we get

$$I_{o,s} = (1 - \tau D)I_i = I_i \qquad ...(1.29)$$

Total response = Transient response + steady state response

or $$I_o = Ae^{-t/\tau} + SI_i \qquad ...(1.30)$$

The constant $A$ is evaluated from the initial conditions as follows :

At $t = 0$, $I_o = 0$

$\therefore$ $0 = A + SI_i$ or, $A = -SI_i$

$$\therefore \quad I_o = \underbrace{-I_i\,e^{-t/\tau}}_{\text{Transient}} + \underbrace{I_i}_{\text{Steady state}}$$

or, $$I_o = I_i(1 - e^{-t/\tau}) \qquad ...(1.31)$$

or, $$\frac{I_o}{I_i} = (1 - e^{-t/\tau}) \qquad ...(1.32)$$

... in *non-dimensional form.*

**Salient features (with step input) :**

Following are the salient features of first-order system with step input :

(*i*) The transient response of the first-order system is *time dependent;* as the time passes, its value decreases (Refer to eqn. 1.32) and after a very long time the value becomes zero approximately. Thus magnitude of output ($I_o$) will be same as input ($I_i$) when the time is very large.

(*ii*) The speed of response relates to the time constant $\tau$. A large $\tau$ indicates that response of the system is slow, whereas a small $\tau$ represents a fast system response. Thus in order to get good fidelty (*i.e.,* for accurate dynamic measurements) efforts should be made to minimise the value of $\tau$.

(*iii*) Refer to Fig. 1.23, which shows the time response of a first-order system to a step-input; when $t = \tau$, $\frac{I_o}{I_i} = (1 - e^{-1}) = 0.632$. Thus, the *time constant* ($\tau$), *for a rising exponential function, is defined as the time to reach* 63.2% *of its steady state value*. The time constant, for a *decaying function* would correspond to the time taken to fall to *36.8% of its initial value.*

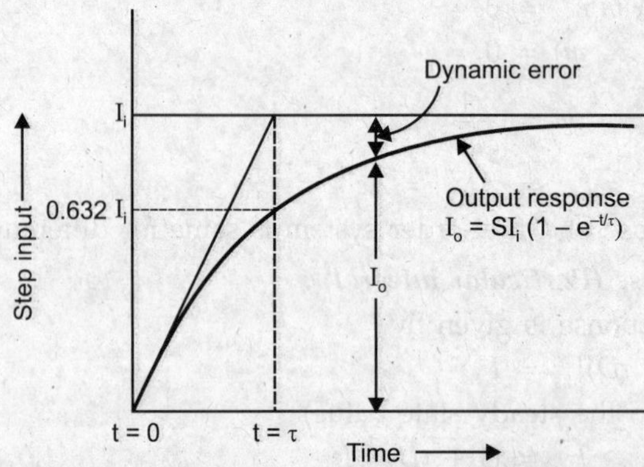


**Fig. 1.23.** Time response of a first-order system to step input.

(*iv*) Dynamic error (*i.e.,* vertical difference between the input and output response curve),

$$\begin{aligned} E_{dy.} &= I_i - I_o \\ &= I_i - I_i(1 - e^{-t/\tau}) \\ &= -I_i\, e^{-t/\tau} \qquad \text{...(1.33)} \end{aligned}$$

or,
$$\frac{E_{dy.}}{I_i} = -e^{-t/\tau} \qquad \text{...(1.34)}$$

... (*in dimensionless form*)

(*v*) In case the measurand has an initial value of $I_{\text{initial}}$ at $t = 0$, then the output $I_o$ at any instant $t$ is given by

$$I_o = I_i[1 - e^{-t/\tau}] + I_{\text{initial}}\, e^{-t/\tau}$$

or,
$$I_o = I_i + (I_{\text{initial}} - I_i)\, e^{-t/\tau} \qquad \text{...(1.35)}$$

(*vi*) The speed response of a system is defined in terms of *settling time* (it is time taken by the system to reach and remain within a certain percentage tolerance band of the final state value). Smaller the settling time, faster is the response. Typical values of tolerance band are 2% and 5% settling times.

For unit step input, the output reaches the values given in Table 1.1 at various intervals of time. This table also shows the per unit dynamic error at different intervals of time.

**Table 1.1. Per Unit Output and Dynamic Error for a Step Input for a First-order System**

| $t/\tau$ | Per unit output $(I_o/I_i)$ | Per unit dynamic error $(E_{dy}/I_i)$ |
|---|---|---|
| 0 | 0 | 1.000 |
| 1 | 0.632 | 0.368 |
| 2 | 0.865 | 0.135 |
| 3 | 0.950 | 0.050 |
| 4 | 0.982 | 0.018 |
| 5 | 0.993 | 0.007 |
| ∞ | 1.000 | 0 |

Thus, 5% settling time means that the system has reached its specified value after a time which is thrice the time constant (Table 1.1).

**2. Ramp Input :**

Consider that the input signal varies linearly with time so that ramp input $I_i$ is prescribed by the relation $I_i + \psi t$, where $\psi$ is constant. Then the governing differential equation is given by

$$(1 + \tau D)I_o = \psi t \qquad ...(1.36)$$

Now, *Transient response* (complimentary function) is given as:

$$I_{o,t} = A\, e^{-t/\tau} \qquad \text{... as before}$$

and, *steady state response* (particular integral) will be;

$$I_{o,s} = (1 - \tau D + \text{terms in } D^2 \text{ and higher})\, \psi t = \psi t - \tau \frac{d}{dt}(\psi \tau)$$

or,
$$I_{o,s} = \psi t - \psi \tau \qquad ...(1.37)$$

∴ Complete response = Transient response + steady state response

$$I_o = A\, e^{-t/\tau} + (\psi t - \psi \tau) \qquad ...(1.38)$$

The value of constant $A$ can be evaluated by applying the initial condition,

At, $t = 0 \qquad I_0 = 0$

or, $0 = A - \psi\tau \quad \therefore \quad A = \psi\tau$

$$\therefore \quad I_o = \psi t - \psi\tau + \psi\tau \times e^{-t/\tau} = \psi(t - \tau) + \psi\tau\, e^{-t/\tau} \qquad ...(1.39)$$

or,
$$I_o = \psi[t - \tau(1 - e^{-t/\tau})] \qquad ...(1.39a)$$

Fig. 1.24 shows the time response of a first-order system to a ramp input.

*The dynamic error,*

$$E_{dy.} = I_i - I_o$$
$$= \psi t - [\psi t - \psi\tau + \psi\tau\, e^{-t/\tau}]$$
$$= \underbrace{\psi\tau}_{\text{steady}} \; \underbrace{- \psi\tau\, e^{-t/\tau}}_{\text{Transient}} \qquad ...(1.40)$$

or,
$$\frac{E_{dy.}}{\psi\tau} = 1 - e^{-t/\tau} \qquad ...(1.41)$$

... (in dimensionless form)

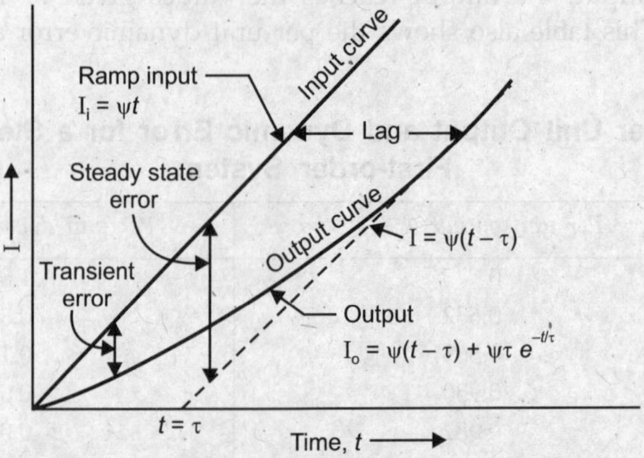


Fig. 1.24. Time response of a first-order system to a ramp input.

*Salient features (with ramp input ) :*

(*i*) The term $\psi\tau$ being independent of time continues to exist and so it is called the steady state error. The term $\psi\tau\, e^{-t/\tau}$ gradually decreases with time and hence is called the *transient error.*

— Since the steady state error is directly proportional to $\tau$ (time constant), therefore, the larger the value of $\tau$ the larger will be the magnitude of the error.

— When $\tau$ is made small the transient error decreases rapidly; this implies, that the system attains the steady state at a faster pace.

(*ii*) The output response curve always lags behind the input curve by a constant amount known as **lag.**

**3. Sinusoidal (Harmonic) input :**

The frequency analysis of a system pertains to the steady state response of the system to a sinusoidal input. In this analysis, the system is subjected to a sinusoidal input and the system response is studied with frequency as the independent variable. The sinusoid is a unique input signal, and the resulting output signal for a linear system is sinusoidal in the steady state. However, the output signal differs from the input waveform in amplitude and phase.

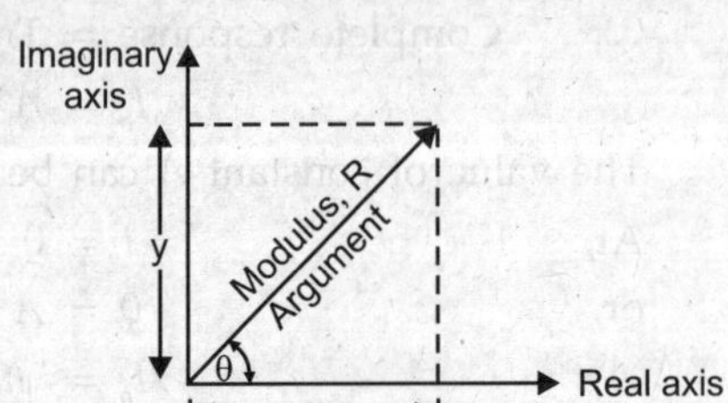


Fig. 1.25.

In order to determine the frequency response of sinusoidal input to a first-order system, let us replace the transfer operator $D$ by a factor $j\omega$ in the input/output relationship; then we get,

$$\frac{I_o}{I_i} = \frac{1}{1+D\tau} = \frac{1}{1+j\omega\tau} \qquad ...(1.42)$$

where, $\omega$ = Input frequency, rad/s, and

$j = \sqrt{(-1)}$.

In a frequency response the following two quantities are of interest : Refer to Fig. 1.25.

(*i*) *Amplitude ratio or modulus* $\left(\frac{I_o}{I_i}\right)$. It prescribes the size of the output amplitude relative to the input amplitude.

(*ii*) *Phase shift* of output relative to input.

For the first-order system represented by the equation (1.42),

$$\text{Modulus} = \sqrt{1+(\omega\tau)^2}$$

$$\text{Argument/Phase angle} = \tan^{-1}(\omega\tau) \quad ...(1.43)$$

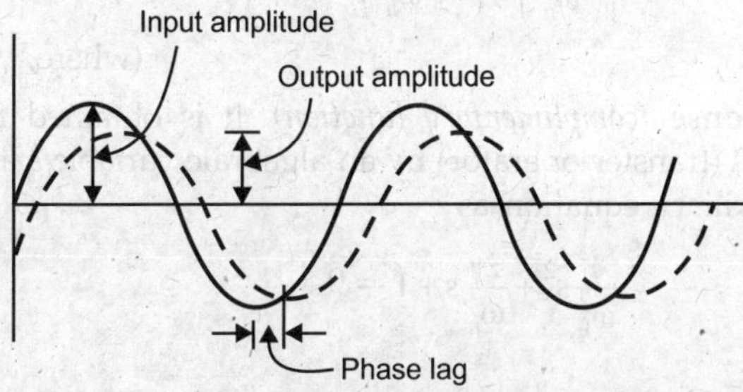


Fig. 1.26. Relationship between an input frequency and corresponding output frequency.

***Salient features (with sinusoidal input)***: Refer to Fig. 1.26.

(*i*) When a system is subjected to a sinusoidal input with frequency ω, its output will also be sinusoidal, but the magnitude of the output amplitude necessarily may not be the same (as the input one). *The ratio of the amplitude* (often called *attenuation*) is given as :

$$\frac{I_o}{I_i} = \frac{1}{\sqrt{1+(\omega\tau)^2}} \quad ...(1.44)$$

*Thus, with the increase in input frequency, the amplitude ratio decreases.*

(*ii*) The output from the system may not necessarily be in phase with the input; and the phase difference is given by

$$\phi \text{ (phase angle)} = -\tan^{-1}(\omega\tau) \quad ...(1.45)$$

–ve indicates that output *lags* behind the input. When $\omega = \frac{1}{\tau}$ the phase lag is $\frac{\pi}{4}$ or 45°.

As the accuracy of an instrument measuring dynamic input depends upon the time constant, therefore, smaller the time constant, greater the accuracy; for phase shift to be small, the time period τ should be small.

(*iii*) When the input and output signals are given by the relations :

$$I_i = A \sin \omega t, \text{ and } I_o = B \sin(\omega t + \phi) = zA \sin(\omega t + \phi),$$

Then the amplitude ratio may be represented as follows :

$$K = \left|\frac{I_o}{I_i}\right| = \frac{z}{\sqrt{1+(\omega\tau)^2}} \quad ...(1.46)$$

In order to produce amplitude of sinewave without any attenuation ($K = 1$) we must use an instrument whose time constant, $\tau = \frac{\sqrt{z^2-1}}{\omega}$.

### 1.11.6. Second-order System Responses

In case of typical second-order system having unit static sensitivity, the homogeneous equation is given by :

$$\frac{I_o}{I_i} = \frac{1}{\left(\frac{1}{\omega_n^2}\right)D^2 + \left(\frac{2\gamma}{\omega_n}\right)D + 1}$$

or, $$\left[\left(\frac{1}{\omega_n^2}\right)D^2 + \left(\frac{2\gamma}{\omega_n}\right)D + 1\right]I_o = I_i \quad \text{...(1.47)}$$

(where, $\gamma$ = damping ratio)

**(a) Transient response** *(complimentary function).* It is obtained from the auxiliary equation by replacing $D$ (transfer operator) by an algebraic variable $s$ and putting $I_i$ equal to zero; we get the auxiliary equation as :

$$\frac{1}{\omega_n^2}s^2 + \frac{2\gamma}{\omega_n}s + 1 = 0$$

The roots are :

$$s_1, s_2 = \frac{\frac{-2\gamma}{\omega_n} \pm \sqrt{\left(\frac{2\gamma}{\omega_n}\right)^2 - \frac{4}{\omega_n^2}}}{\frac{2}{\omega_n^2}}$$

$$= -\gamma\omega_n \pm \sqrt{\gamma^2\omega_n^2 - \omega_n^2}$$

or, $$s_1, s_2 = -\gamma\omega_n \pm \omega_n\sqrt{\gamma^2 - 1} \quad \text{...(1.48)}$$

The *transient solution* has the accepted form,

$$I_{o,t} = A\,e^{s_1 t} + B\,e^{s_2 t}$$

where, $A$ and $B$ = Arbitrary constants to be determined from initial conditions, and

$s_1, s_2$ = Roots of the auxiliary equation (The roots may be real and different, real and equal or imaginary and *that determines* the *nature of transient response of the system*).

The response of the system is of the following three types depending upon the roots of the characteristic equations :

(*i*) Overdamped systems.
(*ii*) Critically damped. systems.
(*iii*) Under-damped systems.

**(*i*) *Over-damped systems.*** In this case $\gamma > 1$ and the roots are *real and unequal.*

- There is *heavy damping* and the system responds to the final steady-state value without any oscillations but in a *sluggish manner.*
- *No overshoot* in step response and no *"resonance"* (resonance refers to the output signal *greater* in magnitude than the ideal output) in the *frequency response.*
- The overdamped systems, owing to their sluggish response, are usually unsuitable for several control applications.

**(*ii*) *Critically-damped systems.*** In this case $\gamma = 1$ and the roots are *real and equal.*

- The system has a *quick and smooth response to the final steady state value without any oscillations.*
- *No overshoot in the step response and no resonance in the frequency response.*

***(iii) Under-damped systems.*** For an undamped system $\gamma < 1$ and roots of the characteristic equation are a complex conjugate pair; these are given as

$$s_1, s_2 = -\gamma\omega_n \pm j\omega_n\sqrt{(1-\gamma^2)}$$

$$= -\gamma\omega_n \pm j\omega_d \qquad ...(1.49)$$

where, $\omega_d = \omega_n\sqrt{(1-\gamma^2)}$ ; this quantity is called *"damped natural frequency"* of the system (This is the frequency at which the damped system freely oscillates when disturbed).

- Such systems *take a long time to reach steady state, but have quick initial response.*
- In these systems, there are oscillations in the step response and resonance effects in the frequency response for values of $\gamma > 0.707$.
- Majority of instruments and control are generally underdamped (*light damping*).

**(*b*) The steady state response** (*particular integral*). It is given by :

$$\left(\frac{1}{\omega_n^2}D^2 + \frac{2\gamma}{\omega_n}D + 1\right)I_{o,s} = I_i$$

or,

$$I_{o,s} = \left(1 + \frac{2\gamma}{\omega_n}D + \frac{1}{\omega_n^2}D^2\right)^{-1} I_i$$

$$= \left(1 - \frac{2\gamma}{\omega_n}D + \text{terms in } D^2 \text{and higher}\right)I_i \qquad ...(1.50)$$

**1. Step input :** Fig. 1.27 shows the time response of a second-order system to a step input.

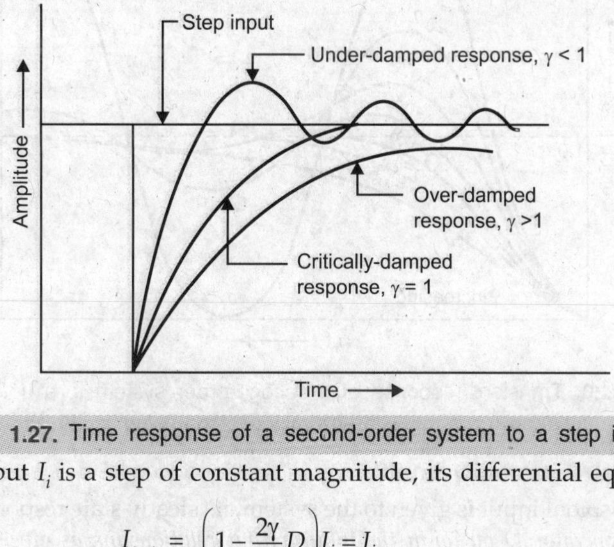


**Fig. 1.27.** Time response of a second-order system to a step input.

Since the input $I_i$ is a step of constant magnitude, its differential equals zero and we have,

$$I_{o,s} = \left(1 - \frac{2\gamma}{\omega_n}D\right)I_i = I_i$$

∴ The complete response

$$I_o = I_{o,s} + I_{o,t}$$

or, $$I_o = I_i A e^{s_1 t} + B e^{s_2 t} \qquad ...(1.51)$$

For the under-damped system, the complex conjugate pair of roots are given by

$$s_1, s_2 = -\gamma\omega_n \pm j\omega_d$$

∴ $$I_o = I_i + Ae^{-(\gamma\omega_n + j\omega_d)t} + Be^{-(\gamma\omega_n - j\omega_d)t} \qquad ...(1.52)$$

Replacing the complex exponentials by sines and cosines, we get

$$I_o = I_i + e^{-\gamma\omega_n t}(A \cos w_d t + B \sin w_d t) \qquad ...(1.53)$$

By applying the initial conditions,

At $t = 0$, $\quad I_o = 0$ and $\frac{dI_o}{dt} = 0,$

we get the values of the constants $A$ and $B$ as :

$$A = -I_i, \text{ and}$$

$$B = \frac{-I_i \gamma}{\sqrt{1-\gamma^2}}$$

Inserting these values in equation (1.53), we have

$$I_o = I_i\left[1 - e^{-\gamma\omega_n t}\left\{\cos \omega_d t + \frac{\gamma}{\sqrt{1-\gamma^2}} \sin \omega_d t\right\}\right] \qquad ...(1.54)$$

Fig. 1.28 shows the transient response of a second-order system to a unit step function for different values of damping factor γ; *the curves indicate the overshoot and oscillations increase with a reduced damping in the system.*

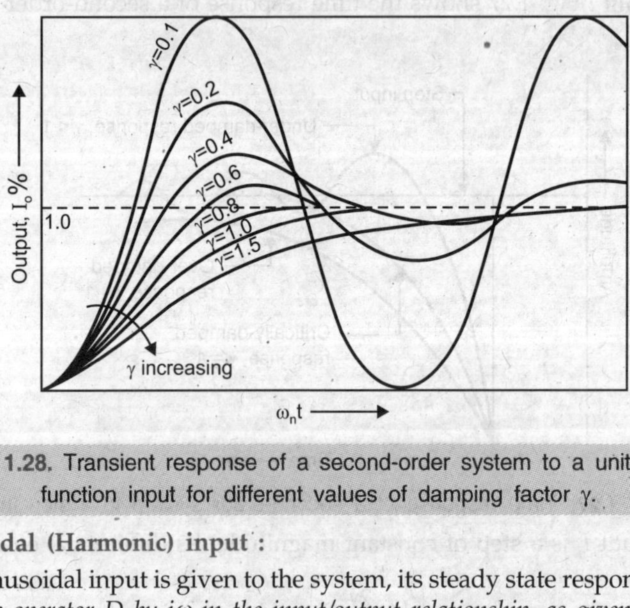


**Fig. 1.28.** Transient response of a second-order system to a unit step function input for different values of damping factor γ.

**2. Sinusoidal (Harmonic) input :**

When a sinusoidal input is given to the system, its steady state response is determined by replacing *the operator D by jω in the input/output relationship, as given below :*

$$\frac{I_o}{I_i} = \frac{1}{\frac{1}{\omega_n^2}D^2 + \frac{2\gamma}{\omega_n}D + 1} = \frac{\omega_n^2}{D^2 + 2\gamma\omega_n D + \omega_n^2}$$

$$= \frac{\omega_n^2}{(j\omega)^2 + 2\gamma\omega_n(j\omega) + \omega_n^2} = \frac{\omega_n^2}{(\omega_n^2 - \omega^2) + j(2\gamma\omega_n\omega)} \quad ...(1.55)$$

where, $\omega$ = Input frequency in rad/s, and

$j = \sqrt{-1}$

The denominator is a complex number having :

$$\text{Modulus} = \sqrt{\left[\left(\omega_n^2 - \omega^2\right)^2 + (2\gamma\omega_n\omega)^2\right]}$$

$$\text{Argument} = \tan^{-1}\left\{\frac{2\gamma\omega_n\omega}{\omega_n^2 - \omega^2}\right\} \quad ...(1.56)$$

$\therefore$ The amplitude ratio,

$$\left|\frac{I_o}{I_i}\right| = \frac{\omega_n^2}{\sqrt{\left[\left(\omega_n^2 - \omega^2\right)^2 + (2\gamma\omega_n\omega)^2\right]}} \quad ...(1.57)$$

and, the system phase lag,

$$\phi = \tan^{-1}\left\{\frac{2\gamma\omega_n\omega}{\omega_n^2 - \omega^2}\right\} \quad ...(1.58)$$

...[using eqn. (1.56)

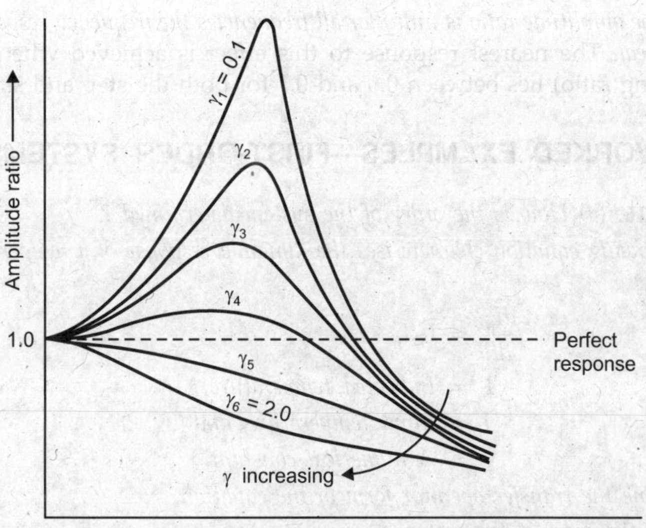


**Fig. 1.29.** Step response of a second-order system.

Fig. 1.29 and Fig. 1.30 show the variation of amplitude ratio and phase lag verses frequency ratio ($\omega/\omega_n$) for various values of damping ratio ($\gamma$). From these graphs we observe that the *salient features* of the *steady state response* of a second-order system are :

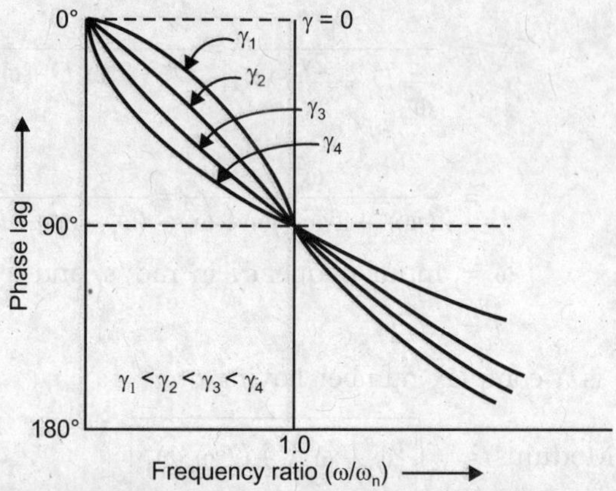


**Fig. 1.30.** Plot between phase lag and frequency ratio for a second-order system.

(*i*) As *frequency ratio* $\rightarrow 0$ :
Amplitude ratio $\rightarrow 1$;
Phase lag $\rightarrow 0°$

(*ii*) As frequency ratio $\rightarrow \infty$ :
Amplitude ratio $\rightarrow 0$;
Phase lag $\rightarrow 180°$

(*iii*) When frequency ratio = 1 :
Amplitude ratio $\rightarrow \infty$ in undamped systems ($\gamma = 0$)
Phase lag $\rightarrow -90°$ in all the systems.
This condition is known as "*resonance*" and can *result in destructive oscillations in lightly damped system.*

(*iv*) *When the amplitude ratio is unity for all frequencies the frequency response is considered to be ideal.* The nearest response to this effect is achieved when the value of $\gamma$ (damping ratio) lies between 0.6 and 0.7 for both the step and sinusoidal inputs.

## WORKED EXAMPLES—FIRST-ORDER SYSTEMS

**Example 1.16.** *(a) How is the order of the system determined ?*

*(b) The following equation characterises the dynamic response of a temperature measuring instrument :*

$$\frac{dI_o}{dt} = C(I_i - I_o)$$

*where,* $I_o$ = *Indicated temperature,*
$I_i$ = *Input temperature, and*
$C$ = *A numerical constant.*

*(i) Determine the transfer operator form of the equation.*

*(ii) What is the order of the system ?*

**Solution.** (*b*) *Given equation :* $\frac{dI_o}{dt} = C(I_i - I_o)$

**(*i*) Transfer operator form :**

The given equation can be rewritten as :

$$\frac{1}{C}\cdot\frac{dI_o}{dt} = I_i - I_o$$

or,
$$\frac{1}{C}\cdot\frac{dI_o}{dt} + I_o = I_i$$

or,
$$I\cdot\frac{dI_o}{dt} + I_o = I_i \qquad \text{(where, } \tau = \text{time constant} = \frac{1}{C}\text{)}$$

or,
$$(\tau D + 1)I_o = I_i$$

∴ The transfer operator form of the equation is given by :

$$\frac{I_o}{I_i} = \frac{1}{(\tau D + 1)} \quad \textbf{(Ans.)}$$

**(ii) Order of the system :**

Since the highest differential in the denominator of the transfer operator is *unity*, therefore, the temperature measuring instrument has a *first-order system.*

The grouping of the measurement and control systems is done according to the *order of the highest differential in the denominator* of the system transfer operation.

*Examples :*

- First-order system : $\frac{I_o}{I_i} = \frac{2}{1+3D}$
- Second-order system : $\frac{I_o}{I_i} = \frac{6}{D^2+3D+4}$, or

$$\frac{I_o}{I_i} = \frac{14}{(1+0.3D)(1+0.2D)}.$$

**Example 1.17.** *Formulate the governing equation for a first-order system-temperature measurement by a thermal measuring element (say a thermometer or thermocouple).*

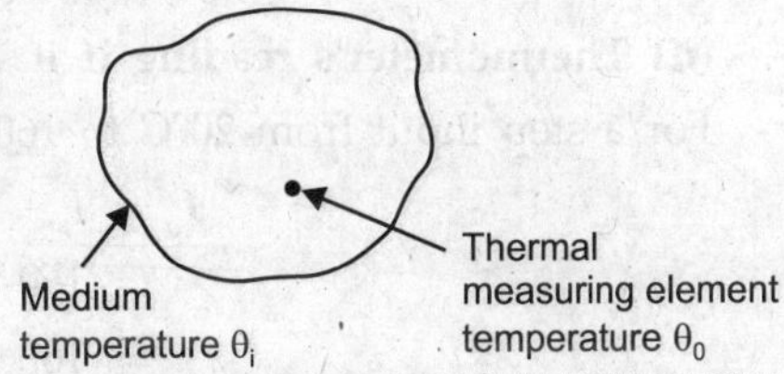


**Fig. 1.31.** Thermal element.

**Solution.** Refer to Fig. 1.31.

Let $I_i$ = Temperature of the medium,

$I_o$ = Temperature indicated by the thermal measuring instrument (say a thermometer or thermocouple),

$A$ = Exposed area of the thermal measuring element,

$h$ = Convective heat transfer coefficient,

$m$ = Mass of thermal element, and

$c$ = Specific heat of the element.

Then, the rate of heat flux into the element is,

$$Q = hA(\theta_i - \theta_0) \qquad ...(i)$$

The rate of enthalpy gain by the element

$$= mc\frac{d\theta_0}{dt} \qquad ...(ii)$$

Since the rate of heat flow equals the rate of enthalpy gain by the element, therefore, equating (*i*) and (*ii*) we get :

$$mc\frac{d\theta_0}{dt} = hA(\theta_i - \theta_o)$$

or, $$\frac{mc}{hA}\cdot\frac{d\theta_o}{dt}+\theta_o = \theta_i$$

$$\tau\frac{d\theta_o}{dt}+\theta_o = \theta_i$$

where, $\tau = \frac{mc}{hA}$ is known as *time constant* of the system.

In terms of *D*-operator (where $D = \frac{d}{dt}$), we have

$$(\tau D + 1)\theta_0 = \theta_i$$

or, $$\frac{\theta_o}{\theta_i} = \frac{1}{\tau D+1}$$ ...*Required equation.*

which is an equation of *first-order*. **(Ans.)**

**Example 1.18.** *A thermometer, idealised as a first-order system with a time constant of 2.2 seconds, is suddenly given an input of 160°C from 0°C.*

*(i) What will be the reading of the thermometer after 1.2 seconds ?*

*(ii) Determine its reading if it is initially held at 20°C.*

**Solution.** *Given* : $I_i$ = 160°C; *t* = 1.2*s*; *t* = 2.2*s*; $I_{initial}$ = 20°C.

**(i) Thermometer's reading after 1.2*s* :**

We know that, $$I_o = I_i(1 - e^{-t/\tau})$$ ...[Eqn. (1.31)]

$$= 160[1 - e^{-(1.2/2.2)}] = 67.27°C$$ **(Ans.)**

**(ii) Thermometer's reading if it was initially held at 20°C :**

For a step input from 20°C to 160°C, we have

$$I_o = I_i + (I_{initial} - I_i)\,e^{-t/\tau}$$ ...[Eqn. (1.35)]

$$= 160 + (20 - 160)\,e^{-(1.2/2.2)}$$

$$= 160 + (20 - 160) \times 0.5796 = 78.86°C$$ **(Ans.)**

**Example 1.19.** *A temperature sensing device can be modelled as a first-order system with a time constant of 5 seconds. It is suddenly subjected to a step input of 30°C – 160°C. Calculate the temperature indicated by the device after 10 seconds after the start of the process.*

**Solution.** *Given* : τ = 5*s*; $I_{initial}$ = 30°C, $I_i$ = 160°C; *t* = 10*s*.

Temperature after 10 seconds is calculated as follows :

$$I_o = I_i + (I_{initial} - I_i)\,e^{-t/\tau}$$ ...[Eqn. (1.35)]

$$= 160 + (30 - 160)\,e^{-10/5}$$

$$= 160 - 130 \times 0.1353 \doteq 142.4°C$$ **(Ans.)**

**Example 1.20.** *A temperature sensitive transducer when subjected to sudden temperature change takes 9 seconds to reach equilibrium conditions (Three time constants). Calculate the time taken by the transducer to read half of the temperature difference.*

**Solution.** Time taken to reach equilibrium condition = 3τ = 9*s* (*Given*)

∴ Time constant, $$\tau = \frac{9}{3} = 3s$$

Time taken by the transducer to read half of the temperature difference is calculated as follows :

$$I_o = I_i(1 - e^{t/\tau}) \qquad ...[\text{Eqn. (1.31)}]$$

or, $$\frac{I_o}{I_i} = 1 - e^{-t/\tau}$$

or, $$0.5 = 1 - e^{-t/3}$$

or, $$e^{-t/3} = 0.5 \quad \text{or,} \quad e^{t/3} = 2$$

$$\therefore \quad t = \mathbf{2.08s} \quad \textbf{(Ans.)}$$

**Example 1.21.** *A resistance temperature detector has a steady state gain of 2.96 W/°C and a time constant of 4.8 seconds. It is subjected to a step change of 80°C in temperature. If its stable resistance before the step change is 90W, write down the time-domain equation for resistance and find its value after 12 seconds of application of step input.*

**Solution.** Step input in terms of resistance

= Steady stage gain × change of temperature

= 0.296 × 80 = 23.68 Ω step input in terms of resistance.

In order to determine the time domain equation for resistance, the value of initial resistance must be added to it.

*i.e.,* $$R_t = 23.68(1 - e^{-t/4.8}) + 90 \quad \textbf{(Ans.)}$$

*...Time domain equation*

∴ The value of resistance at $t$ = 12s.

$$R_{12} = 23.68(1 - e^{-12/4.8}) + 90 = \mathbf{111.74\ \Omega} \quad \textbf{(Ans.)}$$

**Example 1.22.** *A thermometer initially at a temperature of 15°C is suddenly plunged into a liquid bath maintained at 140°C. After a time interval of 4 seconds, the thermometer indicated a reading of 75°C. Determine :*

*(i) The time constant for the thermometer.*

*(ii) The indicated temperature after five times constant; comment upon this result.*

**Solution.** *Given :* $I_{initial}$ = 15°C; $I_i$ = 140°C; $t$ = 4s; $I_o$ = 75°C.

**(i) The time constant for the thermometer, τ :**

When a first-order system is subjected to a step input from $I_{initial}$ to $I_i$, the indicated temperature is given by the relation :

$$I_o = I_i + (I_{initial} - I_i)e^{-t/\tau} \qquad ...[\text{Eqn. (1.35)}]$$

$$75 = 140 + (15 - 140)e^{-4/\tau}$$

or, $$e^{-4/\tau} = \frac{75-140}{15-140} = 0.52$$

or, $$e^{4/\tau} = 1.923$$

Taking logs on both sides, we get

$$\frac{4}{\tau} = 0.6539$$

$$\therefore \quad \tau = \mathbf{6.12\ s} \quad \textbf{(Ans.)}$$

**(ii) Indicated temperature after five times constant, $I_o$ :**

Substituting the value in the above equation, we get

$$I_o = 140 + (15 - 140)\, e^{-5\tau/\tau}$$

$$= 140 + (15 - 140)e^{-5} = \mathbf{139.16°C} \quad \textbf{(Ans.)}$$

*Comments* : After a time interval equivalent to five times constants, the thermometer reaches the *equilibrium condition* (*i.e.*, it indicates a temperature which is approximately equal to the maximum step input).

**Example 1.23.** *A temperature sensitive transducer used to measure the temperature of a furnace has been idealised as a first-order system subjected to ramp input. If the maximum permissible error in temperature measurement is limited to 3.9°C, calculate the time constant of the transducer when the furnace temperature increases at the rate of 0.2°C/s.*

**Solution.** *Given* : Dynamic error, $E_{dy.}$ = 3.9°C; slope, $\psi$ = 0.2°C/s.

For the ramp input, the input signal varies linearly with time ($I_i - \psi t$) and the dynamic or measurement error,

$$E_{dy.} = \underbrace{\psi\tau}_{\text{Steady}} - \underbrace{\psi\tau e^{-t/\tau}}_{\text{Transient}} \quad \text{...[Eqn. (1.40)]}$$

Since the transient error gradually vanishes with time, we have

$$E_{dy.} = \psi\tau$$

$$\therefore \quad \text{Time constant,} \quad \tau = \frac{E_{dy.}}{\psi} = \frac{3.9}{0.2} = \mathbf{19.5\ s} \quad \textbf{(Ans.)}$$

**Example 1.24.** *A balloon (weather) carrying a temperature sensing device (of first-order, time constant 8 seconds) rises through the atmosphere at 5 m/s. It supplies information about temperature and altitude through radio signals. At 2500 m height, a temperature indication of 30°C has been received. If the temperature varies with altitude at a uniform rate of 0.011°C/m, determine the time altitude at which 30°C temperature occurs.*

**Solution.** Time constant of temperature sensing device, $\tau$ = 8s

Rate of rise of the balloon = 5 m/s

Temperature indicated at an altitude of 2500 m = 30°C

Rate of temperature variation with altitude = 0.011°C/m

**True altitude at 30°C :**

Rate of change of temperature with time $y$ = 0.011 × 5 = 0.055°C/s

For a ramp signal applied to a first-order system, the error, when transient conditions die away,

$$E_{dy.} = \psi\tau = 0.055 \times 8 = 0.44°C$$

$$\therefore \quad \text{Error in amplitude} = \frac{0.44}{0.011} = 40 \text{ m}$$

$$\therefore \quad \textit{Actual altitude} = 2500 - 40 = \mathbf{2460\ m} \quad \textbf{(Ans.)}$$

**Example 1.25.** *A thermometer with time constant of 50 seconds is used for measuring temperature cycling with 500s time period. Calculate :*

*(i) The ratio of output to input.*

*(ii) The time lag.*

**Solution.** Time constant of Thermometer, $\tau$ = 50 s

Time period, $t$ = 500 s

Now, frequency of temperature variation,

$$\omega = \frac{2\pi}{t} = \frac{2\pi}{500} = 0.01257 \text{ rad/s}$$

**(i) The ratio of output to input, $\frac{I_o}{I_i}$ :**

$$\frac{I_o}{I_i} = \frac{1}{\sqrt{1+(\omega\tau)^2}}$$

$$= \frac{1}{\sqrt{1+(0.01257\times 50)^2}} = \mathbf{0.8467} \quad \textbf{(Ans.)}$$

**(ii) Time lag :**

Phase shift = $\tan^{-1}(\omega\tau) = \tan^{-1}(0.01257 \times 50) = 32.15°$ or 0.561 rad.

$\therefore$ Time lag = $\frac{1}{\omega}\times$phase shift in radian

$$= \frac{1}{0.01257}\times 0.561 = \mathbf{44.63\ s} \quad \textbf{(Ans.)}$$

**Example 1.26.** *During a particular process, the air temperature cycle at the rate of 1 cycle every 4 minutes. The time constant of the temperature measuring device is 20 seconds. Determine :*

*(i) The variation in the indicated temperature if true (input) temperature has a sinusoidal variation of ± 25°C.*

*(ii) The time by which the maximum reading of the thermometer lags the true maximum value.*

**Solution.** *Given:* Air temperature variation= 1 cycle every 4 minutes

Time constant of the device, $\tau = 20$ s

Sinusoidal variation of the input, $I_i = \pm 25°C$

Now, $t$ = 4 minutes

$$\therefore \quad f = \frac{1}{t} = \frac{1}{4\times 60} = \frac{1}{240}\text{ cycles/s}$$

and,
$$\omega = 2\pi f = 2\pi\times\frac{1}{240} = 0.0262\text{ rad./s}$$

**(i) The variation in the indicated temperature ($I_o$) :**

The steady state (when transient errors are omitted) solution for the sinusoidal input is given as :

$$\frac{I_o}{I_i} = \frac{1}{\sqrt{(1+\omega\tau)^2}}\sin(\omega t-\phi) \quad \text{...[Eqn. (1.44)]}$$

where, $I_o$ = Output temperature as indicated by the thermometer,

$I_i$ = True temperature, and

$\phi$ = Phase lag = $\tan^{-1}(\omega\tau)$

Substituting the value of $\omega$ and $\tau$ (calculated above) in the above equation, we get

$$\frac{I_o}{I_i} = \frac{I}{\sqrt{1+(0.0262\times 20)^2}} = 0.886$$

$$I_o = \pm 25 \times 0.886 = \mathbf{\pm 22.15°C} \quad \textbf{(Ans.)}$$

**(ii) The lag (in seconds) :**

Phase lag = $\tan^{-1}(\omega\tau)$

or, $\phi = \tan^{-1}(0.0262 \times 20) = 27.65°$ or 0.4826 rad.

*The lag in seconds*

$$= \frac{1}{\omega} \times \text{phase lag in radian}$$

$$= \frac{1}{0.0262} \times 0.4826 = \mathbf{18.42\ seconds} \quad \textbf{(Ans.)}$$

**Example 1.27.** *A first-order instrument having an input of sine-wave pattern is required to measure signals with frequency response of 120 cycles per second. If an amplitude inaccuracy of 4 percent it permissible, calculate the maximum time constant and time lag at 90 cycles per second.*

**Solution.** *Given :* Input frequency = 120 cycles per second ;

Amplitude inaccuracy permissible = 4%.

Now, $\omega = 2\pi f = 2\pi \times 120 = 754$ rad/s

**Maximum time constant τ :**

Corresponding to 4% amplitude inaccuracy, the amplitude ratio of output and input signals is $\left(1 - \frac{4}{100}\right) = 0.96$.

For a first-order instrument, the amplitude ratio is given as :

$$\frac{I_o}{I_i} = \frac{1}{\sqrt{1 + (\omega\tau)^2}} \quad \text{...[Eqn. (1.44)]}$$

or, $$0.96 = \frac{1}{\sqrt{(1 + 754\tau)^2}}$$

or, $$(1 + 754\tau)^2 = \left(\frac{1}{0.96}\right)^2 = 1.085$$

or, $$1 + 754\tau = 1.0416$$

∴ $$\tau = \mathbf{5.517 \times 10^{-5}\ s} \quad \textbf{(Ans.)}$$

**Time lag at 90 cycles per second :**

We know that, for a sinusoidal input,

$$\phi = \tan^{-1}(\omega\tau) = \tan^{-1}(2\pi \times 90 \times 5.517 \times 10^{-5})$$

$$= 1.787° \quad \text{or} \quad 0.03119$$

∴ $$\text{Time lag} = \frac{1}{\omega} \times \text{phase lag in radian}$$

$$= \frac{1}{2\pi \times 90} \times 0.03119 = \mathbf{5.515 \times 10^{-5}\ s} \quad \textbf{(Ans.)}$$

**Example 1.28.** *The temperature of a furnace is found to vary sinusoidally between 520°C and 580°C with a periodic time of 50 seconds. A thermocouple system with a time constant of 10 seconds is employed to measure the furnace temperature. Determine :*

*(i) The maximum and minimum values that will be indicated by the thermocouple.*

*(ii) The phase shift and the corresponding time lag between the temperature signals and the thermocouple input signals.*

**Solution.** *Given :* Range of variation of temperature = 520° to 580°C

Periodic time, $t = 50$ seconds

Time constant of the thermocouple system, $\tau = 10$ seconds.

**(i) The maximum and minimum values of the temperature :**

The temperature would oscillate about the mean value of $\frac{580+520}{2} = 550°C$, with an amplitude of 30°C.

Angular frequency of oscillation,

$$\omega = 2\pi f = \frac{2\pi}{t} = \frac{2\pi}{50} = 0.1256 \text{ rad/s}$$

The amplitude ratio, after the transient effects die away, is given by :

$$\frac{I_o}{I_i} = \frac{1}{\sqrt{1+(\omega\tau)^2}} = \frac{1}{\sqrt{1+(0.1256\times10)^2}} = 0.6229$$

$$\therefore \quad I_o = 0.6229 \times 30 = 18.7°C$$

$\therefore$ *Maximum temperature* = 550 + 18.7 = **568.7°C** **(Ans.)**

*Minimum temperature* = 550 – 18.7 = **531.3°C** **(Ans.)**

**(ii) Phase shift and time lag :**

$$\text{Phase shift} = \tan^{-1}(\omega\tau)$$

$$= \tan^{-1}(0.1256 \times 10) = 51.47°, \text{ or, } 0.898 \text{ rad.}$$ **(Ans.)**

$$\text{Time lag} = \frac{1}{\omega} \times \text{phase shift in radian}$$

$$= \frac{1}{0.1256} \times 0.898 = 7.15 \text{ s}$$ **(Ans.)**

Fig. 1.32 shows the variation of actual and indicated temperatures of furnace.

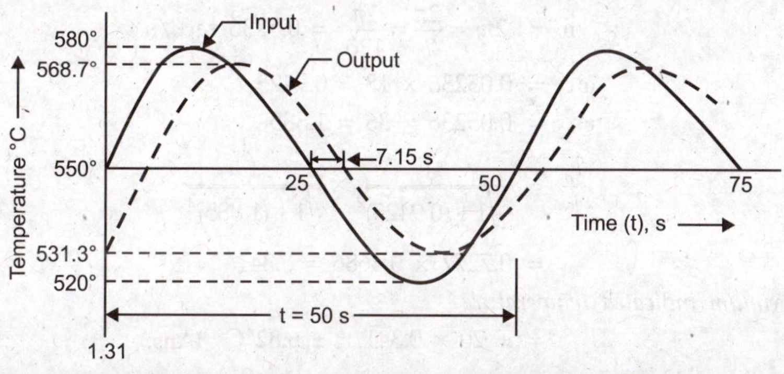


**Fig. 1.32.** Variation of actual and indicated temperatures of furnace.

**Example 1.29.** *A first-order instrument is subjected to a sinusoidal input given by relation : $I_i = 0.35 \sin 25t$. If the instrument has a time constant of 0.3 second, develop an expression for the corresponding output.*

**Solution.** *Given* : Sinusoidal input relation, $I_i = 0.35 \sin 25\,t$

Time constant of the instrument, $\tau = 0.3$ s.

**Expression for the output :**

The amplitude ratio, after the transisent effects die away, is given as :

$$\frac{I_o}{I_i} = \frac{1}{\sqrt{1+(\omega\tau)^2}} = \frac{1}{\sqrt{1+(25\times0.3)^2}} = 0.132$$

(Here, $\omega$ = 25 ..... from the given relation)

∴ Magnitude of output,

$$I_o = 0.132 \times \text{magnitude of input}$$
$$= 0.132 \times {}^*0.35 = 0.0462 \quad \text{(* from the given relation)}$$

Phase shift, $\phi = \tan^{-1}(\omega\tau)$

$$= \tan^{-1}(25 \times 0.3) = 82.4°$$

∴ $I_o =$ **$0.0462 \sin (25t - 82.4°)$** ...(*Required expression*) **(Ans.)**

**Example 1.30.** *A thermometer (its glass bulb protected by a well) is subjected to a temperature which is changing ± 20°C every 120 seconds. The measurement system may be idealised by a double capacity system with time constants of 18 seconds for the bulb and 36 seconds for the well.*

*Determine the maximum value of temperature indicated by the thermometer. After how much time of the actual occurrence of temperature, it will be indicated by the instrument (i.e., time lag) ?*

**Solution.** *Given :* Rate of change of temperature : ± 20°C every 120 seconds

Time constant for the bulb = 18 seconds

Time constant for the well = 36 seconds.

**Maximum value of the temperature indicated :**

The amplitude ratio ($I_o/I_i$) for a double capacity system is given as :

$$\frac{I_o}{I_i} = \frac{1}{\sqrt{1+(\omega\tau_1)^2}} \times \frac{1}{\sqrt{1+(\omega\tau_2)^2}}$$

and, the angle of lag, $\phi = \tan^{-1}(\omega\tau_1) + \tan^{-1}(\omega\tau_2)$

From the given data :

$$\omega = 2\pi f = \frac{2\pi}{t} = \frac{2\pi}{120} = 0.5236 \text{ rad./s}$$

$$\omega\tau_1 = 0.05236 \times 18 = 0.9425$$

$$\omega\tau_2 = 0.05236 \times 36 = 1.885$$

$$\frac{I_o}{I_i} = \frac{1}{\sqrt{1+(0.9425)^2}} \times \frac{1}{\sqrt{1+(1.885)^2}}$$

$$= 0.7277 \times 0.4686 = 0.341$$

∴ *Maximum indicated temperature*

$$= \pm 20 \times 0.341 = \pm \mathbf{6.82°C} \quad \textbf{(Ans.)}$$

**Time lag :**

$$\text{Angle of lag} = \tan^{-1}(0.9425) + \tan^{-1}(1.885)$$
$$= 43.30° + 62.05° = 105.35° \text{ or } 1.839 \text{ rad.}$$

$$\therefore \quad \text{Time lag (delay)} = \frac{1}{\omega} \times \text{angle of lag (in rad.)}$$

$$= \frac{1}{0.05236} \times 1.839 = \mathbf{35.12 \text{ seconds}} \quad \textbf{(Ans.)}$$

**Example 1.31.** *An input of 2 sin2t + 0.5 sin10t is applied to a first-order instrument having a time constant of 0.3 second. Find the output.*

**Solution.** *Given :* Input equation :

$$I_i = 2 \sin 2t + 0.5 \sin 10t$$

Time constant, $\tau = 0.3$ s

**Output :**

The output of the instrument to an input 2 sin2*t* + 0.5 sin10*t* when time constant is 0.3s, is given as:

$$I_o = \frac{1}{\sqrt{1+(2\times0.3)^2}}\sin[2t-\tan^{-1}(2\times0.3)]+\frac{1}{\sqrt{1+(10\times0.3)^2}}\sin[10t-\tan^{-1}(10\times0.3)]$$

$$= 0.857 \sin(2t - 30.96°) + 0.316 \sin(10t - 71.56°) \quad \textbf{(Ans.)}$$

**Example 1.32.** *A first-order system having a time constant of 0.15 second is used to measure a signal prescribed by the relation: $I_i$ = 2 sin2t + 0.2 cos8t. Develop an expression for the corresponding output.*

**Solution.** *Given :* Relation followed by the signal : $I_i = 2 \sin 2t + 0.2 \cos 8t$

Time constant of the system, $\tau = 0.15$ s

**Expression for the output :**

The concept of sinusoidal transfer function can be applied for the frequency response only if input, $I_i$ is either a sinewave or sum of sinewaves of different frequencies and amplitudes. In order to satisfy this condition, the given input need to be rewritten in the following form :

$$I_i = 2 \sin 2t - 0.2 \sin [8t + \pi \text{ (or } 180°)]$$

Further, the first-order measuring systems are *linear one,* and therefore the output can be obtained by using the *superposition principle.*

- For the sinusoidal input : $I_i = 2 \sin 2t$

$$\frac{I_o}{I_i} = \frac{1}{\sqrt{1+(\omega\tau)^2}} = \frac{1}{\sqrt{1+(2\times0.15)^2}} = 0.9578$$

Magnitude of output = 0.9578 × 2 = 1.9156

Phase shift, $\phi = \tan^{-1}(\omega\tau) = \tan^{-1}(2 \times 0.15) = 16.7°$

∴ Output, $I_o = 1.9156 \sin(2t - 16.7°)$

- For the sinusoidal input : $I_i = -0.2 \sin(8t + \pi)$

$$\frac{I_o}{I_i} = \frac{1}{\sqrt{1+(\omega\tau)^2}} = \frac{1}{\sqrt{1+(8\times0.15)^2}} = 0.6402$$

Magnitude of output = 0.2 × 0.6402 = 0.128

Phase shift, $\phi = \tan^{-1}(\omega\tau) = \tan^{-1}(8 \times 0.15) = 50.19°$

∴ Output, $I_o = -0.128 \sin(8t + \pi - 50.19°)$

Hence, *overall output,* $\boldsymbol{I_o} = 1.9156 \sin(2t - 16.7°) - 0.128 \sin(8t + \pi - 50.19°)$ **(Ans.)**

## SECOND-ORDER SYSTEMS

**Example 1.33.** *Formulate the governing equation for a second-order system–spring mass system with damping.*

**Solution.** Refer to Fig. 1.33.

Let, $x_i$ = Input displacement,

$x_o$ = Output displacement,

$k$ = Stiffess of the spring,

$C_d$ = Viscous damping coefficient, and

$\gamma$ = Damping ratio.

The *forces* acting on the mass are :

(*i*) As both ends of the springs are free to move, therefore

*Spring force* = Spring stiffness × displacement of one end of the spring relative to other

$= k(x_i - x_o)$, acting *downward.*

(*ii*) One end of the dashpot fixed; there is a reaction force acting in the *upward* direction.

*Damping force* = Damping coefficient × velocity

$$= C_d \times \frac{dx_o}{dt}$$

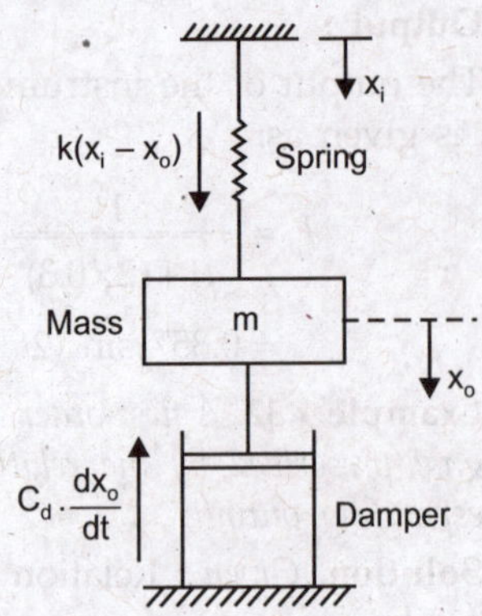


**Fig. 1.33.** Spring-mass system.

For translational systems, the Newton's law states that,

$$\Sigma \text{ Force} = \text{Mass} \times \text{acceleration} = m\frac{d^2x_o}{dt^2}$$

or,
$$m \cdot \frac{d^2x_o}{dt^2} = k(x_i - x_o) - C_d\frac{dx_o}{dt}$$

or,
$$mD^2x_o + C_dDx_o + kx_o = kx_i \qquad \left(\text{where } D = \frac{d}{dt}, \text{ and } D^2 = \frac{d^2}{dt^2}\right)$$

or,
$$x_i = \left(\frac{m}{k}D^2 + \frac{C_d}{k}D + 1\right)x_o \qquad \text{...Required equation}$$

which is an equation of *second-order* type. **(Ans.)**

Comparing the above expression with the standard second-order form, we have :

*Undamped* natural frequency,

$$\omega_n = \sqrt{\frac{k}{m}}$$

and
$$\frac{2\gamma}{\omega_n} = \frac{C_d}{k}$$

∴ Damping ratio,
$$\gamma = \frac{C_d\omega_n}{2k} = \frac{C_d}{2k}\sqrt{\frac{k}{m}}$$

or,
$$\gamma = \frac{C_d}{2\sqrt{km}} \qquad \text{...Required equation.}$$

**Example 1.34.** *(a) Write down the expression describing the motion of linear and rotational displacement systems of the second-order.*

*(b) The pen arrangement of U.V. recorder (second-order system) has a mass of 4.5g. Calculate the percentage reduction in mass if it is desired to have 15 percent increase in natural frequency of the recorder.*

**Solution.** (*a*) • The expression for *linear displacement* (spring-mass-damper) *system* is given by :

$$m\frac{d^2I_o}{dt^2}+C_d\frac{dI_o}{dt}+k\,I_o = k\,I_i \quad ...(i)$$

where, $m$ = Mass (kg),
$C_d$ = Viscous damping force (Ns/m),
$k$ = Spring stiffness (N/m),
$I_o$ = Output reading, and
$I_i$ = Input reading.

• For the *rotational system*, the expression may be written as :

$$J\frac{d^2I_o}{dt^2}+C_d\frac{dI_o}{dt}+q\,I_o = q\,I_i.$$

where, $J$ = Inertia (kg m$^2$), and
$q$ = Torsional stiffness.

Comparing these expressions with differential equation in the standard form,

$$\frac{1}{\omega_n^2}\cdot\frac{d^2I_o}{dt^2}+\frac{2\gamma}{\omega_n}\cdot\frac{d\,I_o}{dt}+I_o = I_i, \text{ we get}$$

Natural frequency, $\omega_n = \sqrt{\frac{k}{m}} = \sqrt{\frac{q}{J}}$, as the case may be. **(Ans.)**

(*b*) *Given* : $m$ = 4.5 g; Percentage increase required = 15%.

Using subscripts 1 and 2 for initial and final values respectively, we have :

$$\omega_{n1}^2 = \frac{k}{m_1}, \text{ and, } \omega_{n2}^2 = \frac{k}{m_2}$$

$$\therefore \quad m_2 = m_1\times\left(\frac{\omega_{n1}}{\omega_{n2}}\right)^2$$

But, $\omega_{n2} = 1.15\,\omega_{n1}$ ...(*Given*)

$$\therefore \quad m_2 = m_1\times\left(\frac{\omega_{n1}}{1.15\,\omega_{n1}}\right)^2 = 0.756\,m_1$$

∴ *Percentage reduction in mass*

$$= \left(\frac{m_1-m_2}{m_1}\right) = \left(\frac{m_1-0.756m_1}{m_1}\right)\times 100 = \mathbf{24.4\%} \quad \textbf{(Ans.)}$$

**Example 1.35.** *A second-order system follows the differential equation given below :*

$$\frac{d^2I_o}{dt^2}+3\frac{dI_o}{dt}+30\,I_o = 30\,I_i$$

*where, $I_o$ and $I_i$ are output and input quantities respectively. Determine the following :*

(*i*) *Damping ratio.*
(*ii*) *Damped natural frequency.*
(*iii*) *Static sensitivity.*
(*iv*) *Time constant.*

**Solution.** The standard form of the differential equation of a second-order system is given as :

$$\frac{1}{\omega_n^2}\cdot\frac{d^2I_o}{dt^2}+\frac{2\gamma}{\omega_n}\cdot\frac{dI_o}{dt}+I_o = k\,I_i \qquad ...(i)$$

Since the term $I_o$ in eqn. (*i*) has a unit coefficient, therefore to recast the given equation in the standard form, let us divide it throughout by 30; we get

$$\frac{1}{30}\cdot\frac{d^2I_o}{dt^2}+\frac{1}{10}\cdot\frac{dI_o}{dt}+I_o = I_i \qquad ...(ii)$$

Comparing eqns. (*i*) and (*ii*), we get

$$\omega_n^2 = 30;\ \frac{2\gamma}{\omega_n}=\frac{1}{10}=0.1;\ k = 1$$

∴ Natural frequency, $\omega_n = \sqrt{30} = 5.477$ rad./s

**(*i*) Damping ratio, γ :**

$$\frac{2\gamma}{\omega_n} = 0.1, \quad \text{or} \quad \gamma = \frac{\omega_n}{2}\times 0.1$$

or,

$$\gamma = \frac{5.477}{2}\times 0.1 = \mathbf{0.274} \quad \textbf{(Ans.)}$$

**(*ii*) Damped natural frequency, $\omega_d$ :**

$$\omega_d = \omega_n\sqrt{1-\gamma^2} = 5.477\sqrt{1-0.274^2} = \mathbf{5.267\ rad./s} \quad \textbf{(Ans.)}$$

**(*iii*) Static sensitivity :**

*Static sensitivity*, $k$ = **1** **(Ans.)**

**(*iv*) Time constant : τ :**

$$\tau = \frac{1}{\omega_n}=\frac{1}{5.477} = \mathbf{0.1826\ s} \quad \textbf{(Ans.)}$$

**Example 1.36.** *The following data pertain to a torque sensing transducer connected to the shaft of an electric motor which drives a load :*

*Input torque = Sinusoidal with frequency varying at 1.95 rad./s; Maximum permissible error = ± 8 per cent; Moment of inertia of load, J = 0.14 kg m²; Torsional constant of the shaft, q = 1.22 Nm/rad. Calculate the following :*

(*i*) *Natural frequency.*
(*ii*) *Damping ratio.*
(*iii*) *Damped natural frequency.*
(*iv*) *Time constant of the system.*

**Solution.** *Given* : $\omega$ = 1.95 rad/s; Max. permissible error = ± 8%; $J$ = 0.14 kg m²; $q$ = 1.22 Nm/rad.

**(*i*) Natural frequency, $\omega_n$ :**

$$\omega_n = \sqrt{\frac{q}{J}}=\sqrt{\frac{1.22}{0.14}} = \mathbf{2.95\ rad./s} \quad \textbf{(Ans.)}$$

**(*ii*) Damping ratio, γ :**

Also, normalised frequency ratio, $r = \frac{\omega}{\omega_n}=\frac{1.95}{2.95}=0.661$

Corresponding to maximum permissible error of ± 8%, the amplitude ratio is 1.08. Further the amplitude ratio ($M$) is prescribed by the relation :

$$M = \frac{1}{\sqrt{(1-r^2)^2 + (2\gamma r)^2}}$$

or,
$$1.08 = \frac{1}{\sqrt{(1-0.661^2)^2 + (2\gamma\times 0.661)^2}}$$

$$= \frac{1}{\sqrt{0.317 + 1.747\gamma^2}}$$

or,
$$0.317 + 1.747\gamma^2 = \left(\frac{1}{1.08}\right)^2 = 0.857$$

$\therefore$
$$\gamma = \mathbf{0.555} \quad \textbf{(Ans.)}$$

**(*iii*) Damped natural frequency, $\omega_d$ :**

$$\omega_d = \omega_n\sqrt{1-\gamma^2}$$

$$= 2.95\sqrt{1-(0.555)^2} = \mathbf{2.454\ rad./s} \quad \textbf{(Ans.)}$$

**(*iv*) Time constant, $\tau$ :**

$$\tau = \frac{1}{\omega_n} = \frac{1}{2.95} = \mathbf{0.339\ s} \quad \textbf{(Ans.)}$$

**Example 1.37.** (*a*) *Discuss the time domain parameters for the response of step input to second-order system.*

(*b*) *When a step input was given to a second-order system, the measurements revealed that the system had an overshoot of 12 per cent in a rise time of 0.22 second. Determine :*

(*i*) *Effective damping ratio.*

(*ii*) *Undamped natural frequency of the system.*

**Solution.** (*a*) In an under-damped system, the time domain parameters of interest are (Refer to Fig. 1.34) :

(*i*) How long does the system take to practically reach its final steady value ?

(*ii*) How fast the system moves to follow the applied input ?

(*iii*) How oscillatory is the system ?

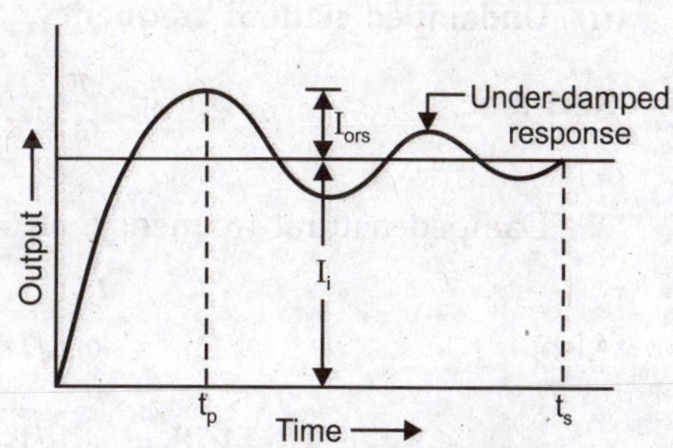


**Fig. 1.34.** Step response illustrating response, rise and settling times.

- **Settling time ($t_s$)** is defined as the *time required for the system to reach a steady state.*
- **Rise time ($t_r$)** is the *time required for the system to rise from 0 to 100% of its final value.*
- **Peak time ($t_p$)** is the *time taken to reach the first peak (overshoot).*

The percentage overshoot $= \frac{I_{ors}}{I_i} \times 100$.

Mathematical analysis of the under-damped system would show that :

$$\textit{Rise time } (t_r) = \frac{\pi}{\omega_d}$$

$$\textit{Peak time } (t_p) = \frac{\pi}{\omega_n\sqrt{1-\gamma^2}}$$

$$\textit{Overshoot} = \exp\left[\frac{-\pi\gamma}{\sqrt{(1-\gamma^2)}}\right]$$

(*b*) *Given* : Percentage overshoot = 12%; Rise time = 0.22 s.

**(*i*) Effective damping ratio, γ :**

$$\text{Percentage overshoot} = \exp\left[\frac{-\pi\gamma}{\sqrt{1-\gamma^2}}\right]\times 100$$

or,
$$12 = \exp\left[\frac{-\pi\gamma}{\sqrt{1-\gamma^2}}\right]\times 100$$

Rearranging and taking logs on both sides, we get

$$\log_e\left(\frac{12}{100}\right) = \frac{-\pi\gamma}{\sqrt{1-\gamma^2}}$$

$$-2.12 = \frac{-\pi\gamma}{\sqrt{1-\gamma^2}}$$

or,
$$0.675\sqrt{1-\gamma^2} = \gamma$$

$$0.456 - 0.456\,\gamma^2 = \gamma^2 \quad \text{...(Squaring both sides)}$$

∴
$$\gamma = \sqrt{\frac{0.456}{1.456}} = \mathbf{0.56} \quad \textbf{(Ans.)}$$

**(*ii*) Undamped natural frequency, $f_n$ :**

Rise time,
$$t_r = \frac{\pi}{\omega_d}$$

∴ Damped natural frequency,
$$\omega_d = \frac{\pi}{t_r} = \frac{\pi}{0.22} = 14.28 \text{ rad./s}$$

Also,
$$\omega_d = \omega_n\sqrt{1-\gamma^2} \quad (\text{where, } \omega_n = \text{undamped natural frequency})$$

$$14.28 = \omega_n\sqrt{1-0.56^2}$$

or,
$$\omega_n = \frac{14.28}{\sqrt{1-0.56^2}} = 17.24 \text{ rad./s}$$

∴
$$f_n = \frac{\omega_n}{2\pi} = \frac{17.24}{2\pi} = \mathbf{2.74 \text{ Hz}} \quad \textbf{(Ans.)}$$

**Example 1.38.** *A second-order instrument has a natural frequency of 5 Hz and a damping ratio of 0.62. If the excitation frequency of the system is 7 Hz, determine the error due to the proximity of excitation frequency with the natural frequency of the system.*

**Solution.** *Given :* Natural frequency of the instrument, $f_n$ = 5 Hz.

Damping ratio, $\gamma$ = 0.62

Excitation frequency of the system, $f$ = 7 Hz

Now, the ratio of excitation frequency to natural frequency

$$r = \frac{\omega}{\omega_n} = \frac{f}{f_n} = \frac{7}{5} = 1.4$$

$$\text{Amplitude ratio, } M = \frac{1}{\sqrt{(1-r^2)^2 + (2\gamma r)^2}} = \frac{1}{\sqrt{(1-1.4^2)^2 + (2 \times 0.62 \times 1.4)^2}} = 0.504$$

∴ The "*error*" due to proximity of excitation frequency with the natural frequency of the system

$$= (1 - 0.504) \times 100 = \mathbf{49.6\%} \quad \textbf{(Ans.)}$$

**Example 1.39.** *While measuring force by a diaphragm type pressure transducer (second-order system), the system is found to have a natural frequency of 800 cycles per second (cps) and a damping ratio of 62 per cent.*

*Determine the frequency range over which the amplitude ratio corresponding to sinusoidal input deviates by a maximum amount of 12 per cent.*

**Solution.** *Given :* Natural frequency of the system = 800 cps

Damping ratio, $\gamma$ = 62 per cent

Maximum amount of deviation of amplitude ratio = 12 per cent.

Corresponding to a maximum deviation of 12%, the amplitude ratio ($M$) is either 0.88 or 1.12.

Now,
$$M = \frac{1}{\sqrt{(1-r^2)^2 + (2\gamma r)^2}}$$

where, $r$ = Frequency ratio = $\frac{f}{f_n} = \frac{\omega}{\omega_n}$

(*i*) *When M = 1.12 :*

$$1.12 = \frac{1}{\sqrt{(1-r^2)^2 + (2 \times 0.62 \times r)^2}}$$

or, $(1 - r^2)^2 + (2 \times 0.62\, r)^2 = \left(\frac{1}{1.12}\right)^2 = 0.797$

or, $1 + r^4 - 2r^2 + 1.538\, r^2 = 0.797$

or, $r^4 - 0.462\, r^2 + 0.203 = 0$

The solution has imaginary roots and this implies that with $\gamma$ = 0.62, the *output would never be 1.12 times the output for any frequency.*

(*ii*) *When M = 0.88 :*

$$0.88 = \frac{1}{\sqrt{(1-r^2)^2 + (2 \times 0.62 \times r)^2}}$$

or, $(1 - r^2)^2 + (2 \times 0.62\, r)^2 = \left(\frac{1}{0.88}\right)^2 = 1.29$

or, $1 + r^4 - 2r^2 + 1.538\, r^2 = 1.29$

or, $r^4 - 0.462\, r^2 - 0.29 = 0$

or, $$r^2 = \frac{0.462 \pm \sqrt{(0.462)^2 + 4 \times 0.29}}{2} = 0.817$$

or, $r = 0.904$

$\therefore$ $f = f_n \times r = 800 \times 0.904 = 723$ cps

Thus, the deviation remains within 12 per cent of output for the *frequency range* **0 – 723 cps** **(Ans.)**

**Example 1.40.** *A unit amplitude sinusoidal input signal having a frequency of 0.6 Hz is given to a second-order control system of transfer function :* $\frac{I_o}{I_i} = \frac{8}{D^2 + 4D + 20}$.

*Develop an expression for the steady state response of the system and determine the corresponding output amplitude, output frequency and phase lag.*

**Solution.** Frequency, $\omega = 2\pi f = 2\pi \times 0.6 = 3.77$ rad/s

As the sinusoidal input signal has a *unit* amplitude, therefore

$$I_i = 1 \sin (3.77\, t)$$

Putting $D = j\omega$ in the given relation, we get

$$\frac{I_o}{I_i} = \frac{8}{(j\omega)^2 + 4j\omega + 20} = \frac{8}{(3.77j)^2 + 4 \times 3.77j + 20}$$

$$= \frac{8}{-14.21 + 15.08j + 20} = \frac{8}{5.79 + 15.08j} = \frac{1}{0.724 + 1.885j}$$

The denominator has a magnitude of

$$\sqrt{(0.724)^2 + (1.885)^2} = 2.02$$

$$\text{Phase lag, } \phi = \tan^{-1}\frac{1.885}{0.724} = 69° = 1.204 \text{ rad}$$

$$\text{Magnitude of output} = \frac{1}{2.02} \times \text{magnitude of input}$$

$$= \frac{1}{2.02} \times 1 = 0.495$$

$\therefore$ The required expression for the output in the sinusoidal form is :

$$I_o = 0.495 \sin (3.77\, t - 69°) \quad \textbf{(Ans.)}$$

Hence, *corresponding output amplitude* = **0.495** (*as compared with unit amplitude of input*) **(Ans.)**

*Output frequency* = **3.77** rad./s (*same as that of input signal*) **(Ans.)**

*Phase lag* = **69°** (*between input and output*) **(Ans.)**

## 1.12 ERRORS IN MEASUREMENTS

### 1.12.1. Introduction

Errors and uncertainties are *inherent* in the process of making any measurement and in the instrument with which the measurements are made. The study of error is important as a step in finding ways of reducing them, and also as means of estimating the reliability of final results.

The accuracy and precision of a measuring instrument depends upon its design, proper selection of material for each and every part, and skill of manufacturing.

### 1.12.2. Limiting (or Guarantee) Errors

In most instruments the accuracy is generated to be *within a certain percentage of full scale reading.* The manufacturer has to specify the deviations from the *nominal value* of a particular quantity. *The limits of these deviations from the specified value are defined as* **Limiting errors** *or* **Guarantee errors.**

The **relative error** is *defined as the ratio of the error to the specified (nominal) magnitude of a quantity :*

It, $Q_s$ = The specified quantity or true quantity of a component, and

$Q_a$ = The actual or erroneous quantity ;

Then, error $= Q_a - Q_s \ (= \delta Q)$

and, Relative limiting error (also called accuracy),

$$E_r = \frac{Q_a - Q_s}{Q_s} = \frac{\delta Q}{Q_s}$$

Percentage limiting error,

$$\% \ E_r = E_r \times 100$$

If a component has a relative limiting error $\pm E_r$, then limiting values of the component, $Q_s = (1 \pm E_r)$. Thus if the specified magnitude of a resistor is 100 Ω with a limiting error of ± 4 ohms, then the magnitude of the resistor will lie between the limits 100 ± 4 *i.e.,* 96 ohms and 104 ohms. In other words the manufacturer of the resistor guarantees its value to lie between 96 and 104 ohms.

**Example 1.41.** *A wattmeter having a range of 500 W has an error of ± 1.5 per cent of full scale deflection. If the true power is 50 W, what would be the range of the readings ?*

*If the error is specified as percentage of true value, what would be the range of the readings ?*

**Solution.** *Given :* Range of the wattmeter = 500 W

Error (% age of full scale deflection ) = ± 1.5%

True or specified power, $Q_s$ = 50 W

- When the error is specified as a *percentage of full scale deflection*, the magnitude of limiting error at full scale

$$= \pm\left(\frac{1.5}{100}\right)\times 500 = \pm 7.5 \text{ W}$$

Thus the wattmeter reading when the true reading is 50 W may be 50 ± 7.5 *i.e.,* between **42.5 to 57.5 W** **(Ans.)**

- Relative error, $E_r = \dfrac{7.5}{50}\times 100 = 15\%$

Now suppose the error is specified as percentage of true value, then,

$$\text{Magnitude of error} = \pm \frac{1.5}{100} \times 50 = \pm 0.75 \text{ W, and}$$

The meter may read between 50 ± 0.75 or **49.25 and 50.75 W. (Ans.)**

**Example 1.42.** *The guaranteed accuracy of a flow meter working on thermal principles is ± 3 per cent of full scale reading of 2.5 × $10^{-6}$ $m^3$/s. If the flow measured by this meter is 1.25 × $10^{-6}$ $m^3$/s, calculate the limiting error in percent. Comment upon the results.*

**Solution.** *Given :* Guaranteed accuracy of flow meter,

$$E_r = \pm 3\% \text{ (of full scale reading)}$$

Full scale reading, $Q_s = 2.5 \times 10^{-6} \text{ m}^3/\text{s}$

Flow measured by the meter = $1.25 \times 10^{-3}$ $m^3$/s.

**Limiting error (Percent) :**

The magnitude of limiting error at full scale of the meter is,

$$\delta Q = E_r \times Q_s$$
$$= \pm 0.03 \times 2.5 \times 10^{-6}$$
$$= \pm 0.075 \times 10^{-6} \text{ m}^3/\text{s}$$

The magnitude of the flow being measured is $1.25 \times 10^{-6}$ $m^3$/s;

∴ The relative error at this flow,

$$E_r' = \frac{\delta Q}{Q_s} = \frac{0.075 \times 10^{-6}}{1.25 \times 10^{-6}} = 0.06$$

∴ The value of flow being measured is,

$$Q = Q_s(1 \pm E_r') = 1.25 \times 10^{-6} (1 \pm 0.06)$$
$$= 1.25 \times 10^{-6} \pm 0.075 \times 10^{-6} \text{ m}^3/\text{s}$$

Hence, *Percentage limiting error,*

$$\% E_r = \pm \frac{0.075 \times 10^{-6}}{1.25 \times 10^{-6}} \times 100 = \pm 6\% \quad \textbf{(Ans.)}$$

***Comments.*** From above results we observe that when the meter reading is full scale value of 2.5 × $10^{-6}$ $m^3$/s the flow meter is guaranteed to have a limiting error of ± 3%, but when the meter is reading its half scale value of 1.25 × $10^{-6}$ $m^3$/s, the limiting error is ±6%. Thus when reading is halved, the limiting error is doubled. *This indicates that the percent limiting error will be greater if smaller values are read by the meter.* Therefore, while selecting instruments, special care should be taken as regards the range; the values to be measured *should not lie in the lower third of the range.* Hence as far as possible meter which will read up their scale should be used.

**Resultant error of a device composed of several components :**

A measuring instrument/device/system may be formed by combining several components, each having its limiting error. The limiting error of the combination can computed in the following ways depending on how the various components are connected in the total system:

When two or more quantities, each having a limiting error, are combined, it is advantageous to be able to compute the limiting error of the combination. The limiting error can be easily found by *considering the relative increment of the function if the final result is in the form of an algebraic equation.*

**(i) When the final result is either sum or difference of outputs of various components :**

Let, $q_1, q_2, q_3, \ldots\ldots\ldots\ldots q_n$ = Outputs of each device, and

$Q$ = The final result;

Then, $Q$ (sum or difference) $= q_1 \pm q_2 \pm q_3 \pm \ldots\ldots\ldots\ldots\ldots\ldots \pm q_n$,

and,
$$\frac{\delta Q}{Q} = \pm\left[\frac{q_1}{Q}\cdot\frac{\delta q_1}{q_1} + \frac{q_2}{Q}\cdot\frac{\delta q_2}{q_2} + \frac{q_3}{Q}\cdot\frac{\delta q_3}{q_3} + \ldots\right] \quad \ldots(1.59)$$

Equation (1.59) shows that the *resultant limiting error in this case is equal to the sum of the products formed by multiplying the individual relative errors by the ratio of each term to the function.*

**(ii) When the result is either product or quotient of output of various components :**

In this case, $Q = q_1 \times q_2 \times q_3 \times \ldots q_n$

or, $Q = q_1 \div q_2 \div q_3 \div \ldots q_n$

or, $Q = q_1 \times q_2 \div q_3 \times q_4 \div \ldots q_n$

Then,
$$\frac{\delta Q}{Q} = \pm\left(\frac{\delta q_1}{q_1} + \frac{\delta q_2}{q_2} + \frac{\delta q_3}{q_3} + \ldots\ldots\frac{\delta q_n}{q_n}\right) \quad \ldots(1.60)$$

In other words the *relative limiting error of the system is the sum of relative errors of all devices/components.*

**(iii) When the final output is composite factor of output of various components :**

In this case,
$$Q = q_1^{\pm n_1} \times q_2^{\pm n_2} \times q_3^{\pm n_3} \ldots \times q_n^{\pm n_n}$$

Then,
$$\frac{\delta Q}{Q} = \pm\left[n_1\frac{\delta q_1}{q_1} + n_2\frac{\delta q_2}{q_2} + n_3\frac{\delta q_3}{q_3} + \ldots\right] \quad \ldots(1.61)$$

**Example 1.43.** *The three resistors $R_1$, $R_2$ and $R_3$ have the following ratings :*

$$R_1 = 25\ \Omega \pm 4\%,\ R_2 = 65\ \Omega \pm 4\%;\ R_3 = 45\ \Omega \pm 4\%$$

*Determine the following :*

*(i) Limiting value of the resultant resistance.*

*(ii) Percent limiting error of series combination of resistance.*

**Solution.** *Given :*
$R_1 = 25\ \Omega \pm 4\%$
$R_2 = 65\ \Omega \pm 4\%$
$R_3 = 45\ \Omega \pm 4\%$.

**(i) Limiting value of the resultant resistance :**

The values of the resistances are :

$$R_1 = 25 \pm \frac{4}{100} \times 25 = 25 \pm 1.0\ \Omega$$

$$R_2 = 65 \pm \frac{4}{100} \times 65 = 65 \pm 2.6\ \Omega$$

$$R_3 = 45 \pm \frac{4}{100} \times 45 = 45 \pm 1.8\ \Omega$$

∴ The limiting value of resultant resistance,

$$R = (25 + 65 + 45) \pm (1.0 + 2.6 + 1.8) = (135 \pm 5.4)\ \Omega$$

*Magnitude of resistance* = **135 Ω** and, *error* = **± 5.4 Ω (Ans.)**

***(ii)* Percent limiting error of the series combination of resistance :**

Percent limiting error of the combination

$$= \pm\frac{5.4}{135}\times 100 = \pm \mathbf{4\%} \quad \textbf{(Ans.)}$$

**Example 1.44.** *A power meter and an ammeter are used to measure resistance of an unknown resistance. Find the limiting error in the measurement of resistance when the limiting errors in the measurement of power and current are ± 1.2% and 0.8% respectively.*

**Solution.** *Given* : Limiting error in the measurement of power ($P$) = $\frac{\delta P}{P} = 1.2\%$

Limiting error in the measurement of current ($I$) = $\frac{\delta I}{I} = 0.8\%$

We know that, $R = \frac{P}{I^2}$

∴ Limiting error in the measurement of resistance,

$$\frac{\delta R}{R} = \pm\left(\frac{\delta P}{P} + 2\frac{\delta I}{I}\right)$$

$$= \pm (1.2 + 2 \times 0.8) = \pm \mathbf{2.8\%} \quad \textbf{(Ans.)}$$

**Example 1.45.** *The unknown resistance in a Wheatstone bridge is measured by using three known resistance such that :*

$$R_4 = \frac{R_2 R_3}{R_1}$$

*where,* $R_1 = 50 \pm 0.5\%\ \Omega;\ R_2 = 500 \pm 0.5\%\ \Omega;\ R_3 = 440 \pm 0.5\%\ \Omega$

*Determine the magnitude of the unknown resistance* ($R_4$) *and the limiting error in percent and in ohms (for the unknown resistance).*

**Solution.** *Given :* $R_1 = 50 \pm 0.5\%\ \Omega;\ R_2 = 500 \pm 0.5\%\ \Omega;\ R_3 = 440 \pm 0.5\%\ \Omega$

The unknown resistance is given by

$$R_4 = \frac{R_2 R_3}{R_1} = \frac{500\times 440}{50} = \mathbf{4400\ \Omega} \quad \textbf{(Ans.)}$$

The relative limiting error of the unknown resistance is,

$$\frac{\delta R_4}{R_4} = \pm\left(\frac{\delta R_2}{R_2} + \frac{\delta R_3}{R_3} + \frac{\delta R_1}{R_1}\right)$$

$$= \pm (0.5 + 0.5 + 0.5) = \pm \mathbf{1.5\%} \quad \textbf{(Ans.)}$$

*Limiting error in ohms* $= \frac{1.5}{100}\times 4400 = \pm \mathbf{66\ \Omega}$ **(Ans.)**

*i.e., The guaranteed values of resistance are between* **4334** (*i.e.,* 4400 – 66) **Ω** and **4466** (*i.e.,* 4400 + 66) **Ω** **(Ans.)**

**Example 1.46.** *A cantilever beam of length l, width b and depth d is used to measure the force F. The deflection is measured using an LVDT of accuracy ± 0.1 mm. The length, width and depth are measured to an accuracy of ± 0.01 mm and their values are 45 mm, 4.5 mm and 0.9 mm respectively. The modulus of rigidity is known accurate to 0.2 %. If the deflection noted was 1.8 mm, determine the error in force measured.*

*Given force,* $F = \frac{3Ebd^3}{12l^3}\cdot y$

*where,* $E$ = *Modulus of elasticity;* $b$ = *width,* $d$ = *depth, and*

$l$ = *length of the beam ;* $y$ = *deflection.*

**Solution.** The *limiting error in force F is,*

$$\frac{\delta F}{F} = \pm\left(\frac{\delta E}{E}+\frac{\delta b}{b}+3\frac{\delta d}{d}+3\frac{\delta l}{l}+\frac{\delta y}{y}\right)$$

$$= \pm\left(0.2+\frac{0.01}{4.5}+3\times\frac{0.01}{0.9}+3\times\frac{0.01}{45}+\frac{0.1}{1.8}\right)$$

$$= \pm\ (0.2 + 0.00222 + 0.03333 + 0.00067 + 0.05555)$$

$$= \pm\ \mathbf{0.2917\%}\quad \textbf{(Ans.)}$$

**Example 1.47.** *A dynamometer is used to measure the output power of a rotating shaft. The output power is given by the following relation :*

$$P = \frac{2\pi\times 9.81\, FLR}{t\times 10^6}\ kW$$

*where,* $F$ = *Force at the end of torque arm (kg),*

$L$ = *Length of torque arm (mm),*

$R$ = *Number of revolutions during time t, and*

$t$ = *Time for test run (s).*

*The given test data are*

$F = 4.26 \pm 0.02$ *kg;* $L = 382 \pm 1.2$ *mm;*

$R = 1192 \pm 1.0$ *revolutions;* $t = 60 + 0.50$ *s*

*The errors are limiting (absolute) errors.*

*(i) The magnitude of power.*

*(ii) The magnitude of the limiting error in the computed power.*

**Solution.** *Given :* $F = 4.26 \pm 0.02$ kg; $L = 382 \pm 1.2$ mm; $R = 1192 \pm 1.0$ revolutions; $t = 60 + 0.50$ s

**(i) The magnitude of power *P* :**

$$P = \frac{2\pi\times 9.81\times FLR}{t\times 10^6}\ \text{kW} \qquad ...(Given)$$

Inserting the various values in the above expression, we get

$$P = \frac{2\pi\times 9.81\times 4.26\times 382\times 1192}{60\times 10^6} = \mathbf{1.993\ kW}\quad \textbf{(Ans.)}$$

**(ii) Limiting errors in the computed power :**

Total per unit error in computed power,

$$\frac{\delta P}{P} = \pm\left(\frac{\delta F}{F}+\frac{\delta L}{L}+\frac{\delta R}{R}+\frac{\delta t}{t}\right)$$

$$= \pm\left(\frac{0.02}{4.26}+\frac{1.2}{382}+\frac{1.0}{1192}+\frac{0.50}{60}\right)$$

$$= \pm\ (0.00469 + 0.00314 + 0.00084 + 0.00833) = \pm\ 0.017$$

*Limiting error in computer power*

$$\delta P = \pm 0.017 \times P = \pm\ 0.017 \times 1.993 = \mathbf{0.0339\ kW}$$

The result is written as: $P$ = **1.993 ± 0.0339 kW** **(Ans.)**

**Example 1.48.** *(a) How are "known errors" treated ?*

*(b) During a test the current was measured as 26.5 A flowing in a resistor of 0.12 Ω. Later on it was discovered that the ammeter reading was low by 1.1% and the marked resistance was high by 0.25%.*

*Determine the true power as a percentage of the power that was originally calculated.*

**Solution.** (*a*) **Known errors.** When the error of quantity of an instrument is known the effect of this error when combined with errors, can be computed in a manner similar to the combinations of the limiting errors. However, the difference is that in case of known errors the *signs of relative errors are given and must be presented in the calculations.*

(*b*)

$$\text{True value of current } I = 26.5(1 + 0.011) = 26.791 \text{ A}$$

$$\text{True value of resistance } R = 0.12(1 - 0.0025) = 0.1197 \ \Omega$$

$$\text{True value of power} = I^2R = (26.791)^2 \times 0.1197 = 85.916 \text{ W}$$

$$\text{Measured value of power} = (26.5)^2 \times 0.12 = 84.27 \text{ W}$$

$$\therefore \quad \frac{\text{True power}}{\text{Measured power}} \times 100 = \frac{85.916}{84.27} \times 100 = \mathbf{101.9\%} \quad \textbf{(Ans.)}$$

*Alternatively :*

$$\text{Power, } P = I^2R$$

$$\text{Total relative error} = \frac{\delta P}{P} = 2 \cdot \frac{\delta I}{I} + \frac{\delta R}{R}$$

$$= 2(-0.011) + 0.0025 = -0.0195$$

$$\frac{\text{True power}}{\text{Measured power}} = \frac{1}{1 + \text{relative error}}$$

$$= \frac{1}{1 - 0.0195} = \mathbf{101.9\%} \quad \textbf{(Ans.)}$$

### 1.12.3. Types of Errors

In measurement systems, the errors could originate from several sources. Broadly these may be *classified* as follow :

1. *Gross errors*
2. *Systematic errors :*
   (*i*) Instrumental errors;
   (*ii*) Environmental errors;
   (*iii*) Observational errors.
3. *Random errors.*

#### 1.12.3.1. Gross errors

These errors occur due to *human mistakes* in reading instruments and recording and calculating results of measurement. Although it is probably impossible to eliminate the gross error completely, yet one should try to anticipate and correct them.

The mathematical analysis of gross errors is impossible since these may occur in different amounts. While some gross errors may be easily detected, others may go unnoticed.

These errors can be *avoided* by adopting *two means* :

(*i*) Immense care should be taken while taking the reading and recording the data.

(*ii*) Two, three, or even more readings should be taken for the quantity being measured.

● Designers of instruments to some extent can take care to avoid gross errors but much would depend upon the interest and competence of observer.

### 1.12.3.2. Systematic errors

The systematic errors are repeated consistently with the repetition of the experiment and are caused by *such effects as* :

— Sensitivity shift;

— Zero off-set;

— Known non-linearity.

● These errors *cannot* be determined by direct and repetitive observations of the measurand made each time with same technique.

*These errors can be located only by having repeated measurements under different conditions or with different equipment and where possible by an entirely different method.*

These errors may be *instrumental, environmental or observational errors.*

**1. Instrumental errors :**

These errors arise *due to* :

— Inherent shortcomings in the instruments.

— Misuse of the instruments.

— Loading effects of instruments.

Instrument errors are inherent in instruments due to construction, calibration or operation of the instruments. *These errors may cause the instrument to read too low or too high.*

Some *"examples"* of such errors are :

(*i*) Mechanical friction and wear, backlash, yielding of supports, pen or pointer drag and hysteresis of elastic members due to aging.

(*ii*) Constructional faults resulting from finite width of knife edges, excessive friction at the mating parts, lost motion due to necessary clearance in gear teeth etc.

(*iii*) Improper selection and poor maintenance of the instrument.

(*iv*) Unavoidable physical phenomenon due to friction, capillary attraction and imperfect rarefaction.

***Assembly errors.*** These errors may be due to the following *reasons* :

(*i*) Displaced scale *i.e.,* incorrect fitting of the scale zero with respect to the actual zero position of the movement.

(*ii*) Non-uniform division of the scale.

(*iii*) Bent or distorted pointer.

The assembly errors do not change with time, so these can be easily discovered and rectified.

**Misuse of instruments.** A good instrument used in an unintelligent way may give erroneous results.

● *Loading effects.* In several measurement techniques, the detector/sensor utilizes the energy of the medium, thus altering slightly the parameter to be measured, resulting in errors.

**2. Environmental errors :**

The environmental errors are due to conditions external to the measuring device (*e.g.,* effects of pressure, temperature, humidity, dust, vibrations, external magnetic or electrostatic fields). These errors are much more troublesome than assembly errors.

These errors can be *eliminated or reduced* by taking the following measures :

(i) Using the instrument in controlled conditions of pressure, temperature, humidity in which it was originally assembled and calibrated.

(ii) Measure deviations in the local conditions from the calibrated ones and then apply a suitable corrections to the instrument readings.

(iii) Automatic compensation for departures from the calibrated conditions by using sophisticated devices.

(iv) Make a complete new calibration under the local conditions.

(v) In case it is suspected that external magnetic or electrostatic fields can effect the readings of the instruments, magnetic or electrostatic shields may be provided.

(vi) Apply computed corrections. Normally, efforts are made to avoid the use of applications of computed corrections, but where these corrections are needed and are necessary, they are incorporated for the computation of the results.

**3. Observational errors :**

These errors occur due to *carelessness of operators*. Even when an instrument has been properly selected, carefully installed and faithfully calibrated, shortcoming in the measurement occurs due to certain failings on the part of the observer.

The observational errors may be due to the following *reasons :*

(i) *Parallax*—Since the parallax errors arise on account of pointer and the scale not being in the same plane, we can eliminate this error by having the pointer and the scale in the same plane.

(ii) Wrong scale reading, and wrong recording of data.

(iii) Inaccurate estimates of average reading, lack of ability to interpolate properly between graduations.

(iv) Incorrect conversion of units in between consecutive readings, and non-simultaneous observation of interdependent quantities.

(v) Tendency to read high or low, anticipate a signal and read too soon—Personal bias.

- Modern electrical instruments have *digital display of output* which completely eliminates the errors on account of human observational or sensing power as the output is in the *form of digits.*

**1.12.3.3. Random errors**

- The random errors are *accidental, small and independent.*
- *They vary in an unpredictable manner.*
- The magnitude and direction of these errors *cannot* be predicted from a knowledge of measurement system; however these errors are *assumed to follow the law of probabilities.*

*The most common causes of these errors are :*

(i) Friction in instrument movement.

(ii) Backlash in the movement.

(iii) Parallax errors between pointer and scale.

(iv) Finite dimensions of the pointer and scale divisions.

(v) Hysteresis in elastic members.

(vi) Mechanical vibrations.

## 1.12.4. Sources of Errors

Besides the errors so far enumerated (of the physical and mechanical nature) and the errors which necessarily result from the faulty calibration of a measuring system, the errors arise from the *following sources also :*

(*i*) **Noise.** *It is defined as any signal that does not convey useful information.*

— Noise signals/errors can be reduced to a minimum level through filtering, careful selection of components, shielding and isolation of the entire measuring system.

(*ii*) **Response time.** *It is defined as the time taken by the instrument to show 63.2% change in a reading to a step input.* This factor contributes to the uncertainty of the measurement.

(*iii*) **Design limitations.** In the design of the instrument, certain inevitable factors such as friction and resolving power lead to uncertainty of measurements.

(*iv*) **Energy exchanged by interaction.** As earlier pointed out, whenever the energy required for operating the measuring system is extracted from the measurand, the value of the latter is altered to a greater or leaser extent. This alternation is dependent upon the capacity of the system.

(*v*) **Transmission.** During the transmission of information from the primary sensing element to the indicator, the signal may be *attenuated* due to any of the following *reasons* :

— It may suffer loss through leakage ;

— It may be absorbed or otherwise consumed in the communication channel;

— It may be distorted by resonance, attenuation or delay phenomenon whose actions are selective on various signal components.

(*vi*) **Deterioration of measuring system.** The examples of some types of deterioration which occur in a measuring system and result in as a source of error in measurement are :

— Change in resistance of a circuit element through strain relief ;

— Alteration of thermocouple characteristics through oxidizing or reducing atmosphere;

(*vii*) **Ambient influences on measuring systems.**

(*viii*) **Errors of observation and interpretation.**

### 1.12.5. Statistical Analysis of Test Data

The *systematic and random errors* are evaluated and studied by *statistical procedures* which make it possible to state from a limited group of data the most probable value of a quantity, the probable uncertainty of a single observation, and the probable limits of uncertainty of the best value that can be derived from the data.

— The object of the statistical methods based on laws of chance which operate only on random errors and not on systematic errors, is to achieve consistency (precision) of value and not their accuracy.

The experimental data is obtained in the following *two forms of tests* :

(*i*) Single-sample test;

(*ii*) Multi-sample test.

(*i*) **Single-sample test.** *A single measurement (or succession of measurements) done under identical conditions excepting for time is known as "single-sample test".*

(*ii*) **Multi-sample test.** In a multi-sample test, *repeated measurements of a given quantity are taken under different test conditions such as using different instruments, different ways of measurement and by employing different observers.* This type of test does not materialise result by simply making measurements with the same equipment, procedure, technique and same observer.

• For getting the exact value of the quantity under measurement, tests should be conducted using as many different procedures, techniques as experiments as practicable. It should be understood that the *statistical means which help us to arrive at correct results are only valid for multi-sample tests.*

**1.12.5.1. Statistical averages**

When several multi-sample observations are taken experimentally there is a scatter of the data about some central value, one of the methods, presenting the result in the form of a histogram (also called a *frequency distribution curve*). Thus, histogram (Fig. 1.35) is a pictorial representation of the test data where the ordinates indicate the number of observed readings of a particular value. With more and more data taken at smaller and smaller increments, the histogram finally changes into a *smooth curve.* The smooth curve is symmetrical with respect to the central value.

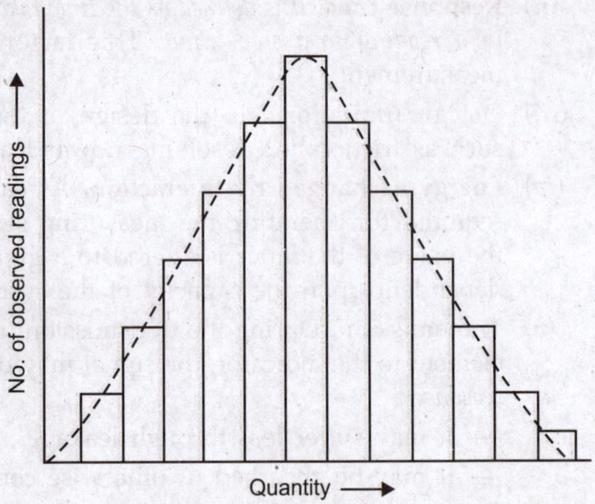


**Fig. 1.35.** Histogram.

(*i*) **Arithmetic mean (AM).** *The most probable value of measured variable (variate) is the arithmetic mean of the number of readings taken.* We arrive at the best approximation if the number of readings of the same quantity are very large. The mean is computed by summing all the values and dividing by the number of measurements. If $q_1, q_2, q_3 \cdots q_n$ be the set of measured values of a *quantity*, then the arithmetic mean $\bar{q}$ will be given by,

$$\bar{q} = \frac{q_1 + q_2 + q_3 + ....q_n}{n} = \frac{1}{n}\sum_{i=1}^{n} q_i \qquad ...(1.62)$$

(*ii*) **Geometric mean (G.M.).** It is defined as the nth root of the product of *n* terms.

$$\text{Geometric mean (G.M)} = \sqrt[n]{q_1 \times q_2 \times q_3 \times .....\times q_n} \qquad ...(1.63)$$

(*iii*) **Median.** *The middle value of a set of an "odd" number of readings, if variables are arranged in numerical order, is called the "median".* For an *"even"* number of readings, the median is the *arithmetic average of the two central readings.*

(*iv*) **Mode.** *The value of the variable that occurs most frequently is termed as "mode".*

• For a symmetrical distribution curve, the three measures of central value *viz., 'mean', 'median' and 'mode' are equal.*

**1.12.5.2. Dispersion from mean**

*The property which denotes the extent to which the values are dispersed about the central value is termed as* **dispersion** *or* **stored** *or* **scatter**.

A large dispersion indicates that some factors involved in the measurement process are not under close control and therefore it becomes difficult to estimate the measured quantity with confidence and definiteness.

*The simplest possible measure of dispersion is the* **Range** *which is the difference between greatest and least values of data.*

(*i*) **Deviation.** *The departure of the observed reading from the arithmetic mean of the group of readings is termed as* ***deviation***. Let the deviation of reading $q_1$ be $d_1$ and that of reading $q_2$ be $d_2$ etc.

Then,

$$d_1 = q_1 - \bar{q}$$
$$d_2 = q_2 - \bar{q}$$
$$\ldots \quad \ldots$$
$$d_n = q_n - \bar{q} \qquad \ldots(1.64)$$

and,

$$\bar{q} = \frac{\Sigma(q_n - d_n)}{n} \qquad \ldots(1.65)$$

$$\text{Algebraic sum of deviations} = d_1 + d_2 + d_3 + \ldots\ldots d_n$$
$$= (q_1 - \bar{q}) + (q_2 - \bar{q}) + (q_3 - \bar{q}) + \ldots.(q_n - \bar{q})$$
$$= (q_1 + q_2 + q_3 + \ldots. + q_n) - n\bar{q} = 0$$
$$(\because q_1 + q_2 + q_3 + \ldots + q_n) = n\bar{q}$$

Thus, the *algebraic sum of deviations is zero.*

The above mathematical result implies that the deviations can be *positive as well as negative.*

(*ii*) **Average deviation.** *It is defined as the sum of the absolute values of deviations divided by the number of readings.*

Average deviation may be expressed as

$$\bar{d} = \frac{|d_1| + |d_2| + |d_3| + \ldots\ldots.. + |d_n|}{n}$$

$$= \frac{\Sigma|d|}{n}, \text{ or, } \sum_{i=1}^{n} \frac{|q_i - \bar{q}|}{n} \qquad \ldots(1.66)$$

The average deviation is taken as a measure of accuracy of measurement.

- The average deviation is an *indication of the precision of the instruments* used in making the measurements. *Highly precise instruments yield a low average deviation between readings.*

(*iii*) **Standard deviation.** "*Standard deviation*" (or "*mean square deviation*") of an infinite number of data is defined *as the square root of the sum of individual deviations squared divided by the number of readings.*

Thus, standard deviation

$$\sigma = \sqrt{\frac{d_1^2 + d_2^2 + d_3^2 + \ldots\ldots + d_n^2}{n}}$$

$$= \sqrt{\frac{\Sigma d^2}{n}} = \sqrt{\sum_{i=1}^{n} \frac{(q_i - \bar{q})^2}{n}} \qquad \ldots(1.67)$$

- In practice, however the number of observations is *finite*. When the number of observations is *greater than 20*, standard deviation is denoted by the symbol σ while if it is *less than* 20 the symbol used is *s*. The standard deviation of a finite number of data is given by :

$$s = \sqrt{\frac{d_1^2 + d_2^2 + d_3^2 + ...... + d_n^2}{n-1}} = \sqrt{\frac{\Sigma d^2}{n-1}} \quad ...(1.68)$$

**Variance**. *The square of the standard deviation is called* ***variance***.

Variance, $V = (\text{Standard deviation})^2$

$$= \sigma^2 = \frac{d_1^2 + d_2^2 + d_3^2 ...... d_n^2}{n} = \frac{\Sigma d^2}{n} \quad ...(1.69)$$

But when the number of observation is less than 20,

$$\text{Variance,} \quad V = s^2 = \frac{\Sigma d^2}{n-1} \quad ...(1.70)$$

**1.12.5.3. Best value from a sample of readings**

Let, $q_b$ = The best value (approximately the true value) from a set of readings/data,

Then, the deviation or individual component error,

$$d_i = q_i - q_b \qquad \text{where, } i = 1, 2, 3, ....n$$

This deviation may be *+ve*, *zero* or *–ve*.

(*i*) **Sum of deviations :**

$$\sum_{i=1}^{n} d_i = \sum_{i=1}^{n} (q_i - q_b) \quad ...(1.71)$$

In case of a *best value*, the sum of deviations is equal to *zero*.

*i.e.,*

$$\sum_{i=1}^{n} (q_i - q_b) = 0 \quad ...(1.72)$$

or,

$$q_b = \frac{1}{n}\sum_{i=1}^{n} q_i \text{, which is the arithmetic mean.}$$

Thus, the arithmetic mean may be interpreted as the best value around which the chances of error being *+ve* or *–ve* likely to be equal.

(*ii*) $d_i = q_i - q_b$

Squaring both sides, we have

$$d_i^2 = q_i^2 - 2q_iq_b + q_b^2$$

Sum of the square of deviations,

$$\sum_{i=1}^{n} d_i^2 = \sum_{i=1}^{n} q_i^2 - 2q_b\sum_{i=1}^{n} q_i + nq_b^2 \quad ...(1.73)$$

The sum of the square of deviations, corresponding to the *best* would be *minimum*. This condition can be obtained by setting the deviation w.r.t., $q_b$ equal to zero;

*i.e.,*

$$\frac{d}{dq_b}\left[\sum_{i=1}^{n} d_i^2\right] = 0 \quad ...(1.74)$$

or, $$\frac{d}{dq_b}\left[\sum_{i=1}^{n} q_i^2 - 2q_b \sum_{i=1}^{n} q_i + nq_b^2\right] = 0$$

As all the $q_i^2$s are constant, the differentiation gives,

$$-2\sum_{i=1}^{n} q_i + 2nq_b = 0$$

or, $$q_b = \frac{1}{n}\sum_{i=1}^{n} q_i \qquad ...(1.75)$$

which is again the arithmetic mean. Obviously, *the arithmetic mean is the most probable value or the best value which one can obtain from a set of scattered measurement data.*

**Example 1.49.** *By using a micrometer screw the following readings were taken of a certain physical length :*

*1.34, 1.38, 1.56, 1.47, 1.42, 1.44, 1.53, 1.48, 1.40, 1.59 mm*

*Assuming that only random errors are present, calculate the following :*

*(i) Arithmetic mean,*

*(ii) Average deviation,*

*(iii) Standard deviation, and*

*(iv) Variance.*

**Solution.** *Given* : Readings of the micrometer screw: 1.34, 1.38, 1.56, 1.47, 1.42, 1.44, 1.53, 1.48, 1.40, 1.59 mm.

*(i)* **Arithmetic mean :**

We know that, arithmetic mean,

$$\bar{q} = \frac{1}{n}\sum_{i=1}^{n} q_i \qquad ...(\text{Eqn. } 1.62)$$

$$= \frac{q_1 + q_2 + q_3 + q_4 + q_5 + q_6 + q_7 + q_8 + q_9 + q_{10}}{10}$$

$$= \frac{1.34 + 1.38 + 1.56 + 1.47 + 1.42 + 1.44 + 1.53 + 1.48 + 1.40 + 1.59}{10}$$

= **1.461 mm (Ans.)**

*(ii)* **Average deviation :**

From eqn. (1.64), the *deviations* are :

$$d_1 = q_1 - \bar{q} = 1.34 - 1.461 = -0.121$$

$$d_2 = q_2 - \bar{q} = 1.38 - 1.461 = -0.081$$

$$d_3 = q_3 - \bar{q} = 1.56 - 1.461 = +0.099$$

$$d_4 = q_4 - \bar{q} = 1.47 - 1.461 = +0.009$$

$$d_5 = q_5 - \bar{q} = 1.42 - 1.461 = -0.041$$

$$d_6 = q_6 - \bar{q} = 1.44 - 1.461 = -0.021$$

$$d_7 = q_7 - \bar{q} = 1.53 - 1.461 = +0.069$$

$$d_8 = q_8 - \bar{q} = 1.48 - 1.461 = +0.019$$

$$d_9 = q_9 - \bar{q} = 1.40 - 1.461 = -0.061$$

$$d_{10} = q_{10} - \bar{q} = 1.59 - 1.461 = 0.129$$

∴ *Average deviation,*

$$\bar{d} = \frac{\Sigma|d|}{n}\left(\text{or } \sum_{i=1}^{n} \frac{|q_i - \bar{q}|}{n}\right)$$

$$= \frac{0.121 + 0.081 + 0.099 + 0.009 + 0.041 + 0.021 + 0.069 + 0.019 + 0.061 + 0.129}{10}$$

$= \mathbf{0.065}$ **(Ans.)**

It may be noted that for calculation of average deviations, the *signs of deviations are not considered.*

(*iii*) **Standard deviation :**

Since the number of readings is 10, which is *less than 20*, therefore equation (1.68) is used for calculation of standard deviation.

$$\text{Standard deviation, } s = \sqrt{\frac{\Sigma d^2}{n-1}} \qquad \text{...[Eqn. (1.68)]}$$

$$\text{or, } s = \sqrt{\frac{(-0.121)^2 + (-0.081)^2 + (+0.099)^2 + (0.009)^2 + (-0.041)^2 + (-0.021)^2 + (0.069)^2 + (0.019)^2 + (-0.061)^2 + (0.129)^2}{(10-1)}}$$

$$= \sqrt{\frac{0.05869}{9}} = \mathbf{0.0807\ mm} \quad \textbf{(Ans.)}$$

(*iv*) **Variance :**

$$\textit{Variance, } V = s^2 = (0.0807)^2 = \mathbf{0.00651\ mm^2} \quad \textbf{(Ans.)}$$

**Example 1.50.** *Eight different students tuned in the circuit for resonance and the values of resonant frequency in kHz were recorded as :*

*412, 428, 423, 415, 426, 411, 423, 416*

*Calculate the following :*

*(i) Arithmetic mean,*

*(ii) Average deviation,*

*(iii) Standard deviation, and*

*(iv) Variance,*

**Solution.** *Given* : Values of resonant frequency : 412, 428, 423, 415, 426, 411, 423, 416

(*i*) **Arithmetic mean :**

*Arithmetic mean,*

$$\bar{q} = \frac{\Sigma q}{n} = \frac{412 + 428 + 423 + 415 + 426 + 411 + 423 + 416}{8}$$

$= \mathbf{419.25\ kHz}$ **(Ans.)**

(*ii*) **Average deviation :**

The deviations are :

$$d_1 = q_1 - \bar{q} = 412 - 419.25 = -7.25 \text{ kHz}$$

$$d_2 = q_2 - \bar{q} = 428 - 419.25 = +8.75 \text{ kHz}$$

$$d_3 = q_3 - \bar{q} = 423 - 419.25 = 3.75 \text{ kHz}$$

$$d_4 = q_4 - \bar{q} = 415 - 419.25 = -4.25 \text{ kHz}$$

$$d_5 = q_5 - \bar{q} = 426 - 419.25 = +6.75 \text{ kHz}$$

$$d_6 = q_6 - \bar{q} = 411 - 419.25 = -8.25 \text{ kHz}$$

$$d_7 = q_7 - \bar{q} = 423 - 419.25 = +3.75 \text{ kHz}$$

$$d_8 = q_8 - \bar{q} = 416 - 419.25 = -3.25 \text{ kHz}$$

∴ *Average deviation,*

$$\bar{d} = \frac{\Sigma|d|}{n}$$

$$= \frac{7.25 + 8.75 + 3.75 + 4.25 + 6.75 + 8.25 + 3.75 + 3.25}{8}$$

$$= \mathbf{5.75 \text{ kHz}} \quad \textbf{(Ans.)}$$

(For the calculation of average deviation, the *signs of deviations are not considered*)

(*iii*) **Standard deviation :**

Since the number of readings is 8, which is less than 20, therefore eqn. (1.68) is used for calculation of standard deviation.

$$\text{Standard deviation, } s = \sqrt{\frac{\Sigma d^2}{n-1}} \qquad \text{...[Eqn (1.68)]}$$

$$\text{or, } s = \sqrt{\frac{(-7.25)^2 + (+8.75)^2 + (+3.75)^2 + (-4.25)^2 + (+6.75)^2 + (-8.25)^2 + (+3.75)^2 + (-3.25)^2}{(8-1)}}$$

$$= \mathbf{6.54 \text{ kHz}} \quad \textbf{(Ans.)}$$

(*iv*) **Variance :**

$$\textit{Variance, } V = s^2 = (6.54)^2 = \mathbf{42.77 \ (kHz)^2} \quad \textbf{(Ans.)}$$

**1.12.5.4. Normal or Gaussian curve of errors**

In experimental work the random effects are normally studied with reference to the Gaussian or normal error distribution. Here, the following *assumptions* are made :

(*i*) All observations include a large number of small random disturbing effects, and these small random errors may be both +*ve* and –*ve*. Thus, there is a strong central tendency, *i.e., there is a greater probability of small errors and large errors of both +ve and –ve values.*

(*ii*) There is equal probability of +*ve* and –*ve* disturbing effects. As such the probability curve of errors plotted against the magnitude of error will be *symmetrical about the zero value.*

The *law of probability states that the normal occurrence of deviations from average value of an infinite number of measurements or observations can be expressed as :*

$$P = \frac{1}{\sigma\sqrt{2\pi}} e^{-x^2/2\sigma^2} \qquad \text{...(1.76)}$$

where, $P$ = Number of readings at any deviation '$d$' *i.e.*, probability of occurrence of deviation $d$,

$\sigma$ = Standard deviation, and

$x$ = Magnitude of deviation from mean.

Fig. 1.36 shows a Gaussian distribution curve in which $P$ has been plotted as a function of $x$. The curve is symmetrical about the mean value and may be regarded as the limiting form of a histogram. The two values of standard deviation $\sigma_1$ and $\sigma_2$ ($\sigma_1 > \sigma_2$) have been illustrated in the given figure. *Larger the value of σ, the flatter the curve and hence larger the expected error of all measurements.*

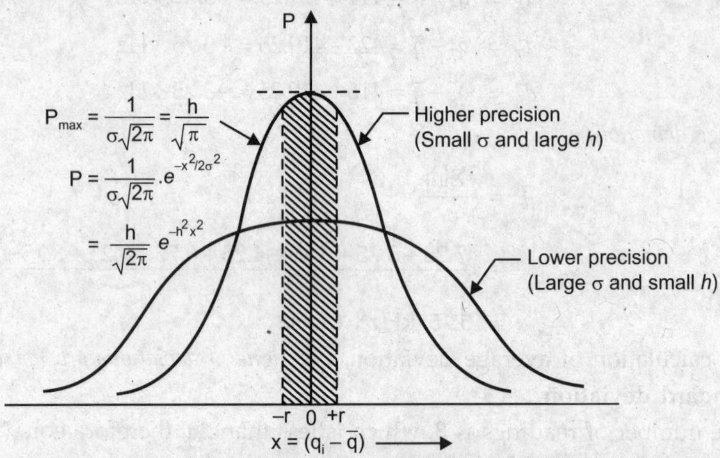


**Fig. 1.36.** Gaussian distribution curve.

If we integrate $P$ between the limits $x = \pm x$, we shall obtain the area under the curve between these two values of $x$. Because the curve is related to the probability that the measured lies within same range, we have

$$\int_{-x}^{+x} P dx = 1.0 \qquad ...(1.77)$$

*i.e.*, area under the probability curve is *unity*.

If the total number of readings are represented by unity, then the fraction of the total number of readings lying between $x_1$ and $x_2$ will be represented by the area under the curve between these two values of deviation ;

*i.e.*, $$n_{1-2} = \int_{-x}^{+x} P dx$$

If the area between $x_1$ and $x_2$ is 0.6, then 60 per cent of the deviations fall between $x_1$ and $x_2$.

**Precision index.** It may be observed from the normal distribution function that the maximum probability occurs when deviation is *zero, i.e.*, $q = \overline{q}$ and the corresponding value of probability is,

$$P_{max} = \frac{1}{\sigma\sqrt{2\pi}} = \frac{h}{\sqrt{\pi}} \qquad ...(1.78)$$

where, $h = \frac{1}{\sigma\sqrt{2}}$ is called the **precision index.**

A large value of $h$ represents high precision of the data because the probability of occurrence of variates in a given range falls off *rapidly* as the deviation increases because the variates tend to cluster (become closer) into a narrow range. On other hand, a small value of $h$ represents low precision because the probability of occurrences of variates in a given range falls off *gradually* as the deviation increases; this is because the variates are spread over a wide range.

**Probable error :** Let us consider the two points $-r$ and $+r$ marked in Fig. 1.35. These points are so located that the area bounded by the curve, the X-axis and the ordinates erected at $x = -r$ and $x = +r$ is equal to half of the total area under the curve. That is half the deviations lie between $x = \pm r$. The location of the points $r$ can be worked out as follows :

$$\int_{-r}^{+r} P.dx = \frac{1}{2}$$

or,
$$\frac{h}{\sqrt{\pi}} \int_{-r}^{r} e^{-h^2x^2} dx = \frac{1}{2}$$

This gives *probable error,*

$$r = \frac{0.4769}{h} \qquad \text{...(1.79)}$$

**Average deviation for the normal curve :**

In case of normal curve,

Average deviation,

$$\bar{d} = \int_{-\infty}^{+\infty} |x| P\, dx$$

(|$x$| represents the numerical value of deviation)

$$= \frac{h}{\sqrt{\pi}} \int_{-\infty}^{+\infty} e^{-h^2x^2} x.dx = \frac{2h}{\sqrt{\pi}} \int_{0}^{+\infty} e^{-h^2x^2} x.dx$$

$$= \frac{2h}{\sqrt{\pi}} \int_{0}^{+\infty} e^{-h^2 z} dz \qquad \left[\begin{array}{l}\text{where, } z = x^2 \\ dz = 2x.dx \\ x.dx = \dfrac{dz}{2}, \text{and} \\ \displaystyle\int e^{ax} dx = \frac{e^{ax}}{a}\end{array}\right]$$

$$= \frac{2h}{\sqrt{\pi}} \times \frac{1}{2} \left[\frac{e^{-h^2 z}}{-h^2}\right]_0^{+\infty}$$

$$= \frac{h}{\sqrt{\pi}} \left[\frac{0-1}{-h^2}\right] = \frac{h}{\sqrt{\pi}} \times \frac{1}{h^2} = \frac{1}{h\sqrt{\pi}}$$

$$= \frac{r}{0.4769\sqrt{\pi}} = \frac{r}{0.8453}$$

*i.e.,*
$$\bar{d} = \frac{r}{0.8453} \qquad \text{...(1.80)}$$

**Standard deviation for the normal curve :**

We know that, standard deviation $\sigma$ is given by :

$$\sigma = \sqrt{\frac{\Sigma d^2}{n}} \qquad \text{...[Eqn. (1.67)]}$$

Following a method similar to that followed above, we have

$$\sigma^2 = \frac{2h}{\sqrt{\pi}} \int_0^{+\infty} e^{(-h^2x^2)} x^2 . dx = \frac{1}{2h^2}$$

or, $$\sigma = \frac{1}{\sqrt{2}h} \quad ...(1.81)$$

$$= \frac{r}{0.4769 \times \sqrt{2}}$$

$$= \frac{r}{0.6745} \quad ...(1.82)$$

From eqns. (1.80) and (1.82), we have *probable error*,

$$r = 0.8453 \bar{d} \quad ...(1.83)$$

and, $$r = 0.6745\ \sigma \quad ...(1.84)$$

Also, *Standard deviation of mean*,

$$\sigma_m = \frac{\sigma}{\sqrt{n}} \quad ...(1.85)$$

and, *Standard deviation of standard deviation*,

$$\sigma_\sigma = \frac{\sigma}{\sqrt{2n}} = \frac{\sigma_m}{\sqrt{2}} \quad ...(1.86)$$

**1.12.5.5. Probability tables**

The probability of an individual measurement to fall within the deviations $x_1$ and $x_2$ from the mean is given by,

$$P(x_1, x_2) = \frac{h}{\sqrt{\pi}} \int_{x_1}^{x_2} e^{(-h^2x^2)} dx$$

for a normal Gaussian curve.

The parameter $h = \frac{1}{\sigma\sqrt{2}}$ and the mean value $\bar{q}$ may be known from the raw data. Since the evaluation of the integral is a cumbersome and a tedious task, hence, **probability tables** have been prepared from which the value of probability can be directly read off.

Let $$y = \sqrt{2}hx = \frac{x}{\sigma},$$

or, $$x = \sigma \cdot y$$

Now, $$P(x_1, x_2) = P(y_1, y_2)$$

$$= \frac{1}{\sqrt{2\pi}} \int_{y_1}^{y_2} e^{(-y^2/2)} dy \quad ...(1.87)$$

where, $$y_1 = \sqrt{2}hx_1$$

and, $$y_2 = \sqrt{2}hx_2$$

Eqn. (1.87) may be written as :

$$P(y_1, y_2) = \frac{1}{\sqrt{2\pi}} \int_0^{y_2} (e)^{(-y^2/2)} dy - \frac{1}{\sqrt{2\pi}} \int_0^{y_1} e^{(-y^2/2)}.dy \quad \text{...(1.88)}$$

Each of the two integrals in equation (1.88) is a function of a single parameter $y_1$ or $y_2$. Therefore, probability,

$$P(0, y) = \frac{1}{\sqrt{2\pi}} \int_0^{y} (e)^{(-y^2/2)} dy \quad \text{...(1.89)}$$

The results of this integral are given in Table 1.2 and can be used to compute the probability between the limits $x_1$ and $x_2$. Tabulated values of the integral represent the area under the normalised Gaussian error curve for $y = 0$ and $y = y$ as shown in Fig. 1.36.

As the curve is *symmetrical* the same table may be used for *–ve* deviations replacing $y$ by $-y$ in equation (1.89) since the value of probability is not affected by this substitution.

— Suppose we are required to find the probability that the deviation lies between $\pm\sigma$. For the *+ve* deviation $\sigma$,

$$y = \frac{q-\bar{q}}{\sigma} = \frac{\sigma}{\sigma} = 1$$

The table 1.2 indicates the probability value $P = 0.3413$ for $y = 1.0$. Due to symmetrical nature of the error curve, an equal probability would exist for the deviations to lie between $-\sigma$ and 0. Hence the probability that deviation lies between $\pm\sigma$ works out to be $2 \times 0.3413 = 0.6826$.

**Table 1.2. Probability Tables**

| y | 0 | 0.01 | 0.02 | 0.03 | 0.04 | 0.05 | 0.06 | 0.07 | 0.08 | 0.09 |
|---|---|---|---|---|---|---|---|---|---|---|
| 0.0 | 0.0000 | 0.0040 | 0.0080 | 0.0120 | 0.0160 | 0.0199 | 0.0239 | 0.0279 | 0.0319 | 0.0359 |
| 0.1 | 0.0398 | 0.0438 | 0.0478 | 0.0517 | 0.0557 | 0.0596 | 0.0636 | 0.0675 | 0.0714 | 0.0754 |
| 0.2 | 0.0793 | 0.0832 | 0.0871 | 0.0910 | 0.0948 | 0.0987 | 0.1026 | 0.1064 | 0.1103 | 0.1141 |
| 0.3 | 0.1179 | 0.1217 | 0.1255 | 0.1293 | 0.1331 | 0.1368 | 0.1406 | 0.1443 | 0.1480 | 0.1517 |
| 0.4 | 0.1554 | 0.1591 | 0.1628 | 0.1664 | 0.1700 | 0.1736 | 0.1772 | 0.1808 | 0.1844 | 0.1879 |
| 0.5 | 0.1915 | 0.1950 | 0.1985 | 0.2019 | 0.2054 | 0.2088 | 0.2123 | 0.2157 | 0.2190 | 0.2224 |
| 0.6 | 0.2258 | 0.2291 | 0.2324 | 0.2357 | 0.2389 | 0.2422 | 0.2454 | 0.2486 | 0.2518 | 0.2549 |
| 0.7 | 0.2580 | 0.2612 | 0.2642 | 0.2673 | 0.2704 | 0.2734 | 0.2764 | 0.2794 | 0.2823 | 0.2852 |
| 0.8 | 0.2881 | 0.2910 | 0.2939 | 0.2967 | 0.2996 | 0.3023 | 0.3051 | 0.3079 | 0.3106 | 0.3133 |
| 0.9 | 0.3159 | 0.3186 | 0.3212 | 0.3238 | 0.3264 | 0.3289 | 0.3315 | 0.3340 | 0.3365 | 0.3389 |
| 1.0 | 0.3413 | 0.3438 | 0.3461 | 0.3485 | 0.3508 | 0.3531 | 0.3554 | 0.3577 | 0.3599 | 0.3621 |
| 1.1 | 0.3643 | 0.3665 | 0.3686 | 0.3708 | 0.3729 | 0.3749 | 0.3770 | 0.3790 | 0.3810 | 0.3830 |
| 1.2 | 0.3849 | 0.3869 | 0.3888 | 0.3907 | 0.3925 | 0.3944 | 0.3962 | 0.3980 | 0.3997 | 0.4015 |
| 1.3 | 0.4032 | 0.4049 | 0.4066 | 0.4082 | 0.4099 | 0.4115 | 0.4131 | 0.4147 | 0.4162 | 0.4177 |
| 1.4 | 0.4192 | 0.4207 | 0.4222 | 0.4236 | 0.4251 | 0.4265 | 0.4279 | 0.4292 | 0.4306 | 0.4319 |
| 1.5 | 0.4332 | 0.4345 | 0.4357 | 0.4370 | 0.4382 | 0.4394 | 0.4406 | 0.4418 | 0.4430 | 0.4441 |
| 1.6 | 0.4452 | 0.4463 | 0.4474 | 0.4485 | 0.4495 | 0.4505 | 0.4515 | 0.4525 | 0.4535 | 0.4545 |
| 1.7 | 0.4554 | 0.4564 | 0.4573 | 0.4582 | 0.4591 | 0.4599 | 0.4608 | 0.4616 | 0.4625 | 0.4633 |
| 1.8 | 0.4641 | 0.4649 | 0.4656 | 0.4664 | 0.4671 | 0.4678 | 0.4686 | 0.4693 | 0.4700 | 0.4706 |

| 1.9 | 0.4713 | 0.4719 | 0.4726 | 0.4732 | 0.4738 | 0.4744 | 0.4750 | 0.4756 | 0.4762 | 0.4767 |
|---|---|---|---|---|---|---|---|---|---|---|
| 2.0 | 0.4773 | 0.4778 | 0.4783 | 0.4788 | 0.4793 | 0.4798 | 0.4803 | 0.4808 | 0.4812 | 0.4817 |
| 2.1 | 0.4821 | 0.4826 | 0.4830 | 0.4834 | 0.4838 | 0.4842 | 0.4846 | 0.4850 | 0.4854 | 0.4857 |
| 2.2 | 0.4861 | 0.4865 | 0.4868 | 0.4871 | 0.4875 | 0.4878 | 0.4881 | 0.4884 | 0.4887 | 0.4890 |
| 2.3 | 0.4893 | 0.4896 | 0.4898 | 0.4901 | 0.4904 | 0.4906 | 0.4909 | 0.4911 | 0.4913 | 0.4916 |
| 2.4 | 0.4918 | 0.4920 | 0.4922 | 0.4925 | 0.4927 | 0.4929 | 0.4931 | 0.4932 | 0.4934 | 0.4936 |
| 2.5 | 0.4938 | 0.4940 | 0.4941 | 0.4943 | 0.4945 | 0.4946 | 0.4948 | 0.4949 | 0.4951 | 0.4952 |
| 2.6 | 0.4953 | 0.4955 | 0.4956 | 0.4957 | 0.4959 | 0.4960 | 0.4961 | 0.4962 | 0.4963 | 0.4964 |
| 2.7 | 0.4965 | 0.4966 | 0.4967 | 0.4968 | 0.4969 | 0.4970 | 0.4971 | 0.4972 | 0.4973 | 0.4974 |
| 2.8 | 0.4974 | 0.4975 | 0.4976 | 0.4977 | 0.4977 | 0.4978 | 0.4979 | 0.4980 | 0.4980 | 0.4981 |
| 2.9 | 0.4981 | 0.4982 | 0.4983 | 0.4983 | 0.4984 | 0.4984 | 0.4985 | 0.4985 | 0.4986 | 0.4986 |
| 3.0 | 0.4987 | 0.4987 | 0.4987 | 0.4988 | 0.4988 | 0.4989 | 0.4989 | 0.4989 | 0.4990 | 0.4990 |
| 3.1 | 0.4990 | 0.4991 | 0.4991 | 0.4991 | 0.4992 | 0.4992 | 0.4992 | 0.4992 | 0.4993 | 0.4993 |
| 3.2 | 0.4993 | 0.4993 | 0.4994 | 0.4994 | 0.4994 | 0.4994 | 0.4994 | 0.4995 | 0.4995 | 0.4995 |
| 3.3 | 0.4995 | 0.4995 | 0.4996 | 0.4996 | 0.4996 | 0.4996 | 0.4996 | 0.4996 | 0.4996 | 0.4997 |
| 3.4 | 0.4997 | 0.4997 | 0.4997 | 0.4997 | 0.4997 | 0.4997 | 0.4997 | 0.4997 | 0.4998 | 0.4998 |
| 3.5 | 0.4998 | 0.4998 | 0.4998 | 0.4998 | 0.4998 | 0.4998 | 0.4998 | 0.4998 | 0.4998 | 0.4998 |
| 3.6 | 0.4998 | 0.4999 | 0.4999 | 0.4999 | 0.4999 | 0.4999 | 0.4999 | 0.4999 | 0.4999 | 0.4999 |
| 3.7 | 0.4999 | 0.4999 | 0.4999 | 0.4999 | 0.4999 | 0.4999 | 0.4999 | 0.4999 | 0.4999 | 0.4999 |
| 3.8 | 0.4999 | 0.4999 | 0.4999 | 0.4999 | 0.4999 | 0.4999 | 0.4999 | 0.5000 | 0.5000 | 0.5000 |

– Further, let it be required to find the probable error which represents the interval about the mean in which lies half of the data. This corresponds to symmetrical area about the mean which has a value of 0.5. Only half of this area lies to the right $y = 0$ and accordingly we look for the value of $y$ in the probability table that corresponds to 0.25. The corresponding value of $y$ = **0.6745** and accordingly the probable error is given by : *Probable error* = $\pm$ 0.6745 $\sigma$.

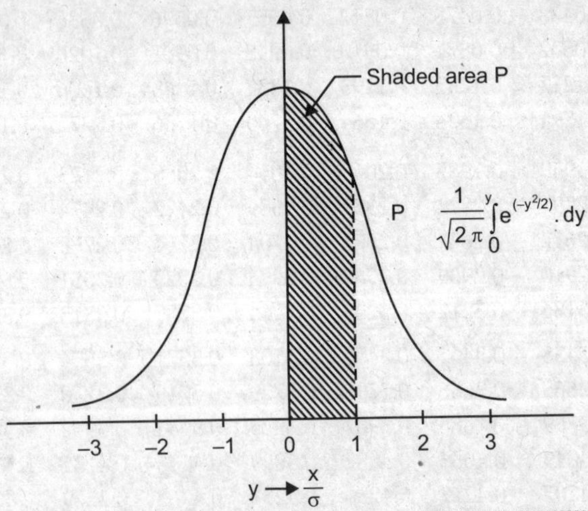


**Fig. 1.37.** Graphical representation of probability.

**Example 1.51.** *While measuring a temperature the following ten readings were recorded : 39.6, 39.9, 39.7, 39.9, 40.0, 39.8, 39.9, 39.8, 40.4 and 39.7°C.*

*Calculate the following :*

*(i) The mean,*

*(ii) The standard deviation,*

*(iii) The probable error of one reading,*

*(iv) the probable error of mean, and*

*(v) The range.*

**Solution.** The statistical values of the given date are computed with the help of the following table :

| *Observation number* | *Temperature reading, °C* $(q)$ | *Deviation,* $d = q - \bar{q}$ | $d^2$ |
|---|---|---|---|
| 1 | 39.6 | − 0.27 | 0.0729 |
| 2 | 39.9 | + 0.03 | 0.0009 |
| 3 | 39.7 | − 0.17 | 0.0289 |
| 4 | 39.9 | + 0.03 | 0.0009 |
| 5 | 40.0 | + 0.13 | 0.0169 |
| 6 | 39.8 | − 0.07 | 0.0049 |
| 7 | 39.9 | + 0.03 | 0.0009 |
| 8 | 39.8 | − 0.07 | 0.0049 |
| 9 | 40.4 | + 0.53 | 0.2809 |
| 10 | 39.7 | − 0.17 | 0.0289 |
| $n = 10$ | $\Sigma q = 398.7$ <br> $\bar{q} = \frac{\Sigma q}{n} = \frac{398.7}{10} = 39.87$ | | $\Sigma d^2 = 0.441$ |

**(*i*) The mean value of the reading :**

$$\bar{q} = \frac{\Sigma q}{n} = \frac{398.7}{10} = \mathbf{39.87°C} \quad \textbf{(Ans.)}$$

**(*ii*) The standard deviation :**

$$s = \sqrt{\frac{\Sigma d^2}{(n-1)}} \quad \text{...since the number of observations is 10 (\textit{i.e.}, less than 20).}$$

$$= \sqrt{\frac{0.441}{10-1}} = \mathbf{0.22°C} \quad \textbf{(Ans.)}$$

**(*iii*) Probable error of one reading :**

$$r_1 = 0.6745s = 0.6745 \times 0.22 = \mathbf{0.15°C} \quad \textbf{(Ans.)}$$

**(*iv*) Probable error of mean :**

$$r_m = \frac{r_1}{\sqrt{n-1}} = \frac{0.15}{\sqrt{10-1}} = \mathbf{0.05°C} \quad \textbf{(Ans.)}$$

**(*v*) Range :**

$$Range = 40.4 - 39.6 = \mathbf{0.8°C} \quad \textbf{(Ans.)}$$

**Example 1.52.** *In a test temperature, measurements of temperature were made 100 times with variations in apparatus and procedures. After applying the corrections the following results were obtained :*

| *Temperature °C* | 197 | 198 | 199 | 200 | 201 | 202 | 203 | 204 | 205 |
|---|---|---|---|---|---|---|---|---|---|
| *Frequency of Occurrence* | 2 | 4 | 10 | 24 | 36 | 14 | 5 | 3 | 2 |

*Calculate the following :*

*(i) Arithmetic mean,*

*(ii) Average deviation,*

*(iii) Standard deviation,*

*(iv) Variance,*

*(v) Probable error of one reading,*

*(vi) Probable error of the mean, and*

*(vii) Standard deviation of the standard deviation.*

**Solution.** The computations in tabular form are given below :

| *Temperature °C* $T$ | *Frequency of occurrence, f* | $T \times f$ | *Deviation* $d\left(=T-\frac{\Sigma T \times f}{n}\right)$ | $f \times d$ | $d^2$ | $f \times d^2$ |
|---|---|---|---|---|---|---|
| 197 | 2 | 394 | – 3.77 | – 7.54 | 14.213 | 28.426 |
| 198 | 4 | 792 | – 2.77 | – 11.08 | 7.673 | 30.692 |
| 199 | 10 | 1990 | – 1.77 | – 17.7 | 3.133 | 31.330 |
| 200 | 24 | 4800 | – 0.77 | – 18.48 | 0.593 | 14.232 |
| 201 | 36 | 7236 | + 0.23 | + 8.28 | 0.053 | 1.908 |
| 202 | 14 | 2828 | + 1.23 | + 17.22 | 1.513 | 21.182 |
| 203 | 5 | 1015 | + 2.23 | + 11.15 | 4.973 | 24.865 |
| 204 | 3 | 612 | + 3.23 | + 9.69 | 10.433 | 31.299 |
| 205 | 2 | 410 | + 4.23 | + 8.46 | 17.893 | 35.786 |
| Total | $n = 100$ | $\Sigma T \times f =$ 20077 | | $\Sigma\lvert f \times d\rvert$ = 109.6 | | $\Sigma f \times d^2$ = 219.72 |

**(*i*) Arithmetic mean :**

$$\text{Arithmetic mean} = \frac{\Sigma T \times f}{n} = \frac{20077}{100} = \mathbf{200.77°C} \quad \textbf{(Ans.)}$$

**(*ii*) Average deviation :**

$$\sigma = \frac{\Sigma|f \times d|}{n} = \frac{109.6}{100} = \mathbf{1.096°C} \quad \textbf{(Ans.)}$$

**(*iii*) Standard deviation :**

$$\textit{Standard deviation, } \sigma = \sqrt{\frac{\Sigma f \times d^2}{n}} = \sqrt{\frac{219.72}{100}} = \mathbf{1.482°C} \quad \textbf{(Ans.)}$$

**(*iv*) Variance :**

$$\text{Variance, } V = \sigma^2 = (1.482)^2 = \mathbf{2.196°C^2} \quad \textbf{(Ans.)}$$

**(*v*) Probable error of one reading :**

*Probable error of one reading,*

$$r_1 = 0.6745\ \sigma = 0.6745 \times 1.482 \simeq \mathbf{1°C} \quad \textbf{(Ans.)}$$

**(*vi*) Probable error of the mean :**

$$r_m = \frac{r_1}{\sqrt{n}} = \frac{1(°C)}{\sqrt{100}} = \mathbf{0.1°C} \quad \textbf{(Ans.)}$$

(*vii*) **Standard deviation of the standard deviation :**

Standard deviation of the mean,

$$\sigma_m = \frac{\sigma}{\sqrt{n}} = \frac{1.482}{\sqrt{100}} = 0.148°C$$

∴ Standard deviation of the standard deviation,

$$\sigma_\sigma = \frac{\sigma_m}{\sqrt{2}} = \frac{0.148}{\sqrt{2}} = \mathbf{0.1046°C} \quad \textbf{(Ans.)}$$

**Example 1.53.** *A known current of 80A is measured by an ammeter. If 40% of the readings are within 0.8A of true value, determine the following :*

*(i) The standard deviation for the meter.*

*(ii) The probability of an error of 1.2 A.*

**Solution.** (*i*) **Standard deviation for the meter :**

Considering a normal distribution curve, half the readings, *i.e.*, 20% of the total readings have *a +ve* error and the remaining 20% have a *–ve* error.

From probability tables, it is found that corresponding to a probability of $P(y) = 0.2$,

$$y = \frac{x}{\sigma} = 0.5248$$

∴ Standard deviation,

$$\sigma = \frac{x}{y} = \frac{0.8}{0.5248} = \mathbf{1.524} \quad \textbf{(Ans.)}$$

(*ii*) **Probability of an error of 1.2 A :**

Now,
$$y = \frac{x}{\sigma} = \frac{1.2}{1.524} = 0.787$$

Corresponding to $y = 0.787$, probability is 0.2842

∴ Probability of an error of 1.2*A* = 2 × 0.2842 = 0.5684

Thus about 57% of the readings are *within ± 1.2A of the true value.* **(Ans.)**

**Example 1.54.** *An underdamped galvanometer was energized 100 times under the same carefully controlled experimental conditions and maximum deflection was read in each case. The readings were distributed about a mean value of 21.9 mm and had a probable error of 2.1 mm. How many of 100 readings would you estimate exceeded 25 mm ?*

**Solution.** Deviation, $x = 25 - 21.9 = 3.1$ mm

Probable error, $r = 2.1$ mm

From eqn., 1.82, standard deviation,

$$\sigma = \frac{r}{0.6745} = \frac{2.1}{0.6745} = 3.1$$

$$\text{Ratio, } y = \frac{x}{\sigma} = \frac{3.1}{3.1} = 1.0$$

Corresponding to ratio $y = \frac{x}{\sigma} = 1.0$, the area under the curve is 0.3413 (from table 1.2).

∴ Number of readings having a deviation within ± 3.1 mm

= 2 × 0.3413 × 100 = 68

Number of readings exceeding a deviation of ± 3.1 mm

$$= 100 - 68 = 32$$

Hence, the number of readings having a deviation of + 3.1 mm

$$= \frac{32}{2} = 16.$$

Thus the *number of readings exceeding a maximum deflection of 25 mm i.e.,* (21.9 + 3.1) *mm is 16.* **(Ans.)**

**Example 1.55.** *A machine shop was assigned the task to manufacture 10,000 aluminium rods of a nominal length of 20 mm; the rod length was stipulated neither to exceed 20.25 mm and nor to be smaller than 19.5 mm. When inspected for quality control, it was found that 1000 of the rods were too long to fit into a gauge set at 20.25 mm. Determine the number of remaining 9000 rods which will conform to the specifications. Assume that the measurement data conforms to Gaussian normal distribution curve.*

**Solution.** In case of normal distribution, there is equal probability of the +*ve* and –*ve* errors about the mean. Therefore 5000 rods have lengths greater than 20 mm and 5000 rods have lengths smaller than 20 mm. Since 1000 rods have length greater than 20.25 mm, therefore,

The number of rods where length lies between 20 mm and 20.25 mm

$$= 5000 - 1000 = 4000$$

∴ Probability that 4000 rods have a value greater than 20 mm and less than 20.25 mm,

$$P(y) = \frac{4000}{10,000} = 0.4$$

From probability tables, it is found that corresponding to a probability of $P(y) = 0.4$,

$$y = \frac{x}{\sigma} \simeq 1.3 \qquad \text{...(From table 1.2)}$$

∴ Standard deviation,

$$\sigma = \frac{x}{1.3} = \frac{20.25 - 20.00}{1.3} = \frac{0.25}{1.3} = 0.1923$$

Now, for rods with nominal length of 19.5 mm and 20 mm,

$$y = \frac{x}{\sigma} = \frac{(20 - 19.5)}{0.1923} = 2.60$$

Corresponding probability, $P(y) = 0.4953$ ...(From table 1.2)

∴ Number of rods that have lengths between 19.5 mm and 20 mm

$$= 10000 \times 0.4953 = 4953$$

Hence, *total number of rods whose lengths lie between specified limits of 19.5 mm to 20.25 mm,*

$$= 4000 + 4953 = \mathbf{8953} \quad \textbf{(Ans.)}$$

**Example 1.56.** *The speed of a hydraulic turbine model, which is running at 1500 rpm, is measured by a tachometer. The hydraulic turbine is subject to variations in speed. For a sample of 20 readings at this speed, how many readings would you expect between 1485 and 1515. Assume that the tachometer gives a normal set of deviations with precision index of 0.04 at 1500 rpm.*

**Solution.** *Given* : Deviation, $d = q - \bar{q} = \pm 15$ r.p.m.

and, precision index, $h = 0.04$

Standard deviation, $\sigma = \dfrac{1}{\sqrt{2}h}$ ...[Eqn. (1.81)]

$$= \frac{1}{\sqrt{2} \times 0.04} = 17.68$$

$$y = \frac{x}{\sigma} = \frac{q - \bar{q}}{\sigma} = \frac{15}{17.68} = 0.8484$$

Corresponding to $y = 0.8484$, as read from the probability tables

$P(y) = 0.3015$ ...(From table 1.2)

Probability of an error ± 15 r.p.m. = 2 × 0.3015 = **0.603**

Thus we expect 60% of 0.6 × 20 = **12 readings** *to lie between 1485 to 1515 r.p.m.* **(Ans.)**

**Example 1.57.** *The depth of water which had a nominal value of 15 cm, was measured with a hook. Measurements were taken 40 times and 10 readings were found to lie outside a particular range. Assuming that the measurements conform to normal distribution with precision index 9 cm$^{-1}$, determine the prescribed range.*

**Solution.** *Given* : Nominal value of depth of water = 15 cm

Total number of times the measurements taken = 40

No. of measurement readings found to lie outside a particular range = 10

Precision index, $h = 9$ cm$^{-1}$

**Prescribe range :**

Probability of falling within a particular range

$$= \frac{(40 - 10)}{40} = 0.75$$

Half of these measurements have a *+ve* error and half have a *–ve* error.

$$P(y) = \frac{0.75}{2} = 0.375$$

From probability tables, it is found that corresponding to a probability

$P(y) = 0.375,$

$y = \dfrac{x}{\sigma} = \dfrac{q - \bar{q}}{\sigma} = 1.15$ ...(From table 1.2)

Now, standard deviation, $\sigma = \dfrac{1}{\sqrt{2}h}$ ...[Eqn. (1.81)]

$$= \frac{1}{\sqrt{2} \times 9} = 0.0786$$

$\therefore$ $x = (q - \bar{q}) = 1.15 \times 0.0786 = 0.0904$ cm

*Thus 75 percent of depth measurements lie within the* **range (15 ± 0.0904) cm. (Ans.)**

**Example 1.58.** *(a) A mass flow meter, using electromagnetic principle is set to record a large number of readings. The fixed mass flow rate is 100 kg/s; half of the readings lie between 104.8 and 95.2 kg/s. The sample measurements give a straight line between the (0, 50%) points on the probability coordinates. Calculate the precision index of the instrument.*

*(b) An alarm rings when the flow meter gives a reading of 88 kg/s.*

*(i) How many false alarm are expected in the month of November if the flow is checked automatically 4 times a day ?*

*(ii) What will be the value of precision index if one wants to reduce the false alarm rate by a factor of 2 ?*

**Solution.** (*a*) The overall probability is 0.5. Considering normal distribution, 0.25 of the measurements have a *+ve* error and the rest 0.25 of the measurements have a *–ve* error.

Corresponding to a probability of $P(y) = 0.25$, from probability tables, we have

$$y = \frac{x}{\sigma} = 0.675 \quad \text{...(From table 1.2)}$$

$$\therefore \text{ Standard deviation, } \quad \sigma = \frac{x}{0.675} = \frac{4.8}{0.675} = 7.11$$

$$\textbf{Precision index,} \quad h = \frac{1}{\sqrt{2}\sigma} = \frac{1}{\sqrt{2} \times 7.11} = \mathbf{0.099} \quad \textbf{(Ans.)}$$

(*b*) (*i*) For a mass flow rate of 88 kg/s.

$$y = \frac{x}{\sigma} = \frac{100-88}{7.11} = 1.69$$

Corresponding probability $P(y) = 0.45$ ...(From table 1.1)

Thus it is expected that (0.5 – 0.45) = 0.05 or 5% of the flow measurements will fall in the range of false alarms.

Number of measurements in the month of November = 30 × 4 = 120

∴ *Expected false alarms* = 120 × 0.05 = **6** **(Ans.)**

(*ii*) Reduced number of false alarms $= \frac{1}{2} \times 6 = 3$

Probability of false alarms $= \frac{3}{120} = 0.025$ (or 2.5%)

Probability of data to lie in the tolerant band,

$$P(y) = 0.5 - 0.025 = 0.475$$

Corresponding to $P(y) = 0.475$

$$y = \left(\frac{x}{\sigma}\right) = 1.96 \quad \text{...(From table 1.2)}$$

$$\text{or, Standard deviation, } \quad \sigma = \frac{x}{1.96} = \frac{100-88}{1.96} = 6.12$$

$$\therefore \textit{ Precision index,} \quad h = \frac{1}{\sqrt{2}\sigma} = \frac{1}{\sqrt{2} \times 6.12} = \mathbf{0.1155} \quad \textbf{(Ans.)}$$

### 1.12.5.6. Odds

***Odds*** *is the number of chances that a particular reading will occur when error limit is specified.* Thus the probability of occurrence can be stated in terms of *odds*.

*Example.* When the error limits are specified as ± 0.6745σ, the chances are that 50 percent of observations will lie between the above limits or in other words we can say odds are 1 to 1.

The odds can be calculated as under :

$$\text{Probability of occurrence} = \frac{\text{Odds}}{\text{Odds} + 1} \quad \text{...(1.90)}$$

Certain deviations, their probability of occurrence and the corresponding odds are given in the table 1.3.

**Table 1.3. Deviations, Probability and Odds**

| *Deviation* | *Probability* | *Odds* |
|---|---|---|
| $\pm$ 0.6745 $\sigma$ | 0.5000 | 1 to 1 |
| $\pm$ $\sigma$ | 0.6826 | 2.15 to 1 |
| $\pm$ 2$\sigma$ | 0.9546 | 21 to 1 |
| $\pm$ 3$\sigma$ | 0.9974 | 369 to 1 |

#### 1.12.5.7. Specification of measurement data

The results of statistical analysis of multi-sample data are normally expressed as deviations about a mean value; the deviations are expressed as follows :

(*i*) **Probable error.** The result is expressed as $\bar{q} \pm 0.6745\,\sigma$, which implies that 0.50 (50%) of the readings are within this limit and odds are 1 to 1. Obviously there is an even probability that a new reading will lie within with these limits.

(*ii*) **Standard deviation.** The result is expressed as $\bar{q} \pm \sigma$. In this case, the error limit is the standard deviation. It implies that 0.6828 (about 68%) of the readings are within the limits $\sigma = \pm$ 1 and the odds are 2.15 to 1. Thus there is approximately a 2 to 1 possibility that a new reading will fall within this limit.

(*iii*) **$\pm$ 2$\sigma$ limit.** The result is express as $\bar{q} \pm 2\sigma$, which implies that 0.9546 (about 95%) of the readings are within the given limits and the odds are 21 to 1.

(*iv*) **$\pm$ 3$\sigma$ limit.** The result is expressed as $\bar{q} \pm 3\sigma$, which implies that 0.9974 (about 99%) of the readings are within the given limits. The odds of any observation falling within this limit are 369 to 1.

#### 1.12.5.8. Confidence interval and confidence level

*The range of deviation from the mean value within which a certain fraction of all values are expected to lie is called* **confidence interval.**

*The probability that the value of a randomly selected observation will lie in the range of confidence interval is termed as* **confidence level.**

When the number of observations is large and their errors are random and follow Gaussian distribution, the various confidence intervals about the mean value $\bar{q}$ are given in table 1.4.

**Table 1.4. Confidence Intervals**

| *Confidence level* | *Confidence interval* | *Values lying outside confidence interval* |
|---|---|---|
| 0.500 | $\bar{q} \pm 0.674\sigma$ | 1 in 2 |
| 0.800 | $\bar{q} \pm 1.282\sigma$ | 1 in 5 |
| 0.900 | $\bar{q} \pm 1.645\sigma$ | 1 in 10 |
| 0.950 | $\bar{q} \pm 1.960\sigma$ | 1 in 20 |
| 0.990 | $\bar{q} \pm 2.576\sigma$ | 1 in 100 |
| 0.999 | $\bar{q} \pm 3.291\sigma$ | 1 in 1000 |

If the number of observations is small (say *less* than 20) and the standard deviation is not accurately known, the confidence interval must be *broadened*. The standard deviation, in this case, is computed as :

$s = \sqrt{\frac{\Sigma d^2}{n-1}}$; this standard is multiplied by a suitable factor to establish the confidence interval.

**1.12.5.9. Probability graph paper**

Prior to undertaking statical analysis of the measurement data, the experimenter needs to check that the data is in conformity with the normal distribution. To accomplish this, the data is plotted on a probability graph paper wherein the abscissa is the value of a particular reading (on a linear scale) and the ordinate has a percentage of readings at or below the value of abscissa (on a logarithmic scale). A straight line is then drawn through the data points; however, while fitting the curve much weightage is not given to the points at low and high ends of the plot. *If the 50 percent reading at the ordinate exactly corresponds to averages of all the data points, the data is considered to follow the Gaussian distribution.*

**1.12.5.10. Rejection of test data–Chauvenet's criterion**

During experimentation, it is observed that some points are noticeably different from the majority of the data. Such points can be determined and subsequently eliminated by applying Chauvenet's criterion.

**Chauvenet's criterion** states as follows :

*"An observed reading may be rejected if the probability of obtaining the particular deviation from the mean is less than* $\frac{1}{2n}$". The quantity $n$ refers to the number of observations which are large enough to presume that the data follows the Gaussian error distribution.

Let us consider that during a measurement act ten readings have been made. Then,

$$\frac{1}{2n} = \frac{1}{2 \times 10} = 0.05 \text{ or } 5\%$$

Thus readings with 95 percent of probability are to be accepted, and less than that can be rejected. Corresponding to probability

$$P(y) = \frac{0.95}{2} = 0.475;\ y = 1.96$$

Results based on such calculations for different number of readings are given in table 1.5.

**Table 1.5. Chauvenet's Criterian for Rejecting Data**

| *Number of readings* | *Ratio of maximum acceptable deviation to standard-deviation,* $d_{max}/\sigma$ |
|---|---|
| 2 | 1.15 |
| 3 | 1.38 |
| 4 | 1.54 |
| 5 | 1.65 |
| 6 | 1.73 |
| 7 | 1.80 |
| 10 | 1.96 |
| 15 | 2.13 |

| | |
|---|---|
| 25 | 2.33 |
| 50 | 2.57 |
| 100 | 2.81 |
| 300 | 3.14 |
| 500 | 3.29 |
| 1000 | 3.48 |

When applying Chauvenet's criterion, in order to eliminate any dubious data, the following *procedure* is adopted :

(*i*) Using all data points calculate the standard deviation.

(*ii*) Compare the deviation of the individual readings with standard deviation.

(*iii*) If the ratio of deviation of a reading to the standard deviation exceeds the limits given in table 1.5, reject the readings.

(*iv*) Calculate again the mean value and the standard deviations by excluding the rejected readings from the data.

- A criterion (*based upon "confidence intervals"*) used for discarding a data point is that its deviation from the mean exceeds four times the probable error of a single reading. This results in discarding a data outside a confidence interval for a single reading at a confidence level of 0.993.

— A better criterion which does not involve the evaluation of probable error when set of data points is small and standard deviation is not accurately known, *is to discard a reading that lies outside the interval corresponding to confidence level of 0.99 for a single observation* (Refer to table 1.4 and 1.5). On this basis not more than one reading in 100 would lie outside their range.

— A still better method is to use confidence interval corresponding to a confidence level of 0.95 in order to scrutinize the measurement procedure adopted.

**Example 1.59.** *10 copper rods, selected at random, were found to a have the following lengths in meters : 5.30, 5.73, 6.77, 5.26, 4.33, 5.45, 6.09, 5.64, 5.81, 5.75.*

*Determine any reading that can be rejected by applying Chauvenet's criterion. The ratio of maximum deviation to standard deviation should not exceed 1.96.*

**Solution.** The computations in a tabular form are given below :

| *S.No.* | *Length q, m* | Deviation $d = q - \bar{q}$ | $d^2$ | $\frac{\lvert d \rvert}{s}$ |
|---|---|---|---|---|
| 1 | 5.30 | − 0.313 | 0.09797 | 0.5 |
| 2 | 5.73 | + 0.117 | 0.01369 | 0.19 |
| 3 | 6.77 | + 1.157 | 1.33865 | 1.85 |
| 4 | 5.26 | − 0.353 | 0.12461 | 0.56 |
| 5 | 4.33 | − 1.283 | 1.64609 | 2.05 |
| 6 | 5.45 | − 0.163 | 0.02657 | 0.26 |
| 7 | 6.09 | + 0.477 | 0.22753 | 0.76 |
| 8 | 5.64 | + 0.027 | 0.00073 | 0.04 |
| 9 | 5.81 | + 0.197 | 0.03881 | 0.31 |
| 10 | 5.75 | + 0.137 | 0.01877 | 0.22 |
| $n = 10$ | $\Sigma q = 56.13$ <br> $\bar{q} = \frac{56.13}{10} = 5.613$ | | $\Sigma d^2 = 3.53342$ | |

Arithmetic mean, $\bar{q} = \frac{\Sigma q}{n} = \frac{56.13}{10} = 5.613$

Standard deviation, $s = \sqrt{\frac{\Sigma d^2}{n-1}}$ ... since the number of readings is less than 20.

$$= \sqrt{\frac{3.53342}{10}} = 0.626$$

It is given that for 10 readings the ratio of deviation to standard deviation is not to exceed 1.96 and therefore **reading number 5** *i.e.,* **4.33 m** *should be rejected.* **(Ans.)**

### 1.12.5.11. Method of least squares

- Let us consider a set of $n$ readings $q_1, q_2, q_3, .... q_n$. The sum of squares of their deviations for some mean value $Q_m$ is,

$$S = \sum_{i=1}^{n} (q_1 - Q_m)^2$$

For minimizing $S$ with respect to mean value $Q_m$, we have

$$\frac{\partial S}{\partial Q_m} = 2\left(\sum_{i=1}^{n} q_1 - nQ_m\right) = 0$$

or, $$Q_m = \frac{1}{n}\sum_{i=1}^{n} q_i$$

But, $$\frac{1}{n}\sum_{i=1}^{n} q_i = \bar{q}$$

$$\therefore \quad Q_m = \bar{q} \quad ...(1.91)$$

Thus the *mean value which minimizes the sum of the squares of the deviation is the arithmetic mean.*

- Now consider two variables $u$ and $v$ measured over a range of values and let us find the linear function connecting $y$ and $x$.

Let the linear function is :

$$v = au + b$$

Now let us minimize the quantity,

$$S = \sum_{i=1}^{n} [v_i - (au_i + b)]^2$$

This is accomplished by taking derivatives of $S$ w.r.t. $a$ and $b$ and setting them equal to zero. This gives

$$nb + a\Sigma u_i = \Sigma v_i \quad ...(i)$$

and, $$b\Sigma u_i + a\Sigma u_i^2 = \Sigma u_i\, v_i \quad ...(ii)$$

Solving, the above eqns., (*i*) and (*ii*) we get

$$a = \frac{n\Sigma u_i v_i - (\Sigma u_i)(\Sigma v_i)}{n\Sigma u_i^2 - (\Sigma u_i)^2} \quad ...(1.92)$$

$$b = \frac{(\Sigma v_i)(\Sigma u_i^2) - (\Sigma u_i v_i)(\Sigma u_i)}{n\Sigma u_i^2 - (\Sigma u_i)^2} \quad ...(1.93)$$

The standard deviations may be found as :

$$s_a = \sqrt{\frac{n}{n\Sigma u_i^2 - (\Sigma u_i)^2}} s_v \quad ...(1.94)$$

and

$$s_b = \sqrt{\frac{\Sigma u_i^2}{n\Sigma u_i^2 - (\Sigma u_i)^2}} s_v \quad ...(1.95)$$

where, $s_v = \sqrt{\frac{1}{n}\sum (au_i + b - v_i)^2}$

Standard deviation of $u$ is,

$$s_u = \sqrt{\frac{1}{n}\sum\left(\frac{v_i - b}{a} - u_i\right)^2} = \frac{s_v}{a} \quad ...(1.96)$$

**Example 1.60.** *The data given below are expected to follow a linear relationship :*

$$v = au + b$$

| $u$ | 1.8 | 4.6 | 6.6 | 9.0 | 11.4 | 13.4 |
|---|---|---|---|---|---|---|
| $v$ | 2.2 | 3.2 | 5.2 | 6.4 | 8.0 | 10.0 |

*(i) Obtain the best linear relation, and*

*(ii) Calculate the standard deviation.*

**Solution.** *Given* : $v = au + b$ .... Linear relationship ;

Number of reading, $n = 6$

The computations in a tabular form are given below :

| $u_i$ | $v_i$ | $u_i v_i$ | $u_i^2$ |
|---|---|---|---|
| 1.8 | 2.2 | 3.96 | 3.24 |
| 4.6 | 3.2 | 14.72 | 21.16 |
| 6.6 | 5.2 | 34.32 | 43.56 |
| 9.0 | 6.4 | 57.60 | 81.0 |
| 11.4 | 8.0 | 91.20 | 129.96 |
| 13.4 | 10.0 | 134.00 | 179.56 |
| $\Sigma u_i = 46.8$ | $\Sigma v_i = 35$ | $\Sigma u_i v_i = 335.8$ | $\Sigma u_i^2 = 458.48$ |

**(*i*) Best linear relation :**

Using equations (1.92) and (1.93), we have

$$a = \frac{(6)\times(335.8) - (46.8)\times(35)}{(6)\times(458.48) - (46.8)^2} = \frac{376.8}{560.64} = 0.672$$

$$b = \frac{(35)\times(458.48) - (335.8)(46.8)}{(6)(458.48) - (46.8)^2} = \frac{331.36}{560.64} \doteq 0.591$$

*The best linear relation is* : $\boldsymbol{v = 0.672\,u + 0.591}$ **(Ans.)**

(*ii*) **Standard deviation :**

The computations for finding standard deviation, in a tabular form, are given below :

| $u_i$ | $v_i$ | $au_i + b$ | $au_i + b - v_i$ | $(au_i + b - v_i)^2$ |
|---|---|---|---|---|
| 1.8 | 2.2 | 1.8006 | – 0.3994 | 0.1595 |
| 4.6 | 3.2 | 3.6822 | + 0.4822 | 0.2325 |
| 6.6 | 5.2 | 5.0262 | – 0.1738 | 0.0302 |
| 9.0 | 6.4 | 6.6390 | + 0.239 | -0.0571 |
| 11.4 | 8.0 | 8.2518 | + 0.2518 | 0.0634 |
| 13.4 | 10.0 | 9.5958 | – 0.4074 | 0.1660 |
| | | | | $\Sigma(au_i + b - v_i)^2 = 0.7087$ |

∴ Standard deviation of $v$,

$$s_v = \sqrt{\frac{1}{n}\Sigma(au_i + b - v_i)^2} = \sqrt{\frac{0.7087}{6}} = \pm \mathbf{0.3436} \quad \textbf{(Ans.)}$$

From eqn. (1.96), standard deviation of $u$,

$$s_u = \frac{s_v}{a} = \frac{\pm 0.3436}{0.672} = \pm \mathbf{0.511} \quad \textbf{(Ans.)}$$

From eqns. (1.94 and 1.95) the standard deviations in $a$ and $b$ are :

$$s_a = \sqrt{\frac{n}{n\Sigma u_i^2 - (\Sigma u_i)^2}}\, s_v$$

$$= \sqrt{\frac{6}{6 \times 458.48 - (46.8)^2}} \times (\pm 0.3436) = \pm \mathbf{0.0355} \quad \textbf{(Ans.)}$$

$$s_b = \sqrt{\frac{\Sigma u_i^2}{n\Sigma(u_i^2) - (\Sigma u_i)^2}}\, s_v$$

$$= \sqrt{\frac{458.48}{6 \times 458.48 - (46.8)^2}} \times (\pm 0.3436) = \pm \mathbf{0.3107} \quad \textbf{(Ans.)}$$

**Example 1.61.** *(a) Enumerate the factors which should be considered while selecting an instrument for a specific measurement application.*

*(b) The iron losses in a ferromagnetic material used in a transformer vary with frequency f of the supply given to the transformer. For a particular transformer these iron losses were determined at various frequencies with a constant flux density in the ferromagnetic material; the results obtained are given below :*

| *Frequency, f (Hz) :* | 550 | 700 | 850 | 1000 |
|---|---|---|---|---|
| *Iron losses, P (mW) :* | 23 | 31 | 37 | 61 |

*Assuming the iron losses to have general form* $P \doteq af^2 + bf$, *determine the constants a and b to achieve best fit for the four measured values, using method of least squares.*

**Solution.** (*a*) The following *factors* need to be considered for the selection of an instrument to be used for a specific measurement application :

1. Easy calibration.
2. Convenience and ease in reading the instrument.
3. The maximum and minimum values the input variables are expected to assume.

4. Nature of the input signal.
5. Non-interference with measuring system.
6. Accuracy expected from the instrument.
7. Cost criterion.
8. Specific application of the data.
9. Required form of data.
10. Negligible or well-known environmental errors.

(*b*) *Given* : General form $P = af^2 + bf$

or, $$\frac{P}{f} = af + b$$

The above relation may be expressed as

$$v = au + b$$

which is a straight line relationship between, $v = \frac{P}{f}$ and $u = f$,

*i.e.*, $u = f$ : 550 700 850 1000

$v = \frac{P}{f}$ : 0.04182 0.04429 0.05529 0.061

The computations are tabulated below :

| $u_i$ | $v_i$ | $u_i v_i$ | $u_i^2$ |
|---|---|---|---|
| 550 | 0.04182 | 23 | $0.302 \times 10^6$ |
| 700 | 0.04429 | 31 | $0.49 \times 10^6$ |
| 850 | 0.05529 | 47 | $0.722 \times 10^6$ |
| 1000 | 0.0610 | 61 | $1 \times 10^6$ |
| $\Sigma u_i = 3100$ | $\Sigma v_i = 0.2024$ | $\Sigma u_i v_i = 162$ | $\Sigma u_i^2 = 2.514 \times 10^6$ |

In order to find the values of $a$ and $b$ for the best straight line fit using the method of least squares, let us use the equations (1.92) and (1.93) as follows :

$$a = \frac{n\Sigma u_i v_i - (\Sigma u_i)(\Sigma v_i)}{n\Sigma u_i^2 - (\Sigma u_i)^2} \quad \text{...[Eqn (1.92)]}$$

$$= \frac{4 \times 162 - (3100)(0.2024)}{4 \times 2.514 \times 10^6 - (3100)^2} = \frac{20.56}{446000} = 46.1 \times 10^{-6}$$

$$b = \frac{(\Sigma v_i)(\Sigma u_i^2) - (\Sigma u_i v_i)(\Sigma u_i)}{n\Sigma u_i^2 - (\Sigma u_i)^2} \quad \text{...[Eqn. (1.93)]}$$

$$= \frac{0.2024 \times 2.514 \times 10^6 - 162 \times 3100}{4 \times 2.514 \times 10^6 - (3100)^2}$$

$$= \frac{6633.6}{446000} = 14.87 \times 10^{-3}$$

Hence, *the best fit relationship is :*

$$P = 46.1 \times 10^{-6} f^2 + 14.87 \times 10^{-3} f \, mW \quad \textbf{(Ans.)}$$

**Example 1.62.** *The final result of a measurement depends on two measurements in series, each having values of 50 and 100 units with errors of ± 2% and ± 1% such that final result is sum of both. Determine the error of final result if errors are considered as :*

*(i) Limiting errors, and*

*(ii) Standard deviations.*

**Solution. Error of final result :**

*(i) Limiting errors :*

When individual errors are limiting errors, then fractional errors of combination

$$= \pm\left(\frac{q_1}{q_1+q_2} \times \frac{\delta q_1}{q_1} + \frac{q_2}{q_1+q_2} \times \frac{\delta q_2}{q_2}\right)$$

$$= \pm\left(\frac{50}{50+100} \times 0.02 + \frac{100}{50+100} \times 0.01\right)$$

$$= 0.0133 \text{ or } \mathbf{1.33\%} \quad \textbf{(Ans.)}$$

*(ii) Standard deviations :*

When individual errors are standard deviations, then errors in individual measurements are 2% of 50 and 1% of 100 *i.e.,* 1 and 1

$$\text{Thus, resultant error} = \sqrt{1^2 + 1^2} = \sqrt{2}$$

$$= \frac{\sqrt{2}}{(50+100)} \times 100\% = \mathbf{0.943\%} \quad \textbf{(Ans.)}$$

## 1.13 ELECTROMECHANICAL INSTRUMENTS

### 1.13.1. General Aspects

- *The deflection type instruments with a scale and movable pointer are called* ***analog*** *instruments.*
- *The instruments which use logic circuits and techniques to obtain a measurement and then display it in numerical reading (digital) form are called* ***digital*** *instruments.*

The digital instruments claim the following *advantages* over analog instruments :

*(i)* Greater accuracy.

*(ii)* Better resolution.

*(iii)* Automatic polarity and zeroing.

*(iv)* Easy readability.

Table 1.6. gives the comparison between analog and digital type instruments.

**Table 1.6. Comparison between Analog and Digital type Instruments.**

| S.No. | Aspects | Analog type instruments | Digital type instruments |
|---|---|---|---|
| 1. | *Information form* | As the position of a pointer against a calibrated scale or dial. | As a number. |
| 2. | *Possibility of human error* | Exists | Does not exist. |
| 3. | *Best possible accuracy* | ± 0.25% | ± 0.005% or better. |
| 4. | *Resolution* | One in several hundreds. | One part in several hundred thousands. |
| 5. | *Presence of moving parts* | Moving parts involved. | These instruments can be made without moving parts. |
| 6. | *Construction* | Simple in construction and direct reading type; can perform under unfavourable conditions. | Since these instruments involve electronics, proper environmental conditions are essential. |
| 7. | *Rate of change of parameter* | These instruments enable the operator to judge the rate of change of parameter by seeing the needle movement. | Change of digital reading does not give any knowledge of rate of change of parameter. |
| 8. | *Time required to observe the reading* | If exact reading is required operator takes more time as he has to guess the approximate tenths of small division. | Reading of digital meters is very fast. |
| 9. | *Auxiliary power requirement* | These instruments require no auxiliary source of power for actuation but derive driving power for indicating system from the process. | These instruments require auxiliary source of power. |
| 10. | *Mobility* | Can be portable also. | Usually stationary type. |
|  | *Examples :* | Examples of pointer-dial output devices are :<br>• Micrometer and platform scales<br>• Manometers and Bourdon-tube pressure gauge;<br>• Mercury in glass and filled system thermometers;<br>• Speedometer of an automobile;<br>• Common voltmeters and ammeters etc. | Examples of Digital output devices are :<br>• Digital ammeters and voltmeters;<br>• Electronic and mechanical counters;<br>• Odometers;<br>• Yes-No light (On or Off);<br>• Time on a scoreboard etc. |

• *Analogue* ammeters and voltmeters are classified together, since there is no basic difference in their operating principles. The action of all ammeters and voltmeters, except those of the electrostatic variety, depends upon a deflecting torque principles by an electric current. In an ammeter this torque is produced by the current to be measured, or by a definite fraction of it. In a voltmeter, it is produced by a current that is proportional to the voltage to be measured. Hence both *voltmeters and ammeters are essentially current measuring devices.*

— The *"ammeter"* has a *low resistance* so that when it is connected in ***series*** with any circuit, it does not change the current.

The *"voltmeter"* has a *high resistance* and it is so designed that when connected in *parallel* to the circuit for measuring voltages it does *not* take appreciable current.

— *An ammeter of low range can be used as a voltmeter by connecting an external resistance in series with it.*

• ***Essential requirements of a measuring instrument :***

The essential requirements of a measuring instrument are :

1. Its introduction into the circuit where measurements are to be made, should not change the circuit conditions.
2. Its power consumption should be small.

### 1.13.2. Types of Instruments mainly used as Ammeters and Voltmeters

The following types of instruments are mainly used as *ammeters* and *voltmeters.*

1. **PMMC** (Permanent Magnet Moving Coil) meter movement; also called *D' Arsonval* meter movement :
   - Can be used for *D.C. measurements* only (most accurate type instrument).
   - This type of instrument is frequently constructed to have substandard accuracy.
   - It depends upon the magnitude effect of current.
2. **Moving iron :**
   - It is the *most commonly used* form of indicating instrument, as well as the *cheapest.*
   - It can be used for both A.C. and D.C. measurements and is very accurate, if properly designed.
3. **Electrodynamometer :**
   - The calibration of this type of instrument is the same for A.C. and D.C.
4. **Hot wire and Thermocouple :**
   - For these instruments also the above situation prevails.
   - These instruments are particularly suitable for A.C. measurements since their deflection depends directly upon the heating effect of the A.C. *i.e.,* upon the r.m.s. value of the current. Their readings are therefore *independent* of the frequency.
5. **Electrostatic :**
   - These instruments used as voltmeters have the advantage that *their power consumption is exceedingly small.*
   - They can be made to cover a large range of voltage and can be constructed to have sub-standard accuracy.
6. **Induction type :**
   - The induction principle is most generally used for watt-hour meters. This principle is not preferred for use in ammeters and voltmeters because of the comparatively high cost and inaccuracy of the instrument.
7. **Rectifier.**

### 1.13.3. Permanent Magnet Moving Coil (PMMC) Meter Movement

The basic PMMC meter movement (*widely used in electronic instruments*) is often called the D'Arsonval movement, after its inventor. It can be used for *D.C. measurements only*.

#### 1.13.3.1. Construction and working

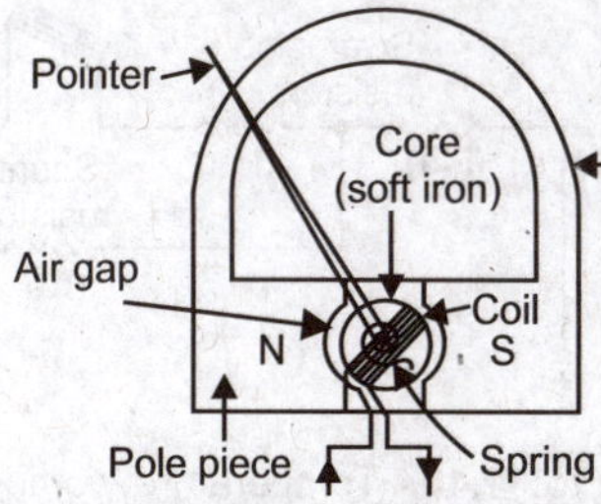


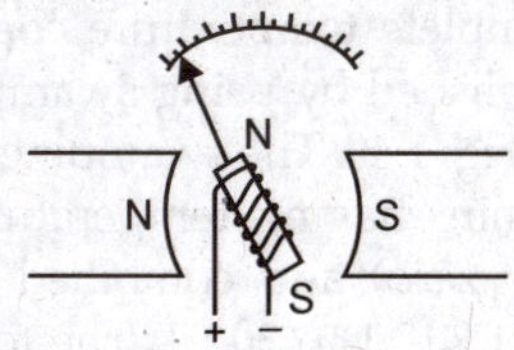


**Fig. 1.38.** PMMC meter. **Fig. 1.39.** D'Arsonval principle.

***Construction :*** Refer to Fig. 1.38

— It consists of a permanent horse-shoe magnet with soft iron pole pieces attached to it.

— Between the two pole-pieces is situated a cylinder-shaped soft iron core around which moves a coil of fine wire wound on a light metal frame. The metal frame is mounted in jewel bearings so that it can rotate freely.

— A light pointer attached to the moving coil moves up-scale as the coil rotates when the current is passed through it. The rotating coil is prevented from continuous rotation by a *spring which provides restoring torque.*

● The moving coil movement described above has now been replaced by ***"taut-band"*** movement in which the moving coil and the pointer are suspended between bands of spring metal so that the *restoring force is torsional*. The band performs two functions : (*i*) They support the coil, and (*ii*) they provide restoring torque thereby *eliminating the pivots and jewels* used with coil spring movement.

— Generally speaking, taut-band suspension instruments can be made with *higher sensitivities* than those using pivots and jewels, and they can be used in almost every application served by pivoted instruments.

— Furthermore, taut-band instruments are *relatively insensitive to shock and temperature* and are *capable of withstanding greater overloads comparatively.*

***Working :***

— This meter movement (Fig. 1.39) works on the "***motor principle*** ("*when a current carrying conductor is placed in a magnetic field, it is acted upon by a force which tends to move it to one side and out of the field.*") and is a *current-responding device*. The deflection of the pointer is directly proportional to the amount of current passing through the coil.

— When D.C. current flows through the coil, the magnetic field so produced reacts with the field of the permanent magnet. The resultant force turns the coil along with its pointer.

— The amount of deflection is *directly proportional* to the amount of current in the coil. Hence the scale is *linear*.

● The *power requirements* of the D'Arsonval movement are surprisingly small : Typical values range from *25 μW to 200 μW*. *Accuracy* of the instrument is generally of the order of *2 to 5 percent of full-scale reading*.

• The PMMC instrument is *unsuitable for A.C. measurements, unless the current is rectified* before application to the coil.

**1.13.3.2. Temperature compensation**

The PMMC basic movement is not inherently insensitive to temperature, but it may be *temperature-compensated* by the appropriate use of series and shunt resistors.

• Complete temperature compensation may be accomplished by using swamping resistors as shown in Fig. 1.40. The swamping resistor is made of manganin (having temperature coefficient of practically zero) and combined with copper in the ratio of 20 : 1 to 30 : 1. The total resistance of the coil and swamping resistor increases slightly with rise in temperature but only just enough to counteract the change of springs and magnet, so that the overall temperature effect is zero.

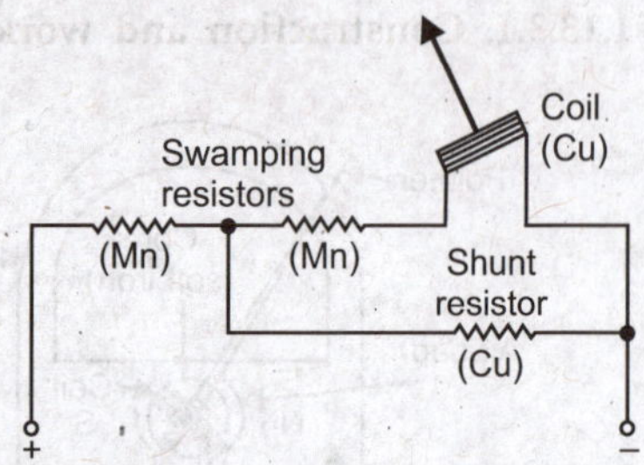


**Fig. 1.40.** Temperature compensation using series and shunt resistors.

— The main "*disadvantage*" of the use of swamping resistors is a reduction in the full-scale sensitivity of the movement; because a higher applied voltage is necessary to sustain the full-scale current.

**1.13.3.3. Characteristics of PMMC meter movement**

The following *three* characteristics of PMMC meter movement will be discussed :

1. ***Full-scale deflection current ($I_m$) :***

• *It is the current needed to deflect the pointer all the way to the right to the last mark on the calibrated scale.*

• Typical valves of the $I_m$ for D'Arsonval movement vary from 2μA to 30 mA.

— It may be remembered that for smaller currents, the number of turns in the moving coil should be *more* so that the magnetic field produced by the coil is strong enough to react with the field of permanent magnet for producing reasonable deflection of the pointers. *Fine wire* has to be used for reducing the weight of the moving coil but this increases its resistance.

— For heavy currents, thick wire with lesser number of turns is needed so that resistance of the moving coil is comparatively less.

2. ***Internal resistance ($R_m$) :***

• *It is the D.C. ohmic resistance of the wire of the moving coil.*

• Typical values of $R_m$ range from 1.2Ω for a 30 mA movement to 2 kΩ for a 50 μA movement.

3. ***Sensitivity (S) :***

• It is also known as *current-sensitivity* or *sensitivity factor*.

• It is given by the reciprocal of full scale reflection current $I_m$.

$$\therefore \qquad S = \frac{1}{I_m} \text{ ohm/volt}$$

— The sensitivity of a meter movement depends on the *strength of the permanent magnet and number of turns in the coil. Larger the number of turns, smaller the amount of current required to produce full-scale deflection and hence, higher the sensitivity.*

#### 1.13.3.4. Variations of PMMC meter movement

The PMMC system can be converted into an instrument to measure D.C. as well as A.C. quantities like current, voltage and resistance etc. It can carry a maximum current of $I_m$, without any modification and can withstand a maximum D.C. voltage, $V = I_m R_m$.

The various modifications of PMMC meter movement may be summed up as follows :

1. ***D.C. instruments :***
   (*i*) D.C. ammeter ....... By using a *shunt resistor.*
   (*ii*) D.C. voltmeter ....... By using *series multiplier resistor.*
   (*iii*) Ohmmeter ....... By using *battery and series resistor.*
2. ***A.C. instruments :***
   (*i*) Audio-frequency (AF) A.C. ammeter or voltmeter ....... By using a *rectifier.*
   (*ii*) Radio-frequency (RF) ammeter or voltmeter ....... By using a *thermocouple.*

#### 1.13.3.5. Advantages and disadvantages

The moving-coil permanent-magnet type instruments have the following *advantages* and *disadvantages* :

**Advantages :**

(*i*) Low power consumption.
(*ii*) Uniform scales.
(*iii*) No hysteresis loss.
(*iv*) High torque/weight ratio.
(*v*) They have very effective and efficient eddy-current damping.
(*vi*) Range can be extended with shunts or multipliers.
(*vii*) No effect of stray magnetic field as intense polarised or unidirectional field is employed.

**Disadvantages :**

(*i*) Somewhat costlier as compared to moving-iron instruments.
(*ii*) Cannot be used for A.C. measurements.
(*iii*) Friction and temperature might introduce errors as in case of other instruments.
(*iv*) Some errors are set in due to the ageing of control springs and the permanent magnets.

- Moving-coil permanent-magnet instrument can be *used as* :
  (*i*) Ammeters — By using a *low resistance shunt.*
  (*ii*) Voltmeters — By using a *high series resistance.*
  (*iii*) Flux meters — By *eliminating the control spring.*
  (*iv*) Ballistic galvanometers — By *making control springs of large moment of inertia.*

### 1.13.4. D.C. Ammeters

***Shunt resistor :***

Since the coil winding of a basic movement of a D.C. ammeter (PMMC galvanometer) is small and light, it *can carry only very small currents.* When it is required to measure large current, it is necessary to bypass the major part of the current through a resistance, called a *shunt*, as shown in Fig. 1.41. The resistance of the shunt can be calculated as follows :

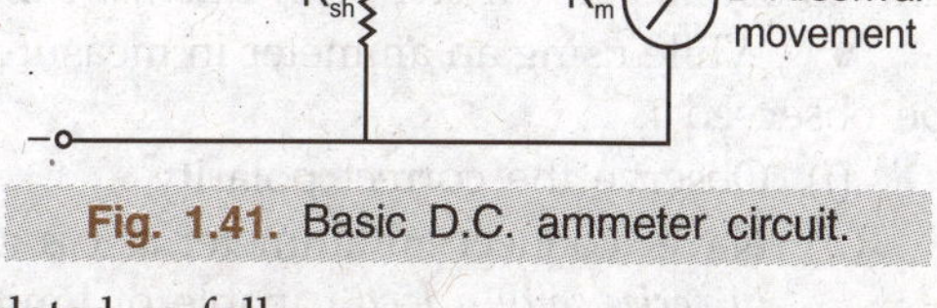


**Fig. 1.41.** Basic D.C. ammeter circuit.

Let, $R_m$ = Internal resistance of the meter,
$R_{sh}$ = Resistance of the shunt,
$I_m$ = Current through the meter,
$I_{sh}$ = Shunt current, and
$I$ = Current of the circuit to be measured.

Thus, $$I = I_m + I_{sh}$$

$$(\text{or } I_{sh} = I - I_m).$$

Also, since the voltage drop across the shunt and the meter/instrument is same,

$$I_{sh}R_{sh} = I_m R_m \quad \text{...(1.97)}$$

or, $$R_{sh} = \frac{I_m R_m}{I_{sh}}$$

or, $$R_{sh} = \frac{I_m R_m}{(I - I_m)} \quad \text{...(1.98)}$$

The shunt resistance with PMMC meter movement may consist of a length of constant-temperature resistance wire within the case of the instrument or it may be an external (manganin or constantan) shunt having a very low resistance.

— The ***external shunt*** consists of evenly spaced sheets of resistance material welded into a large block of heavy copper on each end of the sheets.

— The resistance material has a very low temperature coefficient, and a low thermoelectric effect exists between the resistance material and copper.

— External shunts of this type are normally used for measuring very large currents.

— ***Fig. 1.42*** shows a simple ***multirange ammeter*** in which the current range of the D.C. ammeter may be further extended by a number of shunts, selected by a *range switch*.

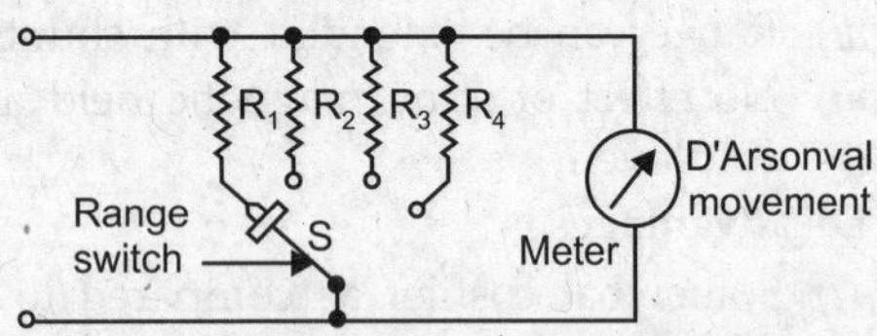


**Fig. 1.42.** Simple multirange ammeter.

— ***Fig. 1.43*** shows a ***universal or Ayrton shunt***, the use of which eliminates the possibility of having the meter in the circuit without a shunt. This *advantage* is gained at the cost of a slightly higher overall meter resistance.

— The Ayrton shunt provides an excellent opportunity *to apply basic network theory to a practical circuit.*

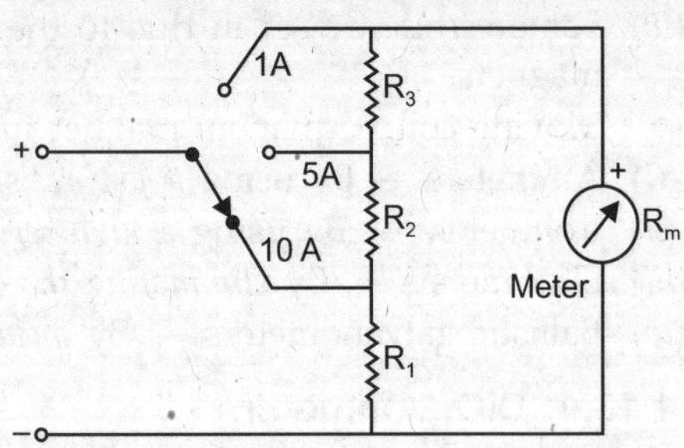


**Fig. 1.43.** Universal or Ayrton shunt.

● D.C. ammeters are commercially available in a large number of ranges, from 20μA to 50A full-scale for a *self-contained* meter and to 500 A for a meter with *external shunt*.

● While using an ammeter in measurement work, the following ***precautions*** should be observed :

(*i*) Observe the correct polarity.

(*ii*) Never connect an ammeter *across* an e.m.f. source. Always connect an ammeter *in series with a load* capable of limiting the current.

(*iii*) While using a multirange meter, first use the higher current range; then decrease the current range until substantial deflection is obtained.

### 1.13.5. D.C. Voltmeters

The basic D'Arsonval movement can be converted into a D.C. voltmeter by adding a series resistor or *multiplier*, as shown in Fig. 1.44. The multiplier *limits* the current through the meter so as not to *exceed* the value of the full-scale deflection current ($I_{fsd}$). The value of multiplier, required to extend the voltage range, is calculated from Fig. 1.44 as follows :

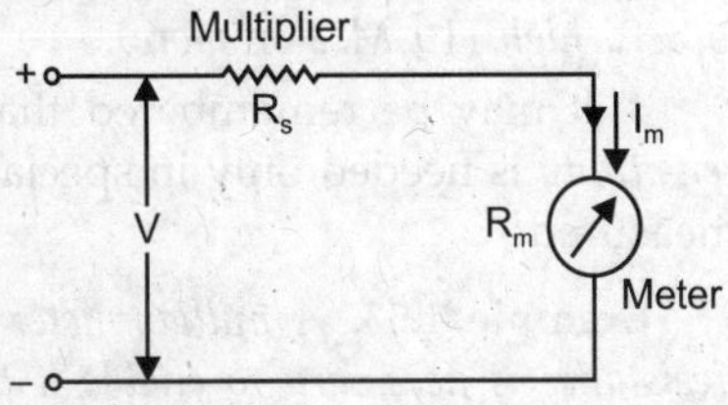


**Fig. 1.44.** Basic D.C. voltmeter circuit.

Let, $I_m$ = Full scale deflection current of the meter ($I_{fsd}$)

$R_m$ = Internal resistance of the meter,

$R_s$ = Multiplier resistance, and

$V$ = Voltage of the circuit to be measured.

Now, $V = I_m (R_s + R_m)$

or, $$R_s = \frac{V - I_m R_m}{I_m} = \frac{V}{I_m} - R_m \quad ...(1.99)$$

— For moderate ranges upto 500 V, the multiplier is usually mounted *inside* the case of the voltmeter. However, for *higher voltages*, the multiplier may be mounted separately *outside the case* on a pair of binding posts *to avoid excessive heating inside the case.*

• ***Fig. 1.45*** shows a ***multirange voltmeter*** where the multipliers are connected in a series string and range selector switches the appropriate amount of resistance in series with the meter.

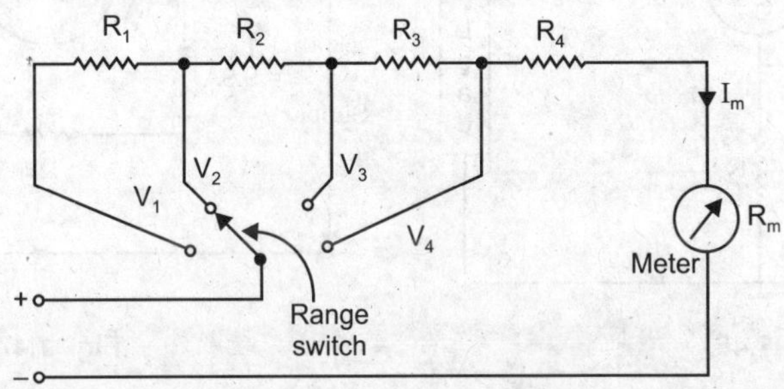


**Fig. 1.45.** Multirange voltmeter.

— The *advantage* of this system is that all multipliers except the first have standard resistance values and can be obtained commercially in precision tolerances. The low-range multiplier, $R_4$, is the only special resistor that must be manufactured to meet the specific circuit requirements.

**Loading effect of a voltmeter :**

If the resistance of a voltmeter is *not high* as compared to the resistance of the circuit across which it is connected, the measured voltage will be *less* than the actual one. The decrease in voltage may be negligible or it may be appreciable depending on the sensitivity and input resistance of the voltmeter. It is called ***loading effect*** because the voltmeter *loads down the circuit* across which it is connected.

*Loading effect can be minimised by using a voltmeter whose resistance is as high as possible as compared to that of the circuit across which it is connected.*

— In case of an *electronic voltmeter "there is no loading problem because its input resistance is very high (10 MΩ or more).*

• It may be remembered that whereas *accuracy* is always required in instruments, *sensitivity* is needed only in special applications where *loading* distorts that which is being measured.

**Example 1.63.** *A milliammeter of 2.5 ohms resistance reads upto 100 milliamperes. What resistance is necessary to enable it to be used as :*

*(i) A voltmeter reading upto 10 V.*

*(ii) An ammeter reading upto 10 A.*

*Draw the connection diagram in each case.*

**Solution.** Resistance of the milliammeter, $R_m = 2.5\ \Omega$

Maximum current of the milliammeter, $I_m = 100$ mA = 0.1 A.

(*i*) Voltage to be measured, $V = 10$ volts

Resistance to be connected in series.

$$R_s = \frac{V}{I_m} - R_m = \frac{10}{0.1} - 2.5 = \mathbf{97.5\ \Omega} \quad \textbf{(Ans.)}$$

Connection diagram is shown in Fig. 1.46.

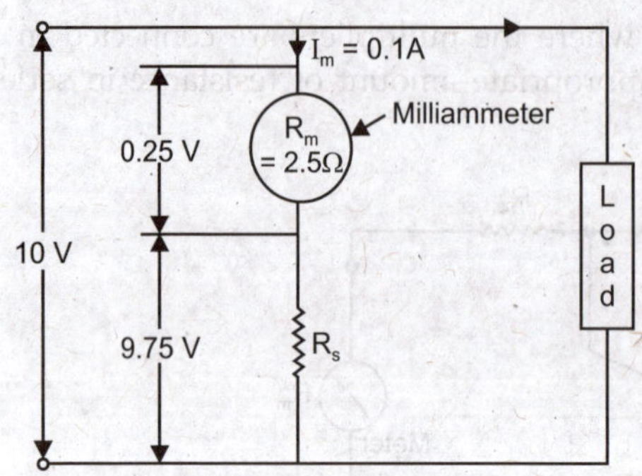


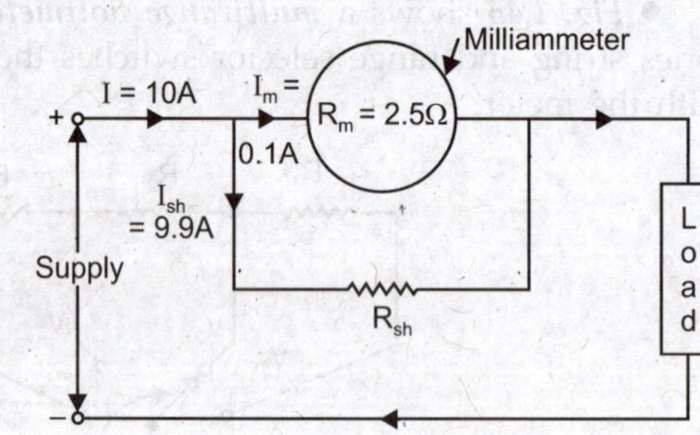


**Fig. 1.46.** **Fig. 1.47.**

(*ii*) Current to be measured, $I = 10$ A

∴ Resistance to be connected in parallel,

$$R_{sh} = \frac{I_m R_m}{I - I_m} = \frac{0.1 \times 2.5}{(10 - 0.1)} = \mathbf{0.02525\ \Omega} \quad \textbf{(Ans.)}$$

Connection diagram is shown in Fig. 1.46.

**Example 1.64.** *A moving-coil milliammeter having a resistance of 10 ohms gives full scale deflection when a current of 5 mA is passed through it. Explain how this instrument can be used for measurement of :*

*(i) Current upto 1 A.*

*(ii) Voltage upto 5 V.*

**Solution.** Resistance of the milliammeter, $R_m = 10\ \Omega$

Full scale deflection current, $I_m = 5$ mA = 0.005 A.

*(i) To measure current upto 1 A :*

**Resistance of the shunt, $R_{sh}$ :** Refer to Fig. 1.48.

Since voltage drop across the milliammeter and the shunt are equal

$$\therefore \quad I_m R_m = I_{sh} R_{sh} = (I - I_m)\, R_{sh}$$

$$\therefore \quad R_{sh} = \frac{I_m R_m}{(I - I_m)} = \frac{0.005 \times 10}{(1 - 0.005)} = \mathbf{0.05025\ \Omega} \quad \textbf{(Ans.)}$$

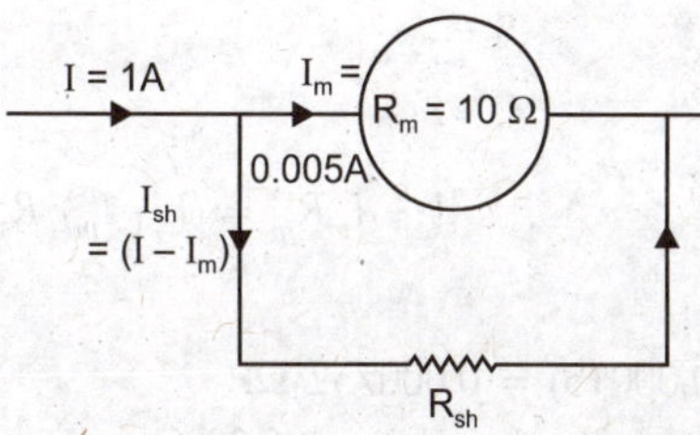


Fig. 1.48.

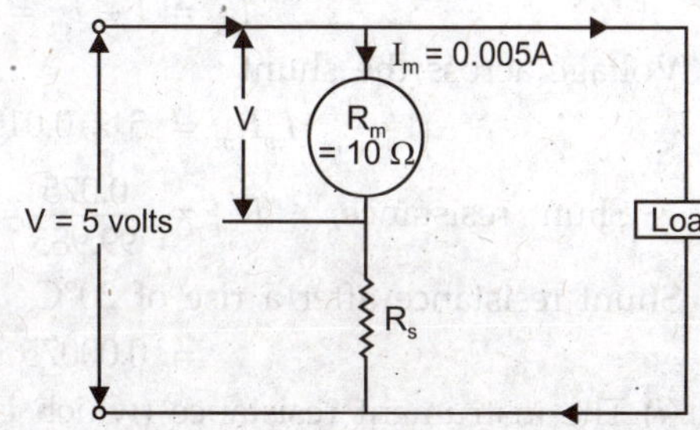


Fig. 1.49.

*(ii) To measure voltage upto 5 V :*

**The value of external series resistance, R :** Refer to Fig. 1.49.

Now, voltage across supply loads

= Voltage drop across the milliammeter + voltage drop across external series resistance $R_s$

$$\therefore \quad V = I_m R_m + I_m R_s$$

$$\text{or,} \quad 5 = 0.005 \times 10 + 0.005\, R_s$$

$$\text{or,} \quad R_s = \frac{5 - 0.005 \times 10}{0.005} = \mathbf{990\ \Omega} \quad \textbf{(Ans.)}$$

**Example 1.65.** *If the moving coil of a voltmeter consists of 100 turns wound on a square former which has a length of 30 mm and the flux density in the air gap is 0.09 Wb/m², calculate the turning moment on the coil when it is carrying a current of 10 mA.*

**Solution.** Number of turns, $N = 100$

Length of each side, $l = 30$ mm = 0.03 m

Flux density, $B = 0.09$ Wb/m²

Current through the coil, $I = 10$ mA = 0.01 A

The force $F$ on each side of the coil is given by :

$$F = NBIl \text{ newton}$$

∴ Turning moment (*i.e.,* deflecting torque),

$$T = F \times \text{breadth} = F \times l = NBIl^2 \text{ N-m}$$

$$= 100 \times 0.09 \times 0.01 \times (0.03)^2 \text{ N-m}$$

$$= \mathbf{8.1 \times 10^{-5}\ N\text{-}m.} \quad \textbf{(Ans.)}$$

**Example 1.66.** *A moving-coil instrument has a resistance of 5 Ω between terminals and full-scale deflection is obtained with a current of 0.015 A. This instrument is to be used with a manganin shunt to measure 100 A full scale. Calculate the error caused by a 20°C rise in temperature.*

*(i) When the internal resistance of 5 Ω is due to copper only.*
*(ii) When a 4 Ω manganin swamping resistor is used in series with a copper resistor of l Ω.*
*The temperature-resistance co-efficients are :*
*Copper : $\alpha_c$ = 0.4% per °C, Manganin : $\alpha_m$ = 0.015% per °C.*

**Solution.** Resistance of the instrument, $R_m = 5\Omega$

Current through the instrument, $I_m = 0.015$ A

Current to be measured, $I = 100$ A

Current through the shunt,

$$I_{sh} = I - I_m = 100 - 0.015 = 99.985 \text{ A}$$

Voltage across the shunt

$$I_m R_m = 5 \times 0.015 = 0.075 \text{ V}$$

∴ Shunt resistance, $R_{sh} = \frac{0.075}{99.985} = 0.00075\ \Omega$   $[\because I_m R_m = (I - I_m) R_{sh}]$

Shunt resistance after a rise of 20°C

$$= 0.00075 (1 + 20 \times 0.00015) = 0.000752\ \Omega.$$

(*i*) The instrument resistance (which is wholly copper) after a rise of 20°C

$$= 5 (1 + 20 \times 0.004) = 5.4\ \Omega$$

Hence, current through the instrument corresponding to 100 A in the line

$$= \frac{0.000752}{(5.4 + 0.000752)} \times 100 = 0.01392 \text{ A}$$

Reading of the instrument = 0.01392 × 100/0.015 = 92.8 A

∴ **Percentage error** = 100 – 92.8 = **7.2** (*low*) **(Ans.)**

(*ii*) The instrument resistance after a rise of 20°C

$$= 1 (1 + 20 \times 0.004) + 4 (1 + 20 \times 0.00015)$$
$$= 1.08 + 4.012 = 5.092\ \Omega$$

Instrument current with a line current of 100 A

$$= \frac{0.000752}{(5.092 + 0.000752)} \times 100 = 0.01476 \text{ A}$$

∴ Instrument reading = $0.01476 \times \frac{100}{0.015} = 98.4 \text{ A}$

**Percentage error** = 100 – 98.4 = **1.6** (*low*) **(Ans.)**

**Example 1.67.** *The coil of a 250 V moving iron voltmeter has a resistance of 500 Ω and inductance of 1 H. The current taken by the instrument when placed on 250 V, D.C. supply is 0.05 A. Determine the percentage error when the instrument is placed on 250 V, A.C. supply at 100 Hz.*

**Solution.** Total ohmic resistance, $R = \frac{250}{0.05} = 5000\ \Omega$

(Original calibration of the instrument is with direct current)

Reactance of the coil, $X_L = 2\pi f L = 2\pi \times 100 \times 1 = 628\ \Omega$

Coil impedance, $Z = \sqrt{R^2 + X_L^2} = \sqrt{(5000)^2 + (628)^2} = 5039\ \Omega$

$\therefore$ Voltage reading on A.C. $= \frac{250 \times 5000}{5039} = 248\text{ V}$

$\therefore$ Error $= 248 - 250 = -2\text{V}$

**Percentage error** $= \frac{2}{250} \times 100 = \mathbf{0.8}$ **(Ans.)**

**Example 1.68.** *A 15-volt moving iron voltmeter has a resistance of 300 Ω and an inductance of 0.12 H. Assuming that this instrument reads correctly on D.C., what will be its readings on A.C. at 15 volts when frequency is (i) 25 Hz and (ii) 100 Hz ?*

**Solution.** On *D.C.*, *only ohmic resistance* is involved and the voltmeter reads correctly. But on *A.C.*, it is the *impedance* of the instrument which has to be taken into account.

(*i*) **When frequency is 25 Hz :**

Impedance at 25 Hz,

$$Z = \sqrt{R^2 + X_L^2} = \sqrt{R^2 + (2\pi f L)^2}$$

$$= \sqrt{(300)^2 + (2\pi \times 25 \times 0.12)^2} = 300.6\,\Omega$$

$\therefore$ **Voltmeter reading** $= 15 \times \frac{300}{300.6} = \mathbf{14.97\ V}$ **(Ans.)**

(*ii*) **When frequency is 100 Hz :**

Impedance at 100 Hz,

$$Z = \sqrt{(300)^2 + (2\pi \times 100 \times 0.12)^2} = 309.3\,\Omega$$

$\therefore$ **Voltmeter reading** $= 15 \times \frac{300}{309.3} = \mathbf{14.55\ V}$ **(Ans.)**

Incidently, it may be noted that as the frequency is increased, the impedance of the voltmeter is also increased. Hence, the current is decreased and, therefore, the voltmeter readings are lower.

## 1.13.6. Ohmmeters

The ohmmeters may be of the following two types :

1. Series-type ohmmeter.
2. Shunt-type ohmmeter.

1. **Series-type ohmmeter :**

Fig. 1.50 shows the elements of a simple single-range series ohmmeter :

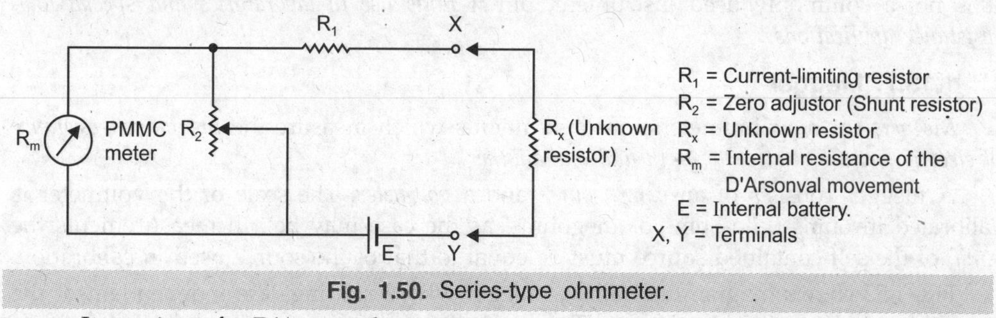


**Fig. 1.50.** Series-type ohmmeter.

— It consists of a D'Arsonval movement connected in series with a resistance and a battery to a pair of terminals (*X*, *Y*) to which the unknown resistance ($R_x$) is connected.

The current through the meter then depends on the magnitude of unknown resistor, and the meter indication is proportional to the value of the unknown, provided that the calibration problems are taken into account,

— When $R_x = 0$ (*i.e.*, terminals $X$ and $Y$ shorted), *maximum* current flows in the circuit. Under this condition, shunt resistor $R_2$ is adjusted until the meter indicates full-scale current ($I_{fsd}$); this position of the pointer is marked "$0\Omega$" on this scale.

— When $R_x = \infty$ (*i.e.*, terminals $X$ and $Y$ are open), the current in the circuit drops to *zero* and the meter indicates zero current, which is then marked "$\infty$" on the scale.

*Intermediate markings* may be placed on the scale by connecting different known values of $R_x$ to the instrument.

— The accuracy of these scale markings depends on the repeating accuracy of the meter and the tolerances of the calibrating resistors.

**2. Shunt-type Ohmmeter :**

Fig. 1.51 shows the circuit diagram of a shunt-type ohmmeter :

— It consists of a battery in series with an adjustable resistor $R_1$ and PMMC meter movement. The unknown resistor $R_x$ is connected across terminals $X$ and $Y$ in parallel with the meter. An *off-on switch* is provided to disconnect the battery from the circuit when the instrument is not used.

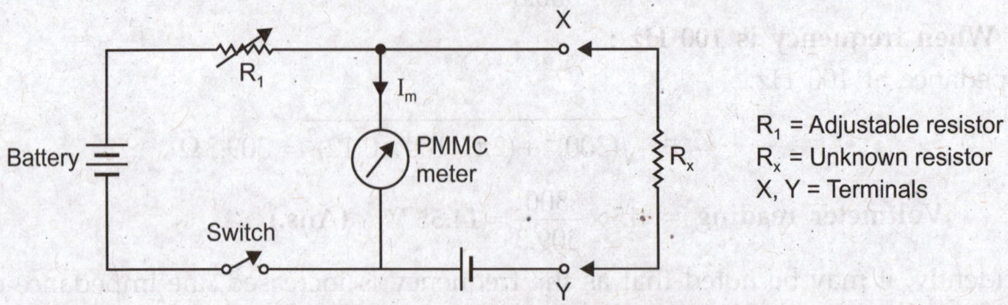


**Fig. 1.51.** Shunt-type ohmmeter.

— When $R_x = 0\ \Omega$ ($X$ and $Y$ terminals shorted), the meter current is zero.

— When $R_x = \infty$ ($X$ and $Y$ terminals open), the current finds a path only through the meter, and by the appropriate selection of the value $R_1$, the pointer can be made to read full scale.

The ohmmeter therefore has the 'zero' mark at the left hand side of the scale (no current) and the "infinite" mark on the right hand side of the scale (full-scale deflection).

- This type of ohmmeter is particularly suited to the measurement of *low-value resistors*. It is not a commonly used instrument, but it *finds use in laboratories and special low-resistance applications*.

### 1.13.7. Megger

*Meggers* (or megohmmeters) are instruments which measure the *insulation resistance of electric circuits relative to earth and one another*.

A megger consists of an *e.m.f. source* and a *voltmeter*. The scale of the voltmeter is calibrated in ohms (kilo-ohms or megohms, as the case may be). In measurements the *e.m.f.* of the self-contained source must be equal to that of the source used in calibration.

Fig. 1.52 shows diagrammatically a megger whose readings are independent of the speed of the self-contained generator. The moving system incorporates two coils 1 (current coil) and 2 (pressure coil) mounted on the same shaft and *placed in the field of a permanent magnet* (not shown) 90° apart. The generator energizes the two coils over separate wires. Connected in series with one coil is a fixed resistance $R_1$ (or several different resistances in order to extend the range of the instrument). The unknown resistance $R_x$ is connected in series with the other coil. *The currents in the coils interact with the magnetic field and produce opposing torques.*

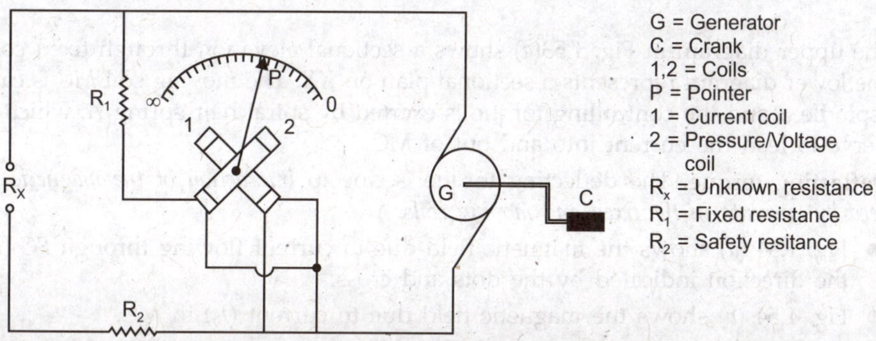


**Fig. 1.52.** Circuit diagram of megger.

*The deflection of the moving system depends on the ratio of the currents in the coils and is independent of the applied voltage.* The unknown resistance is read *directly* from the scale of the instrument. (The accuracy of measurement is unaffected by variations in the speed of the generator between 60 and 180 r.p.m.).

## 1.13.8. Electrodynamic or Dynamometer Instruments

In an electrodynamic instrument the *operating field is produced by another fixed coil and not by permanent magnet.* This instrument can be used as an ammeter or as voltmeter but is *generally used as a wattmeter.*

Refer Fig. 1.53 (*a*), (*b*).

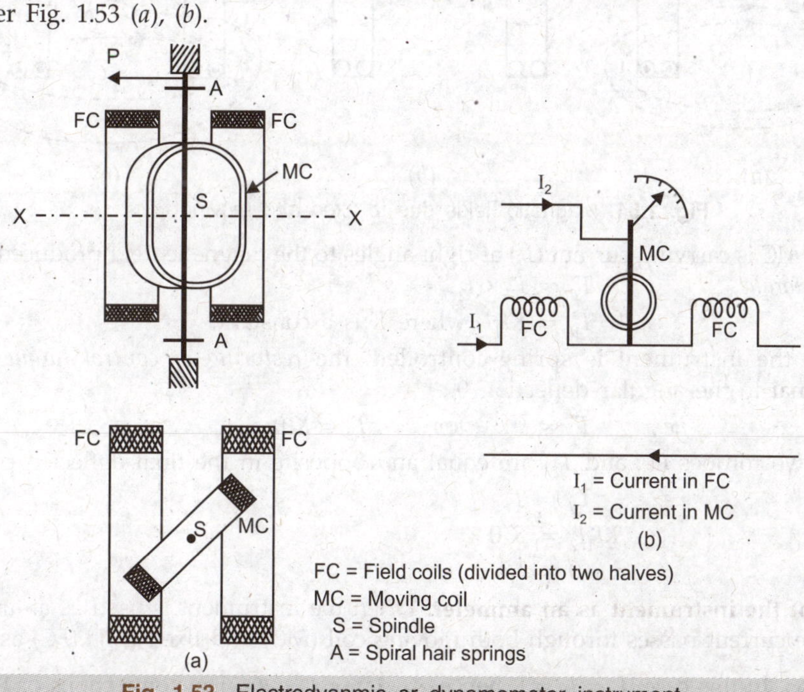


**Fig. 1.53.** Electrodyanmic or dynamometer instrument.

These instruments essentially consist of fine wire *moving coil* placed in the magnetic field produced by *another fixed coil* when carrying currents. The coils are usually *air cored* to *avoid hysteresis, eddy currents and other errors* when the instrument is used on A.C. The *fixed coil FC* is divided into *two halves* placed close together and *parallel* to each other in order to provide a *fairly uniform field* within the range of the movement of the moving coil.

The upper diagram in Fig. 1.53(*a*) shows a sectional elevation through fixed coil *FC* and the lower diagram represents a sectional plan on *XX*. The moving coil *MC* is carried by a spindle *S* and the controlling torque is exerted by spiral hair spring *A*, which may also serve to lead the current into and out of MC.

**Deflecting torque.** The deflecting torque is due to *interaction of the magnetic fields produced by currents in the fixed and moving coils.*

- Fig. 1.54 (*a*) shows the magnetic field due to current flowing through *FC* ($I_1$) in the direction indicated by the dots and cross.
- Fig. 1.54 (*b*) shows the magnetic field due to current ($I_2$) in *MC*.
- Fig. 1.54 (*c*) shows the combined effect of the above magnetic fields. By combining these magnetic fields it will be seen that when currents ($I_1$ and $I_2$) flow simultaneously through *FC* and *MC*, the resultant magnetic field is *distorted* and effect is to exert a clockwise torque on *MC*.

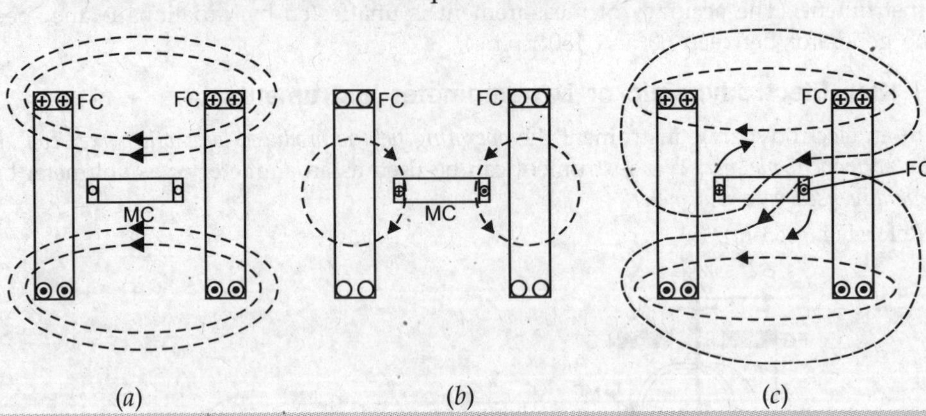


**Fig. 1.54.** Magnetic fields due to fixed and moving coils.

Since *MC* is carrying current ($I_2$) at right angles to the magnetic field produced by *FC*,

*deflecting torque,* $T_d \propto I_1 \times I_2$

or, $T_d = KI_1I_2$ where *K* is a constant.

Since the instrument is spring-controlled, the *restoring or control torque* ($T_c$) is proportional to the angular deflection θ.

*i.e.,* $T_c \propto \theta$ or $T_c = K'\theta$

The two torques ($T_d$ and $T_c$) are equal and opposite in the final deflected position.

*i.e.,* $T_d = T_c$

or, $KI_1I_2 = K'\theta$

or, $\theta \propto I_1I_2$

**Use of the instrument as an ammeter.** When the instrument is used as an ammeter then same current passes through both moving coil (*MC*) and fixed coils (*FC*) as shown in Fig. 1.55. In this case, $I_1 = I_2 = I$, hence $\theta \propto I^2$ or $I \propto \sqrt{\theta}$. The connections of Fig. 1.55

are used when *small currents* are to be measured.

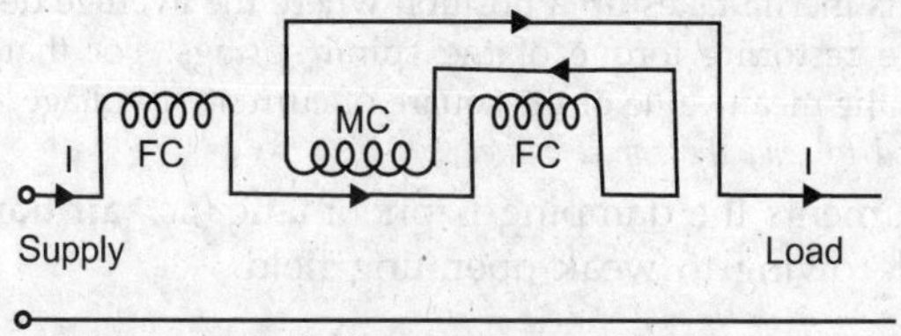


**Fig. 1.55.** Measurement of small currents.

In the case of heavy currents, a shunt is used to limit current through the moving coil as shown in Fig. 1.56.

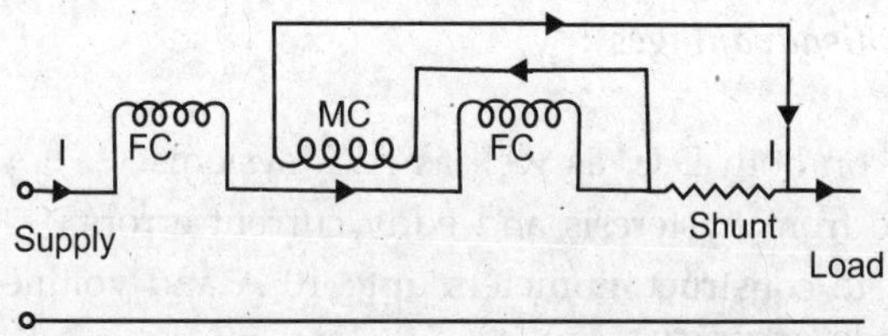


**Fig. 1.56.** Measurement of heavy currents.

**Use of the instrument as voltmeter.** When the instrument is used as a voltmeter, the fixed and moving coils are used in series along with high resistance as shown in Fig. 1.56.

Here again, $I_1 = I_2 = I$,

where, $I = \frac{V}{R}$ ......in D.C. circuits

and, $I = \frac{V}{Z}$ ......in A.C. circuits

$\therefore$ $\theta \propto V \times V$ or $\theta \propto V^2$

or, $V = \sqrt{\theta}$

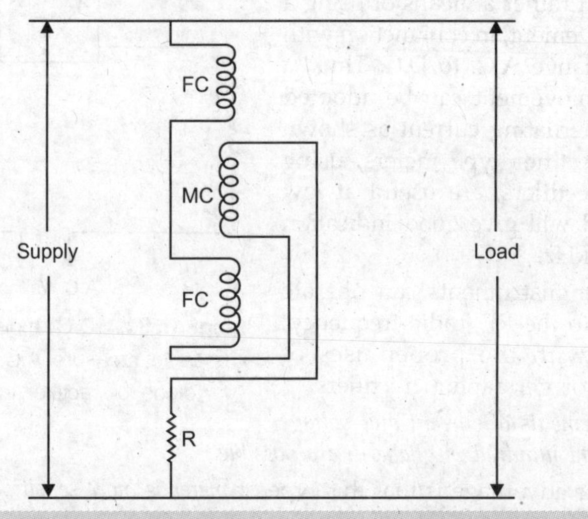


**Fig. 1.57.** Use of the instrument as a voltmeter.

Thus whether the instrument is used as an ammeter or voltmeter, its scale is *uneven* through the whole of its range and is cramped or crowded near the zero in particular.

*Note:* When the dynamometer instrument is used to measure an alternating current or voltage, the moving coil due to its inertia takes up a position where the average deflecting torque over one cycle is balanced by the restoring torque of the spiral springs. For that position, the deflecting torque is proportional to the mean value of the square of current or voltage, and the instrument scale can therefore be *calibrated to read the r.m.s. value.*

- In these instruments the damping is pneumatic (*i.e.,* **air damping**). Eddy current damping is admissible owing to weak operating field.

**Ranges :**

***Ammeter.*** (*i*) With fixed and moving coils in series ............. 0/0.01 A – 0/0.05 A

(*ii*) With moving coil shunted or parallel connections ........... upto 0/30 A.

***Voltmeters.*** Upto 0–750 volts.

***Advantages and disadvantages :***

***Advantages :***

(*i*) Can be used on both D.C. as well as A.C. systems.

(*ii*) They are free from hysteresis and eddy current errors.

(*iii*) It is possible to construct ammeters upto 10 A and voltmeters upto 600 V with precision grade accuracy.

***Disadvantages :***

(*i*) Since torque/weight ratio is small, such instruments have low sensitivity.

(*ii*) The scale is not uniform because $\theta \propto \sqrt{I}$

(*iii*) Cost of these instruments is higher in comparison to those of moving iron instruments. So, these are *only used as voltmeters and ammeters for precision measurements.*

(*iv*) Higher friction losses.

### 1.13.9. Rectifier Instruments

These are not separate types of instruments, but rather a means of using a D'Arsonval movement, in conjunction with a *rectifier,* to change A.C. to D.C. Thus, a direct current movement can be adopted for use with alternating current as shown in Fig. 1.58. Rectifier type meters, using copper oxide rectifiers, are useful at low frequencies and will give good indication upto about 20 kHz.

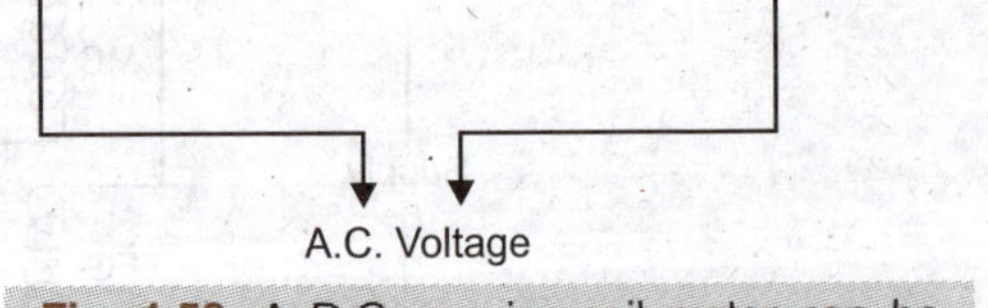


**Fig. 1.58.** A D.C. moving-coil meter can be used to measure A.C. voltage by putting a diode or rectifier in the meter circuit.

- Rectifier instruments can operate well into the RF (radio-frequency) range with the proper uses of silicon or germanium rectifiers.
- *Measurements of current and voltage at several hundred meghahertz are possible.*
- Another advantage is that this type of *meter is more sensitive* than any other type of A.C. meter.

**Characteristics of Rectifier instruments :**

The *characteristics* of rectifier instruments are given below :

1. They provide an economical and practical means of A.C. measurements in radio and communication circuits at audio frequencies (1000 to 10000 Hz) and in other A.C. circuits where small available power makes it necessary to use sensitive low-energy instruments.
2. A rectifier instrument is, in general, approximately *fifty times* more sensitive than either an electrodynamometer or a moving-iron instrument.
3. In most instances rectifier instruments have *essentially linear scales*.
4. They commonly possess sensitivity of the order of 1000 to 2000 ohms per volt and more.
5. The power consumed by a rectifier instrument is generally several times as high as its permanent magnet moving coil (PMMC) mechanism because of the resistance of its rectifier.
6. Rectifier instruments are manufactured as micrometers and milliammeters in ratings from 100 μA to 15 mA for full scale deflection. The higher rating of 15 mA is more or less governed by the special size of the rectifier in comparison with the mechanism with which it is to be used.
7. Shunting of rectifier instruments is not practical because of the change in resistance of the rectifier with both temperature and the amount of the current.

**Rectifier ammeters.** The *rectifier ammeters usually* consists of four rectifier elements L, M, N and P arranged in the form of a bridge (as shown in Fig. 1.59) and A represents a moving-coil ammeter. The apex of the black triangle (shown in the bridge) indicates the direction in which the resistance is low.

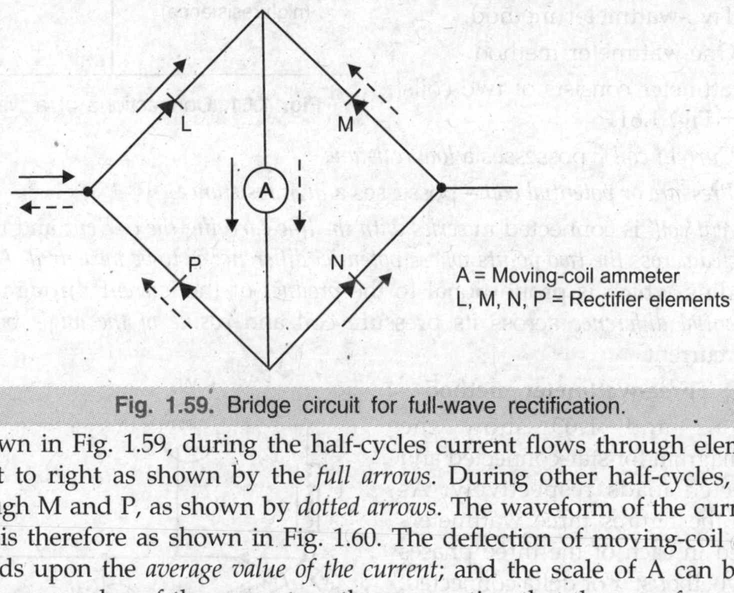


**Fig. 1.59.** Bridge circuit for full-wave rectification.

As shown in Fig. 1.59, during the half-cycles current flows through elements *L* and *N* from left to right as shown by the *full arrows*. During other half-cycles, the current flows through M and P, as shown by *dotted arrows*. The waveform of the current flowing through A is therefore as shown in Fig. 1.60. The deflection of moving-coil ammeter A, thus depends upon the *average value of the current*; and the scale of A can be calibrated to read the r.m.s. value of the current on the assumption that the waveform of the latter is sinusoidal with a form factor of 1.11.

**Rectifier voltmeters.** In a *rectifier voltmeter*, A is a milliammeter and the bridge circuit of Fig. 1.59 is connected in *series with a suitable resistor*.

**Advantage of rectifier instrument :**

1. The primary advantage of the rectifier *voltmeter* is that it is *far more sensitive* as compared to other types of voltmeter suitable for measuring A.C. voltages.

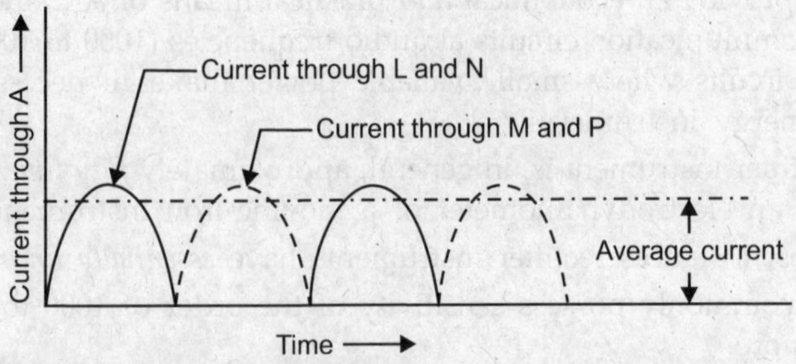


**Fig. 1.60.** Waveform of current through moving-coil ammeter.

2. Metal rectifiers can be incorporated in universal instruments, such as ammeter, thereby enabling a moving-coil milliammeter to be used in combination with shunt and series resistances to measure various ranges of D.C. and D.C. *voltage*, and in combination with a *bridge rectifier and suitable resistors* to measure various ranges of A.C. and A.C. *voltage.*

### 1.13.10. Measurement of Power in 3-Phase Circuit

- The power in 3-phase load can be measured by using the following methods :
  1. Three-wattmeter method.
  2. Two-wattmeter method.
  3. One-wattmeter method.
- A wattmeter consists of two coils : Refer Fig. 1.61.
  1. *Current coil* – possesses a *low resistance.*
  2. *Pressure or potential coil* – possesses a *high resistance.*

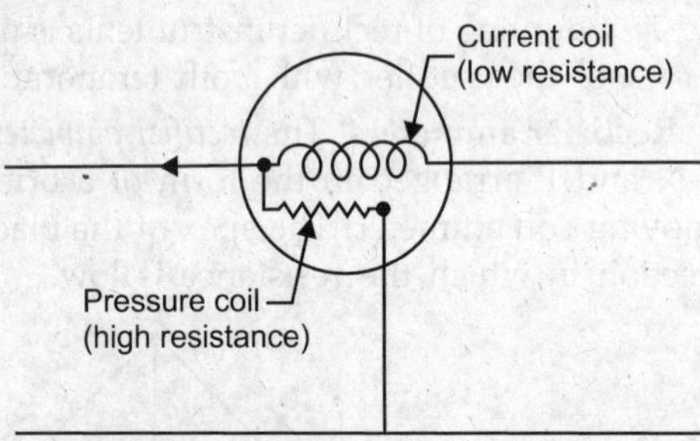


**Fig. 1.61.** Connections of a wattmeter.

The '*current coil*' is connected in *series with the line carrying the current* and the '*pressure coil' is connected across the two points whose potential difference is to be measured.* A wattmeter shows a reading which is proportional to the *product* of the *current* through its current coil, the *potential difference* across its pressure coil and *cosine of the angle* between this voltage and current.

#### 1.13.10.1. Three-wattmeter method

Figs. 1.62 and 1.63 show the connection diagram for star-connected and delta-connected loads respectively. As indicated in the figures three wattmeters are connected in each of the three phases of the load whether star or delta connected. The current coil of each wattmeter carries the current of one phase only and the pressure coil measures the phase-voltage of the phase. Hence, *each wattmeter measures the power in a single phase. The total power in the load is given by the algebraic sum of the readings of the three wattmeters.*

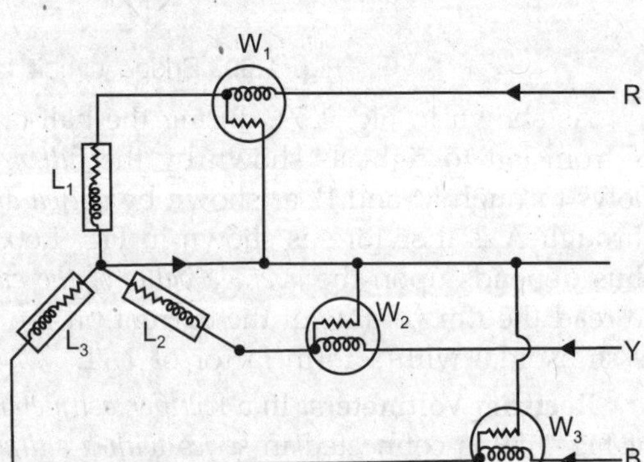


**Fig. 1.62.** Star-connected load.

While using this method following *difficulty* is met with :

- In case of star-connected load it is not always possible to get a neutral point which is required for connections (Fig. 1.62).
- In case of delta-connected load, under ordinary conditions it is not generally feasible to break into the phases of the load.

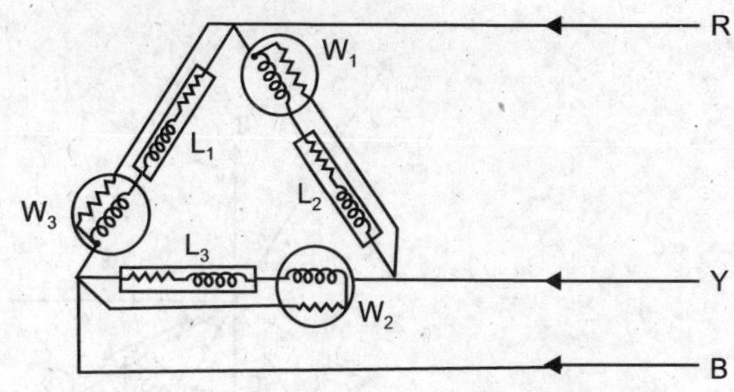


Fig. 1.63. Delta-connected load.

To measure power it is not necessary to use three wattmeters, two wattmeters can be used for the purpose as explained in the Article 1.13.10.2.

**1.13.10.2. Two-wattmeter method – balanced load**

The total power consumed by a *balanced load* can be found by using two wattmeters (Figs. 1.61 and 1.62). When load is assumed inductive in Fig. 1.65 the vector diagram for such a balanced star-connected load is shown in Fig. 1.67.

Let us consider the problem in terms of r.m.s. values (instead of instantaneous values).

Let $E_R, E_Y, E_B$ = r.m.s. values of the three phase voltages,

and $I_R, I_Y, I_B$ = r.m.s. values of the currents.

Since these voltages and currents are assumed sinusoidal, they can be represented by vectors, the currents lagging behind their respective phase voltages by $\phi$.

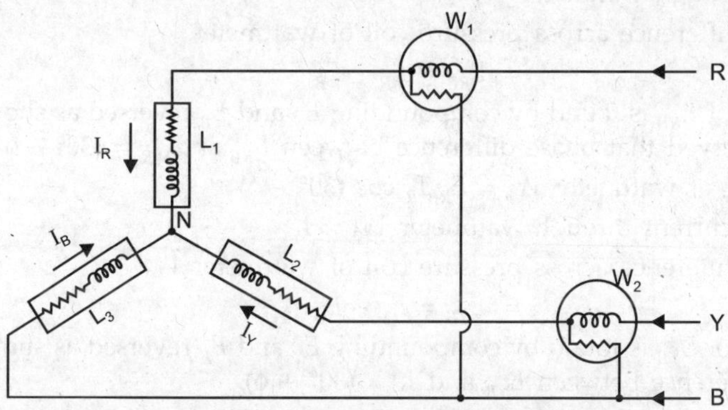


Fig. 1.64. Star-connected load.

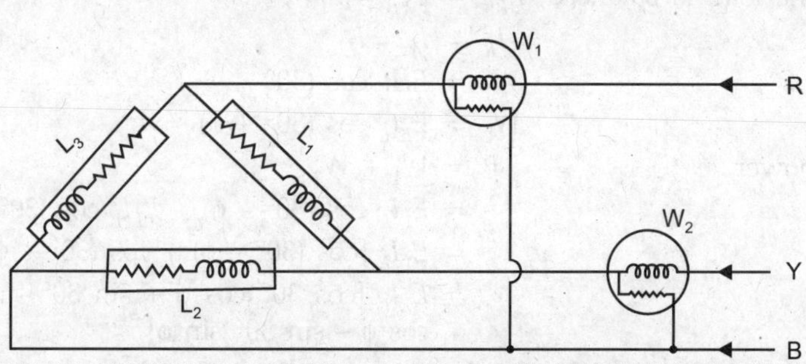


Fig. 1.65. Delta-connected load.

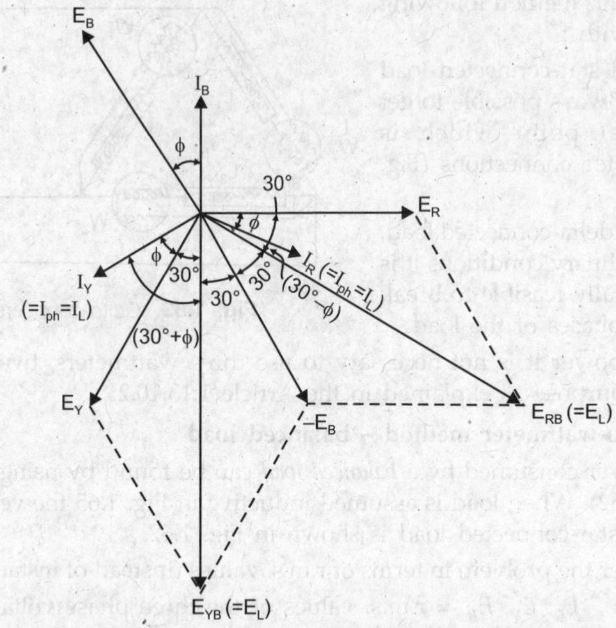


**Fig. 1.66.** Vector diagram – Two-wattmeter method.

Refer to Fig. 1.66.

Current through wattmeter $W_1 = I_R$

Potential difference across pressure coil of wattmeter $W_1$

$$= E_{RB} = E_R - E_B \qquad \text{(Vectorially)}$$

The value of $E_{RB}$ is found by compounding $E_R$ and $E_B$ reversed as shown in Fig. 1.66. It may be observed that phase difference between $E_{RB}$ and $I_R = (30° - \phi)$.

$\therefore$ Reading of wattmeter $W_1 = E_{RB}I_R \cos(30° - \phi)$ ...(*i*)

Similarly, current through wattmeter $W_2 = I_Y$

Potential difference across pressure coil of wattmeter $W_2$

$$= E_{YB} = E_Y - E_B \qquad \text{(Vectorially)}$$

The value of $E_{YB}$ is found by compounding $E_Y$ and $E_B$ reversed as shown in Fig. 1.66. The phase difference between $E_{YB}$ and $I_Y = (30° + \phi)$.

$\therefore$ Reading of wattmeter $W_2 = E_{YB}I_Y \cos(30° + \phi)$ ...(*ii*)

Since the load is balanced, $E_{RB} = E_{YB} = E_L$ (Line voltage)

and, $I_R = I_Y = I_L$ (Line current)

$\therefore$ $W_1 = E_LI_L \cos(30° - \phi)$

and, $W_2 = E_LI_L \cos(30° + \phi)$

$\therefore$ Total power,

$$P = W_1 + W_2$$
$$= E_LI_L \cos(30° - \phi) + E_LI_L \cos(30° + \phi)$$
$$= E_LI_L [\cos(30° - \phi) + \cos(30° + \phi)]$$
$$= E_LI_L [\cos 30° \cos\phi + \sin 30° \sin\phi + \cos 30° \cos\phi - \sin 30° \sin\phi]$$
$$= E_LI_L (2 \cos 30° \cos\phi) = E_LI_L \times 2 \times \frac{\sqrt{3}}{2} \cos\phi$$

$$= \sqrt{3}E_L I_L \cos\phi$$

*i.e.,* $$P = \sqrt{3}E_L I_L \cos\phi$$

Hence the *sum of the readings of the two wattmeters give the total power consumption in the 3-phase load.*

It is worth noting that in the above case the phase sequence of *RYB* has been assumed, the readings of the two wattmeters will change if the phase sequence is reversed.

**Variations in wattmeter readings.** As shown above that for a *lagging power factor*

$$W_1 = E_L I_L \cos(30° - \phi)$$

and, $$W_2 = E_L I_L \cos(30° + \phi)$$

From above it is evident that individual readings of the wattmeters *not only depend on the load but also upon its power factor.* Let us take up the following cases :

(*i*) When $\phi = 0$

*i.e.,* power factor is *unity* (*i.e.,* load is *resistive*)

Then, $$W_1 = W_2 = E_L I_L \cos 30°$$

The reading of each wattmeter will be equal and opposite (*i.e.,* up-scale reading).

(*ii*) When $\phi = 60°$

*i.e.,* power factor = 0.5 (lagging)

Then, $$W_2 = E_L I_L \cos(30° + 60°) = 0$$

Hence, the power is measured by $W_1$ above.

(*iii*) When $90° > \phi > 60°$

*i.e.,* $0.5 > \text{p.f.} > 0$

Then $W_1$ is still *positive* but reading of $W_2$ is *reversed*. For a leading p.f., conditions are just the opposite of this. In that case, $W_1$ will read *negative* because the phase angle between the current and voltage is *more than* 90°. For getting the total power, the reading of $W_2$ is to be *subtracted* from that of $W_1$. Under this condition, $W_2$ will read *'down scale' i.e., backwards.* Hence, *to obtain a reading on* $W_2$, *it is necessary to reverse either its pressure coil or current coil, usually the former.*

*All readings taken after reversal of pressure oil are to be taken as negative.*

(*iv*) When $\phi = 90°$

(*i.e.,*p.f. = 0 *i.e.,* pure inductive or capacitive load)

Then $$W_1 = E_L I_L \cos(30° - 90°) = E_L I_L \sin 30°$$

and $$W_2 = E_L I_L \cos(30° + 90°) = -E_L I_L \sin 30°$$

These two readings are equal in magnitude but opposite in sign

$$\therefore \quad W_1 + W_2 = 0$$

So far we have considered *lagging angles* (taken as *positive*). Now let us discuss how the readings of wattmeters change when the power factor is *leading* one.

— For $\phi = +60°$ (lag) : $W_2 = 0$

— For $\phi = -60°$ (lead) : $W_1 = 0$

Thus we find that for angles of lead the readings of the two wattmeters are interchanged.

Hence, when the power is *leading* :

$$W_1 = E_L I_L \cos(30° + \phi)$$

$$W_2 = E_L I_L \cos(30° - \phi).$$

**Power Factor – When the load is 'balanced'.** When load is balanced with a *lagging power factor* and the voltage and currents are sinusoidal :

$$W_1 + W_2 = E_L I_L \cos(30° - \phi) + E_L I_L \cos(30° + \phi) = \sqrt{3} E_L I_L \cos\phi \quad ...(i)$$

Similarly,
$$W_1 - W_2 = E_L I_L \cos(30° - \phi) - E_L I_L \cos(30° + \phi)$$
$$= E_L I_L (2 \sin\phi \sin 30°) = E_L I_L \sin\phi \quad ...(ii)$$

Dividing (*ii*) by (*i*), we get

$$\tan\phi = \frac{\sqrt{3}(W_1 - W_2)}{(W_1 + W_2)} \quad ...(1.100)$$

For a *leading power*, this expression becomes

$$\tan\phi = -\frac{\sqrt{3}(W_1 - W_2)}{(W_1 + W_2)} \quad ...(1.101)$$

After finding tan $\phi$, hence $\phi$, the value of power factor cos $\phi$ can be found (from trigonometrical tables).

One important point which must be kept in mind is that if $W_2$ reading has been taken after reversing the pressure coil *i.e.*, if $W_2$ is *negative*, then the eqn., (1.13) becomes

$$\tan\phi = \sqrt{3}\frac{W_1 - (-W_2)}{W_1 + (-W_2)}$$

or,
$$\tan\phi = \sqrt{3}\frac{W_1 + W_2}{W_1 - W_2} \quad ...(1.102)$$

### 1.13.10.3. One-wattmeter method

In this method the current coil is connected in any one line and the pressure coil is connected alternately between this and the other two lines as shown in Fig. 1.67. The two readings so obtained, for a balanced load, correspond to those obtained by normal two-wattmeter method.

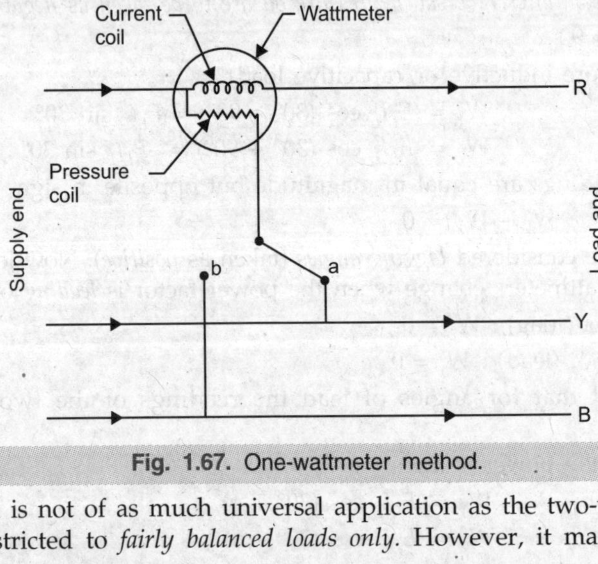


Fig. 1.67. One-wattmeter method.

This method is not of as much universal application as the two-wattmeter method because it is restricted to *fairly balanced loads only*. However, it may be conveniently applied, for instance, when it is desired to find the power input to a factory motor *in order to check the load upon the motor*.

**1.13.10.4. Measurement of reactive volt amperes**

In order to measure reactive power in a single phase circuit, a *compensated wattmeter* is used. In this wattmeter, the voltage applied to the pressure coil is 90° *out of phase with the actual voltage* and hence it will read VI cos (90° – φ) *i.e.*, VI sin φ, reactive power.

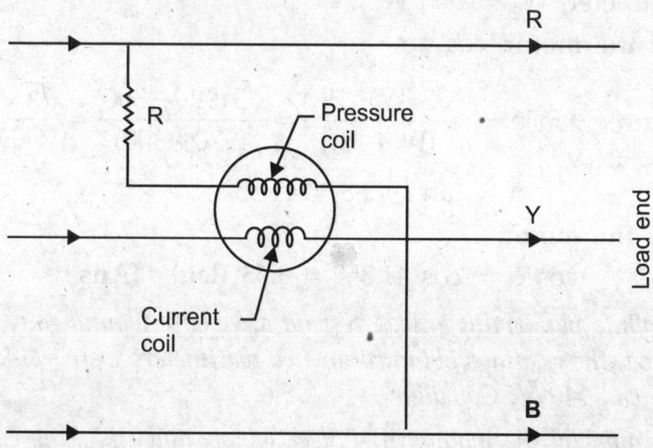


Fig. 1.68. Measurement of reactive volt-amperes.

In balanced three-phase circuit the reactive power can be determined by using one wattmeter. The necessary connections are shown in Fig. 1.68. The current coil is inserted in one line and the pressure coil is connected across the other two lines.

The current flowing through the current coil of the wattmeter.

$$= I_Y = I_{ph} \text{ (say)}$$

The potential difference across the potential coil of wattmeter,

$$E = E_R - E_B$$

$$= \sqrt{3}E_{ph} \text{ leading vector } E_Y$$

by 90°

$$= \sqrt{3}E_{ph} \text{ leading the vector}$$

of current $I_Y$ by (90° + φ)

∴ Reading of the wattmeter (Fig. 1.69)

$$= \sqrt{3}E_{ph}I_{ph}\cos(90° + \phi)$$

$$= -\sqrt{3}E_{ph}I_{ph}\sin\phi = -W_r$$

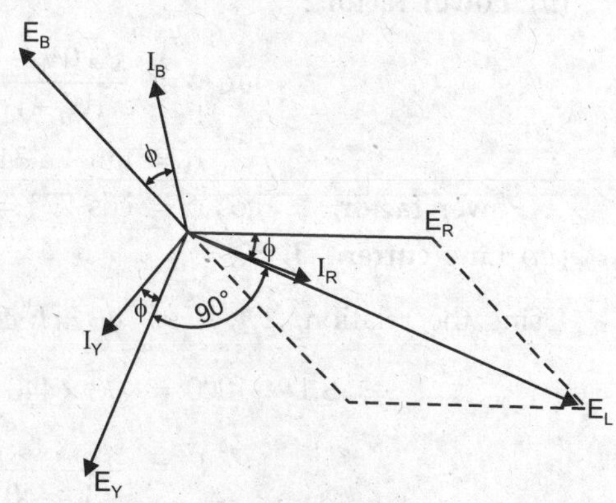


Fig. 1.69.

Total reactive power of the circuit,

$$= 3\,E_{ph}I_{ph}\sin\phi = -\sqrt{3}\,W_r$$

Reactive power (as earlier started) can also be determined from two wattmeter readings connected for measurement of power.

$$W_1 - W_2 = E_LI_L \cos(30° - \phi) - E_LI_L \cos(30° + \phi)$$

or $$W_1 - W_2 = E_LI_L \times 2 \sin 30° \sin \phi$$

or $$E_LI_L \sin\phi = W_1 - W_2$$

Reactive power of load circuit.

$$W_r = \sqrt{3}E_L I_L \sin\phi = \sqrt{3}(W_1 - W_2). \quad ...(1.103)$$

**Example 1.69.** *The power input to a 3-phase induction motor is read by two wattmeters. The readings are 920 W and 300 W. Calculate the power factor of the motor.*

**Solution.** Reading of wattmeter, $W_1 = 920$ W

Reading of wattmeter, $W_2 = 300$ W

**Power factor of the motor, cos φ :**

Using the relation, $\tan\phi = \dfrac{\sqrt{3}(W_1 - W_2)}{(W_1 + W_2)} = \dfrac{\sqrt{3}(920 - 300)}{(920 + 300)} = \dfrac{\sqrt{3} \times 620}{1220} = 0.88$

$$\therefore \quad \phi = \tan^{-1} 0.88 = 41.35°$$

**Power factor of the motor,**

$$\cos\phi = \cos 41.35° = \mathbf{0.75\ (lag)} \quad \textbf{(Ans.)}$$

**Example 1.70.** *While performing a load test on a 3-phase wound-rotor induction motor by two wattmeters method, the readings obtained on two wattmeters were + 14.2 kW and – 6.1 kW and the line voltage was 440 V. Calculate :*

*(i) True power drawn by the motor, (ii) Power factor, and (iii) Line current.*

**Solution.** Reading of wattmeter, $W_1 = 14.2$ kW

Reading of wattmeter, $W_2 = -6.1$ kW

Line voltage, $E_L = 440$ V

(*i*) **True power drawn by the motor :**

True power drawn by the motor = 14.2 – 6.1 = **8.1 kW** **(Ans.)**

(*ii*) **Power factor :**

$$\tan\phi = \frac{\sqrt{3}(W_1 - W_2)}{(W_1 + W_2)} = \frac{\sqrt{3}(14.2 - 6.1)}{[14.2 + (-6.1)]} = \frac{\sqrt{3} \times 20.3}{8.1} = 4.34$$

$$\therefore \quad \phi = \tan^{-1} 4.34 = 77°$$

∴ **Power factor,** $\cos\phi = \cos 77° = \mathbf{0.2249\ (lag).}$ **(Ans.)**

(*iii*) **Line current, $I_L$ :**

Using the relation, $P = \sqrt{3}E_L I_L \cos\phi$, we have

$$8.1 \times 1000 = \sqrt{3} \times 440 \times I_L \times 0.2249$$

$$\therefore \quad I_L = \frac{8.1 \times 1000}{\sqrt{3} \times 440 \times 0.2249} = 47.26 \text{ A}$$

Hence, *line current* = **47.26 A** **(Ans.)**

**Example 1.71.** *A 3-phase, 440 V motor load has a power factor of 0.6. Two wattmeters connected to measure the power show the input to be 25 kW. Find the reading on each instrument.*

**Solution.** Input power, = 25 kW

Line voltage, $E_L = 440$ V

Power factor of the motor load, cos φ = 0.6

**$W_1$ ; $W_2$ :**

Using the relation : $\tan\phi = \dfrac{\sqrt{3}(W_1 - W_2)}{W_1 + W_2}$ ...(*i*)

Now, $\cos\phi = 0.6$, $\therefore \phi = 53.13°$

$\therefore$ $\tan\phi = \tan 53.13° = 1.333$

Also $W_1 + W_2 = 25$ kW (*given*) ...(*ii*)

Substituting these values in (*i*), we get

$$1.333 = \frac{\sqrt{3}(W_1 - W_2)}{25}$$

$$\therefore \quad W_1 - W_2 = \frac{1.333 \times 25}{\sqrt{3}} = 19.24 \text{ kW} \quad ...(iii)$$

From (*ii*) and (*iii*), we get

$$\mathbf{W_1 = 22.12\ kW}$$

and $\mathbf{W_2 = 2.88\ kW}$ **(Ans.)**

### 1.13.11 Induction Type Single Phase Energy Meters

These are, by far, the most common form of A.C. meters met with in every-day *domestic* and *industrial installations.* These meters measure electric energy in kWh.

The principle of these meters is practically the same as that of induction wattmeters. Instead of the control spring and the pointer of the wattmeter, the watt-hour meter, (energy meter) employs a *brake magnet,and a counter attached to the spindle.* Just like other watt-hour meters, the eddy currents induced in the aluminium disc by the brake magnet due to the revolution of the disc, are utilised to control the continuously rotating disc.

**Construction.** The construction of a typical meter of this type is shown in Fig. 1.70. The brake-magnet and recording wheel-train being omitted for clearances. It consists of the following :

(*i*) Series magnet $M_1$

(*ii*) Shunt magnet $M_2$

(*iii*) Brake magnet

(*iv*) A rotating disc.

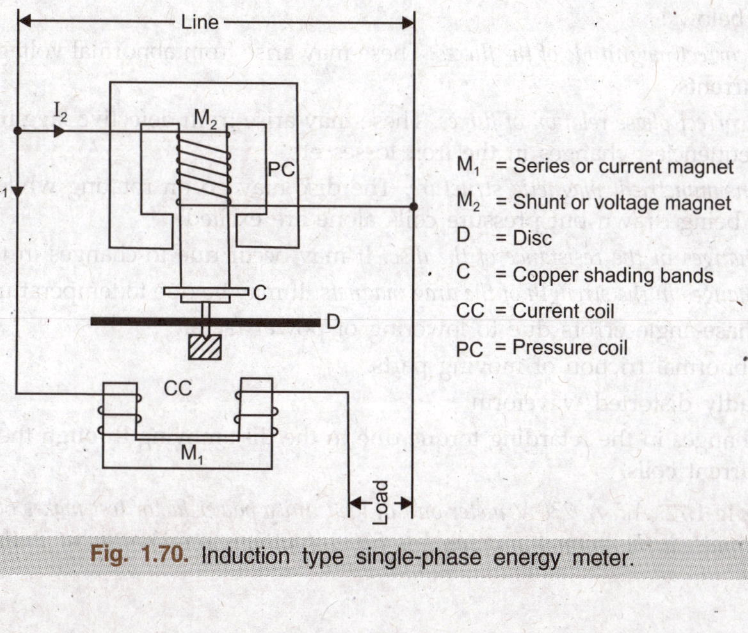


**Fig. 1.70.** Induction type single-phase energy meter.

The *series electro-magnet* $M_1$ consists of a number of U-shaped iron laminations assembled together to form a core, wound with a *few turns of a heavy gauge wire*. This wound coil is known as *current coil* and is connected in one of the lines and in series with the load to be metered. The series electromagnet is energized and sets up a magnetic field cutting through the rotating disc mounted on a vertical spindle and supported on a sapphire cup contained in a bottom screw. The bottom pivot, which is usually removable, is of hardened steel, and the end, which is hemispherical in shape, rests in the sapphire cup. The top pivot (not shown) merely serves to maintain the spindle in a vertical position under working condition and does not support any weight or exert any appreciable thrust in any direction.

The *shunt magnet* $M_2$ consists of a number of *M* shaped iron laminations assembled together to form a core. A core having *large number of turns of fine wire* is fitted on the middle limb of the shunt magnet, this coil is known as *pressure coil PC* and is connected across the supply mains.

The *brake magnet* consists of C-shaped piece of alloy steel bent round to form a complete magnetic circuit, with the exception of a narrow gap between the poles. This magnet is mounted so that the disc revolves in the air gap between the polar extremities. The movement of the rotating disc through the magnetic field crossing the air gap *sets up eddy currents in the disc which react with the field and exerts a braking effect*. The speed of the rotating disc may be adjusted by changing the position of the brake magnet or by diverting some of the flux there from.

**Working.** The shunt electromagnet produces a magnetic field which is of pulsating character; it cuts through the rotating disc and induces eddy currents there in, but normally *does not in itself produce any driving force*. Similarly series electromagnet induces eddy currents in the rotating disc, but *does not in itself produce any driving force*. In order *to obtain driving force in this type of meter, phase displacement of 90° between the magnetic field set up by shunt electromagnet and applied voltage V is achieved by adjustment of copper shading band* C (also known as power factor compensator or compensating loop). *The reaction between these magnetic fields and eddy currents set up a driving torque in the disc.*

**Sources of errors.** The various sources of errors in an induction-type energy meter are given below :

(*i*) *Incorrect magnitude of the fluxes.* These may arise from abnormal voltages and load currents.

(*ii*) *Incorrect phase relation of fluxes.* These may arise from defective lagging, abnormal frequencies, changes in the iron losses etc.

(*iii*) *Unsymmetrical magnetic structure.* The disc may go on rotating while no current is being drawn but pressure coils alone are excited.

(*iv*) *Changes in the resistance of the disc.* It may occur due to changes in temperature.

(*v*) *Changes in the strength of the drag magnets.* It may be due to temperature or ageing.

(*vi*) Phase-angle errors due to lowering of power factor.

(*vii*) Abnormal friction of moving parts.

(*viii*) Badly distorted waveform.

(*ix*) Changes in the retarding torque due to the disc moving through the field of the current coils.

**Example 1.72.** *A 5 A, 230 V meter on full load unity power factor test makes 60 revolutions in 360 seconds. If the normal disc speed is 520 revolutions per kWh, what is the percentage error ?*

**Solution.** Energy consumed in 360 seconds

$$= \frac{VI\cos\phi \times t}{3600 \times 1000} \text{ kWh}$$

$$= \frac{230 \times 5 \times 1 \times 360}{3600 \times 1000} = 0.115 \text{ kWh}$$

where, $t$ is in seconds.

Energy recorded by the meter

$$= \frac{60}{520} = 0.11538 \text{ kWh}$$

$\therefore$ % age error $= \frac{0.11538 - 0.115}{0.115} \times 100 = \mathbf{0.33\%\ (fast).}$ **(Ans.)**

**Example 1.73.** *The constant of a 230 V, 50 Hz, single-phase energy meter is 185 revolutions per kWh. The meter takes 190 seconds for 10 revolutions while supplying a non-inductive load of 4.5 A at normal voltage. What is the percentage error of the instrument ?*

**Solution.** Energy consumed in 190 seconds

$$= \frac{VI\cos\phi}{1000} \times t = \frac{230 \times 4.5 \times 1}{1000} \times \frac{190}{3600} = 0.0546 \text{ kWh}$$

[cos $\phi$ = 1, since load supplied in non-inductive]

Energy registered by the meter $= \frac{10}{185} = 0.054$ kWh

$\therefore$ % age error $= \frac{0.054 - 0.0546}{(0.054)} \times 100 = \mathbf{0.06\%\ (slow).}$

**Example 1.74.** *The name plate of a meter reads "1 kWh = 15000 revolutions". In a check up, the meter completed 150 revolutions during 45 seconds. Calculate the power in the circuit.*

**Solution.** Power metered in 150 revolutions

$$= 1 \times \frac{150}{15000} = 0.01 \text{ kWh}$$

If $P$ kilowatt is the power in the circuit, then energy consumed in 45 seconds

$$= \frac{P \times 45}{3600} \text{ kWh} = 0.0125\, P \text{ kWh}$$

Equating the two amounts of energy, we have

$$0.0125\, P = 0.01$$

$$P = 0.8 \text{ kW} = \mathbf{800\ W.} \quad \textbf{(Ans.)}$$

**Example 1.75.** *A 230 V ampere-hour type meter is connected to a 230 V D.C. supply. If the meter completes 225 revolutions in 10 minutes when carrying 14 A, calculate : (i) The kWh registered by the meter, and (ii) The percentage error of the meter above or below the original calibration.*

*The timing constant of the meter is 40 A-s/revolution.*

**Solution.** During 225 revolutions the meter would register 40 × 225 A-s or columbs.

Since time taken is 10 minutes or 600 seconds it corresponds to a current of $\frac{40 \times 225}{600}$ = 15 A.

(*i*) Energy recorded by the meter $= \dfrac{VIt}{1000}$ kWh

where, $t$ is in hour

$$= \frac{230 \times 15}{1000} \times \frac{10}{60} = 0.575 \text{ kWh.}$$

(*ii*) Actual energy consumed $= \dfrac{230 \times 14}{1000} \times \dfrac{10}{60} = 0.5367$ kWh

**Percentage error** $= \dfrac{0.575 - 0.5367}{0.575} \times 100 = \mathbf{6.66\%}$ **(Ans.)**

### 1.13.12. Power Factor Meters

By definition, the *"power factor" is the cosine of the phase angle between voltage and current;* the power-factor measurements usually involve the determination of this phase angle.

Fig. 1.71 shows the connections for a *single-phase crossed-coil power-factor meter.*

- This instrument is basically an electrodynamometer movement, where the moving element consists of two coils, mounted on the same shaft but at right angles to each other. The moving coils rotate in magnetic field provided by the field coil that carries the line current.

— As shown in Fig. 1.71, the field coil is connected as usual in series with the line and carries the line current.

— *One coil* of the movable element is connected in series with a *resistor* across the lines and receives its current from the applied potential difference. *Second coil* of the movable element is connected in series with the *inductor* across the lines.

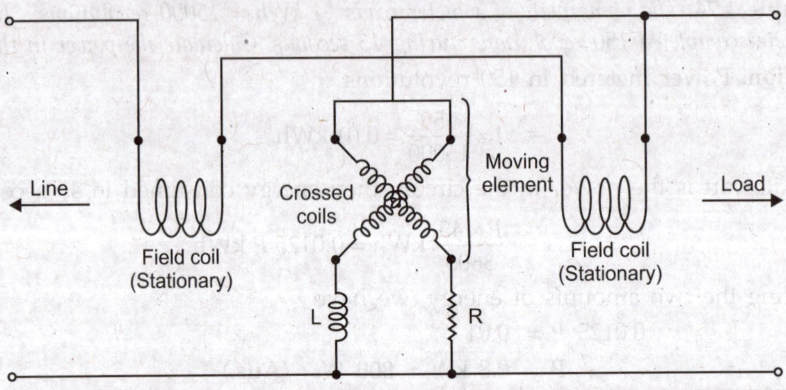


**Fig. 1.71.** Connections for a single-phase crossed-coil power-factor meter.

— Because no control springs are used, the *balance position of the movable element depends on the resulting torque developed by the two crossed coils.*

— When the movable element is in a balanced position, the contribution to the total torque by each element must be equal but of opposite sign. The developed torque in each coil is a function of the current through the coil and therefore depends on the impedance of that coil circuit. The torque is also proportional to the mutual inductance between each part of the crossed coil and the stationary field coil. The mutual inductance depends on the angular position of the cross-coil elements w.r.t. the position of the stationary field coil.

— *When the movable element is at balance, it can be shown that its angular displacement is a function of the phase angle between line current (field coil) and line voltage (crossed coils).*

The *indication of the pointer,* which is connected to the movable element, *is calibrated directly in terms of the phase angle or power factor.*

• For the measurement of phase angle in B-phase power systems, the *"polarized-vane power-factor meter" is used.*

The above mentioned instruments are limited to measurement at comparatively low frequencies and are typically used at the power line frequency (50 Hz). *Phase measurements at higher frequencies often are more accurately performed by special electronic instruments or techniques.*

## HIGHLIGHTS

1. *Measurement* is the act, or the result of a quantitative comparison between a predetermined standard and an unknown magnitude. Measurements provide us with a means of describing various phenomena in quantitative terms.
2. The physical quantity or the characteristic condition which is the object of measurement is an instrumentation system is variously termed as *"measurand"*, *"measurement variable"*, *"instrumentation variable"* or *"process variable"*.
3. The technology of using instruments to measure and control the physical and chemical properties of materials is called *"instrumentation"*.
4. The *"measurement standard"* is the physical embodiment of the unit of measurement as well as that of its submultiple value.
5. The highest standard of either a base unit or a derived unit is called a *"primary standard"*. The "secondary standards" are the reference calibrated standards designated and calibrated from the primary standards. *"Working standards"* are the normal standards which are used by the workers and technicians who actually carry out the measurements.
6. The *static characteristics* pertain to a system where the quantities to be measured are constant or vary slowly with time. The main static characteristics are : *Accuracy, sensitivity, reproducibility, drift, static error and dead zone.*
7. The *relative static error* is defined as the ratio of absolute static error to the true value of the quantity under measurement.
8. The difference between the true value and the measured value of a quantity is called *static correction.*
9. *Static calibration* is a process by which all the static performance characteristics are obtained in one form or another.
10. The difference between the largest and the smaller reading of the instrument is called the *range* of the instrument.
11. *"Dynamic characteristics"* quantify the dynamic relation between the input and output.
12. The limits of the deviations from the specified value are defined as *limiting errors* or *guarantee errors.*
13. The *relative error* is defined as the ratio of the error to the specified (nominal) magnitude of a quantity.

**14.** The errors are classified as :

1. Gross errors
2. Systematic errors :
   (*i*) Instrumental errors.
   (*ii*) Environmental errors.
   (*iii*) Observational errors.

**15.** *Sources of errors are :*

Noise, Response time; Design limitations; Energy exchanged by interaction; Transmission; Determination of measuring system; Ambient influences on measuring systems, Errors of observation and interpretation.

**16.** The deflection type instruments with a scale and movable pointer are called *analog* instruments.

**17.** The instruments which use logic circuit and techniques to obtain a measurement and then display it in numerical reading (digital) form are called *digital* instruments.

## OBJECTIVE TYPE QUESTIONS

**1.** The purpose of instruments is to
(*a*) allow measurements to be made (*b*) transmit the information
(*c*) change signals (*d*) any of the above.

**2.** The measurement refers to which of the following :
(*a*) Primary signal (*b*) Measured variable
(*c*) Output (*d*) All of the above.

**3.** The measurement of a quantity
(*a*) is an act of comparison of an unknown quantity with another quantity.
(*b*) is an act of comparison of an unknown quantity with a known quantity whose accuracy may be known or may not be known.
(*c*) is an act of comparison of an unknown quantity with a predefined acceptable standard which is accurately known.
(*d*) none of the above.

**4.** In present day measurement systems
(*a*) direct methods are commonly used.
(*b*) one of the direct methods is limited but indirect methods are commonly used.
(*c*) both direct and indirect methods are commonly used.
(*d*) all of the above.

**5.** The measurand is
(*a*) output (*b*) measured variable
(*c*) secondary signal (*d*) any of the above.

**6.** The output stage of a generalised measurement system may comprise
(*a*) manipulator (*b*) detector-transducer
(*c*) indicating or recording unit (*d*) all of the above.

**7.** A null type of instrument as compared to a deflection type instrument has
(*a*) a higher accuracy (*b*) a lower sensitivity
(*c*) a faster response (*d*) all of the above.

8. In an intelligent information system used for the measurement of a physical quantity
   (*a*) the digital computer must perform all or nearly all the signal and information processing.
   (*b*) the observer must perform all or nearly all the signal and information processing.
   (*c*) the digital computer is not required.
   (*d*) none of the above.

9. A high-grade of slip gauges preserved in a factory and not put into general use would be a
   (*a*) primary standard (*b*) secondary standard
   (*c*) tertiary standard (*d*) any of the above.

10. The temperature measurement by a thermocouple is
    (*a*) primary measurement (*b*) secondary measurement
    (*c*) tertiary measurement (*d*) any of the above.

11. Purely mechanical instruments cannot be used for dynamic measurements because they have
    (*a*) high inertia (*b*) large time constant
    (*c*) higher response time (*d*) all of the above.

12. The usage of electronic instruments is becoming more extensive because they have
    (*a*) a high sensitivity and reliability.
    (*b*) a fast response and compatibility with digital computers.
    (*c*) the capability to respond to signals from remote places.
    (*d*) all of the above.

13. The intermediate stage of an instrument is used to
    (*a*) sense the primary (input) signal.
    (*b*) indicate or record the measurements.
    (*c*) amplify and transmit the secondary signal (signal after it has been transduced).
    (*d*) any of the above.

14. The function of the signal manipulating element, in a generalised measurement system, is to
    (*a*) change the quantity under measurement to an analogous signal.
    (*b*) change the magnitude of the input signal while retaining its identity.
    (*c*) to perform non-linear operations like filtering, chopping and clipping and clamping.
    (*d*) none of the above.

15. A modifying input to a measurement system can be defined as an input,
    (*a*) which changes the input-output relationship for desired inputs only.
    (*b*) which changes the input-output relationship for desired as well as interfering inputs.
    (*c*) which changes the input-output relationship for interfering inputs only.
    (*d*) none of the above.

16. Which of the following errors can arise, as a result of mistakes in reading, parallax, improper instrument location and inadequate lighting ?
    (*a*) Construction error (*b*) Transmission error
    (*c*) Observation error (*d*) Translation error.

17. Gearing, backlash, friction between moving parts, and scale accuracies are generally known as
    (*a*) instrument errors (*b*) interference errors
    (*c*) calibration errors (*d*) interaction errors.

**18.** Errors which may be variable both in magnitude and nature (positive or negative) are classified as

(*a*) hysteresis errors (*b*) random errors

(*c*) systematic errors (*d*) interaction errors.

**19.** The characteristic of an observer to read either systematically higher or systematically lower than other observers is referred to as

(*a*) insensitivity (*b*) lack of precision

(*c*) lack of discrimination (*d*) operator bias.

## ANSWERS

| | | | | | | |
|---|---|---|---|---|---|---|
| 1. (*a*) | 2. (*c*) | 3. (*c*) | 4. (*b*) | 5. (*b*) | 6. (*c*) | 7. (*a*) |
| 8. (*a*) | 9. (*b*) | 10. (*c*) | 11. (*d*) | 12. (*d*) | 13. (*c*) | 14. (*b*) |
| 15. (*b*) | 16. (*c*) | 17. (*a*) | 18. (*b*) | 19. (*d*). | | |

## THEORETICAL QUESTIONS

1. What do you understand by the term "measurement" ?
2. Describe the significance of measurements.
3. Explain briefly the term "measurand".
4. State the basic requirements for getting the meaningful result of the measurement.
5. Explain the difference between an analog signal and a digital signal.
6. Define the term "instrumentation".
7. How are standards of measurement classified ?
8. Explain clearly the difference between primary, secondary and working standards.
9. Explain briefly the following methods of complete solutions to complex problem in mechanical design :
   (*i*) The empirical method.
   (*ii*) The rational method.
   (*iii*) The experimental methods.
10. Distinguish between direct and indirect methods of measurement.
11. Describe the mechanics and power types of measurements. Explain with suitable examples.
12. What are primary, secondary and tertiary measurements ? Explain with examples.
13. Explain the difference between the following :
    (*i*) Measurand and measurement.
    (*ii*) Measurement and instrument.
    (*iii*) Input and output
14. Draw a block diagram representation of a generalised measurement system. Identify the various elements and point out the functions performed by each element/component.
15. Explain briefly various stages of the general measurement system.
16. Describe briefly the following examples of measurement systems :
    (*i*) Bourdon tube pressure gauge.
    (*ii*) Pressure-actuated thermometer.
17. How are instruments classified ?
18. Explain briefly any three types of the instruments :
    (*i*) Absolute and secondary instruments.
    (*ii*) Analog and digital instruments.

(*iii*) Mechanical, electrical and electronic instruments.
(*iv*) Self-operated and power-operated instruments.
(*v*) Deflection and null-point instruments.

**19.** State the advantages and disadvantages of mechanical instruments over electrical/electronic instruments.

**20.** State the advantages of electrical/electronic techniques of measurements.

**21.** Discuss the factors relating to selection of instruments.

**22.** Describe briefly the following main functions of the instruments :
(*i*) Indicating function.
(*ii*) Recording function.
(*iii*) Controlling function.

**23.** What do you mean by characteristics of a measuring instrument ?

**24.** What are static characteristics ? How do they differ from dynamic characteristics ?

**25.** Define the following terms as applied to the measuring instrument :
(*i*) True value (*ii*) Correction
(*iii*) Range (*iv*) Sensitivity
(*v*) Accuracy (*vi*) Linearity
(*vii*) Dead zone (*viii*) Backlash.

## Static Characteristics :

**26.** Enumerate the main static characteristics of measuring instruments.

**27.** Explain briefly the terms : Accuracy, errors and correction.

**28.** Explain the following terms :
(*i*) Static error (*ii*) Static correction
(*iii*) Relative error , (*iv*) Percentage relative error.

**29.** How can "accuracy" be expressed ? Explain briefly.

**30.** What do you mean by "Static calibration" ? Give the steps which are necessary in performing a calibration.

**31.** Differentiate between "Scale range" and "Scale span" giving suitable examples.

**32.** What is an "Error calibration curve" ?

**33.** What do you mean by "Scale readability" ?

**34.** Differentiate between the terms "Repeatability" and "Reproducibility".

**35.** Explain the term "Drift" using suitable examples.

**36.** How is "drift" classified ?

**37.** Explain briefly the following :
(*i*) Zero drift (*ii*) Span drift (*iii*) Zonal drift.

**38.** Differentiate between the terms "Accuracy" and "Precision" with suitable examples.

**39.** What is "sensitivity" ?

**40.** Define the term "Deflection factor".

**41.** What is "linearity" ? Explain briefly.

**42.** What do you mean by the term "Non-linearity" ? Explain briefly "Terminal linearity" and "Best fit linearity".

**43.** Describe briefly the term "Hysteresis".

**44.** Explain clearly the difference between "Thershold" and "Resolution".

**45.** Explain briefly the following :
(*i*) Dead zone; (*ii*) Dead time.

**46.** What do you mean by the term "loading effect" as applied to measuring instruments ? Explain briefly.

47. Explain briefly the terms "Static stiffness" and "Static compliance".
48. What is "noise" ?
49. What is signal to noise ratio ?
50. Explain briefly the following :
    (*i*) Generated noise; (*ii*) Conducted noise; (*iii*) Radiated noise.

**Dynamic Characteristics :**

51. Define the following terms :
    (*i*) Response; (*ii*) Dynamic response of a measuring system
52. Explain briefly the following types of dynamic inputs :
    (*i*) Transient response; (*ii*) Steady state periodic.
53. Define the following terms :
    (*i*) Transient response; (*ii*) Steady state response.
54. Define the dynamic response of a system and distinguish between steady state response and transient response.
55. Define the following terms :
    (*i*) Speed of response; (*ii*) Measuring lag;
    (*iii*) Fidelity; (*iv*) Dynamic error;
    (*v*) Overshoot.
56. What are the different standard inputs for studying the dynamic response of a system ? Explain them briefly.
57. Derive the equations for time response of a first-order system when subject to :
    (*i*) Unit step input; (*ii*) Unit ramp input.
58. Derive the expression for time response of a second-order underdamped system when subjected to unit ramp input. Show that the nature of the response is same as that for a unit step input.
59. Derive the expression for the magnitude and phase of a first-order system when subjected to a sinusoidal input signal. Draw the magnitude and phase versus frequency plots.
60. Derive the expressions for the frequency response of a second-order system. Draw the magnitude and phase versus frequency plots.
61. Sketch and explain the response of a first-order system when subjected to step input signal.
62. Sketch and explain the response of a first-order system subjected to ramp input signal. Indicate thereon the transient error, steady error and the time lag.
63. Describe the dynamics for sinusoidal input to a first-order system.
64. Develop an expression for the amplitude ratio and phase difference of indicated reading and the input signal which is of sinusoidal form.
65. What is the importance of time constant in appropriate measuring instruments ?

**Errors :**

66. What are limiting errors ?
67. Explain briefly the term *"relative error"*.
68. How are errors classified ?
69. What are "Gross errors " ? How can these be avoided ?
70. What are "Systematic errors" Explain briefly the following systematic errors with examples :
    (*i*) Instrumental errors.
    (*ii*) Environmental errors.
    (*iii*) Observational errors.
71. How can the systematic errors be located ?

72. What are assembly errors ? Can they be easily discovered and rectified ? If yes, why ?
73. State the factors which give rise to observational errors ?
74. What are random errors ? State the common causes of their occurrence.
75. Enumerate the various sources of errors.
76. Explain briefly the following sources of errors :
    (*i*) Noise (*ii*) Response time
    (*iii*) Design limitations (*iv*) Energy exchanged by interaction
    (*v*) Transmission (*vi*) Deterioration of measuring system.
77. What are single sample and multi-sample tests ? Explain briefly.
78. Explain briefly the following terms :
    (*i*) Arithmetic mean (*ii*) Geometric mean
    (*iii*) Median (*iv*) Mode.
79. Describe briefly the following terms :
    (*i*) Deviation (*ii*) Average deviation
    (*iii*) Standard deviation (*iv*) Variance.
80. What is the law of probabilities ? Explain briefly.
81. Explain briefly the term "Precision index".

## UNSOLVED EXAMPLES

1. A meter reads 136.6 V and the true value of the voltage is 136.52 V. Determine :
   (*i*) The static error; (*ii*) The static correction for this instrument
   **[Ans.** (*i*) + 0.08 V (*ii*) – 0.08 V]
2. An analog indicating instrument with a scale range of 0 – 10.0 V shows a voltage of 5.30 V. The true value of a voltage is 5.40 volts.
   (*i*) What are the values of absolute error and correction ?
   (*ii*) Express the error as a function of the true value and full scale deflection.
   **[Ans.** (*i*) – 0.1 V, 0.1 V, (*ii*) 1.85 %, – 1 %]
3. A pressure indicator showed reading as 89 bar on a scale range of 0 – 100 bar, but the true value was 88.5 bar. Determine :
   (*i*) Static error, (*ii*) Static correction, and
   (*iii*) Relative static error. **[Ans.** (*i*) + 0.5 bar, (*ii*) – 0.5 bar, (*iii*) 0.565%]
4. A thermometer is calibrated 100°C to 150°C. The accuracy is specified within ± 0.25 percent of instrument span. What is the maximum static error ? **[Ans.** ± 0.125°C]
5. In a Wheatstone bridge a change of 9 Ω in the unknown arm of the bridge is required to produce a change in the deflection of 3 mm of the galvanometer. Determine (*i*) Sensitivity, and (*ii*) Deflection factor. **[Ans.** (*i*) 0.333 mm/Ω; (*ii*) 3Ω/mm]
6. A pressure measuring system consists of a piezoelectric transducer, a charge amplifier and an ultra violet charge recorder; their sensitivities are 8.5 pC/bar, 0.004 V/pC and 20 mm/V respectively. Determine the deflection on the chart for a pressure change of 25 bar. **[Ans.** 17 mm]
7. Determine the resolution of a moving coil voltmeter having a uniform scale with 100 divisions; the full scale reading is 100 V and $\frac{1}{10}$ of a scale division can be estimated with a fair degree of certainty. **[Ans.** 0.1 V]
8. The calibration range of a certain pyrometer is 200°C to 900°C. If the dead zone in it is 0.12 percent of the span, determine the temperature change which might occur before it is detected. **[Ans.** 0.84°C]

**9.** On the application of a step input of 100 bar to a pressure gauge, the pointer swings to pressure of 102.5 bar and finally comes to rest at 101.3 bar. Determine :

(*i*) The overshoot of the gauge readings and express it as a percentage of the final reading.

(*ii*) The percentage error of the gauge. **[Ans.** (*i*) 1.2 bar, 1.18%, (*ii*) 1.3%]

**10.** A thermometer has a time constant of 2.5 seconds. It is quickly taken from a temperature 0°C to liquid bath having a temperature 200°C. Calculate the temperature indicated by the thermometer after a time of 1.5 seconds.

Determine its reading when it was initially held at 25°C. **[Ans.** 90.24°C, 103.96°C]

**11.** A temperature sensitive device can be modelled as first-order system with a time constant of 6 seconds. It is suddenly subjected to a step input of 25°C – 150°C. Calculate the temperature indicated by the device after 10 seconds after the start of the process.

**[Ans.** 126.4°C]

**12.** A temperature sensitive transducer when subjected to sudden temperature change takes 10 seconds to reach the equilibrium conditions (five times constant). Calculate the time taken by the transducer to read half of the temperature difference. **[Ans.** 1.39s]

**13.** A resistance temperature detector has a steady state gain of 0.3925 Ω/0°C and a time constant of 5.5 seconds. It is subjected to a step change of 75°C in temperature. If its stable resistance before the step change is 100 Ω, write down the time-domain equation for resistance and find its value after 15 seconds of applications of step input.

**[Ans.** $R_t = 29.44\ (1 - e^{-t/5.5}) + 100$, 127.5 Ω]

**14.** A thermometer initially at a temperature of 20°C is suddenly plunged into a liquid bath maintained at 150°C. After a time interval of 3 seconds, the thermometer indicated a temperature of 95°C. Determine.

(*i*) The time constant for the thermometer.

(*ii*) The indicated temperature after five times constant; comment upon this result.

**[Ans.** (*i*) 3.487 s, (*ii*) 149.12°C]

**15.** A temperature sensitive transducer used to measure the temperature of a furnace has been idealised as a first-order system subjected to ramp input. If the maximum permissible error in temperature measurement is limited to 4.2 °C, calculate the time constant of the transducer when the furnace temperature increases at the rate of 0.14/°C.

**[Ans.** 30°C]

**16.** A weather balloon carrying a temperature sensing device (of first-order, time constant 10 seconds) rises through the atmosphere at 6m/s. It supplies information about temperature and altitude through radio signals. At 2800 m height a temperature indication of 35°C has been received. If the temperature varies with altitude at a uniform rate of 0.01°C/m, determine the true altitude at which 35°C temperature occurs. **[Ans.** 2740 m]

**17.** The temperature of a furnace is found to vary sinusoidally between 400° and 450°C with a periodic time of 60 seconds. A thermocouple system with a time constant of 8 seconds is employed to measure the furnace temperature. Determine :

(*i*) The maximum and minimum values that will be indicated by the thermocouple.

(*ii*) The phase shift and corresponding time lag between the temperature signals and the thermocouple input signals. **[Ans.** (*i*) 444.2°C, 405.8 °C, (*ii*) 39.76°, 6.669s]

**18.** A first-order system having a time constant of 0.1 second is used to measure a signal prescribed by the relation; $I_i = 3 \sin 2t + 0.4 \cos 10t$. Develop an expression for the corresponding output. **[Ans.** $I_o = 2.94 \sin (2t - 11.31°) - 0.283 \sin (10t + \pi - 45°]$

**19.** A liquid thermometer has it glass bulb protected by a well. The system can be represented by a double capacity system with time constants of 40s for the well and 20s for the bulb. The thermometer is subjected to a cyclic change of ± 10°C which occurs every 120 seconds. Find the maximum value of the indicated temperature and the delay time.

**[Ans.** ± 2.97°C; 36.8s]

20. An input of 1 sin2$t$ + 9.3 sin20$t$ is applied to a first-order instrument having a time constant of 0.2 second. Find the output.

[**Ans.** 0.93 sin (2$t$ – 21.8°) + 0.073 sin (20$t$ – 76°]

21. The pen arrangement of a U.V. recorder has a mass of 4.8 g. What should be the percentage reduction in mass if it is desired to have 18 percent increase in the natural frequency of the recorder ? Consider the pen motion to conform to a second-order system.

[**Ans.** 28.18%]

22. (*a*) Determine the damping ratio; damped natural frequency, static sensitivity and time constant for a second-order system which follows the differential equation given as :

$\frac{d^2I_o}{dt^2} + 2\frac{dI_o}{dt} + 25I_o = 25I_i$ where $I_o$ and $I_i$ are the output and input quantities respectively. [**Ans.** 0.2; 4.899 rad/s; 1.0; 0.2s]

(*b*) While measuring force by a diagram type pressure transducer (second-order systems), the system is found to have a natural frequency of 900 cycles per second (cps) and damping ratio of 60 percent.

Determine the frequency range over which the amplitude ratio corresponding to sinusoidal input deviates by a maximum amount of 10 percent. [**Ans.** 0 – 824 cps]

23. A unit amplitude sinusoidal input signal having a frequency of 0.5 Hz is given to second-order control system of transfer function : $(I_o/I_i) = 8/(D^2 + 4D + 20)$.

Develop an expression for the steady state response of the system and determine the corresponding output amplitude, output frequency and phase lag.

[**Ans.** $I_o$ = 0.4587 (sin $\pi t$ – 46.04°); 0.4587; $\pi$; 46.04°]

24. A torque sensing transducer is connected to the shaft of an electric motor which drives a load that has a moment of interia of 0.20 kg-m$^2$. If the torsional constant is 1.5 Nm/rad, what should be the damping ratio of the system when a simusoidal torque varying at frequency of 2 rad/s it to be reproduced with a maximum error of + 10 percent ?

[**Ans.** 0.534]

25. A second-order instrument has a natural frequency of 4 Hz and a damping ratio of 0.66. If the excitation frequency of the system is 6Hz, determine the error due to the proximity of excitation frequency with the natural frequency of the system. [**Ans.** 57.3%]

26. Three resistors have the following have the following ratings :

$r_1$ = 74 Ω ± 5%, $R_2$ = 150 Ω ± 5%; $R_3$ = 100 Ω + 5%.

Determine the magnitude and limiting error in ohms and in percent of the resistance of these resistances connected in series. [**Ans.** 324 ± 16.20 ± 5%]

27. The resistance of a circuit is found by measuring current flowing and the power fed into the circuit. Find the limiting error in the measurement of resistance when the limiting errors in the measurement of power and current are 0.9% and ± 0.6% respectively.

28. The unknown resistance in a Wheatstone bridge is measured by using three known resistances such the $R_4 = \frac{R_2R_3}{R_1}$;

where, $R_1$ = 100 ± 0.5% Ω ; $R_2$ = 500 ± 5% Ω ; $R_3$ = 280 ± 0.5% Ω.

Determine the magnitude of the unknown resistance ($R_4$) and the limiting error in percent and in ohms (for the unknown resistance). [**Ans.** 1400 Ω ; 1.5% ± 21 Ω].

29. During a test the current was measured as 30.4A, flowing in a resistor of 0.105 Ω. Later on its was discovered that the ammeter reading was low by 1.2% and the marked resistance was high by 0.3%. Determine the true power as a percentage of the power that was originally calculated.

[**Ans.** 102.1%]

**30.** A circuit was tuned for resonance by eight different students, and the values of resonant frequency in KHz was recorded as 532, 548, 543, 535, 546, 531, 543 and 536. Calculate the following :

(*i*) Arithmetic mean.
(*ii*) Average deviation, and
(*iii*) Variance. [**Ans.** (*i*) 539.25 kHz, (*ii*) 5.75 kHz; (*iii*) 6.54 kHz, (*iv*) 42.77 $(kHz)^2$]

**31.** During a test run, measurements of temperature were made 100 times with variation in apparatus and procedure. After applying corrections for known systematic errors, the following data were obtained :

| Temperature ° C : | 397 | 398 | 399 | 400 | 401 | 402 | 403 | 404 | 405 |
|---|---|---|---|---|---|---|---|---|---|
| Frequency of Occurrence : | 1 | 3 | 12 | 23 | 37 | 16 | 4 | 2 | 2 |

Calculate the following :

(*i*) Arithmetic mean,
(*ii*) Mean deviation,
(*iii*) Standard deviation,
(*iv*) Probable error of one reading,
(*v*) Probable error of the mean, and
(*vi*) Standard deviation of the standard deviation.

[**Ans.** (*i*) 400.78°C, (*ii*) 1.028°C, (*iii*) 1.380°C, (*iv*) 0.93°C, (*v*) 0.093°C, (*vi*) 0.0796°C]

**32.** The speed of a hydraulic turbine mode, which is running at 1000 r.p.m. is measured by a tachometer. The hydraulic turbine is subject to variations in speed. For a sample of 20 readings at this speed, how many readings would you expect between 990 and 1010 r.p.m. Assume that the tachometer gives a normal set of deviation with precision index of 0.04 at 1000 r.p.m. [**Ans.** 9 readings]

**33.** The depth of water which had a norminal value of 18 cm, was measured with a hook. Measurements were taken 24 times and 6 readings were found to lie outside a particular range. Assume that the measurements conform to normal distribution with precision index 8 $cm^{-1}$, determine the prescribed range. [**Ans.** (18 ± 0.1016) cm]

**34.** (*a*) A mass flow meter using electromagnetic principle is set to record a large number of readings. The fixed mass flow rate is 50 kg/s; half of the readings lie between 52.4 and 47.6 kg/s. The sample measurements give a straight line between the (0, 0.50%) points on the probability coordinates. Calculate the precision index of the instrument.

(*b*) An alarm rings when the flow meter gives a reading of 44 kg/s.

(*i*) How many false alarms are expected in the month of June if the flow is checked automatically eight times a day ?

(*ii*) What will be the value of precision index if one wants to reduce the false alarm rate by a factor of two ? [**Ans.** (*a*) 0.199; (*b*) (*i*) 12, (*ii*) 0.231]

**35.** The data given below are expected to follow a linear relationship : $v = u + ab$

| $u$ : | 0.9 | 2.3 | 3.3 | 4.5 | 5.7 | 6.7 |
|---|---|---|---|---|---|---|
| $v$ : | 1.1 | 1.6 | 2.6 | 3.2 | 4.0 | 5.0 |

Using least square analysis : (*i*) Obtain the best linear relation, and (*ii*) Calculate the standard deviation.

[**Ans.** $v = 0.672u + 0.296$, (*ii*) $s_v = \pm 0.1716$; $s_u = \pm 0.255$; $s_a = \pm 0.0355$; $s_b = \pm 0.155$]

**36.** The final result of a measurement depends on two measurements in series, each having values of 25 and 50 units with errors of ±2% and ±1%, such that result is sum of both. Determine the error of final result if errors are considered final as (*i*) Limiting errors, (*ii*) Standard deviations. [**Ans.** (*i*) 1.34%; (*ii*) 0.943%]

CHAPTER

# 2 Electronic Instruments

**2.1. Introduction 2.2. Essentials of an electronic instrument 2.3. Advantages of electronic instruments 2.4. Electronic voltmeters** – Introduction – Advantages of electron voltmeters – Vacuum tube voltmeters (VTVM) – Transistor voltmeters (TVM) – Differential voltmeter (D.C) – Rectifier type A.C. voltmeter – Average responding voltmeter – Peak responding voltmeter – True r.m.s. – reading voltmeter – Solid state voltmeter – Vector voltmeter – Electronic multimeter (or VOM) – Selection of an analog voltmeter. **2.5. Cathode ray oscilloscope** – Introduction – Applications of CRO – Block diagram of an oscilloscope – Cathode ray tube (CRT) – Electrostatic deflection – Magnetic deflection – Vertical amplifier – Time base generator – Types of sweeps – Horizontal amplifier – Delay line – Basic controls of CRO – Dual track oscilloscope – Dual beam oscilloscope – Storage oscilloscope – Sampling oscilloscope – Digital read out oscilloscope – Digital storage oscilloscope – High frequency oscilloscope – *Accessories* of cathode ray oscilloscope – probes – Electronic switch – Calibrators – Cameras – *CRO measurements* – Voltage measurement – Current measurement – Time period measurement – Lissajous figures/patterns – Phase angle measurement – Frequency measurement – Dielectric loss measurement –Delayed sweep – **2.6. Bridge and other measurements** – Introduction – Types of bridges – Wheatstone bridge – Kelvin bridge – A.C. bridge – general form – Maxwell bridge – Hay bridge – Schering bridge – Anderson bridge – Owen bridge – De Sauty bridge – Heaviside mutual inductance bridge – Carey Foster bridge – Campbell bridge – Wien bridge – Resonance bridge – Wagner earthing device – Types of detectors – Precautions to be taken while using a bridge – Q–meter – Impedance (R-L-C) bridge – LCR meter – RX meter – Measurement of resistance (effective), inductance and capacitance – Automatic bridges – Measurement of current using electronic instruments – Measurement of power – Measurement of audio and radio frequencies – *Highlights – Objective Type Questions – Theoretical Questions – Unsolved Examples.*

## 2.1 INTRODUCTION

**"Instrument"** *may be defined as a device for measuring the value or magnitude of a quantity or variable.* **"Measurement"** *involves using an instrument as a physical means of determining a quantity or variable.*

For any measuring instrument, the *primary requirements* are :

1. Power consumption should be as small as possible.
2. No alteration in the circuit conditions with the introduction of instrument in the circuit.
   - Most of the scientific and industrial instruments require *very fast responses*. Such requirements cannot be met with by mechanical and electrical instruments. The "*electronic instruments*" use *semiconductor devices*. In the electronic devices, since the only movement involved is that of electrons, *the response time is*

*extremely small owing to very small inertia of electrons.* With the use of electronic devices *very weak signals can be detected by using pre-amplifiers and amplifiers.*

- The electronic instruments are based on electrical or electronic principles for their measurement.
- These instruments are either indicating or recording type instruments just like the electrical measuring instruments. In addition to indicating and recording a quantity, some instruments also control a quantity e.g. tachogenerator may be used for indicating or reccording the speed as well for controlling the speed in a speed controlled motor.

## 2.2 ESSENTIALS OF AN ELECTRONIC INSTRUMENT

An electronic instrument is made up of the following three elements (See Fig. 2.1). 1. Transducer; 2. Signal modifier; 3. Indicative device.

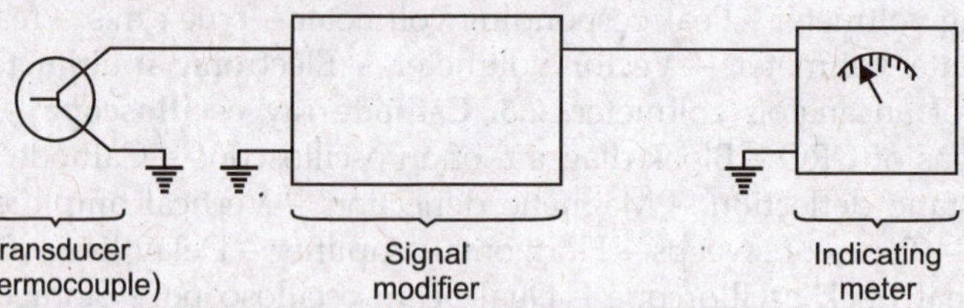


Fig. 2.1. An electronic instrument.

1. **Transducer :**
   - This sensing element is required only for measuring non-electrical quantities e.g. pressure, temperature. Its function is to convert the non-electrical quantity into an electrical signal.
   - It is not required when the quantity being measured is already in the electrical form.
2. **Signal modifier :**
   - The function of this element is to make the incoming signal suitable for application to the indicating device.

   ***Example.*** The signal may need *amplification* before it can be properly displayed.

   — Other types of signal modifiers are : Voltage dividers for reducing the amount of signal applied to the indicating device or wave shaping circuits such as *filters, rectifiers* or *choppers* etc.
3. **Indicating device :**
   - The indicating device is usually a deflection type meter (See F.g. 2.1) for general purpose instruments like voltmeters, ammeters, or ohm meters.
   - For digital readout instruments, the indicating device is of *digital design*.

## 2.3 ADVANTAGES OF ELECTRONIC INSTRUMENTS

Following are the *advantages of electrical/electronic techniques of measurements :*

1. Less power consumption and less load on the system being measured.
2. Non-contact measurements are possible.
3. Light, compact and more reliable instrumentation.
4. Remote indication and recording feasible.
5. The amplification produced is greater than that produced by a mechanical contrivance.
6. The mathematical processing of signals like summation, differentiating and integrating is possible.

7. The frequency and transient response is good.
8. Greater flexibility.
9. Lighter in construction.
   - The electronic instruments are *widely used in detection of electro-magnetically produced signals such as radio, video and microwaves.*
   - The use of these instruments has made it possible to *build analog and digital computers* which are quite instrumental in the modern developments in the fields of science and technology.

## 2.4 ELECTRONIC VOLTMETERS

### 2.4.1 Introduction

*"Electronic voltmeters"* use *rectifiers, amplifiers and other circuits to generate a current proportional to the voltage being measured.* The current so produced is measured by a permanent moving coil instrument (normally abbreviated as PMMC or D'Arsonval movement).

— Any measurement using PMMC movement must draw a current of 50 μA from the measured quantity for its operation for full scale deflection if conventional voltmeters are used. This would produce *great loading effects* specially in electronic and communication circuits. *"Electronic voltmeters"* avoid the loading errors by supplying power required for measurement by using external circuits like *'amplifiers'*. The amplifiers not only supply power for the operation but also make it possible for low level signals, which produce a current less than 50 μA for full scale deflection, to be detected which otherwise cannot be detected in the absence of amplifiers.

### 2.4.2. Advantages of Electronic Voltmeters

The electronic voltmeters claim the following *advantages over conventional ones* :

1. Low power consumption.
2. High accuracy.
3. High frequency range.
4. Compactness and portability.
5. Low level signal detection.
6. Less loading effect.
7. High input impedance.
8. High sensitivity.

The *comparison* between electronic voltmeters and conventional analog meters is given in table 2.1. below :

**Table 2.1. Comparison between Electronic Voltmeters and Conventional Analog Voltmeters**

| S.No. | *Aspects* | *Electronic voltmeters* | *Conventional analog voltmeters* |
|---|---|---|---|
| 1. | *Power consumption* | Low | High |
| 2. | *Type of construction* | Not rugged | Rugged |
| 3. | *Use of amplifiers* | Yes | No |
| 4. | *Voltage range* | Higher | Limited |
| 5. | *Loading effect* | Not present | Present |
| 6. | *Low level signals* | Can be detected | Cannot be detected |
| 7. | *Sensitivity* | High | Low |
| 8. | *Frequency range* | High | Limited |
| 9. | *Input impedance* | Very high | Low |
| 10. | *Instrument range* | More | Less |

### 2.4.3. Types of Electronic Voltmeters

The multimeter, also called volt-ohm-milliammeter (VOM), can be used to measure voltages but it lacks both sensitivity and high input resistances; moreover, its input resistance is different for each range. The *electronic voltmeter* (EVM), on the other hand, has *input resistance ranging from 100 MΩ to 1000 MΩ, thus producing less loading of the circuit under test than the VOM*. Another advantage of EVM is that its *input remains constant over all ranges.*

Following two types of voltmeters are used these days :

1. Analog voltmeters.
2. Digital voltmeters.
   - A "digital instrument" is one which uses internal circuitry of digital design. A *"digital readout instrument" is* one whose measuring circuitry is of analog design but the indicating device is of digital design.

The electronic voltmeters are also *classified according to technology used*, as follows:

1. Vacuum-tube voltmeter (VTVM) :
   - It uses vacuum tubes with deflection meter movement.
2. Digital voltmeters :
   - Transistor voltmeter (TVM).
   - FET voltmeter (FETVM)

### 2.4.4. Vacuum Tube Voltmeters (VTVM)

**Introduction :**

R.A. Heising, in 1915, used a **valve** to measure the voltage. Most of the pioneer work in the united Kingdom was done by E.B. Moullin Oxford University. In fact, at one time, valve voltmeters were usually referred to as Moullin voltmeters. In U.S.A. these meters are known as '**Vacuum Tube Voltmeters**' usually abbreviated as 'V.T.V.M.'.

— '*Valve–voltmeters*' essential consists of a *thermionic valve which has a millimeter connected in its anode circuit*. The voltage which is to be measured is normally applied to its control grid circuit, which imposes very little load on the circuit, even at a high frequency. Although the basic arrangement of a voltmeter has limited range, this can be extended by the use of a potential divider.

— The valve voltmeter *takes practically no power at all from the source under test*; this factor is an important one in the measurement of voltages in a *radio circuit*. An ordinary moving coil meter, no matter how sensitive it may be, always draws some power from the circuit under test. In circuits where a plenty of power is available this is not serious, but when dealing with a circuit in which *even a load of a few microamperes could seriously affect the accuracy of the reading, a valve voltmeter should be used.*

**Description :**

- A **vacuum tube voltmeter** is a useful measuring device to measure *A.C. Voltages over a wide frequency range.*

  — It can work on high frequencies, without losing any efficiency, even if it is calibrated at such a low frequency as 50 Hz.

  — It consumes practically no power.
- It works on the principle that a *detector can be used as a voltmeter.*

The voltage to be measured is first rectified and the rectifier plate current is used to measure the applied voltage, by having a D.C. indicating instrument in the plate circuit of the tube. Usually an *amplifier stage is also added, which in addition to increasing sensitivity of the instrument, also increases the input impedance.* Because of this increase in input impedance, the power drawn by the instrument *reduces* nearly to *nil.* This makes the measurements *more accurate* and the instrument can be used over a wide voltage range with the same high sensitivity.

**Types of vacuum type voltmeter :**

The different types of V.T.V.M. are given below :

1. D.C. vacuum tube voltmeter
2. Diode rectifier-amplifier meter
3. Voltmeter
4. Plate circuit rectification meter
5. Gridh rectification meter
6. Amplifier type meter.

## True R.M.S. voltmeter :

Fig. 2.2. shows one type of commercial true r.m.s. instrument :

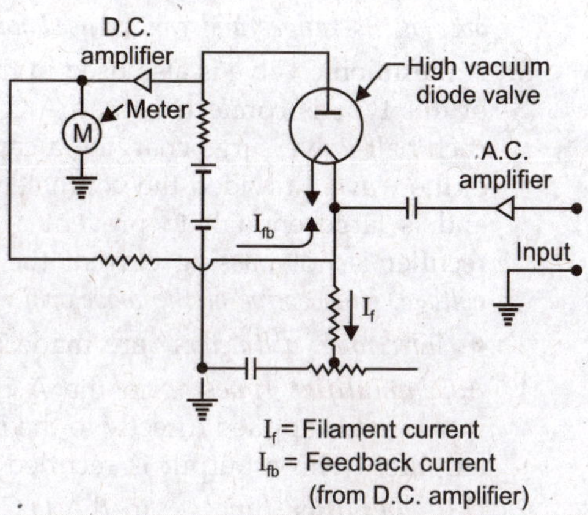


**Fig. 2.2.** True r.m.s. voltmeter.

- In this instrument the input signal is amplified and fed into the directly-heated filament of high vacuum diode valve, thus furnishing part of the filament heating power. The filament must have a small thermal time constant if the instrument is to respond to rapid changes in signal level.
- After amplification of A.C. input signal a cathode follower is used to obtain an impedance match between the amplifier and the low-impedance filament. The signal current will tend to raise the filament temperature, which will cause the anode current to rise and the anode-to-cathode voltage to fall. This drop in voltage is amplified by the D.C. amplifier. The output of this amplifier supplies the component of current $I_{fb}$ (feedback current) to the filament.
- The feedback circuit is adjusted so that the decrease in $I_{fb}^2 R_f$ equals the signal power, $I_s^2 R_f$, in the filament, where $I_s$ is the *r.m.s. signal current.*
- The meter reads the *output voltage of the amplifier, which is proportional to the feedback current* $I_{fb}$. It can be shown that the feedback current $I_{fb}$ *is proportional to the square of the r.m.s. signal current*. The meter reading is, *therefore, proportional to the r.m.s. signal current.*
- The meter will have a *linear power scale.*

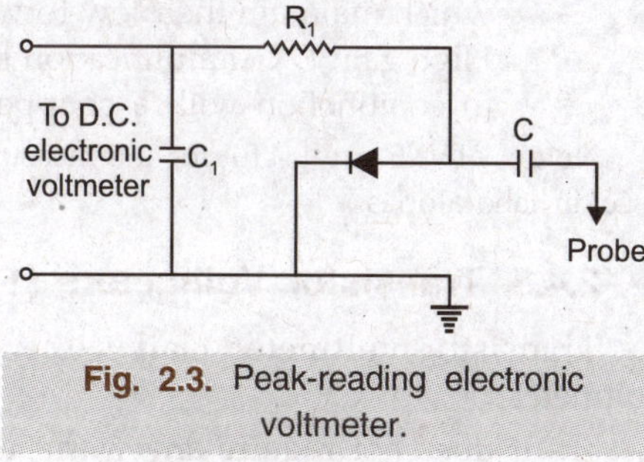


**Fig. 2.3.** Peak-reading electronic voltmeter.

**Peak-reading Electronic Voltmeter :**

In Fig. 2.3. is shown the basic circuit of a peak-reading electronic voltmeter. The diode is placed in a probe for high-frequency response and low-level signals. The diode permits the A.C. signal to chargé C to the peak value of the signal voltage. The network $R_1C_1$ is a simple filter which removes all ripple voltage from the D.C. that is delivered to the D.C. amplifier in the instrument.

**Advantages-Applications of valve voltmeters :**

1. An important advantage of the valve voltmeter is that it may be designed to cover *practically any frequency;* commercially manufactured instruments are suitable for frequencies upto 50 MHz and above.
2. The valve voltmeter will often be used on circuits where the voltage is not very high so that input impedance can only be compared with moving-coil meter which has been adjusted to appropriate range. This means that it is usually impossible to determine accurately a low voltage on the high voltage range of a moving-coil meter. On the other hand, the valve voltmeter *maintains high impedance over all its ranges and can be used for low-voltage R.F. measurement.*
3. A thermionic valve is also used in the diode peak voltmeter. When an instrument of this type is connected to an A.C. source, rectification takes place in the diode each half-wave surge charging a capacitor in the output circuit to the peak value of the wave. Provided the voltmeter in the circuit is of sufficiently high resistance and is large enough to prevent a loss of charge through the meter when the rectifier is not passing current the reading on the meter will indicate the *peak voltage, irrespective of the waveform.*

- *"Electronic voltmeters"* are made in the following ***two basic forms*** :

1. ***A.C. amplifier types*** ... in the A.C. amplifier configuration, the voltage to be measured is applied directly to the input terminals of a wide-band gain-stabilized amplifer, whose output is rectified and applied to a moving-coil meter.
2. ***D.C. amplifier types*** ... in the D.C. amplifer construction, the input voltage is applied directly to a diode detector, and the resulting D.C. is amplified and monitored.

— It is generally accepted that the A.C. *amplifier type is usually the more sensitive of the two, but the usable range is restricted by the amplifier band-width.*

— The D.C. amplifer form permits voltage measurement upto frequencies approaching the resonance of the diode unit. Sensitivity however, is restricted by the linearity of the diode and stability of the D.C. amplifer. The first of these limitations is reduced considerably by the use of semiconductor diodes, which maintain their low forward resistance down to very small voltage levels. High gain D.C. amplification is obtained by the use of an A.C. coupled amplifer in conjunction with a chopper and synchronous detector.

**Note :** VTVMs are useful for the measurement of D.C. and A.C. voltages. They are till today used in laboratories.

### 2.4.5. Transistor Voltmeters (TVM)

Transistor voltmeter claims the following *advantages* over the vacuum tube voltmeter :

1. It does not require any warm up time.
2. It is comparatively compact and is light in weight and as such suited for field work.
3. Being battery operated, it can be used where a source of external power is not available.

- A transistor offers a low input resistance. That is why in early days hybrid (valve + transistor) configuration was used to provide high input resistance. However, now-a-days field effect transistors (FET) which offers as high input resistance as valve are extensively used.

## D.C. VOLTMETERS

### 2.4.5.1 Transistorised voltmeter using cascaded transistors

Fig. 2.4 shows a TVM using cascaded transistors (TVM using a single transistor has *low sensitivity*) :

— It consists of two transistors $Q_1$ and $Q_2$ directly coupled in a complementary configuration (i.e. one is NPN type while other is PNP type); they form a *direct coupled amplifier*.

— The input voltage is applied through a network of resistances. The range switch selects the resistor that is in series with the transistor for the desired range of voltage. It can be seen from Fig. 2.4 that for the 0 to 1V range the value of resistance is 0.2 MΩ *i.e.* 200 K Ω; this means sensitivity is 200 kΩ/V neglecting small resistance offered by the transistor $Q_1$. Thus a current of $\frac{1}{200 \times 1000}$ a.m.p. or 5 μA is drawn from the circuit. The values of the other resistors are so chosen that the same current (*i.e.* 5μA) is drawn from other range.

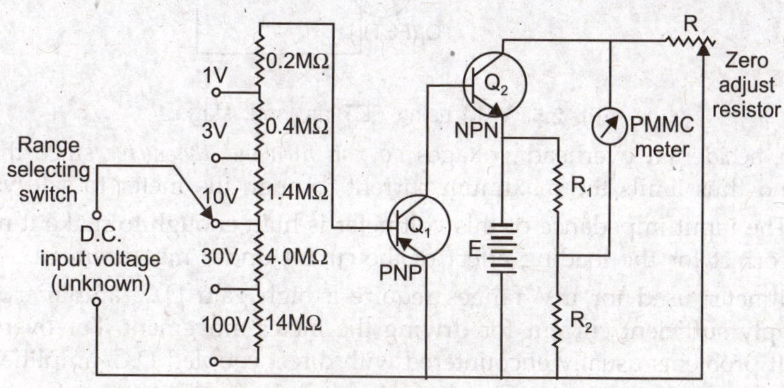


$Q_1$, $Q_2$ = Transistors; E = Battery

**Fig. 2.4.** TVM using cascaded transistors.

— A variable resistance R is put in the circuit for zero adjustment of the PMMC. It controls the bucking current from the supply E to buck out the quiescent current.

- The *drawback* of such a voltmeter is that it has to work under specified ambient temperature to get the required accuracy otherwise excessive drift problem occurs during operation.

### 2.4.5.2. Transistorised voltmeter using FET as input stage

Fig. 2.5. shows a TVM D.C. voltmeter with a *FET* (field effect transistor) forming the *input stage* of a direct coupled amplifier :

— In this voltmeter, voltage to be measured is firstly attenuated with range selection switch to keep the input voltage of amplifier within its level.

— *FET* is used in the input stage of amplifier because of its *high impedance so that it does not load the circuit of which voltage is to be measured and it also keeps the sensitivity of voltmeter very high.* As FET is a voltage controlled device so resistance network of attenuator is put in *shunt* in the circuit.

— The two transistors $Q_1$ and $Q_2$ form a *direct coupled D.C. amplifier* using the PMMC meter movement. Provided that the amplifier operates within the limits of its dynamic range, the deflection of meter is *directly proportional* to the magnitude of the input applied voltage. The gain of this amplifier permits the instrument to be used in the *mV range.*

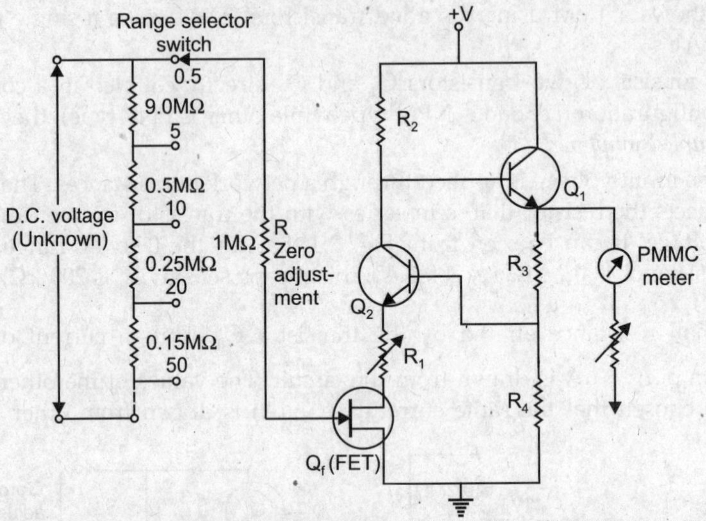


Fig. 2.5. VTM using FET as input stage.

Further, accidental overhead voltages *do not burn out the meter* since the amplifier saturates and thus limits the maximum current through the meter to safe value.

— The input impedance of this voltmeter is high enough to make it necessary to correct for the loading effect of the circuit under measurement.

- Voltmeter used for mV ranges require a high gain D.C. amplifier in order to supply sufficient current for driving the meter movement. For overcoming the drift problems usually encountered with direct coupled D.C. amplifiers, the D.C. input is converted into A.C. voltage, before it is applied to an A.C. amplifier. The amplified A.C. voltage is then converted back into a D.C. voltage.

— In order to avoid the drift problems of D.C. amplifiers, *chopper type D.C. amplifiers* are commonly used in high sensitivity voltmeters.

**Advantages :**

1. Very high sensitivity (as much as 100 times the normal PMMC meter).
2. No damage of the meter due to the overload.
3. No loading effect, since the amount of power drawn is very low.
4. High input impedance (to isolate meter from the measurement circuit).
5. The voltages in the mV range can be measured.

**2.4.5.3. Balanced bridge TVM**

This is also called *"difference-amplifier type TVM".*

A basic configuration of a balanced bridge TVM is shown in Fig. 2.6 :

— The bridge arms consist of resistances $R_{L1}$, $R_{L2}$ and internal collector resistances of two identical transistors $Q_1$ and $Q_2$.

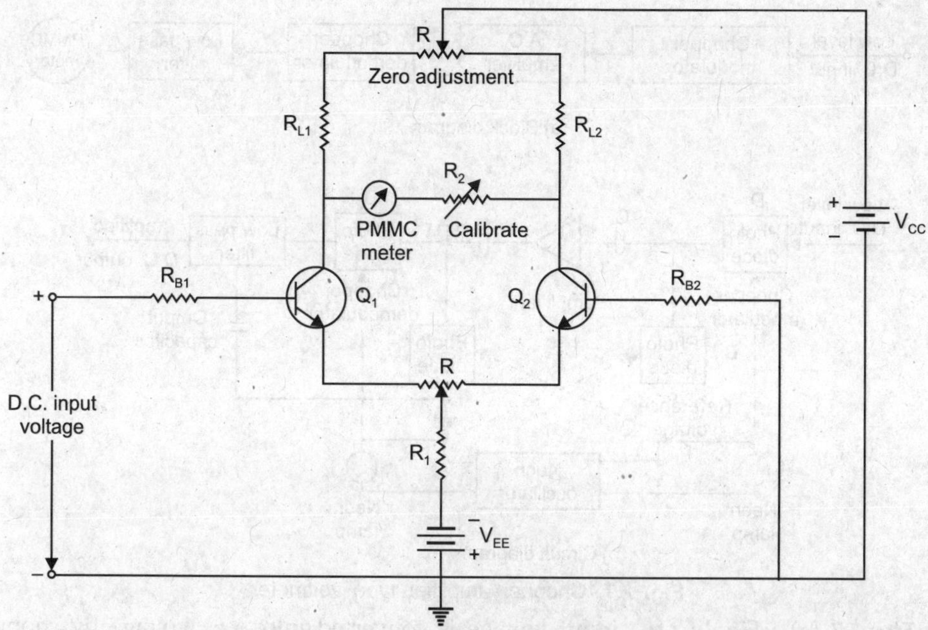


**Fig. 2.6.** Balanced bridge TVM.

— When a positive voltage is applied to the base of $Q_1$, an increased emitter current flows. In turn, $Q_2$ draws decreased emitter current due to increased voltage drop across $R_1$. Thus the bridge is unbalanced and a current proportional to input voltage flows through the meter which can be calibrated to read the voltage supply directly.

— Resistor $R$ is used for *zero adjustment*. The purpose of this adjustment is to equalise both halves of the difference amplifier so that there is no current through the meter under the absence of D.C. input voltage.

— Resistance $R_2$ is used for calibration purpose – It does accurate selling of the full-scale reading of the ammeter reading. In order to get several different full-scale ranges a range switch may be used.

***Advantages :***

1. When transistors $Q_1$ and $Q_2$ are similar then with the change in ambient temperature, β valves of both the transistors will change equally so there will be *no effect on collector current* in the changed ambient temperature condition.

   Also, with change in ambient temperature, saturation current of both the transistors will change equally, and, therefore, the *meter reading will remain unaffected.*

2. Any fluctuation in power supply $V_{CC}$ and $V_{EE}$ does not affect its performance as it *changes the biasing of both transistors equally.*

***Disadvantage :***

The main disadvantage of this type of voltmeter is its comparatively *low input resistance.* However a high resistance comparable to that of VTVM can be achieved by designing the voltmeter with FET as bridge circuit component.

### 2.4.5.4. Chopper amplifier type voltmeter

This type of voltmeter is *highly sensitive* D.C. electronic voltmeter.

Figs. 2.7 (*a*), 2.7 (*b*) show the block diagram and the circuit diagram of a chopper amplifier type voltmeter respectively.

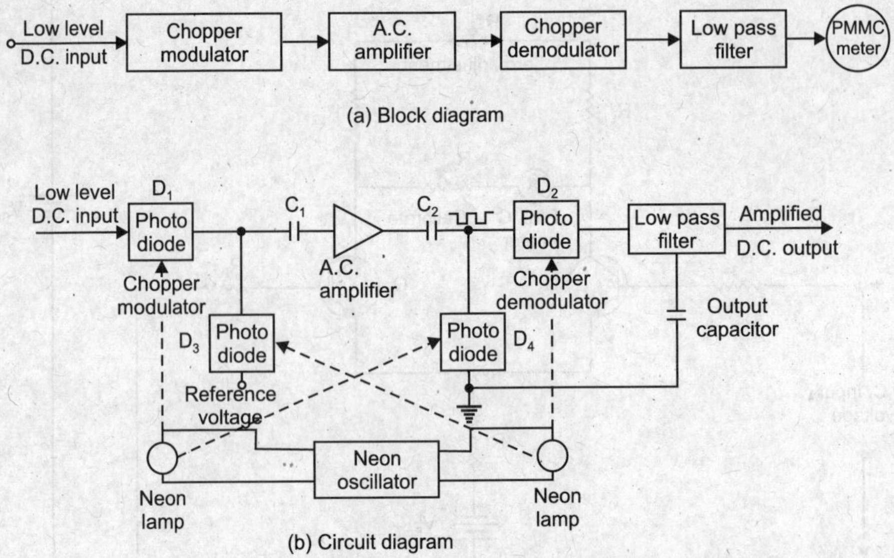


**Fig. 2.7.** Chopper amplifier type voltmeter.

**Fig. 2.7 (*a*) :** Firstly D.C. input voltage is converted into A.C. voltage by chopper modulator and then it is supplied to an A.C. amplifier. Output of amplifier is then demodulated to a D.C. voltage proportional to the original input voltage. Modulator chopper and demodulator chopper act in synchronism.

**Fig. 2.7. (*b*) :**

— The D.C. input voltage is chopped to a low frequency, 200 to 300 Hz A.C.

— It is passed through a blocking capacitor, amplified and then passed through another blocking capacitor so as to remove D.C. drift or offset from the amplified signal. By suitable choice of $C_1$ and $C_2$ it is possible to design a sensitive D.C. voltmeter which is drift free. This has high D.C. input impedance and high degree of accuracy.

— Such sensitive voltmeter employs photodiodes $D_1$, $D_2$, $D_3$ and $D_4$ for modulation (that is conversion from D.C. to A.C.) and demodulation (that is conversion from A.C. to D.C.). A photoconductor has a low resistance (say of the order of few hundred ohms) when it is illuminated by an incandescent or a neon lamp. Its resistance sharply increases to several megaohms when it is not illuminated. An oscillator drives two neon bulbs into illumination on alternate half cycles of oscillation. Each neon lamp illuminates one photodiode in the input circuit of the amplifier and in the output circuit as shown in Fig. 2.7 (b). Together they act like switch across the input to the amplifier alternately opening and closing at a rate determined by the frequency of the neon oscillator.

— The input signal to the amplifier is a square wave voltage whose amplitude is *proportional to the input voltage level and its frequency is equal to the frequency of oscillator*. The A.C. amplifier gives an amplified square wave in magnitude.

— The photodiodes in the amplifier *output circuit*, operating in antisynchronism with the input chopper, produces a *demodulating* action yielding D.C. output voltage.

— This D.C. signal is passed through a *low pass filter to remove any residual A.C. components*. This amplified drift free D.C. output is then fed to PMMC movement for measurement.

- A commercial available instrumentary using a photo chopper amplifier has an input impedance of 100 M Ω, a resolution of 0.1 μV and input ranges from 3 μ V full scale to 1 k V full scale with an accuracy of ±2% of full-scale deflection.

**Note.** The input impedance of chopper amplifier D.C. voltmeter is high except on very low input ranges. To overcome this high impedance problem an arrangement for *nulling* is included in meter circuit. A control in the front panel enables to *null the input voltage with a bucking voltage.* When the bucking voltage is equal to the input voltage a null is indicated and the meter shows an infinite impedance and therefore the loading effects are totally eliminated. The input voltage is then removed, and bucking voltage which is equal to the input voltage is shown by the meter.

**Advantages :**

1. The input impedance of chopper amplifier is usually of the order of 10 M Ω or higher, except on very low input ranges.
2. The drift in an ordinary D.C. amplifier is of the order of m V. The full scale range of an ordinary D.C. amplifier is limited to measuring input signal of 1 – 100 mV. In a chopper modulator system with the use of A.C. amplifier, drift can be cut down by a factor of 100, thus allowing an input signal range of about 0.01 mV = 10 μV full scale to be handled.

### 2.4.6. Differential voltmeter (D.C.)

- *This is one of the most common and accurate methods of measuring unknown voltages.*
- In this voltmeter the unknown voltage is *compared* with a known voltage. The principle of operation of this voltmeter is similar to a potentiometer that is why it is often called a *potentiometric voltmeter*.

Fig. 2.8 shows the *basic circuit of a differential voltmeter* :

— A precision resistance potential divider *P* is used to divide down an accurately known reference voltage $E_{ref}$.

— The adjustments of *P* are done till the output voltage *equals* the input/unknown voltage $E_{unk}$.

— Under balance conditions, neither the input source nor the reference delivers any current to the null detector, and thus voltmeter offers an infinite impedance to the source under test.

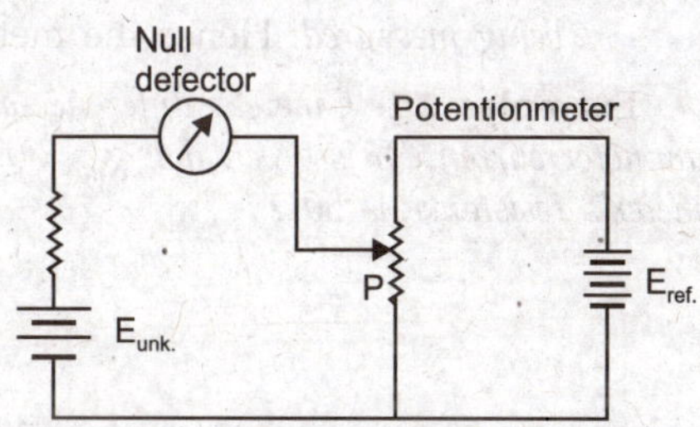


**Fig. 2.8.** Basic circuit of a differenital voltmeter.

— The reference source usually consists of a low voltage D.C. standard or a low voltage zener-controlled D.C. source. In order to measure high voltages, a high voltage reference source may be used.

- The ***drawback*** of this system is that due to use of potential divider, the input resistance of the voltmeter is no longer infinite and this meter has a *loading effect which is undesirable.*

**Direct current FETVM using difference amplifer :**

Fig. 2.9. shows the schematic diagram of FETVM using difference amplifier.

— The two FETs ($Q_1$ and $Q_2$) are identical so that increase in the current of one FET is offset by corresponding decrease in source current of the other. The two FETs form the lower arms of the balanced bridge circuit whereas the two drain resistors $R_D$ form the upper arms. The meter movement is connected across the drain terminals of the FETs.

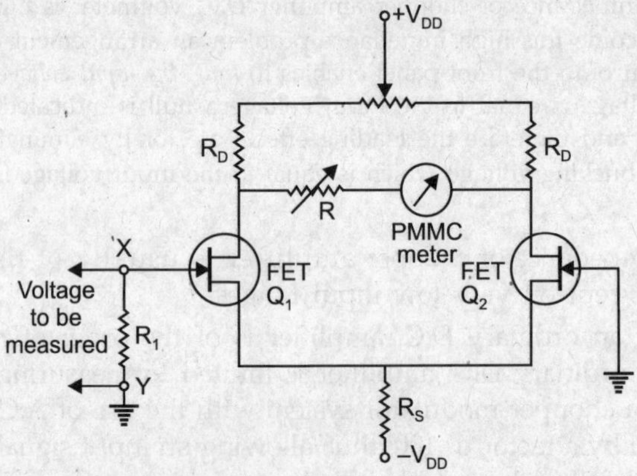


**Fig. 2.9.** FETVM using difference amplifier.

— The circuit is balanced under zero-input-voltage condition provided the two FFTs are identical. In that case there would no current through the PMMC meter. Zero-adjustment potentiometer is used to get null deflection in case there is a small current through the meter under zero-signal condition.

— The variable resistor $R$ is used to adjust the full-scale calibration.

— When positive voltage is applied to the gate of $Q_1$, some current flows through the meter. The magnitude of this current is found to be *proportional to the voltage being measured*. Hence the meter is calibrated is volts to indicate input voltage.

**Examples 2.1.** *Given a difference-amplifier type FETVM shown in Fig. 2.10, find the ammeter current. If is given that $g_m$ = 0.005 siemens, FETs have resistance of 100 kΩ and the meter's resistance is 50Ω.* **(M.S. University, Baroda)**

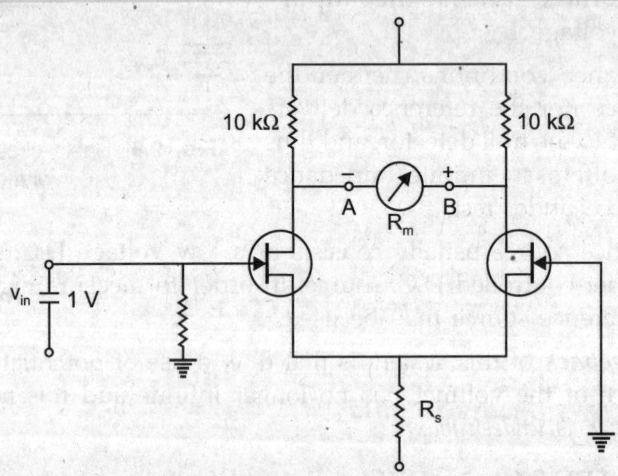


**Fig. 2.10.**

**Solution.** *Given* : $g_m$ (Transconductance) = 0.005 siemens : $R_{Q_1} = R_{Q_2} = R_Q = 100$ kΩ; $R_m = 50$ Ω; $R_D$ (Drain resistance) = 10 kΩ

Let us use Thevenin's theorem to solve the problem.

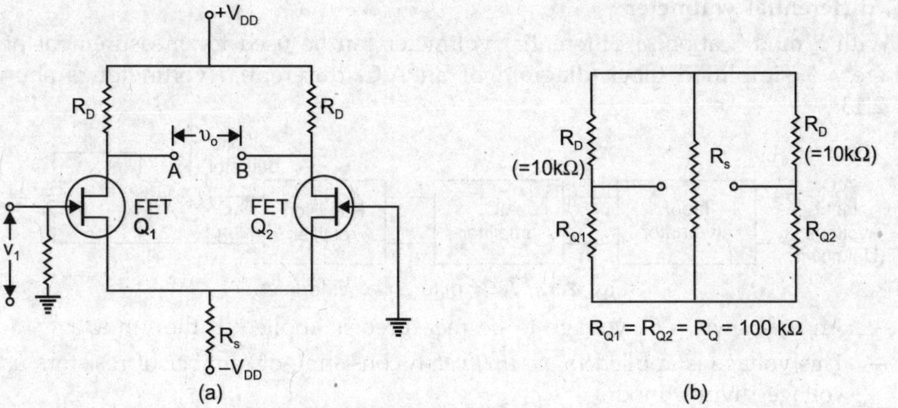


**Fig. 2.11.** Thevenin's equivalent circuit.

When we remove the ammeter (considering it as the load) the circuit will become as shown in Fig. 2.11 (a). The output voltage is the voltage gain of single FET times the difference between the input of $Q_1$ and input of $Q_2$. Since the input voltage of $Q_2$ is zero as shown in Fig. 2.11 (a) the output voltage under open circuit conditions is,

$$v_o = g_m \left( \frac{R_Q R_D}{R_Q + R_D} \right) v_1 \qquad \text{...(1)}$$

Where, $R_D$ = Drain resistor,

$R_Q$ = Resistance of FET.

In order to find the resistance of Thevenin source generator looking into terminals A and $B$, $v_1$ and $V_{DD}$ are made zero. Under this condition both FETs have a resistance of $R_Q$ as shown in Fig. 2.11 (b). From this balance circuit and assuming $R_S$ to be large, the resistance looking into terminals $A$ and $B$ is

$$R_{AB} = 2\left( \frac{R_Q \times R_D}{R_Q + R_D} \right) \qquad \text{...(2)}$$

The Thevenin's equivalent circuit with the ammeter connected as load is shown in Fig. 2.12. The current through the PMMC meter is

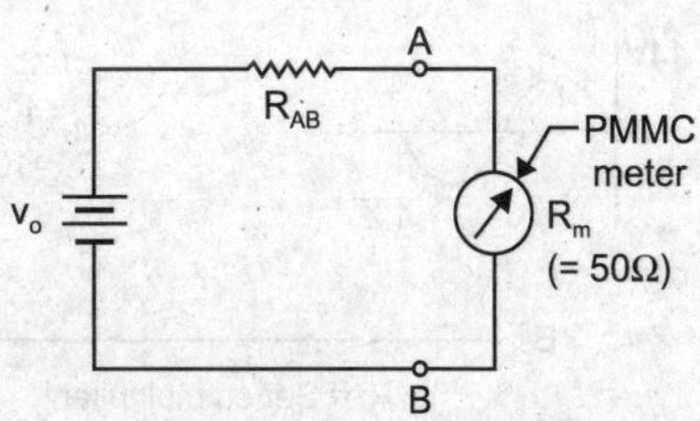


**Fig. 2.12.**

$$i = \frac{v_o}{R_{AB} + R_m}$$

or,
$$i = \frac{[g_m R_Q\, R_D / (R_Q + R_D)]\, v_1}{[2R_Q R_D / (R_Q + R_D)] + R_m}$$

...using eqs. (1) and (2)

$$= \frac{[0.005 \times (100 \times 10^3) \times (10 \times 10^3)/(100 \times 10^3 + 10 \times 10^3)] \times 1}{[2 \times (100 \times 10^3)\,(10 \times 10^3)/(100 \times 10^3 + 10 \times 10^3)] + 50}$$

$$= \frac{[0.005 \times 10^9 / 110{,}000] \times 1}{[2 \times 10^9 / 110{,}000] + 50} = \frac{45.45}{18231.8} = 2.5 \times 10^{-3}\,\text{A}$$

or, $i$ = **2.5 mA (Ans.)**

**A.C. differential voltmeter :**

With a modification, a differential voltmeter can be used for measurement of A.C. voltages. A simplified block diagram of an A.C. differential voltmeter is shown in Fig. 2.13:

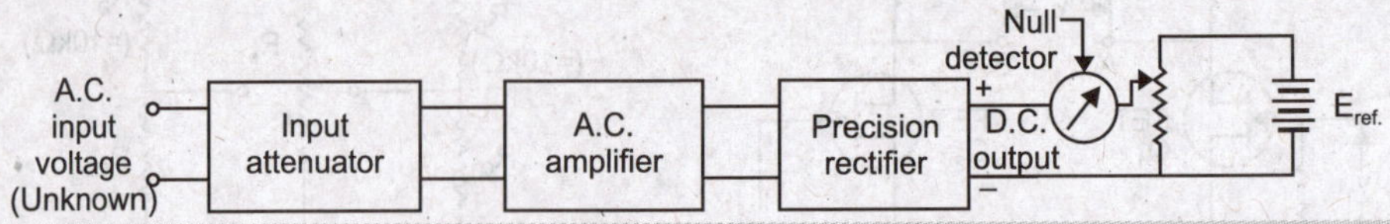


**Fig. 2.13.** A.C. differential voltmeter.

— An unknown A.C. voltage to be measured is applied at the input.

— This voltage is applied to an *attenuator*, consisting of number of resistors used in voltage divider mode.

— The output of attenuator is given to an A.C. amplifier which amplifies the small signals.

— The amplified A.C. voltage is then converted to D.C. voltage, by using a precision *rectifier*. Once D.C. output is proportional to the A.C. input voltage obtained, the remaining function of the voltmeter is exactly same as D.C. differential voltmeter as discussed above.

## 2.4.7. Rectifier Type A.C. Voltmeter

The PMMC (or D' Arsonval) meter movement can be used for measuring alternating voltage (and current) by inserting a rectifier in the measuring circuit. Such meters are more widely used than either (costly but more accurate) dynamometer type or more delicate thermal and hotwire type.

**1. A.C. voltmeter using half-wave rectifier :**

Fig. 2.14 shows an A.C. voltmeter using half-wave rectifier. Here, a half-wave diode rectifier has been combined in series with a D.C. meter movement.

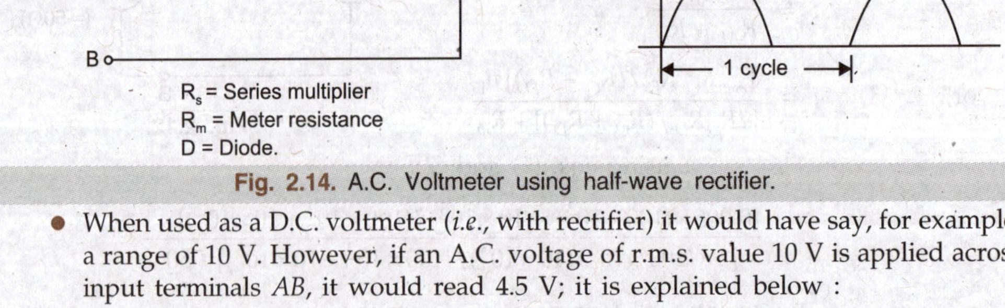


**Fig. 2.14.** A.C. Voltmeter using half-wave rectifier.

- When used as a D.C. voltmeter (*i.e.*, with rectifier) it would have say, for example, a range of 10 V. However, if an A.C. voltage of r.m.s. value 10 V is applied across input terminals *AB*, it would read 4.5 V; it is explained below :

— An A.C. voltage of r.m.s. value 10 V has a peak valve of $10 \times \sqrt{2} = 14$ V and an average value of 0.636 × 14 = 9 V. Since in the half-wave rectified output, one half-cycle is absent, the average for full cycle = $\frac{9}{2} = 4.5$ V. The meter movement will, therefore, read 4.5 V *i.e.* 45% of the D.C. value.

— It may also be noted that *A.C. sensitivity of a half-wave A.C. meter is only 45 percent* of *the D.C. sensitivity* (*i.e.* A. C. voltage is *less sensitive* than D.C. voltmeter).

- The value of series multiplier ($R_S$) for A.C. voltmeters way be obtained as follows:

$$R_S = \frac{E_{D.C.}}{I_{D.C.}} - R_m$$

or,

$$R_S = \frac{0.45\, E_{r.m.s}}{I_{D.C.}} - R_m$$

where, $I_{DC}$ = Full scale deflection current.

**2. A.C. voltmeter using full-wave rectifier :**

Fig. 2.15 shows an A.C. voltmeter using full-wave rectifier.

- In this case, obviously, the meter reading would be 90% of r.m.s. input voltage *i.e.*, 90% of the D.C. value.

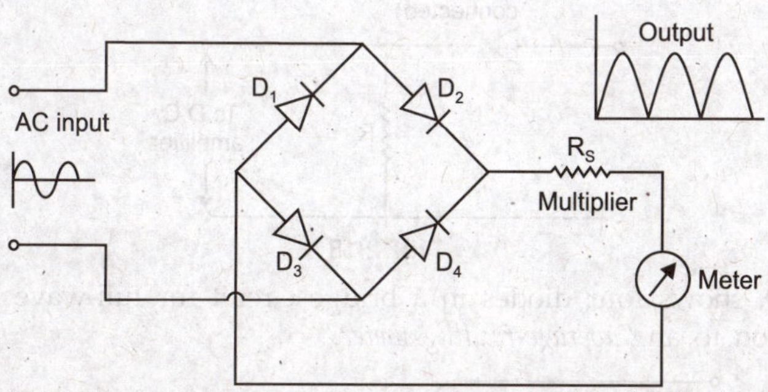


**Fig. 2.15.** A.C. voltmeter using full-wave rectifier.

- ***Electronic A.C. voltmeters*** are basically identical to D.C. voltmeters except that the A.C. input voltage must be *rectified* before it can be applied to the D.C. meter circuit. Basic A.C. voltmeter *circuits* are :

— In one approach, the rectification takes place *before* amplification, in which case a simple diode rectifier circuit precedes the amplifier and meter, as shown in Fig. 2.16. This approach ideally requires a D.C. amplifier with zero drift characteristics and *unity voltage gain*, and a D.C. meter movement with adequate sensitivity.

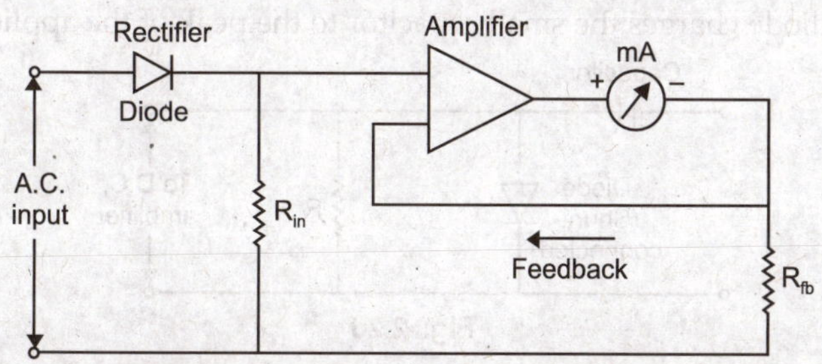


**Fig. 2.16.** D.C. mode of operation.

— In another approach the A.C. signal is rectified after amplification (See fig. 2.17) where full-wave rectification takes place in the meter circuit connected to the output terminals of the A.C. amplifier. This approach generally requires an A.C. amplifier with *high open-loop gain* and *large amounts of negative feedback* to overcome the nonlinearity of the rectifier diodes.

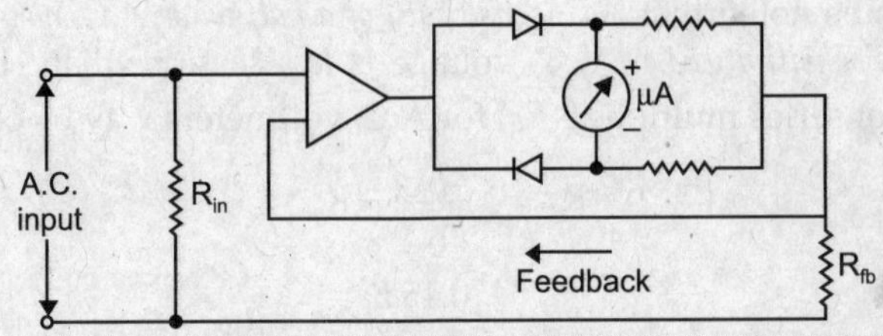


**Fig. 2.17.** A.C. mode of operation.

- The various *rectifier circuits used in A.C. voltmeter* are shown in Figs. 2.18, 2.19, and 2.20.

— **Fig. 2.18.** shows a series connected diode, providing *half-wave rectification* for an *"average-reading voltmeter"*.

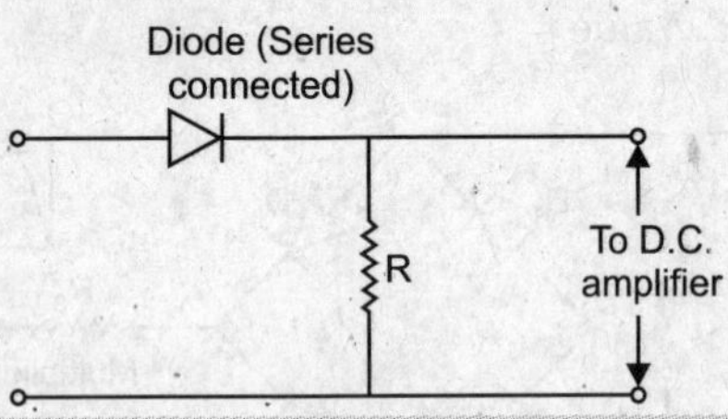


**Fig. 2.18**

— **Fig. 2.19.** shows four diodes in a bridge circuit for full-wave rectification and application to an *"average-reading meter."*

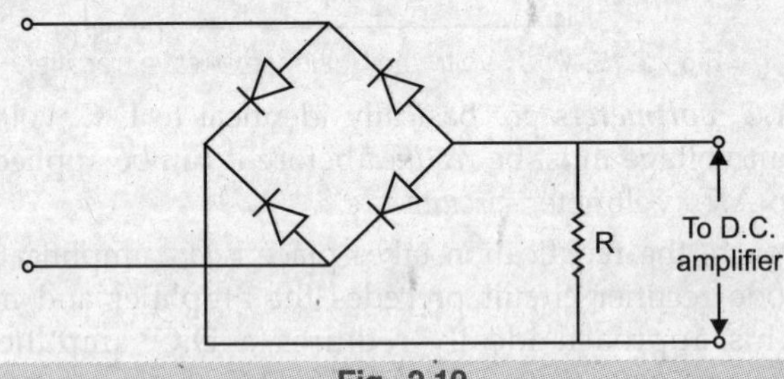


**Fig. 2.19**

— **Fig. 2.20** shows shunt-connected diode used in a *"peak-reading voltmeter"*. The rectifier diode charges the small capacitor to the peak of the applied input voltage.

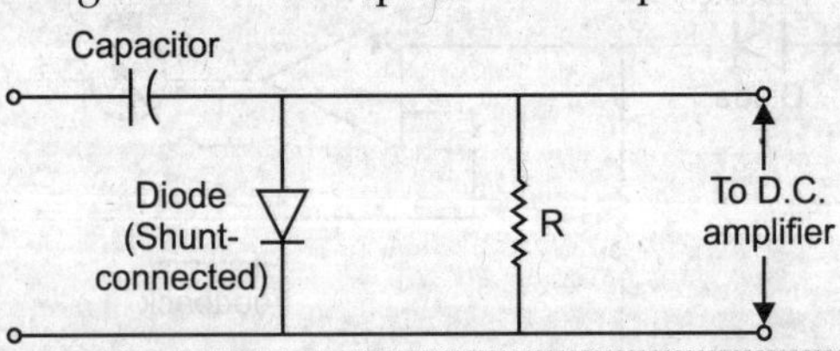


**Fig. 2.20**

- Usually A.C. voltmeters are of **average-responding** *type, with the meter scale calibrated in terms of the r.m.s. value of a sine wave*. In electronics since so many wave forms are sinusoidal, this is an entirely satisfactory solution and certainly *much less expensive* than a true-responding voltmeter. However, this meter will read *high or low* depending on the form factor of the wave, in case of nonsinusoidal wave-forms (Refer to examples 2.2 and 2.3).

## 2.4.8. Average Responding Voltmeter

Fig. 2.21 shows the block diagram of a typical average responding voltmeter :

— The input/applied wave form is *amplified* in a high gain stabilised amplifier and then *rectified* and fed to a D.C. mA calibrated in terms of r.m.s. input voltage. In this meter instrument, the rectified current is *averaged by a filter* to produce a steady deflection of the pointer of the meter.

— The input *blocking capacitor* (preceding the high gain amplifier) *excludes* the D.C. component (in the applied voltage) from the measurement.

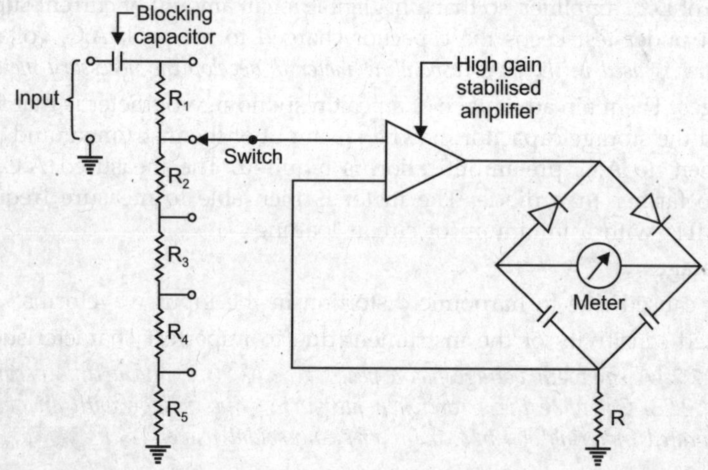


**Fig. 2.21.** Average responding voltmeter.

— The large amount of negative feedback from the A.C. amplifier ensures gain stability for measurement accuracy, and an increased frequency range of the instrument. The inclusion of the meter in the feedback path minimises the effect of diode non-linearity and meter impedance variations on performance of the circuit.

— The *capacitors* in the meter circuit tend to act as storage or filter capacitors for the rectifier diodes as well as coupling capacitors for the feedback signal.

— The *diodes* in the meter circuit act as switches to maintain unidirectional meter current despite changes in the instantaneous polarity of the input voltage.

- The *accuracy* with which an average responding voltmeter indicates the r.m.s. value of a wave with harmonic content *depends not only on the amplitude of the harmonic but also on the phase*

  — The *errors* in the reading of an average responding voltmeter may be *due to the application of complex waveforms.*

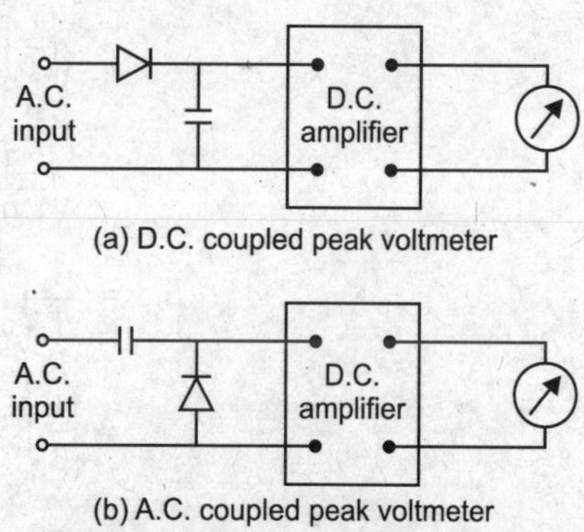


**Fig. 2.22.** Peak responding voltmeter.

## 2.4.9. Peak Responding Voltmeter

The main difference between peak responding voltmeters and average responding voltmeters is the *use of storage capacitors with the rectifying diode in the former case*. The capacitor charges through the diode to the peak value of the applied voltage and the meter circuit then responds to voltage of the capacitor.

Fig. 2.22 (*a*) and (*b*) shows the *two* most common types of peak responding voltmeters:

— Fig. 2.22 (a) shows a *D.C. coupled peak voltmeter;* here the capacitor charges to the total peak voltage above ground reference.

In this type of meter, the reading of the meter is affected by the presence of D.C. with A.C. voltage.

— Fig. 2.22 (b) shows an *A.C. coupled peak voltmeter.*

In both the circuits, the capacitor discharges very slowly through the high impedance input of D.C. amplifier, so that a negligible small amount of current supplied by the circuit under test keeps the capacitor charged to the peak A.C. voltage. *The D.C. amplifier is used in the peak responding meter to develop the necessary meter current.*

**Advantages.** The main advantage of a peak responding voltmeter is that the rectifying diode and the storage capacitor may be taken out of the instrument and placed in the probe when no A.C. pre-amplification is required. The measured A.C. signal then travels no farther than diode. The meter is then able to measure frequencies upto 100s of MHz with a minimum of circuit loading.

**Disadvantages :**

1. Error caused due to harnomic distortion in the input waveforms.
2. Limited sensitivity of the instrument due to imperfect characteristics of diode.

**Example 2.2.** *A sawtooth voltage has a peak value of 80 V and a time period of 2.0 s as shown in Fig. 2.23. Calculate the error when measuring this voltage with an average reading voltmeter calibrated in terms of r.m.s. value of a sinusoidal wave.*

**Solution.** The given sawtooth wavefrom can be mathematically expressed as :

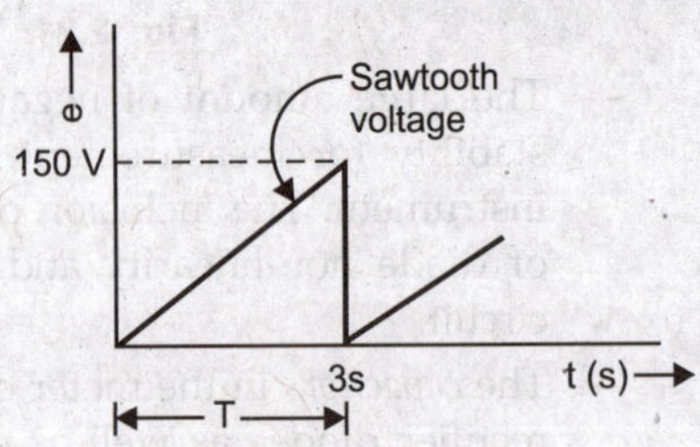

Fig. 2.23.

$$e = mt, \text{ where } m \text{ is the slope of ramp}$$

$$\therefore \quad m = \frac{e}{t} = \frac{150}{3} = 50$$

or $$e = 50t \qquad 0 \le t \le T$$

$$\therefore \quad E_{r.m.s} = \sqrt{\frac{1}{T}\int_o^T e^2 . dt}$$

$$= \sqrt{\frac{1}{3}\int_0^3 (50t)^2 dt} = \sqrt{\frac{2500}{3}\left|\frac{t^3}{3}\right|_0^3}$$

$$= \sqrt{\frac{2500}{9}(3^3 - 0)} = 50\sqrt{3} \text{ V}$$

$$E_{av} = \frac{1}{T}\int_0^T e . dt$$

$$= \frac{1}{3}\int_0^3 50\, t\, dt = \frac{50}{3}\left|\frac{t^2}{2}\right|_0^3 = \frac{50}{6}(3^2 - 0) = 75 \text{ V}$$

$$\therefore \quad \text{Form factor, } k_f = \frac{E_{r.m.s}}{E_{av}} = \frac{50\sqrt{3}}{75} = 1.155$$

The ratio of the two form factors is,

$$\frac{(k_f)_{\text{sine wave}}}{(k_f)_{\text{sawtooth}}} = \frac{1.11}{1.155} = 0.961$$

Thus, the meter indication is **low** by 0.961.

$$\therefore \quad \text{Percentage error} = \frac{0.961 - 1}{1} \times 100 = -\mathbf{3.9\%} \textbf{ (Ans.)}$$

**Example 2.3.** *Calculate the error when a symmetrical square-wave voltage as shown in Fig. 2.24, is applied to an average reading A.C. voltmeter with scale calibrated in terms of r.m.s. value of a sinusoidal wave.*

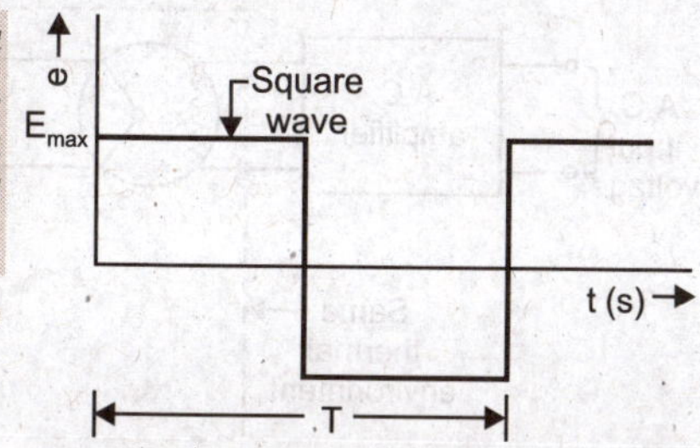


Fig. 2.24

**Solution.** The r.m.s. value of the square voltage is given as :

$$E_{r.m.s.} = \sqrt{\frac{1}{T}\int_0^T e^2\, dt}$$

$$= \sqrt{\frac{1}{T}\int_0^T E_{\max}^2\, dt} = \sqrt{\frac{1}{T} E_{\max}^2 \, |t|_0^T}$$

$$= \sqrt{\frac{1}{T} \times E_{\max}^2\,(T-0)} = E_{\max}$$

$$E_{av.} = \frac{1}{(T/2)}\int_0^{T/2} e.\, dt$$

$$= \frac{2}{T} E_{\max} . |t|^{T/2} = \frac{2}{T} E_{\max}\left(\frac{T}{2} - 0\right) = E_{\max}$$

$$\therefore \quad \text{Form factor, } k_f = \frac{E_{r.m.s}}{E_{av}} = \frac{E_{max}}{E_{max}} = 1$$

The ratio of the two form factors is,

$$\frac{(k_f)_{sine\ wave}}{(k_f)_{square\ wave}} = \frac{1.11}{1} = 1.11$$

Thus, the meter indication is **high** by a factor 1.11.

$$\therefore \quad \text{Percentage error} = \frac{1.11 - 1}{1} \times 100 = \mathbf{11\%} \quad \textbf{(Ans.)}$$

### 2.4.10. True R.M.S.-reading Voltmeter

The average reading voltmeter can be converted into an r.m.s. reading voltmeter by calibrating the meter scale in terms of the r.m.s. value of a *sine wave*. However, if the waveform is not sinusoidal the reading causes considerable error. *Complex waveforms can be accurately measured by r.m.s.-reading voltmeter.*

This type of voltmeter provides an output by sensing the heating power of the waveform. *Since the heating power of the waveform is proportional to the square of the r.m.s. value of the voltage, the scale of the instrument can be calibrated to read the r.m.s. directly.* In practice, the heating power is measured by measuring an amplified version of the input waveform to the heater element of thermocouple whose output voltage is proportional to the square of the r.m.s. value of input voltage.

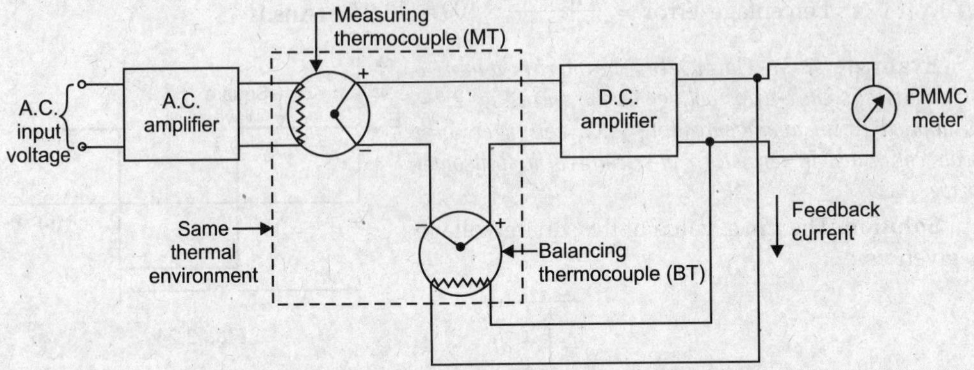


**Fig. 2.25.** Block diagram of a true r.m.s.-reading voltmeter.

Fig. 2.25. shows the block diagram of a true r.m.s.-reading voltmeter :

— The r.m.s.-reading voltmeter consists of two thermocouples MT and BT as shown in Fig. 2.25; MT is called measuring thermocouple and BT the balancing thermocouple. BT is used in the feedback loop to cancel out the non-linear effect of the MT. These two thermocouples MT and BT form part of the bridge circuits connected to the input of a D.C. amplifier.

— The unknown A.C. voltage is amplified and applied to the heating element of MT. As already stated, heating produces an output voltage which *upsets the balance of bridge*.

— This *unbalanced voltage* is *amplified by D.C. amplifier* and is fed back to the heater element of the BT.

— Balance is re-established when the voltage output of the thermocouple currents in the heater element of BT is equivalent to the A.C. current in the heating element of MT. D.C. current is directly proportional to the r.m.s. value of input voltage.

— The PMMC meter is calibrated to read the r.m.s. voltage directly.

- A typical laboratory-type r.m.s.-reading voltmeter provides accurate r.m.s.reading of complex waveforms having a *crest factor* (ratio of peak value to r.m.s. value) of 10/1.
  — At 10 per cent of full-scale meter deflection, where there is less chance of amplifier saturation, waveforms with crest factors as high as 100/1 could be accommodated.
  — With the use of good instruments, voltages throughout a range of 100μV to 300V within a frequency range of 10 Hz to 10 MHz may be measured.

***Advantages :***

1. Sensitivities in the mV region is possible.
2. The non-linear behaviour is avoided by using two thermocouples (MT and BT) in the same thermal environment.

3. The true r.m.s. value measured is independent of the A.C. input waveform, if the peak amplitude of A.C. input is within the dynamic range of the A.C. amplifier.

***Limitations :***

1. The overall meter response is sluggish (since the response of thermocouples is slow).
2. In case of highly non-linear waveforms, the crest factor puts the limitation on the meter reading.
3. The cost of meter is high as compared to average and peak-responding meters.

**Non-thermocouple type true r.m.s. voltmeter :**

As the overall response of a r.m.s. voltmeter using thermocouples is sluggish, a non-thermal type of true r.m.s. voltmeter is used.

This type of voltmeter *works on the principle that when a small A.C. voltage is applied between the bases of the two transistors connected in parallel, the A.C. component of the voltage developed across the common collector resistors of the two transistors is proportional to the square of the applied input voltage.*

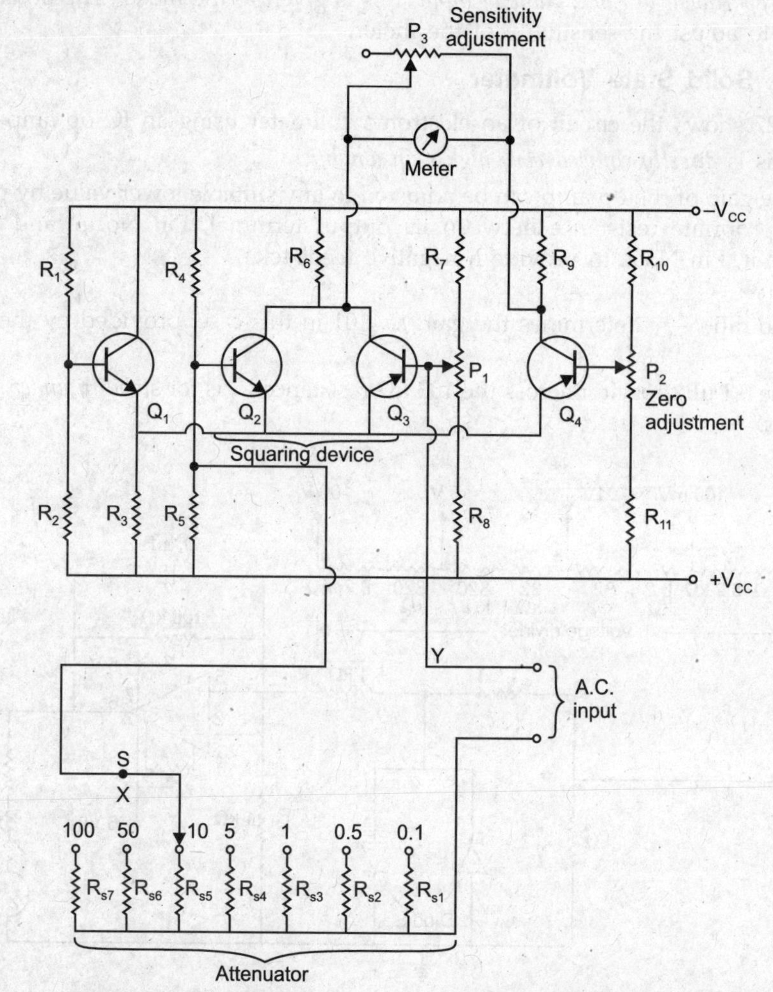


$Q_1, Q_2, Q_3, Q_4$ = Transistors; $P_1, P_2, P_3$ = Potentiometers; $S$ = Switch

**Fig. 2.26.** Non-thermocouple type true r.m.s. voltmeter.

Fig. 2.26 shows a non-thermocouple type transistorised true R.M.S. voltmeter, using a squaring device and other components :

— The transistors $Q_2$ and $Q_3$ constitute the *squaring device*. The resistance $R_6$ is common collector resistance. The potentiometer $P_1$ is used for *bias setting*.

— The transistor $Q_4$ constitutes the other side of the bridge. The potentiometer $P_2$ is used for bias setting which is nothing but *zero adjustment*. It adjusts zero for zero A.C. input *i.e.* shorting the terminals $X$ and $Y$. Under this condition the bridge is balanced and drops across $R_6$ and $R_9$ are equal.

— The function of transistor $Q_1$ is to improve temperature stability through emitter resistance $R_3$. The resistances $R_1$ and $R_2$ are used to adjust the biasing to $Q_1$ such that drop across $R_3$ is 0.7V for silicon transistors.

— A.C. input is applied through an attenuator consisting of resistances $R_{S1}$ and $R_{S7}$. The proper range is selected through the switch $S$ (a range selector).

On the application of input, the bridge is unbalanced and *drop between resistances $R_6$ and $R_9$ is proportional to r.m.s. value of input;* this is given to the meter. The potentiometer $P_3$ is used to adjust the sensitivity of the meter.

### 2.4.11. Solid State Voltmeter

Fig. 2.27 shows the circuit of an electronic voltmeter using an IC op amp 741C :

— This is *directly coupled very high gain amplifier*.

— The gain of the op-amp can be adjusted to any suitable lower value by providing appropriate resistance between its output terminal, Pin No. 6, and inverting input, Pin No.2, to provide a negative feedback.

— The ratio $\frac{R_2}{R_1}$ determines the gain *i.e.* 101 in this case, provided by the op-amp.

— The 0.1 μF capacitor across the 100 kΩ resistance $R_2$ is for *stability under stray pick-ups*.

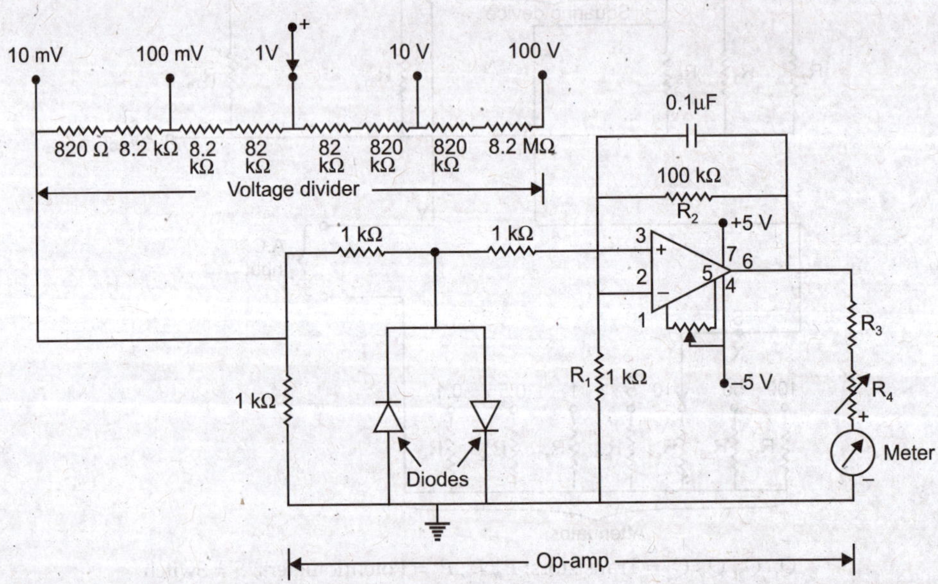


**Fig. 2.27.** Solid state mV voltmeter using opamp.

— Terminals 1 and 5 are called *offset null terminals*. A 10 kΩ potentiometer is connected between these two offset null terminals with its centre tap connected to a –5V supply. This potentiometer is called "*zero set* " and is used for adjusting zero output for zero input conditions.

— The two diodes used are for *IC protection*. Under normal conditions, they are non-conducting, as the maximum voltage across them is 10 mV. If an excessive voltage, say more than 100 mV appears across them, then depending upon the polarity of the voltage, one of the diode conducts and protects the IC. A μA scale of 50-1000μA full scale deflection can be used as an indicator. $R_4$ is adjusted to get maximum full scale deflection.

### 2.4.12. Vector Voltmeter

A ***vector voltmeter*** *measures the amplitude of a signal at two points in a circuit and simultaneously measures the phase difference between the voltage waveforms at these two points.*

This voltmeter has wide applications especially in situations where other methods are very difficult or time consuming. The vector voltmeter is useful in VHF applications and can be used successfully in measurements such as :

(*i*) Complex insertion loss.
(*ii*) Amplifier gain and phase shift.
(*iii*) Two-port network parameters.
(*iv*) Filter transfer functions.
(*v*) Radio frequency distortion.

- This voltmeter basically converts two RF signals of the same fundamental frequency (from 1MHz to 16Hz) to two IF signals with 20 kHz fundamental frequencies. These I.F. signals have the same amplitudes, waveforms and phase relationships as the original RF signals. Consequently, the fundamental components of the IF signals have the same amplitude and phase relationships as the fundamental components of the RF signals. *These fundamental components are filtered from the IF signals and are measured by a voltmeter and a phase meter.*

  A vector voltmeter consists of *five* major sections :

  (*i*) RF-to-IF converter for channel-1.
  (*ii*) RF-to-IF converter for channel-2.
  (*iii*) Automatic phase control section.
  (*iv*) A phase meter circuit.
  (*iv*) A voltmeter circuit.

The power supply section for the instrument generates all the necessary supply voltages for the various sections of the instrument.

- The RF-to-IF converters and the phase control section produce two 20 kHZ sine waves with the same amplitudes and the same phase relationships as the fundamental components of the RF signals applied to channels 1 and 2. The phase meter section continuously *monitors* these two 20 kHz sine waves and indicates the phase angle between them. The voltmeter section can be switched to channel - 1 or channel -2 to *provide a meter display of the amplitude.*

### 2.4.13. Electronic Multimeter (or VOM)

The solid-state electronic multimeter or VOM (Voltage-Ohm Meter) is one of the most versatile general-purpose shop instruments *capable of measuring D.C. and A.C. voltages as well as current and resistance.*

An electronic multimeter generally consists of the following *elements* :

1. Balanced-bridge D.C. amplifier and indicating meter.
2. Input attenuator or Range switch...*To limit the magnitude of the input voltage to the desired value.*
3. Rectifier section... *To convert an A.C. input voltage to a proportional D.C. value.*
4. Internal battery and additional circuitry... *To provide the capability of resistance measurement.*
5. "Function switch" ... *To select the various measurement functions of the instrument such as voltage, current or resistance.*

The instrument is also usually provided with a built-in power supply for operation on A.C. mains and, in most cases, one or more batteries for operation as *portable test instrument*.

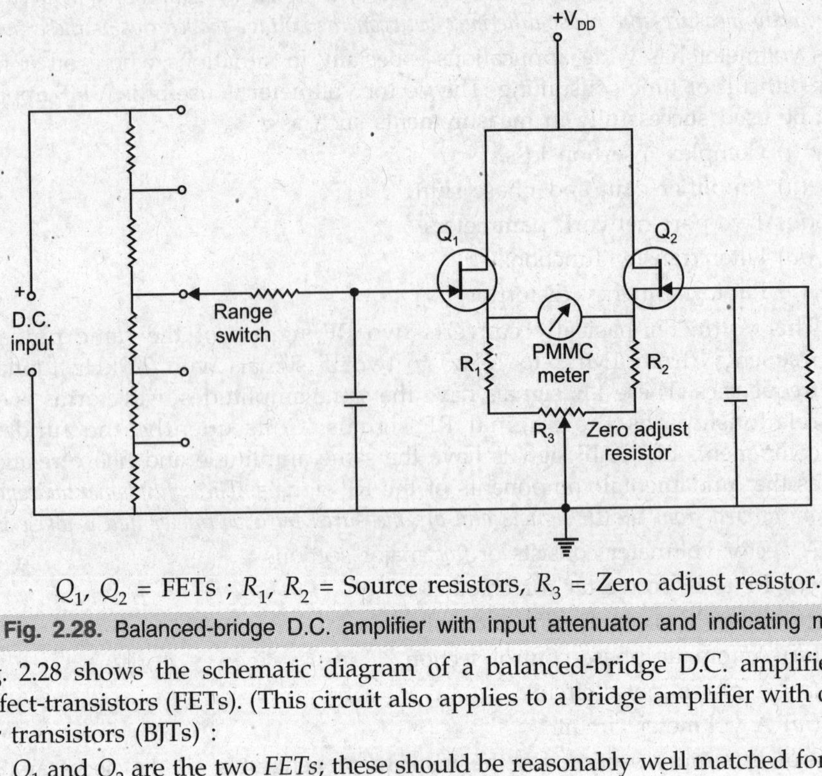


$Q_1$, $Q_2$ = FETs ; $R_1$, $R_2$ = Source resistors, $R_3$ = Zero adjust resistor.

**Fig. 2.28.** Balanced-bridge D.C. amplifier with input attenuator and indicating meter.

Fig. 2.28 shows the schematic diagram of a balanced-bridge D.C. amplifier using field effect-transistors (FETs). (This circuit also applies to a bridge amplifier with ordinary bipolar transistors (BJTs) :

— $Q_1$ and $Q_2$ are the two *FETs*; these should be reasonably well matched for current gain to ensure thermal stability of the circuit.

— The two *FETs*, $Q_1$ and $Q_2$ constitute the *upper* arms of a bridge circuit, whereas, source resistors $R_1$ and $R_2$, together with *"Zero"* adjust resistor $R_3$ form the *lower* bridge arms,

— The PMMC meter is connected between the source terminals of the FETs, representing two opposite corners of the bridge.

**Working :**

***Voltage measurement :***

- Where there is no input signal, the gate terminals of the FETs are at ground potential and the transistors operate under identical quiescent conditions. Under such a condition the bridge is balanced and meter reads zero. However, in practice, small differences in the operating characteristics of the transistors, and slight tolerance differences in the various resistors, cause a certain amount of unbalance in the drain currents, and the meter shows a small deflection from zero. This current is reduced to zero by adjust control resistor $R_3$ and now the bridge is balanced.

- On the application of the voltage to the gate of input FET $Q_1$, its drain current increases which causes the voltage at the source terminal to rise. The resulting unbalance between $Q_1$ and $Q_2$ source voltages is indicated by the PMMC meter, whose scale is calibrated to agree with the magnitude of the applied input voltage.
- The maximum voltage which can be applied to the gate of $Q_1$ is determined by the operating range of FET and is usually on the order of a few volts. The range of input voltages can easily be extended by an input attenuator a *"Range switch"*, as shown in Fig. 2.29. *The unknown D.C. input voltage is applied through a large resistor in the probe body to a resistive voltage* divider. Thus as shown in Fig. 2.29 :
  - With the *"Range switch"* in 3V position, the voltage at the gate of the input FET is developed across 8 MΩ of the total resistance of 11.3 MΩ and the circuit is so arranged that the meter deflects full scale with 3 V applied to the tip of the probe.
  - With the *"Range switch"* in the 12V position, the gate voltage is developed across 2 MΩ of the total divider resistance of 11.3 MΩ and an input voltage of 12V is required to cause the same full-scale meter deflection.

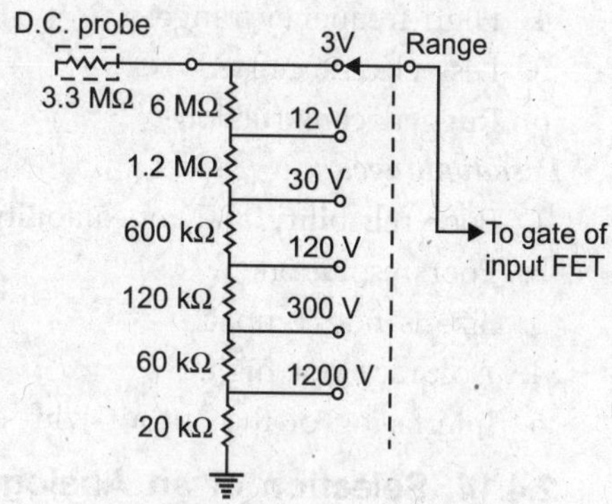


**Fig. 2.29.** Typical input voltage attenuator for a multimeter.

***Resistance measurement :***

In an electronic multimeter, when the *function switch* is placed in the 'OHMS' position the unknown resistor is connected in series with an internal battery, and the PMMC meter simply measures the voltage drop across the unknown.

Fig. 2.30 shows a typical circuit, where a separate divider network, used only for resistance measurements, provides for a number of differents resistance ranges.

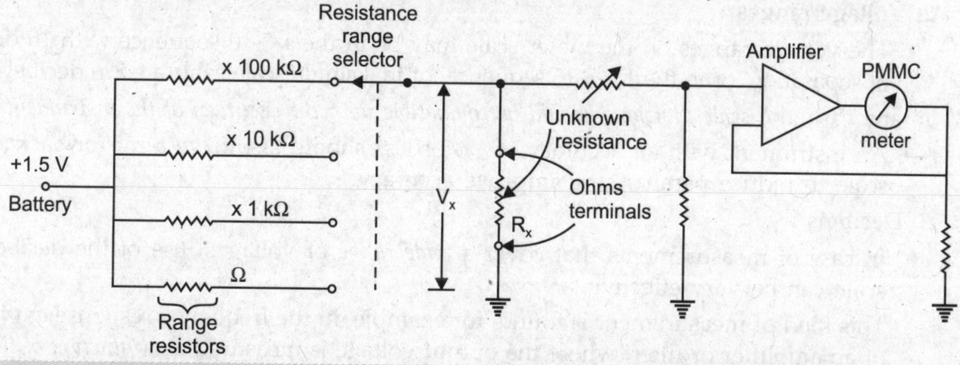


**Fig. 2.30.** Resistance range selector circuit of an electronic multimeter.

— When an unknown resistor $R_x$ is connected to the OHMS terminals of the multimeter, the 1.5 V battery supplies current through one of the range resistors and unknown resistor to ground.

— Voltage drop $V_x$ (across $R_x$) is applied to the input of the bridge amplifier which causes a deflection on the meter. Since the voltage drop across $R_x$ is *directly proportional to its resistance, the meter scale can be calibrated in terms of resistance.*

If may be noted that the resistance scale of a multimeter reads increasing resistance from left to right, opposite the way resistance scale reads on conventional meters. It is expected because electronic multimeter reads a *larger resistance as a higher voltage, whereas the ordinary multimeter indicates a higher resistance as a smaller current.*

**Advantages and disadvantages :**

Following are the *advantages* and *disadvantages* of an electronic multimeter :

***Advantages :***

1. Simple circuit.
2. Less costly.
3. High input impedance.
4. High frequency range.
5. Less electric noise.
6. Rugged construction.

***Disadvantages :***

1. Poor reliability and repeatability.
2. Poor resolution.
3. Size is not compact.
4. Accuracy not high.
5. Interfacing of the output with external devices is difficult.

### 2.4.14. Selection of an Analog Voltmeter

The selection of the most appropriate instrument for a particular voltage measurement depends on the performance required in a given situation. Some important considerations needed in selecting an analog voltmeter (*an analog voltmeter always measures average value; it is calibrated to read r.m.s. or peak values*) are as follows :

**1. Input impedance :**

- The input resistance/impedance should be at least of an order of magnitude higher than the impedance of the circuit under measurement to avoid loading effects.

**2. Voltage ranges :**

- The voltage ranges on the meter scale may be in the 1-3-10 sequence with 10 db of separation, or in the 1.5-5-15 sequence, or in a single scale calibrated in decibels.

In any case, the *scale divisions should be compatible with the accuracy of the instrument :*

— An instrument with an accuracy of 1% or less should also have a mirror-backed scale to reduce parallax and improve accuracy.

**3. Decibels :**

- In case of measurements that cover a *wide range* of voltages, use of the decibel scale can be very effective.

— This kind of measurement is found, for example, in the frequency response curve of an amplifier or filter, where the output voltage is measured as a *function of the frequency of the applied input voltage.*

— All voltmeters with *db* scales are calibrated in *dbm*, referenced to some particular impedance.

**4. Sensitivity versus bandwidth :**

- Noise is a function of bandwidth.

— A voltmeter with a broad bandwidth will pick up and generate more noise than one operating over a narrow range of frequencies.

**5. Battery operation :**

- A voltmeter powered by an internal battery is essential for field work.

**6. A.C. current measurements :**

- A sensitive A.C. voltmeter and a series resistance may be used to measure currents.
- In the usual case, however, an A.C. *current probe* is used which enables the operator to measure an A.C. current without disturbing the circuit under test.

In view of the above considerations, the following *general guidelines* can be stated :

(*i*) *For D.C. measurements* : ............... Select the meter with the broadest capability meeting the requirements of the circuit.

(*ii*) *For A.C. measurements* : [involving sine waves with only modest amounts of distortion (less than 10%)] ............... The average-responding voltmeter provides the best accuracy and most sensitivity per unit investment.

(*iii*) *For high frequency measurements* : (greater than 10 MHz) ............... The peak-responding voltmeter with a diode-probe input is the most economical choice.

(*iv*) *For measurements involving the effective power of waveforms which depart from the true sinusoidal form* : ............... The r.m.s. responding voltmeter is the appropriate choice.

## 2.5 CATHODE RAY OSCILLOSCOPES (CROs)

### 2.5.1. Introduction

**Cathode ray oscilloscope** *(CRO) is a device that allows the amplitude of electrical signals, whether they be voltage, current, power, etc., to be displayed primarily as a function of time.*

It is a very useful and versatile laboratory instrument used for display, measurement and analysis of waveforms and other phenomena in electrical and electronic circuits.

— It is basically an *electron beam voltmeter*.

- The oscilloscope depends on the movement of an *electron beam* to impinge on a phosphor surface, which produces a *visible spot*. If the electron beam is deflected in either of two orthogonal axes, such as the familiar *X* and *Y* axes used in conventional graph construction, the luminous spot can be used to create *two-dimensional displays*. Typically, the *X*-axis of the oscilloscope is deflected at a constant rate, relative to time, and the vertical or *Y*-axis is deflected in response to an input stimulus such as a voltage. This produces the time-dependence variation of the input voltage, which is very important to the design and development of electronic circuits.
- The oscilloscope is *capable of displaying events that take place over periods of microseconds and nanoseconds.*

- CROs *operate on voltages*. However it is possible to convert current, strain, acceleration, pressure and other physical quantities into voltages with the help of transducers and thus to present visual representations of a wide variety of dynamic phenomena on CROs.

These oscilloscopes have been evolved continuously, and they are now available which can measure frequencies upto 1GHz and observe events as small as 20 Hz in duration.

- With some oscilloscopes several additional features, such as digital multimeters and counters, are available. Several oscilloscopes are *microprocessor controlled*. They have the ability to calculate several features, such as rise time or pulse width of the measured waveform, and to display these values along with the display of waveforms. They are *easier to use and internal routines often act as a guide for the user, and display a warning if there is any error in setting*.

Although, most oscilloscopes are monochromatic, *colour oscilloscopes* are finding increasing applications in computers and in television.

- Most present day oscilloscopes are *capable of accepting two or more inputs displaying them simultaneously*. This may be achieved through using a *split beam* or *by using a multiple beam tube*.

### 2.5.2. Applications of CRO

There are variety of applications in which CRO is used. Some of these applications are:

1. Tracing of an actual waveform of current or voltage.
2. Determination of amplitude of a variable quantity.
3. Comparison of phase and frequency.
4. Measurement of capacitance and inductance.
5. In televisions.
6. In radar.
7. For finding B-H curves for hysteresis loop.
8. For engine pressure analysis.
9. For studying the heart beats, nervous reactions etc.
10. For tracing transistor curves.
11. For determining the modulation characteristics and to detect the standing waves in transmission lines.
12. It can be used to check the diodes and the faulty components in the various circuits.

### 2.5.3. Block Diagram of an Oscilloscope

Fig. 2.31. shows the block diagram of general-purpose oscilloscope. Its consists of the following major systems :

1. ***Cathode ray tube :***
   — It generates the electron beam, accelerates the beam to a high velocity, deflects the beam to create the image, and contains a phosphor screen where the electron beam eventually becomes visible.
   — It displays the quantity being measured.
2. ***Vertical amplifier.*** It amplifies the signal waveform to be viewed.
3. ***Horizontal amplifier.*** It is fed with a sawtooth voltage which is then applied to horizontal deflection plates.

4. ***Sweep or time base generator.*** It produces sawtooth voltage waveform used for horizontal deflection of the electron beam.
5. ***Trigger circuit.*** It produces trigger pulses to start horizontal sweep.
6. ***High and low-voltage supply :***
   — The power supply block provides the voltages required by the cathode ray tube to generate and accelerate the beam, as well as to supply the required operating voltages for the other circuits of the oscilloscope.
   — Relatively high voltages are required by cathode ray tubes, on the order of a few thousand volts, for acceleration, as well as a low voltage for the heater of the electron gun, which emits the electrons. Supply voltages for other circuits are various values, usually not more than a few hundred volts.
7. ***Delay line.*** High frequency oscilloscopes always include delay line in the vertical amplifiers. *The purpose of such lines is to delay the vertical signal enough to keep it from reaching the CRT deflection plates before the horizontal sweep circuits are running.*

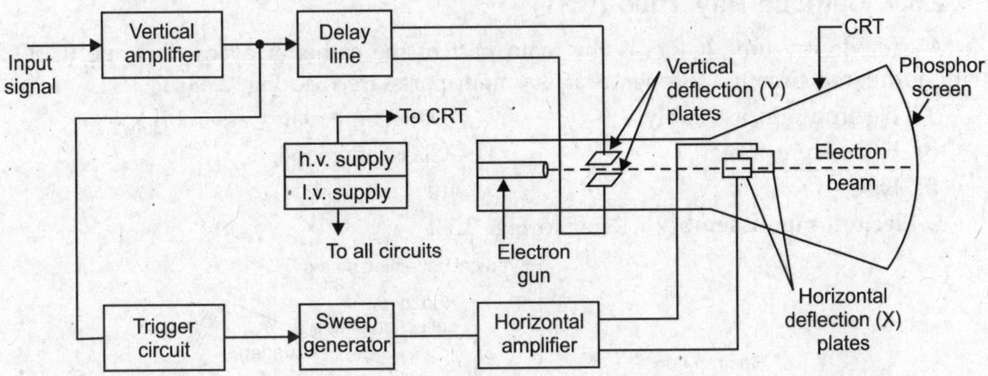


**Fig. 2.31.** Block diagram of a general purpose CRO.

**Working of a CRO :**

- When signal is to be displayed or viewed on the screen it is applied across the Y-plates of a cathode ray tube (CRT). But to see its waveform or pattern, it is essential to spread it to horizontally from left to right. This is achieved by applying a sawtooth voltage wave to X-plates.

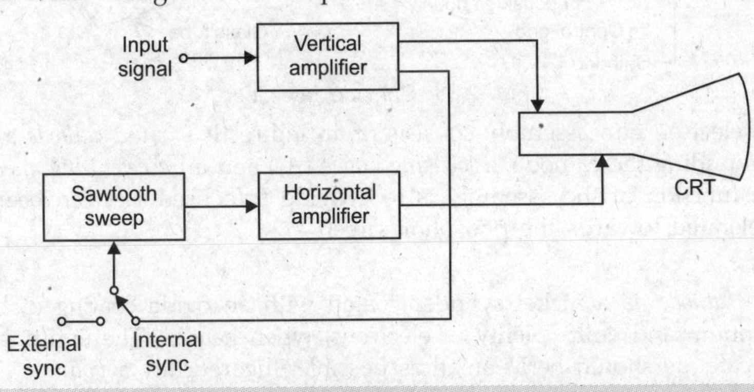


**Fig. 2.32**

Under these conditions the electron beam would move uniformly from left to right thereby graphing vertical vibrations of the input signal versus time. Due to repetitive tracing of the viewed waveform, we get a continuous display because of persistence of vision.

- However, to get a stable stationary display on the screen it is essential to *synchronize the horizontal sweeping of the beam* (sync) with the input signal across the Y-plates. The signal will be properly synced only when *its frequency equals the sweep generator frequency*.

The usual method of synchronizing the input signal is to *use a portion of the input signal to trigger the sweep generator so that the frequency of the sweep signal is locked or synchronized to the input signal*. It is called internal sync because the synchronization is obtained by internal wiring connection as shown in Fig. 2.32.

— In the *external synchronization*, an external trigger is used to trigger the signal being measured.

— In the *line synchronization*, the trigger is obtained from the power supply to CRO which is normally 230 V, 50Hz.

### 2.5.4. Cathode Ray Tube (CRT)

A cathode ray tube (CRT) is the main part of the cathode ray oscilloscope (CRO), with additional circuitry to operate it. Its main parts are (See Fig. 2.33) :

1. Electron gun assembly;
2. Deflection plates assembly;
3. Fluorescent screen;
4. Glass envelope;
5. Base.

**1. Electron gun assembly :** Refer to Fig. 2.33.

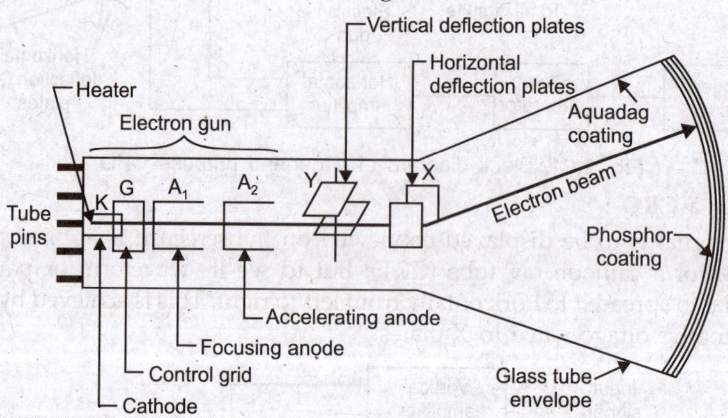


**Fig. 2.33.** Cathode ray tube.

- The electron gun assembly consists of an indirectly heated *cathode* a *control grid* surrounding the cathode, a *focusing anode* ($A_1$) and an *accelerating anode* ($A_2$). The sole function of the assembly is to provide a focused electron beam which is accelerated towards the phosphor screen.

*Cathode :*

- The '*cathode*' is a nickel cylinder coated with an oxide coating of barium and strontium and emits plenty of electrons, when heated. The emitting surface of the cathode should be as small as possible, theoretically a *point*.
- Rate of emission of electrons (or say the intensity of electron beam) depends on the magnitude of cathode current, which can be controlled by the control grid in a manner similar to a conventional vacuum tube.

- The typical values of current and voltage required by an indirectly heated cathode are 600 mA at 6.3 V (A.C. or D.C.). The special low power designs use 140 mA at 1.5 V.

***Control grid :***

- The *'control grid'* is usually a metal cylinder covered at one end but with a small hole in the cover. This is usually a metal cap of low permeability steel, about 15 mm in diameter and 15 mm long. An aperture of approximate 0.25 mm is drilled in the grid cap for the passage of electrons through it.
- The grid is kept at *negative potential* (variable) w.r.t. cathode and its function is to vary the electron emission and so the brilliancy of the spot on the phosphor screen. The hole in the grid is provided to allow passage for electrons through it and concentrate the beam of electrons along the axis of tube.

***Focusing and accelerating anodes :***

- Electron beam comes out from the control grid through a small hole in it and enters a *pre-accelerating anode*, which is a hollow cylinder in shape and is at a potential of few hundred volts more positive than the cathode so as to accelerate the electron beam in the electric field.
- The electron beam is then focused on the screen by an electrostatic lens consisting of two more cylindrical anodes called the *'focusing anode'* and *'accelerating anode'* apart from the pre-accelerating anode. The function of these anodes is to concentrate and focus the beam on the screen and also to accelerate the speed of electrons.
- The pre-accelerating and accelerating anodes are connected to a common positive high voltage of about 1500V; the focusing anode is connected to a lower adjustment voltage of 500 V.

**2. Deflection plates assembly :**

- The electron beam, after leaving the electron gun, passes through two pairs of deflection plates. One pair of plates is mounted horizontally and produces an electric field in the vertical plane. This pair produces a vertical deflection and is thus called *Vertical deflection plates* or *Y-plates*. The other pair of plates is mounted vertically and produces a horizontal deflection. This pair of plates is called *Horizontal deflection plates or X-plates*.

  The plates are flared so as to allow the beam to pass through them without striking the plates.
- When a sinusoidal voltage is applied to Y-plates, the beam will be moved up and down according to the variation of plate potential. If one frequency of variation is more than 16 Hz the deflection will be a vertical line in the centre of the screen. In case the sinusoidal voltage is applied to X-plates and frequency of variation is more than 16 Hz the deflection will be a horizontal line. If potentials are applied to both sets of plates simultaneously, the deflection will be an oblique line. The amount of deflection is in proportion to the voltage applied to the pair of plates.

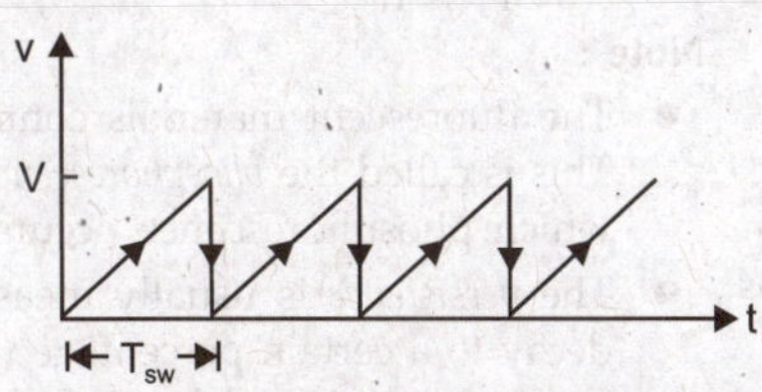


**Fig. 2.34**

It must be remembered that :

— The signal to be displayed on the screen is always applied across the Y-plates.

— The voltage applied across X-plates is a ramp voltage *i.e.* a voltage which increases linearly with time. It has a *'sawtooth waveform'* as shown in Fig. 2.34. It is also called *horizontal time-base or sweep voltage*. It has a sweep time of $T_{sw}$.

**3. Fluorescent screen :**

- The end wall of the CRT, called *'screen'*, is coated with *phosphor*. When struck by high-energy electrons, it glows. In other words, it absorbs the kinetic energy of the electrons and converts it into light–the process being known as *fluorescence*. That is why the screen is called *fluorescent screen*. The colour of the emitted light depends on the type of phosphor used.

The intensity of light emitted from the CRT screen, called *luminance*, depends upon the following *factors* :

(*i*) The number of bombarding electrons striking the screen per second.

(*ii*) The energy with which the bombarding electrons strike the screen which is determined by the accelerating potential (more the accelerating potential more the luminance).

(*iii*) The time the beam strikes a given area of the phosphor *i.e.,* sweep speed.

(*iv*) The physical characteristics of the phosphor itself.

- Among the fluorescent materials used are :
  (*i*) *Zinc or orthosilicate* ... for giving a green trace very suitable for visual observations;
  (*ii*) *Calcium tungstate* ... For giving blue and ultra-violet radiations very suitable for photography;
  (*iii*) *Zinc sulphide* (with other materials) ... For giving a white light suitable for T.V.;
  (*iv*) *Zinc phosphate* ... It gives a pronounced after glow and is useful when studying transient phenomena because the trace persists for short while after the transient has disappeared. .
- The phosphor of the oscilloscope is designated as follows :

$P_1$ = Green medium; $P_2$ = Blue green medium;
$P_5$ = Blue very short; $P_{11}$ = Blue short ect.

— The material $P_{31}$ is used commonly for general purpose oscilloscopes due to the following *characteristics* :

(*i*) High illumination level. (*ii*) High writing speed.

(*iii*) High *burn resistance (to avoid accidental damage)

(*iv*) It gives colour to which human eye response is maximum.

(*v*) It gives short persistence required to avoid multiple image display.

* When an electron beam strikes the screen its K.E. is absorbed by the phosphor material. This creates heating. The heat produced gives rise to *"phosphor burn"* which is damaging and may cause phosphor destruction. So the phosphor material should have a *high burn resistance*.

**Note :**

- The fluorescent materials continue to emit light even after radiation exposure is cut off. This is called the *phosphorescence* characteristics of the materials. The length of time during which phosphorescence occurs is called the *persistence* of the phosphor.
- The *persistence* is usually measured in terms of the time required for the CRT image to decay to a certain percentage (usually 10%) of the original light output. It is expressed as short, medium and long.

(*i*) *"Short persistence"*. It lasts for *microseconds*. It is required for extremely high speed phenomena, to avoid smearing and interference caused when one image persists and overlaps with next one.

(*ii*) "*Medium persistence*". It lasts for *milliseconds*. These traces are mostly used for general purpose applications.

(*iii*) "*Long persistence*". It lasts for seconds to several minutes. These traces are used for transients as they keep the fast transient on the screen for observation after the transient has disappeared.

**4. Glass envelope :**

- The glass envelope is funnel-shaped having a phosphor-coated screen at its flared end. It is *highly-evacuated in order to permit the electron beam to traverse the tube easily.*
- The inside part of the flared part of the tube is coated with a conducting graphite layer called "*aquadag*" which is maintained at the same potential as the second anode. This layer performs the two functions : (*i*) It accelerates the electron beam after it passes between the deflecting plates, and (*ii*) It collects the electrons produced by secondary emission when electron beam strikes the screen. Hence it prevents the formation of negative charge on the screen.

— In some tubes, particularly CRTs with magnetic focusing (such as TV picture tubes), the accelerating anode is dispersed with entirely and the conductive coating is used as the final accelerating anode.

- *Horizontal and vertical marks* an marked on the screen of the CRT to provide user a correct measurement. These marks, *usually in rectangular form*, are called *graticule.*

— The graticule may be *external*, *internal* and *projected*.

— The graticules are generally laid out in an 8 × 10 pattern as shown in Fig. 2.35. with 8 major divisions along the vertical axis and 10 major divisions along the horizontal axis. The tick marks on the centre graticule represent minor divisions or subdivisions.

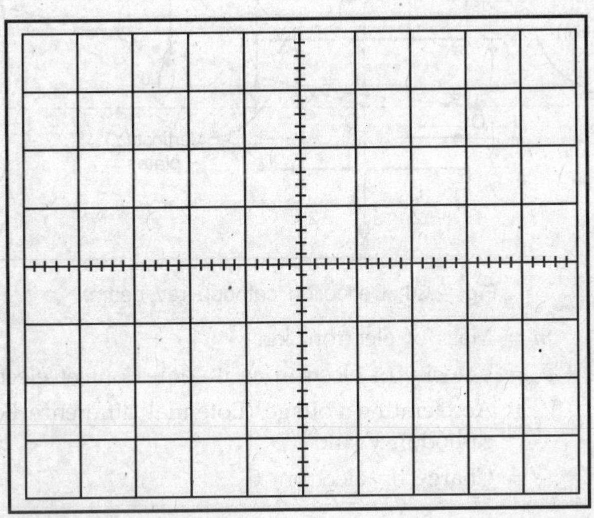

**Fig. 2.35.** The CRT graticule.

**Size :** Size refers to *screen diameter*. CRTs for Oscilloscopes are available in sizes 1, 2, 3, 5 and 7 inches. 3 inch tube is most common for portable instruments.

**Example :** *A CRT having a number $5GP_1$ is 5 inch tube with a medium persistence green trace. The first number indicates that it is a 5 inch Tube and $P_1$ indicates a medium persistence green trace.*

— Both round and rectangular CRTs are found is oscilloscopes. Vertical viewing size is 8 cms and horizontal is 10 cms (See Fig. 2.35).

- **Operating voltages :**

The CRT requires a heating voltage of 6.3 V A.C. or D.C. at 600 mA. Several D.C. voltages are listed below; the voltages vary with the types of tube used :

(*i*) Negative grid (control) voltage : –14 V to –200 V

(*ii*) Positive anode $A_1$ (Focusing anode) : –100 V to –1100 V

(*iii*) Positive anode $A_2$ (Accelerating anode) : 600 V to 6000 V

$A_3$ : 200 V to 20000 V in some cases

**5. Base.** The base is provided to the CRT through which the connections are made to the various parts.

### 2.5.5. Electrostatic Deflection

Fig. 2.36. shows the upward deflection of an electron beam when it passes between the vertical or *Y*-plates of a CRT. The beam deflects upwards because the upper *Y*-plate has been made positive w.r.t. the lower plate. *Reversing* the polarity of the *"applied voltage"* would, obviously, cause the beam to deflect downwards.

The velocity of an electron coming out of the electron gun can be found by equating the kinetic and potential energies.

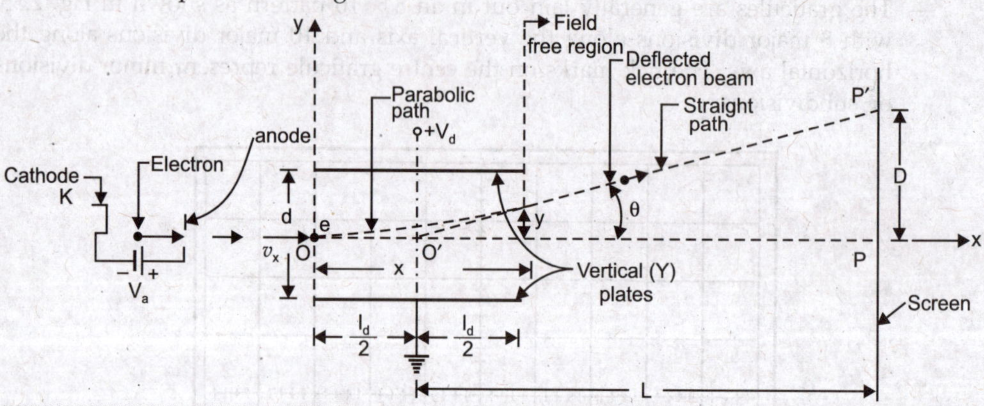


**Fig. 2.36.** Deflection cathode ray beam.

If, $m$ = Mass of electron, kg,

$v_x$ = Velocity of electron as it comes out of electron gun, $m/s$,

$V_a$ = Accelerating voltage (Potential difference between anode and cathode), $V$ and

$e$ = Charge on electron, $C$ ;

Then,

$$\frac{1}{2} m v_x^2 = eV_a$$

or,
$$v_x = \sqrt{\frac{2eV_a}{m}} \qquad ...(2.1)$$

This velocity $v_x$ *remains constant* because the fields due to X-and Y-plates act at right angles to each other.

Fig. 2.36. shows the *electrostatic deflection* due to the Y-plates.

Let, $d$ = Distance between Y-plates,

$l_d$ = Length of each Y-plate,

$L$ = Distance from $O'$ to screen,

$V_d$ = Deflecting voltage (*i.e.*, potential difference between Y-plates), and

$D$ = Deflection on the screen.

(The straight line of travel of the electron is tangent to the parabola at $x = l_d$)

Now, electric field between Y-plates

$$= \frac{V_d}{d} \text{ volts/m} \qquad ...(2.2)$$

$$\therefore \quad \text{Force on electron} = \frac{eV_d}{d} \text{ newtons} \qquad ...(2.3)$$

Also, force on electron = $m \times$ acceleration in $y$-direction (*i.e.* $a_y$)

$$\therefore \quad a_y = \frac{eV_d}{md} m/s^2 \qquad ...(2.4)$$

The time taken by the electron to traverse the length $l$ is $\frac{l_d}{v_x}$. Since the initial velocity in the $y$-direction is zero, $v_y$, *i.e.* velocity in $y$-direction when electron leaves the field between plates, is

$$v_y = 0 + a_y.t$$

$$= \frac{eV_d}{md} \cdot \frac{l_d}{v_x}$$

$$i.e. \quad v_y = \frac{eV_d}{md} \cdot \frac{l_d}{v_x} \qquad ...(2.5)$$

After the electron leaves the field of the Y-plates, it moves in a *straight line* because it is in field-free space/region. *During the motion between the deflecting plates, a uniform acceleration is acting on the electron and therefore, its path is parabolic.* If the straight line is extended backward, it can be shown that it passes through O′, midway between the plates.

The deflection of the electron when it leaves the field is,

$$y = (\text{Average velocity in } y\text{-direction}) \times \frac{l_d}{v_x}$$

$$= \left(\frac{1}{2} \cdot \frac{eV_d}{md} \cdot \frac{l_d}{v_x}\right) x \frac{l_d}{v_x}$$

$$\text{or,} \quad y = \frac{1}{2}\left(\frac{eV_d}{md}\right)\left(\frac{l_d}{V_x}\right)^2 \qquad ...(2.6)$$

From Fig. 2.36; $\frac{y}{D} = \frac{(l_d/2)}{L}$ *or* $\frac{l_d}{2L}$

or, $$D = \frac{2L}{l_d} . y = \frac{2L}{l_d} \times \frac{1}{2} \times \frac{eV_d}{md} \times \frac{l_d^2}{v_x^2}$$

i.e. $$D = \frac{L\,e\,V_d\,l_d}{m\,d\,v_x^2} \qquad ...(2.7)$$

Substituting the values of $v_x$ from eqn. (2.1), we get

$$D = \frac{L\,eV_d l_d}{m\,d} \times \frac{m}{2eV_a} = \frac{LV_d l_d}{2dV_a}$$

i.e. $$D = \frac{LV_d\,l_d}{2\,dV_a} \qquad ...(2.8)$$

It is seen from eqn. (2.8) that $D$ (vertical deflection on screen) is proportional to $V_d$ (deflection voltage). This direct proportionality indicates that CRT may be used as a *linear voltage-indicating device*. The deflection voltage, although *assumed fixed* in the above discussion, usually varies and the image on the screen follows the variations of the deflection voltage in a linear manner.

- **Electrostatic deflection sensitivity. (S)** *is defined as the deflection per volt of the deflection voltage.*

  i.e., $$S = \frac{D}{V_d} = \frac{L\,l_d}{2\,d\,V_a} \qquad ...(2.9)$$

  Its units are $m/V$.

- **Deflection factor (G)** of a cathode ray tube is given by the *reciprocal of the sensitivity*.

  i.e. $$G = \frac{1}{S} = \frac{2dV_a}{L\,l_d} \qquad ...(2.10)$$

  Its units are $V/m$.

— It is seen that $S$ is independent of the deflection voltage $V_d$ and the ratio $\frac{e}{m}$.

Moreover, S varies inversely with $V_a$ (acceleration voltage); thus by decreasing $V_a$, sensitivity can be increased.

On the other hand high $V_a$ value, produces a bright spot. But for high $V_a$, high $V_d$ is required and such a beam which is highly accelerated is *difficult to deflect*. Such a beam is called *"hard beam"*.

— The typical values of deflection sensitivities are 1.0 mm/V to 0.1 mm/V corresponding to the deflection factors of 10 V/cm to 100 V/cm respectively.

### 2.5.6. Magnetic Deflection

When a conductor of length $l$ carrying a current $I$ is placed in a magnetic field of flux density $B$, a force acts on the conductor. This force $F$ is given by (assuming the directions of $I$ and $B$ to be perpendicular),

$$F = BIl \text{ newtons} \qquad ...(2.11)$$

The direction $F$ is perpendicular to the directions of $I$ and $B$. If $I$ and $B$ are not perpendicular, only the component of $I$ perpendicular to $B$ contributes to the force.

Let the length $l$ contains $n$ electrons; if these electrons take $t$ seconds to travel length

$l$ then the charge per second *i.e.* current is given as :

$$I = \frac{ne}{t} \quad ...(2.12)$$

where $e$ is the charge on an electron, C.

∴ Force on the $n$ electrons is,

$$F = \frac{Bnel}{t} \quad ...(2.13)$$

$\frac{l}{t}$ is the drift speed (*i.e.* $v$) of the electrons.

∴ Force *per electron* $= Bev$ ...(2.14)

Equ. (2.14) indicates that if $v$ is zero, F is zero. This means that if electron is at rest, it does not experience any force and it remains at rest. If the velocity is along the direction of magnetic flux, there is no force on the electron and it continues to move at the same velocity.

*Refer to* **Fig. 2.37.** When an electron enters a magnetic field, having flux density directed into the page, with a speed $v$ the force $F$ is perpendicular to $v$ and therefore the speed $v$ remains constant. This forces causes the electron to move in a circular motion with constant speed $v$. If $r$ is the radius of the circular path, the acceleration towards the centre is $\frac{v^2}{r}$.

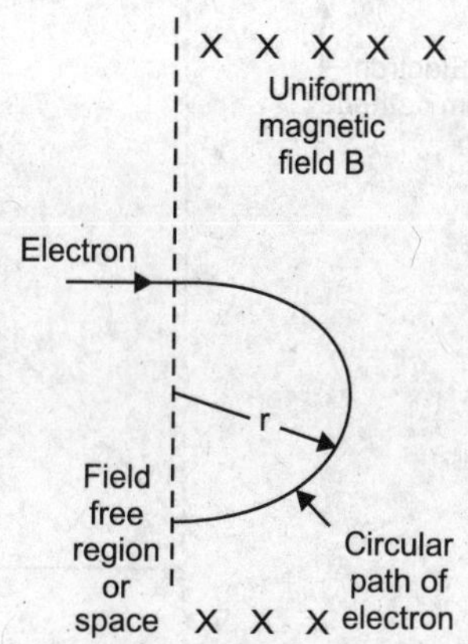


**Fig. 2.37.** Electron moving in a magnetic field.

Then, $$F = \frac{mv^2}{r} = Bev$$

or, $$r = \frac{mv}{eB} \quad ...(2.15)$$

Also, angular velocity,

$$\omega = \frac{v}{r} = \frac{v}{(mv/eB)} = \frac{eB}{m} \text{ rad./sec.} \quad ...(2.16)$$

and, Time period $T$ (*i.e.*, to complete one revolution) is given as :

$$T = \frac{2\pi}{\omega} = \frac{2\pi}{(eB/m)} = \frac{2\pi m}{eB} \text{ rad./sec.} \quad ...(2.17)$$

From above discussion, we find that the radius of the circular path is directly proportional to $v$. $\omega$ and $T$ do not depend on $r$ or $v$. Thus, the fast moving particles travel in a bigger circle while the slow moving particles travel in smaller circle.

The *time period for both the slow and the fast particles is the same*. This result forms the basis of operation of *cyclotron, magnetic focusing apparatus and many other apparatus*.

- A magnetic field can be used for deflection of electron beam in a cathode ray tube.

Fig. 2.38 shows a transverse magnetic field of flux density $B$ for detecting an electron beam. The field is pointing into page. The electron moves in a straight line from cathode to boundary $A$ of the magnetic field. In the magnetic field region, the electron experiences a force $F$ (eqn. 2.14) and moves in a circular path $AB$. The radius of the circular path is $r$ (eqn. 2.15) and the centre is at point $C$. The angle of deflection $\theta$ is very small and is given by,

$$\theta \simeq \frac{l}{r} \qquad \text{...(2.18)}$$

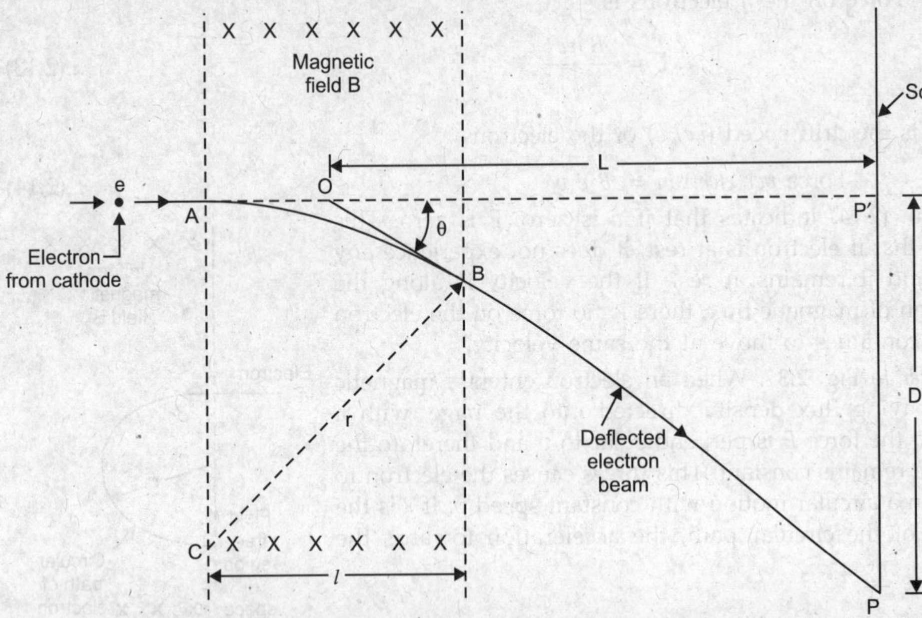


Fig. 2.38. Deflection of electron beam by a magnetic field.

As distance $L$ is very large as conspired to $l$, therefore, the straight line $PB$ when projected backward passes through point $O$ at the centre of the magnetic field. The deflection $D$ is given as :

$$D = L\tan\theta \simeq L\theta = \frac{Ll}{r} = \frac{LleB}{mv} \qquad \text{...(2.19)}$$

The velocity $v$ is $v_x$ as given by equ. (2.1). Substituting this value in eqn. (2.19), we get :

$$D = \frac{LleB}{\sqrt{\dfrac{2eV_a}{m}}} = BLl\left(\sqrt{\frac{e}{2mV_a}}\right) \qquad \text{...(2.20)}$$

**Magnetic deflection sensitivity (S)** *is defined as the deflection on the screen in meters per tesla of magnetic flux density.*

*i.e.*
$$S = \frac{D}{B} = Ll\sqrt{\frac{e}{2mV_a}} \qquad \text{...(2.21)}$$

- It is seen that magnetic deflection and magnetic deflection sensitivity are inversely proportional to $\sqrt{V_a}$. Therefore, magnetic deflection is preferable in tubes with high accelerating voltage like *TV tube* and *radar* (while *electrostatic deflection is prefered for general purpose oscilloscope*).

**Comparison between deflection methods :**

The comparison between 'electrostatic deflection' and 'magnetic deflection' method is given below :

| S.N. | *Aspects* | *Electrostatic deflection* | *Magnetic deflection* |
|---|---|---|---|
| 1. | *Method used to obtain deflection.* | By the application of voltage to the plates. | By controlling the magnetic field by varying the current through the coils. |
| 2. | *Relationship between the deflection and $V_a$.* | Inversely proportional to $V_a$. | Invesely proportional to $\sqrt{V_a}$. |
| 3. | *Size of the beam for given $V_a$.* | Small | Large (wide). |
| 4. | *Sensitivity (S) formula.* | $S = \frac{Ll_d}{2dV_a}$ | $S = Li\sqrt{\frac{e}{2mV_a}}$ |
| 5. | *Dependance of deflection on the ratio e/m.* | No | Yes |
| 6. | *Size of tubes necessary for given display area.* | Longer tubes. | Shorter tubes (can be built for the given display area). |
| 7. | *Effect on the spot brightness with reduction in $V_a$ (for greater sensitivity)* | Reduced brightness. | Increased brightness. |
| 8. | *Applications.* | General propose oscilloscopes. | T.V. and radar. |

- ***Differences between CRT and television picture tube.***
  1. The CRT used in CRO uses electrostatic method of focussing as compared to a TV picture tube which employs electromagnetic envelope.
  2. In CRT used in CRO, connection to accelerating anode is brought out through pin in the base along with other electrode connections as shown in Fig. 2.33. However in TV picture tube the lead in from accelerating anode is brought out through the sides of glass envelope.
  3. In CRT used in CRO, EHT, (*i.e.*, extra high tension transformer) to accelerating anode is given from mains supply whereas in TV picture tube EHT supplied to accelerating anode is generated in the line output stage during flyback. During flyback the anode current of line output stage is suddenly cut off and anode voltage generates a spike of 2 to 3 kV amplitude. By increasing the number of turns on the primary winding to form an auto-transformer of say 1 : 5 turns pulses of 15 kV or more is obtained. After rectification and smoothing, EHT is obtained and applied to accelerating anode.

This method of applying EHT has an advantage because if input to line and valve fails, no EHT is produced and thus the tube is saved.

**Example 2.4 :** *A CRT has an anode voltage of 2 kV and parallel deflecting plates 2 cm long and 5 mm apart. The screen is 30 cm from the centre of the plates. Find the input voltage required to deflect the beam through 3 cm. The input voltage is applied to the deflecting plates through amplifiers having an overall gain of 100.* **(Nagpur University S - 2005)**

**Solution.** *Given* : $V_a$ = 2 kV or 2000 V; $l_d$ = 2 cm or 0.02 m; $d$ = 5 mm or 0.005 m; L = 30 cm or 0.3 m; D = 3 cm or 0.03 m; Overall gain = 100.

We know that, $$D = \frac{Ll_d V_d}{2dV_a} \quad \text{...[Eqn. (2.1)]}$$

or, $$V_d = \frac{2dV_a D}{Ll_d}$$

$$= \frac{2\times0.005\times2000\times0.03}{0.3\times0.02} = 100\text{V}$$

∴ Input voltage required for deflection of 3 cm

$$= \frac{V_d}{gain} = \frac{100}{100} = \mathbf{1\ V} \ \textbf{(Ans.)}$$

**Example 2.5.** *An electrostatically deflected cathode ray tube has plane parallel deflecting plates which are 2.5 cm long and 0.5 cm apart, and the distance from their centre to the screen is 20 cm. The electron beam is accelerated by a potential difference of 2500 V and is projected centrally between the plates.*

*Calculate the deflection voltage required to cause the beam to deflect through 3 cm and find the corresponding deflection sensitivity of the screen.* **(PTU Dec. - 2005)**

**Solution.** *Given* : $l_d$ = 2.5 cm = 0.025 m; $d$ = 0.5 cm = 0.005 m; $L$ = 20 cm = 0.2 m; $V_a$ = 2500 V ; $D$ = 3 cm = 0.03 m

$\mathbf{V_d}$ **;** $\mathbf{S}$ **:**

We know that, $$D = \frac{Ll_d V_d}{2dV_a} \quad \text{...[Eqn. (2.8)]}$$

or, *Deflection voltage*, $$V_d = \frac{D\times2dV_a}{Ll_d} = \frac{0.03\times2\times0.005\times2500}{0.2\times0.025} = \mathbf{150\ V}\ \textbf{(Ans.)}$$

∴ *Deflection sensitivity*, $$S = \frac{D}{V_d} = \frac{0.03}{150}\ \text{m/V or } \mathbf{0.2\ mm/V}\ \textbf{(Ans.)}$$

**Example 2.6.** *The horizontal deflection plates in a CRT are 20 mm long and 5 mm apart. The centre of the plates is 20 cm from the screen. Accelerating voltage is 2500 V. Determine the deflection sensitivity.* **(UPTU - 2004)**

**Solution.** *Given*: $l_d$ = 20 mm = 0.02 m; $d$ = 5 mm = 0.005 m; $L$ = 20 cm = 0.2 m; $V_a$ = 2500 V.

We know that, deflection sensitivity,

$$S = \frac{Ll_d}{2dV_a} \quad \text{...[Eqn. (2.9)]}$$

or, $$S = \frac{0.2\times0.02}{2\times0.005\times2500}\ \text{m/V or } \mathbf{0.16\ mm/V}\ \textbf{(Ans.)}$$

**Example 2.7.** *In a CRT, the anode to cathode voltage is 2500 V. The parallel deflection plates are 1.5 cm long and spaced 5 mm. The screen is 50 cm from the centre of the deflecting plates. If mass of electron is 9.109 × $10^{-31}$ kg and charge on electron 1.602 × $10^{-19}$C, find :*

(i) *The beam speed.*

(ii) *The deflection sensitivity of the tube.*

**Solution.** *Given :* $V_a = 2500$ V; $l_d = 1.5$ cm $= 0.015$ m; $d = 5$ mm $= 0.005$ m;
$L = 50$ cm $= 0.5$ m; $m = 9.109 \times 10^{-31}$ kg; $e = 1.602 \times 10^{-9}$ C.

(*i*) **The beam speed, $v$ :**

We know that,
$$v = \sqrt{\frac{2eV_a}{m}} \qquad \text{...(Eqn. (2.1)]}$$
$$= \sqrt{\frac{2 \times 1.602 \times 10^{-19} \times 2500}{9.109 \times 10^{-31}}}$$
$$= \mathbf{29.65 \times 10^6\ m/s} \quad \textbf{(Ans.)}$$

(*ii*) **Deflection sensitivity, $S$ :**

$$S = \frac{Ll_d}{2dV_a} \qquad \text{...[Eqn. (2.2)]}$$
$$= \frac{0.5 \times 0.015}{2 \times 0.005 \times 2500}\ m/V$$

or,
$$= \mathbf{0.3\ mm/V} \quad \textbf{(Ans.)}$$

**Example 2.8.** *A cathode-ray tube, with magnetic deflection, has its screen 22 cm from the centre of the magnetic field. The width of the uniform magnetic field is 3.3 cm and the final anode potential is 6 kV. Calculate the density of the magnetic field to produce a deflection of 4.4 cm on the screen.*

*Take : Mass of electron = $9.107 \times 10^{-31}$ kg; charge on electron = $1.6 \times 10^{-19}$ C.*

**Solution.** *Given :* $L = 22$ cm $= 0.22$ m; $l = 3.3$ cm $= 0.033$ m; $V_a = 6$ kV;
$D = 4.4$ cm $= 0.044$ m; $m = 9.107 \times 10^{-31}$ kg; $e = 1.6 \times 10^{-19}$ C.

**Density of magnetic field, $B$ :**

We know that,
$$D = BLl\left(\sqrt{\frac{e}{2mV_a}}\right) \qquad \text{...[(Eqn. (2.20)]}$$

Substituting the values we get,
$$0.044 = B \times 0.22 \times 0.033\left(\sqrt{\frac{1.6 \times 10^{-19}}{2 \times 9.107 \times 10^{-31} \times (6 \times 10^3)}}\right)$$

or,
$$0.044 = B \times 27.779$$

or,
$$B = 1.584 \times 10^{-3}\ \text{Wb/m}^2 \quad \text{or} \quad \mathbf{1.584\ m\ Wb/m^2}\ \textbf{(Ans.)}$$

**Example 2.9.** *In a cathode-ray tube, the electron beam is displaced vertically by a magnetic field of flux density $1.8 \times 10^{-4}$ Wb/m². The length of the magnetic field along the tube axis is same as that of the electrostatic deflection plates. The final anode voltage is 800 V.*

*Calculate the voltage which should be applied to the Y-deflection plates 10 mm apart, to return the spot back to the centre of the screen.*

*Take : Mass of electron = $9.107 \times 10^{-31}$ kg; charge on electron = $1.6 \times 10^{-19}$ C.*

**Solution.** *Given :* $B = 1.8 \times 10^{-4}$ Wb/m²; $V_a = 800$ V; $d = 10$ mm $= 0.01$ m.

**Deflection voltage, $V_d$ :**

**Case I.** *Magnetic deflection :*

In this case, deflection, $D = BLl\left(\sqrt{\dfrac{e}{2mV_a}}\right)$ ...[Eqn. (2.20)]

$$= 1.8\times10^{-4}\, L\, l\left(\sqrt{\frac{1.6\times10^{-19}}{2\times9.107\times10^{-31}\times800}}\right)$$

or, $D = 1.886\ lL$ metres ...(1)

**Case II.** *Electrostatic deflection* :

In this case, $D = \dfrac{LV_d\, l}{2dV_a}$ ...[Eqn. (2.8)]

or $D = \dfrac{lL\,V_d}{2\times0.01\times800} = 0.0625\ l\ M\ Vd$ metres ...(2)

Equating (1) and (2), we get

$$1.886\ lL = 0.0625\ lLV_d$$

or, $V_d = \dfrac{1.886}{0.0625}$

$= $ **30.176 V (Ans.)**

### 2.5.7. Vertical Amplifier

- The input signal is fed to the $Y$-deflecting plates through a vertical amplifier.
- The function of vertical amplifier is to *amplify the input signal.*
- It may consist of several stages in cascade and generally, operates with fixed gain. The output stage of this amplifier is invariably a push-pull amplifier. It is necessary to adjust the magnitude of the input signal so that a large signal may not cause the display to go out of the screen. Therefore, the input signal is fed to the vertical amplifier through an *attenuator*.
- The *"attenuator"* is a network of resistances and capacitances and it *reduces the input signal* so that the display of large input signals *may not go out of the screen.*

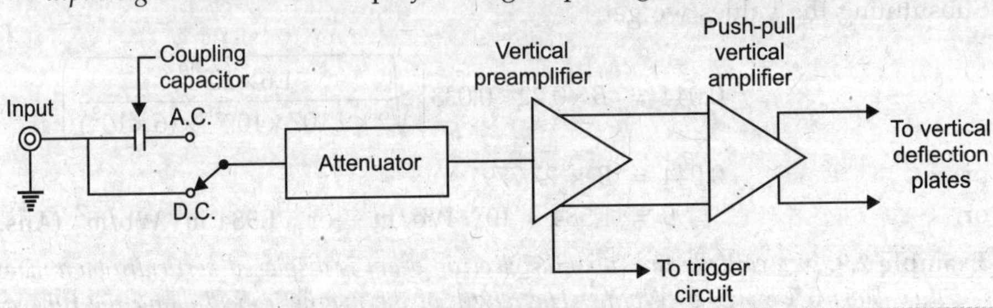


**Fig. 2.39.** Block diagram of the vertical section of an oscilloscope.

Fig. 2.39 shows the block diagram of a complete *vertical deflection system.*

— The input convector feeds an input attenuator, after which follows the vertical amplifier. The input impedance of an oscilloscope is rather high, being on the order of 1 MΩ, which is desirable for measuring voltages in high impedance circuits.

— The attenuator sets the sensitivity of the oscilloscope in the common 1-2-5 sequence. As an example, the input attenuator could provide for 10, 20, 50, 100, 200 mV, etc. per centimetre.

— Practically all oscilloscopes provide a switchable input coupling capacitor. This is provided so that measurements of A.C. signals may be viewed in the presence of high D.C. voltages by including the coupling capacitor. When D.C. measurements are to be made, the capacitor may be removed. The value of the capacitor is chosen so that the frequency response of the oscilloscope is preserved down to a few Hz.

- *The vertical amplifier is the principal factor in determining the sensitivity and bandwidth of an oscilloscope.*
  - — The gain of the vertical amplifier determines the smallest signal that the oscilloscope can satisfactorily reproduce on the CRT screen. The sensitivity of an oscilloscope is directly proportional to gain of the vertical amplifier that is, as gain increases sensitivity increases, which allows us to observe smaller-amplitude signal.
  - — The vertical sensitivity of an oscilloscope is the smallest deflection factor that can be selected with the rotary switch.
  - — The *bandwidth* of an oscilloscope determines the range of frequencies that can be observed with the instrument. The bandwidth of an oscilloscope is the range of frequencies over which the gain of the vertical amplifier is within 3 dB of the midband frequency gain.

### 2.5.8. Time Base (Sweep or ramp) Generator

If the horizontal deflecting plates X-plates are not energised, the electron beam would draw a vertical line on the screen.

- The *function of the time base generator* is to drive the beam at a steady speed across the screen and, when it reaches the right-hand side of the screen, the beam is made to fly back to the starting position on the left-hand side of the screen. The time base generator produces a *saw-tooth wave* (Fig. 2.40) of *the same frequency as the* input signal to the *Y*-plates.

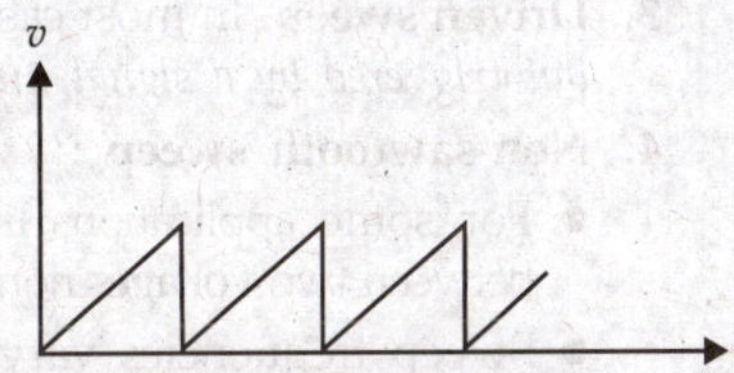

**Fig. 2.40.** Saw-tooth voltage wave.

- The horizontal deflection of the beam is proportional to the instantaneous voltage of the saw-tooth wave, therefore, the beam sweeps at a uniform rate across the screen in the horizontal direction, running to the starting position almost instantaneously and sweeping again.
- The time base generator gives an output sweep signal of adjustable frequency so that the *Y* input signals of a wide range of frequencies may be displayed on the screen.

There are many applications of a CRO in which a time base generator is *not* needed and it is necessary to apply an external input signal to the X-plates (through a switch).

### 2.5.9. Types of Sweeps

There are *four* types of sweeps :

**1. Free running sweep :**

- In very basic oscilloscopes the sweep generator is continuously charging and discharging a capacitor. One ramp voltage is followed *immediately* by another, hence the sawtooth pattern appears. A sweep generator operating in this manner is said to be *"free running"*.

- In order to present a stationary display on the screen, the sweep generator signal must be forced to run in *synchronization* with a vertical input signal. In basic oscilloscopes this is accomplished by carefully adjusting the sweep frequency to a value very close to the exact frequency of the vertical input signal, or submultiple of this frequency. With signals at the same frequency, an *internal synchronization pulse will lock the sweep generator into the vertical input signal*. This method of synchronization has some serious limitations when an attempt is made to observe *low-amplitude* signals. However, the most serious limitation is probably the inability of the instrument to maintain synchronization when the amplitude or frequency of the vertical signal is not constant, such as voice or music signals.

2. **Triggered sweep :**
   - A waveform to be observed on the CRO may not be periodic but may perhaps occur at irregular intervals. In this case it is desirable that the sweep circuit remain inoperative and the sweep be initiated by the waveform under examination. In some cases the waveform may be periodic, but it may be that the interesting part of the waveform is of very short duration compared to the period of the waveform. Under such cases a *triggered switch is used.*
   - In triggered sweep or single sweep, the spot is swept once across the screen in response to a trigger signal.
   - The *triggered sweep is used for examination of transient or one time signals and waveform is photographed for record.* The trigger can be obtained from the signal under investigation or by an external source.
3. **Driven sweep.** In most cases, a *'driven sweep'* is used where the sweep is *recurrent but triggered by a signal under test.*
4. **Non-sawtooth sweep :**
   - For some applications like comparison of two frequencies or finding phase shift between two voltages non-sawtooth sweep voltages are utilised for the sweep circuit.
   - Sweep frequencies vary with the type of the oscilloscope.

A laboratory oscilloscope may have sweep frequencies upto several MHz; a simple oscilloscope for audio work has an upper limit 100 kHz. Most T.V. services require a sweep voltage frequency upto 1 MHz.

### 2.5.10. Horizontal Amplifier

- The function of the horizontal amplifier is to *amplify the signal applied to the X-plates.*
- This amplifier consists of several stages in the cascade.
- For measurement of *voltages and currents,* the input to the horizontal amplifier is a *sawtooth wave.* For *frequency and phase* measurements, the input is an *external signal.* The horizontal amplifier should be capable of amplifying both these inputs without distortion.

The input impedance of most general purpose oscilloscopes is about 1 MΩ shunted by a capacitance of about 20 pF.

### 2.5.11. Delay Line

The *purpose of a "delay line" is to delay the vertical signal enough to keep it from reaching the CRT deflection plates before the horizontal sweep circuits are running.*

- In an oscilloscope all electronic circuitry (attenuators, amplifiers, pulse shapers, etc.) causes a certain amount of line delay in the transmission of signal voltages to the deflection plates. Almost all of this delay is created in circuits that *switch*, *shape* or *generate*.

Referring to Fig. 2.41, we observe that the horizontal signal (time base, or sweep voltage) is initiated or *triggered*, by a portion of the output signal applied to the vertical plates of the CRT. In horizontal channel, the signal processing consists of generating and shaping a trigger pulse (trigger pick off) that starts the sweep generator, whose output is fed to the horizontal amplifier and then to the horizontal deflection plates. This whole process consumes time on of the order of 80 ns or so. Thus, to allow the operator to observe the leading edge of the signal waveform, the signal drive for the vertical CRT plates must therefore be *delayed* by at least the same amount of time. This is the function of the vertical delay line. In the Fig. 2.41 is shown a 200 ns delay line which has been added to the vertical channel. It may be remembered that a trigger pick off must *precede* the delay line, although the delay line can appear almost anywhere along the vertical signal path.

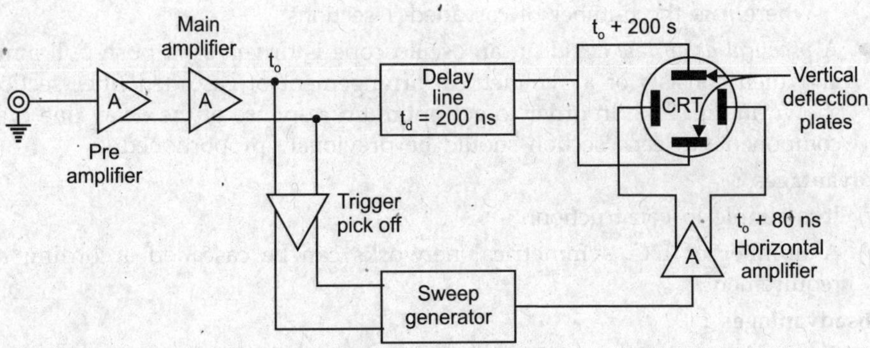


**Fig. 2.41.** Delay line circuit.

In CRO the following two types of delay lines are used :

1. Lumped-parameter delay line. 2. Distributed-parameter delay line.

**1. Lumped-parameter delay line :**

- This type of delay line consists of a cascaded symmetrical LC networks, such as the so called T-section of Fig. 2.42.

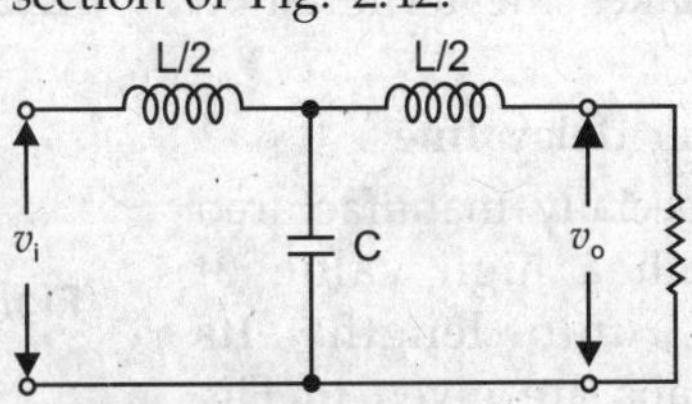


**Fig. 2.42.** T-filter section.

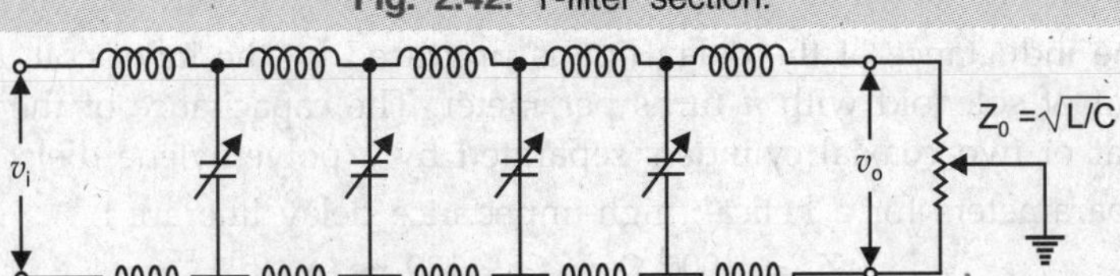


**Fig. 2.43.** Practical delay line.

- Each section is capable of delaying the signal by 3 to 6 ns.
- The T-section is terminated into its characteristic impedance $Z_o$. So, it acts as

a low-pass filter, its cut-off frequency is given as :

$$f_c = \frac{1}{\pi\sqrt{LC}} \quad ...(2.22)$$

If the spectrum of input signal $v_i$ consists of frequencies much less than the cut-off frequency, output signal $v_o$ will be a faithful reproduction of $v_i$ but delayed by a time,

$$t_s \simeq \frac{1}{\pi f_c} = \sqrt{LC} \quad ...(2.23)$$

where, $t_s$ is the delay time for a single T-section. However, when a number of T-sections are cascaded into a so-called *lumped-parameter* delay line, the total delay time,

$$t_d = n\, t_s \quad ...(2.24)$$

where $n$ is the number of cascaded T-sections.

- A *practical delay line circuit* in an oscilloscope is driven by a push-pull amplifier and then consists of a symmetrical arrangement of cascaded filter sections as shown in Fig. 2.43. In order to get optimum response of the delay line $L$ and $C$ components in each section should be previously proportioned.

**Advantages :**

(*i*) It is simple in construction.

(*ii*) A number of L.C. symmetrical networks can be cascaded according to the requirement.

**Disadvantages :**

(*i*) Since it has a sharp cut-off frequency therefore, as the input frequency increases the amplitude and phase *distortion* takes place.

(*ii*) In order to get better transient response the delay line should be matched to the characteristic impedance at the input and output side. This makes the termination circury *complex*.

**2. Distributed-parameter delay line :**

- It consists of a specially manufactured coaxial cable with a high value of inductance per unit length. Its constructional details are given in Fig. 2.44.

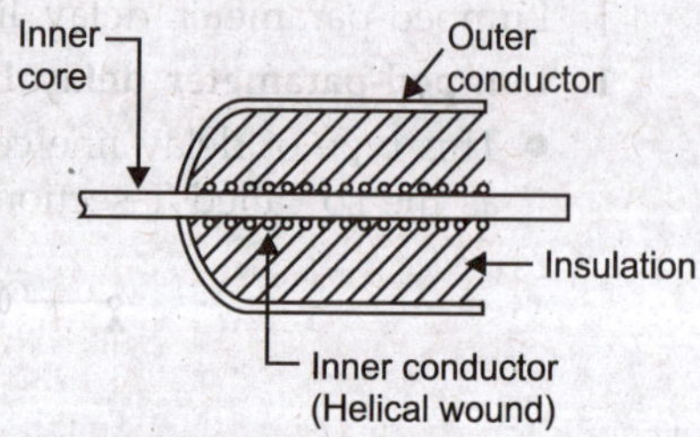


**Fig. 2.44.** Distributed parameter delay line.

- The inductance of the delay line is produced by the inner coil, and it equals that of solenoid with $n$ turns per meter. The capacitance of the delay line is that of two coaxial cylinders separated by a polyethylene dielectric.

Typical parameters for a helical, high-impedance delay line are :

$$Z_o = 1000\ \Omega\ ;\ t_d = 180\ \text{ns/m}.$$

**Advantages :**

(*i*) Occupies less space.

(*ii*) It does not require careful adjustment of a lumped-parameter delay line.

### 2.5.12. Basic Controls of CRO

Fig. 2.45. shows the detailed block diagram of a general purpose oscilloscope, exhibiting a number of controls required for the proper functioning of CRO.

**1. Vertical (Y) deflection system :**

- The signals to be examined are usually applied to the vertical or Y-deflection plates through an attenuator and a number of amplifier stages. Vertical amplifier is required because the signals are not strong enough to produce measurable deflection on CRT screen.

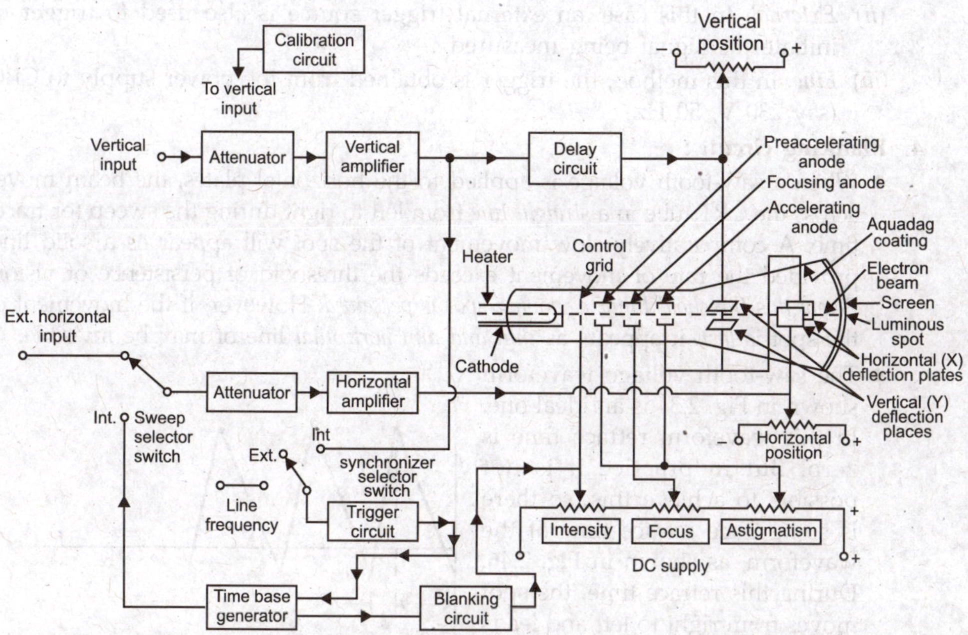


**Fig. 2.45.** Detailed block diagram of a general purpose oscilloscope.

— *The amplifier response must be wide enough to pass faithfully the entire band of frequencies to be measured.*

- The high voltage signals (to be examined) must be attenuated to bring them as within the range of vertical amplifies. The output of the vertical amplifier is also applied to the synchronizing amplifier through the *synchronizer selector switch in internal position (Int.)*. This permits the horizontal sweep circuit to be triggered by the signal being investigated.

**2. Horizontal (X) deflection system :**

- The horizontal or X-deflection plates are fed by a sweep voltage that provides a *time base*. The horizontal plates are supplied through an amplifier, but they can be fed directly when voltages are of sufficient magnitudes.
- When external signals are to be applied to the horizontal deflection system, they can also be fed through the horizontal amplifier, *via* the sweep selector switch which is in the external position (Ext.). When the sweep selector switch is in the internal position, the horizontal amplifier receives an input from the saw tooth sweep generator which is triggered by the synchronizing amplifier.

3. **Synchronization :**
   - In order to obtain a *stationary pattern* synchronization [irrespective of the type of sweep (e.g. free running, triggered sweep, driven sweep, non saw-tooth sweep] has to be done. This requires that the time base be operated at a submultiple frequency of the signal under measurement (applied to *Y*-plates).

   The following three usual sources for synchronization can be selected by synchronizing selector :

   (*i*) *Internal*. Here the trigger is obtained from the signal being measured through the vertical amplifier.

   (*ii*) *External*. In this case, an external trigger source is also used to trigger or initiate the signal being measured.

   (*iii*) *Line*. In this method, the trigger is obtained from the power supply to CRO (say 230 V, 50 Hz)

4. **Blanking circuit :**
   - When a saw-tooth voltage is applied to the horizontal plates, the beam moves across the CRT tube in a *straight line* from left to right during the sweep (or trace) time. A comparatively slow movement of the spot will appear as a solid line, provided the rate of movement exceeds the threshold of persistence of vision. *Below this threshold limit, a moving spot is perceived*. However, if the movement of the spot is *fast*, it appears as *thin and dim horizontal* line or may be invisible.
   - The saw-tooth voltage waveform shown in Fig. 2.34 is an ideal one. In this waveform retrace time is zero. But in practice it is not possible to achieve this, so there is some trace retrace time of the waveform, as shown in Fig. 2.46. During this retrace time, the spot moves from right to left and leads to confusion. *So this trace is blanked out by supplying a high -ve voltage to the grid during the retrace time*. The blanking voltage is usually developed/triggered by time-base generator.

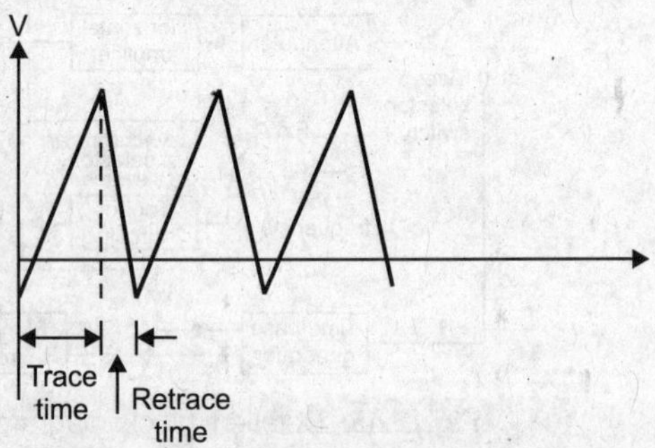


**Fig 2.46.** Sawtooth voltage waveform.

5. **Positioning controls.** These controls involve two knobs–one for controlling the horizontal position and another for controlling the vertical position.
   — The spot can be moved to *left or right i.e.*, horizontally with the help of a knob, which regulates the D.C. potential applied to the *X*-deflection plates in addition to the usual saw-tooth wave. Similarly the spot can be moved *up and down i.e.* vertically with the help of another knob, which regulates the D.C. potential applied to the *Y*-deflection plates in addition to the signal.

6. **Intensity (Z-axis) modulation :**
   - The intensity modulation is done by inserting a signal between the ground and the cathode/control grid.
   - It is applied during normally visible portion of the trace.
   - It can be used for *brightening the display*.
     — The periodically brightened spots may be used as markers for time calibration of the main waveform.

**7. Intensity control :**

- The electron-beam intensity is varied by the *intensity control potentiometer* which changes the grid potential w.r.t. cathode. The grid potential determines the amount of electrons leaving the cathode and thus controls the intensity of the beam.
- It must be ensured that the electron beam does not form burning spots on the screen. A stationary spot should be kept on very low intensity. If the intensity is kept high, the spot must be kept moving. If a *"halo"* appears around the spot, the intensity is too high; the intensity should be turned before turning the oscilloscope on.

**8. Focus control :**

- The middle anode, in the electron gun of a CRT, is kept at a lower potential w.r.t. other two anodes; it acts like an electrostatic lens and focal length of this lens can be varied by varying the potential of the middle anode w.r.t. other two anodes. So the focusing of an electron beam is done by varying the potential of the middle anode with the help of a potentiometer, as shown in Fig. 2.45.
- *By increasing the +ve potential applied to the focusing anode the electron beam can be narrowed and the spot on the screen can be made a pin point.*

**9. Astigmatism:**

- *Astigmatism* is an additional focusing control (in most modern oscilloscopes) and is analogous to as astigmatism in optical lenses.
- In order to focus the spot correctly, it is necessary to stop it near the centre of the screen by switching off the time base and adjusting the $X$ and $Y$ positioning controls. The spot is then made as *sharp as possible by successive adjustment of focus and astigmatism controls.*

**10. Calibration circuit :**

- Laboratory oscilloscopes normally have an internally generated and stablised voltage of known amplitude which is used for calibration purposes.
- The calibrating voltage usually has a square waveform.

## 2.5.13. Dual Trace Oscilloscope

The comparison of two or more voltages is invariably necessary in the analysis and study of many electronic circuits and systems. This is possible by using more than one oscilloscope but in such a case it is difficult to trigger the sweep of each oscilloscope precisely at the same time. A common and less costly method to solve this problem is to use dual trace or multitrace oscilloscopes. In this method, the same electron beam is *used to generate "two traces" which can be deflected from two independent* vertical sources.

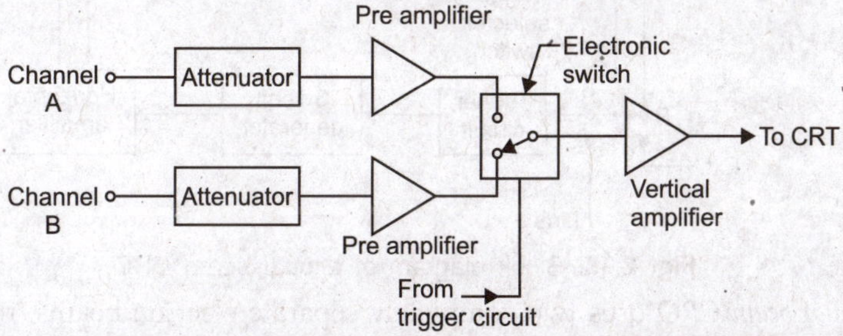


**Fig. 2.47.** Block diagram of a dual trace oscilloscopes.

Dual trace oscilloscopes are used extensively by *industrial firms and research laboratories.*

These oscilloscopes produce a dual-trace display by means of electronic switching of two separate input signals.

Fig. 2.47 shows the block diagram of a dual trace oscilloscope :

- There are two vertical input circuits marked channels *A* and *B* with identical preamplifiers.
- The outputs of these preamplifiers are fed to an electronic switch which alternately connects them to the main vertical amplifier of the oscilloscope.
- In this type of oscilloscope there is *only one electron beam*.
- There are two common operating modes for the electronic switch, called *alternate* and *chop*; these are selected from the instrument's front panel.
  - — In ***alternate mode*** the electronic switch alternates between channels A and B, letting each through for one cycle of the horizontal sweep. The display is blanked during the flyback and hold-off periods, as in conventional oscilloscopes provided the sweep speed is much greater than the decay time of the CRT phosphor, the screen will show a steady display of both the waveforms at channels *A* and *B*. *This mode can not be used for displaying very low frequency signals.*
  - — In the ***chop mode***, the electronic switch free runs at a high frequency of the order of 100 kHz to 300 kHz. The result is that small segments from channels *A* and *B* are connected alternately to the vertical amplifier, and displayed on the screen. Provided the chopping rate is much faster than the horizontal sweep rate, the display will show a continuous line for each channel. If the sweep rate approaches the chopping rate then the individual segments will be visible and the alternate mode should now be used.

## 2.5.14. Dual Beam Oscilloscope

Fig. 2.48 illustrates a dual beam oscilloscope.

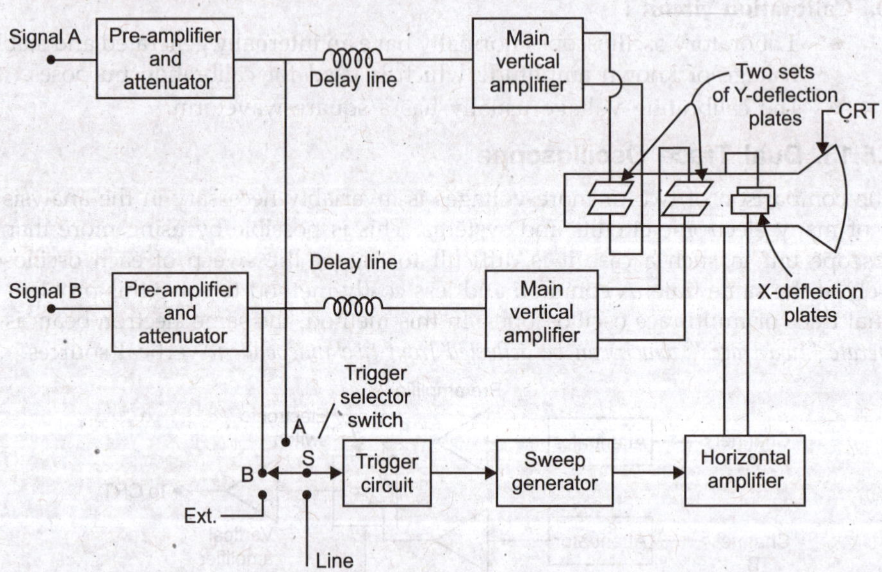


**Fig. 2.48.** Block diagram of a dual beam CRO.

- The *dual beam CRO* uses two completely separate electron beams, two sets of *Y*-deflection plates and a single set of *X*-deflection plates (whereas the dual trace oscilloscope has one cathode ray gun, and an electronic switch which switches two signals to a single vertical amplifier).

- Only one beam can be synchronized at one time, since the sweep is the same for both signals, *i.e.*, a common time base is used for both beams. Therefore, the signals must have the same frequency or must be related harmonically, in order to obtain both beams locked on the CRT screen, *e.g.*, the input signal of an amplifier can be used as signal *A* and its output signal as signal *B*.
- ***Multiple beam oscilloscope*** has a *single tube but several beam producing systems inside. Each system has separate vertical deflecting pair of plates and generally a common time base system*. The triggering can be done internally using either of the multiple inputs or externally by an external signal or line voltages.

**Comparison of dual trace and dual beam oscilloscopes :**

| *S. No.* | *Aspects* | *Dual trace CRO* | *Dual beam CRO* |
|---|---|---|---|
| 1. | *No. of electron beams used.* | One (to generate two traces) | Two |
| 2. | *No. of vertical amplifiers used.* | One | Two |
| 3. | *Display of two signals.* | Not simultaneous (the two signals are not displayed simultaneously in real time but appears to be displayed simultaneously. | Simultaneous |
| 4. | *Switching between traces.* | Difficult (since beam is shared between the two signals) | Easy |
| 5. | *Frequencies of signals.* | Different | Same frequency or their frequencies must be integer multiples of each other |
| 6. | *Size and weight.* | Less | More |
| 7. | *Modes of operation.* | Alternate and chop | The two different types are using double gun tube or split beam using single electron gun |

## 2.5.15. Storage Oscilloscope

In the conventional CRT the persistence of the phosphor ranges from a *few milliseconds to several seconds*, so that an event that occurs only once will disappear from the screen after a relatively short period of time.

- A "*storage CRT can retain the display much longer, upto several hours after the image was first written on the phosphor*. This *retention* feature can also be useful when displaying the waveform of a very low-frequency signal. In the conventional (non-storage) oscilloscope, the start of such a display would fade before the end is written.
- A storage oscilloscope uses the phenomenon of "*secondary electron emission" to build up and store electrostatic charges on the surface of an insulated target*.
  — Storage CRT tubes can be classified as : *Bistable tubes* and *halftone tubes*. The "*bistable tube*" will either store or not store an event and produces only one

level of image brightness. The *"halftone tube"* can retain an image for varying lengths of time (variable persistence) and at different levels of image brightness.

- The storage oscilloscopes are specially useful in the following cases :
  (*i*) For real-time observation of events that occur only once.
  (*ii*) For displaying the waveform of a very low-frequency signal.

### 2.5.16. Sampling Oscilloscope

Above the range of 50-300 MHz, *sampling techniques* have to be employed to obtain suitable display and CRO employing such sampling methods are called *"sampling oscilloscopes"*. The display is made up from as many 1000 dots of luminescence. The vertical deflection for each dot is obtained from progressively later points in each successive cycles of input waveform as shown in Fig. 2.49. The horizontal deflection of the electron beam is obtained by applications of a staircase waveform to X-deflection plates. Fig. 2.50 shows the display of dots.

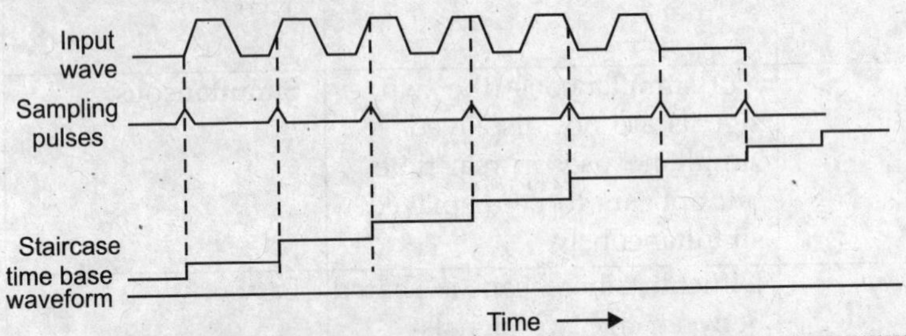


**Fig. 2.49.** Sampling principle.

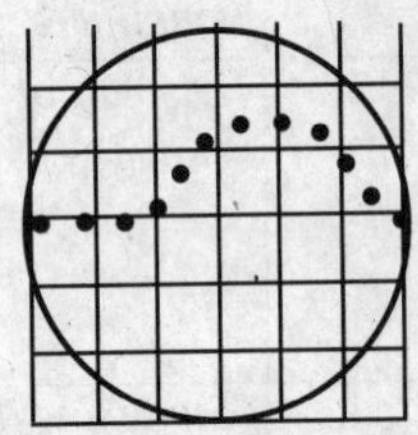

**Fig. 2.50.** Display of dots.

Fig. 2.51 shows the block diagram of a sampling oscilloscope :

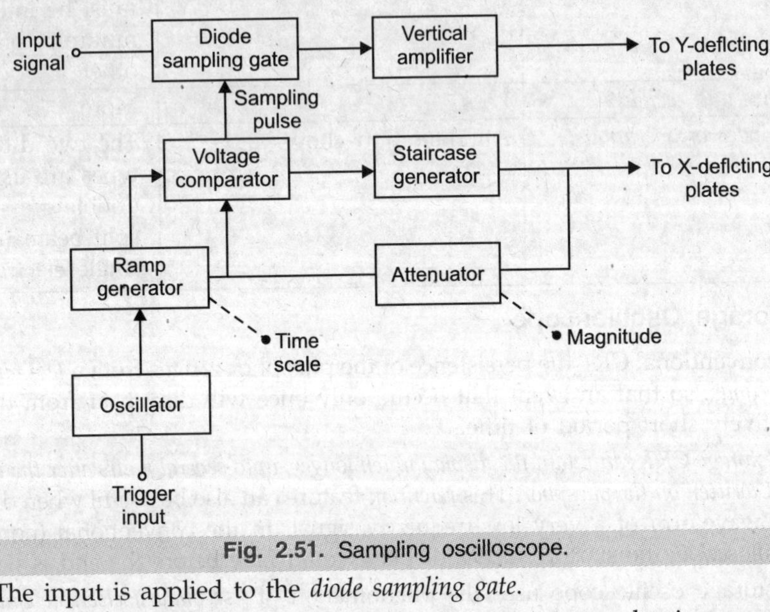


**Fig. 2.51.** Sampling oscilloscope.

— The input is applied to the *diode sampling gate*.

— At the start of each sampling cycle a trigger input pulse is generated which activates the blocking *oscillator*. The oscillator output is given to the *ramp generator* which generates the linear ramp signal.

— Since the sampling must be synchronized with the input signal frequency, the signal is delayed in the *vertical amplifier*.

— The *staircase generator* produces a staircase waveform which is applied to an *attenuator*. The attenuator controls the magnitude of the staircase signal and then it is applied to a *voltage comparator*. Another input to the voltage comparator is the output of the ramp generator. The voltage comparator compares the two signals and produces the output pulse when the two voltages are equal. This is nothing but a *sampling pulse* which is applied to sampling gate through the gate control circuitry. This pulse opens the diode gate and the sample is taken in. This sampled signal is then applied to the vertical amplifier and the *Y*-deflecting plates.

— The output of the staircase generator is also applied to the *X*-deflecting plates. During each step of staircase the spot moves on the screen. *The comparator output advances the staircase output through one step*. After the certain number of pulses about 100 or so, the staircase generator resets. The smaller the size of the steps of the staircase generator, larger is the number of samples and higher is the resolution of image.

— The sampling oscilloscope responds and stores rapid bits of information and displays them continuously. It is this quantity of sampling oscilloscopes which has limited sensitivity, bandwidth and small display size. The *sampling method transforms the input signal into lower frequency range, where low frequency circuitry is then capable of producing an effective display*.

**Advantages :**

1. The display produced is clear.
2. High speed electrical signals can be produced.
3. Very high frequency performance can be achieved.
4. By controlling the size of the steps of the staircase generator, the number of samples and hence the resolution can be controlled.
5. The sampling technique allows the design of the oscilloscope with wide bandwidth, high sensitivity even for low duty cycle pulses.

**Limitation.** This oscilloscope *cannot be used to display* the *transient waveforms*.

### 2.5.17. Digital Read Out Oscilloscope

This oscilloscope provides readout of the signal information such as voltage, time etc., in addition to the conventional CRT display.

It consists of a *high-speed laboratory CRO and an electronic counter, both in one cabinet*.

Fig. 2.52. shows the block diagram of a digital CRO for voltage measurement :

— The input signal is first sampled using sampling circuit.

— The sampled signal is given as one output to each start and step comparators. When input is sampled, the sampling circuit advances the sampling position by a fixed amount. This process is known as *strobing*. The selected sweep time per cm control and number of samples taken per cm decide the equivalent time between two samples.

— The CRT trace is used to identify 0% and 100% zone positions. The positions can be shifted anywhere on the display. The potential divider is used which taps the voltage between 0% and 100% levels. The 0% level is used to produce a pulse for opening of gate while the 100% is used to produce a pulse of closing of gate.

— The coincidence of any of the input waveforms with the selected percentage point is sensed by the voltage comparator. When 0% level gives start pulse, clock gate opens and counter starts counting the pulses. When 100% level gives stop pulse, clock gate closes and counter stops counting.

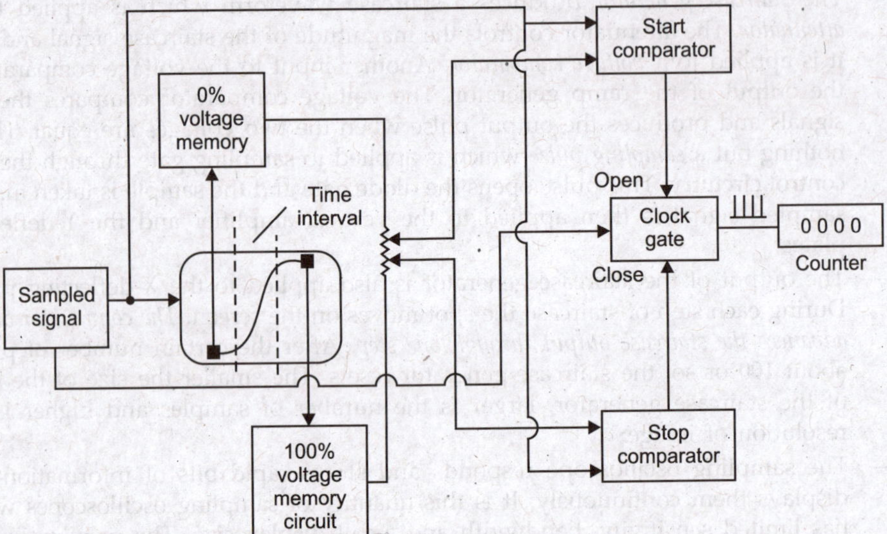


**Fig. 2.52.** Block diagram of a digital CRO for voltage measurement.

— *The number of pulses counted by a counter are proportional to actual sample taken*. The digital readout is obtained by using Nixie display tube.

### 2.5.18. Digital Storage Oscilloscope

Following are the distinct *disadvantages of the storage cathode ray tube* :

(*i*) The storage tube can preserve a stored waveform for a finite amount of time; eventually, waveform is lost. The power to the storage tube must be present as long as the image is to be stored.

(*ii*) The trace of a storage tube is, generally, not as fine as a normal cathode ray tube.

(*iii*) The writing rate of the storage tube is less than a conventional cathode ray tube, which limits the speed of the storage oscilloscope.

(*iv*) The storage cathode ray tube is considerably more expensive than a conventional tube and requires additional power supplies.

(*v*) Only one image can be stored. If two traces are to be compared, they must be superimposed on the same screen and displayed together.

The "*digital storage oscilloscope*" is a superior method of trace storage. In this technique, the waveform to be stored is *digitized, stored in a digital memory, and retrieved for display on the storage oscilloscope*. The stored display can be displayed indefinitely as long as power is applied to the memory, which can be supplied with a small battery.

Fig. 2.53 shows the block diagram of a digial storage oscilloscope :

— The input is amplified and attenuated with input amplifiers (as in any oscilloscope).

— The digital storage oscilloscope uses the same types of input circuitry and oscilloscope probes as a conventional oscilloscope, and many digital storage oscilloscopes can operate in a conventional mode, bypassing the digitizing and storing features.

— The output of the input signal amplifiers feeds an A/D (analog-to-digital) converter. The main requirement of the A/D converter is its speed, while accuracy and resolution are of secondary impedance.

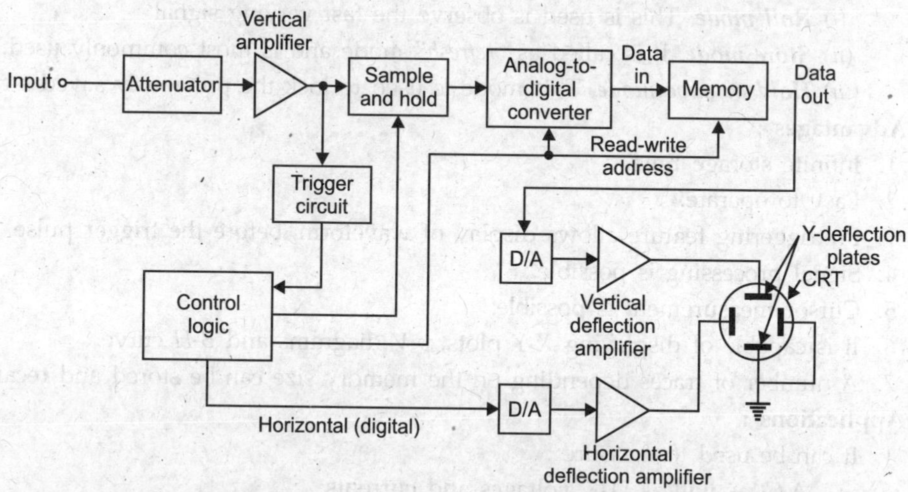


**Fig. 2.53.** Digital storage oscilloscope.

— The digitized output is in the binary form and not BCD form. Generally, successive approximation type of ADC is used in digital storage oscilloscope. If a 12 bit converter is used then accuracy of about 0.1% and resolution of about 0.025% is obtained, while if a 10 bit ADC is used then the resolution of about 0.1% is obtained.

— Digitizing the analog signal means to take samples of the input signal at periodic intervals of time. The sampling rate should be greater than twice the highest frequency present in the input signal, so that the sampling theory is satisfied. This ensures that aliasing effect will not occur and there is no loss of information. The sampling rate may be as high as 100,000 samples per second. For such high sampling rates, a fast conversion A/D convertor is required. *'Flash' A/D converters* are used at the cost of decreased resolution with increase in sampling rate.

— The size of the memory is related to the amount of horizontal segments of the trace that can be divided into one sweep of the time base, *e.g.*, if memory size is 1K then the trace will be divided into 1000 horizontal segments. If a 10 bit ADC is used then the frequency response of 25 kHz is obtained; the total memoery storage capacity is 4096 for a single channel, 2048 for two channels, 1024 for four channels each.

— The selection of the sampling rate and memory size depends on the type of waveform being recorded. As the analog signal is sampled at suitable rates, the data points are readable on to the memory. Once the signal is stored, the required manipulations can be done.

— A *continuous storage oscilloscope* consists of a feature called ***pretrigger view***. This indicates that the events that occurred before the trigger input signal was applied, are displayed. This selection is a percentage selection.

This mode is useful when failure occurs.

— *Single shot events* one sometimes transient in nature and hence they are quickly lost. Unless the wavefrom is stored the observer cannot see such events. *Such events are stored in the memory of the digital storage oscilloscope and memory can be rapidly read and on the screen a continuous waveform can be displayed.*

- The digital storage oscilloscope has the following threee modes of operation:
  - (*i*) ***Roll mode.*** This is used is observe the fast varying signal.
  - (*ii*) ***Store mode.*** It is called as "*refresh*" mode and is most commonly used.
  - (*iii*) ***Hold or save mode.*** This mode is used to lock the previously saved signal.

**Advantages :**

1. Infinite storage time.
2. Easy to operate.
3. Pretriggering feature allows display of waveform, before the trigger pulse.
4. Signal processing is possible.
5. Cursor measurement is possible.
6. It is capable of displaying *X-Y* plots, *p-V* diagrams and *B-H* curve.
7. A number of traces depending on the memory size can be stored and recalled.

**Applications :**

1. It can be used to measure :
   — A.C. as will as D.C. voltages and currents;
   — Frequency, time period, time interval between two signals etc.,
   — Inductance, capacitance.
2. It is used to give the visual representation for a target of radar such as aeroplane, ship etc.
3. In medical fields, it is used to display cardiograms that are useful for diagnosis of heart of the patient.
4. It is used to observe the radiation pattern generated by the transmitting antenna.
5. It can be used to check the faulty components in various circuits. It can also be used to analyse TV waveforms.
6. It can be used to determine the modulation characteristics and detect the standing waves in transmission lines.

## 2.5.19. High Frequency Oscilloscope

A general purpose oscilloscope has an upper frequency limits of about *300 kHz*. However, there are many applications which involve measurement or examination of signals whose frequencies are much higher than this upper limit. For such high-frequency (HF) applications conventional HF oscilloscopes are built. These HF oscilloscopes can display signals with frequencies upto *500 MHz*. They differ in their internal construction from low-frequency (LF) oscilloscopes by their special HF CRTs and high-frequency amplifiers.

Fig. 2.54 shows a high frequency CRT.

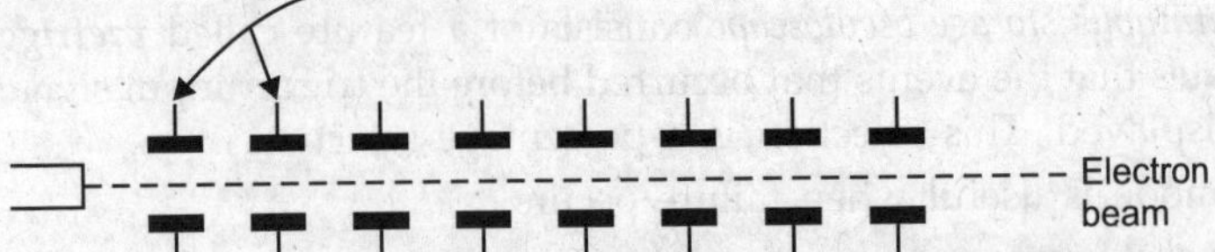


Fig. 2.54. High-frequency CRT.

— In an ordinary cathode ray oscilloscope, there is only one pair of vertical deflection plates. When the HF signal is to be displayed, the electron beam does not get sufficient time to pick up the instantaneous level of the signal. Also, at high frequencies the number of electrons striking the screen in a given time and the intensity of the beam is reduced. Hence, instead of one set of vertical deflection plates, a *series of vertical deflection plates* are employed.

— The vertical deflection plates are so *shaped* and *spaced that an electron moving along CRT receives from each set of plates an additional deflection force in proper time sequence.* This synchronization is obtained by making the signal travel from one plate to the next plate at the *same speed as the transit time of electrons.*

— The signal is applied to each pair of plates, and as the electron beam moves the signal also moves through the delay times. The time delays are so arranged that the same electrons are deflected by the input signal. Thus the electron beam picks up the level of the input signal. *The time delays between the plates correspond exactly to the electron transit times.* In addition, new fluorescent materials have now been developed to increase the brightness at high frequencies.

### 2.5.20. Accessories of Cathode Ray Oscilloscopes

The CRO is one of the most useful instruments in the electronic industry. Its usefulness is further extended by the provision of accessories or auxiliary equipment. Source of the accessories are enumerated and described below :

1. Probes
2. Electronic switch
3. Calibrators
4. Cameras.

#### 2.5.20.1. Probes

The *function of a probe is to connect the test circuit to the oscilloscope without altering, loading, or otherwise disturbing the test circuit.*

Probes are of the following types :

1. Direct probe;
2. Isolation probe;
3. High impedance or 10:1 probe;
4. Active probes;
5. Current probe.

**1. *Direct probe* :**

- It is the simplest of all the probes and uses a shielded co-axial cable.
- It avoids stray pick-ups which may create problems when low level signals are being measured.
- It is usually used for low frequency or low impedance circuits.
- While using the shielded probe, the shunt capacitance of the probe and cable is added to the input impedance and capacity of the oscilloscope and acts to lower response of the oscilloscope to high impedance and high frequency circuits.

**2. *Isolation probe* :**

- An isolation probe is employed *to avoid the undesirable circuit loading effect of the shielded probe.*

- It is made by placing a carbon resistor in series with the test lead, as shown in Fig. 2.55.

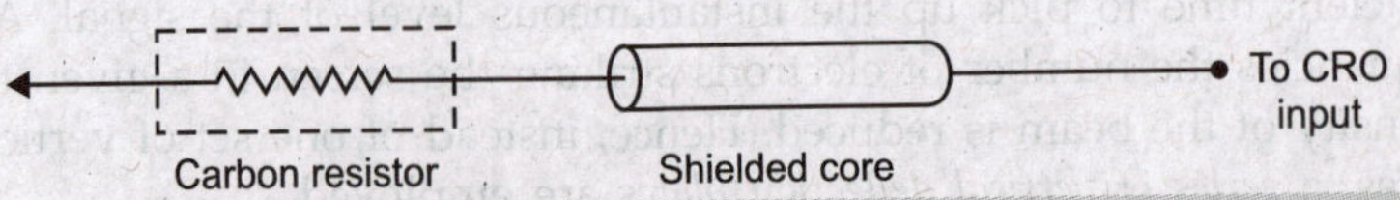


**Fig. 2.55.** Isolation probe.

- This probe causes a slight change in the amplitude of the waveform and a slight change in the wave shape. To avoid this possibility, a high impedance compensated probe, called a *low capacitance* or a 10 : 1 probe is employed.

**3. *High Impedance or 10:1 Probe :***

These probes are used *to increase the input resistance and reduce the effective input capacitance of an oscilloscope.*

Fig. 2.56 (*a, b*) shows the layout of a high impedance or 10:1 probe and its equivalent circuit.

The capacitor is adjusted so that the elements of the bridge are balanced. Under conditions of balance we have :

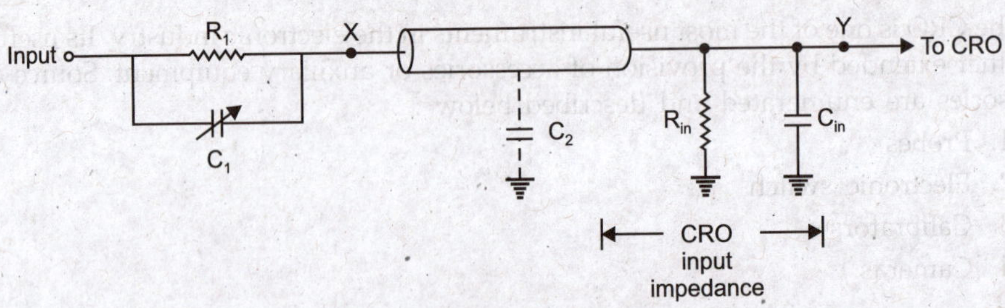


(a) Lay out of a 10:1 probe.

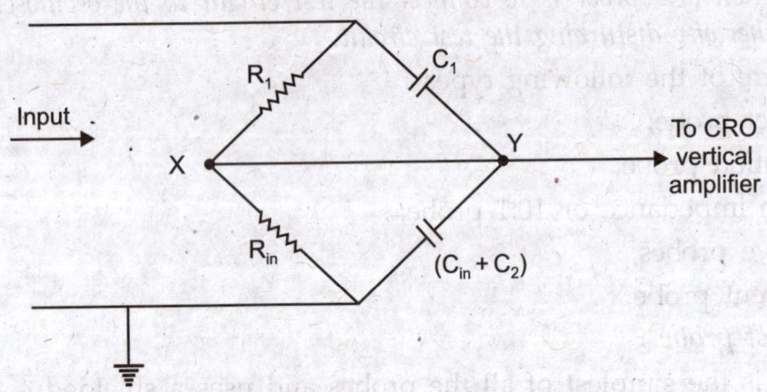


(b) Equivalent circuit of 10:1 probe.

**Fig. 2.56.** High impedance or 10:1 probe.

$$R_1 X_{(C_{in}+C_2)} = R_{in} X_{(C_1)}$$

or,
$$\frac{R_1}{\omega (C_{in}+C_2)} = \frac{R_{in}}{\omega C_1}$$

or,
$$R_1 C_1 = R_{in} (C_{in} + C_2) \quad ...(2.25)$$

Therefore, $X$ and $Y$ are equipotential and the effect of the probe is equivalent to placing a potential divider consisting of $R_1$ and $R_{in}$ across, the input circuit.

- The attenuation of the signal is 10:1, *i.e.*, $\frac{R_1 + R_{in}}{R_{in}} = 10:1$ over a wide frequency range. Hence, it is called a compensated 10:1 probe.
- Capacitor $C_1$ is adjusted by connecting the probe tip to a square-wave of 1 kHz and observing the CRT display. When the CRT display has *optimum response* the value of $C_1$ is deemed to be appropriate.

Then,
$$V_{out} = 0.1\ V_{in} = \frac{V_{in} \times R_{in}}{R_1 + R_{in}} \quad ...(2.26)$$

**4. *Active probes :***

- These probes are designed *to provide an efficient method of coupling high frequency, fast rise time signals to the CRO input.*
- Usually active probes have *very high input impedance with less attenuation than passive probes.*
- Active devices may be diodes, FETs, BJTs, etc.
- These probes are *more expensive and bulky than passive probes, but they are useful for small signal measurements, because their attenuation is less.*

Fig. 2.57 shows a basic circuit of an active probe using a FET and BJT.

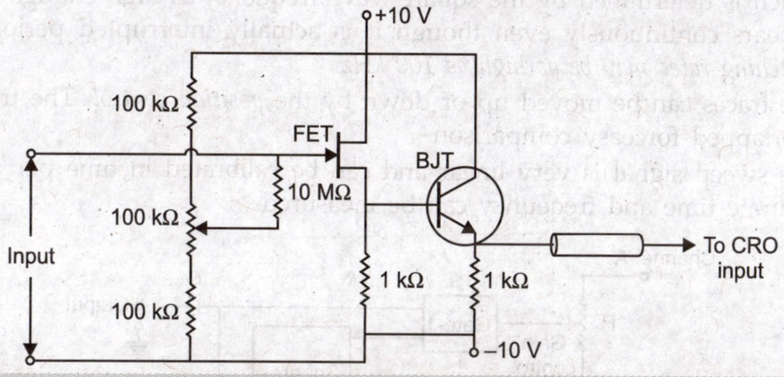


**Fig. 2.57.** FET probe.

- The FET is used as the active element to amplify the input signal. Although voltage gain of FET follower circuit shown is unity, the follower circuit provides a power gain so that input impedance can be increased.
- To be effective the FET must be mounted directly in the voltage probe tip, so that the capacitance of the interconnecting cable can be eliminated. This requires that the power for the FET be supplied from the oscilloscope to the FET in the probe tip. The FET voltage follower drives a coaxial cable but instead of the cable connecting directly to the high input impedance of the oscilloscope it is terminated in its characteristic impedance.
- Because there is no signal attenuation between FET amplifier and probe tip, the range of the signals that can be handled by the FET probe is limited to the dynamic range of the FET amplifier and is typically *less than a few volts*. This is a *significant drawback of the FET probe.* Therefore, for *handling a large dynamic range, external attenuators are added at the probe tip.*
- *Active probes have limited use because the FET probe effectively becomes an attenuator probe. So the oscilloscopes are typically used with 10:1 attenuator probe.*

5. ***Current probe :***
- It is a very valuable probe and can be clamped around a wire carrying current *without any physical contact to the probe,* allowing the oscilloscope to be employed to measure the magnitude of the current with a frequency from 0(D.C.) to 50 Hz.
- The current sensor consits of two parts : A *conventional transformer* for transforming alternating current to voltage, and a *Hall effect device* for converting direct current to a voltage.

### 2.5.20.2. Electronic switch

The electronic switch is an extremely useful accessory for displaying the two signals simultaneously on single trace oscilloscopes. Fig. 2.58 shows the block diagram of an electronic switch :

— Basically, the electronic switch is a *square-wave generator.*

— Two signals which are to be displayed are each applied to the separate gain control and gate stage. The gate stages are alternately biased to cut-off by square-wave signals from the square-wave generator. Thus only one gate stage is in a condition to pass its signal at any given time.

— The output of both stages are applied directly to the oscilloscope input. Amplitudes of the two signals are adjusted by their gain controls. When the switching rate, which is determined by the square-wave frequency is high enough each signal appears continuously even though it is actually interrupted periodically. *The switching rates may be as high as 100 kHz.*

— The traces can be moved up or down by the *position control.* The traces can be overlapped for easy comparison.

— The sweep signal is very linear and can be calibrated in time per mm so that accurate time and freqeuncy can be measured.

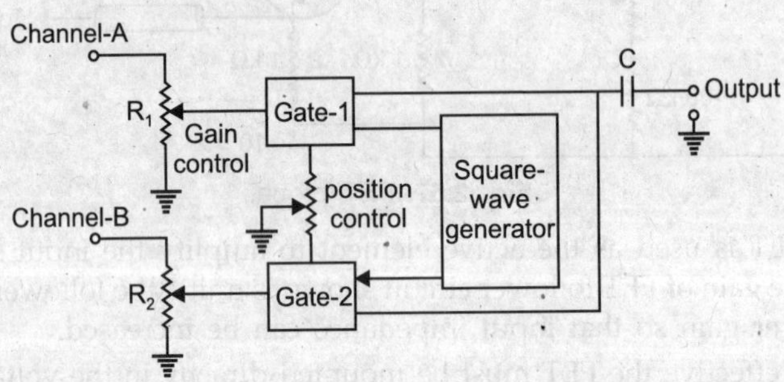


**Fig. 2.58.** Block diagram of electronic switch.

### 2.5.20.3. Calibrators

Several oscilloscopes have a *built-in reference source of voltage* which has usually a square waveform with a frequency of 1 kHz. The voltage may have a single magnitude or may be of several selectable magnitudes. *Provision of this facility enables the oscilloscope amplifiers and time base to be calibrated and checked for accuracy.*

### 2.5.20.4. Cameras

The use of photography is the best method of permanent records of oscilloscope traces for analysis.

- The special cameras used for keeping the records are of the following two types :

(*i*) 35 mm film; (*ii*) Polaroid film.

- *35 mm unit* is usually prefocused and bolted over the tube face. Some cameras have a fixed lens aperture, the exposure is simply controlled by hand operation of flapper shutter.
- *Polaroid film* photographic technique approaches *instant photography*. A permanent record is obtained with 10–20s. However, this process is comparatively *expensive*.

## 2.5.21. CRO Measurements

CRO can be used to measure various characteristics of an input signal and the properties of the output signal. The various parameters which can be measured by using CRO are

(*i*) Voltage; (*ii*) Current;
(*iii*) Period; (*iv*) Frequency;
(*v*) Phase; (*vi*) Amplitude;
(*vii*) Peak-to peak value (*viii*) Duty cycle etc.

### 2.5.21.1. Voltage measurement

The procedure for voltage measurement is as follows :

— The voltage to be measured is applied to Y-deflection plates through the vertical amplifier. Generally, a coaxial cable is used for connection.
— The X-deflection plates are excited by the time-base generator.
— To aid the measurement, a set of squares is marked on the transparent screen cover. Using these markings and the gain of the vertical amplifier, the peak-to-peak (p-p) value (and hence r-m.s. value) of the voltage wave can be measured.
— After noting down the selection in *volts/division* from the front panel (selected for measurement) the peak-to-peak value, amplitude and r.m.s value of sinusoidal signal can be obtained as given below :

(*i*) Peak-to-peak value, $V_{p-p} = \left(\frac{\text{Volts}}{\text{div.}}\right) \times \text{(number of divisions)}$

(The volts/div. is nothing but *deflection sensitivity* of CRO)

(*ii*) Amplitude, $V_{max} = \frac{V_{p-p}}{2}$

(*iii*) R.M.S. value, $V_{r.m.s.} = \frac{V_{p-p}}{2\sqrt{2}}$ (for sinusoidal signals only)

- It must be remembered that an oscilloscope is very complex device and that the accuracy of measurements is not better than ± 3%. It is basically a *display device so that we can observe the waveform in detail and find out the extent of harmonics in the wave.*

**Example 2.10.** *For the waveform shown in Fig. 2.59, if vertical attenuation is 3 mV/div., determine :*

(*i*) *Peak-to-peak value,*
(*ii*) *Amplitude, and*
(*iii*) *R.M.S. value of the signal.*

**Solution.** Vertical attenuation = 3 mV/div.

From the figure, it is evident that the screen is divided such that one part is subdivided into 5 units.

$\therefore$ 1 subdivision $= \frac{1}{5} = 0.2$ units (see Fig. 2.59)

(*i*) **Peak-to-peak value-$V_{p-p}$ :**

It can be observed that positive peak of the signal corresponds to two full divisions and three subdivisions. Hence positive peak is $2 + 3 \times 0.2 = 2.6$ units. While the negative peak also corresponds to 2.6 units.

$$\therefore \quad V_{p-p} = 2.6 + 2.6 = 5.2 \text{ divisions}$$

$$\text{Also,} \quad V_{p-p} = \frac{\text{Volts}}{\text{division}} \times \text{No. of divisions} = 3 \times 5.2 = \mathbf{15.6\ mV\ (Ans.)}$$

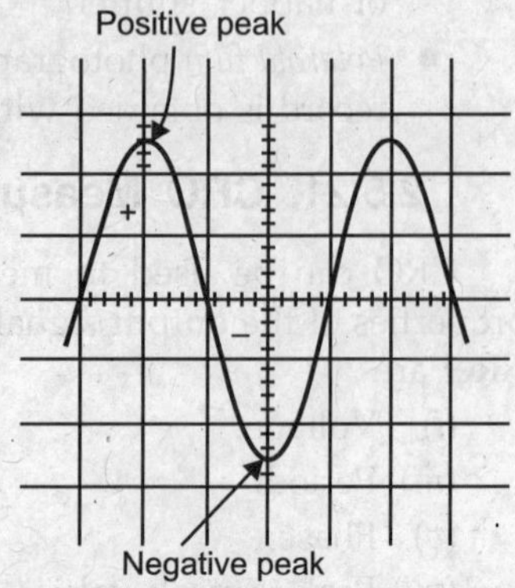


Fig. 2.59.

(*ii*) **Amplitude, $V_{max}$ :** $V_{max} = \frac{V_{p-p}}{2} = \frac{15.6}{2} = \mathbf{7.8\ mV\ (Ans.)}$

(*iii*) **R.m.s. value, $V_{r.m.s.}$ :** $V_{r.m.s.} = \frac{V_{max}}{\sqrt{2}} = \frac{7.8}{\sqrt{2}} = \mathbf{5.515\ mV\ (Ans.)}$

### 2.5.21.2 Current measurement

A CRO has a very high input impedance and cannot be used for direct measurement of current. However, the current can be measured in terms of voltage drop across a standard resistance as shown in Fig. 2.60; *i.e.*, $I = \frac{V \text{ (measured on CRO)}}{R}$.

- For measurement of *high frequency currents, the shunt should be non-inductive.*

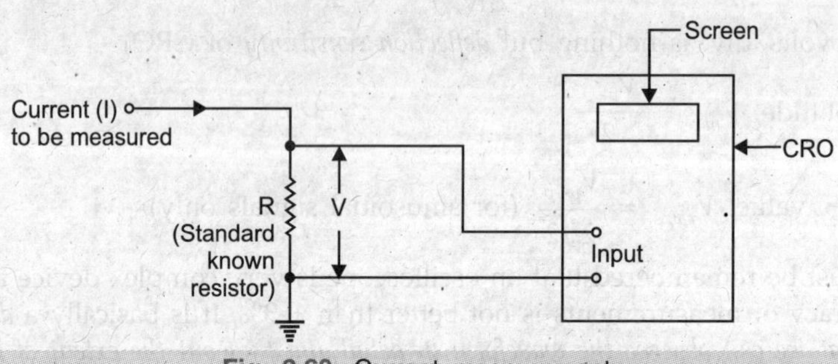


Fig. 2.60. Current measurement.

### 2.5.21.3. Time period measurement

For the measurement of time period (*T*) of the waveform, it is displayed on the screen such that *one complete cycle is visible on the screen*. The accuracy increases if a single cycle occupies as much as the horizontal distance on the screen.

After noting the time/division selected on the front panel, the time period of the waveform can be obtained as :

$$\text{Time period,} \quad T = \frac{\text{Time}}{\text{Division}} \times \text{number of division occupied by } \textit{one cycle}.$$

The frequency ($f$) is then given as :

$$f = \frac{1}{T}$$ (This is the method of frequency measurement without Lissajous pattern)

### 2.5.21.4. Lissajous figures/patterns

***Lissajous patterns*** (named in the honour of French scientist who first obtained them geometrically and optically) *are formed when two sine waves are applied simultaneously to the vertical and horizontal deflecting plates of CRO*. The two sine waves may be obtained from two audio oscillators. Obviously, in this case, a *sine wave sweeps a sine-wave input signal.*

- The shape of the Lissajous pattern depends on the frequency and phase relationship of the two sine waves.
- Two sine waves of the *same* frequency and amplitude may produce a line, an ellipse or a circle depending on their *phase difference* as shown in Fig. 2.61.

  In general, the shape of Lissajous figures depend on :

(*i*) Amplitude ;
(*ii*) Phase difference ;
(*iii*) Ratio of frequency of the two waves.

**Uses.** The Lissajous figures are used for the following purposes :

1. To determine an unknown frequency by comparing it with a known frequency.
2. To check audio oscillator with a known-frequency signal.
3. To check audio amplifiers and feedback networks for phase shift.

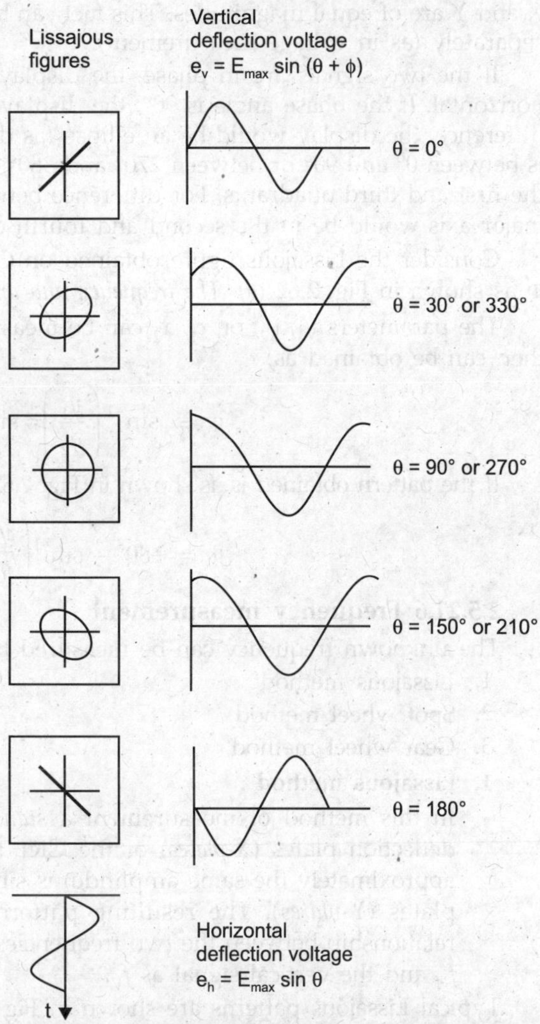


**Fig. 2.61.** Lissajous patterns for same frequency, different phase shift.

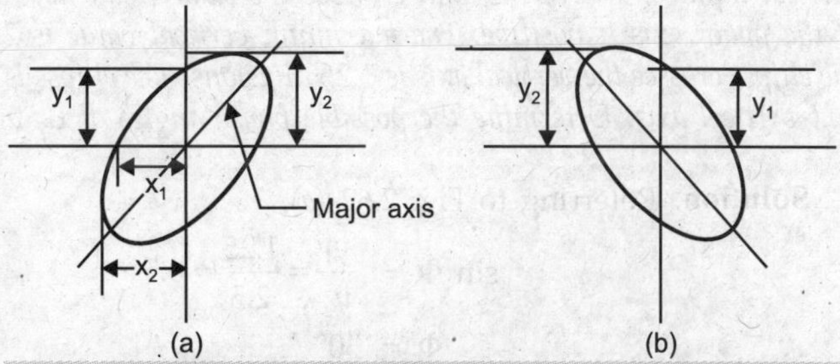


**Fig. 2.62.** Measurement of phase angles by using CRO.

### 2.5.21.5. Phase-angle measurement

An oscilloscope can be used to find the phase angle between the two sinusoidal quantities of the same frequency. One of the signals is fed to the *Y*-plates. The time-base generator is switched out and the second signal is fed to the *X*-plates. It is necessary that *X* and *Y* are of equal magnitudes. This fact can be checked by measuring the amplitudes separately (as in voltage measurement).

If the two signals are in phase, the display would be a straight line at 45° to the horizontal. If the phase angle is 90°, the display would be a circle. For any other phase difference, the display would be an ellipse (as discussed earlier). If the phase difference is between 0° and 90° or between 270° and 360°, the ellipse would have its major axis in the first and third quadrants. For difference between 90° and 180° or 180° and 270°, the major axis would be in the second and fourth quadrants.

Consider the Lissajous figure obtained on CRO with an unknown phase difference Φ as shown in Fig. 2.62 (*a*). *The frequency and amplitude of the two waves are same.*

The parameters $y_1$, $y_2$, or $x_1$, $x_2$ can be measured (See Fig. 2.62 (*a*); the phase angle then can be obtained as,

$$\Phi = \sin^{-1}\left(\frac{y_1}{y_2}\right) = \sin^{-1}\left(\frac{x_1}{x_2}\right)$$

If the pattern obtained is as shown in Fig. 2.62 (*b*), then the phase angle Φ is given by:

$$\Phi = 180° - \sin^{-1}\left(\frac{y_1}{y_2}\right)$$

### 2.5.21.6 Frequency measurement

The unknown frequency can be measured by the following methods :

1. Lissajous method.
2. Spot wheel method.
3. Gear wheel method.

**1. Lissajous method :**

In this method of measurement a *standard frequency* is applied to one set of deflection plates (*X-plates*) of the CRT tube while the unknown frequency (of approximately the same amplifide) is simultaneously applied to the other set of plates (*Y-plates*). The resulting patterns depend on the integral and phase relationship between the two frequencies. The horizontal signal is designated as $f_H$ and the vertical signal as $f_V$.

Typical Lissajous patterns are shown in Fig. 2.63 (*a* & *b*) for sinusoidal frequencies which are equal, integral and in ratio.

**Example 2.11.** *Voltage $E_1$ is applied to the horizontal input and voltage $E_2$ is applied to the vertical input of a CRO. $E_1$ and $E_2$ have the same frequency. The trace is an ellipse. The slope of the major axis is positive. The maximum vertical value is 2.5 divisions and the point where the ellipse crosses the vertical axis is 1.25 divisions. The ellipse is symmetrical about the horizontal and vertical axis. Determine the possible phase angles of $E_2$ with rasped to $E_1$.*

**(Nagpur University, W-2005)**

**Solution.** Referring to Fig. 2.62 (*a*),

$$\sin \Phi = \frac{y_1}{y_2} = \frac{1.25}{2.5} = 0.5$$

$$\therefore \quad \Phi = 30°$$

Thus possible phases are **30°** or **330° (Ans.)**

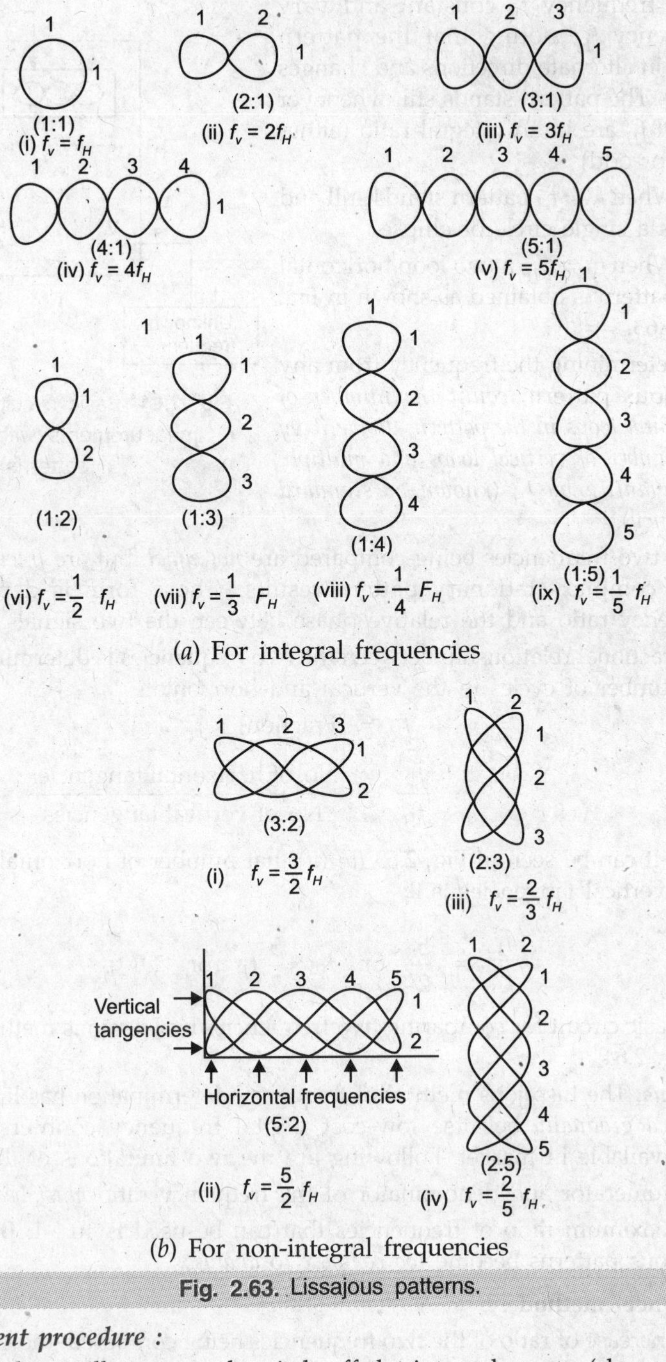


**Fig. 2.63.** Lissajous patterns.

***Measurement procedure :***

- Set up the oscilloscope and switch off the internal sweep (change to Ext.)
- Switch off sync. control.
- Connect the signal source as shown in Fig. 2.64.
- Set the horizontal and vertical gain control for the desired width and height of the pattern.

- Keep frequency $f_V$ constant and vary frequency $f_H$, noting that the pattern spins in alternate directions and changes shape. The pattern stands still whenever $f_V$ and $f_H$ are in an integral ratio (either even or odd).
  - When $f_V = f_H$ pattern stands still and is a single circle or ellipse.
  - When $f_V = 2f_H$, a two loop horizontal pattern is obtained as shown in Fig. 2.63.
- For determining the frequency from any Lissajous pattern, *count the number of horizontal loops in the pattern, divide it by the number of vertical loops and multiply this quantity by $f_H$ (known or standard frequency).*

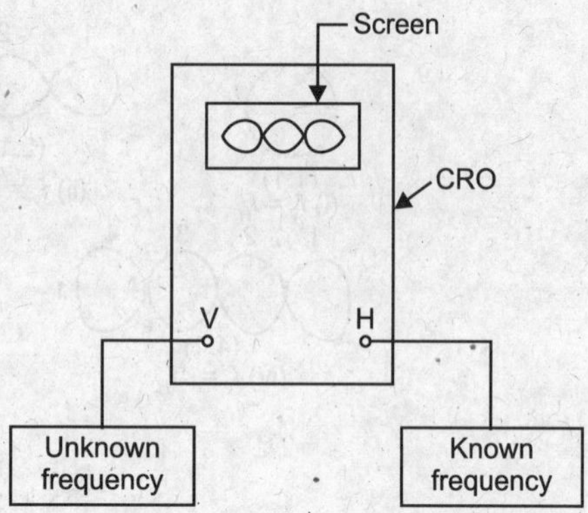


**Fig. 2.64.** Basic circuit for frequency measurements with Lissajous patterns.

- If the two frequencies being compared are *not equal,* but are *fractionally related,* a more complex stationary pattern results, whose form is dependent on the frequency ratio and the relative phase between the two signals.

  The fractional relationship between the two frequencies is determined by counting the number of cycles in the vertical and horizontal.

$$f_V = (\text{Fraction}) \times f_H$$

or,

$$\frac{f_V}{f_H} = \frac{\text{No. of horizontal tangencies}}{\text{No. of vertical tangencies}}$$

**Example.** It can be seen in Fig. 2.63 (*b*) (*ii*) that number of horizontal tangencies is 5 and those of vertical tangencies is 2;

Hence,

$$\frac{f_V}{f_H} = \frac{5}{2}, \text{ or } f_V = \frac{5}{2} f_H \text{ or } 2.5\, f_H.$$

- The basic circuit for comparing two frequencies by Lissajous method is illustrated in Fig. 2.64.

***Limitations.*** The Lissajous method of frequency determination has limitations and is being *discarded gradually* because low-cost digital frequency counters are becoming increasingly available in market. Following are the two limitations of this method :

1. The numerator and denominator of the frequency ratio *must be whole number.*
2. The maximum ratio of frequencies that can be used is 10 : 1. Beyond that, the Lissajous patterns become *too complex* to *analyse.*

**2. Spot wheel method :**

With the increase of ratio of the two frequencies being compared, the Lissajous pattern becomes more complicated hence more difficult to interpret. In such cases, it is advantageous to use spot wheel method of display.

**Fig. 2.65** illustrates the *spot wheel method of frequency determination* :

— *Fig. 2.65 (a)* shows the *basic circuit* for frequency measurement, whereby an intensity modulated circular or elliptical pattern is produced on the face of the CRT.

— The *lesser* of the two frequencies being compared is applied to the deflecting plates of the CRT through resistance-capacitance phase shifter as illuspraed, thereby producing the circular or elliptical sweep.

— The higher frequency is applied to the control grid of the CRT (biased slightly below cut off), as shown in *Fig. 2.65* (*b*), thus modulating the intensity of beam. As a consequence, the pattern appears as a series of alternate bright and dark spots, as shown in *Fig. 2.65* (*c*), where the ratio of high to low frequencies is given by the *number of bright spots*.

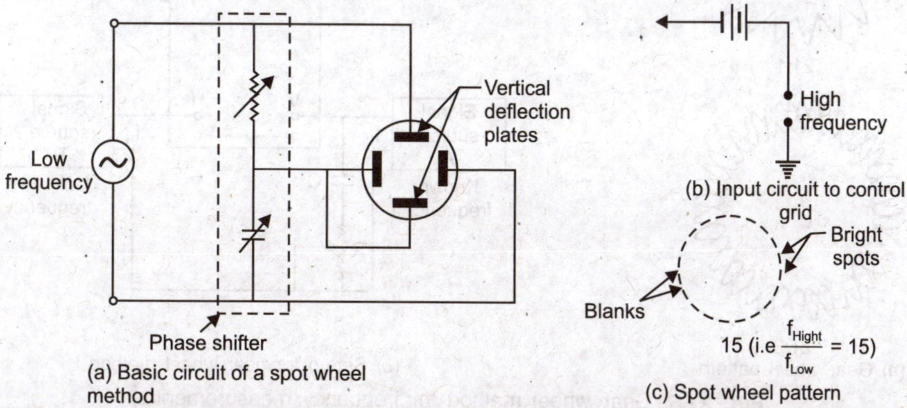


**Fig. 2.65.** Spot wheel method for frequency measurements.

— When the voltages across *R* and *C* are equal, the pattern is circle; if they are unequal, the pattern is elliptical.

— The variation of the grid-cathode potential changes the density of the electron stream within the CRT and *determines the intensity*. When the grid bias of the CRT reaches cut-off potential, there is no electron stream and fluorescence at the screen cannot occur. *This property is used for the comparison of two frequencies, provided their ratio is an integer. The high frequency signal is applied to the grid.* Sufficient grid must be present so that the negative peaks of the sine wave cut off the electron beam.

— The high frequency signal need not be a pure sine wave, it can be a square wave.

Fig. 2.65(*c*) shows that the frequency ratio is 15 : 1 $\left(i.e. \frac{f_{High}}{f_{Low}} = 15\right)$ and there are 15 blanks in the pattern :

**3. Gear wheel method :**

When the number of loops in a Lissajous figure is large, the accurate counting becomes difficult.

- A test method that uses a *modulated ring pattern* in place of looped figure and permits a higher count is shown in Fig. 2.66 (a). This pattern is also called a 'gear wheel' or 'toothed wheel', because of its shape. The unknown frequency is determined by multiplying the known frequency by the number of teeth in the pattern.
- Fig. 2.66 (b) shows the set-up for gear wheel method for frequency measurements :
  — In this method, a phase shift network (RC) introduces a 90° phase shift

between the horizontal and vertical channels of the oscilloscope, which is needed to produce a ring or circle pattern with a known frequency $f_V$. This ring is *modulated* by the voltage from the unknown frequency.

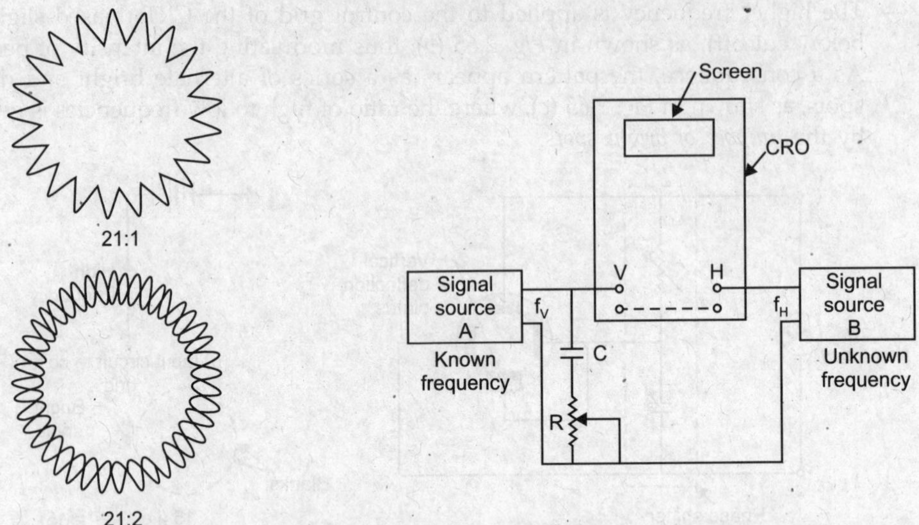


(*a*) Gear wheel pattern (*b*) Set-up for gear wheel method

**Fig. 2.66.** Gear wheel method for frequency measurements.

— When the *voltages across* and $R$ and $C$ are unequal, the pattern is elliptical. If they are equal, the pattern is circle.

— *The unknown frequency must be higher than the known frequency and the amplitude of the unknown must be below that of the known to prevent distortion.*

***Measurement procedure :***

- After setting up the oscilloscope, switch off the and Sync–control and connect the circuit as shown in Fig. 2.66 (b). Switch off the signal $B$ temporarily.
- *Switch on the signal source A* and adjust $R$ to obtain a ring pattern on the screen. Adjust the horizontal and vertical gain controls to spread the ring over the screen as much as possible.
- *Switch on the signal source B*, noting that the ring becomes *toothed by the signal.* Adjust the known frequency $f_V$ to stop the ring from rotating. Adjust the amplitude of $f_H$ to obtain a distinct teeth pattern. The unknown frequency is then given by :

$$f_H = nf_V$$

where $f_H$ is an integral multiple (odd or even) of $f_V$.

— *The wheel rotates or spins if $f_H$ is not exact multiple of $f_V$, therefore adjust the known frequency to stop the wheel from rotating.*

### 2.5.21.7. Dielectric loss measurement

Fig. 2.67 (*a*) shows the circuit diagram or measurement of dielectric loss by cathode ray oscilloscope :

- C is capacitor formed by dielectric material under test and $C_s$ is the *standard capacitor of negligible loss;* these capacitors C and $C_s$ are connected in series across A.C. supply source.
- The voltage across the capacitor C or a known fraction of it applied to the X-deflection plates and the voltage across the standard capacitor $C_s$ is applied to

the $Y$-deflection plates of CRO. There will be phase difference between the voltages across capacitors $C$ and $C_s$, because voltage across standard capacitor $C_s$ will lag 90° behind the screen flowing through it but the voltage across capacitor $C$ will not *lag exactly 90°* behind the current flowing through it *owing to its losses*.

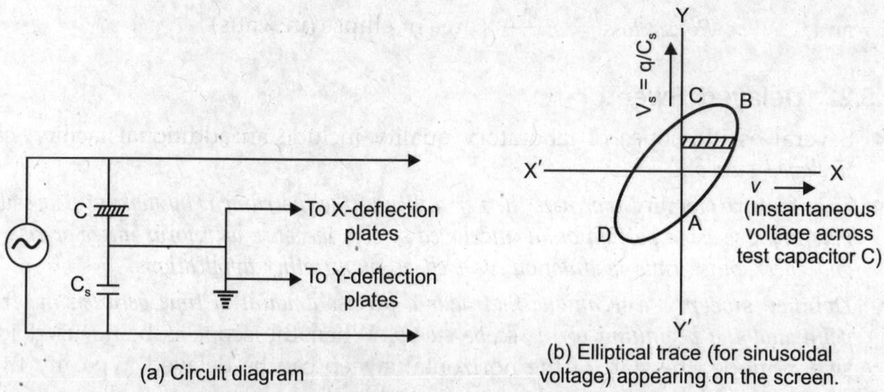


**Fig. 2.67.** Dielectric loss measurement by CRO.

- When the voltage is sinusoidal then the trace appearing on the screen of CRT will be an ellipse, as shown in Fig. 2.67 (*b*), and the *area of this trace will represent the power loss per cycle of the capacitor C under test*. More the power loss in the test capacitor $C$, more will be the phase difference of the two voltages and so more will be the area of the trace.

The work done on the test capacitor $C$ over any period of time is given as :

$$dw = \int vidt = v\int \frac{dq}{dt}.dt = \int v.\,dq$$

where, $v$ = Instantaneous voltage across the test capacitor $C$ and

$q$ = Instantaneous charge on the standard capacitor $C_s$.

If the deflection sensitivity CRO is $x$ and $y$ mm/V in $X$-and $Y$-axes respectively, then the area of the hatched section shown in the figure will be :

$$x.\,v.\,y.\frac{dq}{C_s}\,mm^2$$

— When the *spot moves from point A to point B*, the area between curve $AB$ and the $Y$-axis is

$$\int x.v.y.\frac{dq}{C_s} = \frac{xy}{C_s}\int v.dq = \frac{xy}{C_s} \times \text{work done on capacitor } C$$

— When the *spot moves from point B to point C*, both the voltages across the capacitor $C$ and charge on the capacitor $C_s$ are being reduced so that *energy is being restored* to the circuit and this amount of energy is represented by the curve $BC$ and the $Y$-axis. But the area enclosed by curve $ABC$ and the $Y$-axis represents the work done on the dielectric and *not returned to the circuit*. Similarly the area enclosed by $CDA$ represents work done during the remaining half cycle, so that the *area of the ellipse represents the work done on the dielectric per cycle*.

Since area of ellipse $= \dfrac{xy}{C_x} \times$ work done on capacitor $C$, therefore:

*Work done or dilectric loss per cycle*

$$= \frac{C_s}{x.y} \times \text{area of ellipse}$$

and, *Power loss* $= \dfrac{f.C_s}{x.y} \times \text{area of ellipse (in watts)}.$

### 2.5.22. Delayed Sweep

- Several oscilloscopes of laboratory quality include an additional facility, called *'Delayed sweep'*.
- This feature/facility *increases the versatility of the instrument by making it possible to magnify a selected portion of an undelayed sweep, measure waveform jitter or rise time, and check pulse time modulation, as well as many other applications.*
- ***Delayed sweep*** *is a technique that adds a precise amount of time between the trigger point and the beginning of the scope sweep.* When the scope is being used in the sweep mode, the start of the horizontal sweep can be delayed, typically from a few μs to perhaps 10 seconds or more. Delayed sweep operation allows the users to view a small segment of the waveform, e.g. an oscillation or ringing that occurs during a small portion of a low frequency waveform.

The following are the most common approaches used by the oscilloscope manufacturers for delayed sweep operations :

(*i*) Intensified sweep, where the delayed sweep acts as a positional magnifier.

(*ii*) Normal triggering sweep after the desired time delay, which is set from the panel controls.

(*iii*) A Delay Plus Trigger mode, where a visual indication, such as light, indicates that the delay time has elapsed and sweep is ready to be triggered.

## 2.6. BRIDGE AND OTHER MEASUREMENTS

### 2.6.1. Introduction

For several years the precision measurements of component values have been made by using various forms of bridges.

- Bridge circuits are extensively used for measuring component values such as R, L and C.
- A "bridge circuit" in its simplest form consists of a network of four resistance arms forming a closed circuit, with a D.C. source of current applied to two opposite junctions and a current detector connected to the other two junctions.
- Since the bridge circuit merely compares the value of an unknown component with that of an accurately known component (a standard), its measurement accuracy can be very high. This is because the readout of this comparison is based on the null indication at bridge balance, and is essentially independent of the characteristics of the null detector. The *measurement accuracy is, therefore, directly related to the accuracy of the bridge component and not to that of the null detector used.*
- *Advantages* of the bridge circuits are :

(*i*) High measurement accuracy (as the measurement is done by comparing the unknown value with the standard value).

(*ii*) The accuracy is independent of null detector's characteristics (it depends on the values of the components).

(*iii*) The balance equation is independent of the magnitude of the input voltage or its source impedance, the sensitivity and impedance of the null detector or any impedance shunting the detector.

(*iv*) The interchange of the source and detctor does not affect the balance condition.

(*v*) The bridge circuit can be used in control circuits.

### 2.6.2. Types of Bridges

Bridges are of the following two types :

1. **D.C. bridges :**
   - These bridges are used to measure the *"resistances."*
   - They use the *D.C. voltage* as the excitation voltage.

   ***Examples :*** (*i*) Wheatstone bridges;
   (*ii*) Kelvin bridges.

2. **A.C. bridges :**
   - These bridges are used to measure the *"impedances"* consisting of inductances and capacitances.
   - They use the *alternating voltage* as the excitation voltage.

   ***Examples :*** (*i*) Maxwell's bridge
   (*ii*) Hay's bridge
   (*iii*) Schering bridge
   (*iv*) Anderson bridge etc.

## D.C. BRIDGES

### 2.6.3. Wheatstone Bridge (*for measurement of* "resistance")

The *simplest form of bridge is for the purpose of measuring resistance and is called the "Wheatstone bridge."*

- The Wheatstone bridge is widely used for precision measurement of resistance from *approximately 1Ω to the low megaohm range.*

— The *"upper limit"* is set by the reduction in sensitivity to unbalance, caused by high the resistance values, thus reducing the galvanometer current.

— The *"lower limit"* is set by the resistance of the connecting leads and the contact resistance at the binding posts. The resistance of the leads could be calculated or measured, and the final result modified, but *contact resistance is very hard to compute or measure. For low resistance measurements, therefore, the Kelvin bridge is generally the preferred instrument.*

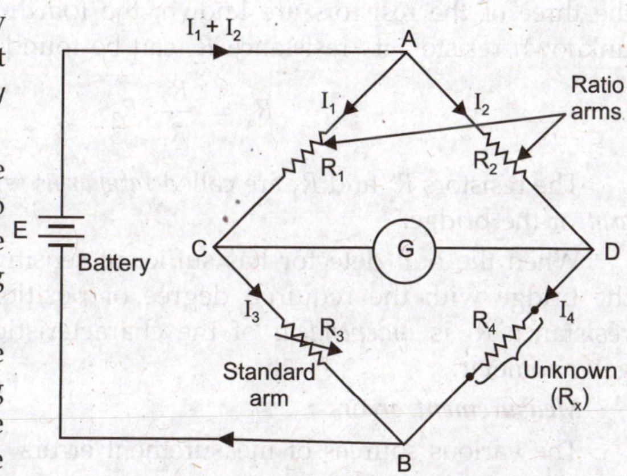


**Fig. 2.68.** Wheatstone bridge.

- *It is the most accurate method available for measuring resistances and is popular for laboratory use.*

**Basic operation :**

Fig. 2.68 shows the basic circuit diagram of a Wheatstone bridge :

— The bridge has four resistive arms, together with a source of e.m.f. (a battery) and null detector, usually a galvanometer (G) or other sensitive current meter.

— The current through the galvanometer depends on the potential difference between the points $C$ and $D$. The bridge is said to be *balanced* when the potential difference across the galvanometer is *zero volt* so that there is no current through the galvanometer. Hence the bridge is balanced when potential difference between points $C$ and $D$ is equal,

*i.e.* $$I_1R_1 = I_2R_2 \quad ...(2.27)$$

When the current through the galvanometer is zero, the following conditions should be satisfied :

$$I_1 = I_3 = \frac{E}{R_1 + R_3} \quad ...(2.28)$$

and, $$I_2 = I_4 = \frac{E}{R_2 + R_4} \quad ...(2.29$$

Substituting the values of $I_1$ and $I_2$ in eqn. (2.27), we get

$$\frac{E}{R_1 + R_3} \times R_1 = \frac{E}{R_2 + R_4} \times R_2$$

or, $$\frac{R_1}{R_1 + R_3} = \frac{R_2}{R_2 + R_4}$$

or, $$R_1(R_2 + R_4) = R_2\,(R_1 + R_3)$$

or, $$R_1R_2 + R_1R_4 = R_1R_2 + R_2R_3$$

or $$R_1R_4 = R_2R_3 \quad ...(2.30)$$

Eqn. (2.30) is the well known expression for balance of the wheatstone bridge. When the three of the resistors are known, the fourth may be determined. Thus if $R_4$ is the unknown resistor, its resistance $R_x$ can be found as follows :

$$R_x = \frac{R_2}{R_1} . R_3 \quad ...(2.31)$$

The resistors $R_2$ and $R_1$ are called *ratio arms* while the resistor $R_3$ is called the *standard arm* of the bridge.

When the null detector has sufficient sensitivity to indicate the balance position of the bridge with the required degree of position, the measurement of the unknown resistance $R_x$ is *independent* of the characteristics or calibration of the null-detecting galvanometer.

***Measurement errors :***

The various sources of measurement errors are :

(*i*) The limiting errors of three known resistances–main measurement error.

(*ii*) Insufficient sensitivity of the null detector.

(*iii*) Changes in resistance of the bridge arms due to heating effect of the current through the resistors.

— In order to check this error the power dissipation in the bridge arms must be computed in advance, particularly when low-resistance values are to be measured, and the *current must be limited to a safe value.*

(*iv*) Thermal e.m.fs. in the bridge circuit or the galvanometer circuit.

— To prevent thermal effects, the more sensitive galvanometers sometimes have copper coils and copper suspension systems to avoid having dissimilar metals in contact with one another and generating thermal e.m.fs.

(*v*) Errors due to the resistance of leads and contacts exterior to the actual bridge circuit. These errors can be *reduced* by using a Kelvin bridge.

**Applications of Wheatstone bridge :**

Following are the applications of Wheatstone bridge :

1. To measure the D.C. resistance of various types of wire either for the purpose of quality control of the wire itself, or of some assembly in which it is used.

***Example*** : *Measurement of the resistance of motor windings, transformers, solenoids, and relay coils.*

2. This bridge is also used extensively by telephone companies and others to locate cable faults.

**Limitations :**

The Wheatstone bridge entails the following *limitations* :

1. In case of measurement of low resistance the resistance of the leads and contacts becomes significant, resulting in introduction of an error. This error can be *eliminated* by using Kelvin's double bridge.
2. While measuring high resistances the resistance of the bridge becomes so large that the galvanometer becomes *insensitive to imbalance.* Consequently, a power supply has to replace the battery and a D.C. VTVM replaces the galvanometer. The Wheatstone bridge *cannot* be used for measuring high resistances in megaohms.
3. The *change in resistance* of the bridge arms due to the heating effect of current through the resistance–the temperature rise causes a change in the value of the resistance, and *excessive current may cause a permanent change in value.*

**Example 2.12.** *Fig. 2.68 consists of the following parameters :*
*$R_1 = 20$ k Ω; $R_2 = 30$ k Ω; $R_3 = 80$ k Ω.*
*Determine the unknown resistance $R_x$.*

**Solution.** Using the bridge balance equation, we have :

$$R_1R_x = R_2R_3$$

or,

$$R_x = \frac{R_2R_3}{R_1} = \frac{30 \times 80}{20}$$

$$= \mathbf{120\ k\ \Omega\ (Ans.)}$$

## 2.6.4. Kelvin Bridge

This bridge is a modified version of Wheatstone bridge and provides greatly increased accuracy in the measurement of *low-value resistances*, generally below 1 Ω.

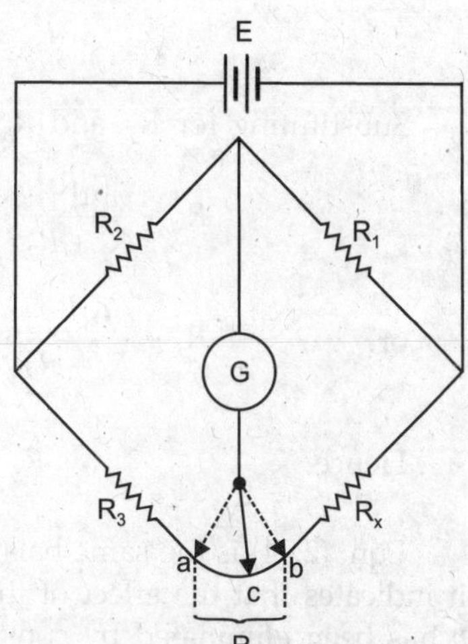


**Fig. 2.69.** Kelving bridge.

Fig. 2.69 shows the schematic of Kelvin bridge :

— $R_y$ represents the *resistance of connecting lead* from $R_3$ to $R_x$

— Two galvanometer connections are available, to point *a* or to point *b*. When the galvanometer is connected to point *a*, the resistance $R_y$

(of connecting lead) is added to the unknown resistor $R_x$, resulting in too high an indication for $R_x$. When connection is made to $b$, $R_y$ is added to bridge arm $R_3$ and the resulting measurement of $R_x$ will be lower than it should be, because now the actual value of $R_3$ is higher than its normal value by resistance $R_y$. When the galvanometer connection is made to a point $c$, in between the two points $a$ and $b$, in such a way that the ratio of the resistances from $b$ to $c$ and from $a$ to $c$ equals the ratio of resistors $R_1$ and $R_2$, then

$$\frac{R_{bc}}{R_{ac}} = \frac{R_1}{R_2} \quad \text{...(2.32)}$$

The balance equation for the bridge yields

$$R_x + R_{bc} = \frac{R_1}{R_2}(R_3 + R_{ac}) \quad \text{...(2.33)}$$

Also, $$R_{ac} + R_{bc} = R_y \quad \text{and} \quad \frac{R_{bc}}{R_{ac}} = \frac{R_1}{R_2}$$

Now, $$\frac{R_{bc}}{R_{ac}} + 1 = \frac{R_1}{R_2} + 1$$

or, $$\frac{R_{bc} + R_{ac}}{R_{ac}} = \frac{R_1 + R_2}{R_2}$$

or, $$\frac{R_y}{R_{ac}} = \frac{R_1 + R_2}{R_2}$$

or, $$R_{ac} = \frac{R_2\, R_y}{R_1 + R_2} \quad \text{...}(i)$$

and, $$R_{bc} = R_y - R_{ac}$$

$$= R_y - \frac{R_2\, R_y}{R_1 + R_2}$$

$\therefore$ $$R_{bc} = \frac{R_1\, R_y + R_2 R_y - R_2 R_y}{(R_1 + R_2)} = \frac{R_1 R_y}{R_1 + R_2} \quad \text{....}(ii)$$

Substituting for $R_{ac}$ and $R_{bc}$ in eqn. (2.33), we get

$$R_x + \frac{R_1 R_y}{R_1 + R_2} = \frac{R_1}{R_2}\left(R_3 + \frac{R_2\, R_y}{R_1 + R_2}\right)$$

or, $$R_x + \frac{R_1 R_y}{R_1 + R_2} = \frac{R_1 R_3}{R_2} + \frac{R_1 R_y}{(R_1 + R_2)}$$

Hence, $$R_x = \frac{R_1 R_3}{R_2} \quad \text{...(2.34)}$$

Eqn. (2.34) is the same balance equation as developed for the Wheatstone bridge and it indicates that the effect of the resistance of the connecting lead from point $a$ to point $b$ has been eliminated by connecting the galvanometer to the intermediate position $c$.

The above principle forms the basis for construction of the Kelvin double bridge, commonly known as *Kelvin bridge*.

**Kelvin double bridge :**

This bridge is so called (*i.e.* double bridge) since the circuit contains a *second set of ratio arms* ($l$ and $m$) as shown in Fig. 2.70. This second set of ratio arms connects the galvanometer to a point $f$ at the appropriate potential between $c$ and $d$ and it eliminates the effect of the yoke resistance $R_y$. The ratio of the resistances of arm $l$ and $m$ is the same as the ratio of $R_1$ and $R_2$.

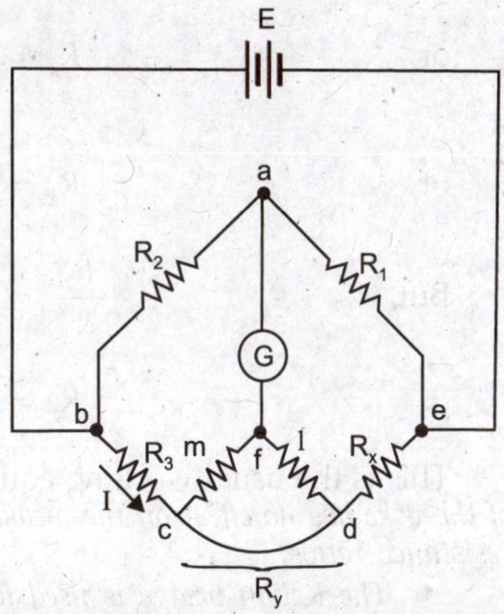


**Fig. 2.70.** Kelvin double bridge.

The galvanometer will read zero when the potential at $a$ equals the potential at $f$.

*i.e.,* $$E_{ab} = E_{bcf}$$

But, $$E_{ab} = \frac{R_2}{R_1 + R_2} \times E \quad ...(2.35)$$

and, $$E = I\left[R_3 + R_x + \frac{(l+m)\,R_y}{(l+m)+R_y}\right]$$

Substituting for $E$ in eqn. (2.35), we get

$$E_{ab} = \frac{R_2}{R_1+R_2} \times I\left[R_3 + R_x + \frac{(l+m)\,R_y}{(l+m)+R_y}\right] \quad ...(2.36)$$

Similarly, $$E_{bcf} = I\left[R_3 + \frac{m}{l+m}\left\{\frac{(l+m)\,R_y}{(l+m)+R_y}\right\}\right] \quad ...(2.37)$$

But, $$E_{ab} = E_{bcf}$$

$$\therefore \quad \frac{IR_2}{R_1+R_2}\left[R_3 + R_x + \frac{(l+m)\,R_y}{(l+m)+R_y}\right] = I\left[R_3 + \frac{m}{l+m}\left\{\frac{(l+m)\,R_y}{(l+m)+R_y}\right\}\right]$$

or, $$R_3 + R_x + \frac{(l+m)\,R_y}{(l+m)+R_y} = \frac{R_1+R_2}{R_2}\left[R_3 + \frac{m\,R_y}{(l+m)+R_y}\right]$$

or, $$R_3 + R_x + \frac{(l+m)\,R_y}{(l+m)+R_y} = \left(\frac{R_1}{R_2}+1\right)\left(R_3 + \frac{mR_y}{l+m+R_y}\right)$$

or, $$R_x + \frac{(l+m)\,R_y}{l+m+R_y} + R_3 = \frac{R_1R_3}{R_2} + R_3 + \frac{m\,R_1R_y}{R_2\,(l+m+R_y)} + \frac{m\,R_y}{l+m+R_y}$$

or, $$R_x = \frac{R_1R_3}{R_2} + \frac{m\,R_1R_y}{R_2\,(l+m+R_y)} + \frac{m\,R_y}{l+m+R_y} - \frac{(l+m)\,R_y}{l+m+R_y}$$

or, $$R_x = \frac{R_1R_3}{R_2} + \frac{m\,R_1R_y}{R_2\,(l+m+R_y)} + \frac{m\,R_y - l\,R_y - m\,R_y}{l+m+R_y}$$

or, $$R_x = \frac{R_1 R_3}{R_2} + \frac{m R_1 R_y}{R_2 (a + b + R_y)} - \frac{l R_y}{l + m + R_y}$$

or, $$R_x = \frac{R_1 R_3}{R_2} + \frac{m R_y}{l + m + R_y}\left(\frac{R_1}{R_2} - \frac{l}{m}\right) \quad \text{...(2.38)}$$

But, $$\frac{R_1}{R_2} = \frac{l}{m}$$

∴ $$R_x = \frac{R_1 R_3}{R_2}$$

This is the usual working equation for the Kelvin bridge. *It indicates that the resistance of the yoke has no effect on the measurement, provided that the two sets of ratio arms have equal resistance ratios.*

- *The Kelvin bridge is used for measuring very low resistances, from approximately 1 Ω to as low as 0.00001 Ω* (accuracy varying from ± 0.05 % to 0.2%).

**Example 2.13.** *In a Kelvin double bridge, there is error due to mismatch between the ratios of outer and inner arm resistances. The following data relate to this bridge :*

*Standard resistance = 100.03μ Ω; Inner ratio arms = 100.31Ω and 200Ω; Outer ratio arms = 100.24Ω and 200Ω; The resistance of connecting leads from standard to unknown resistor is 680μ Ω.*

*Calculate the unknown resistance.*

**Solution.** Refer to Fig. 2.70. *Given* : $R_3$ = 100.03 μ Ω; $l$ = 100.31Ω; $m$ = 200Ω; Ω ; $R_1$ = 100.24 Ω; $R_2$ = 200 Ω; $R_y$ = 680μ Ω

**Unknown resistance, $R_x$ :**

We known that, $$R_x = \frac{R_1 R_3}{R_2} + \frac{m R_y}{l + m + R_y}\left(\frac{R_1}{R_2} - \frac{l}{m}\right) \quad \text{...(2.38)}$$

Inserting the various values, we get

$$R_x = \frac{100.24 \times 100.03 \times 10^{-6}}{200} + \frac{200 \times 680 \times 10^{-6}}{100.31 + 200 + 680 \times 10^{-6}}\left(\frac{100.24}{200} - \frac{100.31}{200}\right)$$

$$= 50.135 \times 10^{-6} + 4.528 \times 10^{-4} (-3.5 \times 10^{-4})$$

$$= 49.97 \times 10^{-6}\ \Omega \quad \text{or} \quad \mathbf{49.97\ \mu\ \Omega\ (Ans.)}$$

**Wheatstone bridge with digital readout :**

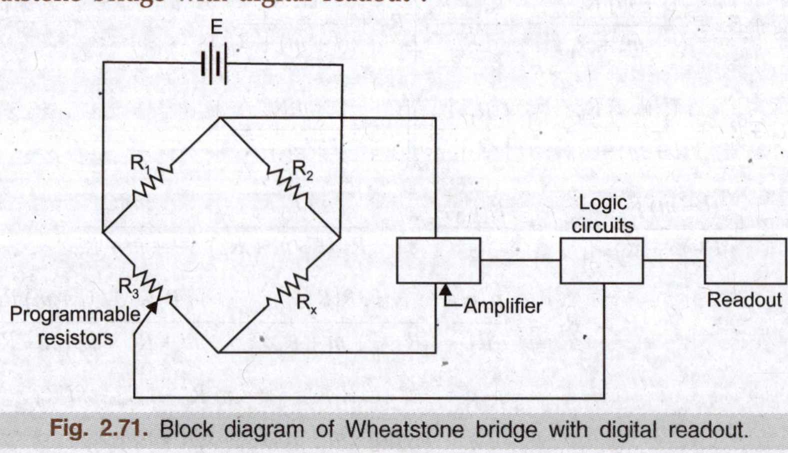


**Fig. 2.71.** Block diagram of Wheatstone bridge with digital readout.

Fig. 2.71 shows the block diagram of Wheatstone bridge with digital readout. Here a logic circuit is used to provide a signal to $R_3$, reuse the null, and provide a digital readout representing the value of $R_x$.

## A.C. BRIDGES

### 2.6.5. A.C. Bridge—General form

An A.C. bridge in its basic form consists of the following elements/components :

1. ***Four bridge arms.***
2. ***A source of excitation.*** The power source supplies an A.C. voltage to the bridge at the desired frequency.

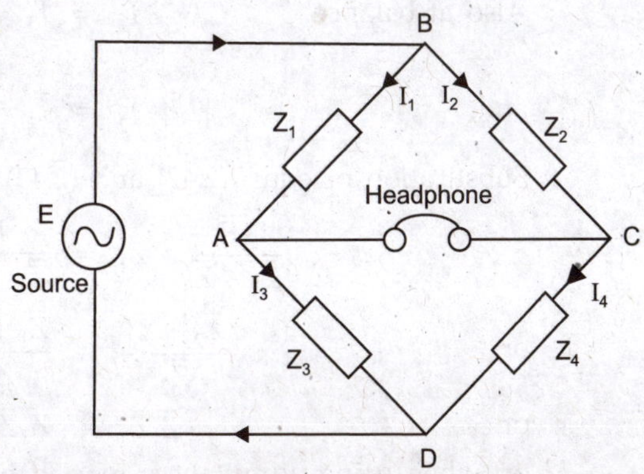


**Fig. 2.72.** General form of the A.C. bridge.

- For power measurements at low frequencies, the power line may serve as the source of excitation.
- At higher frequencies, the electronic oscillators are universally used as the source of supply due to the following *reasons* :

(*i*) The output waveform is very close to sine wave.
(*ii*) The output frequency is very stable.
(*iii*) The output frequency can be determined with accuracy and is also easily adjustable.
(*iv*) The output power is adequate to drive the logic circuits.

— A typical oscillator has a frequency range of 40 Hz to 125 kHz with power output of 7W.

3. ***Null detector.*** It must respond to A.C. unbalance currents. The null/balance detectors commonly used for A.C. bridges are :

(*i*) *Headphones.* These are used as detectors at the frequencies of 250 Hz to 3 to 4 kHz.
(*ii*) *Tunable amplifier circuits* :

— A tuned detector is the most sensitive detector while working with single frequency.

— Tunable amplifier detectors are used for frequency range of 10 Hz to 100 Hz.

(*iii*) *Vibration galvanometers.* These are useful for low audio frequency range from 5 Hz to 1000 kHz but are commonly used below 200 Hz.

**General equation for bridge balance :**

Fig. 2.72 shows the general form of an A.C. bridge.

- The four bridge arms $\bar{Z}_1$ $\bar{Z}_2$ $\bar{Z}_3$ and $\bar{Z}_4$ (Bar *i.e.*—indicates complex notation) are indicated as unspecified impedances and detector is represented by headphones. (cheapest and effective from).
- The balance condition is reached when the detector response is zero, or indicates a null. Balance adjustment to obtain a null response is made by varying one or more of the bridge arms.
- The *condition for bridge balance* requires that potential difference from *A* to *C* in Fig. 2.72 be zero. This will be the case when the voltage drop from *B* to *A* equals

the voltage drop from $B$ to $C$ in both *magnitude* and *phase*. In complex notation we can write :

$$\bar{E}_{BA} = \bar{E}_{BC} \quad ...(2.40)$$

or,

$$\bar{I}_1\bar{Z}_1 = \bar{I}_2\bar{Z}_2 \quad ...(2.41)$$

Also at balance,

$$\bar{I}_1 = \bar{I}_3 = \frac{\bar{E}}{\bar{Z}_1+\bar{Z}_3} \quad ...(2.42)$$

and,

$$\bar{I}_2 = \bar{I}_4 = \frac{\bar{E}}{\bar{Z}_2+\bar{Z}_4} \quad ...(2.43)$$

Substitution of eqns. (2.42) and (2.43) into eqn. (2.41) yields :

$$\frac{\bar{E}}{\bar{Z}_1+\bar{Z}_3}\times\bar{Z}_1 = \frac{\bar{E}}{\bar{Z}_2+\bar{Z}_4}\times\bar{Z}_2$$

or,

$$\frac{\bar{Z}_2+\bar{Z}_4}{\bar{Z}_2} = \frac{\bar{Z}_1+\bar{Z}_3}{\bar{Z}_1} \quad \text{or} \quad 1+\frac{\bar{Z}_4}{\bar{Z}_2} = 1+\frac{\bar{Z}_3}{\bar{Z}_1}$$

or,

$$\bar{Z}_1\bar{Z}_4 = \bar{Z}_2\bar{Z}_3 \quad ...(2.44)$$

or, when using admittances instead of impedance, we have

$$\bar{Y}_1\bar{Y}_4 = \bar{Y}_2\bar{Y}_3 \quad ...(2.45)$$

Eqn. (2.44) is the most convenient form in most cases and is the *general equation for balance* of the A.C. bridge. Eqn. (2.45) is useful when dealing wih parallel components in bridge arms.

— Eqn. (2.44) states the product of the impedances of one pair of opposite arms must be equal to the product of impedances of the other pair of opposite arms, with the impedances expressed in complex notation. If the impedance is written in the form $\bar{Z} = Z\angle\theta$, where $Z$ represents the magnitude and $\theta$ the phase angle of the complex impedance, Eqn. (2.44) can be written in the form

$$(Z_1\angle\theta_1)(Z_4\angle\theta_4) = (Z_2\angle\theta_2)(Z_3\angle\theta_3) \quad ...(2.46)$$

or,

$$Z_1Z_4\angle(\theta_1+\theta_4) = Z_2Z_3\angle(\theta_2+\theta_3) \quad ...(2.47)$$

— Eqn. (2.47) shows that following *two conditions* must be met *simultaneously* when balancing an A.C. bridge :

(*i*) The magnitudes of the impedances should satisfy the relationship :

$$Z_1Z_4 = Z_2Z_3 \quad ...(2.48)$$

(*ii*) The phase angles of the impedances should satisfy the relationship :

$$\angle\theta_1 + \angle\theta_4 = \angle\theta_2 + \angle\theta_3 \quad ...(2.49)$$

**Example 2.14.** *The following data relate to the basic A.C. bridge of Fig. 2.72 :*

$\bar{Z}_1$ = *50∠80° (inductive impedance);*

$\bar{Z}_2$ = *125Ω (pure resistance);*

$\bar{Z}_3$ = *200∠30° (inductive impedance);*

$\bar{Z}_4$ = *Unknown*

*Determine the constants of the unknown arm :*

**Solution.** *Given* : $\bar{Z}_1 = 50\angle 80°$; $\bar{Z}_2 = 125\ \Omega$, $\bar{Z}_3 = 200\angle 30°$

$\bar{Z}_4$ :

We know that :

$$Z_1Z_4 = Z_2Z_3 \quad ...(2.48)$$

or, $$Z_4 = \frac{Z_2 Z_3}{Z_1} = \frac{125 \times 200}{50} = 500\Omega$$

Also, $$\angle\theta_1 + \angle\theta_4 = \angle\theta_2 + \angle\,\theta_3 \quad ...(2.49)$$

or, $$80° + \theta_4 = 0° + 30°$$

or, $$\theta_4 = 0° + 30° - 80° = -50°$$

Hence, $$\bar{Z}_4 = \mathbf{500\angle{-50°}\ \Omega},$$

indicating that we are dealing with a *capacitive* element, possibly consisting of a series combination of a resistor and a capacitor. **(Ans.)**

**Example 2.15.** *The following data relate to the balanced A.C. bridge shown in Fig. 2.73.*

*Arm AB, $R_1 = 225\ \Omega$; Arm BC, $R_2 = 150\ \Omega$ in series with $C_2 = 0.53\ \mu F$; arm CD, unknown; arm DA, $R_3 = 100\ \Omega$ in series with $L = 7.95$ mH. The oscillator frequency = 1 kHz.*

*Calculate the constants of arm CD.*

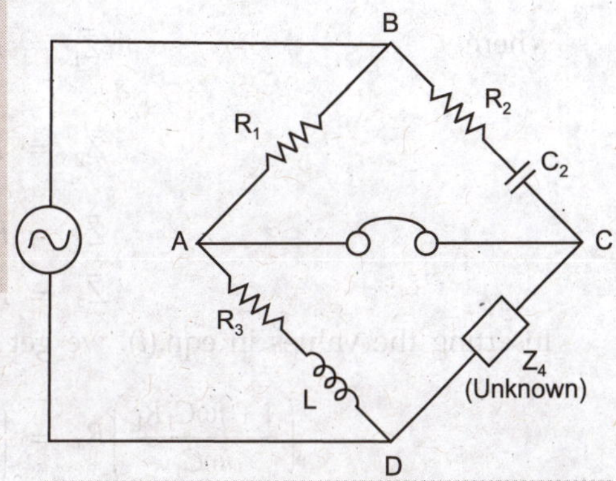


Fig. 2.73.

**Solution.** *Given* : $f = 1$ kHz, $R_1 = 225\ \Omega$; $R_2 = 150\ \Omega$, $C = 0.53\ \mu F$ or $\omega C_2 = 2\pi f \times 0.53 \times 10^{-6} = 2\pi \times 1000 \times 0.53 \times 10^{-6}\ \Omega = 3.33 \times 10^{-3}\ \Omega$; $R_3 = 100\ \Omega$, $L = 7.95$ mH or $\omega L = 2\pi f L = 2\pi \times 1000 \times 7.95 \times 10^{-3} \simeq 50\Omega$.

**Arm *CD*, $\bar{Z}_4$ :**

Now, $$\bar{Z}_1 = R_1 = 225\ \Omega$$

$$\bar{Z}_2 = R_2 - \frac{j}{\omega C_2} = 150 - \frac{j}{3.33 \times 10^{-3}} = (150 - j\,300)\ \Omega$$

$$\bar{Z}_3 = R_3 + j\omega L = (100 + j\,50)\ \Omega$$

$$\bar{Z}_4 = \text{Unknown}$$

The general equation for bridge balance states that :

$$\bar{Z}_1\,\bar{Z}_4 = \bar{Z}_2\,\bar{Z}_3$$

or, $$\bar{Z}_4 = \frac{\bar{Z}_2\,\bar{Z}_3}{\bar{Z}_1} = \frac{(150 - j300)(100 + j\,50)}{225}$$

$$= \frac{(15000 + j\,7500 - j\,30000 + 15000)}{225}$$

$$= \frac{30000 - j\,22500}{225} = \mathbf{133.33 - j\,100}$$

*i.e.* $$R_4 = 133.33\ \Omega \text{ and } X_{C_4} = \frac{1}{\omega C_4} = \frac{1}{2\pi f C_4} = 100$$

or, $$C_4 = \frac{1}{2\pi \times 1000 \times 100} = 1.59 \times 10^{-6}\ F \text{ or } 1.59\ \mu F$$

Hence, $$\mathbf{R_4 = 133.33\ \Omega,\quad C_4 = 1.59\ \mu\ F}\ \textbf{(Ans.)}$$

**Example 2.16.** *In a four-arm bridge network (Fig. 2.74), the arm AB consists of an imperfect condenser, BC and CD are non-reactive resistances of 1000 Ω each and DA is a standard capacitor of 0.0115 μF capacitance in series with a resistance of 140 Ω. If the bridge is balanced for frequency ω = 7500 radians/sec., find the shunt-less resistance and capacitance of the imperfect condenser.* **(UPTU, 2004-05)**

**Solution.** *Given* : ω = 7500 rad/s ; $R_2$ = 140 Ω; $R_3 = R_4$ = 1000 Ω ; $C_2$ = 0.0115 μ F.

Refer to Fig. 2.74 :

We know that for balance,

$$\bar{Z}_1 \bar{Z}_4 = \bar{Z}_2 \bar{Z}_3 \quad ...(i)$$

where,

$$\bar{Z}_1 = R_1 + \frac{1}{j\omega C_1} = \frac{1 + j\omega C_1 R_1}{j\omega C_1}$$

$$\bar{Z}_2 = R_2 + \frac{1}{j\omega C_2} = \frac{1 + j\omega C_2 R_2}{j\omega C_2}$$

$$\bar{Z}_3 = R_3$$

$$\bar{Z}_4 = R_4$$

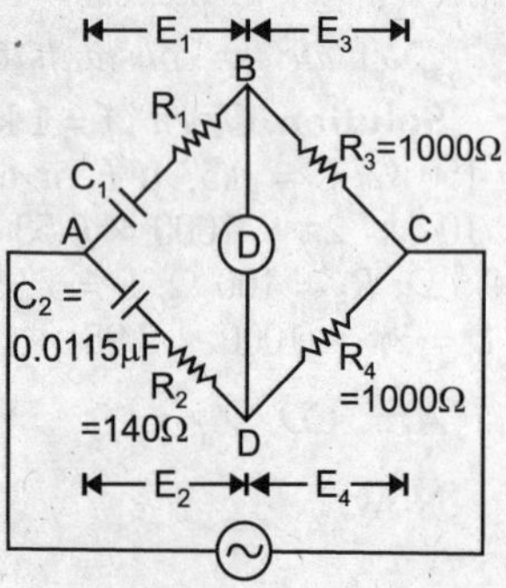


Fig. 2.74

Inserting the values in eqn.(*i*), we get

$$\left(\frac{1 + j\omega C_1 R_1}{j\omega C_1}\right) R_4 = \left(\frac{1 + j\omega C_2 R_2}{j\omega C_2}\right) R_3$$

or,

$$j\omega C_2 R_4 - \omega^2 R_1 C_1 C_2 R_4 = j\omega C_1 R_3 - \omega^2 C_1 C_2 R_2 R_3$$

Comparing the *real part*, we have :

$$\omega^2 R_1 C_1 C_2 R_4 = \omega^2 C_1 C_2 R_2 R_3$$

$$R_1 = \frac{R_2 R_3}{R_4} \quad ...(ii)$$

or,

$$\mathbf{R_1} = \frac{140 \times 1000}{1000} = \mathbf{140\ \Omega\ (Ans.)}$$

Comparing the *imaginary part*, we get :

$$\omega C_2 R_4 = \omega C_1 R_3$$

$$C_2 R_4 = C_1 R_3 \quad ...(iii)$$

or,

$$\mathbf{C_1} = \frac{C_2 R_4}{R_3} = \frac{(0.0115 \times 10^{-6}) \times 1000}{1000}$$

$$= 0.0115 \times 10^{-6} \text{ F} \quad \text{or} \quad \mathbf{0.0115\ \mu\ F\ (Ans.)}$$

### 2.6.6. Maxwell Bridge (*for measurement of* "inductance")

Maxwell bridge, shown in Fig. 2.75 *measures an unknown inductance in terms of a known capacitor* :

— The use of standard arm offers the advantage of compactness and easy shielding.

— The capacitor is almost a loss-less component.

— One arm has a resistance $R_1$ in parallel with $C_1$, and hence it is easier to write the balance equation using the admittance of arm 1 instead of the impedance.

The general equation for bridge balance is,

$$\bar{Z}_1 \bar{Z}_x = \bar{Z}_2 \bar{Z}_3$$

or,
$$\bar{Z}_x = \frac{\bar{Z}_2 \bar{Z}_3}{\bar{Z}_1} = \bar{Z}_2 \bar{Z}_3 \bar{Y}_1 \quad ...(2.50)$$

where, $\bar{Z}_1 = R_1$ in parallel with $C_1$ *i.e.* $\bar{Y}_1 = \frac{1}{\bar{Z}_1}$

$$\bar{Y}_1 = \frac{1}{R_1} + j\omega C_1$$

$$\bar{Z}_2 = R_2$$

$$\bar{Z}_3 = R_3$$

$\bar{Z}_x = R_x$ in series with $L_x$ $(= R_x + j\omega L_x)$

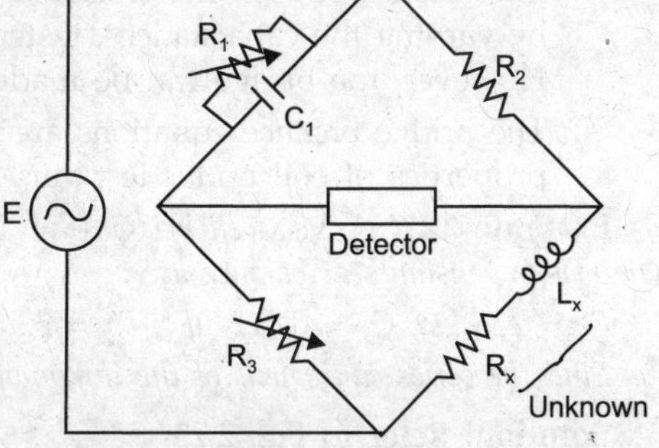


**Fig. 2.75.** Maxwell bridge for inductance measurements.

Substituting these values in eqn. (2.50), we get

$$R_x + j\omega L_x = R_2 R_3 \left( \frac{1}{R_1} + j\omega C_1 \right)$$

or,
$$R_x + j\omega L_x = \frac{R_2 R_3}{R_1} + j\omega C_1 R_2 R_3$$

Equating real terms and imaginary terms, we have

$$\left. \begin{aligned} R_x &= \frac{R_2 R_3}{R_1}; \\ L_x &= C_1 R_2 R_3 \end{aligned} \right\} \quad ...(2.51)$$

To obtain the bridge balance, first $R_3$ is adjusted for inductive balance and $R_1$ is adjusted for resistive balance.

The *"quality factor"* of the coil is given by,

$$Q = \frac{\omega L_x}{R_x} = \frac{\omega . C_1 R_2 R_3}{(R_2 R_3 / R_1)} = \frac{\omega . C . R_2 R_3 R_1}{R_2 R_3}$$

*i.e.*
$$Q = \omega C_1 R_1 \quad ...(2.52)$$

- Maxwell bridge is limited to the measurement of *low Q values (1-10)*.
- The bridge is *particularly suited for inductance measurements*, since comparison with a capacitor is more ideal than with another inductance. Commercial bridges measure from *1-1000 H, with ± 2 % error* (If $Q$ is very large, $R_1$ becomes excessively large and it is impractical to obtain a satisfactory variable standard resistance in the range of values required).

***Advantages :***

1. The balance equation is independent of losses associated with inductance.
2. The measurement is independent of the excitation *frequency*.
3. The scale of the resistance can be calibrated to read inductance directly.
4. A wide range of inductance at power and audio frequencies can be measured.

***Disadvantages :***

1. High $Q$ values cannot be measured.
2. It cannot be used for measurement of very low $Q$ values, because of balance coverage problem.
3. The Maxwell bridge using a fixed capacitor has the disadvantage that there is an interaction between the resistance and reactance balances. This can be avoided by varying the capacitances, instead of $R_2$ and $R_3$, to obtain a reactance balance. However, the bridge can be made to read directly in $Q$.
4. The bridge balance equations are independent of frequency. But practically, the properties of coil under test vary with frequency which can cause error.

**Example 2.17.** *A Maxwell bridge (Fig. 2.75) is used to measure an inductive impedance. The bridge constants at balance are :*

$R_1 = 235\ k\Omega,\ C_1 = 0.012\ \mu F\ ;\ R_2 = 2.5\ k\Omega;\ R_3 = 50\ k\Omega$

*Find the series equivalent of the unknown impedance.*

**Solution.** Refer to Fig. 2.75.

$R_x$ : ; $L_x$ :

We know that :

$$R_x = \frac{R_2 R_3}{R_1} = \frac{2.5 \times 50}{235}$$

$$= \mathbf{0.53\ k\ \Omega\ (Ans.)}$$

$$L_x = C_1 R_2 R_3 = 0.012 \times 10^{-6} \times (2.5 \times 10^3) \times (50 \times 10^3)$$

$$= \mathbf{1.5\ H\ (Ans.)}$$

The *equivalent series circuit* is shown in Fig. 2.76.

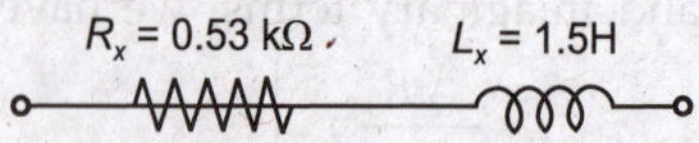


**Fig. 2.76.**

## 2.6.7. Hay Bridge (*For measurement of* "inductance")

The Hay bridge shown in Fig. 2.77 differs from the Maxwell bridge by having resistor $R_1$ in series with standard capacitor $C_1$ instead of in parallel. It is apparent that for large phase angles, $R_1$ should have a very low value. This bridge is, thus, *more convenient for measuring high–Q coils.*

The general equation for bridge balance is,

$$\bar{Z}_1 \bar{Z}_x = \bar{Z}_2 \bar{Z}_3 \quad ...(i)$$

where,

$$\bar{Z}_1 = R_1 - \frac{j}{\omega C_1};$$

$$\bar{Z}_2 = R_2 ;$$

$$\bar{Z}_3 = R_3 ;$$

$$\bar{Z}_x = R_x + j\omega L_x$$

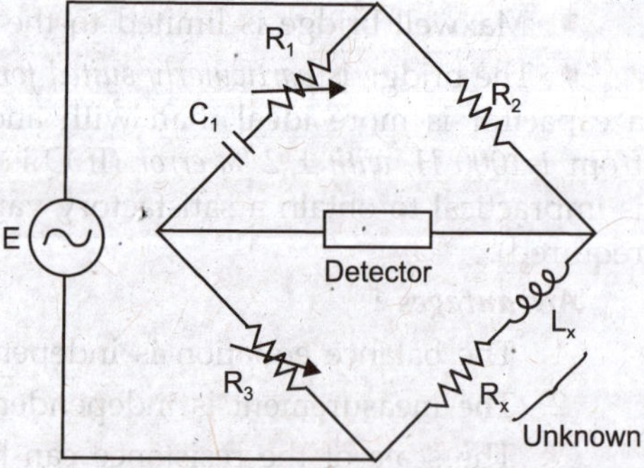


**Fig. 2.77.** Hay bridge for inductance measurements.

Substituting these values in eqn. (*i*), we get :

$$\left(R_1 - \frac{j}{\omega C_1}\right)(R_x + j\omega L_x) = R_2 R_3$$

$$R_1R_x + j\omega L_x R_1 - \frac{jR_x}{\omega C_1} + \frac{L_x}{C_1} = R_2 R_3$$

Equating the *real* and *imaginary* terms, we get :

$$R_1 R_x + \frac{L_x}{C_1} = R_2 R_3 \quad \text{...(2.53)}$$

and,

$$\frac{R_x}{\omega C_1} = \omega L_x R_1 \quad \text{...(2.54)}$$

or,

$$R_x = \omega^2 L_x C_1 R_1 \quad \text{...[2.54 (a)]}$$

Substituting for $R_x$ in eqn. (2.53), we have

$$R_1(\omega^2 L_x C_1 R_1) + \frac{L_x}{C_1} = R_2 R_3$$

or,

$$\omega^2 R_1^2 C_1 L_x + \frac{L_x}{C_1} = R_2 R_3$$

Multiplying both sides by $C_1$, we get

$$\omega^2 R_1^2 C_1^2 L_x + L_x = R_2 R_3 C_1$$

or,

$$L_x (1 + \omega^2 R_1^2 C_1^2 L_x) = R_2 R_3 C_1$$

or,

$$L_x = \frac{R_2 R_3 C_1}{1 + \omega^2 R_1^2 C_1^2} \quad \text{...(2.55)}$$

Substituting for $L_x$ in eqn. (2.54*a*), we have

$$R_x = \frac{\omega^2 C_1^2 R_1 R_2 R_3}{1 + \omega^2 R_1^2 C_1^2} \quad \text{...(2.56)}$$

The term $\omega$ appears in the expressions for both $L_x$ and $R_x$. This indicates that the bridge is *frequency sensitive.*

— This bridge is also used in the *measurement of incremental inductance.* The inductance balance equation depends on the losses of the inductor (or $Q$) and also on the operating frequency.

- *A commercial bridge measures from 1 μH–100H with ± 2 % error.*
- An inconvenient feature of this bridge is that equation giving the balance condition for inductance contains the multiplier $\frac{1}{(1 + 1/Q^2)}$ (where $Q = \frac{1}{\omega R_1 C_1}$). The inductance balance thus depends on its $Q$ and frequency.

$$\therefore \quad L_x = \frac{R_2 R_3 C_1}{1 + (1/Q^2)} \quad \text{...(2.57)}$$

For a value of $Q$ greater than 10, the term $\frac{1}{Q^2}$ will be smaller than $\frac{1}{100}$ and can therefore be *neglected*. Then,

$L_x = R_2 R_3 C_1$, which is the same as Maxwell's equation. But for inductors with a $Q$ less than 10, the $\frac{1}{Q^2}$ term cannot be neglected. Hence *this bridge is not suited for measurements of coils having $Q$ < 10.*

**Example 2.18.** *Find the series equivalent inductance and resistance of the network that causes an opposite angle (Hay bridge) to null the following bridge arms (Fig. 2.78)*
$\omega = 3000$ *rad/s ;* $R_2 = 9k\Omega$ *;* $R_1 = 1.8\ k\Omega$ *;* $C_1 = 0.9\ \mu F$*,* $R_3 = 0.9\ k\Omega$*.*

**Solution.** We know that,

$$R_x = \frac{\omega^2 C_1^2 R_1 R_2 R_3}{1+\omega^2 R_1^2 C_1^2} \quad ...(2.56)$$

$$= \frac{(3000)^2 (0.9\times10^{-6})^2 \times (1.8\times1000)(9\times1000)(0.9\times1000)}{1+(3000)^2 (1.8\times1000)^2 (0.9\times10^{-6})^2}$$

$$= \frac{106.27\times10^3}{1+14.58}$$

$$= 6822\ \Omega \text{ or } 6.822\ k\Omega$$

Also, $$L_x = \frac{R_2 R_3 C_1}{1+\omega^2 R_1^2 C_1^2} \quad ...(2.57)]$$

$$= \frac{(9\times1000)(0.9\times1000)\times(0.9\times10^{-6})}{1+(3000)^2 (1.8\times1000)^2 (0.9\times10^{-6})^2}$$

$$= \frac{7.29}{1+14.58} = 0.468\text{H}$$

*i.e.* $R_x = 6.822\ k\Omega$; $L_x =$ **0.468 H (Ans.)**

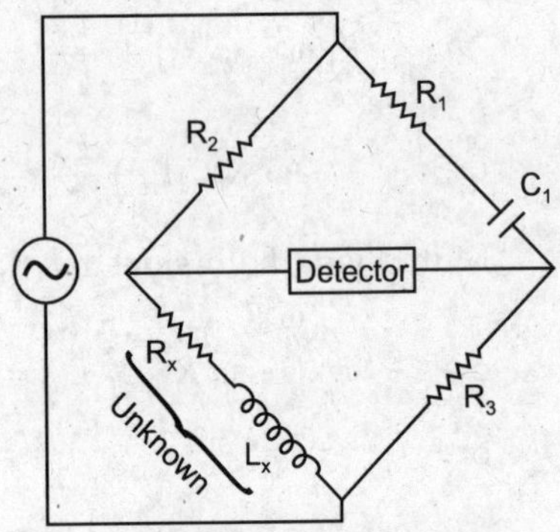


Fig. 2.78.

### 2.6.8. Schering Bridge (*For measurement of* "Capacitance")

It is one of the most important A.C. bridges, widely used for the *measurement of unknown capacitors, dielectric loss and power factor.*

Fig. 2.79 shows the basic circuit arrangement of Schering bridge. The capacitor $C_3$ is a high quality mica capacitor (low-loss) for general measurements, or an air capacitor (having a very stable value and a very small electric field) for insulation measurement.

The general equation for balance is,

$$\bar{Z}_1 \bar{Z}_x = \bar{Z}_2 \bar{Z}_3$$

or, $$\bar{Z}_x = \frac{\bar{Z}_2 \bar{Z}_3}{\bar{Z}_1} = \bar{Z}_2 \bar{Z}_3 \bar{Y}_1 \quad ...(i)$$

where, $$\bar{Z}_x = R_x - \frac{j}{\omega C_x}$$

$$\bar{Z}_2 = R_2$$

$$\bar{Z}_3 = -\frac{j}{\omega C_3}$$

$$\bar{Y}_1 = \frac{1}{R_1} + j\omega C_1$$

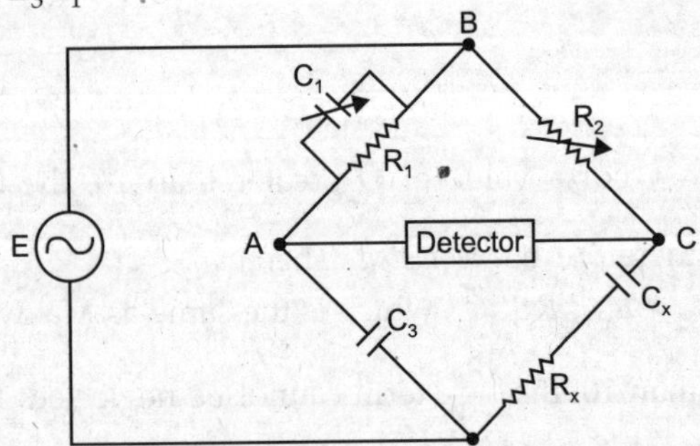


Fig. 2.79. Schering bridge for the measurement of capacitance.

Substituting the values in eqn. (*i*), we get

$$R_x - \frac{j}{\omega C_x} = R_2\left(-\frac{j}{\omega C_3}\right)\times\left(\frac{1}{R_1} + j\,\omega\,C_1\right)$$

or $$\left(R_x - \frac{j}{\omega C_x}\right) = \frac{R_x\,(-j)}{R_1\,\omega\,C_3} + \frac{R_2\,C_1}{C_3}$$

Equating the *real* and *imaginary* terms, we get :

$$R_x = \frac{R_2\,C_1}{C_3} \quad \text{...(2.58)}$$

and, $$C_x = \frac{R_1\,C_3}{R_2} \quad \text{...(2.59)}$$

The dial of capacitor $C_1$ can be calibrated directly to give the dissipation factor at a particular frequency.

- The lower junction of the bridge is grounded. At the frequency normally used on this bridge, the reactances of capacitors $C_3$ and $C_x$ are much higher than the resistances of $R_1$ and $R_2$. Hence if the junction of $R_1$ and $R_2$ is grounded, the detector is effectively at ground potential. This reduces any stray-capacitance effect, and makes bridge more stable.

**Power factor.** The *power factor* (p.f.) of a series R.C. combination is defined as *the cosine of the phase angle of the circuit.*

$\therefore$ The power factor of the unknown, p.f. $= \dfrac{R_x}{Z_x}$

For phase angles very close to 90°, the reactance is almost equal to the impedance and we can approximate the power factor to,

$$\text{p.f.} \simeq \frac{R_x}{X_x} = \frac{R_x}{\dfrac{1}{\omega C_x}} = \omega\,R_x\,C_x \quad \text{...(2.60)}$$

**Dissipation factor (D).** The *dissipation factor* of a series *R-C* circuit is defined *as the cotangent of the phase angle.*

*i.e.,* $$D = \frac{R_x}{X_x} = \omega\,C_x\,R_x \quad \text{...(2.61)}$$

Also, since the quality of a coil is defined by $Q = \dfrac{X_L}{R_L}$, we find that the dissipation factor, *D*, is the *reciprocal of the quality factor, Q,*

*i.e.* $$D = \frac{1}{Q} \quad \text{...(2.62)}$$

The dissipation factor tells us something about the quality of the capacitor; *i.e.,* how close the phase angle of the capacitor is to the ideal value of 90°.

Substituting the value of $C_x$ in eqn. (2.59) and of $R_x$ in eqn. (2.58) in the eqn. (2.61), we get

$$D = \omega\times\frac{R_1\,C_3}{R_2}\times\frac{R_2\,C_1}{C_3} = \omega R_1\,C_1 \quad \text{...(2.63)}$$

— When resistor $R_1$ in the Schering bridge of Fig. (2.79) has a fixed value, the dial of capacitor $C_1$ may be calibrated directly in dissipation factor *D*. This is the usual practice in a Schering bridge. The appearance of the term ω in the expression for the dissipation factor eqn. (2.63) means, of course, that the calibration of the $C_1$ dial holds for only one particular frequency at which dial is calibrated. A different frequency can be used, provided that a correction is made by multiplying the $C_1$ dial reading by the ratio of the two frequencies.

- Commerical units *measure from 100 pF- 1μF, with ± 2% accuracy.*
- The Schering bridge is *widely used for testing small capacitors at low voltages with very high precision.*

**Example 2.19.** *The Schering bridge shown in Fig. 2.79 has the following constants :*
*Arm AB ... Capacitor of 0.4 μF in parallel with 1.5 kΩ resistance;*
*Arm BC ... Resistance of 3 kΩ;*
*Arm CD ... Unknown capacitor $C_x$ and $R_x$ in series*
*Arm DA ... Capacitor of 0.4 μ F;*
*Frequency ... 1 kHz.*
*Determine the following :*
*(i) $R_x$ and $C_x$.* *(ii) Dissipation factor.*

**Solution.** *Given:* $R_1$ = 1.5 kΩ; $C_1$ = 0.4 μF ; $R_2$ = 3 kΩ; $C_3$ = 0.4 μF.

*(i)* **$R_x$, $C_x$ :**

We know, that $$R_x = \frac{R_2 C_1}{C_3} = \frac{3k\Omega \times 0.4\mu F}{0.4\mu F} = \mathbf{3\ k\Omega\ (Ans.)}$$

and, $$C_x = \frac{R_1 C_3}{R_2} = \frac{1.5\,k\Omega \times 0.4\,\mu F}{3\,k\Omega} = \mathbf{0.2\ \mu F\ (Ans.)}$$

*(ii)* **Dissipation factor, *D* :**

$$D = \omega\, C_x\, R_x \quad ...(2.61)$$
$$= 2\,\pi\, f.\, C_x.\, R_x$$
$$= 2\,\pi \times 1000 \times 0.2 \times 10^{-6} \times (3 \times 1000) = 3.77$$

*i.e.* $D$ = **3.77 (Ans.)**

### 2.6.9 ANDERSON BRIDGE (*For measurement of "Inductance"*)

*This method is one of the commonest and best bridge methods for precise measurement of inductance over a wide range.* Here, the self-inductance is *measured in terms of a standard capacitor.*

Fig. 2.80 shows the circuit and the phasor diagrams of the bridge for *balanced conditions.*

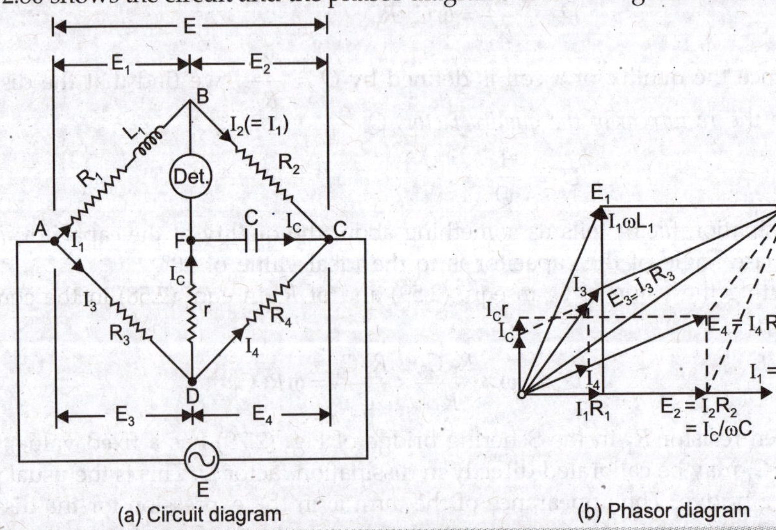


(a) Circuit diagram (b) Phasor diagram

**Fig. 2.80.** Anderson bridge.

Let, $L_1$ = Self-inductance to be measured,

$R_1$ = Resistance of the coil under test,

$R_2, R_3, R_4, r$ = Known non-inductive resistances, and

$C$ = Standard known capacitor.

*At. balance* :

$$\bar{I}_1 = \bar{I}_2 \quad \text{and} \quad \bar{I}_3 = \bar{I}_c + \bar{I}_4$$

Now,
$$\bar{I}_1 R_2 = I_c \times \frac{1}{j\omega C}$$

or,
$$I_c = \bar{I}_1 j\omega C R_2 \qquad ...(i)$$

Other balance equations are :

$$\bar{I}_1 (R_1 + j\omega L_1) = \bar{I}_3 R_3 + I_c.r \qquad ...(ii)$$

and,
$$\bar{I}_c.r + \bar{I}_c.\frac{1}{j\omega C} = (\bar{I}_3 - \bar{I}_c) R_4$$

or,
$$I_c\left(r + \frac{1}{j\omega C}\right) = (\bar{I}_3 - \bar{I}_c) R_4 \qquad ...(iii)$$

Substituting the value of $\bar{I}_c$ from eqn. (*i*) in eqns. (*ii*) and (*iii*), we get :

$$\bar{I}_1 (R_1 + j\omega L_1) = \bar{I}_3 R_3 + \bar{I}_1 j\omega C R_2.r$$

or,
$$\bar{I}_1 (R_1 + j\omega L_1 - j\omega C R_2.r) = \bar{I}_3 R_3 \qquad ...(iv)$$

and,
$$\bar{I}_1 j\omega C R_2 \left(r + \frac{1}{j\omega C}\right) = (\bar{I}_3 - \bar{I}_1 j\omega C R_2) R_4$$

or,
$$\bar{I}_1 j\omega C R_2 r + \bar{I}_1 R_2 = \bar{I}_3 R_4 - \bar{I} j\omega C R_2 R_4$$

or,
$$\bar{I}_1 (j\omega C R_2 r + j\omega C R_2 R_4 + R_2) = \bar{I}_3 R_4 \qquad ...(v)$$

Substituting the value of $R_4$ from (*v*) in (*iv*), we get :

$$\bar{I}_1 (R_1 + j\omega L_1 - j\omega C R_2.r) = \bar{I}_1 \left(\frac{j\omega C R_2 R_3 r}{R_4} + j\omega C R_2 R_3 + \frac{R_2 R_3}{R_4}\right)$$

Equating the *real* and *imaginary* terms, we get :

$$\boldsymbol{R_1} = \frac{R_2 R_3}{R_4} \qquad ...(2.62)$$

$$L_1 - CR_2 r = \frac{CR_2 R_3 r}{R_4} + CR_2 R_3$$

or,
$$L_1 = CR_2 r + \frac{CR_2 R_3 r}{R_4} + CR_2 R_3$$

or,
$$L_1 = CR_2 \left[ r\left(1 + \frac{R_3}{R_4}\right) + R_3 \right]$$

or,
$$\boldsymbol{L_1} = \frac{CR_2}{R_4}[r(R_3 + R_4) + R_3 R_4] \qquad ...(2.63)$$

***Advantages :***

1. A fixed capacitor can be used instead of a variable capacitor as in the case of a Maxwell bridge.
2. It is much easier to obtain balance in case of Anderson bridge than a Maxwell bridge for low Q-coils.
3. This bridge may be used for accurate determination of capacitance in terms of inductance.

***Disadvantages/Limitations :***

1. It is more complex than its prototype Maxwell bridge.
2. The shielding of the bridge becomes more difficult due to an additional junction point.

- In view of the above complications of the Anderson bridge, in all the cases where a variable capacitor is permissible the more simple Maxwell bridge is used (instead of Anderson bridge).

**Example 2.20.** *Fig. 2.81 shows the connection of an Anderson bridge for measuring the inductance L and resistance R of the coil. Find R and L, if balance is obtained when Q = S = 1,000 Ω, P = 500 Ω, r = 100 Ω and C = 0.5 μF.* **(AMIE)**

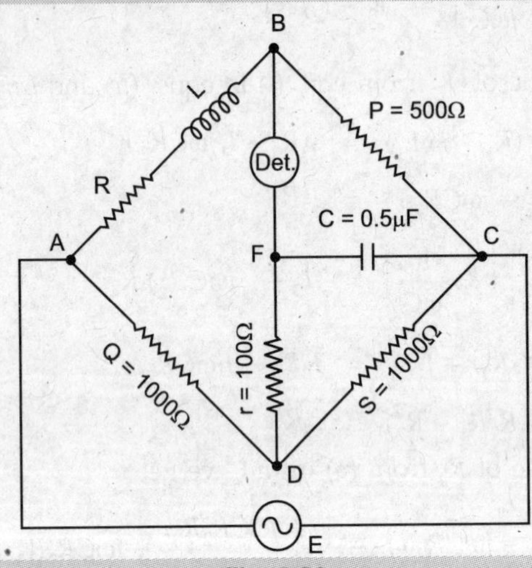


**Fig. 2.81**

**Solution.** *Given* : $Q = S = 1{,}000\ \Omega$, $P = 500\ \Omega$, $r = 100\ \Omega$, $C = 0.5\ \mu F$.

**R, L :**

In the balance condition of bridge :

$$R = \frac{P \times Q}{S} \quad \text{...[Eqn. (2.62)]}$$

$$= \frac{500 \times 1000}{1000} = \mathbf{500\ \Omega}\ \textbf{(Ans.)}$$

$$L = \frac{CP}{S}[r(Q+S)+QS] \quad \text{...[Eqn. (2.63)]}$$

$$= \frac{0.5 \times 10^{-6} \times 500}{1000}[100(1000 + 1000) + 1000 \times 1000]$$

$$= \mathbf{0.3\ H}\ \textbf{(Ans.)}$$

**Example 2.21.** *The following data relate to Anderson bridge shown in Fig. 2.82 :*

*The arms BC, CD and DA consist of resistances having values 1000 Ω, 1000 Ω and 500 Ω respectively.*

*A resistance of 100 Ω and a capacitance of 3 μF are connected respectively to the arms DF and CF.*

*An A.C. supply of 100 Hz is applied between the terminals A and C and a detector is connected between the terminals B and F.*

*The detector indicates null under the above conditions. Determine the values of R and L connected to the arm AB.*

**Solution.** *Given :* $R_2 = 1000\ \Omega$; $R_3 = 500\ \Omega$; $R_4 = 1000\ \Omega$; $r = 100\ \Omega$; $C = 3\mu F$

**R, L :**

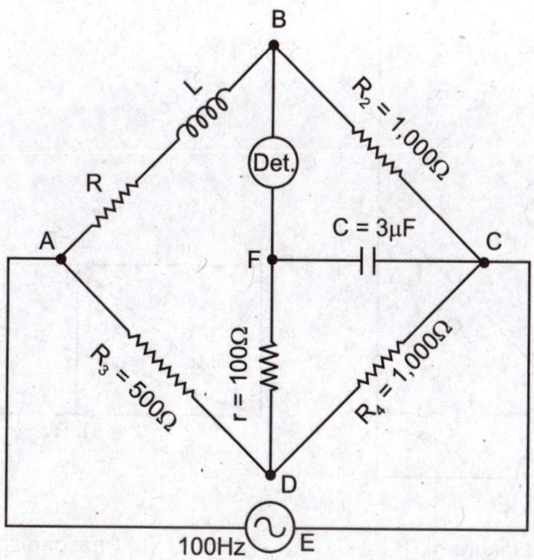


Fig. 2.82

At balance condition of the bridge :

We know that :
$$R = \frac{R_2 R_3}{R_4} \quad \text{...[Eqn. (2.62)]}$$
$$= \frac{1000 \times 500}{1000} = \mathbf{500\ \Omega} \quad \textbf{(Ans.)}$$
$$L = \frac{CR_2}{R_4}\,[r\,(R_3 + R_4) + R_3R_4] \quad \text{...[Eqn. (2.63)]}$$
$$= \frac{3 \times 10^{-6} \times 1000}{1000}\,[100(500 + 1000) + 500 \times 1000]$$
$$= \mathbf{1.95\ H} \quad \textbf{(Ans.)}$$

## 2.6.10 OWEN BRIDGE (*For measurement of "Inductance"*)

This bridge may be used *for measurement of an inductance in terms of capacitance.*

Fig. 2.83 shows the circuit and phasor diagrams for this bridge, under balance condition.

Let, $L_1$ = Unknown self-inductance of resistance $R_1$,

$R_2$ = Fixed non-inductive resistance,

$R_3$ = Variable non-inductive resistance,

$C_3$ = Variable standard capacitor, and

$C_4$ = Fixed standard capacitor.

When the bridge is balanced :

$$\bar{Z}_1 . \bar{Z}_4 = \bar{Z}_2 . \bar{Z}_3$$

$$(R_1 + j\omega L_1)\left(\frac{1}{j\omega C_4}\right) = R_2 . \left(R_3 + \frac{1}{j\omega C_3}\right)$$

or,

$$\frac{R_1}{j\omega C_4} + \frac{L_1}{C_4} = R_2 R_3 + \frac{R_2}{j\omega C_3}$$

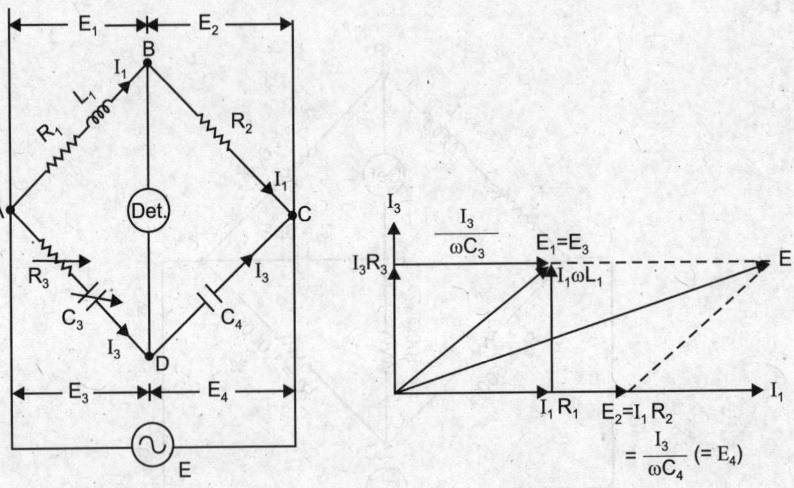


(a) Circuit diagram (b) Phasor diagram

**Fig. 2.83.** Owen bridge.

Separating the *real* and *imaginary* terms, we have :

$$\frac{L_1}{C_4} = R_2 R_3$$

or, $$\boldsymbol{L_1} = R_2 R_3 C_4 \quad ...(2.64)$$

and, $$\frac{R_1}{C_4} = \frac{R_2}{C_3}$$

or, $$\boldsymbol{R_1} = \frac{R_2 C_4}{C_3} \quad ...(2.65)$$

***Advantages :***

1. The balance equations are quite simple and do not contain any frequency component.
2. Since $R_3$ and $C_3$, the variable elements, are in the same arm, convergence to balance conditions is much easier.
3. The bridge can be used over a wide range of measurement of inductances.

***Disadvantages :***

1. While measuring high $Q$ coils, the value of capacitance $C_3$ tends to become rather large.

2. The bridge requires a variable capacitor which is an expensive item and also its accuracy is about 1%.

**Example 2.22.** *Fig. 2.84 shows an Owen bridge to measure the properties of a sample of sheet at 2 kHz. At balance, arm AB is the test specimen; arm BC is $R_2 = 100\ \Omega$; arm CD is $C_4 = 0.1\ \mu F$ and arm DA is $R_3 = 834\ \Omega$ in series with $C_3 = 0.124\ \mu F$.*

*Calculate the effective impedance of the specimen under test conditions.*

**(Punjab University)**

**Solution.** Let $R_1$ and $L_1$ be the effective resistance and inductance of the specimen respectively.

When the bridge is *balanced* :

$$\bar{Z}_1\bar{Z}_4 = \bar{Z}_2\bar{Z}_3$$

or, $$(R_1 + j\omega L_1)\left(\frac{1}{j\omega C_4}\right) = R_2\left(R_3 + \frac{1}{j\omega C_3}\right)$$

or, $$R_1 \cdot \frac{1}{j\omega C_4} + \frac{L_1}{C_4} = R_2R_3 + R_2 \cdot \frac{1}{j\omega C_3}$$

Equating real and imaginary terms, we get :

$$\frac{L_1}{C_4} = R_2R_3$$

or, $$\mathbf{L_1} = R_2R_3C_4 = 100 \times 834 \times 0.1 \times 10^{-6}$$
$$= 8.34 \times 10^{-3}\ \text{H or } \mathbf{8.34\ mH}\ \textbf{(Ans.)}$$

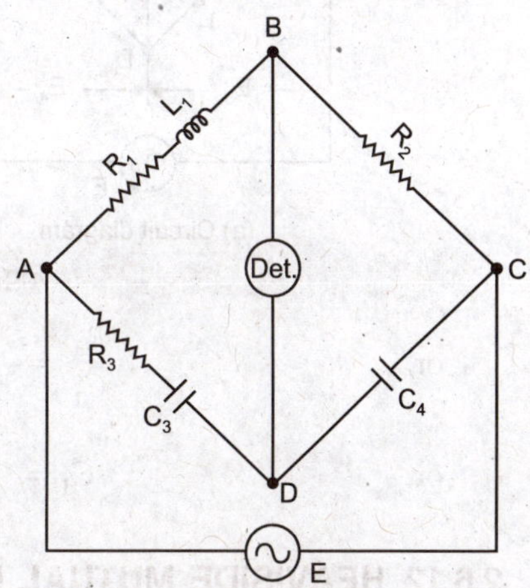

Fig. 2.84

and, $$\frac{R_1}{C_4} = \frac{R_2}{C_3}$$

or, $$\mathbf{R_1} = R_2 \times \frac{C_4}{C_3} = 100 \times \frac{0.1}{0.124} = \mathbf{80.6\ \Omega}\ \textbf{(Ans.)}$$

Reactance of specimen at 2 kHz,

$$X_1 = 2\pi f L_1 = 2\pi \times (2 \times 1000) \times 8.34 \times 10^{-3} = 104.8\ \Omega$$

$\therefore$ Impedance of the specimen, $Z_1 = \sqrt{R_1^2 + X_1^2} = \sqrt{(80.6)^2 + (104.8)^2} = \mathbf{132.2\ \Omega}$ **(Ans.)**

## 2.6.11 DE SAUTY BRIDGE (*For measurement of "Capacitance"*)

This bridge is the simplest method of *comparing two capacitances.*

Fig. 2.85 (*a* & *b*) shows the circuit and phasor diagrams of the bridge.

Let, $C_1$ = Capacitor whose capacitance is to be measured,

$C_3$ = A standard capacitor, and

$R_2, R_4$ = Non-inductive resistors

At balance : $I_1 = I_2$ ; $I_3 = I_4$

and, $$\left(\frac{1}{j\omega C_1}\right)R_4 = \left(\frac{1}{j\omega C_3}\right)R_2$$

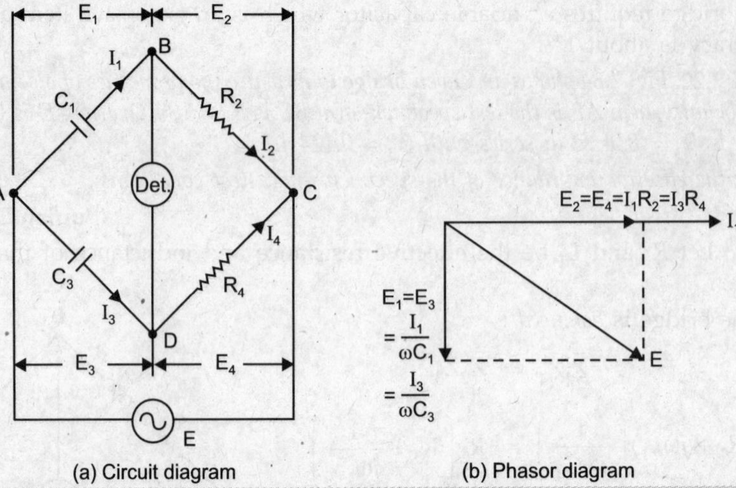


**Fig. 2.85.** De Sauty bridge.

or, $$\frac{R_4}{C_1} = \frac{R_2}{C_3}$$

or, $$C_1 = \frac{C_3R_4}{R_2} \qquad \text{...(2.66)}$$

## 2.6.12 HEAVISIDE MUTUAL INDUCTANCE BRIDGE

**Measurement of mutual inductance.** If the terminals of two coils whose mutual inductance is to be measured are available for series connections, measurement may be made by one of the self-inductance bridges discussed earlier. If the connections are made so that the magnetic fields of two coils are additive as shown in Fig. 2.86(*a*), the effective inductance of the two coils in series is :

$$L_{e1} = L_1 + L_2 + 2M \qquad \text{...(2.67)}$$

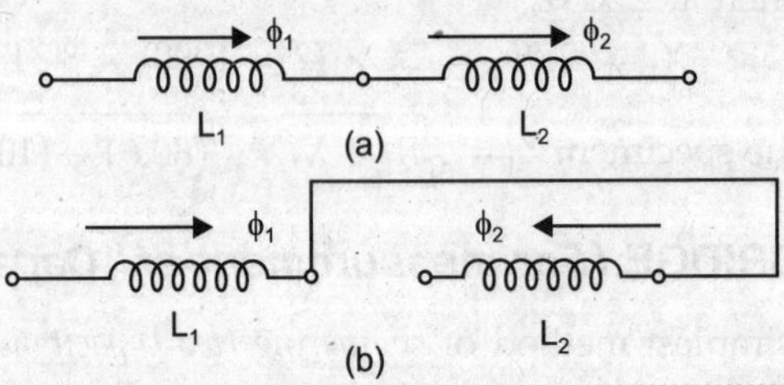


**Fig. 2.86.** Positive and negative couplings for mutual inductors.

If the connections of one coil are *reversed* as shown in Fig. 2.86(*b*), *then*

$$L_{e2} = L_1 + L_2 - 2M \qquad \text{...(2.68)}$$

Subtracting eqn. (2.69) from eqn. (2.68), we get :

Mutual inductance, $$M = \frac{1}{4}(L_{e1} - L_{e2}) \qquad \text{...(2.69)}$$

Hence, mutual inductance is obtained as one-fourth of difference of self-inductance measured with *series additive* and *series subtractive* connections.

*This method is of advantage only when fairly high coupling between the two coils is obtained otherwise it results in poor accuracy due to nearly equal terms of eqn. (2.69).*

*Fig. 2.87 shows the circuit and phasor diagrams of Heaviside mutual inductance bridge, which measures mutual inductance in terms of a known self-inductance.*

Let, $M$ = Unknown mutual inductance,

$L_1$ = Self-inductance of secondary of mutual inductance,

$L_3$ = Known self-inductance, and

$R_1, R_2, R_3, R_4$ = Non-inductive resistors.

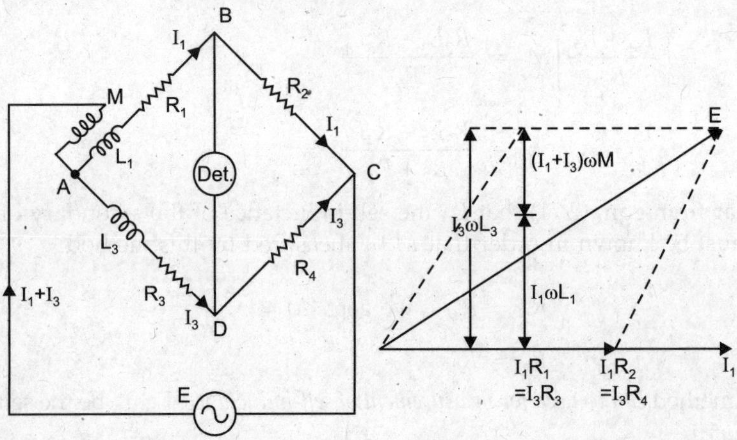


(a) Circuit diagram (b) Phasor diagram

**Fig. 2.87.** Heaviside mutual inductance bridge.

At balance voltage drop between *B* and *C* must equal the voltage drop between *D* and *C*. Also voltage drop across *A-B-C* equals the voltage drop across *A-D-C*. Thus we have the following equations at balance :

$$\bar{I}_1 R_2 = \bar{I}_2 R_4 \quad \text{...(i)}$$

and, $$(\bar{I}_1 + \bar{I}_2)(j\omega M) + \bar{I}_1(R_1 + R_2 + j\omega L_1) = \bar{I}_2 (R_3 + R_4 + j\omega L_3) \quad \text{...(ii)}$$

Inserting the value of $\bar{I}_1 = \dfrac{\bar{I}_2 R_4}{R_2}$ from (*i*) in (*ii*), we get :

$$\left(\frac{\bar{I}_2 R_4}{R_2} + \bar{I}_2\right)(j\omega M) + \frac{\bar{I}_2 R_4}{R_2}(R_1 + R_2 + j\omega L_1) = \bar{I}_2 (R_3 + R_4 + j\omega L_3)$$

or, $$\bar{I}_2\left(\frac{R_4}{R_2} + 1\right)(j\omega M) + \frac{\bar{I}_2 R_4}{R_2}(R_1 + R_2 + j\omega L_1) = \bar{I}_2 (R_3 + R_4 + j\omega L_3)$$

or, $$j\omega M\left(\frac{R_4}{R_2} + 1\right) + \frac{R_4 R_1}{R_2} + R_4 + j\omega L_1 \frac{R_4}{R_2} = R_3 + R_4 + j\omega L_3$$

Separating the real and imaginary terms, we have :

$$\frac{R_4 R_1}{R_2} + R_4 = R_3 + R_4$$

or, $$\frac{R_4 R_1}{R_2} = R_3$$

or, $$R_1 = \frac{R_2 R_3}{R_4} \quad \text{...(2.70)}$$

and, $$\left(\frac{R_4}{R_2}+1\right)M + L_1\frac{R_4}{R_2} = L_3$$

$$\left(\frac{R_4}{R_2}+1\right)M = L_3 - \frac{L_1 R_4}{R_2}$$

or, $$\left(\frac{R_4+R_2}{R_2}\right)M = \frac{R_2L_3 - R_4L_1}{R_2}$$

or, $$M = \frac{R_2L_3 - R_4L_1}{R_2 + R_4} \quad \text{...(2.71)}$$

It is clear from eqn. (2.71) that $L_1$, the self-inductance of the secondary of the mutual inductor, must be known in order that $M$ be measured by this method.

In case, $R_2 = R_4$, we get, $$M = \frac{L_2 - L_1}{2} \quad \text{...(2.72)}$$

and, $$R_1 = R_3 \quad \text{...(2.73)}$$

- This method *can be used for measurement of self-inductance.* Let $L_3$ be the self-inductance to *be determined.*

Then, from eqns. (2.70) and (2.71), we get :

$$L_3 = \frac{M(R_2+R_4) + R_4L_1}{R_2}$$

$$= M\left(1+\frac{R_4}{R_2}\right) + \frac{R_4}{R_2}L_1 \quad \text{...(2.74)}$$

and, $$R_3 = \frac{R_1R_4}{R_2} \quad \text{...(2.75)}$$

In case, $R_4 = R_2$, we have :

$$L_3 = 2M + L_1$$

and, $$R_3 = R_1$$

## Campbell's modification of Heaviside bridge :

*This bridge is used to measure self-inductance in terms of a mutual inductance.*

Fig. 2.89 shows a modified Heaviside bridge; this modification is due to Campbell :

— In this case an additional balancing coil $L$, $R$ is included in arm $AD$ in series with the inductor under test.

— An additional resistance $r$ is put in arm $AB$. Balance is obtained by varying $M$ and $r$.

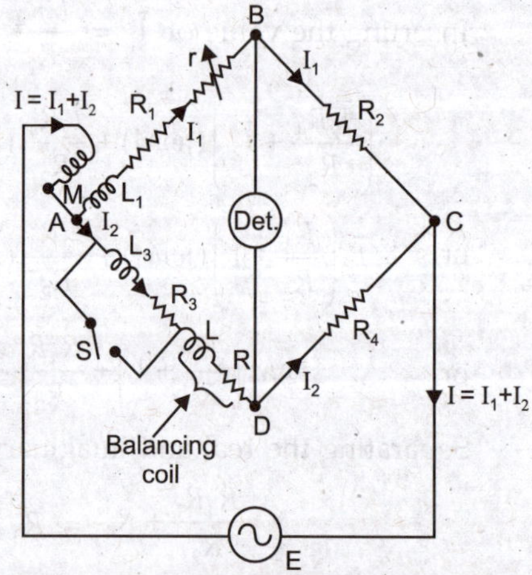


**Fig. 2.89.** Campbell's modification of Heaviside bridge.

— A short-circuiting switch $S$ is placed across the coil $R_3$, $L_3$ *under measurement.*

— Two sets of readings are taken–one with switch being open and the other with switch being closed. Let values of $M$ and $r$ be $M_1$ and $r_1$ with switch open, and $M_2$, $r_2$ with switch closed.

∴ With switch *open*, from eqn. (2.74), we have :

$$L_3 + L = \frac{M_1 (R_2 + R_4) + R_4 L_1}{R_2} \quad ...(i)$$

and, with switch *closed*,

$$L = \frac{M_2 (R_2 + R_4) + R_4 L_1}{R_2} \quad ...(ii)$$

Subtracting (*ii*) from (*i*), we get :

$$L_3 = (M_1 - M_2)\left(1 + \frac{R_4}{R_2}\right) \quad ...(2.76)$$

With switch *open*, from eqn. (2.75), we have :

$$R_3 + R = (R_1 + r_1) \times \frac{R_4}{R_2} \quad ...(iii)$$

and, with switch *closed*, $R = (R_1 + r_2) \times \frac{R_4}{R_2}$ ...(*iv*)

Subtracting eqn. (*iv*) from eqn. (*iii*), we get :

$$R_3 = (r_1 - r_2) \times \frac{R_4}{R_2} \quad ...(2.77)$$

- When we have "*equal ratio arms*", $R_2 = R_4$

and, therefore from eqns. (2.76) and (2.77), we get :

$$L_3 = 2(M_1 - M_2) \quad ...(2.78)$$

and, $$R_3 = r_1 - r_2 \quad ...(2.79)$$

## 2.6.13 CAREY FOSTER (Heydweiller) BRIDGE

Basically Carey Foster used this bridge, but it was subsequently modified by Heydweiller for use on A.C.

This bridge is *used* for the following *purposes* :

(*i*) *To measure the capacitance in terms of a standard mutual inductance.* The bridge in this case is known as *Carey Foster bridge.*

(*ii*) *To measure the mutual inductance in terms of standard capacitance.* The bridge in this case is known as *Heydweiller bridge.*

Fig. 2.90, shows the circuit and phasor diagrams for the bridge under balance conditions.

The unusual feature of this bridge is that one of its arms, *AD*, is *short-circuited* and therefore the potential drop across this arm is *zero*. For obtaining the balance, the potential drop across arm AB *should also be equal to zero* and for this reason *negative coupling is needed for the mutual inductance.*

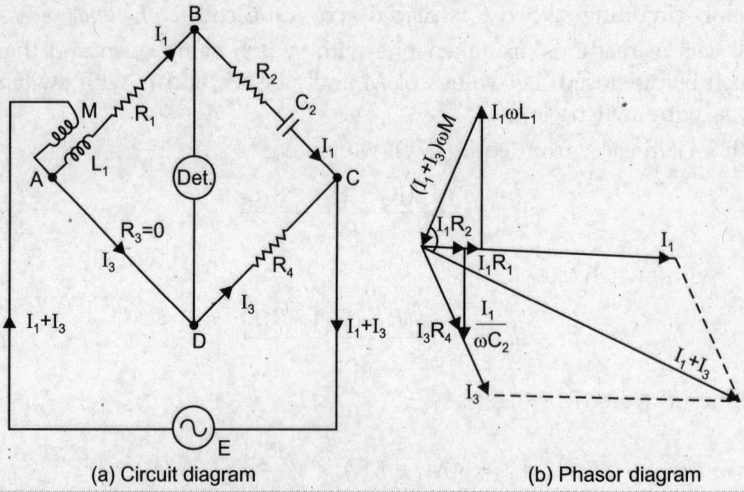


**Fig. 2.90.** Carey Foster (Heydweiller) bridge.

When the bridge is balanced :

$$\bar{I}_1 (R_1 + j\omega L_1) - (\bar{I}_1 + \bar{I}_3)\, j\omega M = 0$$

or, $$\bar{I}_1 (R_1 + j\omega L_1 - j\omega M) = \bar{I}_3 j\omega M \qquad ...(i)$$

and, $$\bar{I}_1 \left( R_2 + \frac{1}{j\omega C_2} \right) = \bar{I}_3 R_4 \qquad ...(ii)$$

Substituting the value of $\bar{I}_3$ from (*ii*) in (*i*), we get :

$$\bar{I}_1 (R_1 + j\omega L_1 - j\omega M) = \frac{\bar{I}_1}{R_4} \left( R_2 + \frac{1}{j\omega C_2} \right) j\omega M$$

or, $$R_1 + j\omega L_1 - j\omega M = \frac{R_2}{R_4} j\omega M + \frac{M}{R_4 C_2}$$

Equating the *real* and *imaginary* terms, we get :

$$R_1 = \frac{M}{R_4 C_2}$$

or, $$\boldsymbol{M} = R_1 R_4 C_2 \qquad ...(2.80)$$

and, $$L_1 = M \left( 1 + \frac{R_2}{R_4} \right)$$

or, $$\boldsymbol{L_1} = \frac{M(R_2 + R_4)}{R_4} = R_1 C_2 (R_2 + R_4) \qquad ...(2.81)$$

- If the bridge is used for *measurement of capacitance*, eqns. (2.80) and (2.81) may be written as :

$$\boldsymbol{C_2} = \frac{M}{R_1 R_4} \qquad ...(2.82)$$

and, $$\boldsymbol{R_2} = \frac{R_4 (L_1 - M)}{M} \qquad ...(2.83)$$

**Example 2.23.** *The following data relate to a modified Carey Foster bridge [Fig. 2.90(a)], used for measurement of capacitance :*

*Arm AB contains the secondary winding of mutual inductance of 18.35 mH and a total non-reactive resistance of 200 Ω. The inductance of secondary of mutual inductor in arm AB is 40.6 mH. Arm AD is short-circuited. Arm BC contains the unknown capacitor in series with a resistance of 119.5 Ω. Arm CD comprises a resistor of 100 Ω resistance. The detector is across BD.*

*Calculate the capacitance and its equivalent series resistance.* **(Roorkee University)**

**Solution.** Refer to Fig. 2.90(*a*).

*Given* : $M = 18.35$ mH; $R_1 = 200\ \Omega$, $L_1 = 40.6$ mH; $R_2 = 119.5\ \Omega$; $R_4 = 100\ \Omega$

We know that :
$$C_2 = \frac{M}{R_1 R_4} \quad \text{...[Eqn. (2.82)]}$$

$$= \frac{18.35 \times 10^{-3}}{200 \times 100} F = \mathbf{0.9175\ \mu F} \quad \textbf{(Ans.)}$$

and,
$$R_2 = \frac{R_4 (L_1 - M)}{M} = \frac{100\,(40.6 - 18.35)}{18.35} = 121.25\ \Omega$$

∴ *Series resistance of capacitor* = 121.25 – 119.5 = **1.75 Ω (Ans.)**

### 2.6.14 CAMPBELL BRIDGE

This bridge *measures an unknown mutual inductance in terms of a standard mutual inductance.*

Fig. 2.91 shows the circuit diagram of Campbell bridge.

Let, $M_1$ = *Unknown* mutual inductance;

$L_1$ = Self-inductance of secondary of mutual inductance $M_1$,

$M_2$ = Variable standard mutual inductance,

$L_2$ = Self-inductance of secondary of mutual inductance $M_2$, and

$R_1, R_2, R_3, R_4$ = Non-inductive resistances.

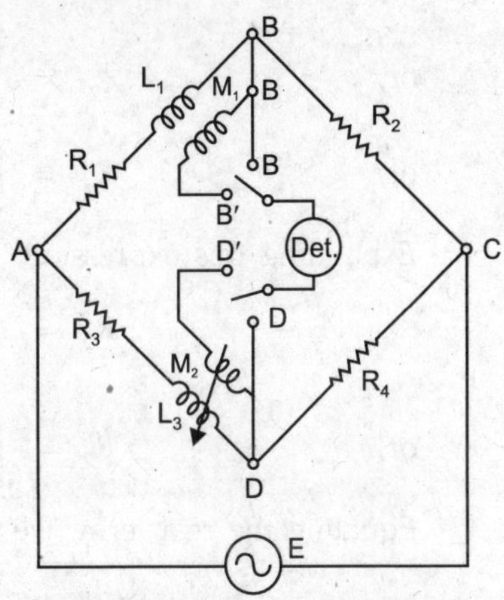


**Fig. 2.91.** Campbell's bridge.

For the balancing process, the following two steps are required :

**Step 1.** Detector is connected between B and D. The circuit now becomes a simple self-inductance comparison bridge. Then, condition for balance is :

$$\frac{L_1}{L_2} = \frac{R_1}{R_3} = \frac{R_2}{R_4} \quad \text{...(2.84)}$$

The bridge may be balanced by adjustment of $R_2$ (or $R_4$) and $R_1$ (or $R_3$).

**Step 2.** Detector is connected between $B'$ and $D'$. Keeping the adjustments as in step 1 above, the variable inductance $M_2$ is varied to get balance. Then,

$$\frac{M_1}{M_2} = \frac{R_2}{R_4}$$

or,
$$M_1 = M_2 \cdot \frac{R_2}{R_4} \quad \text{..(2.85)}$$

## 2.6.15. Wien Bridge (*For measurement of* "Frequency")

The Wien bridge is *primarily known as a "frequency" determining bridge*. It can also be used for the *measurement of an unknown capacitor with great accuracy*.

Fig. 2.92 shows a Wien bridge. It has a series *R-C* combination in one arm and a parallel combination in the adjoining arm.

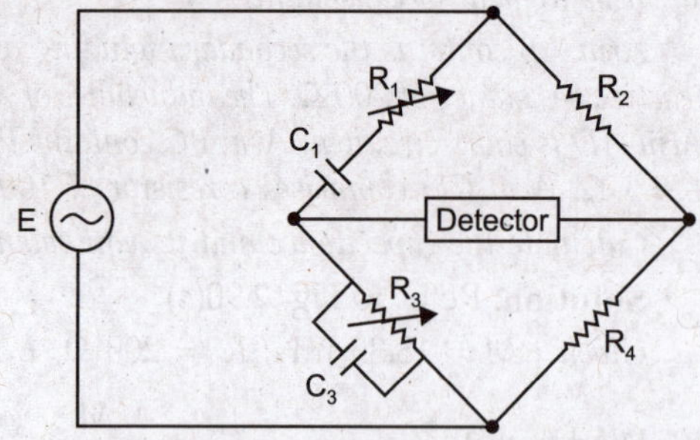


**Fig. 2.92.** Wien bridge for measuring frequency.

The impedance of one arm is,

$$\bar{Z}_1 = R_1 - \frac{j}{\omega C_1}$$

The admittance of parallel arm is,

$$\bar{Y}_3 = \frac{1}{R_3} + j\omega C_3$$

Using the bridge balance equation, we have :

$$\bar{Z}_1 \bar{Z}_4 = \bar{Z}_2 \bar{Z}_3$$

or, $$\bar{Z}_1 \bar{Z}_4 = \bar{Z}_2 \times \frac{1}{\bar{Y}_3}$$

or, $$\bar{Z}_2 = \bar{Z}_1 \bar{Z}_4 \bar{Y}_3$$

or, $$R_2 = \left(R_1 - \frac{j}{\omega C_1}\right) R_4 \left(\frac{1}{R_3} + j\omega C_3\right) \quad ...(2.86)$$

Expanding this expression, we get :

$$R_2 = \frac{R_1 R_4}{R_3} + j\omega R_1 R_4 C_3 - \frac{jR_4}{\omega C_1 R_3} + \frac{R_4 C_3}{C_1}$$

or, $$R_2 = \left(\frac{R_1 R_4}{R_3} + \frac{R_4 C_3}{C_1}\right) - j\left(\frac{R_4}{\omega C_1 R_3} - \omega C_3 R_1 R_4\right) \quad ...(2.87)$$

Equating the real term, we obtain :

$$R_2 = \frac{R_1 R_4}{R_3} + \frac{R_4 C_3}{C_1}$$

or, $$\frac{R_2}{R_4} = \frac{R_1}{R_3} + \frac{C_3}{C_1} \quad ...(2.88)$$

Equating the imaginary terms, we have :

$$\omega C_3 R_1 R_4 = \frac{R_4}{\omega C_1 R_3} \quad ...(2.89)$$

or, $$\omega^2 = \frac{1}{C_1 C_3 R_1 R_3} \quad ...(2.89\ a)$$

or, $$\omega = \frac{1}{\sqrt{C_1 C_3 R_1 R_3}}$$

Since, $\omega = 2\pi f$, therefore,

$$2\pi f = \frac{1}{\sqrt{C_1 C_3 R_1 R_3}}$$

or, $$f = \frac{1}{2\pi\sqrt{C_1 C_3 R_1 R_3}} \quad ...(2.90)$$

We observe that the two conditions for bridge balance, (eqns. 2.88, 2.90), result in an expression determining the required resistance ratio $\frac{R_2}{R_4}$, and another expression determining the frequency of the applied voltage. In other words, if we satisfy eqn. (2.88), and also excite the bridge with a frequency described by eqn. (2.88), the bridge will be in balance.

In most Wien bridge circuits, the components are chosen such that $R_1 = R_3$ and $C_1 = C_3$; this reduces eqn. (2.88) to $\frac{R_2}{R_4} = 2$ and eqn. (2.90) to

$$f = \frac{1}{2\pi RC} \quad ...(2.69)$$

which is *general expression for the frequency of the Wien bridge.*

— In a practical bridge, capacitors $C_1$ and $C_3$ are *fixed* capacitors, and resistors $R_1$ and $R_3$ are variable resistors *controlled by a common shaft*. Provided now that $R_2 = 2R_4$, the bridge may be used as a frequency-determining device balanced by a single *control. This control may be calibrated directly in terms of frequency.*

- By the use of this bridge an *accuracy of 0.5% - 1% can be easily obtained.*
- Because it is frequency sensitive, it is difficult to balance unless the waveform of the applied voltage is purely sinusoidal.

**Applications :**

1. The bridge is employed for *measuring frequency in the audio range.*
   — The audio range is normally divided into 20Hz - 200Hz - 2kHz - 20kHz ranges. In this case, the resistances can be used for range changing and capacitors $C_1$ and $C_3$ for fine frequency control within the range.
2. The bridge can also be used for *measuring capacitances;* in this case operational frequency must be known.
3. The bridge is also used in a harmonic distortion analyser, as a notch filler, and in audio frequency and radio frequency oscillators as a frequency determining element.

**Example 2.24.** *Determine the equivalent parallel resistance and capacitance that causes a Wien bridge (Fig. 2.80) to null with the following component values :*

*$R_1 = 2.8k\Omega$ ; $C_1 = 4.8\mu F$ ; $R_2 = 20k\Omega$; $R_4 = 80k\Omega$; $f = 2$ kHz;*

**Solution.** *Given* : $f$ = 2kHz or $\omega = 2\pi \times (2 \times 1000) = 12.57 \times 10^3$ rad/s.

$$\mathbf{R_3 = ?, C_3 = ?}$$

We know that : $$\frac{R_2}{R_4} = \frac{R_1}{R_3} + \frac{C_3}{C_1} \quad ...(\text{Eqn. } 2.88)$$

and, $$\omega^2 = \frac{1}{C_1 C_3 R_1 R_3} \quad ...(\text{Eqn. } 2.89a)$$

or, $$C_3 = \frac{1}{w^2 C_1 R_1 R_3}$$

Substituting for $C_3$ in eqn. (2.88), we get

$$\frac{R_2}{R_4} = \frac{R_1}{R_3} + \frac{1}{\omega^2 C_1^2 R_1 R_3}$$

$$\frac{R_2}{R_4} = \frac{1}{R_3}\left(R_1 + \frac{1}{\omega^2 C_1^2 R_1}\right)$$

or, $$R_3 = \frac{R_4}{R_2}\left(R_1 + \frac{1}{\omega^2 C_1^2 R_1}\right) \quad ...(i)$$

$$= \frac{80\times10^3}{20\times10^3}\left[2.8\times10^3 + \frac{1}{(12.57\times10^3)^2(4.8\times10^{-6})^2(2.8\times10^3)}\right]$$

$$= 4\,(2.8\times10^3 + 0.098) = \mathbf{11.2k\Omega\ (Ans.)}$$

Now, $$C_3 = \frac{1}{\omega^2 C_1 R_1 R_3} \quad \text{Rearranging (Eqn. 2.89}a)$$

$$= \frac{1}{(12.57\times10^3)^2\times(4.8\times10^{-6})\times(2.8\times10^3)\times11.2\times10^3}$$

$$= \mathbf{42.04\ pF\ (Ans.)}$$

## Additional Worked Examples on Bridges

**Example 2.25.** *In a bridge shown in Fig. 2.93 the balance was obtained with $L_1$ = 52.6 mH and $R_2$ = 1.68Ω. Find the resistance and inductance of the coil.*

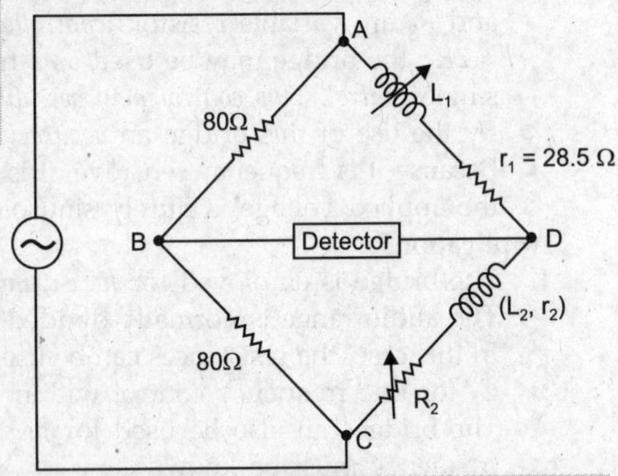


Fig. 2.93

**Solution :** *Given* : $L_1$ = 52.6 mH; $R_2$ = 1.68Ω ; $r_1$ = 28.5Ω.

Refer to Fig. 2.93.

*$r_2$; $L_2$* :

Balance of the bridge was obtained when $L_1$ = 52.6 mH and $R_2$ = 1.68Ω

At balance :

$80\,(r_1 + j\,\omega L_1) = 80\,[(R_2 + r_2) + j\omega L_2]$

or, $r_1 + j\omega L_1 = (R_2 + r_2) + j\omega L_2$

Equating the *real terms*, we get,

$r_1 = R_2 + r_2$

$\therefore\ r_2 = r_1 - R_2 = 28.5 - 1.68 =$ **26.82 Ω (Ans.)**

Equating *imaginary terms*, we have :

$$\omega L_1 = \omega L_2$$

or, $$L_1 = L_2 = \mathbf{52.6\ mH\ (Ans.)}$$

**Example 2.26.** *The bridge shown in the Fig. 2.94 is balanced at 1 kHz. It has the following components :*

*Arm AB = 0.2μ F pure capacitance; Arm BC = 500Ω pure resistance; Arm DA = 300Ω parallel with 0.1μF.*

*Find the constants of arm CD considering it as a series circuit.*

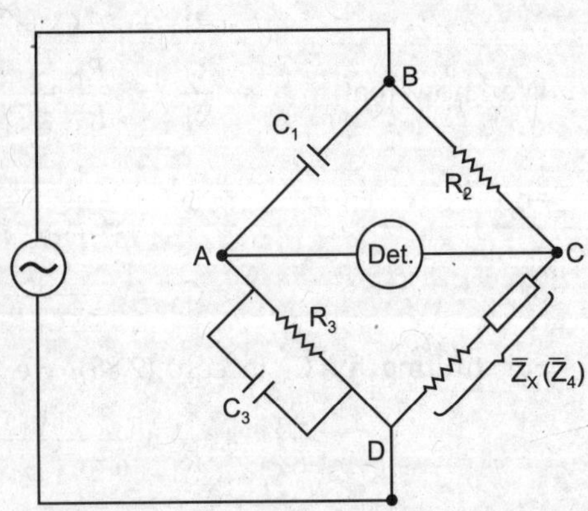


Fig. 2.94

**Solution.** *Given* : $C_1 = 0.2\mu F$ : $R_2$ = 500Ω; $R_3$ = 300Ω, $C_3$ = 0.1μF; $f$ = 1kHz or 1000 Hz.

Now, $$\bar{Z}_1 = 0 - j\left(\frac{1}{\omega C_1}\right)$$

$$\bar{Z}_2 = R_2$$

$$\bar{Z}_3 = R_3 \parallel X_{C3}$$

or,
$$\bar{Y}_3 = \frac{1}{R_3} + j\left(\frac{1}{X_{C3}}\right) = \frac{1}{R_3} + j(\omega C_3)$$

$$\bar{Z}_4 = \bar{Z}_x$$

The basic balance equation is given by :

$$\bar{Z}_1 \bar{Z}_4 = \bar{Z}_2 \bar{Z}_3$$

$\therefore$
$$\bar{Z}_4 = \frac{\bar{Z}_2 \bar{Z}_3}{\bar{Z}_1} = \frac{\bar{Z}_2}{\bar{Z}_1 \bar{Y}_3} = \frac{R_2}{-\frac{j}{\omega C_1}\left[\frac{1}{R_3} + j(\omega C_3)\right]}$$

Now,
$$-\frac{1}{j} = j$$

$\therefore$
$$\bar{Z}_4 (= \bar{Z}_x) = \frac{j\,\omega C_1 R_2}{\frac{1}{R_3} + j\,\omega C_3} = \frac{0 + j\,(2\pi \times 1000) \times 0.2 \times 10^{-6} \times 500}{\frac{1}{300} + j\,(2\pi \times 1000) \times 0.1 \times 10^{-6}}$$

$$= \frac{0 + j\,0.6283}{3.33 \times 10^{-3} + 6.283 \times 10^{-4}}$$

$$= \frac{0.6283 \angle 90^\circ}{3.3888 \times 10^{-3} \angle +10.68^\circ} = 185.4 \angle 79.32^\circ$$

$$= 34.36 + j\,182.19\ \Omega$$

$\therefore$
$$\bar{Z}_x = (34.36 + j\,182.19)\ \Omega$$

or, $\mathbf{R_x = 34.36\ \Omega}$ **(Ans.)**

and, $X_{L_x} = 2\,\pi\,f\,L_x = 182.19$

$\therefore$
$$L_x = \frac{182.19}{2\pi \times 1000};\ 0.029 \text{ or } \mathbf{29\ mH}$$ **(Ans.)**

**Example 2.27.** *The following data relate to the bridge shown in Fig. 2.95.*

*$R_1 = 200\ \Omega$ ; $R_2 = 200\ \Omega$, $C_2 = 5\mu F$; $R_3 = 500\Omega$; $C_3 = 0.2\mu F$.*

*Find the constants of $Z_x$ considering it as a series circuit.*

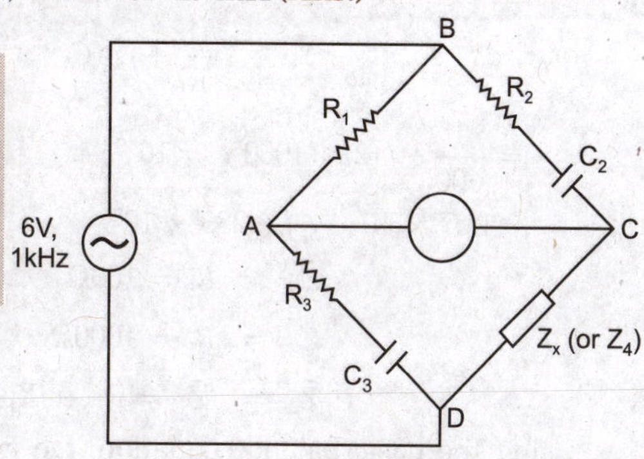

Fig. 2.95

**Solution.** Refer to Fig. 2.95

$$\bar{Z}_1 = R_1$$
$$= 200\Omega$$

$$\bar{Z}_2 = R_2 - j\left(\frac{1}{\omega C_2}\right)$$

$$= 200 - j\left[\frac{1}{2\pi \times 1 \times 10^3 \times 5 \times 10^{-6}}\right] = 200 - j\,31.83\ \Omega$$

$$\bar{Z}_3 = R_3 - j\left(\frac{1}{\omega C_3}\right)$$

$$= 500 - j\left[\frac{1}{2\pi \times 1 \times 10^3 \times 0.2 \times 10^{-6}}\right] = 500 - j\,795.77\ \Omega$$

When the bridge is balanced,

$$\bar{Z}_1\,\bar{Z}_4 = \bar{Z}_2\,\bar{Z}_3$$

$$\therefore \quad \bar{Z}_4 \text{ (or } \bar{Z}_x) = \frac{\bar{Z}_2\,\bar{Z}_3}{\bar{Z}_1} = \frac{(200 - j\,31.83)(500 - j\,795.77)}{200}$$

$$= \frac{(202.517\,\angle -9.04°)(939.81\,\angle -57.86°)}{200}$$

or, $$\bar{Z}_x = 951.64\,\angle -66.9° = (373.36 - j\,875.3)\ \Omega$$

$\therefore$ $R_x$ = **373.36 Ω (Ans.)**

and, $$X_c = \frac{1}{2\pi f C_x} = 875.3$$

or, $C_x = \dfrac{1}{2\pi \times 1 \times 10^3 \times 875.3}$ = **0.1818μF (Ans.)**

**Example 2.28.** *The following data relate to the bridge shown in Fig. 2.96.*

*$R_1 = 600\Omega$; $C_1 = 1\mu F$; $R_2 = 100\Omega$ ; $R_3 = 1000\Omega$*

*Find the values of $R_x$ and $L_x$ at balance.*

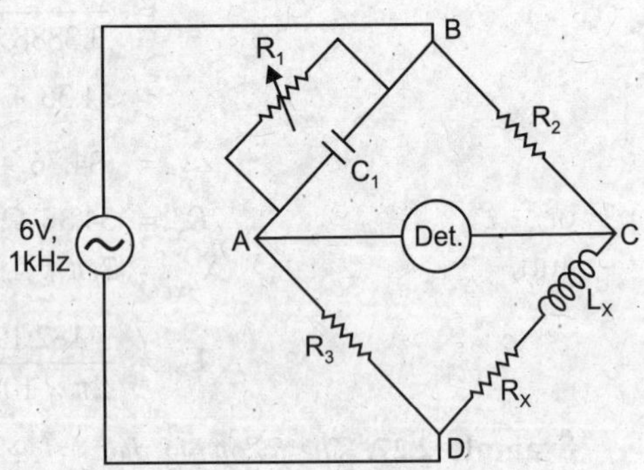


Fig. 2.96

**Solution.** Refer to Fig. 2.96

$$\bar{Z}_1 = R_1 \parallel X_{C_1}$$

or, $$\bar{Y}_1 = \frac{1}{R_1} + j\,\omega C_1$$

$$= \frac{1}{600} + j\,(2\pi \times 1000 \times 1 \times 10^{-6})$$

$$= 1.67 \times 10^{-3} + j\,6.283 \times 10^{-3}$$

$$\bar{Z}_2 = R_2 = 100\Omega$$

$$\bar{Z}_3 = R_3 = 1000\Omega$$

$$\bar{Z}_4 = \bar{Z}_x = R_x + jX_L$$

Using the basic balance equation, we get :

$$\bar{Z}_1\,\bar{Z}_4 = \bar{Z}_2\,\bar{Z}_3$$

or, $$\bar{Z}_4 (= \bar{Z}_x) = \frac{\bar{Z}_2\,\bar{Z}_3}{\bar{Z}_1} = \bar{Z}_2\,\bar{Z}_3\,\bar{Y}_1$$

$$= 100 \times 1000 \times [1.67 \times 10^{-3} + j\,6.283 \times 10^{-3}]$$

$$= (167 + j\,628.3)\ \Omega = (R_x + j\,X_L)\ \Omega$$

$$\therefore \quad R_x = \mathbf{167\ \Omega\ (Ans.)}$$

$$X_{L_x} = 628.3\ \Omega = 2\pi f L_x$$

$$\therefore \quad L_x = \frac{628.3}{2\pi \times 1000};\ \mathbf{0.1\ H\ (Ans.)}$$

**Example 2.29.** *A bakelite sheet of 5 mm thickness is tested at 50 Hz between the electrodes 12 cm in diameter. The Schering bridge (Fig. 2.97) used has an air capacitor $C_2$ of 106 pF, a non-reactive resistance of $R_4$ of $\left(\frac{1000}{\pi}\right)$ $\Omega$ in parallel with a variable capacitor $C_4$ and a non-reactive variable resistance $R_3$. Balance is obtained with $C_4 = 0.55\mu F$ and $R_3 = 270\Omega$.*

*Determine the following :*

*(i) Capacitance ;*

*(ii) Power factor, and*

*(iii) Relative permittivity of the sheet.*

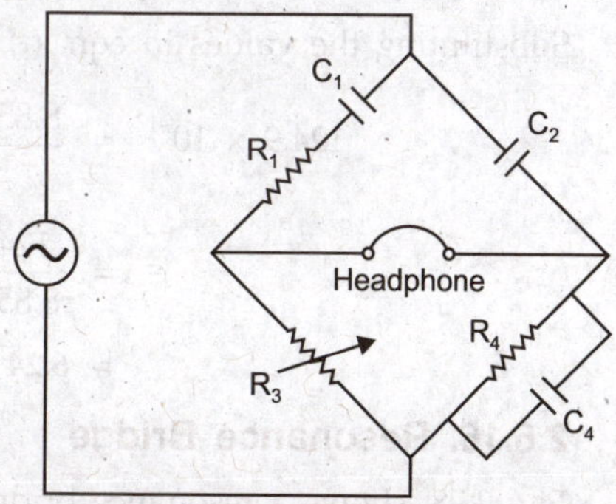


Fig. 2.97

**Solution.** Refer to Fig. 2.97.

*Given :*

$$C_2 = 106\text{pF} = 106 \times 10^{-12}\ \text{F};$$

$$R_4 = \frac{1000}{\pi}\ \Omega;$$

$$C_4 = 0.55\mu\ \text{F} = 0.55 \times 10^{-6}\ \text{F};$$

$$R_3 = 270\ \Omega$$

*(i)* **Capacitance *C* :**

When the bridge is balanced, we have :

$$R_1 = \frac{R_3 C_4}{C_2} \quad \text{...[Eqn. (2.58)]}$$

$$= \frac{270 \times 0.55 \times 10^{-6}}{106 \times 10^{-12}} = 1.4\ \text{M}\ \Omega$$

and,

$$C_1 = \frac{R_4 . C_2}{R_3} \quad \text{...[Eqn. (2.59)]}$$

$$= \frac{(1000/\pi) \times 106 \times 10^{-12}}{270}$$

$$= \mathbf{124.9\ pF\ (Ans.)}$$

*(ii)* **Power factor :**

$$\text{Power factor} = \omega R_1 C_1 \quad \text{...[(Eqn. (2.60)]}$$

$$= 2\pi \times 50 \times (1.4 \times 10^6) \times (124.9 \times 10^{-12})$$

$$= \mathbf{0.0549\ (Ans.)}$$

*(iii)* **Relative permittivity $\epsilon_r$ :**

We know that

$$C = \frac{\epsilon_0 \epsilon_r A}{t} \quad \text{...}(i)$$

where, $C\ (= C_1)$ = Capacitance,

$\varepsilon_o = 8.854 \times 10^{-12}$,

$A$ = Area of the electrodes $= \frac{\pi}{4} \times \left(\frac{12}{100}\right)^2 = 0.0113\ m^2$, and

$t$ = Thickness of bakelite sheet = 5 mm or 0.005 m.

Substituting the values in eqn. (*i*), we get :

$$124.9 \times 10^{-12} = \frac{8.854 \times 10^{-12} \times \epsilon_r \times 0.0113}{0.005}$$

or,

$$\epsilon_r = \frac{124.9 \times 10^{-12} \times 0.005}{8.854 \times 10^{-12} \times 0.0113}$$

$$= \mathbf{6.24\ (Ans.)}$$

### 2.6.16. Resonance Bridge

Fig. 2.98. shows a resonance bridge, consisting of a series resonance circuit :

— The series resonance circuit is formed by $R_4$, $C_4$ and $L_4$ in series.

— All the other arms consist of resistors only.

Using the bridge balance equation, we have :

$$\bar{Z}_1 \bar{Z}_4 = \bar{Z}_2 \bar{Z}_3$$

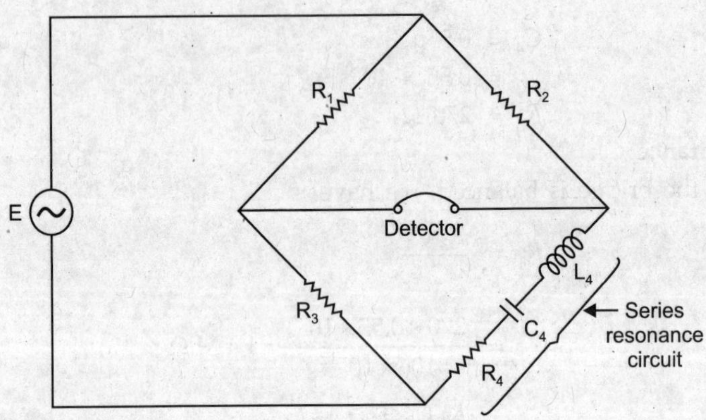


**Fig. 2.98.** Resonance bridge.

Here, $\bar{Z}_1 = R_1$ ; $\bar{Z}_2 = R_2$ ; $\bar{Z}_3 = R_3$ ; $\bar{Z}_4 = R_4 + j\omega L_4 - \frac{j}{wC_4}$

Substituting the values in the balance equation, we get :

$$R_1 \left( R_4 + j\omega L_4 - \frac{j}{\omega C_4} \right) = R_2 R_3$$

or,

$$R_1 R_4 + j\,\omega\, L_4 R_1 - \frac{jR_1}{\omega C_4} = R_2 R_3$$

Equating the real and imaginary terms, we have :

$$R_1R_4 = R_2R_3$$

$$\therefore \qquad R_4 = \frac{R_2R_3}{R_1}$$

and, $$\omega L_4 - \frac{1}{\omega C_4} = 0$$

or, $$\omega^2 = \frac{1}{L_4\,C_4}$$

or, $$(2\pi f)^2 = \frac{1}{L_4\,C_4} \quad \text{or} \quad 2\pi f = \frac{1}{\sqrt{L_4\,C_4}}$$

or, *Resonance frequency,* $$f = \frac{1}{2\pi\sqrt{L_4\,C_4}} \qquad \text{...(2.70)}$$

- This bridge can be *used to measure unknown inductances or capacitances*. The resistor $R_4$ can be measured by keeping a fixed ratio $\frac{R_3}{R_1}$ and using a standard variable resistance to obtain balance.
  — If an *"inductance"* is *being measured, a standard capacitor is varied unitl balance is obtained.*
  — If a *"capacitance"* is *being measured, a standard inductor is varied until balance is obtained.*

The operating frequency of the generator must be known in order to calculate the unknown quantity. *Balance is indicated by the minimisation of sound in the headphones.*

## 2.6.17. Wagner Earthing Device

In the earlier discussion it has been assumed that the four bridge arms consist of simple lumped impedances which do not interact in any way. However, in practice, *stray capacitances* exist between the various bridge elements and ground, and also between bridge arms themselves. *These stray capacitances shunt the bridge arms and cause measurement error, particularly at the higher frequencies or when small capacitors or large inductors are measured.*

One way to control stray capacitances is by *shielding the arms and connecting the shields to the ground*. This does not eliminate the capacitances but at least makes them constant in value, and they can therefore be *compensated*.

- One of the most widely used methods for eliminating some of the effects of stray capacitance in a bridge circuit is the *Wagner earth (ground) connection*. This circuit *eliminates the troublesome capacitance which exists between the detector terminals and the ground.*

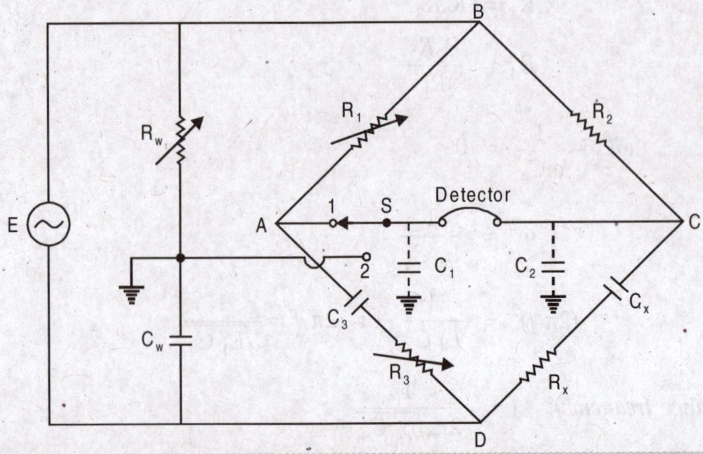


**Fig. 2.99.** Wagner earthing device.

Fig. 2.99. shows the circuit of a capacitance bridge. $C_1$ and $C_2$ represent the stray capacitances. The oscillator is removed from its usual ground connection and bridged by a series combination of resistor $R_w$ and capacitor $C_w$. The junction of $R_w$ and $C_w$ is grounded and is called *Wagner ground connection.*

The initial adjustment of the bridge involves the following *procedure* :

— The detector is connected to point 1, and $R_1$ is adjusted for null or minimum sound in the headphones.

— The switch is then thrown to position 2, which connects the detector to wagner ground point. Resistor $R_w$ is now adjusted for minimum sound.

— When the switch is thrown to position 1 again, some unbalance will probably exist. Resistors $R_1$ and $R_3$ are then adjusted for minimum detector response, and the switch is again thrown to position 2.

This procedure is repeated till a null is reached on both switch positions; under this condition points 1 and 2 are at the same potential and this is *ground potential.* Stray capacitances $C_1$ and $C_2$ are then effectively shorted out and have no effect on normal bridge balance.

The capacitances from point *B* to *D* to ground are also eliminated by the addition of Wagner ground connection, since current through these capacitances will enter the Wagner ground connection.

The capacitances across the bridge arms are *not* eliminated by this earthing device and they will still *affect* the accuracy of the measurement.

- Because the addition of Wagner earthing device does not affect the balance conditions, the *measurement procedure* remains *unchanged.*

### 2.6.18. Types of Detectors

Following types of detectors are used in various applications :

1. *For low frequency ...* Vibration galvanometer.
   — A moving magnet vibration galvanometer has a range of 1.5 k Hz.
2. *For ordinary laboratory work at frequencies up to a few 100 Hz...* Moving coil type of instrument; it has a high sensitivity.

3. *For higher frequencies (> 800 Hz) ...* Telephone (headphone)-best detector (Vibration galvanometers and headphones have no phase selectivity, *i.e.*, they do not indicate whether it is resistance or reactance adjustments that are required).
4. *For frequencies above about 3 kHZ, and particularly at high AF or RF...* ..... Heterodyne or beat tone detector

Following points are worth noting :

- The A.C. galvanometer and separately excited dynamometer are phase selective, and are best suited at low frequencies.
- An electrodynamometer can also be used as an A.C. detector.
- Electrometers are used as detectors because small capacitances possess a very large impedance when used with A.C. circuits at low frequency, and when measured in the bridge they form a high impedance branch. Hence, this detector is used to *increase sensitivity*.
- In several cases, especially in bridges for routine use, a pointer instrument is used. It is advantageous if it can be made phase selective. These pointer instruments are generally moving coil milliammeters operated with some arrangement of copper oxide rectifiers (frequency range 40 Hz - 1 kHz)
- *Modern bridge techniques employ the amplifier as a regular feature.*
- With all detectors, the impedance should be selected to suit that of the bridge. A higher sensitivity can be obtained by using an interbridge transformer.
  — When *headphones* are used as detectors, precations should be taken *to eluiminate capacitance effects* between the observer and the headphones.

### 2.6.19. Precautions to be taken when using a Bridge

When using a bridge, the following *precautions* should be taken to obtain accurate readings :

1. The layout of the leads should be such that no loops or long lengths enclosing magnetic flux are produced, with consequent stray inductance errors.
2. The self-capacitance of the leads, with a large L, is more important than their inductance, so they should be spaced relatively far apart.
3. When measuring a capacitor, it is important to keep capacitance as low as possible. Owing to this reason the leads should *not be too close together* and should be made of fine wire.
4. In case of very precise inductive and capacitive measurements, leads are encased in metal tubes to shield them from mutual electromagnetic action, and are so designed as to completely shield the bridge.

### 2.6.20. Q-Meter

The overall efficiency of the coils and capacitors intended for radio frequency (RF) applications is best evaluated using the Q value.

**Q-meter** *is an instrument designed to measure some electrical properties of coils and capacitors.* The *principle* of the Q-meter is based on *'series resonance'*, the voltage drop across the coil or capacitor is Q times the applied voltage (where *Q is the ratio of reactance to resistance,* $\frac{X_L}{R}$). If a fixed voltage is applied to the circuit, a *voltmeter across the capacitor can be* *calibrated to read Q directly.*

At resonance : $X_L = X_C$ ; $V_L = IX_L$ ; $V_C = IX_C$, $V = IR$

where,

$X_L$ = Inductive reactance,
$X_C$ = Capacitive reactance,
$R$ = Coil resistance,
$I$ = Circuit current,
$V_L$ = Inductive voltage,
$V_C$ = Capacitor voltage, and
$V$ = Applied voltage.

Therefore,
$$Q = \frac{X_L}{R} = \frac{X_C}{R}\left(=\frac{1}{\omega CR}\right) = \frac{V_C}{V} \qquad ...(2.93)$$

In the above eqn. (2.93), if *V is kept constant, the voltage across the capacitor can be measured by a voltmeter calibrated to read directly in terms of Q.*

Fig. 2.100 shows the basic Q-meter circuit :

- A wide-range *oscillator* with frequency range from 50 kHz to 50 MHz is used as a power supply to the circuit. The output of the oscillator is shorted by a low-value resistance, Rsh usually of the order of 0.02Ω. So it introduces almost no resistance into the oscillatory circuit and represents a voltage source with a very small or almost negligible internal resistance. The voltage across the low-value shunt resistance $R_{sh}$ $V$ is measured by a *thermocouple meter* and the voltage across the capacitor $V_C$ is measured by an electronic voltmeter.

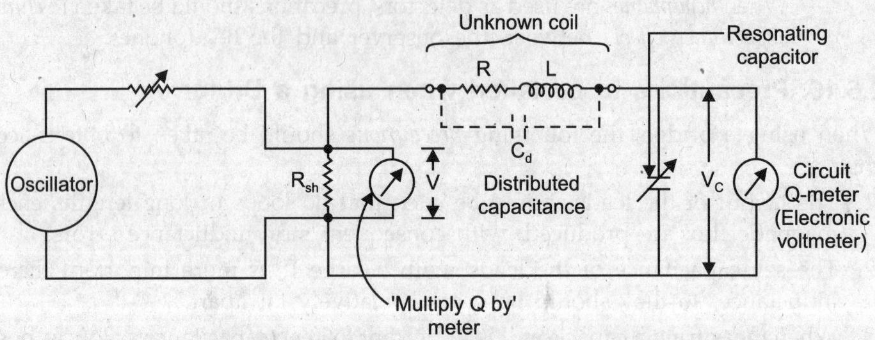


**Fig. 2.100.** Circuit diagram of a Q-meter.

- The inductance (*L*) of the coil can be determined by connecting it to the test terminals of the instrument. The circuit is tuned to resonance by varying either the capacitance or the oscillatory frequency. If the capacitance is varied, the oscillator frequency is adjusted to a given frequency and resonance is obtained. If the capacitance is pre-set to a desired value, the oscillator frequency is varied until resonance occurs. The *Q reading on the output meter must be multiplied by index setting or the "Multiply Q by" switch to obtain the actual value.*

  The inductance of the coil can be calculated from known values of the coil frequency and resonating capacitor (C).

$$X_L = X_C$$

or,
$$2\pi fL = \frac{1}{2\pi fC} \quad \text{or} \quad 4\pi^2 f^2 LC = 1$$

or $$L = \frac{1}{(2\pi f)^2 C} \quad \text{...(2.94)}$$

- The $Q$ indicated is *not the actual* $Q$, because the losses of the resonating capacitor, voltmeter and inserted resistance are all included in the measuring circuit. The actual $Q$ of the measured coil is *somewhat greater than the indicated* $Q$. *This difference is negligible except where the resistance of the coil is relatively small compared to the inserted resistance* $R_{sh}$.

**Measurement of low impedance components :**

Low-impedance components, such as low-value resistors, small coils, and large capacitors, are measured in series with the measuring circuit, as shown in Fig. 2.101. The component to be measured, $Z_x$, is connected in series with a stable *work coil* usually provided with the instrument across the test terminals.

*Two measurements* are made :

(*i*) In the first measurement the unknown is short-circuited by a small *shorting-strap* and the circuit is resonated, establishing a reference condition. The values of the tuning capacitor ($C_1$) and the indicated $Q$ ($Q_1$) are noted.

(*ii*) In the *second* measurement, the shorting strap is removed and the circuit is retuned, giving a new value for the tuning capacitor ($C_2$) and a change in the $Q$ value from $Q_1$ to $Q_2$.

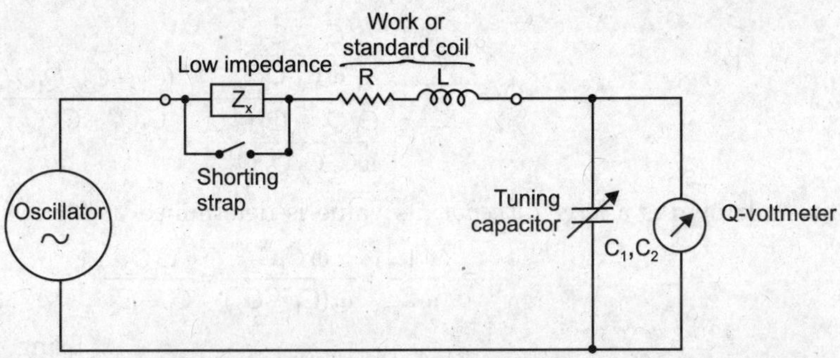


**Fig. 2.101.** Q-meter measurement of a low-impedance component in the *series* connection.

For the second measurement the reactance of the unknown ($X_x$) can be expressed in terms of the *new* value of the tuning capacitor ($C_2$) and the *in-circuit* value of the inductor ($L$), this yields

$$X_x = X_{C2} - X_L$$

or, $$X_x = X_{C2} - X_{C1} \quad \text{...(2.95)}$$

$$(\because X_L = X_{C1})$$

$$= \frac{1}{\omega C_2} - \frac{1}{\omega C_1}$$

or, $$X_x = \frac{C_1 - C_2}{\omega C_1 C_2} \quad \text{...(2.96)}$$

$X_x$ is inductive if $C_1 > C_2$ and capacitive if $C_1 < C_2$.

The resistive component of the unknown impedance $R_x$ can be found in terms of reactance $X_x$ and the indicated values of circuit $Q$, since,

$$R_1 = \frac{X_1}{Q_1}, \text{ and } R_2 = \frac{X_2}{Q_2}$$

Also,
$$R_x = R_2 - R_1 = \frac{1}{\omega C_2 Q_2} - \frac{1}{\omega C_1 Q_1}$$

or,
$$R_x = \frac{C_1 Q_1 - C_2 Q_2}{\omega C_1 C_2 Q_1 Q_2} \qquad ...(2.97)$$

— If the unknown is *purely resistive*, the setting of the tuning capacitor would *not* have changed in the measuring process, and $C_1 = C_2$. The equation for resistance, then, reduces to,

$$R_x = \frac{Q_1 - Q_2}{\omega C_1 Q_1 Q_2} = \frac{\Delta Q}{\omega C_1 Q_1 Q_2} \qquad ...(2.98)$$

— If the unknown is a *small inductor*, the value of the inductance is found from eqn. (2.98) as follows :

$$L_x = \frac{C_1 - C_2}{\omega^2 C_1 C_2} \qquad ...(2.99)$$

$$(\because X_x = \omega L_x)$$

- The $Q$ of the coil can be determined by the definition,

*i.e.*,
$$Q_x = \frac{X_x}{R_x} = \frac{\left(\dfrac{C_1 - C_2}{\omega C_1 C_2}\right)}{\left(\dfrac{C_1 Q_1 - C_2 Q_2}{\omega C_1 C_2 Q_1 Q_2}\right)} = \frac{(C_1 - C_2) Q_1 Q_2}{C_1 Q_1 - C_2 Q_2} \qquad ...(2.100)$$

— If the unknown is a *large capacitor*, its value is determined as follows :

$$C_x = \frac{1}{\omega X_x} = \frac{\omega C_1 C_2}{\omega (C_1 - C_2)} = \frac{C_1 C_2}{C_1 - C_2} \qquad ...(2.101)$$

... using eqn. (2.99)

- The $Q$ of the capacitor can be found by using eqn. (2.100).

**Measurement of high impedance components :**

High impedance components, such as high-value resistors, certain inductors, and small capacitors, are measured by connecting them in *parallel* with the measuring circuits, as shown in Fig. 2.102.

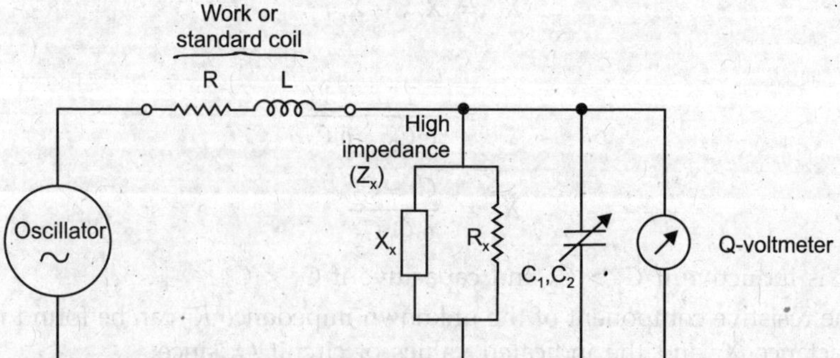


**Fig. 2.102.** Q-measurement of a high-impedance component in the *parallel* connection.

Prior to connecting the unknown, the circuit is resonated, by using a suitable work coil, to establish reference values for $Q$ and $C$ ($Q_1$ and $C_1$). Then, when the unknown component is connected to the circuit, the capacitor is readjusted for resonance, and a *new value* for the tuning capacitance ($C_2$) is obtained and a change in the value of circuit $Q$ ($\Delta Q$) from $Q_1$ to $Q_2$.

The unknown impedance is measured in terms of its parallel components $X_x$ and $R_x$. At the initial resonance condition, when the unknown is *not yet* connected into the circuit, the working coil ($L$) is tuned by the capacitor ($C_1$), and we have

$$\omega L = \frac{1}{\omega C_1} \quad \text{...(2.102)}$$

and,

$$Q_1 = \frac{\omega L}{R} = \frac{1}{\omega C_1 R} \quad \text{...(2.103)}$$

When the unknown impedance ($Z_x$) is connected into the circuit and the capacitor is tuned for resonance, the reactance of the working coil ($X_L$) equals the parallel reactances of the tuning capacitor ($X_{C_2}$) and the unknown ($X_x$), and we have,

$$X_L = \frac{X_{C_2} X_x}{X_{C_2} + X_x}$$

or,
$$X_L (X_{C_2} + X_x) = X_{C_2} X_x$$

or,
$$X_L X_{C_2} + X_L X_x = X_{C_2} X_x$$

or,
$$X_x (X_{C_2} - X_L) = X_L X_{C_2}$$

or,
$$X_x = \frac{X_L X_{C_2}}{(X_{C_2} - X_L)} = \frac{\frac{1}{\omega C_1}\,\frac{1}{\omega C_2}}{\frac{1}{\omega C_2} - \frac{1}{\omega C_1}} = \frac{1}{\omega (C_1 - C_2)} \quad \text{...(2.104)}$$

— If the unknown is *inductive*, $X_x = \omega L_x$, eqn. (2.104) yields the value of the unknown inductance :

$$L_x = \frac{1}{\omega^2 (C_1 - C_2)} \quad \text{...(2.105)}$$

— If the unknown is *capacitive*, $X_x = \dfrac{1}{\omega C_x}$ the eqn. (2.104) yields the value of the unknown capacitor.

$$C_x = C_1 - C_2$$

In a *parallel resonant circuit* the total resistance ($R_t$) at resonance is equal to the product of the circuit $Q$ and the reactance of the coil.

*i.e.*
$$R_t = Q_2 X_L = Q_2 X_{C_1} = \frac{Q_2}{\omega C_1} \quad \text{...(2.106)}$$

[$\because X_L = X_{C_1}$ ... eqn. (2.102)]

The *resistance* ($R_x$) of the unknown impedance can be determined by comaring the conductances in the circuit (see Fig. 2.102).

Let, $G_t$ = Total conductance of the resonant circuit,

$G_x$ = Conductance of the unknown impedance, and

$G_L$ = Conductance of the working coil.

Then, $G_t = G_X + G_L$

or $G_x = G_t - G_L$ ...(2.107)

Also, $G_t = \frac{1}{R_t} = \frac{\omega C_1}{Q_2}$ ... using eqn. (2.106)

$$\therefore \quad \frac{1}{R_x} = \frac{\omega C_1}{Q_2} - \frac{R}{R^2 + \omega^2 L^2}$$

[∵ Conductance of the working coil, $G_L = R/(R^2 + \omega^2 L^2)$]

$$= \frac{\omega C_1}{Q_2} - \frac{1}{R}\left[\frac{1}{1 + \frac{\omega^2 L^2}{R^2}}\right] = \frac{\omega C_1}{Q_2} - \frac{1}{R}\left(\frac{1}{1 + Q_1^2}\right)$$

Normally value of $Q_1$ is very high so $(1 + Q_1^2)$ can be taken equal to $Q_1^2$.

$$\therefore \quad \frac{1}{R_x} = \frac{\omega C_1}{Q_2} - \frac{1}{RQ_1^2}$$

$$= \frac{\omega C_1}{Q_2} - \frac{1}{RQ_1 \times \frac{1}{\omega C_1 R}}$$

$$= \frac{\omega C_1}{Q_2} - \frac{\omega C_1}{Q_1} \quad \text{...[using eqn. (2.103)]}$$

$$= \frac{\omega C_1 Q_1 - \omega_2 C_1 Q_1}{Q_1 Q_2}$$

or
$$R_x = \frac{Q_1 Q_2}{\omega C_1 (Q_1 - Q_2)} \quad \text{...(2.107)}$$

The $Q$ of the unknown impedance,

$$Q_x = \frac{R_x}{X_x} = \frac{\frac{Q_1 Q_2}{\omega C_1 (Q_1 - Q_2)}}{\frac{1}{\omega (C_1 - C_2)}} = \frac{(C_1 - C_2) Q_1 Q_2}{C_1 (Q_1 - Q_2)} \quad \text{...(2.108)}$$

...(using eqn. 2.107 and 2.104, respectively)

**Sources of error :**

The various errors occurring in measurements are discussed below :

1. ***Error due to distributed capacitance :***

The *distributed capacitance or self-capacitance* of the measuring circuit is the most important factor which affects accuracy of the measurement. This source of error affects the actual value of $Q$ and the inductance of the coil.

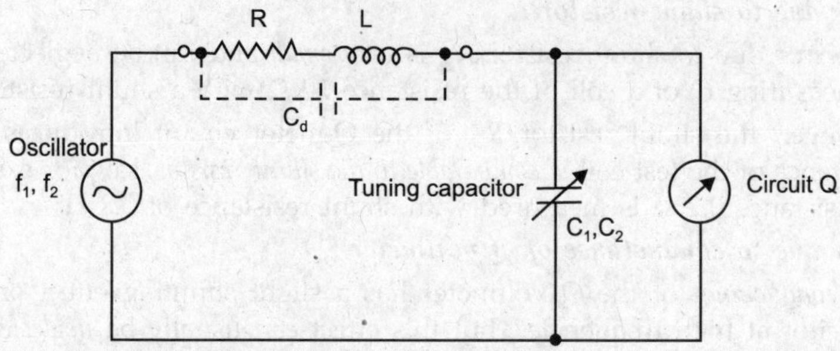


**Fig. 2.103.** Determination of the distributed capacitance of an inductor.

Distributed or self-capacitance $C_d$ of a coil can be measured by making two measurements at different frequencies. The circuit diagram is shown in Fig. 2.103. The coil under test in connected directly to the test terminals of the measuring circuit (See Fig. 2.103). The tuning capacitor is set to a high value, preferably to its maximum position, and the circuit is resonated by adjusting the oscillatory frequency. Resonance is indicated by maximum deflection on the "*Circuit Q*" meter. The values of the tuning capacitor ($C_1$) and the oscillatory frequency ($f_1$) are noted. The frequency is then increased to twice its original value ($f_2 = 2f_1$) and the circuit is retuned by adjusting the resonating capacitor ($C_2$). For these two resonant conditions, we have :

$$f_1 = \frac{1}{2\pi\sqrt{L(C_1 + C_d)}} \qquad \text{...(2.109)}$$

and,

$$f_2 = \frac{1}{2\pi\sqrt{L(C_2 + C_d)}} \qquad \text{...(2.110)}$$

Since,

$$f_2 = 2f_1$$

$$\therefore \quad \frac{1}{2\pi\sqrt{L(C_2 + C_d)}} = 2 \times \frac{1}{2\pi\sqrt{L(C_1 + C_d)}}$$

Squaring both sides and simplifying, we get

$$\frac{1}{C_2 + C_d} = \frac{4}{C_1 + C_d}$$

or,

$$(C_1 + C_d) = 4(C_2 + C_d) \quad \text{or} \quad (4C_2 + 4C_d)$$

or,

$$C_d = \frac{C_1 - 4C_2}{3} \qquad \text{...(2.111)}$$

The *distributed capacitance reduces the effective value of Q* of a coil and the difference between the measured value and actual value depends on the value of the distributed capacitance and the resonating capacitor. It can be shown that,

$$\text{True } Q = Q_e\left(\frac{C + C_d}{C}\right) \qquad \text{...(2.112)}$$

where, $Q_e$ = Effective $Q$ of the coil,

$C$ = Resonating capacitance, and

$C_d$ = Distributed capacitance.

**2.** ***Error due to residual inductance of the instrument :***

- This type of error is very small, of the order of 0.015μH.
- It affects the measurement of only very small inductors, say inductors of value lesser than 0.5μ H.

3. ***Error due to shunt resistor :***

- The error due to shunt resistor $R_{sh}$ is *very small* and can be neglected as in case of measuring $Q$ of a coil of the resistance 20 Ω with a shunt resistor of 0.02 Ω.
- However, the shunt resistor $R_{sh}$ of the Q-meter circuit introduces error when resistance of the test coil is *comparable to the shunt resistor*, say, for example, a coil of resistance 0.2 Ω is measured with shunt resistance of 0.2 Ω.

4. ***Error due to conductance of Q voltmeter :***

- The *conductance* of the $Q$ voltmeter has a slight shunting effect on the tuning capacitor at high frequencies, but this effect can usually be *neglected*.

**Applications :** Some of the specialized uses of this instrument are *to measure* :

(*i*) $Q$ of a coil;
(*ii*) Inductance and capacitance;
(*iii*) Distributed capacitance of a coil;
(*iv*) $Q$ and power factor of a dielectric material;
(*v*) Mutual inductance of coupled circuits;
(*vi*) Coefficient of coupling;
(*vii*) Critical coupling;
(*viii*) Reactance and effective resistance of an inductor at operating frequency;
(*ix*) Bandwidth of a tuned circuit etc.

- It has been very rightly said that *if ever an RF problem arises, a Q meter can always provide the answer.*

**Example 2.30.** *The following data were obtained during the measurement of the distributed or self-capacitance of a coil :*

*First measurement :* $f_1 = 2$ *MHz and* $C_1 = 420$ *pF;*

*Second measurement :* $f_2 = 4$ *MHz and* $C_2 = 90$ *pF.*

*Find the distributed capacitance,* $C_d$.

**Solution.** *Given* : $f_1 = 2$ MHz ; $C_1 = 420$ pF; $f_2\ (= 2f_1) = 4$MHz, $C_2 = 90$ pF.

We know that, $$C_d = \frac{C_1 - 4C_2}{3} \quad \text{...(Eqn. 2.111)}$$

or, $$C_d = \frac{420 - 4 \times 90}{3} = \mathbf{20pF} \ \mathbf{(Ans.)}$$

**Example 2.31.** *Calculate the value of distributed capacitance of a coil when the following measurements are made :*

*At frequency* $f_1 = 2$ *MHz, the tuning capacitor is set at 410pF.*

*At frequency,* $f_2 = 5$ *MHz, the tuning capacitor is tuned at 50pF.*

**Solution.** *Given* : $f_1 = 2$MHz, $C_1 = 410$pF; $f_2 = 5$MHz, $C_2 = 50$pF.

**Distributed capacitance $C_d$ :**

We know that : $$f_1 = \frac{1}{2\pi\sqrt{L(C_1 + C_d)}} \quad \text{...[Eqn. (2.109)]}$$

and, $$f_2 = \frac{1}{2\pi\sqrt{L(C_2 + C_d)}} \quad \text{...[Eqn.(2.110)]}$$

Now, $$\frac{f_2}{f_1} = \frac{5}{2} = 2.5 \text{ or } f_2 = 2.5\, f_1$$

$\therefore$ $$\frac{1}{2\pi\sqrt{L(C_2 + C_d)}} = 2.5 \times \frac{1}{2\pi\sqrt{L(C_1 + C_d)}}$$

or, $$\frac{1}{C_2 + C_d} = \frac{6.25}{C_1 + C_d}$$ ...squaring both sides and simplifying

or, $$(C_1 + C_d) = 6.25(C_2 + C_d)$$

or, $$C_d = \frac{C_1 - 6.25\, C_2}{5.25} = \frac{410 - 6.25 \times 50}{5.25}$$

$$= \mathbf{18.57pF} \textbf{ (Ans. )}$$

**Example 2.32.** *In a Q-meter, after connecting a standard coil, the resonance is obtained with a frequency $f_1$ with resonating capacitor set at $C_1$; the indicated Q-factor is $Q_1$. An unknown impedance is connected in series with the standard coil and the resonance is re-established by reseting the resonating capacitor at $C_2$, the corresponding Q-factor being $Q_2$.*

*Determine the resistive and reactive components of the unknown impedance when $C_1$ = 190pF, $Q_1$ = 75 ; $C_2$ = 170pF and $Q_2$ = 45. The frequency is 200 kHz.*

**Solution.** *Given* : $C_1$ = 190pF; $Q_1$ = 75; $C_2$ = 170pF; $Q_2$ = 45; $f$ = 200kHz.

**Resistive ($R_x$) and reactive components ($X_x$) :**

*Resistive component* ($R_x$) of unknown impedance ($Z_x$) is given by :

$$R_x = \frac{C_1Q_1 - C_2Q_2}{\omega C_1C_2Q_1Q_2}$$ ...[Eqn. (2.97)]

or, $$R_x = \frac{190 \times 10^{-12} \times 75 - 170 \times 10^{-12} \times 45}{(2\pi \times 200 \times 1000) \times 190 \times 10^{-12} \times 170 \times 10^{-12} \times 75 \times 45}$$

$$= \mathbf{48.18\Omega} \textbf{ (Ans.)}$$

*Reactive component* ($X_x$) is given by :

$$X_x = \frac{C_1 - C_2}{\omega\, C_1\, C_2}$$ ...[Eqn. (2.96)]

or, $$X_x = \frac{190 \times 10^{-12} - 170 \times 10^{-12}}{(2\pi \times 200 \times 1000) \times 190 \times 10^{-12} \times 170 \times 10^{-12}}$$

$$= \mathbf{492.74\ \Omega} \textbf{ (Ans.)}$$

**Example 2.33.** *A coil of resistance 4Ω is connected in the circuit shown in Fig. 2.100 and resonance occurs when oscillator frequency is 500 kHz and the resonating capacitor is set at 120 pF. Determine the percentage error introduced in the calculated value of Q if a resistance of 0.02 Ω is used across the oscillatory circuit.*

**Solution.** *Given* : $R$ = 4Ω; $f$ = 500kHz; C = 110pF; Resistance used across the oscillatory circuit = 0.02 Ω.

**Percentage error :**

The true or effective $Q$ of the coil,

$$Q_{true} = \frac{1}{\omega CR} = \frac{1}{(2\pi \times 500 \times 10^3) \times 120 \times 10^{-12} \times 4} = 663.14$$

The indicated or calculated $Q$ of the coil,

$$Q_{indicated} = \frac{1}{\omega C\,(R+0.02)} = \frac{1}{(2\pi\times 500\times 10^{3})\times 120\times 10^{-12}\times(4+0.02)}$$

$$= 659.85$$

$$\therefore \quad \text{Percentage error} = \frac{Q_{true}-Q_{indicated}}{Q_{true}}\times 100$$

$$= \frac{663.14-659.85}{663.14}\times 100 \simeq \mathbf{0.5\ \%\ (Ans.)}$$

**Example 2.34.** *The following results were obtained during tests conducted, using a Q-meter, on a radio-tuning coil to establish the distributed/self-capacitance :*

(*i*) *With the radio coil connected normally, resonance was obtained at 600 kHz with the tuning capacitor of the Q-meter set at 100pF.*

(*ii*) *With a standard inductor connected in place of radio coil, resonance was obtained at 2 MHz and this condition was not changed when the radio coil was connected in parallel with the standard inductor.*

*Determine the self-capacitance of the radio coil.*

**Solution.** *Given* : $C_1$ = 100pF ; $f_1$ = 600kHz ; $f_2$ = 2MHz.

**Self-capacitance of the coil, $C_d$ :**

If $L$ and $C_d$ are the inductance and self-capacitance of the coil respectively, then from the *first test* we have :

$$L = \frac{1}{\omega_1^2\,(C_1+C_d)} \qquad ...(i) \quad ...(\text{From eqn. 2.94})$$

where, $\omega_1$ = Angular frequency, and

$C_1$ = Tuning capacitor setting to produce resonance in first setting.

In the *second test* the reason for resonance condition of the standard inductor being unchanged is that a very high impedance has been placed in parallel, this being the resonant circuit of the radio coil and its own self-capacitance, *i.e.*, the radio coil in parallel with its own self-capacitance is resonant at 2 MHz. Thus, we can write,

$$L = \frac{1}{\omega_2^2\,C_d} \qquad ...(ii)$$

Equating expressions (*i*) and (*ii*), we have :

$$\frac{1}{\omega_1^2\,(C_1+C_d)} = \frac{1}{\omega_2^2\,C_d}$$

or,
$$\omega_2^2\,C_d = \omega_1^2 C_1 + \omega_1^2 C_d$$

or,
$$C_d = \frac{\omega_1^2\,C_1}{(\omega_2^2-\omega_1^2)} = \frac{f_1^2\,C_1}{f_2^2-f_1^2} \qquad (\because\ \omega = 2\pi f)$$

Substituting the value, we get :

$$\boldsymbol{C_d} = \frac{(600\times 1000)^2\times 100\times 10^{-12}}{(2\times 10^{6})^2-(600\times 1000)^2} = \mathbf{9.89pF\ (Ans.)}$$

**Example 2.35.** *A circuit consisting of an unknown coil, a resistance and a variable capacitor connected in series is tuned to resonance using a Q-meter. If the frequency is 400 kHz, the resonating capacitor is set at 220 pF, the resistance is 0.8 Ω and the Q-meter indicates 110, determine the effective inductance and resistance of the unknown coil.*

**Solution.** *Given* : $f = 400\text{kHz}$; $C = 200\text{pF}$; $R_{sh} = 0.8\Omega$; $Q = 110$.

**Inductance (*L*) ; resistance (*R*) :**

Under *resonant conditions*, we have :

$$L_{coil} = \frac{1}{\omega^2 C} = \frac{1}{(2\pi \times 400 \times 1000)^2 \times 220 \times 10^{-12}}$$

$$= \mathbf{719.6\ \mu H\ (Ans.)}$$

$$R_{coil} = \frac{\omega L}{Q} - R_{sh} = \frac{2\pi \times 400 \times 1000 \times 719.6 \times 10^{-6}}{110} - 0.8$$

$$= \mathbf{16.44\ \Omega\ (Ans.)}$$

## 2.6.21. Impedance (Or R-L-C) Bridge

The impedance bridge is employed for the measurement of :

(*i*) Resistance; (*ii*) Inductance;
(*iii*) Capcitance; (*iv*) Dissipation factor;
(*v*) Quality factor.

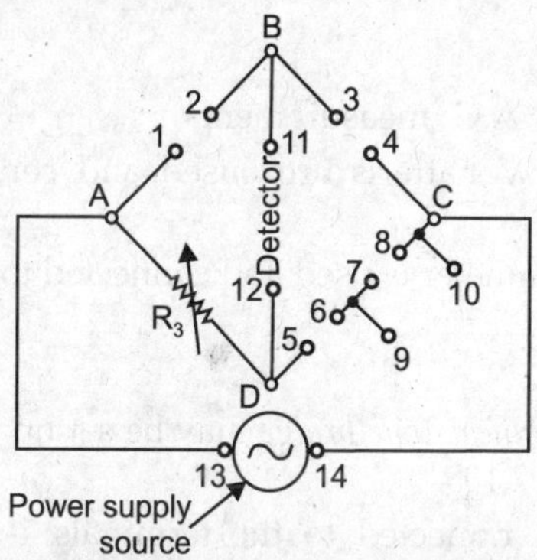


Fig. 2.104. Skelton type R-L-C bridge.

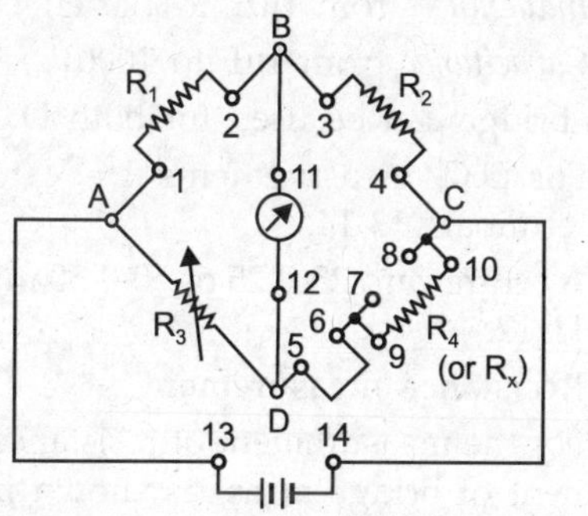


Fig. 2.105. Wheatstone bridge (Resistance measurement)

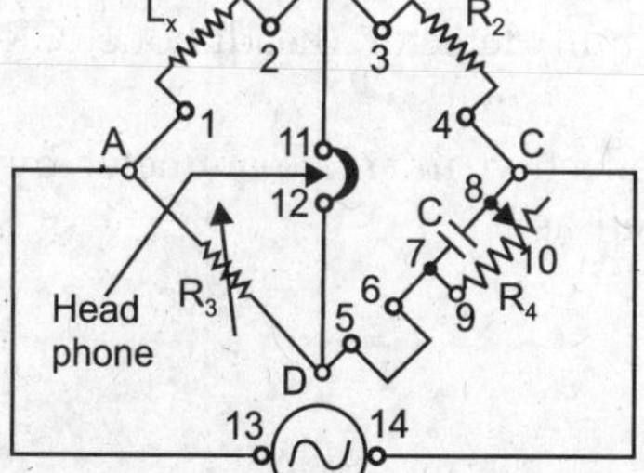


Fig. 2.106. Inductance measurement.

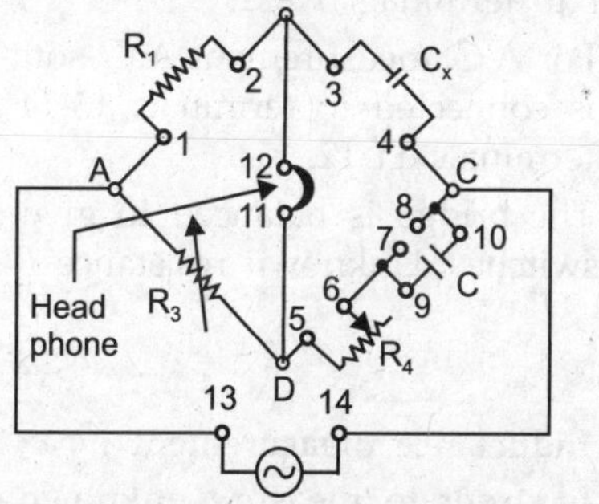


Fig. 2.107. Capacitance measurement.

This bridge consists of the following components/elements :

(i) ***Signal sources;***

(ii) ***Bridges*** :

(a) *Wheatstone bridge* ... For measurement of resistance;

(b) *Maxwell and Hay bridges* ... For measurement of inductance and quality factors;

(c) *De Sauty bridge* ... For measurement of capacitance and dissipation factors.

(iii) ***Imbalance amplifier;***

(iv) ***A metered detector** in addition to power supply.*

— A sensitive centre zero-micro-ammeter (25μA-0.25μA) is used to indicate the balance condition.

— Provision is also made for application of external signal sources and connecting external detectors.

- Fig. 2.104 shows the schematic circuit diagram of a *skelton type bridge* employed for measurement of *R*, *L* and *C* :

— $R_3$ is a decade resistance box.

— The other arms are completed by connecting the unknown and standard components to 1-2, 3-4, 5-6, 7-8 and 9-10, a null detector to 11-12 and a source of supply to 13-14.

This bridge can be used to measure :

(a) *Resistance* from 100mΩ to 10MΩ;

(b) *Inductance* from 1μH to 100 H;

(c) *Capacitance* from 1μF to 100μF.

This bridge can be used for both D.C. and A.C. measurements :

— For D.C. measurements 1.5 V to 1.75 V batteris are2 used and connected to terminals 13-14.

— A centre zero 25-0-25 or 50-0-50 micro-ammeter is used and connected to terminals 11-12.

1. **Resistance measurement :**

- For the measurement of resistance, the *Wheatstone bridge* may be set up by proper arrangement of bridge arms, as shown in Fig. 2.105.

- The standard resistors $R_1$ and $R_2$ are connected to the terminals 1-2 and 3-4 respectively, unknown resistance $R_x$ is connected to terminals 9-10 and a heavy wire jumper is connected to terminals 5-6.

- For D.C. measurement a battery is connected to terminals 13-14 and a micro-ammeter to terminals 11-12.

For A.C. measurement A.C. source (AF oscillator tuned to a frequency of 1kHZ) is connected to terminals 13-14 and a null detector (headphone, CRO etc.) to terminals 11-12.

- The bridge is balanced to give null deflection in micro-ammeter by adjusting decade switches. Unknown resistance $R_x$ is given as :

$$R_x = \frac{R_2 R_3}{R_1} \qquad ...(2.113)$$

2. **Inductance measurement :**

- In order to measure unknown inductance $L_x$, standard resistor $R_2$, standard capacitor *C*, 100Ω rheostat $R_4$, heavy wire jumper, A.C. supply source and detector are connected to terminals 1-2, 3-4, 7-8, 9-10, 5-6, 13-14 and 11-12 respectively as shown in Fig. 2.106.

- The bridge is balanced to give null deflection by adjusting the decade switches and rheostat $R_4$. Then :

Unknown inductance, $L_x = R_2 R_3 C$ henrys

where $R_2$ and $R_3$ are in ohms and $C$ is in farads.

3. **Capacitance measurement :**

- For the measurement of capacitor, standard resistor $R_1$, unknown capacitance $C_x$, a 100Ω rheostat $R_4$, standard capacitor $C$, source of A.C. supply and detector are connected to terminals 1-2, 3-4, 5-6, 9-10, 13-14 and 11-12 respectively as shown in Fig. 2.107.
- The bridge is balanced by adjusting decade switches and rheostat $R_4$. Then :

Unknown capacitance, $C_x = \frac{CR_3}{R_1}$ μF

where $R_1$ and $R_3$ are in ohms and $C$ is in μF.

## 2.6.22. LCR Meter

These meters are used to measure values of inductance (*L*), capacitance (*C*) and resistance (*R*) *directly*.

LCR meters are provided with switching arrangement for the measurement of *L*, *C* and *R*.

Fig. 2.108. illustrates a typical LCR meter with a block diagram and important controls on the face panel of the meter.

— It consists of a permanent magnet moving coil (PMMC) instrument.

— The scale is calibrated in terms of *R*, *L* and *C*.

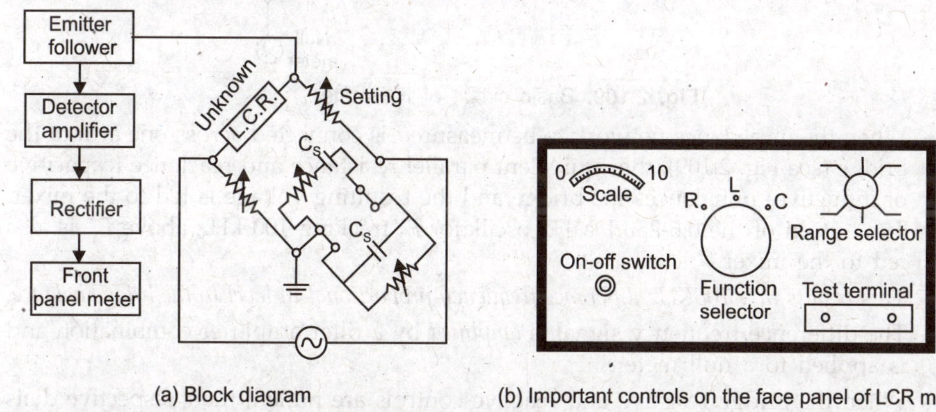


(a) Block diagram (b) Important controls on the face panel of LCR meter

**Fig. 2.108.** Typical LCR meter.

— When the component under test (*R*, *L* or *C*) is connected across the *"test terminals"* of the meter, the pointer of the meter indicates the value of the component *directly*. Basically the meter is an electronic instrument containing various electronic components fabricated on a PCB (printed circuit board) which needs D.C. supply. For this purpose, the meter also has a stepdown transformeter and a rectifier. However, a sinusoidal single-phase supply is given to the meter from outside.

- The *use of this meter* is as follows :

- The *use of this meter* is as follows :

— The meter is connected to the supply and ON/OFF switch is made 'ON'.

— The glow of the indicator confirms that the meter is properly connected with the supply.

— Now the position of function selector is adjusted at L, C or R depending upon the component to be tested. Also suitable range is selected by range selector.

— The unknown component is connected across the test terminals of the meter and the value is read on the scale.

### 2.6.23. RX Meter

This meter is used *to measure resistive and reactive components of a parallel Z network. It is especially useful in the design of RF tank circuits but can be used with any parallel 'Z'.*

Fig. 2.109 shows the basic circuit of a typical *RX* meter :

— It consists of two variable frequency oscillators that track each other at frequency *100 kHz apart*. The output of a 0.5-250MHz oscillator $F_1$ is fed into a bridge.

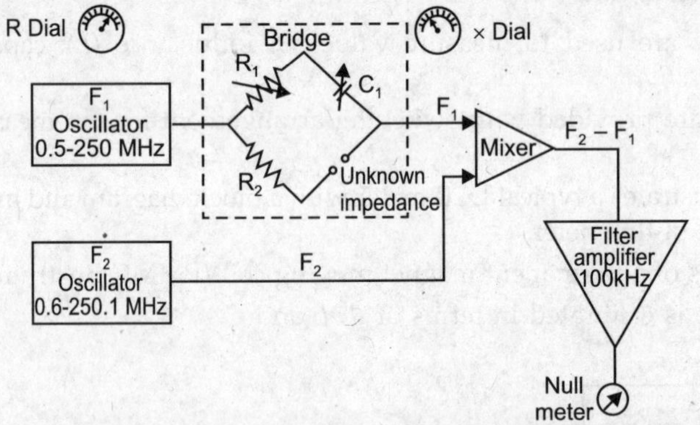


**Fig. 2.109.** Basic circuit of RX meter.

— When the impedance network to be measured is connected across one arm of the bridge (see Fig. 2.109), the equivalent parallel resistance and reactance (capacitive or inductive) unbalances the bridge and the resulting voltage is fed to the mixer. The output of the 0.6-250.1 MHz oscillator $F_2$ tracking 100 kHz above $F_1$, is also fed to the mixer.

This *results in a 100KHz difference frequency proportional in level to the bridge balance.*

— The difference frequency signal is *amplified* by a filter amplifier combination and is applied to a null meter.

— When the bridge resistive and reactive controls are nulled, their respective dials accurately indicate the parallel impedance components of the network under test.

### 2.6.24. Measurement of Resistance (effective), Inductance and Capacitance and Impedance at High Frequencies

- Although some of the methods described earlier *e.g.*, bridge methods can be used to measure *R*, *L* and *C* at high frequencies, say *0.1 to 100 MHz* but these methods do not give correct results owing to the fact that they do not take into consideration the special effects which are inherent in any measurement at high frequencies. Some of the special effects are :

(*i*) High reactance due to stray series inductance and its variation with frequency; (*ii*) Low reactance due to stray shunt capacitance and its variation with frequency; (*iii*) Increase of effective resistance on account of proximity effects ; (*iv*) Increase of effective resistance with frequency on account of skin effects; (*v*) Increase of effective resistance due to production of eddy currents; and (*vi*) Variation of dielectric loss of capacitors with frequency.

- As errors to the tune of 1 to 5% are acceptable in measurements in this frequency range, the methods used involve the use of very simple apparatus which may include :

  (*i*) A set of inductors from 1.0 μ H to 10mH; (*ii*) A variable air capacitor of maximum capacitance 1000 pF; (*iii*) A decade resistance box of resistance up to 100Ω; and (*iv*) A thermoelectric ammeter or an electronic voltmeter in order to indicate resistance conditions.

**2.6.24.1. Measurement of resistance (effective)**

Following are the *two methods* of measurement of effective resistance (at high frequency) of a series R-L-C resonant circuit :

1. Resistance variation method.
2. Reactance variation method.

1. **Resistance variation method :**

The effective series resistance of a resonant circuit is measured by *adding extra resistance from outside.*

Fig. 2.110. illustrates the resistance variation method :

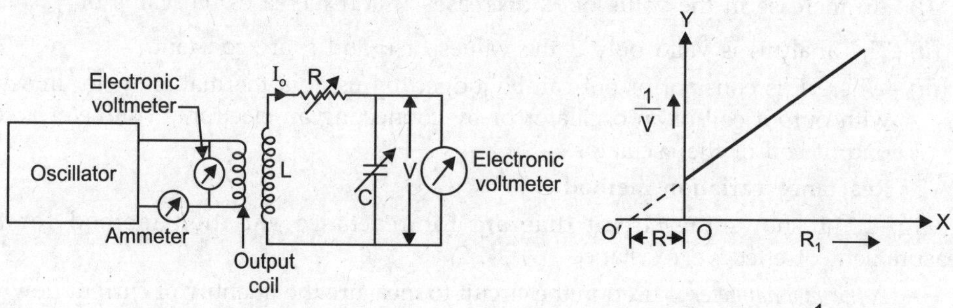


(a) Circuit diagram (b) Plot between $\frac{1}{V}$ and $R_1$

**Fig. 2.110.** Resistance variation method.

Let,
$E$ = Induced e.m.f. in the resonant circuit;
$I_o$ = Current in the resonant circuit,
$R_1$ = Additional resistance inserted in the resonant circuit
$R$ = Effective resistance of resonant circuit excluding resistance $R_1$
$L$ = Inductance of coil,
$C$ = Capacitance of coil,
$V$ = Reading of voltmeter connected across $C$, and
$f_o$ = Resonant frequency.

*At resonance* (achieved by varying capacitor) :

$$X_L = X_C$$

$\therefore$ Total resistance $= R + R_1$

and, $$I_0 = \frac{E}{R + R_1} \quad \text{...(2.114)}$$

$$V = L_o \times X_C = \frac{1}{2\pi f_o C} \quad \text{...(2.115)}$$

or, $$V = \frac{E}{2\pi f_o C (R + R_1)} \quad \text{...[2.115(a)]}$$

or, $$R + R_1 = \frac{E}{2\pi f_o CV} \quad \text{...(2.116)}$$

If the values of $E$, $f_o$ and $C$ are kept constant, we get,

$$R + R_1 = \frac{K}{V}$$

where, $$k = \text{a constant} = \frac{E}{2\pi f_o C} \quad \text{...(2.117)}$$

Thus, under the above conditions, if a curve is plotted between $\frac{1}{V}$ and $R_1$, we get a straight line as shown in Fig. 2.110 (*b*). The effective resistance is given by the intercept 00′ provided that the loss resistance of capacitor and the resistance of the leads is *negligible*.

*Conclusion from the above analysis* :

(*i*) An increase in the value of $R_1$ decreases $V$ and increases the value of $\frac{1}{V}$.

(*ii*) The analysis is valid only if the values of $E$ and $f_o$ are constant.

(*iii*) Where $E$ is constant or not can be judged by using a thermal ammeter in series with output coil of an oscillator or by connecting an electronic voltmeter across output coil of the oscillator.

2. **Reactance variation method :**

Fig. 2.111 shows the circuit diagram for reactance variation method for the measurement of effective resistance :

— A *thermal ammeter* is used in the circuit to measure the quantity of current flowing.

— By varying the circuit capacitance (C) or frequency (*f*), the circuit reactance is varied till the ammeter reads maximum value. This occurs only when the circuit is under resonant conditions.

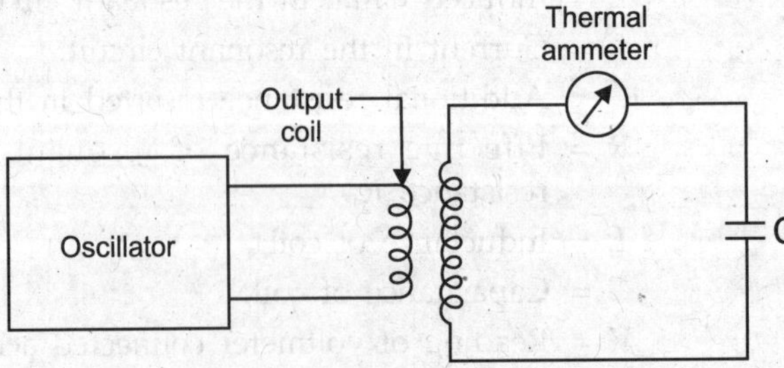


**Fig. 2.111.** Reactance variation method.

The maximum value of current,

$$I_o = \frac{E}{R} \quad ...(2.118)$$

where, $E$ = E.m.f. induced, and

$R$ = Effective resistance.

The reactance is then adjusted so that current $I = \frac{I_o}{\sqrt{2}}$

$$\therefore \quad I = \frac{E}{\sqrt{R^2 + X^2}} = \frac{I_o}{\sqrt{2}} = \frac{E}{\sqrt{2}\,R}$$

$$\text{or,} \quad \sqrt{R^2 + X^2} = \sqrt{2}\,R$$

Squaring boths sides, we get :

$$R^2 + X^2 = 2R^2$$

$$\text{or,} \quad X = \pm R \quad ...(2.119)$$

Thus, there are *two values* of reactance X which give a current, $I = \frac{I_o}{\sqrt{2}}$. The difference of these two values or reactance gives twice the value of effective resistance, *i.e.*, 2R.

Following are the two ways of varying the reactance :

(*i*) Variation of capacitance (C) and

(*ii*) Variation of frequency (*f*).

(*i*) ***Variation of capacitance (C) :***

In this method the value of frequecny (*f*) is kept constant and capacitance (C) is varied.

Let the value of capacitance which gives $I_0$ be $C_0$ and the two values of capacitance which give a current of $\frac{I_0}{\sqrt{2}}$ be $C_1$ and $C_2$.

$$\text{Then,} \quad \frac{1}{\omega C_2} - \frac{1}{\omega C_1} = 2R$$

$$\text{or,} \quad \frac{1}{\omega}\left(\frac{C_1 - C_2}{C_1 C_2}\right) = 2R$$

$$\text{or,} \quad R = \frac{C_1 - C_2}{2\omega\, C_1 C_2} = \frac{C_1 - C_2}{4\pi f C_1 C_2} \quad ...(2.120)$$

*When the circuit loss is small and the resonance is peaked and symmetrical,*

$$C_1 C_2 \simeq C_0^2 \quad ...(2.121)$$

$$\therefore \quad R = \frac{C_1 - C_2}{2\omega C_0^2} = \frac{C_1 - C_2}{4\pi f C_0^2} \quad ...(2.122)$$

Also, for a series resonant circuit,

$$Q = \frac{\omega L}{R} = \frac{1}{\omega C_0 R} \quad ...(2.123)$$

or,
$$Q = \frac{1}{\omega C_0 \left\{\frac{C_1 - C_2}{\omega C_0^2}\right\}}$$

... Substituting for $R$ from eqn. (2.122)

or,
$$Q = \frac{2C_0}{C_1 - C_2} \quad ...(2.124)$$

— The above eqn. (2.124) indicates that the circuit $Q$ can be determined in terms of two values of capacitance.

— The accuracy of this method depends upon the accuracy with which the difference $C_1$–$C_2$ can be known. In order to determine this difference accurately, a micrometer capacitor is put in parallel with the main capacitor.

(ii) ***Variation of frequency :***

In this method the capacitance ($C$) is kept constant and the frequency ($f$) is *varied*. Let $\omega_0$ be the angular frequency to give maximum current $I_0$ and $\omega_1$ and $\omega_2$ be the two angular frequencies which give a value of current $I = \frac{I_0}{\sqrt{2}}$.

Then,
$$\left(\omega_1 L - \frac{1}{\omega_1 C}\right) - \left(\omega_2 L - \frac{1}{\omega_2 C}\right) = 2R$$

or,
$$(\omega_1 - \omega_2) L + \frac{1}{C}\left(\frac{1}{\omega_2} - \frac{1}{\omega_1}\right) = 2R$$

or,
$$(\omega_1 - \omega_2)L + \frac{1}{C}\left(\frac{\omega_1 - \omega_2}{\omega_1 \omega_2}\right) = 2R$$

or,
$$(\omega_1 - \omega_2) L + \omega_0^2 L\left[\frac{\omega_1 - \omega_2}{\omega_1 \omega_2}\right] = 2R \quad \left[\because \frac{1}{C} = \omega_0^2 L, \text{ as } X_L = X_C \text{ at resonance frequency } \omega_0\right]$$

Also, for a peaked and symmetrical resonance curve $\omega_0^2 = \omega_1 \omega_2$

$\therefore$
$$(\omega_1 - \omega_2) L + \omega_0^2 L\left(\frac{\omega_1 - \omega_2}{\omega_0^2}\right) = 2R$$

or,
$$(\omega_1 - \omega_2) L + (\omega_1 - \omega_2) = 2R$$

or,
$$R = (\omega_1 - \omega_2) L \quad \text{or} \quad \frac{(\omega_1 - \omega_2)}{(1/L)} \quad ...(2.125)$$

$$\left(\because \frac{1}{L} = \omega_0^2 C\right)$$

or,
$$R = \frac{\omega_1 - \omega_2}{\omega_0^2 C} \quad ...(2.126)$$

and, $$Q = \frac{\omega_0 L}{R} = \frac{\omega_0}{\omega_1 - \omega_2} \quad ...(2.127)$$

...(using eqn. (2.125)

- The *limitation* of this method is that it is quite difficult to *keep the value of E constant with change of frequency*. However, error introduced on this account is *quite small*.

**Example 2.36.** *A coil of inductance L is tuned by a capacitance C to resonate 2 MHz. A standard capacitor of 210pF is tuned to resonate with the same coil at 4 MHz. It the resonance is indicated by a VTVM of capacitance 6pF in parallel with the circuit, calculate the values of L and C.*

**Solution.** *Given* : Capacitance of standard resonant capacitor, $C_s$ = 210 pF; Capacitance of voltmeter, $C_v$ = 6 pF; Resonant frequencies = 2 MHz and 4 MHz respectively.

**Values of L and C :**

We know that at resonance :

$$X_L = X_C$$

or, $$2\pi f L = \frac{1}{2\pi f C} \quad \text{or} \quad C = \frac{1}{4\pi^2 f^2 L}$$

Now, we have $$C + C_v = \frac{1}{4\pi^2 L (2\times 10^6)^2}$$

or, $$(C + 6) \times 10^{-12} = \frac{1}{16\pi^2 L \times 10^{12}} \quad ...(i)$$

and, $$(C_s + C_v) = \frac{1}{4\pi^2 L (4\times 10^6)^2}$$

or, $$(210 + 6) \times 10^{-12} = \frac{1}{64\pi^2 L \times 10^{12}} \quad ...(ii)$$

or, $$L = \frac{1}{(210+6)\times 10^{-12} \times 64\pi^2 \times 10^{12}}$$

$$= \mathbf{7.33\mu H} \quad \textbf{(Ans.)}$$

Substituting for $L$ in eqn. (*i*), we get :

$$(C + 6) \times 10^{-12} = \frac{1}{16\pi^2 \times 7.33 \times 10^{-6} \times 10^{12}}$$

or, $$C = \mathbf{857.9pF} \quad \textbf{(Ans.)}$$

**Example 2.37.** *A coil of inductance L and resistance R, a calibrated variable capacitor C and a thermojunction, all in series, are put across a constant r.m.s. voltage of frequency 4MHz. The uncalibrated ammeter is found to show the same scale reading I for the two values of C namely 40 and 48pF, one being below and the other above the capacitance $C_0$ needed to produce maximum current. The capacitor is now adjusted to a value $C_0$, and an added series resistance of 60Ω is found to reduce the current previous value of I.*

*Determine the values of L and R.*

**Solution.** *Given* : $C_1$ = 40 pF ; $C_2$ = 48 pF ($C_1$ and $C_2$ are the two values of capacitance (C) which produce the same current);

$f$ = 4MHz; Additional series resistance, $R'$ = 60Ω

**Values of $L$ and $R$ :**

Let the resonance curve be peaked and symmetrical and therefore capacitance required for resonance is :

$$C_o = \frac{C_1 + C_2}{2} = \frac{40 + 48}{2} = 44\text{pF}$$

At resonance,
$$L = \frac{1}{4\pi^2 f^2 C_o} = \frac{1}{4\pi^2 \times (4 \times 10^6)^2 \times 44 \times 10^{-12}} \text{H}$$
$$= \mathbf{35.98MH \ (Ans.)}$$

The value of current when the capacitance in the circuit is $C_1$,

$$I = \frac{E}{\sqrt{R^2 + \left(\omega L - \frac{1}{\omega C_1}\right)^2}}$$

or,
$$I = \frac{E}{\sqrt{R^2 + \left(2\pi \times 4 \times 10^6 \times 35.9 \times 10^{-6} - \frac{1}{2\pi \times 4 \times 10^6 \times 48 \times 10^{-12}}\right)^2}}$$

or,
$$I = \frac{E}{\sqrt{R^2 + [(902.26 - 828.93)]^2}} = \frac{E}{\sqrt{R^2 + 5377}} \quad ...(i)$$

Also,
$$I = \frac{E}{R + R'} = \frac{E}{R + 60} \quad ...(ii)$$

Equating eqns. (*i*) and (*ii*) we get :

$$\frac{E}{\sqrt{R^2 + 5377}} = \frac{E}{R + 60}$$

or,
$$R + 60 = \sqrt{R^2 + 5377}$$

Squaring both sides, we get :

$$R^2 + 120R + 3600 = R^2 + 5377$$

$$\therefore \quad R = \mathbf{14.81\Omega \ (Ans.)}$$

**Example 2.38.** *A circuit having an effective capacitance of 160 pF is tuned to a frequency of 1.2 MHz. In this circuit the current falls to 70.7 percent of its resonant value when the frequency of an e.m.f. of constant magnitude injected in series with the circuit deviates from the resonant frequency by 6 kHz.*

*Calculate the Q factor and effective resistance of the circuit.*

**Solution.** *Given* : C = 160pF ; $f_o$ = 1.2MHz

**Values of $Q$ and $R$ (effective) :**

We know that,
$$Q = \frac{\omega_o}{\omega_1 - \omega_2} = \frac{f_o}{f_1 - f_2} \quad ...[\text{Eqn. } (2.127)]$$

The two frequencies at which current is 70.2% or $\frac{1}{\sqrt{2}}$ time the current at resonant frequency are :

$$f_1 = 1.2 \times 10^6 + 6 \times 10^3$$

and, $$f_2 = 1.2 \times 10^6 - 6 \times 10^3$$

$\therefore$ $$f_1 - f_2 = 12 \times 10^3 \text{ Hz}$$

Hence, $$Q = \frac{1.2 \times 10^6}{12 \times 10^3} = \mathbf{100} \textbf{ (Ans.)}$$

Now, $$R = \frac{\omega_1 - \omega_2}{\omega_0^2\, C} \quad \text{...[Eqn. (2.126)]}$$

or, $$R = \frac{f_1 - f_2}{2\pi\, f_0^2 \times C} = \frac{12 \times 10^3}{2\pi \times (1.2 \times 10^6)^2 \times 160 \times 10^{-12}}$$

$$= \mathbf{8.29\ \Omega} \textbf{ (Ans.)}$$

**2.6.24.2. Measurement of Inductance**

Fig. 2.112 (*a*) shows the circuit diagram for the measurement of inductance of a coil :

— The coil under test is loosely coupled to the output coil of a variable frequency oscillator.

— The oscillator's dial is calibrated in terms of frequency.

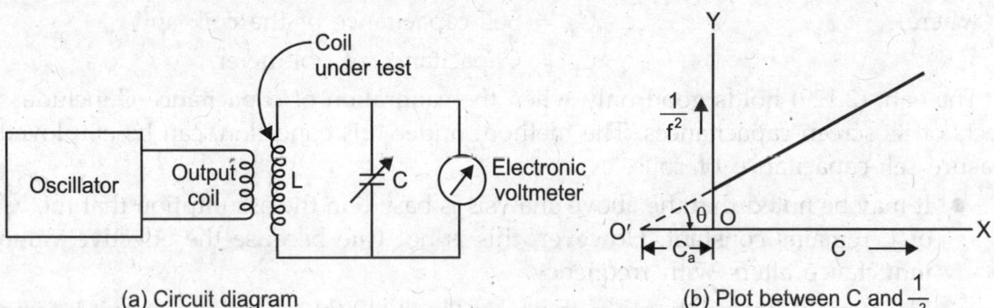


**Fig. 2.112.** Measurement of inductance.

— A variable capacitor, *C*, is used for obtaining resonance conditions which are indicated by an electronic voltmeter connected across it.

At resonance : $$X_L = X_C$$

or, $$\omega L = \frac{1}{\omega C}$$

or, $$LC = \frac{1}{\omega^2} = \frac{1}{4\pi^2 f^2} \quad \text{...(2.128)}$$

Thus knowing the values of $f$ and $C$, the value of $L$ can be calculated.

The value of the inductance (L) obtained by this method is *rarely accurate* due to the following reasons :

— The *additional capacitances* which are introduced into the circuit are *not taken into account*.

— Errors are caused due to the fact that dial setting of capacitor may not be accurately known.

In order to get better accuracy, several readings are take by varying the frequency, $f$.

If $C_a$ is the value of additional capacitance, then for resonance at any frequency, we have :

$$L(C + C_a) = \frac{1}{4\pi^2 f^2} \quad ...(2.129)$$

or,

$$\frac{1}{f^2} = 4\pi^2 L(C + C_a)$$

$$= 4\pi^2 LC + 4\pi^2 LC_a \quad ...[2.129\ (a)]$$

From the eqn. (2.129) it is obvious that if a curve is plotted between $\frac{1}{f^2}$ and C, we get a straight line as shown in Fig. 2.112 (*b*). The line produced back cuts the *x*-axis at *D*. The slope of the line, $\tan\theta = 4\pi^2 L$

$$\therefore \quad L = \frac{\tan\theta}{4\pi^2} \quad ...(2.130)$$

Thus by knowing value of slope of the line, the value of $L$ can be obtained.

Also, the value of additional capacitance $C_a$ is equal to the intercept 00′.

Here $C_a = C_d + C_v$ ...(2.131)

where, $C_d$ = Self-capacitance of the coil; and

$C_v$ = Capacitance of voltmeter.

The eqn. (2.129) holds good only when the calibration of capacitance C includes the effects of its screen capacitances. The method, under this condition, can be employed to measure self-capacitance of coils.

- It may be noted that the above analysis is based on the assumption that the value of L remains constant. However, this is not true because the effective value of inductance alters with frequency.

— The straight line relationship shown in Fig. 2.112 (*b*) is obtained only if L remains constant over the range of frequency used.

— When the value of L changes with frequency, a curve, instead of a straight line, is obtained. Therefore, in order to determine the value of L at a particular frequency a tangent is drawn to the curve at a value of the frequency where the value of L is required. *The value of L is obtained from the slope by using the eqn. (2.130).*

**Example 2.35.** *A capacitance of 200 pF produces resonance with a coil at a frequency of* $\frac{2}{\pi} \times 10^6$ *Hz, while at the second harmonic of this frequency resonance is produced by a capacitance of 40pF. Neglecting the effect of voltmeter capacitance and other stray capacitance, calculate :*

*(i) The self-capacitance of the coil; (ii) The inductance of the coil,*

**Solution.** *Given* : $C_1 = 200\text{pF}$ ; $f = \frac{2}{\pi} \times 10^6 \text{Hz}$; $C_2 = 40\text{pF}$;

$$f_2 \text{ (second harmonic)} = 2f_1 = \frac{4}{\pi} \times 10^6 \text{Hz}$$

**(i) Self-capacitance of the coil, $C_d$ :**

At resonance : $$L = \frac{1}{4\pi^2 f^2 (C + C_d)} \quad \text{...[Eqn. (2.129)]}$$

Therefore, with the given data, we have :

$$L = \frac{1}{4\pi^2 \left(\frac{2}{\pi}\right)^2 \times 10^{12}\,(200 + C_d) \times 10^{-12}}$$

$$= \frac{1}{16\,(200 + C_d)} \quad \text{...}(i)$$

and, $$L = \frac{1}{4\pi^2 \left(\frac{4}{\pi}\right)^2 \times 10^{12}\,(40 + C_d) \times 10^{-12}}$$

$$= \frac{1}{64\,(40 + C_d)} \quad \text{...}(ii)$$

From (*i*) and (*ii*), we get :

$$\frac{1}{16\,(200 + C_d)} = \frac{1}{64\,(40 + C_d)}$$

or, $$64\,(40 + C_d) = 16\,(200 + C_d)$$

$$160 + 4C_d = 200 + C_d$$

or, $$C_d = \mathbf{13.33\ pF\ (Ans.)}$$

Substituting the value of $C_d$ in eqn. (*i*), we get

$$L = \frac{1}{16\,(200 + 13.33)}\ \text{H}$$

$$= \mathbf{292.97\mu\ H\ (Ans.)}$$

### 2.6.24.3. Measurement of capacitance

The capacitance at high frequencies is normally measured by using *"substitution method"*.

Fig. 2.113 shows a simple circuit for measurement of capacitance :

— The unknown capacitance $C$ is connected in parallel with a variable standard capacitor $C_s$, which is of the *same order* as the unknown capacitance.

— Keeping the switch open, resonance is obtained by varying the standard capacitance $C_s$. The resonance condition is indicated by an electronic voltmeter connected across $C_s$.

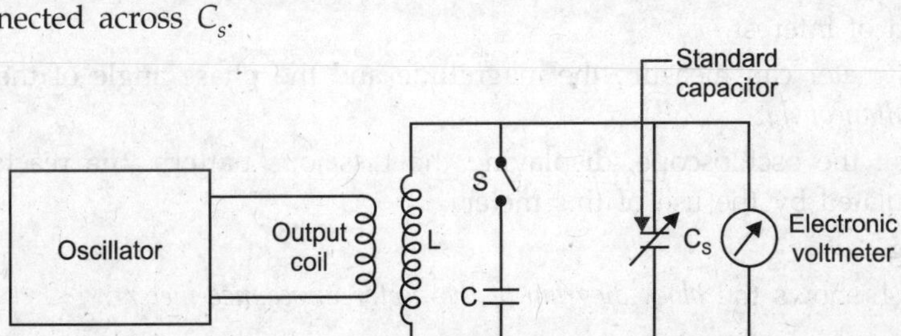


**Fig. 2.113.** Measurement of capacitance.

If $f$ = frequency at resonance (in Hz), $L$ = inductance of coil coupled with the output coil of the oscillator (in H), and $C_s$ = setting of standard capacitor with switch $S$ open (in $F$), then we have :

$$C_{s1} = \frac{1}{4\pi^2 f^2 L} \quad ...(2.132)$$

— Now, close the switch $S$ to put unknown capacitance $C$ in parallel with $C_S$. Keeping the frequency $f$ the same as earlier, obtain the resonant conditions by varying $C_s$. If $C_{se}$ = setting of standard capacitor with switch closed, then we have :

$$C + C_{S2} = \frac{1}{4\pi^2 f^2 L} \quad ...(2.133)$$

From eqn. (2.132) and (2.133), we get :

$$C + C_{S2} = C_{S1}$$

or, Unknown capacitance, $C = C_{S1} - C_{S2}$ ...(2.134)

The eqn. (2.134) indicates that the value of unknown capacitance can be directly known from the *difference of the two settings* of the standard capacitor $C_S$.

### 2.6.24.4. Measurement of impedance

The measuring of impedance means to measure its magnitude and the phase angle. At the frequencies below 100MHz, the impedance magnitude can be determined by voltage and current measurements. The phase difference between the voltage and current waveforms indicates whether the impedance is inductive or capacitive in nature. To specify the component values precisely, many measurements at several different frequencies are necessary. At high frequencies *such measurements become tedious, time consuming and lengthy. For such measurements complex impedance measurement meters are used.*

Two such meters are discussed below :

**1. Vector impedance meter :**

*The meter used for complex impedance measurements at high frequencies is called* **vector impedance meter**.

**Applications :**

(*i*) It is used to measure the impedances over wide frequency range, from 400 kHz to 110 MHz.

(*ii*) It can also be used to obtain *sweep frequency plots, i.e.,* magnitude and phase angle against frequency; these plots provide complete information within the frequency band of interest.

(*iii*) This meter can measure the magnitude and the phase angle of the impedance *simultaneously*.

(*iv*) Using the oscilloscope, displaying the Lissajous pattern, the reactance can be calculated by the use of this meter.

**Working :**

Fig. 2.114. shows the *block diagram of the vector impedance meter* :

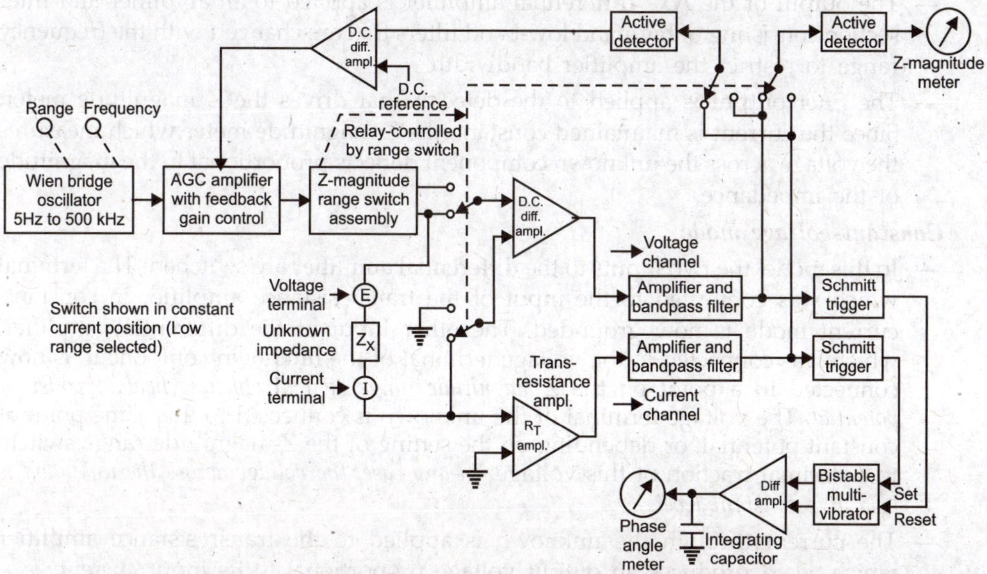


**Fig. 2.114.** Block diagram of the vector impedance meter.

- Two measurements take place.

(*i*) The magnitude of the impedance is determined by measuring the current through the unknown component when a known voltage is applied across it, or by measuring the voltage across the component when a known current is passed through it.

(*ii*) The *phase angle* is found by determining the phase difference between the voltage across the component and the current through the component.

- As shown in Fig. 2.114, the instrument contains a *signal source* (Wien bridge oscillator) with two front panel controls to select the frequency range and to continuously adjust the selected frequency. The output of the oscillator is fed to an *AGC amplifier* which permits accurate gain adjustment by means of its feedback voltage. This gain adjustment is an internal control actuated by the setting of the *impedance range switch,* to which AGC amplifier output is connected.

The impedance range switch allows operation of the instrument in the following two modes :

(*i*) Constant-current mode.

(*ii*) Constant-voltage mode.

***Constant-current mode :***

— In this mode the three lower ranges (×1, ×10, and 100) operate.

— Here, the unknown component is connected across the input of the A.C. differential amplifier; the current supplied (to the unknown) depends on the setting of the impedance range switch. This current is held constant by the action transresistance or $R_T$ amplifier, which converts the current through the unknown to a voltage output equal to the current times its feedback. The output $R_T$ amplifier is fed to a detector circuit and compared to a D.C. reference voltage. The resulting control voltage regulates the gain of the AGC amplifier and hence the voltage applied to the impedance range switch.

— The output of the A.C. differential amplifier is applied to an amplifier and filter section consisting of high-and low-band filters that are changed with the frequency range to restrict the amplifier bandwidth.

— The filter output is applied to the detector that drives the Z magnitude meter. Since the current is maintained constant, the Z magnitude meter which measures the voltage across the unknown component deflects proportional to the magnitude of the impedance.

***Constant-voltage mode :***

— In this mode, the two inputs to the differential amplifier are switched. The terminal which was connected to the input of the transresistance amplifier in constant-current mode is now grounded. The other input of the differential amplifier, which was connected to the voltage terminal of the unknown component, is now connected to a point on the *Z-magnitude range switch which is held at constant potential*. The voltage terminal of the unknown is connected to this same point of constant potential, or depending on the setting of the Z-magnitude range switch, to a decimal fraction of this voltage. *In any case, the voltage across the unknown is held at a constant level.*

— The current through the unknown is applied to the transresistance amplifier which again produces an output voltage proportional to its input current.

— The roles of the A.C. differential amplifier and transresistance amplifier are now *reversed*. The voltage output of the $R_T$ amplifier is applied to the detector and then to the Z-magnitude meter. The output voltage of the differential amplifier controls the gain of the AGC amplifier in the same manner that the $R_T$ amplifier did in the constant current meter.

**Phase angle measurements :**

— The phase angle measurements are carried out simultaneously with the magnitude measurements.

— The outputs of both the voltage channel and the current channel are amplified and each ouput is connected to a schmitt trigger circuit.

— The Schmsitt trigger circuits produce a positive-going spike every time the input sine wave goes through zero crossing. These positive spikes are applied to a *binary phase detector circuit*.

The phase detector consists of a bistable multivibrator, a differential amplifier, and an integrating capacitor.

— The positive-going pulse from the constant-current channel *sets* the multivibrator and the pulse from the constant-voltage channel *resets* the multivibrator. The "set" time of the multivibtrator is, therefore, determined by the *zero crossings of the voltage and current waveforms*.

— The "*set*" and "*reset*" outputs of the multivibrator are applied to the differential amplifier, which applies the difference voltage to an integrating capacitor.

— *The capacitor voltage is directly proportional to the zero-crossing time interval and is applied to the phase-angle meter which then indicates the phase difference in degrees between the voltage and current waveforms.*

- The vector impedance meter is usually *calibrated* by connecting standard components (e.g., standard resistors or capacitors) to the input terminals.

2. **RF vector impedance meters :**

This meter uses the *phase locked sampling system* to obtain and operate in a frequency range of 0.5 MHz to 108 MHz.

Fig. 2.115 shows the block digram of the the RF vector impedance meter :

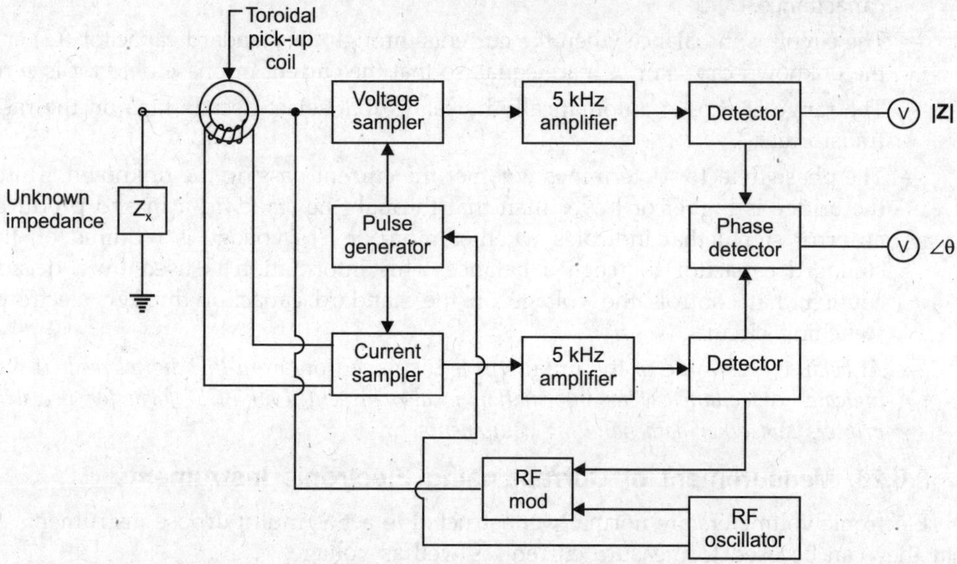


**Fig. 2.115.** RF vector impedance meter.

— The RF current is supplied to the unknown impedance.

— The current is measured by toroidal pick-up coil. It is then sampled, amplified, detected and then applied to the modulator. This forms an *automatic current level control*.

— The voltage is similarly sampled and because the current is held constant, it is a *measure of the magnitude of impedance*.

— The *phase detector* measures the phase angle θ.

- This instrument (commercially available) popularly known as HP 4815 is manufactured by Hewlett-Packard company.

## 2.6.25. Automatic Bridges

*The "automatic bridges" provide an automatic readout without adjustment of balance controls* (as in conventional bridges). In some cases they also provide a *Binary Coded Digital (BCD) readout* to external equipment.

These bridges are similar in operation to digital meters.

Fig. 2.116 shows a typical automatic bridge circuit :

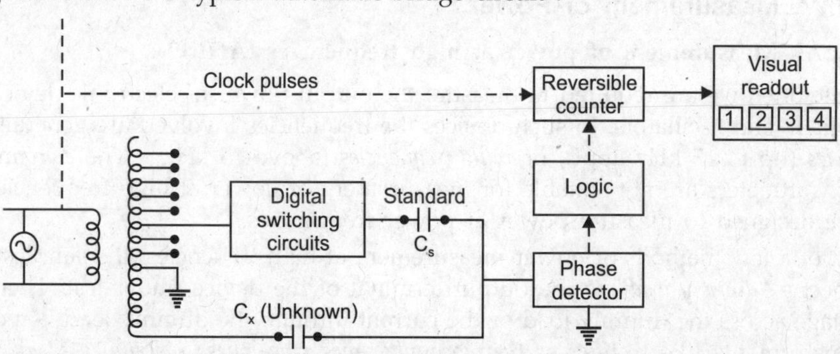


**Fig. 2.116.** Typical automatic bridge circuit.

— The circuit is transformer-ratio-arm bridge for the automatic measurement of capacitance.

— The circuit is in balance when the currents through the standard capacitor ($C_s$) and the unknown capacitor ($C_x$) are equal, so that the current in phase detector is zero.

— The range is chosen automatically by relays that select decade taps on the ratio transformer.

— The phase detector determines whether the current passing the unknown arm of the bridge is higher or lower than that through the standard arm and produces an error signal that indicates whether more or less voltage is required on the standard capacitor to reach a balance. This information is used by a decade counter that controls the voltage on the standard capacitor through electronic switching circuits.

— *At balance*, the value of the unknown is displayed on an *in-line digital readout* that *indicates capacitance. This information is also presented in BCD form for use with printers and other data handling equipments.*

## 2.6.26. Measurement of Current using Electronic Instruments

Electronic voltmeters are normally constructed to act as multipurpose instruments so that they can be used to measure current as well as voltage.

**D.C. current measurements :**

— The unknown current is made to flow through a known standard resistance. The voltage drop across this standard resistor is proportional to the current and is measured by a VTVM or TVM.

— The scale of the meter is *calibrated in terms of currents.*

**A.C. current measurements :**

— For measuring A.C. currents, a rectifier is used to change the alternating current to a corresponding direct current, which is then measured by a VTVM or TVM.

— In another method, an A.C. probe is used which clips around the wire carrying the current and in effect makes the wire a one turn primary of a current transformer (C.T.). The C.T. has finite core and the secondary consists of a large number of turns. The voltage induced in the secondary winding is amplified and the amplifier's output can be measured by any suitable A.C. meter.

— In practice, amplifier is so designed that 1mA current in the wire being measured produces 1mV at the amplifier output. The current is then read directly on the voltmeter using the same scale as for voltage measurements.

## 2.6.27. Measurement of Power

### 2.6.27.1. Measurement of power at high frequencies (AF/RF) :

Quite often we are required to measure the output power of electronic devices such as amplifers and oscillators. In such devices the frequencies involved are generally *audio frequencies* (up to 15 kHz appr.) or *radio frequencies* (above 15 kHz). The dynamometer type of wattmeters are not suitable for measurement of power at high frequencies since they are designed to measure power at power frequencies.

Out of a few methods of power measurement at high freuencies, the simplest one is to connect a *"dummy load"* to the output terminal of the device under test. Then either the voltage across the dummy load or the current through the dummy load is measured using instruments like *rectifier or thermocouple types since these instruments can work at high frequencies*. If $R$ is the resistance of dummy road, then power output can be calculated as follows :

Power, $$P = \frac{V^2}{R}$$

where, $V$ = Voltage across the dummy load, and

$I$ = Current through the dummy load.

Fig. 2.117. shows the basic circuit for power measurement at *"audio frequencies"*.

— The resistance $R$ (non-inductive type) which forms the dummy load is connected across the output terminals of the device under test. The value of $R$ should be nearly equal to that of output impedance of the device.

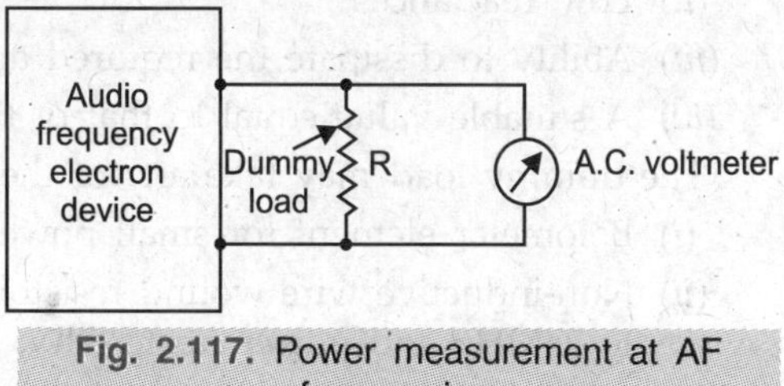


Fig. 2.117. Power measurement at AF frequencies.

— The A.C. voltmeter (a high resistance rectifier type of instrument) when used with a fixed value of $R$ its scale can be directly calibrated in terms of power output at audio frequencies. Such A.C. voltmeters are provided with *decibel scales*.

. Fig. 2.118 shows the basic circuit for power measurement at *radio frequencies* (R.F.)

— The output tank of amplifier dummy or oscillator (consisting of inductance $L_1$ and capacitance $C$) is transmitted to the dummy load consisting of inductance $L_4$, capacitance $C_1$ and resistance $R$ in series with a thermocouple ammeter $A$, by means of an inductive link ($L_2$, $L_3$) that couples $L_1$ and $L_2$ and $L_3$ and $L_4$.

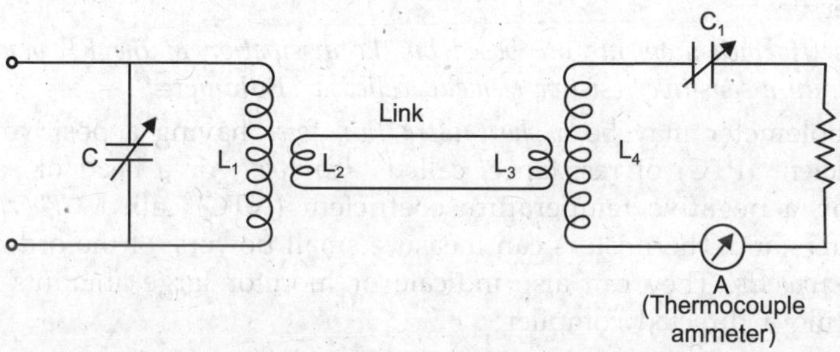


Fig. 2.118. Measurement of power at radio frequencies.

— Using capacitor $C_1$, the dummy load is tuned to resonance at the resonant frequency of the tank circuit. Thus at resonant frequency the dummy load is only $R$.

— A is directly calibrated to read the value of power.

— For the measurement of power at high frequencies, *effective shielding must be used.*

### 2.6.27.2. Bolometer method of power measurement

At *AF* and *RF* it is easier to measure voltage, current and impedance, than to measure power. In this range direct measurement of power is not done. The power is calculated

from the relation : $P = \frac{V^2}{R} = I^2R$ (as stated earlier) .

At *'microwave frequencies', voltage, current and impedance are difficult to determine.* Their values *may vary* at different points in the circuit, and they are affected appreciably by small changes in geometry. Therefore, in this range, *direct measurement of power is more accurate* and the actual load is replaced by a *dummy load.*

The *requirements* of a dummy load are :

(*i*) Low skin effect;

(*ii*) Low reactances;

(*iii*) Ability to dissipate the required amount of power; and

(*iv*) A suitable value equal to that of the actual load.

The dummy load may take any of the following *forms* :

(*i*) Bolometer element for small powers;

(*ii*) Non-inductive wire wound resistors, cooled by air, or water, for frequencies up to 25MHz and power up to 50W; and

(*iii*) Water resistance, for large powers up to 50W etc.

**Method of power measurement :**

— The bolometer is generally incorporated into a bridge network, so that small changes in resistance can be readily detected and the power measurement performed by substitution of low frequency power.

— In one method of power measurement, the bridge is initially balanced with low frequency (or bias) power. When *RF* power is applied to bolometer, *low frequency power is withdrawn until the bridge is balanced again.*

In any case, whatever bridge technique used, the *underlying assumption of measurement is that equal amount of low frequency and RF power produce the same heating effect and resistance changes in the bolometer.*

**Bolometer :**

— *Bolometric measurements are based on the dissipation of the RF power in a small temperature sensitive resistive element, called a "**Bolometer**".*

— The bolometer may be a *short ultra thin wire* having a positive temperature coefficient (PTC) of resistance, called "*Baretter*", or a bead of semiconductor having a negative temperature coefficient (NTC) called "*Thermistor*". Both Baretters and Thermistors can measure small powers, of the order of a *fraction of microwatts.* They can also indicate or monitor large amounts of power by inserting a directed compiler.

— The RF power to be measured heats the bolometer and causes a change in its electrical resistance, which serves as an indication of the magnitude of power.

**Bolometer bridge :**

One of the most commonly used *absorption type meters* is called "*bolometer bridge*".

It is basically a bridge circuit where one of the arms contains a temperature-sensitive resistor ($R_T$), as shown in Fig. 2.119.

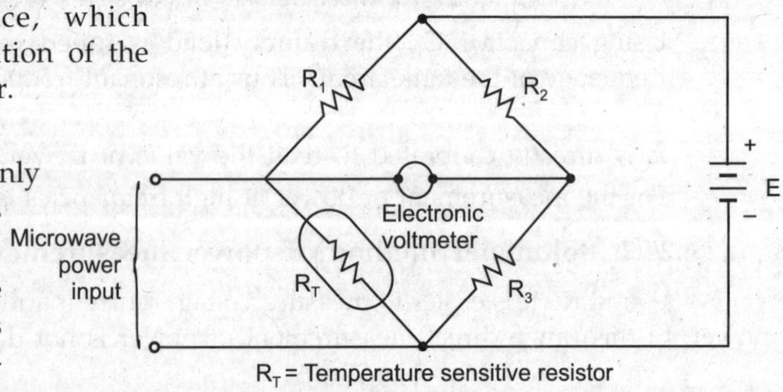


**Fig. 2.119.** Basic bolometer bridge circuit.

The temperature sensitive resistor is placed in the field of a microwave signal whose power is to be measured. *The power is absorbed by the resistor, and the heat generated causes a change in the resistance.* The change of resistance is measured with a bridge circuit.

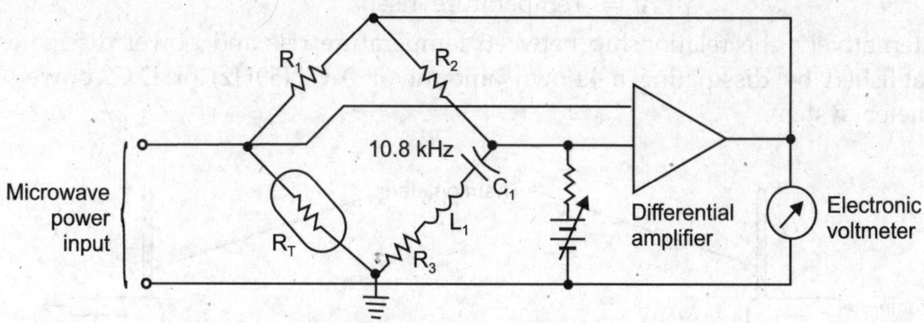


**Fig. 2.120.** Simplified diagram of the measurement circuit of a modern bolometer type power meter.

· Fig. 2.120 shows the *schematic of the measuring circuit* :

— The combination of the differential amplifier and bridge form an oscillator which will *oscillate at particular magnitude when the bridge is unbalanced.*

— The sensitive resistive element ($R_T$) absorbs the power and heats itself until the bridge is very nearly balanced. The D.C. bias is then adjusted until *exact balance is achieved.*

— Now the element is placed in the microwave field. The element absorbs the power and is heated, which causes a change in its resistance, thereby unbalancing the bridge.

— *The imbalance is in a direction opposite to that of the initial cold resistance. This action causes the output from the oscillator to decrease so as to effect a balance.* The electronic voltmeter circuit measures the amount of this power decrease from the oscillator and displays the measurement as a *power increase which is that supplied by the microwave field.*

**Uses of bolomoter :**

(*i*) The bolometer is customarily used for power measurements in the range 0.01 – 10 mW.

(*ii*) A comparison-type bolometer is quite suitable in the range of 10mW to 10W.

(*iii*) Power measurements above 10 W can be made by *using attenuators and a low-power bolometer.*

### 2.6.27.3. Calorimetric method-For measurement of large amount of RF Power

The calorimetric method for measurement of large amount of RF power uses the *principle that a RF power may be directly converted into heat.*

Water acts as a load, and RF power may be used to heat an element such as long transmission line, water being used as a coolant. The RF power may be absorbed directly in the calorimeter fluid.

*The system should have no RF leakage either through lossy joints or by radiation.*

The power radiated (P) is given by the relation :

$$P = m \times c_p \times \Delta t \quad \text{...(2.135)}$$

where, $m$ = Mass flow rate of the fluid (water),

$c_p$ = Specific heat of fluid, and

$\Delta t$ = Temperature rise.

Alternatively, the relationship between temperature rise and power dissipated may be established by dissipating a known amount of A.C. (50Hz) or D.C. power in the calorimeter system.

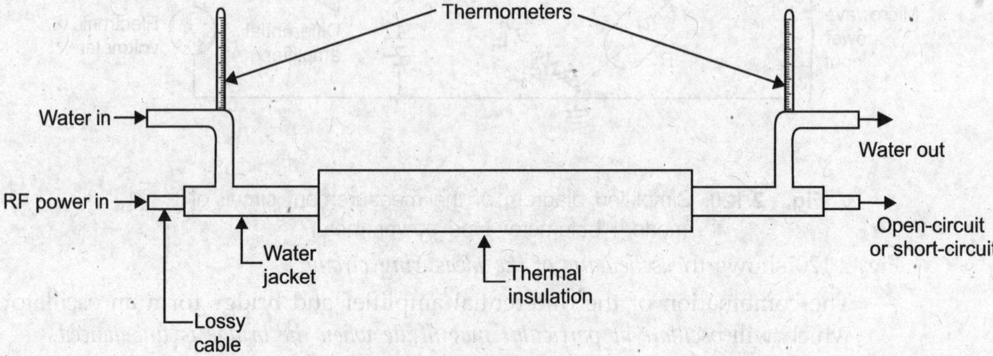


Fig. 2.121. Measurement of power in a lossy cable using the calorimetric method.

- Fig. 2.121 shows an arrangement for the power measurement in a lossy cable using the calorimetric method. Here the difference in temperature is measured by thermometers.

### 2.6.27.4. Ouput power meter

An "*output power wattmeter*" is designed to directly measure the output in an arbitrary load. In this instrument *a set of resistive loads* are provided to be selected for *power measurements*. Besides this, the instrument *can be used to measure impedance and frequency response characteristics.*

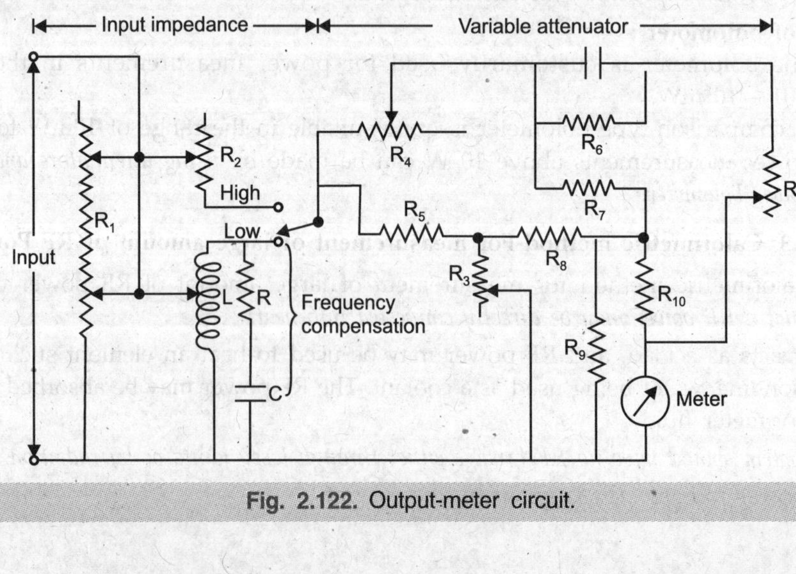


Fig. 2.122. Output-meter circuit.

Fig. 2.122 shows a simple circuit for an output-meter :

- The input impedance network consists of two tapped resistances and a coil. The input impedance can be varied in steps from 2.5Ω to 40kΩ.

— *At low impedance values,* the coil shunts a portion of $R_1$ – an in resistance results in power turns of the coil. This arrangement keeps the meter reading *proportional to the energy dissipated* by the resistor.

— At *high impedance values,* the coil is replaced by another tapped resistance $R_2$.

- The R-C frequency compensator is in parallel with the coil. At low frequency capacitive reactance $X_C$ is high; it decreases as frequency increases. This *compensates* for the losses of the coil at low frequencies.

— The frequency range is generally between 20Hz and 20kHz.

- Resistance $R_3$ is the control in a T-network. This amounts to a variable meter shunt, which is used to *extend the range of the meter.*

— The lowest meter range is normally 5 mW, but it can be extended in decade steps of 50 mW, 500 mW etc.

- The rest of the circuit is a *combination of a calibration and frequency compensation network.*
- This type of meter may have a midscale accuracy of ± 2% at 1 kHz. The accuracy, over the frequency range of 20Hz – 15kHz, may be within ± 2% of the 1kHz value.
- The input meter is subjected to a waveform error when the input is other than sinusoidal.
- *Practical instruments for measuring the power output of oscillators, amplifiers, transformers, transducers and low frequency lines use an input impedance of a tapped transformer with 48 impedance setting.*

*This meter can also be used :*

(*i*) To measure output impedance by adjusting the maximum power.

(*ii*) To check the frequency response characteristics of audio frequency devices.

## 2.6.28. Measurement of Audio and Radio Frequencies

The audio and radio frequencies (AF and RF) can be measured by using the following techniques/methods :

1. Bridge technique.
2. Heterodyne technique.
3. Wavemeters.
4. Interpolation technique.

Some of these methods/techniques are described below :

**1. Bridge technique :**

An unknown frequency can be determined by applying it to bridge in which the balance dependent upon frequency and then calculating the frequency from the circuit constants is required to give balance. Such arrangements find considerable use for measuring AF.

Any type of bridge or null network in which balance depends upon frequency can be used. Typical examples are : *Wien, Hay* and *resonance bridges* and the *null network.*

- Fig. 2.123 shows *"null network"* for measuring AF; the frequency is calculated using the relation :

$$\omega = \frac{1}{R_2}\sqrt{\frac{2}{C_1C_2}} \qquad ...(2.136)$$

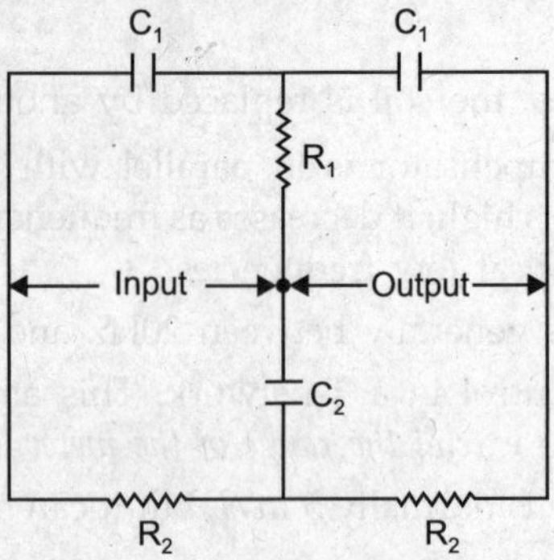


**Fig. 2.123.** Null network for measuring AF.

- The *"Wien bridge"* is particularly satisfactory for measuring AF because it can be brought into balance merely by varying resistive elements containing no inductance, and can be conveniently proportioned to cover a wide frequency range (for details refer to Art. 2.6.15).

  An accuracy in the order of 0.5 to 1.0 % can be readily obtained in frequency bridges. In the frequency range 300 to 500 Hz, balance can be mostly easily made by the use of telephone receivers. At frequencies outside this range some indicating arrangements, such as an amplifier and VTVM combination, is normally required.

**2. Heterodyne technique :**

- The basis of heterodyne frequency meter is a *stable and tunable oscillator*.
- Heterodyne frequency meters are commonly provided with a *detector*, *audio amplifier*, and *headphones* as shown in Fig. 2.124 :

— Energy from the *oscillator* is supplied to the *detector* by internal coupling, and in addition provision is made whereby energy from an external source may be superimposed upon the detector input. The frequency of the heterodyne meter can then be made the same as the unknown frequency by adjusting so that the difference (or beat) frequency of the two oscillators as observed on the headphones is reduced to zero.

— These frequency meters are sometimes provided with a *crystal oscillator calibrator*. This is simple but well-designed crystal oscillator utilizng a *low temperature coefficient crystal to ensure good frequency stability*. The crystal oscillator produces a known frequency which can be used to check one or more points on the frequency calibration of the heterodyne meter with the aid of *detector audio amplifier*.

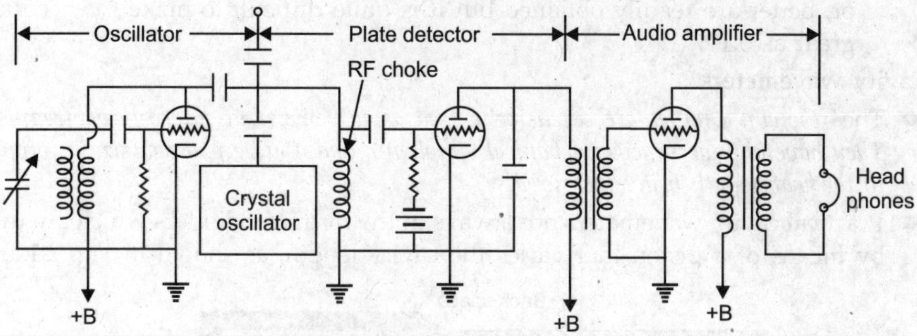


**Fig. 2.124.** Heterodyne frequency meter.

**Uses :** Heterodyne frequency meters are employed for the following *purposes* :

1. To produce known frequencies.
2. To measure an unknown frequency by comparison with the controllable and known frequency of heterodyne frequency meter.
3. To interpolate between fixed frequencies.

**3. Wavemeters :**

Wavemeters are used to measure frequency when high *accuracy is not required* and where *simplicity and portability are important.*

— A wavemeter is an *adjustable resonant circuit provided with a calibration that gives the resonant frequency in terms of setting of the tuning adjustments.*

— A wavemeter may employ any type of resonant circuit that is convenient for the frequency range to be covered; thus resonant circuits based on *lumped constants including butterfly circuits, coaxial two-wire lines, and cavities,* all find use in wavemeter applications.

— Wavemeters can be constructed with lumped circuit elements up to about 1200 MHz. Such arrangements are normally employed either as *absorption or as reaction devices.*

— The response of the wavemeter is commonly indicated by a crystal detector that is loosely coupled to the resonant circuit and associated with a micrometer that indicates the rectified current.

— Arrangements employing a fixed inductance and variable capacitance are suitable for use from the lowest radio frequencies up to about 100 MHz; this entire range can be covered by single variable condenser provided with a suitable set of plug-in coils as shown in Fig. 2.125. In this the coupling between the crystal and resonant circuit is made small by making capacitance $C_1$ small and $C_2$ large.

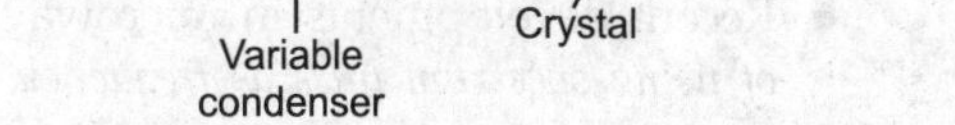


**Fig. 2.125** Tuned-circuit type wavemeter.

— A variation of the inductance-capacitance wavemeter that is useful at *very high frequencies consists in varying both the coil and the condenser simultaneously.* This increases the frequency range that can be covered with a single coil and condenser combination, and is particularly useful at frequencies of the order of 50 to 400 MHz.

- The *accuracy* obtainable with such type of wavemeters depends upon the circuit $Q$ and upon details of construction. In general accuracies of the order of 1%

or, better are readily obtained but it is quite difficult to make the accuracy as great as 0.1 %.

**Cavity wavemeters :**

- The *resonant cavities are extensively used as wavemeter at* ***microwave frequencies****. They have high accuracy, mechanical simplicity, and a large physical size in proportion to the wavelength being measured.*
- Practical cavity wavemeters are always in the form of cylinders which are tunable by means of a piston that varies the cavity length, as shown in Fig. 2.126.

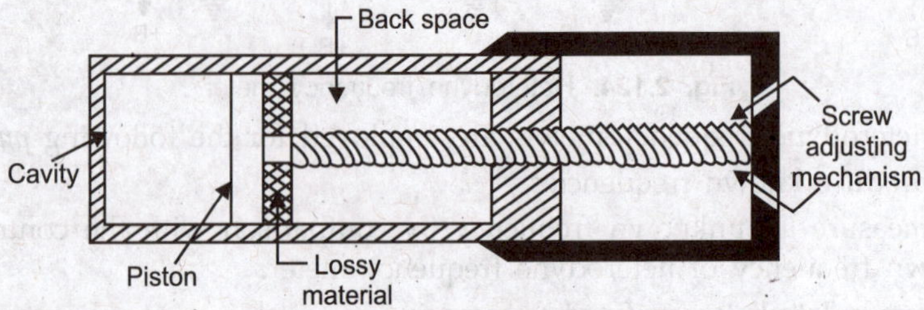


**Fig. 2.126.** Simple cylindrical cavity.

- Coupling to a wavemeter is accomplished by a loop or an orifice. A loop is used when co-axial systems are used to excite a cavity or carry power from it, while orifice is normally used in a waveguides system.
- The accuracy with which frequency can be determined using a well made cavity wavemeter is *quite high* (Precision of the order of 1 part in 1000 is easy to obtain and 1 part in 100,000 can be obtained by careful mechanical design combined with temperature compensation and correction).

**Frequency standards :**

*The fundamental standard of frequencies is the period of rotation of earth.* It can be measured with great accuracy by astronomical methods and is taken as a standard frequency of 1 cycle/day. The error in the measurement is approximately 164 μs. All standards of frequency must ultimately be referred to this fundamental source for calibration purpose.

Practical standards can be *classified* as follows :

(*i*) *Primary standards.* A primary standard of frequency is an oscillator which generates a frequency that is constant over a long period and is checked directly with the earth's rotation.

(*ii*) *Secondary standards.* These standards of frequency are *very stable oscillators* which have their frequencies checked periodically against the primary standards.

- Recent developments in microwave spectroscopy have introduced the *possibility of using spectrum lines as frequency standard.*

## HIGHLIGHTS

1. Essentials of an electronic instrument are :
   (*i*) Transducer; (*ii*) Signal modifier; (*iii*) Indicating device.
2. Valve voltmeter essentially consists of a thermionic valve which has a millimeter connected in its anode circuit.
3. The various electronic voltmeters widely used are :
   (*i*) Transistorised voltmeter using cascaded transistors;

(*ii*) Transistorised voltmeter using FET as input stage;
(*iii*) Balanced bridge T.V.M;
(*iv*) Chopper amplifier type voltmeter;
(*v*) Differential voltmeter (D.C);
(*vi*) Rectifier type A.C. voltmeter;
(*vii*) Averge responding voltmeter;
(*viii*) Peak reading voltmeter;
(*ix*) True R.M.S. reading voltmeter,
(*x*) Solid state voltmeter; and
(*xi*) Vector voltmeter.

4. The *cathode ray oscilloscope* is a device that allows the amplitudes of electrical signals, whether they be voltage, current, power, etc., to be displayed primarily as a function of time.

5. Cathode ray tube (CRT) consists of the following parts :
(*i*) Electron gun assembly;
(*ii*) Deflection plates assembly;
(*iii*) Fluorescent screen;
(*iv*) Glass envelope
(*v*) Base

6. *Electrostatic deflection* :

$$v_x = \sqrt{\frac{2eV_a}{m}} \quad \text{...(i)}$$

$$D = \frac{LV_d l_d}{2dV_a} \quad \text{...(ii)}$$

where, $v_x$ = Velocity of electron as it comes out of electron gun, m/s,
$e$ = Charge on electron, C,
$m$ = Mass of electron, kg,
$V_a$ = Accelerating voltage,
$D$ = Deflection on the screen,
$V_d$ = Deflection voltage,
$l_d$ = Length of each Y-plate,
$d$ = Distance between Y-plates, and
$L$ = Distance of middle of Y-plate to the screen.

Electroatic deflection sensitivity, $S = \frac{D}{V_d}$; its units are *m/V*.

7. Magnetic deflection :

$$D = BLl\left(\sqrt{\frac{e}{2mV_a}}\right)$$

where, $B$ = Flux density.

8. The function of *vertical amplifier* is to amplify the input signal.

9. The function of *time base generator* is to drive the beam at a steady speed across the screen and, when it reaches the right-hand side of the screen the beam is made to fly back to the starting position on the left-hand side of the screen. This generator produces a *saw-tooth wave of the same frequency as the input signal* to the Y-plates.
10. The purpose of *"delay line"* is to delay the vertical signal enough to keep it from reaching the CRT deflection plates before the horizontal sweep circuits are running.
11. Basic controls of CRO are :
    (*i*) Vertical (Y) deflection system; (*ii*) Horizontal (X) deflection system; (*iii*) Synchronization; (*iv*) Blanking circuit; (*v*) Positioning controls; (*vi*) Intensity (Z-axis) modulation; (*vii*) Intensity control; (*viii*) Focus control; (*ix*) Astigmatism; (*x*) Calibration circuit.
12. A *"storage oscilloscope"* uses the phenomenon of *"secondary electron emission"* to build up and store electrostatic charges on the surface of an insulated target.
13. *"Digital readout oscilloscope"* consists of a *high-speed laboratory CRO and electronic counter, both in one cabinet.*
14. The function of a *probe* is to connect the test circuit to the oscilloscope without altering, loading, or otherwise disturbing the test circuit.
15. The *"electronic switch"* is an extremely useful accessory for *displaying the two signals simultaneously on single trace oscilloscopes.*
16. The various parameters which can be measured by using CRO are :
    (*i*) Voltage; (*ii*) Current; (*iii*) Period; (*iv*) Frequency; (*v*) Phase : (*vi*) Amplitude; (*vii*) Peak-to-peak value ; (*viii*) Duty cycle etc.
17. *Lissajous figures/patterns* are formed when two sine waves are applied simultaneously to the vertical and horizontal deflecting plates of CRO.
18. The unknown frequency can be measured by the following methods :
    (*i*) Lissajous method; (*ii*) Spotwheel method : (*iii*) Gearwheel method.
19. *"Delayed sweep"* is a technique that adds a precise amount of time between the trigger point and the beginning of the scope sweep.
20. D.C. bridges are used to measure the *resistance.*
21. A.C. bridges are used to measure the *impedances* consisting of *inductances* and *capacitances.*
22. The simplest form of bridge is for the purpose of measuring resistance and is called the *Wheatstone bridge.*
23. $\bar{Z}_1 \bar{Z}_4 = \bar{Z}_2 \bar{Z}_3$ is the general equation for balance of the A.C. bridge.
24. Maxwell bridge is limited to the measurement of low $Q$ values (1-10).
25. Hay bridge is more convenient for measuring high-$Q$ coils.
26. Schering bridge is widely used for the measurement of unknown capacitors, dielectric loss and power factor.
27. Q-meter is an instrument designed to measure some electrical properties of coils and capacitors.
28. Bolometric measurements are based on the dissipation of the *RF* power in a small temperature sensitive resistive element, called a *bolometer.*
29. An *output power wattmeter* is designed to directly measure the output in an arbitrary load.

## OBJECTIVE TYPE QUESTIONS

### Electronic Voltmeters

1. A true rms reading voltmeter uses two thermocouples in order
   (*a*) to increase sensitivity
   (*b*) that the second thermocouple cancels out the no-linear effects of the first thermocouple
   (*c*) to prevent drift in the d.c. amplifier (*d*) all of the above.
2. In an electronic ohmmeter, an OP-AMP is used as a
   (*a*) Summer (*b*) Multiplier
   (*c*) Buffer amplifier (*d*) Integrator.
3. An average reading VTVM uses one diode connected in series with an external resistance. A high value of series resistance is used so that the instrument should have
   (*a*) a high input impedance (*b*) a linear *v-i* characteristics
   (*c*) low power consumption (*d*) all of the above.
4. In a compensated shunt type peak reading VTVM a bucking battery with a series adjustable resistance is used for
   (*a*) providing power for the circuit (*b*) zero setting
   (*c*) the purpose of amplifiying low level signals
   (*d*) all of the above.
5. The power consumption of a d.c. voltmeter using a direct coupled amplifier when measuring a voltage 0.5 V is of the order of
   (*a*) a few watt (*b*) a few milliwatt
   (*c*) a few microwatt (*d*) a few nanowatt.
6. A chopper stabilized A.C. amplifier may use
   (*a*) an electromechanical chopper (*b*) bipolar transistors as choppers
   (*c*) MOS field effect transistors as choppers (*d*) light activated devices as choppers
   (*e*) all of the above.
7. A voltmeter with a broad bandwidth has
   (*a*) low noise level and high sensitivity (*b*) high noise level and high sensitivity
   (*c*) high noise level and low sensitivity (*d*) low noise level and low sensitivity.
8. In circuits of RF voltmeters
   (*a*) vacuum tube diodes are used
   (*b*) conventional PN junction diodes are used
   (*c*) schottky barrier or point contact type diodes are used
   (*d*) all of the above.
9. In a differential amplifier using two FETs, a resistance has to be used to reduce the meter current to zero even when no voltage is applied to the circuit. This is necessitated on account of
   (*a*) mismatch between characteristics of the FETs
   (*b*) difference between tolerance values of resistors used in the circuit even though they are marked nominally equal.
   (*c*) variations in the operating voltage of the circuit.
   (*d*) both (*a*) and (*b*).

10. A PMMC meter has an internal resistance 200 Ω and the current required for its full scale deflection is 50 μA. The meter is capable of measuring, on its own, a maximum voltage of

(a) 5 mV (b) 10 mV
(c) 5 μV (d) 10 μV.

11. A PMMC meter has an internal resistance of 500 Ω and the current required for its full scale deflection is 100 μA. The power consumed by the meter is

(a) 5 μ W (b) 5 mW
(c) 5 W (d) 0.5 μW.

12. Electronic voltmeters which use rectifiers employ negative feedback. This is done

(a) to increase the overall gain (b) to improve stability
(c) to overcome nonlinearity of diodes (d) none of the above.

13. A voltage having symmetrical square waveshape is applied to an average reading voltmeter with scale calibrated in terms of rms value of a sinusoidal wave which has an error of

(a) –3.9% (b) + 3.9%
(c) – 11% (d) +11%.

14. The resistances, $R_D$, connected in the bridge circuit of a difference amplifier type of Electronic Voltmeter are each 10 kΩ. The value of $R_D$ is much smaller than the value of A.C. drain resistance, $r_d$, of FETs. The transconductance of FETs is 0.005 mho. The value of current through the meter when an input voltage of 1V is applied to the gate of left FET with the gate of right hand FET grounded, and the meter resistance almost negligible is:

(a) 5 mA (b) 2.5 mA
(c) 10 mA (d) 1 mA.

15. In a source follower type electronic voltmeter, the transconductance of FETs is 0.005 mho and the resistance of the meter is 100Ω. The approximate value of current through the meter, when the input voltage is 1V, is :

(a) 2.5 mA (b) 5 mA
(c) 2 mA (d) 1 mA

## Cathode Ray Oscilloscope (CRO)

16. A thin aluminium film is usually deposited on the non-viewing side of the phosphor because

(a) it acts as a heat-sink and prevents phosphor burn
(b) the light scatter from the phosphor is reduced
(c) it does not allow the screen to be negatively charged
(d) all of the above.

17. During the retrace time, the electrons forming the horizontal beam

(a) move from left to right on the screen
(b) move from right to the left on the screen
(c) move from bottom to top of screen
(d) move from top to bottom of screen.

18. The deflection of an electron beam on a CRT screen is 10 mm. Suppose the pre-accelerating anode voltage is halved and the potential between deflecting plates is doubled. The deflection of the electron beam will be :

(a) 80 mm (b) 40 mm
(c) 20 mm (d) 10 mm.

19. If the distance of screen from a CRT to centre of deflection plates is 15 cm, the length of deflection plates is 2 cm, the distance between plates is 1 cm and the accelerating voltage is 500 V, the deflection sensitivity is :
(a) 33.2 V/cm (b) 0.03 cm/V
(c) 66.4 V/cm (d) 0.015 cm/V.

20. Post-acceleration is needed in a CRO if the frequency of the signal is
(a) less than 1 MHz (b) more than 1 MHz
(c) more than 10 MHz (d) more than 10 Hz.

21. In a digital storage oscilloscope, the input signals are
(a) directly applied to the oscilloscope
(b) multiplexed, converted to digitial form and stored and applied to oscilloscope
(c) multiplexed, converted to digital form and stored, converted to analog form and applied to oscilloscope
(d) applied to amplifier, stored as analog signals, multiplexed, converted to digital form, stored in digital form, converted to analog form,and applied to CRO through an amplifier.

22. The source of emission of electrons in a CRT is
(a) PN junction diode
(b) a barium and strontium oxide coated cathode
(c) accelerating anodes (b) post-accelerating anodes.

23. In a CRT the focusing anode is located
(a) between pre-accelerating and accelerating anodes
(b) after accelerating anode
(c) before pre-accelerating anode (d) none of the above.

24. The horizontal amplifier should be designed for
(a) high frequency signals with a fast rise time
(b) high amplitude signals with a slow rise time
(c) high amplitude signals with a fast rise time
(d) low amplitude signals with a fast rise time.

25. A vertical amplifier for a CRO can be designed for
(a) only a high gain (b) Only a broad bandwidth
(c) a constant gain times bandwidth product
(d) all of the above.

26. In an oscilloscope, when the unknown signal applied to the vertical plates is being synchronized with the sweep signal applied to horizontal plates, the pattern seen on the CRO screen moves towards the right. It means that
(a) the frequency of the signal is lower than that of the sweep signal
(b) the frequency of the signal is greater than that of the sweep signal
(c) the frequency of the signal is equal to the frequency of that sweep signal
(d) none of the above.

27. $P_1$ phosphor material is used for display in CRTs for
(a) photographic applications (b) general purpose applications
(c) television applications (d) all of the above.

28. An aquadag is used in a CRO to collect
(a) primary electrons (b) secondary emission electrons
(c) both primary and secondary emission electrons
(d) none of the above.

## Bridge and Other Measurements

**29.** A bridge circuit works at a frequency of 2 kHz. The following can be used as detectors for detection of null conditios in the bridge :

(*a*) Vibration galvanometers and headphones
(*b*) Headphones and tunable amplifiers
(*c*) Vibration galvanometers and tunable amplifiers
(*d*) Vibration galvanometers, headphones and tunable amplifiers.

**30.** Wagner's Earth Devices are used in A.C. bridge circuits for

(*a*) eliminating the effect of earth capacitances
(*b*) eliminating the effect of inter-component capacitances
(*c*) eliminating the effect of stray electrostatic fields
(*d*) shielding the bridge elements.

**31.** Maxwell's inductance-capacitance bridge is used for measurement of inductance of

(*a*) low *Q* coils (*b*) medium *Q* coils
(*c*) high *Q* coils (*d*) low and medium *Q* coils.

**32.** The advantage of Hay's bridge over Maxwell's inductance-capacitance bridge is because

(*a*) its equations for balance do not contain any frequency term
(*b*) it can be used for measurement of inductance of high *Q* coils
(*c*) it can be used for measurement of inductance of low *Q* coils
(*d*) none of the above.

**33.** The equations under balance conditions for a bridge are

$R_1 = R_2R_3/R_4$ and $L_1 = R_2R_3C_4$.

where $R_1$ and $L_1$ are respectively unknown resistance and inductance. In order to achieve converging balance

(*a*) $R_2$ and $R_3$ should be chosen as variables (*b*) $R_2$ and $C_4$ should be chosen as variables
(*c*) $R_4$ and $C_4$ should be chosen as variables (*d*) $R$ and $C_4$ should be chosen as variables.

**34.** Frequency can be measured by using

(*a*) Maxwell's bridge (*b*) Schering bridge
(*c*) Heaviside Campbell bridge (*d*) Wien's bridge.

**35.** The resonant frequency of an RLC series circuit is 1.5 MHz with the tuning capacitor set at 150 pF. The bandwidth is 10 kHz. The effective resistance of the circuit is :

(*a*) 29.5 Ω (*b*) 14.75 Ω
(*c*) 9.4 Ω (*d*) 4.7 Ω

**36.** In a *Q* meter, distributed capacitance of coil is measured by changing the capacitance of the tuning capacitor. The values of tuning capacitor are $C_1$ and $C_2$ for resonant frequencies $f_1$ and $2f_1$ respectively. The value of distributed capacitance is :

(*a*) $\dfrac{C_1 - C_2}{2}$ (*b*) $\dfrac{C_1 - 2C_2}{3}$

(*c*) $\dfrac{C_1 - 4C_2}{3}$ (*d*) $\dfrac{C_1 - 3C_2}{2}$

**37.** The effective reactance of an inductive coil

(*a*) increases because of stray capacitances as the frequency increases.
(*b*) decreases because of stray capcitances as the frequency increases.
(*c*) remains the same irrespective of the increase in frequency even if stray capacitances are present.
(*d*) none of above.

38. A series RLC circuit is used for measurement of inductance of a coil using resonant method. Resonance conditions are produced by varying the capacitance. A thermoelectric type ammeter is used in series with the circuit. Also an electronic voltmeter is connected across the tuning capacitor
   (*a*) The conditions of resonance are more reliably indicated when the ammeter reads maximum
   (*b*) The conditions of resonance are more reliably indicated when the voltmeter connected across the capacitor reads maximum.
   (*c*) The conditions of resonance are the same when both ammeter and voltmeter read maximum.
   (*d*) The conditions of resonance are indicated when both ammeter and voltmeter read minimum.
39. The true value of $Q$ of a coil is 245 and the measured value 244.5. The ratio of distributed capacitance to self-capacitance of the coil is
   (*a*) $2.04 \times 10^{-3}$ (*b*) 1.002
   (*c*) 0.997 (*d*) none of the above.
40. In a $Q$-meter, the value of shunt resistance connected across the oscillator is typically of the order of
   (*a*) Ω (*b*) mΩ
   (*c*) μΩ (*d*) kΩ.
41. If the $Q$ factor of a coil which is $\omega_0 L/R$ is measured by varying the frequency,
   (*a*) the plot between $Q$ and frequency is linear
   (*b*) the value of $Q$ initially decreases with increase of frequency and afterwards increases with increase of frequency.
   (*c*) the value of $Q$ initially increases with increase of frequency and afterwards decreases with increase of frequency
   (*d*) the $Q$ factor remains constant irrespective of the value of frequency.
42. The $Q$ factor of a coil at the resonant frequency 1.5 MHz of an RLC series circuit is 150. The bandwidth is
   (*a*) 225 MHz (*b*) 1.06 MHz
   (*c*) 10 kHz (*d*) None of the above.
43. The effective reactance of an inductive coil
   (*a*) increases due to stray series inductances as frequency increases
   (*b*) decreases due to stray series inductances as frequency increases
   (*c*) remains the same irrespective of the increase in frequency even if stray series inductances are present (*d*) none of the above.
44. The effective resistance of a coil at high frequencies is more than its D.C. resistance on account of
   (*a*) skin effects (*b*) proximity effects
   (*c*) eddy current losses (*d*) all of the above.

## ANSWERS

| | | | | | |
|---|---|---|---|---|---|
| 1. (*b*) | 2. (*c*) | 3. (*d*) | 4. (*b*) | 5. (*d*) | 6. (*e*) |
| 7. (*c*) | 8. (*c*) | 9. (*d*) | 10. (*b*) | 11. (*a*) | 12. (*c*) |
| 13. (*d*) | 14. (*b*) | 15. (*c*) | 16. (*d*) | 17. (*b*) | 18. (*b*) |
| 19. (*b*) | 20. (*c*) | 21. (*d*) | 22. (*b*) | 23. (*a*) | 24. (*b*) |
| 25. (*c*) | 26. (*a*) | 27. (*b*) | 28. (*b*) | 29. (*b*) | 30. (*a*) |
| 31. (*b*) | 32. (*b*) | 33. (*c*) | 34. (*d*) | 35. (*d*) | 36. (*c*) |
| 37. (*b*) | 38. (*b*) | 39. (*a*) | 40. (*b*) | 41. (*c*) | 42. (*c*) |
| 43. (*a*) | 44. (*d*) | | | | |

## THEORETICAL QUESTIONS

### Electronic Voltmeters

1. What is an *"instrument"* ?
2. What do you mean by the term *"Measurement"*?
3. What are the primary requirements for any measuring instrument ?
4. What are the essentials of an electronic instrument ? Explain each of them briefly.
5. Explain briefly the following elements of electronic instrument :
   (*a*) Transducer ; (*b*) Signal modifier;
   (*c*) Indicating device.
6. State the advantages of electronic instruments.
7. What are electronic voltmeters ? Explain briefly.
8. What are the advantages of electronic voltmeters over conventional ones ?
9. Give the comparison between *'Electronic voltmeters'* and *'Conventional analog voltmeters'*.
10. Enumerate the various types of electronic voltmeters.
11. How are electronic voltmeters classified ?
12. Explain briefly *'Vacuum Tube Voltmeters'* (VTVM).
13. List the various types of vacuum type voltmeters.
14. What are the advantages-applications of valve voltmeters ?
15. State the advantages of transistor voltmeter over the vacuum tube voltmeter.
16. Explain briefly any two of the following electronic voltmeters :
    (*a*) Transistorised voltmeter using cascaded transistors.
    (*b*) Transistorised voltmeter using FET as input stage.
    (*c*) Balanced bridge TVM.
17. Describe briefly the following :
    (*a*) Chopper amplifier type voltmeter.
    (*b*) Differential voltmeter (D.C.)
18. Explain briefly 'Rectifier type A.C. voltmeter'.
19. Describe briefly with neat sketches any two of the following :
    (*a*) Average responding voltmeter
    (*b*) Peak responding voltmeter
    (*c*) True r.m.s.-reading voltmeter.
20. What are the advantages and limitations of true r.m.s. reading voltmeter ?
21. What is a *'Solid state voltmeter'* ? Explain briefly.
22. Discuss briefly *'Vector voltmeter'*.
23. Explain with a neat circuit diagram *"Electronic multimeter"* (or VOM).
24. List some important considerations needed in selecting an analog voltmeter.

### Cathode ray Oscilloscope (CRO)

25. What is a 'CRO'? Explain briefly.
26. What are the applications of 'CRO'?
27. Draw the block diagram of an oscilloscope and explain briefly its major systems.
28. Explain briefly the working of a CRO.
29. What is a 'CRT' ? Explain briefly.

30. With the help of a neat diagram, explain briefly the main parts of a cathode ray tube (CRT).
31. Derive an expression for vertical deflection of an electron beam in a 'CRT'.
32. Describe the different types of phosphor materials used in a CRO and list their applications.
33. Explain the functioning of a time-base generator in a CRO.
34. What are the different types of amplifiers used for CROs? Describe the basis on which they are classified.
35. Describe in details the vertical amplifier used in a CRO.
36. Describe the different types of sweeps used in a CRO. Explain their fields of application.
37. Describe the function of attenuators in CROs.
38. Describe the following :
    (*a*) Blanking circuit (*b*) Astigmatism control
    (*c*) Sources of synchronization (*d*) Z-axis modulation.
39. What do you understand by *"Delay line"*? Explain briefly.
40. Enumerate the basic controls of CRO and explain them briefly.
41. Describe briefly the following :
    (*a*) Dual trace oscilloscope (*b*) Dual beam oscilloscope.
42. Give the comparison between *'dual trace'* and *'dual beam'* oscilloscopes.
43. Discuss briefly any two of the oscilloscopes :
    (*a*) Sampling oscilloscope (*b*) Digital readout oscilloscope
    (*c*) High frequency oscilloscope
44. What is a *'probe'* ?
45. Explain briefly high impedance (or 10 : 1) probe.
46. What is an *"Electronic switch"*? Explain its working with the help of a block diagram.
47. Discuss the advantages and disadvantages of analog and digital type of oscilloscopes.
48. Discuss in detail the *'delayed sweep'*.
49. Describe briefly how the following measurements can be made with the use of CRO.
    (*a*) Frequency; (*b*) Phase angle;
    (*c*) Voltage.
50. Explain briefly any two of the following methods of frequency measurement :
    (*a*) Lissajous figures (*b*) Spot wheel method
    (*c*) Gear wheel method.

## Bridge and Other Measurements

51. What is a *"bridge circuit"* ?
52. What are the applications of bridge circuits?
53. What are the advantages of bridge circuits ?
54. What are the types of bridges ?
55. Explain briefly how a 'Wheatstone bridge' is used for measurement of resistance.
56. What are the sources of measurement errors in a Wheatstone bridge?
57. State the limitations of "Wheatstone bridge".
58. Explain briefly the following bridges :
    (*a*) Kelvin bridge; (*b*) Kevin double bridge.
59. Draw the block diagram of Wheatstone bridge with digital readout and explain it briefly.
60. Draw the general form of an A.C. bridge and derive the general equation for its balance.

61. Explain briefly how *"inductance"* can be measured by using a 'Maxwell bridge'.
62. What are the advantages and disadvantages of å Maxwell bridge ?
63. Discuss briefly how "Hay bridge" can be used for the measurement of inductance.
64. With the help of circuit diagram explain how capacitance can be measured by the use of a 'Schering bridge'.
65. Describe briefly how Wien bridge can be used for the measurement of frequency.
66. State the applications of a Wien bridge.
67. Explain briefly *"Resonance bridge"* with the help of a neat circuit diagram. Name the parameters which can be measured by this bridge.
68. Explain with a neat circuit diagram the working of a "Wagner earthing device".
69. How are detectors classified ?
70. What precautions should be taken while using a bridge ?
71. Explain with the help of circuit, the principle of working of a Q-meter.
72. Explain briefly how impedance bridge can be used for measuring resistance, inductance and capacitance.
73. Describe briefly the working of a LCR meter.
74. Draw the basic circuit of RX meter and explain briefly its working.
75. Discuss the problems associated with measurement of resistance, inductance and capacitance at high frequencies.
76. Explain briefly the following two methods of measurement of effective resistanace of a coil at high frequencies :
    (*a*) Resistance variation method (*b*) Reactance variation method.
77. Describe the resonance method for measurement of inductance at high frequencies. Discuss the assumptions made. How is effect of additional capacitances in the tuning capacitor taken care of ?
78. Explain briefly the method for measurement of capacitance at high frequencies using resonance method.
79. Discuss briefly the method of inductance measurement at high frequencies.
80. How is capacitance measured at high frequencies ? Explain.
81. With the help of a block diagram, explain the working of a vector impedance meter.
82. Explain briefly, with a block diagram, the working of a RF vector impedance meter.
83. Discuss briefly the working of a typical automatic bridge, with the help of a neat circuit diagram.
84. Explain briefly the following methods of measuring power :
    (*a*) Bolometer method. (*b*) Calorimeter method.
85. Describe the working of an output-power meter with the help of a neat circuit diagram.
86. List the methods/techniques which are used for the measurement of audio and radio frequencies.
87. Explain briefly any two of the following techniques employed for measurement of AF and RF :
    (*a*) Bridge technique. (*b*) Heterodyne technique.
    (*c*) Wavemeters.
88. Write short notes on the following :
    (*a*) Cavity wavemeters (*b*) Frequency standards.

## UNSOLVED EXAMPLES

1. In a CRT, the anode to cathode voltage is 2000V. The parallel deflector plates are 1.5 cm long and spaced 5 mm. The screen is 50 cm from the centre of the deflection plates. Find the following :
   (*a*) The beam speed.
   (*b*) The deflection sensitivity of the tube.
   Take : Mass of electron = $9.109 \times 10^{-31}$ kg : charge on electron = $1.602 \times 10^{-19}$ C.
   **[Ans.** (*i*) $26.5 \times 10^6$ m/s; (*ii*) 0.325 mm/V]
2. A cathode ray tube, with magnetic deflection, has its screen 20 cm from the centre of the magnetic field. The width of the uniform magnetic field is 3 cm and the final anode potential is 6 kV. Calculate the density of magnetic field to produce a deflection of 4 cm on the screen.
   Take : Mass of electron = $9.107 \times 10^{-31}$ kg : charge on electron = $1.6 \times 10^{-19}$C.
   **[Ans.** 1.7423 Wb/m$^2$]
3. In a cathode-ray tube the electron beam is displaced vertically by a magnetic field of flux density $2 \times 10^{-4}$ wb/m$^2$. The length of the magnetic field along the tube axis is the same as that of the electrostatic deflection plates. The final anode voltage is 800 V. Calculate the voltage which should be applied to the *Y*-deflection plates 1 cm apart, to return the spot back to the centre of the screen. **[Ans.** 33.53 V]
4. Fig. 2.68 consists of the following parameters :
   $R_1$ = 10 kΩ; $R_2$ = 15 kΩ ; $R_3$ = 40 kΩ
   Determine the unknown resistance $R_x$. **[Ans.** 60 kΩ]
5. In a Kelvin double bridge, there is error due to mismatch between the ratios of outer and inner arm resistances. The following data relate to this bridge :
   Standard resistance = 100.03 μΩ; Inner ratio arms = 100.31Ω and 200 Ω; Outer ratio arms = 100.24 Ω and 200 Ω; The resistance of the connecting leads from standard to unknown resistance is 700 μΩ.
   Calculate the unknown resistance. **[Ans.** 49.97 μΩ]
6. The impedances of the basic A.C. bridge of Fig. 2.72 are given follows :
   $\bar{Z}_1$ = 100 Ω∠80° (inductive impedance); $\bar{Z}_2$ = 250 Ω (pure resistance);
   $\bar{Z}_3$ = 400 Ω ∠30° (inductive impedance); $\bar{Z}_4$ = Unknown
   Determine the constants of the unknown arm **[Ans.** 1000 ∠–50°]
7. A Maxwell bridge (Fig. 2.75) is used to measure an inductive impedance. The bridge constants at balance are :
   $R_1$ = 470 kΩ; $C_1$ = 0.01 μF; $R_2$ = 5.1 kΩ ; $R_3$ = 100 kΩ
   Find the series equivalent of the unknown impedance. **[Ans.** $R_x$ = 1.09 kΩ; $L_x$ = 5.1 H]
8. The Schering bridge shown in Fig. 2.79 has the following constants :
   Arm AB ... Capacitor of 0.5 μF in parallel with 1 kΩ resistance;
   Arm BC ... Resistance of 3 kΩ;
   Arm CD ... Unknown capacitor $C_x$ and $R_x$ in series
   Arm DA ... Capacitor of 0.5 μF
   Frequency ... 1 kHz
   Determine the following :
   (*a*) Unknown resistance and capacitance
   (*b*) Dissipation factor. **[Ans.** (*a*) 2Ω, 0.25 μF; (*b*) 3.1416]

9. Determine the equivalent parallel resistance and capacitance that causes a wien bridge (Fig. 2.80) to null with the following component values :
$R_1$ = 3.1 kΩ; $C_1$= 5.2 μF ; $R_2$ = 25 kΩ; $R_4$ = 100 kΩ; $f$ = 2.5 kHz
**[Ans.** 12.4 kΩ ; 20.3 pF]

10. The following data were obtained during the measurement of the distributed or self-capacitance of a coil :
First measurement : $f_1$ = 2 MHz and $C_1$ = 460 pF;
Second measurement : $f_2$ = 4 MHz and $C_2$ = 100 pF.
Find the distributed capacitance, $C_d$. **[Ans.** 20 pF]

11. Calculate the value of distributed capacitance of a coil when the following measurements are made :
At frequency $f_1$ = 2MHz, the tuning capacitor is set at 450 pF;
At frequency $f_2$ = 5MHz, the tuning capacitor is tuned at 60 pF. **[Ans.** 14.28 pF]

12. In a $Q$ meter, after connecting a standard coil, the resoanace is obtained with a frequency $f_1$ with resonating capacitor set at $C_1$; the indicated Q-factor is $Q_1$. An unknown impedance is connected in series with the standard coil and the resonance is re-established by resetting the resonating capacitor at $C_2$, the corresponding Q-factor being $Q_2$. Determine the resistive and reactive components of the unknown impedance when $C_1$ = 208 pF, $Q_1$ = 80, $C_2$ = 184 pF and $Q_2$ = 50. The frequency is 165 kHz. **[Ans.** 46.88 Ω, 605 Ω]

13. A coil of resistance 5 Ω is connected in the circuit shown in Fig. 2.88 and resonance occurs when oscillatory frequency is 500 kHz and resonating capacitor is set at 120 pF. Determine the percentage error introduced in the calculated value of $Q$ if a resistance of 0.02 Ω is used across the oscillatory circuit. **[Ans.** 0.4 %]

14. The following results were obtained during tests conducted, using a $Q$ meter, on a radio-tuning coil to establish the self-capacitance :
(*a*) With the radio coil connected normally resonance was obtained at 800 kHz with the tunning capacitor of the $Q$ meter set at 95 pF.
(*b*) With a standard inductor connected in place of radio coil, resonance was obtained at 2.5 MHz and this condition was not changed when the radio coil was connected in parallel with the standard inductor.
Determine the self-capacitance of the radio coil. **[Ans.** 10.84 pF]

15. A circuit consisting of an unknown coil, a resistance and a variable capacitor connected in series is tuned to resonance using a $Q$ meter. If the frequency is 450 kHz and the resonating capacitor is set at 250 pF, determine the effective inductance and resistance of unknown coil. The $Q$ meter indicates 105 and the resistance is 0.75 Ω.
**[Ans.** 500 μH; 12.76 Ω]

16. A coil of inductance L is tuned by a capacitance C to resontde at 1 MHz. A standard capacitor of 230 pF is tuned to resonate in the same coil at 2 MHz. If the resonance is indicated by a VTVM of capcitance 8 pF in parallel with the circuit, calculate the values of $L$ and $C$. **[Ans.** 26.6 μH; 944 pF]

17. A coil of inductance $L$ and resistance $R$, a calibrated variable capacitor C and a thermojunction, all in series, are put across a constant r.m.s. voltage of frequency 3 MHz. The uncalibrated ammeter is found to show the same scale reading I for the two values of C–namely 37 pF and 45 pF, one being below and the other above capacitance $C_O$ needed to produce maximum current. The capacitor is now adjusted to a value $C_O$, and an added series resistance of of 75 Ω is found to reduce the current to the previous value of $I$. Determine the values of $L$ and $R$. **[Ans.** 69.2 μH ; 70.8 Ω]

18. A circuit having an effective capacitance of 150 pF is tuned to a frequency of 1.5 MHz. In this circuit the current falls to 70.70 per cent of its resonant value when the frequency

of an e.m.f. of constant magnitude injected in series with the circuit deviates from the resonant frequency by 5 kHz. Calculate the $Q$ factor and effective resistance of the circuit.

[**Ans.** 150; 4.7 Ω]

**19.** A capacitance of 250 pF produces resonance with a coil at a frequency $\left(\frac{2}{\pi}\right)\times 10^{6}\ Hz$, while at the second harmonic of this frequency resonance is produced by a capacitance of 50 pF. Calculate : (*i*) The self-capacitance, and (*ii*) The inductance of the coil.

Neglect the effect of voltmeter capacitance and also other stray capacitances.

[**Ans.** 16.67 pF; 234 μH]

CHAPTER

# 3 Signal Generators



## 3.1 GENERAL ASPECTS

*A* **signal generator** *is an instrument that provides a controlled output waveform or signal for use in testing or aligning, or in measurements on other circuits or equipments.*

- The signal generators provide a variety of waveforms for testing electronic circuits, *usually at low powers.*
- The term ***"Oscillator"*** is used to describe an instrument that *provides only a sinusoidal output signal,* and the term ***"Function generator"*** to describe an instrument that provides several output waveforms including *sine wave, square wave, triangular wave and pulse trains, as well as an amplitude modulated waveform.*
  — It may be emphasized that, although we speak of oscillators as generators of signals, no energy is created; it is simply converted from a D.C. source into A.C. energy as some specific frequency.
- In various instrumentation systems, the signal at audio frequency as well as at radio frequency are required. In most of the cases, the signal at particular frequency is generated *using an oscillator which provides fixed frequency signal.* Some other variable frequency oscillators are also available.
- The audio frequency (AF) and radio frequency (RF) signal generators are the equipments which provide signals for general test purposes. These two are *specially designed to cover a wide frequency band from few Hz to several GHz.* The frequency band limits are listed in table 3.1.

**Table 3.1. Frequency Band Limits**

| S. No. | Band | Approximate range |
|---|---|---|
| 1. | Audio frequency (AF) | 20 Hz–20 kHz |
| 2. | Radio frequency (RF) | Above 30 kHz |
| 3. | Very low frequency (VLF) | 15–100 kHz |
| 4. | Low frequency (LF) | 100–500 kHz |
| 5. | Broadcast | 0.5–1.5 MHz |
| 6. | Video | 0–5 MHz |
| 7. | High frequency (HF) | 1.5–30 MHz |
| 8. | Very high frequency (VHF) | 30–300 MHz |
| 9. | Ultra high frequency (UHF) | 300–3000 MHz |
| 10. | Microwave | Beyond 3000 MHz (or 3 GHz) |

**Characteristics of signal generators :**

Although there are various types of signals generators, yet the following *characteristics* are common to all the types.

1. The frequency of the signal should be well known and stable.
2. The amplitude of the output signal should be controllable from very low values to relatively large values.
3. The signal should be free from distortion.

**Applications :**

Signal generators are used for the following *purposes* :

1. To provide appropriate signal for calibration, testing and trouble shooting of the amplifier circuits used in communications, electronics such as radio and television amplifiers.
2. To be employed as power source for the measurement of gain, bandwidth, signal-to-noise (S/N) ratio and standing wave ratio (SWR) and other properties of circuits.
3. To measure the characteristics of antennas and transmission lines.

**Classification of signal generators :**

The signal generators can be *classified* into the following categories :

1. Audio generators.
2. Function generators.
3. Pulse generators.
4. RF generators.
5. Frequency synthesizers.
6. Other signal generators.

The various types of generators are described in the following articles.

## 3.2 CONVENTIONAL STANDARD SIGNAL GENERATOR

A *sine-wave generator* represents the largest single category of signal generators. This instrument covers the frequency range from a few Hz to many GHz. Fig. 3.1 shows the simplest form of the sine-wave generator :

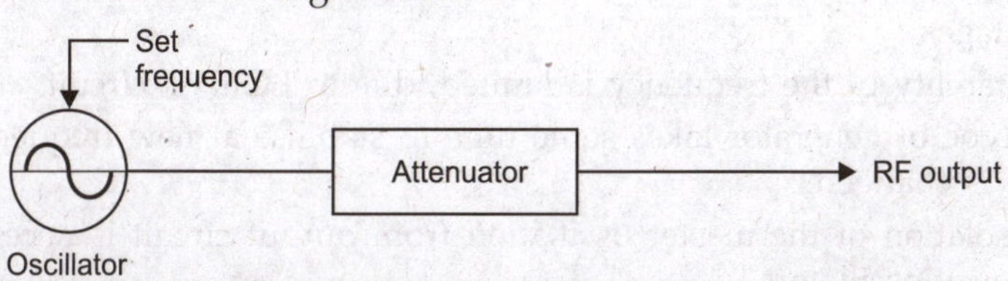


**Fig. 3.1.** Block diagram of a simple sine-wave generator.

— It consists of two basic blocks, an *"oscillator"* and an *"attenuator"*.

— *The frequency accuracy and stability and freedom from distortion depends on the design of the oscillator, while the amplitude accuracy depends on the attenuator's design.*

**Conventional standard signal generator** *is basically a radio frequency (RF) signal generator and produces known and controllable voltages*. It is extensively used in the *testing of radio receivers and transmitters.*

- The generator's output is *frequency modulated or amplitude modulated.*

— The *frequency modulation* is possible using a carrier signal from RF oscillator.

— The *amplitude modulation* can be done by using internal sine-wave oscillator.

The modulation may be done by a sine wave, square wave, triangular wave or a pulse also.

Fig. 3.2, shows the block diagram of a *conventional standard signal generator* :

— The carrier frequency (sine wave) is produced by a *stable RF oscillator* using LC tank circuit. The frequency of oscillation is shown on frequency control and the vernier dial setting.

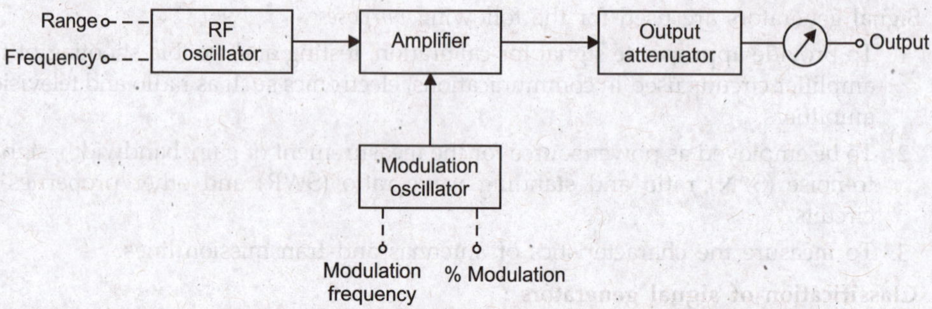


**Fig. 3.2.** Block diagram of a conventional standard signal generator.

— Amplitude modulation is obtained by an *internal or an external fixed frequency oscillators.*

— Modulation is done in output amplifier circuit which delivers its output (*i.e.*, the modulated carrier) to an *attenuator*.

— The output voltage is observed is an *output meter* and the *attenuation output setting*.

- Regulated power supplies are used for frequency stability at all ranges and to maintain it, temperature compensation is used. Buffer amplifiers are employed to isolate the oscillator circuit from the output in high frequency oscillator.

**Advantages :**

1. The output voltage of the conventional standard signal generator can be controlled as per the requirement.
2. The output obtained is stable.

**Disadvantages :**

1. The stability of the frequency is limited, due to LC tank circuit.
2. This type of generator takes some time to stabilise at new frequency when the range is changed.
3. The isolation of the master oscillation from output circuit is necessary in high frequency oscillators.

## 3.3 MODERN SIGNAL GENERATOR

The order to improve the frequency stability, a *single master oscillator is optimally designed for the highest frequency range* and frequency dividers are switched in to produce lower ranges. The *master oscillators is made insensitive to temperature variations and to the influence of the succeeding stages by careful circuit design*. Temperature compensation devices are used for any temperature changes.

Fig. 3.3. shows the block diagram of a modern signal generator (Laboratory type) :

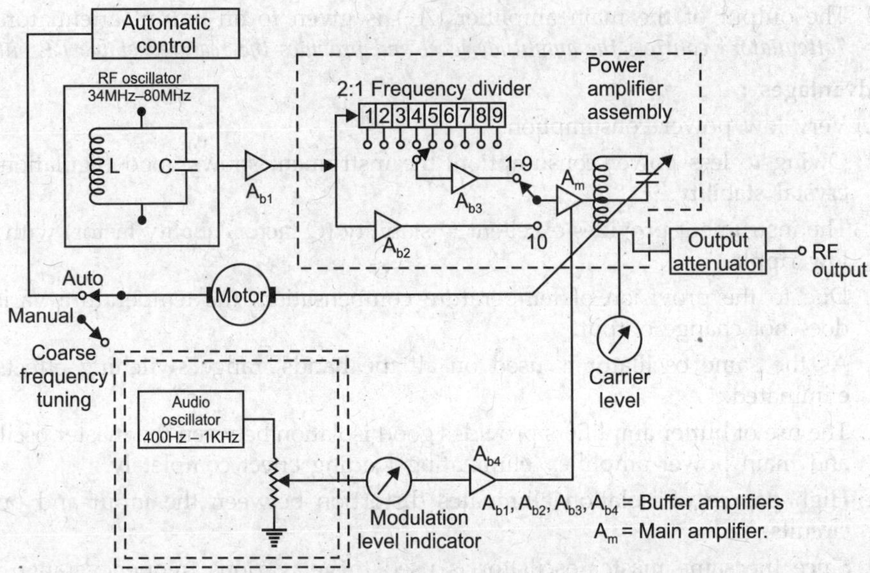


**Fig. 3.3.** Block diagram of a modern signal generator.

— RF oscillator (consisting of LC tank circuit) is used as a *master oscillator;* it gives a frequency range from 34 MHz-80 MHz. This signal is connected to untuned buffer amplifier $A_{b1}$. The output of this amplifier is given to frequency divider circuit (Fig. 3.3); the circuit consists of 9 frequency dividers. The frequency division can be done by using flip-flops to get ratio of 2 : 1.

— The LC tank circuit and frequency divider provides a carrier signal.

— The additional buffer amplifiers $Ab_2$ and $Ab_3$ provide a very high degree of isolation between the master oscillator and the power amplifier, and almost eliminates all the frequency effects (distortion) between the input and output circuits, caused by loading.

— The *modulating signal is provided by an audio oscillator;* the frequency of this oscillator is in the range of 400 Hz – 1 kHz. The modulation is done at the power amplifier stage. The level of modulation can be adjusted upto 95% by using a control device.

— The lowest frequency range produced by the cascaded frequency divider is the highest frequency divided by $2^9$ or 512. Thus, the frequency stability of the highest range is imparted to the lowest frequency ranges.

— The master oscillator is turned automatically or manually. In an automatic controller for tuning master oscillator, a motor driven variable capacitor is used. This system is extensively used in programmable automatic frequency control devices. The oscillator can be fine tuned by means of a large rotary switch with each division corresponding to 0.01% of main dial setting.

— 1 MHz crystal oscillator provides the internal calibration.

— The small power consumption of the instruments makes output with very low ripple.

— *The supply voltage of the master oscillator is regulated by temperature compensated reference circuit.*

— The output of the main amplifier ($A_m$) is given to an output attenuator. The *"attenuator" controls the amplitude level and provides the required stable RF output.*

**Advantages :**

1. Very low power consumption.
2. Owing to less power consumption, the instrument shows good regulation and crystal stability.
3. The instrument provides excellent $Q$ stability ($Q$-factor/quality factor) with very low ripple.
4. Due to the provision of temperature compensation, the temperature variation does not change output.
5. As the same oscillator is used on all the bands, range switching effects are eliminated.
6. The use of buffer amplifiers provides good isolation between the master oscillator and main power amplifier, eliminating loading effect completely.
7. High degree of isolation eliminates distortion between the input and output circuits.
8. Since the same master oscillator is used to get various frequency ranges, the stability is improved even at the highest frequency range.

**Disadvantage :** *Increased cost* (as compared to conventional signal generator) due to complex circuitry.

**Comparison between Standard and Modern signal generators :**

The comparison between standard and modern signal generators is given in the table 3.2 below :

| S. No. | Aspects | Standard signal generator | Modern signal generator |
|---|---|---|---|
| 1. | *Frequency stability* | Limited | Maintained over entire frequency range |
| 2. | *Frequency range* | Small | Wide range possible |
| 3. | *Regulation* | Poor | Excellent |
| 4. | *Construction* | Simple | Complicated |
| 5. | *Cost* | Low | High |
| 6. | *Time taken to stabilise at new frequency* | Takes time | Very quick |
| 7. | *Loading and distortion effects* | More | Less |
| 8. | *Temperature compensation* | Not provided | Provided |
| 9. | *Automatic tuning with motor* | Not available | Possible |
| 10. | *Q-stability* | Less with high ripple | Good with low ripple |

## 3.4 AUDIO FREQUENCY (AF) AND SQUARE WAVE GENERATOR

Fig. 3.4. shows the block diagram of AF sine and square wave generator :

— In this generator, a *Wien bridge oscillator* is used; this oscillator is *best* for the audio frequency range. The frequency of oscillations can be changed by varying the capacitance in the oscillator. The frequency can also be changed in steps by switching in resistors of different values.

— The output of the oscillator goes to the function switch. Depending upon the position of the switch, we get output as sine wave output or square wave output.

— In the square wave generation section, the output of the Wien bridge oscillator is fed to square wave shaper circuit which uses schmitt trigger circuit.

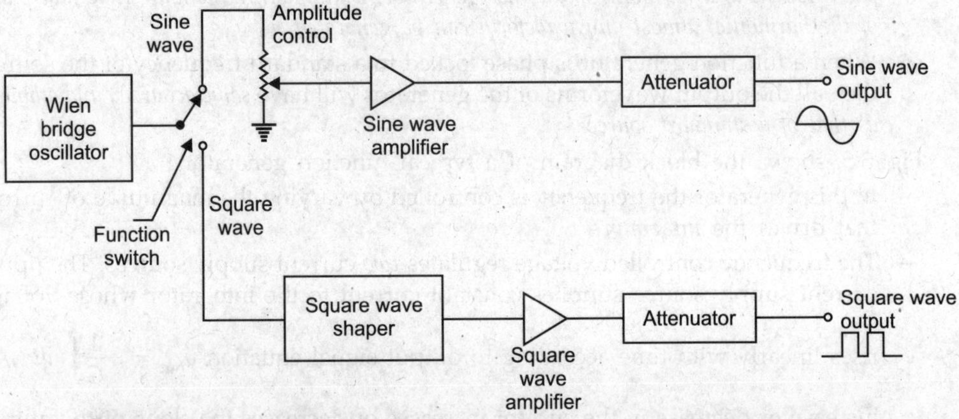


**Fig. 3.4.** Block diagram of AF sine and square wave generator.

— The *attenuators* in both the sections are used to control output signal level. The signal level, before attenuation, is made very high using sine wave amplifier and square wave amplifier.

**Specifications :**

Following are the specifications of a typical AF sine and square wave generator :

(*i*) *Frequency range* : 10 Hz to 1 MHz.

— Frequency is variable almost 5 decades continuously.

(*ii*) *Amplitude (sine wave)* : 5 mV to 5 V (r.m.s.)

— The output is taken through a push-pull amplifier.

— For low output, the impedance is 600 Ω.

(*iii*) *Amplitude (square wave)* : 0 to 20 V (Peak)

— Possible to adjust the symmetry of the square wave from 30 to 70%.

(*iv*) *Power required* : 7 W (at 220 V, 50 Hz).

## 3.5 FUNCTION GENERATOR

A **function generator** *is an instrument that has the capability of producing different types of waveform as its output signal.* The most common output waveforms are :

— Sine waves;

— Triangular waves;

— Square waves;

— Sawtooth waves.

The frequencies of above waves may be adjusted from a fraction of a hertz to several hundred kHz (A very common frequency range for function generators is 0.01 Hz to 3 MHz).

- The *triangular-wave and sawtooth-wave outputs* of function generators are commonly used for those applications which *need a signal that increases or reduces at a specific linear rate.* They are also used in *driving sweep oscillators in oscilloscopes and the X-axis of X-Y recorders.*

- One of the useful features of the function generator is that *it can be phase locked to an external signal source.* One function generator can be phase locked with second function generator; the *two output signals can be displaced in phase by an adjustable amount.* Furthermore, *the fundamental frequency of one generator* can be *phase locked to a harmonic of another generator. By adjusting the amplitude and phase of the harmonic, almost any waveform can be generated.*
- When a function generator is phase locked to a standard frequency of the source, then all the output waveforms of the generator will have *same accuracy and stability as that of a standard source.*

Fig. 3.5. shows the block diagram of a typical function generator :

— In this generator the frequency is controlled by varying the magnitude of current that drives the *integrator.*

— The frequency controlled voltage regulates *two* current supply sources. The upper current supply source supplies constant current to the integrator whose voltage rises linearly with time according to output signal equation $v_{out} = -\frac{1}{C}\int_{o}^{t} i\,dt$. An increase or decrease in the current increases or decreases the slope of the output voltage and thus controls the frequency.

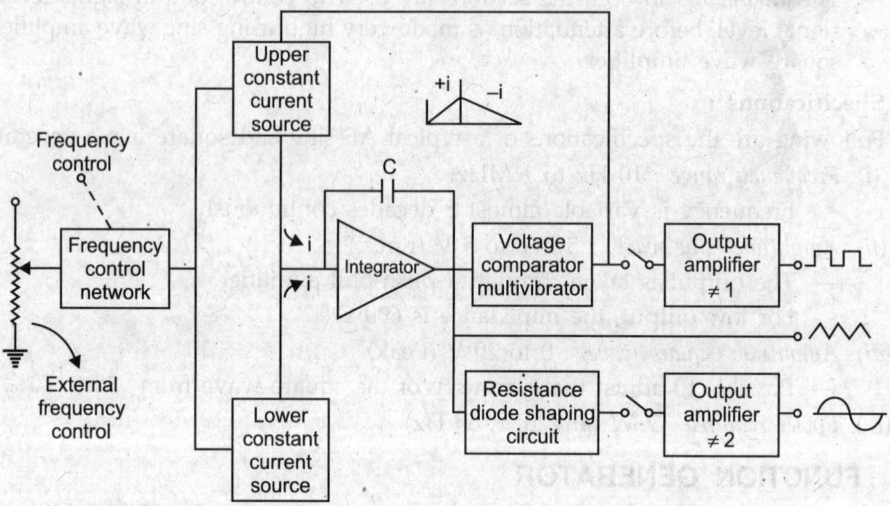


**Fig. 3.5.** Block diagram of a typical function generator.

— The *voltage comparator multivibrator* changes state at a predetermined maximum level of the integrator output voltage. This change cuts-off the current supply from the upper supply source and switches to the lower supply source. The *lower supply source supplies a reverse current to the integrator* so that its *output drops linearly with time.* When the output attains a predetermined level, the voltage comparator again changes state and switch 'ON' to upper supply source.

— *The output of the "integrator" is a "triangular-wave"* whose frequency depends on the upper constant current source. The *"voltage comparator" delivers square wave output voltage of the same frequency as that of input triangular waveform. The triangular wave is synthesised into sine wave using diode resistance network.* In this shaper circuit, the slope of triangular wave is changed as its amplitude changes; this results in a *sine wave with less than 1 per cent distortion.*

**Specifications :**

Following are the typical important specifications of a function generator :

(*i*) *Output waveforms :* Sine, square, triangular, ramp, pulse, AM and FM modulated, arbitrary.

(*ii*) *Frequency range :* 0.001 Hz to 20 MHz.

(*iii*) *Frequency stability :* 0.05%.

(*iv*) *Distortion :* –55 dB below 50 kHz, –40 dB above 50 kHz.

(*v*) *Output amplitude (open circuit) :* 10 $V_{p\text{-}p}$.

(*vi*) *Impedance :* 50 Ω.

**Applications :**

Function generator is used for the following purposes :

1. To test the bandwidth of the audio frequency amplifier—This method is commonly known as ***square wave testing.***
2. To be used as an important instrument in the troubleshooting of different analog and digital circuits and instruments.
3. To act as a source for the alignment of receivers.

**Comparison between "Signal generator" and "Function generator :**

The comparison between 'Signal' and 'Function' generators is given below :

| S. No. | Aspects | Signal generator | Function generator |
|---|---|---|---|
| 1. | *Type of wave generated* | Sine wave | Sine, square, sawtooth, triangular etc. |
| 2. | *Frequency stability* | Limited | High |
| 3. | *Method of frequency control* | By frequency range control and vernier dial setting (by varying capacitor in LC or RC circuit). | By varying the magnitude of the current which drives the integrator. |
| 4. | *Phase locking capability* | No | Yes (It has capability of phase locking with an external source so that two output signals can be displaced in phase with adjustable amount). |

## 3.6 PULSE AND SQUARE GENERATORS

### 3.6.1. Generator Aspects

These generators are *often used with an oscilloscope as the measuring device.*

The primary difference between a pulse generator and a square-wave generator relates to the *"duty cycle"*.

***Duty cycle*** *is defined as the ratio of the average value of the pulse over one cycle to the peak value of the pulse.*

*Or*

It is also defined *as the ratio of the pulse width and the period or pulse repetition time.*

*i.e.,*
$$\text{Duty cycle} = \frac{\text{Pulse width}}{\text{Period}}$$

- The duty cycle of *'square-wave generators'* equals 0.5 or 50% since they produce an output voltage with equal *"on"* and *"off"* times.

- The duty cycle of a *'pulse generator'* may vary.
  - Very short duration pulses give a low duty cycle, and the pulse generator generally can supply *more power* during its *"on"* period than a square-wave generator can.
  - *Short duration pulses reduce the power dissipation in the component under test.*
- Square-wave generators are *used* :
  (*i*) Whenever the low-frequency characteristics of a system are being investigated, *e.g., testing audio systems.*
  (*ii*) When the transient response of a system requires some time to settle down.
- A ***"pulse"*** *differs from a* ***"square wave"*** *in that it needs neither baseline nor left-right symmetry.*

Fig. 3.6 shows one popular approach to constructing a *pulse generator.* A monostable multivibrator (*i.e.,* one-shot) follows a square wave oscillator. The pulse repetition rate is set by the square wave frequency. The one-shot triggers on the leading edge of the square wave and produces one output pulse for each input cycle. The duration of each output pulse is set by the *on-time* of the one-shot and it may be very short or may approach the period of the square wave.

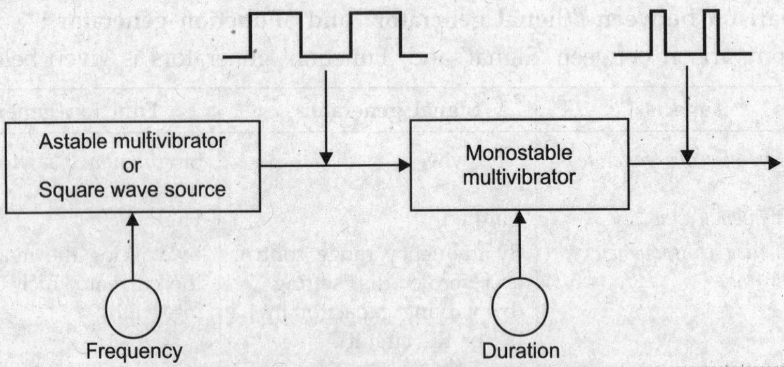


**Fig. 3.6.** Block diagram of a pulse generator.

— A typical pulse generator will allow the user to select the repetition rate, duration, and amplitude, and also the number of pulses to be output in a given burst.

**Circuits used in pulse generation :**

The circuits used in pulse generation fall into the following two categories :

1. *Passive, or pulse-shaping circuits.*
   - In these circuits, a sine-wave oscillator is used as the basic generator and its output is passed through a pulse-shaping circuit to obtain the desired waveform.

     *Example :* An approximate square waveform may be obtained by first amplifying and then clipping a sine wave.
2. *Active, or pulse-generating circuit.*
   - Active generators are usually of the *relaxation type.* The relaxation oscillator uses the charge-and-discharge action of capacitor to control the conduction of a vacuum tube or a transistor.
   - Some common forms of relaxation oscillators are :

     (*i*) *Multivibrators.*

     (*ii*) *Blocking oscillators.*

### 3.6.2. Laboratory Square-wave and Pulse Generator

Fig. 3.7 shows the block diagram of a typical general-purpose generator providing negative pulses of variable frequency, duty cycle, and amplitude.

**Features :**

— The frequency range of the instrument is covered in seven decade steps from 1 Hz to 10 MHz, with a linearly calibrated dial for continuous adjustment on all ranges.

— The duty cycle can be varied from 25% to 75%.

— Two independent outputs are available :

(*i*) A 50 Ω source that supplies pulses with rise-and-fall times of 5 ns at 5 V peak amplitude;

(*ii*) A 600 Ω source that supplies pulses with rise-and-fall times of 70 ns at 30 V peak amplitude.

— The instrument can be operated as a free-running generator or it can be synchronised with external signals.

— Trigger out pulses for synchronisation are also available.

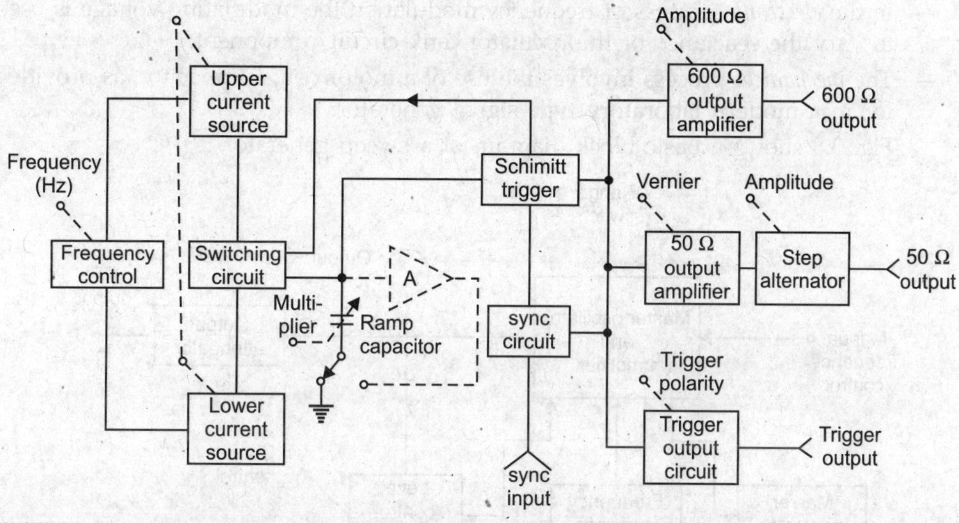


**Fig. 3.7.** Block diagram of a general purpose generator.

**Working :**

- The *upper* current source supplies a constant current to the ramp capacitor. This charges the capacitor at a constant rate, and the ramp voltage increases linearly. When the positive slope of the ramp voltage reaches the upper limit set by internal circuit components, the Schmitt trigger (a bistable vibrator) *changes state.* The trigger circuit output becomes *negative,* reversing the condition of the current control switch, and the capacitor starts discharging. The discharge rate is linear, controlled by *lower* current source. When negative ramp reaches a predetermined lower level, the Schmitt trigger switches back to its original state. This now provides a positive trigger circuit output that reverses the condition of the current switch again, cutting off the lower current source and switching on the upper current source. This gives one cycle of operation. The entire process is repetitive and the Schmitt trigger circuit provides negative pulses at a continuous rate.

- *The Schmitt circuit's output is passed to the trigger output circuit and to the 50 Ω and 600 Ω amplifiers.*
  - — The trigger output circuit differentiates the square-wave output from the Schmitt trigger, inverts the resulting pulse, and *provides a positive triggering pulse.*
  - — The 50 Ω amplifier is provided with an output attenuator to *allow a vernier control of the signal output voltage.*
- Besides its free running mode of operation, the generator can be synchronized or locked into an external signal (by triggering the Schmitt circuit by an external synchronization pulse).

The regulated voltages for all stages of the instrument are provided by an internal supply (by which the unit is powered).

## 3.7 SWEEP GENERATOR

*A* **sweep generator** *provides a sinusoidal output voltage whose frequency varies smoothly and continuously over an entire frequency band, usually at audio rate.*

The frequency modulation process may be accomplished *electronically* or *mechanically.*

— In the *electronic* process of frequency modulation the modulating voltage is used to vary the reactance of the oscillator tank circuit component.

— The *mechanical* process involves the use of a motor driven capacitor, as provided for in a modern laboratory type signal generator.

Fig. 3.8 shows a basic block diagram of a sweep generator :

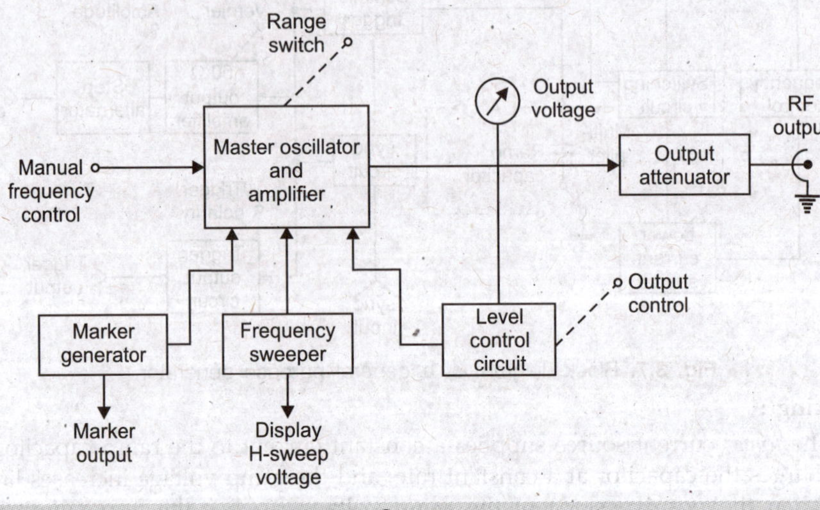


**Fig. 3.8.** Sweep generator.

— The *frequency sweeper* provides a variable modulating voltage which *causes the capacitance of the master oscillator to vary* (may be of the order 20 sweeps/sec). A manual control permits independent adjustment of the oscillator resonant frequency. The frequency sweeper provides a *varying sweep voltage for synchronisation* to drive the horizontal deflection plates of the CRO. Thus the amplitude of the response of a test device will be locked and displayed on the screen.

— For the identification of a frequency interval, a *marker generator* provides half sinusoidal waveforms at any frequency within the sweep range. The marker voltage can be added to the sweep voltage of the CRO during alternate cycles of the sweep voltage, and appears superimposed on the response curve.

— The *automatic level control circuit* is a closed loop feedback system which *monitors the RF level at some point in the measurement system. This circuit holds the power delivered to the load or test circuit constant and independent of frequency and impedance changes.*

A constant power level prevents any source mismatch and also provides a constant readout calibration with frequency.

## 3.8 FREQUENCY SYNTHESIZER

*The* **frequency synthesizer** *is another type of RF generator that uses phase locked loop (PLL) to generate output frequencies over a wide range.* The most common range is from 1 MHz to 160 MHz.

Fig. 3.9 shows a simple block diagram of a frequency synthesizer; its major components are :

(*i*) Voltage controlled oscillator (VCO);
(*ii*) Divide-by-N counter;
(*iii*) Phase detector;
(*iv*) Crystal oscillator;
(*v*) Low-pass filter;
(*vi*) Square-wave circuit.

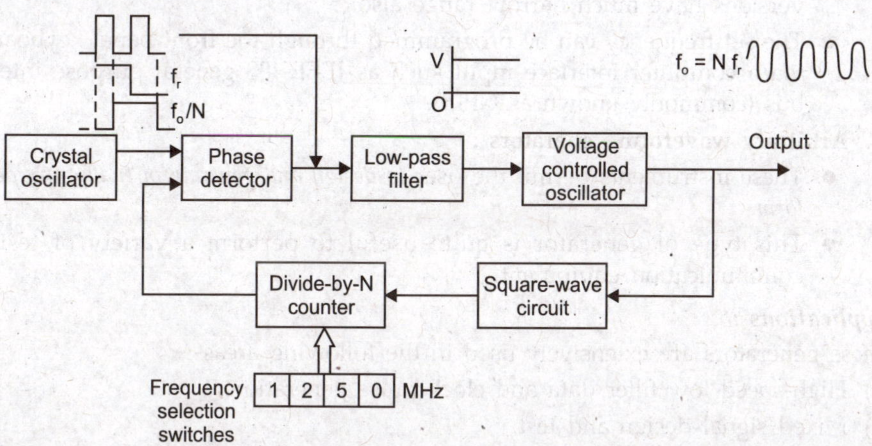


**Fig. 3.9.** Block diagram of a frequency synthesizer.

**Working :**

— The output of the *'crystal oscillator'* (a reference frequency, $f_r$), is fed into one input of a phase detector. The other input of a phase detector has another squarewave applied to it as shown in the figure. *The frequencies of these two square waves are identical but there is a phase difference* ($\phi$) *between them.*

— The output of the *'phase detector'* is a pulse wave waveform with pulse width controlled by the phase difference.

— The output of the phase detector is applied to the low-pass filter which converts it into a D.C. voltage, V.

— The D.C. voltage, V is used as a control voltage for the voltage-controlled oscillator (VCO) and it determines the output frequency of VCO.

— The output of VCO is fed to a circuit that converts it into a square wave for triggering a digital divide-by-N counter. The divide-by-N counter divides the VCO frequency by a number set by a bank of switches. These switches may be push buttons with digital readout or they may be thumb-wheel type which indicate their position numerically. The switches are connected in such a way that the displayed number is the *factor N by which the output frequency is divided before being applied to the phase detector.*

The switches permit the user to obtain frequency which is *any integer multiple of the crystal oscillator frequency.*

- *The frequency synthesizer is used in almost the same areas as the RF signal generators.*

## 3.9 OTHER SIGNAL GENERATORS

Some of the signal generators which do not fit well in various pre-established categories are described below :

**1. RM Makers :**

- These devices are usually crystal controlled and have a fixed output frequency for use as a reference.
- These are used for calibrating TV signals.

**2. Digitally programmable test-oscillators :**

- These instruments can have extremely wide frequency range although some versions have much narrow range also.
- The set frequency can be programmed through the front panel keyboard or via a computer interface input such as IEEE-488 general purpose interface bus (commonly known as GPIB).

**3. Arbitrary waveform generators :**

- These instruments permit the user *to design and generate virtually any desired form.*
- This type of generator is quite useful to perform a variety of tests on communication equipment.

***Applications :***

These generators are extensively used in the following areas :

(*i*) High-speed low filter data and clock pulse generation.

(*ii*) Mixed signal design and test.

(*iii*) Communications design and test for producing *arbitrary IF based signals* and *standard waveforms for communications.*

(*iv*) Read word simulations.

(*v*) Disk drive read/write design and test.

## HIGHLIGHTS

1. Audio signal generators usually produce sine and square waves over the range of at least 20 Hz and 20 kHz. Some will operate down to 1 Hz and upto 100 kHz.

2. The function generator generally produces sine, square, and triangle waveforms at frequencies from fractional Hz to the MHz range.
3. Audio and function generators use 600 Ω output impedance, while RF generators have a 50 Ω output impedance.

   Proper operation of signal generators depends upon using a load matched to the output impedance.
4. RF signal generators operating over 1 GHz are usually called *"microwave generators"*.

## OBJECTIVE TYPE QUESTIONS

1. A triangular waveshape is obtained :
   (*a*) by integrating a square wave  (*b*) by differentiating a sine wave
   (*c*) by differentiating a square wave  (*d*) by integrating a sine wave.
2. An stable multivibrator
   (*a*) gives one output pulse for every two input pulses
   (*b*) gives a timed output pulse for a trigger input
   (*c*) gives a train of output pulses for a trigger input
   (*d*) gives four output pulses for a single input pulse.
3. In signal generators
   (*a*) energy is created
   (*b*) energy is generated
   (*c*) energy is converted from a simple D.C. source into A.C. energy at some specific frequency
   (*d*) all of the above.
4. A video frequency oscillator has a frequency range of :
   (*a*) 20 Hz to 20 kHz  (*b*) 20 kHz to 30 MHz
   (*c*) D.C. to 5 MHz  (*d*) 2 MHz to 30 MHz.
5. The desirable characteristics of a signal source are :
   (*a*) The frequency of the signal should be well known and stable.
   (*b*) The amplitude of a signal source should be controllable from very small to relatively large values.
   (*c*) The output signal should be free from distortion.
   (*d*) All of the above.
6. A voltage controlled oscillator (VCO) is an instrument :
   (*a*) whose frequency is dependent upon the amplitude of input signal.
   (*b*) whose frequency is independent of the amplitude of the input signal.
   (*c*) whose frequency is dependent upon the frequency of the input signal.
   (*d*) none of the above.
7. A random noise generator produces a signal
   (*a*) whose amplitude varies randomly  (*b*) which has no periodic frequency
   (*c*) has an unpredictable power spectrum  (*d*) all of the above.

## ANSWERS

1. (*a*) 2. (*c*) 3. (*c*) 4. (*c*) 5. (*d*) 6. (*a*) 7. (*a*)

## THEORETICAL QUESTIONS

1. What is a signal generator ? Explain briefly.
2. State the common characteristics of signal generators.
3. How are signal generators classified ?
4. Explain briefly, with neat diagrams, the following signal generators :
   (*i*) Conventional standard signal generator.
   (*ii*) Modern signal generator.
5. List the advantages and disadvantages of a modern signal generator.
6. Draw the block diagram of an audio frequency and square wave generator and explain its working.
7. Give the comparison between standard and modern signal generators.
8. What is a function generator ? Explain briefly.
9. Describe the working of 'Function generator' with a neat block diagram.
10. List the typical important specifications of a function generator.
11. Give the comparison between "Signal generator" and "Function generator".
12. What do you mean by "Duty cycle" ? Explain briefly.
13. How does a "pulse" differ from a "square wave" ?
14. Draw a simple diagram of a pulse generator and explain it briefly.
15. Explain briefly circuits used in pulse generation.
16. Explain briefly, with the help of a block diagram, the working of the laboratory square-wave and pulse generator.
17. Explain briefly the following :
    (*i*) Sweep generator.
    (*ii*) Frequency synthesizer.
18. Write short notes on :
    (*i*) RM makers;
    (*ii*) Digitally programmable test oscillators;
    (*iii*) Arbitrary waveform generators.

CHAPTER

# 4 Signal Analysis



## 4.1 INTRODUCTION

It can be shown mathematically that any complex waveform is made up of a *fundamental* and *its harmonics*.

In several applications, the analysis of electrical signals is required. The different instruments which are used for signal analysis are :

(*i*) Wave analysers; (*ii*) Distortion analysers;
(*iii*) Spectrum analysers; (*iv*) Audio analysers;
(*v*) Modulation analysers.

All signal analysis instruments *measure the basic frequency properties of a signal*, but they use different techniques to do so.

(*i*) **A wave analyser.** *It is a voltmeter which can be accurately tuned to measure the amplitude of a single frequency, within a band of about 10 Hz to 40 MHz.*

(*ii*) **Distortion analysers.** These instruments operate over a range of 5 Hz to 1 MHz and *give a measure of the energy present in a signal outside a specified frequency band.* They therefore *tune out the fundamental signal and give an indication of the harmonics.*

(*iii*) **Spectrum analyser.** It sweeps the signal frequency band and *displays a plot of amplitude versus frequency*. It has an operating range of about 0.02 Hz to 250 GHz.

(*iv*) **Audio analyser.** It is similar to a distortion analyser but *can measure additional function, such as noise.*

(*v*) **Modulation analysers.** These instruments tune to the required signal and *recover the whole amplitude modulated (AM), frequency modulated (FM) or phase modulation (PM) envelope for display or analysis.*

## 4.2 WAVE ANALYSERS

### 4.2.1. Introduction

*A* **wave analyser** *is an instrument to measure relative amplitudes of single frequency components in a complex waveform.*

Basically, a wave analyser is a *frequency selective voltmeter* which is tuned to the frequency of one signal while rejecting all other signal components. The desired frequency is selected by a frequency calibrated dial to the input of maximum amplitude. The amplitude is displayed by CRO or a suitable voltmeter.

There are *two* types of wave analysers :

1. Resonant (or Basic) wave analyser.
2. Based upon the frequency ranges.
   (*i*) Frequency selective wave analyser.
   (*ii*) Heterodyne wave analyser.

### 4.2.2. Resonant or Basic Wave Analyser

Fig. 4.1. shows a resonant or basic wave analyser. It consists of the following components :

(*i*) Primary detector—a simple L-C circuit;

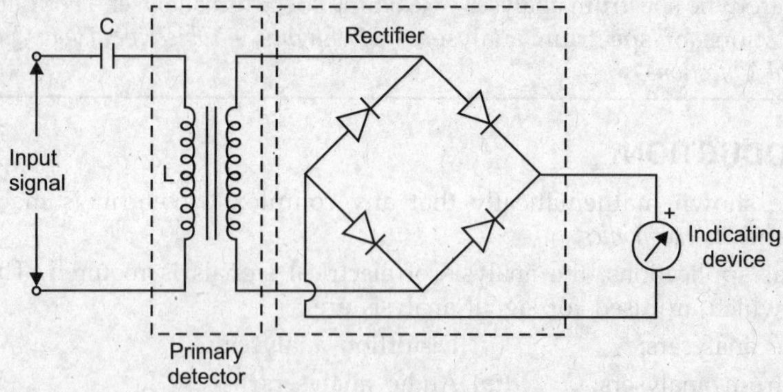


**Fig. 4.1.** Resonant or basic wave analyser.

(*ii*) A full-wave rectifier—intermediate stage;

(*iii*) Indicating device—A simple D.C. voltmeter calibrated to indicate the peak value of the sinusoidal input voltage.

— The *'L-C circuit'* is adjusted for resonance at the frequency of the particular harmonic component to be measured. Since L-C circuit is tuned to a single frequency, *it passes only the frequency to which it is tuned and rejects all other frequencies.*

— The *'full-wave rectifier'* output provides the *average value* of the input signal.

— A number of tuned filters, connected to the *'indicating device'* through a selector switch, would be required for a useful wave analyser.

### 4.2.3. Frequency Selective Wave Analyser

This wave analyser is used for measurements in the *audio frequency range* (*i.e.,* from 20 Hz to 20 kHz). It consists of a *very narrow pass-band filter section that can be tuned to a particular frequency.*

Fig. 4.2. shows the block diagram of frequency selective analyser :

— The waveform to be analysed is passed through an adjustable *'attenuator'*. This acts as a range multiplier and permits a large range of signal amplitudes to be analysed without loading the amplifier.

— A *'driver amplifier'* feeds the attenuated waveform to a *'high-Q active filter'*. The filter consists of a cascaded arrangement of RC resonant sections and filter amplifiers. The passband of the total filter section is covered in decade steps over the entire audio range by switching capacitors in the RC sections. For selecting the frequency ranges, generally close-tolerance polystyrene capacitors are used. Precision potentiometers are used to tune the filter to any desired frequency within the selective band.

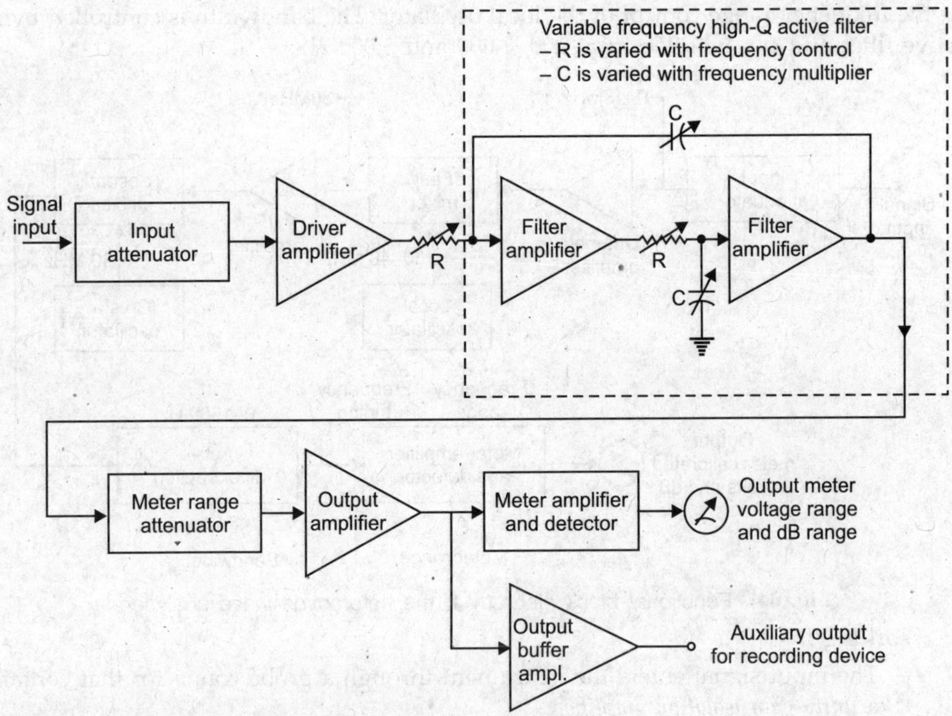


**Fig. 4.2.** Block diagram of frequency selective wave analyser (audio-range).

— A *'final amplifier stage'* supplies the *selected signal* to the meter circuit and to an untuned buffer amplifier.

— The buffer amplifier can be used to drive a recorder or an electronic counter.

— The meter is driven by an average-type detector and usually has several voltage ranges as well as a decibel scale.

- The *bandwidth of the instrument is very narrow, typically about 1% of the selected frequency.*

Fig. 4.3 shows the attenuation characteristics of a frequency selective wave analyser.

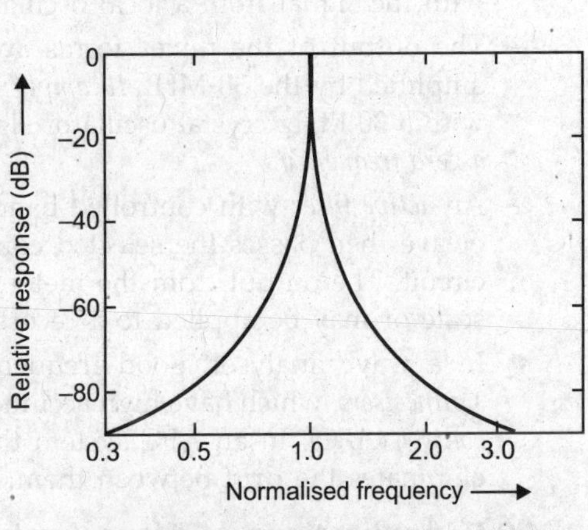


**Fig. 4.3.** Attenuation characteristics of frequency-selective wave analyser.

### 4.2.4. Heterodyne Wave Analyser

An ordinary wave analyser cannot be used for measurements in the RF range and above MHz range. Hence special type of wave analysers working on the principle of, *heterodyning* (mixing) are used. These wave analysers are known as *'Heterodyne wave analysers'*.

Fig. 4.4. shows the functional block diagram of the heterodyne wave analyser. Its operating frequency range is from 10 kHz to 18 MHz in 18 overlapping bands selected by the frequency range control of the local oscillator. The bandwidth is controlled by an active filter and can be selected at 200, 1,000 and 3,000 Hz.

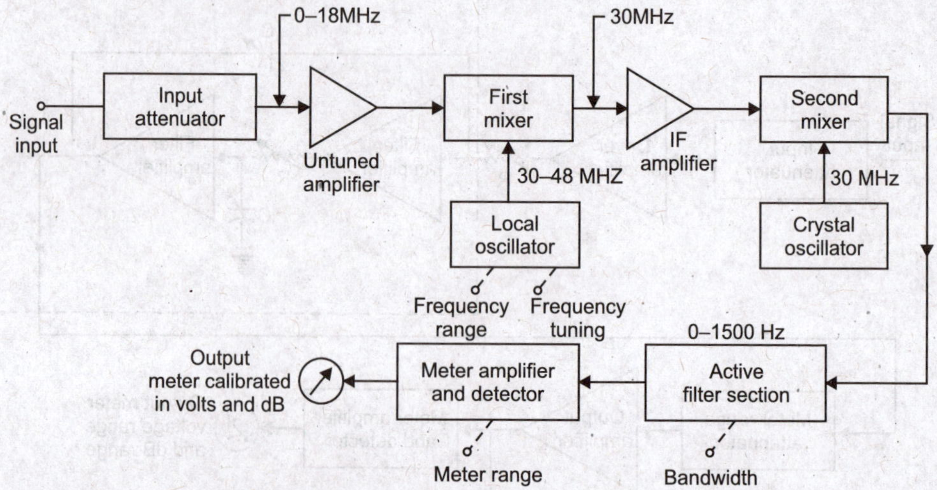


**Fig. 4.4.** Functional block diagram of the heterodyne wave analyser.

**Working :**

— The input signal enters the instrument through a probe connector that contains a *unity-gain isolation amplifier*.

— After appropriate attenuation, the input signal is heterodyned in the mixer stage with the signal from a local oscillator.

— The output of the mixer forms an intermediated frequency that is uniformly amplified by the 30 MHz *IF amplifier*. This amplified signal is then mixed again with a 30 MHz crystal oscillator signal, which results in information *centered on a zero frequency*.

— An *active filter* with controlled bandwidth and symmetrical slopes of 72 dB per octave then passes the selected component to the meter amplifier and detector circuit. The output from the meter detector can be read on a decibel-calibrated scale or may be applied to a recording device.

- In a wave analyser, good frequency stability is achieved by using *frequency synthesisers*, which have high accuracy and resolution, or by an *automatic frequency control (AFC)*. In an AFC system the local oscillator locks to the signal, and so eliminates the drift between them.

### 4.2.5. Applications of Wave Analyser

The various applications of wave analyser are :

1. To carry out complete harmonic analysis.

2. To measure relative amplitudes of single frequency components in a complex waveform.
3. To measure the amplitude in the presence of noise and other interfering signals.
4. To be used to separate and display a large number of harmonics (say 50).
5. To measure the signal energy with well defined bandwidth.
6. To measure the harmonic distortion of an amplifier; the contribution of each harmonic to the total distortion also can be determined.
7. The wave analysers are applied *industrially in the field of reduction of sound and vibrations generated by rotating electrical machines and apparatus.*

## 4.3 HARMONIC DISTORTION ANALYSERS

### 4.3.1. Harmonic Distortion

The sinusoidal input signal when applied to an electronic device, such as an amplifier, should result in generation of a sinusoidal output waveform. However, the output waveform generally is *not exact replica* of the input waveform because of various types of distortions that may occur. ***Distortion*** *may be the result of the inherent non-linear characteristics of different components used in an electronic circuit.*

*Non-linear behaviour circuit elements introduce harmonics in the output waveform and the resultant distortion is often called as* **harmonic distortion** (HD).

The different types of distortions caused by "*amplifiers*" are :

(*i*) Frequency distortion (*ii*) Phase distortion
(*iii*) Amplitude distortion (*iv*) Intermodulation distortion
(*v*) Cross-over distortion.

**Total harmonic distortion :**

A *non-linear system* produces harmonics of an input sinewave, the *harmonics consisting of sine waves with frequencies which are multiples of the fundamental of the input signal. Total harmonic distortion (THD)* is given by,

$$\text{THD} = \frac{\Sigma\,(\text{Harmonics})^{2\;\;1/2}}{\text{Fundamental}} \qquad ...(4.1)$$

A *measure of distortion* represented by a particular harmonic is simply the *ratio of amplitude of harmonic to that of the fundamental harmonic distortion (HD)* and is represented by,

$$D_2 = \frac{E_2}{E_1},\quad D_3 = \frac{E_3}{E_1},\quad D_4 = \frac{E_4}{E_1}$$

where $D_n$ ($n$ = 2, 3, 4, ....) is distortion of $n$th harmonic and $E_n$ represents the amplitude of $n$th harmonic; $E_1$ is the amplitude of *fundamental.*

THD is defined as :

$$\text{THD} = \sqrt{D_2^2 + D_3^2 + D_4^2 + \ldots\ldots D_n^2} \qquad ...(4.2)$$

$$= \sqrt{\left(\frac{E_2}{E_1}\right)^2 + \left(\frac{E_3}{E_1}\right)^2 + \left(\frac{E_4}{E_1}\right)^2 + \ldots\ldots \left(\frac{E_n}{E_1}\right)^2}$$

or, $$\text{THD} = \frac{\sqrt{E_2^2 + E_3^2 + E_4^2 + \ldots.. + E_n^2}}{E_1} \qquad ...(4.3)$$

$\therefore$ $$\text{Percentage harmonic distortion} = \sqrt{D_2^2 + D_3^2 + D_4^2 + \ldots... D_n^2} \times 100 \qquad ...(4.4)$$

$$= \frac{\sqrt{E_2^2 + E_3^2 + E_4^2 + \ldots.. + E_n^2}}{E_1} \times 100 \qquad ...(4.5)$$

Some of the methods used to measure the harmonic distortion factor are described below.

### 4.3.2. Fundamental-Suppression Distortion Analyser

This method of distortion measurement is used when it is important *to measure total harmonic distortion* rather than the distortion caused by each component. Hence, the input waveform is applied to a network that *suppresses or rejects the fundamental frequency but passes all the harmonic-frequency components for subsequent measurement.*

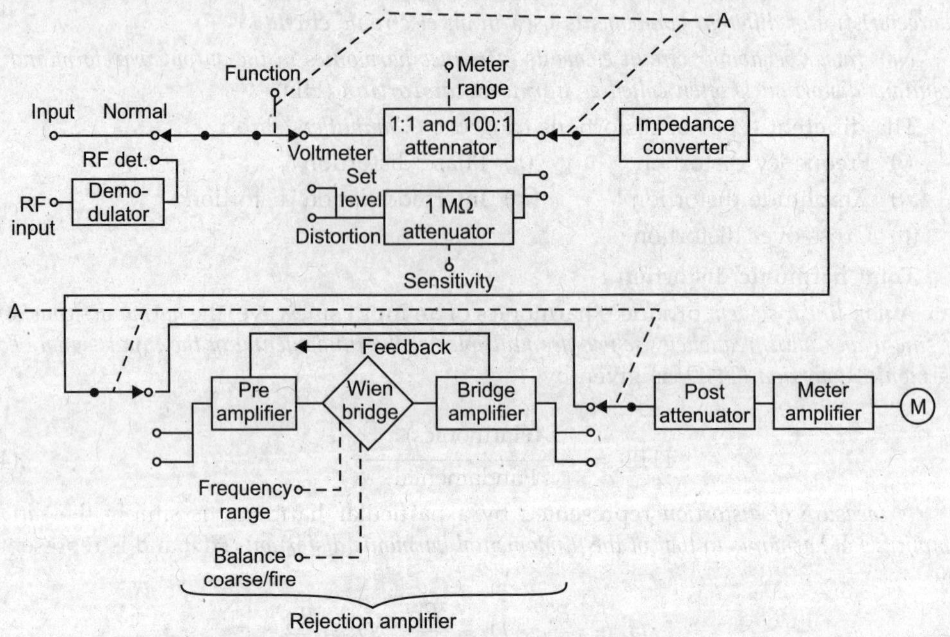


**Fig. 4.5.** Block diagram of a fundamental-suppression distortion analyser.

Fig. 4.5 shows the block diagram of a fundamental-suppression distortion analyser. The instrument consists of the following components:

(*i*) *Impedance converter :* It provides a low-noise, high-impedance input circuit, independent of the signal source at the input terminals to the instrument.

(*ii*) *Rejection amplifier :* It rejects the fundamental frequency of the input signal and passes the remaining frequency components on to the metering circuit where the harmonic distortion is measured.

(*iii*) *Metering circuit :* It provides a visual indication of a total harmonic distortion in terms of a percentage of total input voltage.

(*iv*) *The power supply.*

**Operation/Working.** The following two modes of operation are possible :

1. ***Voltmeter mode :***

- The instrument operates as a conventional A.C. voltmeter (a very convenient feature) when the function switch is in the *"voltmeter" position.*
- In this mode the input signal is applied to the impedance converter circuit through the 1 : 1 and 100 : 1 attenuator, which selects the appropriate meter range.
- The output of the impedance converter then bypasses the rejection amplifier (as shown dotted) and the signal is applied directly to the metering circuit. The meter measures the r.m.s. value of the A.C. input voltage.

2. ***Distortion mode :***

- When the instrument is in distortion mode, the function switch is in the *"distortion"* position and the rejection amplifier is the part of the circuit, and distortion measurements are made.
- In this case the input signal is applied to a 1 MΩ input attenuator that provides 50 dB attenuation in 10 dB steps, controlled by a front panel switch marked *sensitivity.*
- When the desired attenuation is selected, the signal is fed to the impedance converter, which is a low-distortion, high input-impedance amplifier circuit whose gain is independent of the source impedance placed at the input terminals. The overall negative feedback in this amplifier results in unity gain and low distortion. Signals having a high source impedance can be measured accurately and the sensitivity selector can be used in the high-impedance positions without distorting the input signal.
- The preamplifier, Wien bridge, and a bridge amplifier constitute the *"rejection amplifier circuit".*

— The *preamplifier* receives the signal from the impedance converter and provides additional amplification at extremely low distortion levels.

— The *Wien bridge circuit* is used as a *rejection filter for fundamental frequency of the input signal.* With the function switch in the "distortion position", the Wien bridge is connected as an interstage coupling element between the preamplifier and bridge amplifier. The bridge is tuned to the fundamental frequency of the input signal by setting the *frequency range selector* and is *balanced for zero output by the coarse and fine balance controls.*

The Wien bridge, for frequencies other than the fundamental, offers varying degree of phase shift and attenuation, and the resulting output voltage is amplified by the *bridge amplifier.* The output of the bridge amplifier is connected though a post attenuator to the meter circuit and displayed on the front panel meter.

**Advantages :**

This instrument has the following two main *advantages :*

1. The harmonic distortion produced by the instrument itself is negligibly small.
2. Selectivity requirements are not severe because only the fundamental frequency component is to be suppressed.

### 4.3.3. Heterodyne Harmonic Analyser or Wavemeter

This type of analyser makes use of a *highly selective, fixed frequency filter.*

Fig. 4.6. shows the block diagram of a heterodyne harmonic analyser :

— The input signal is mixed with output of variable frequency oscillator in a mixer circuit.

— After the successful mixing, output of mixer supplies the sum and the difference of each harmonic beating against the oscillator frequency. Now for each harmonic component, either the sum or difference of frequencies is made equal to the frequency of the filter by varying the frequency of oscillator.

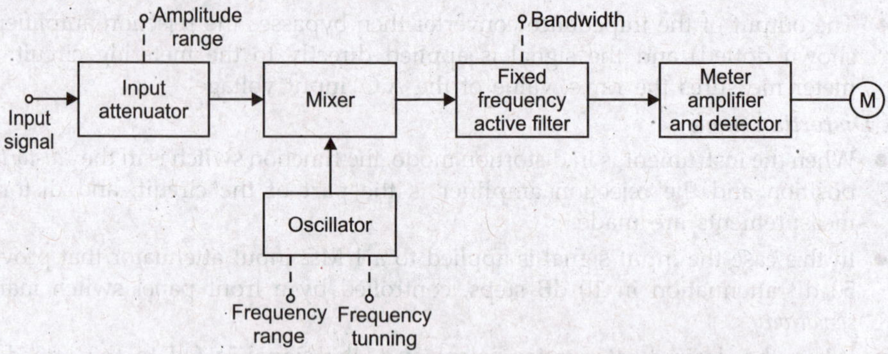


**Fig. 4.6.** Block diagram of heterodyne harmonic analyser or wavemeter.

— Only the mixer output signal that can pass through the filter is that signal whose frequency is equal to the frequency of the filter. *The output of the mixer is the resultant of the variable frequency oscillator output and harmonic component and the frequency of the resultant output from mixer is equal to the filter frequency.*

— The output of mixer has r.m.s. value which is proportional to the r.m.s. value of that harmonic component. So when the output of mixer is supplied to a true r.m.s. reading voltmeter through amplifier, the voltmeter reads the r.m.s. value of that harmonic component. In this way, the instrument *provides r.m.s. value of each harmonic component.*

## 4.4 SPECTRUM ANALYSERS

### 4.4.1. Introduction

**Spectrum analyser** *is an instrument which graphically presents an energy distribution of the signal as a function of frequency on its CRT.*

**Spectrum analysis** *is defined as the study of energy distribution across the frequency spectrum of a given electrical signal.*

The spectrum analysis of a signal provides the information about the following particulars/parameters :

(*i*) Bandwidth.

(*ii*) Effects of different types of modulation and spurious signal generation.

(*iii*) Frequency stability.

(*iv*) The component levels.

(*v*) Spectral purity.

(*vi*) Attenuation.

The spectrum analysis, an account of instrumentation limitations and capabilities, is divided into the following *two* major categories :

(*i*) Audio frequency (AF) analysis.

(*ii*) Radio frequency (RF) analysis.

The *"RF spectrum analysis"* covers a frequency range of 10 MHz to 40 GHz, and hence is more important, because *it includes the vast majority of communication, navigation, radar, and industrial instrumentation frequency bands.*

— In case of *normal oscilloscope* the *signal is plotted as a function of time i.e.,* amplitudes versus time plot is obtained. Such an analysis is called ***time domain analysis***.

— In case of *"spectrum analyser"*, the *signal is plotted as a function of frequency.* This analysis is called ***frequency domain analysis.***

- *Consider a signal which has a fundamental frequency $f_1$ and its second harmonic, $f_2 = 2f_1$* : Refer to Fig. 4.7.

— Fig. 4.7(*a*) shows a three-dimensional representation of a fundamental frequency, $f_1$ and its second harmonic, $f_2 = 2f_1$.

— Fig. 4.7(*b*) shows the two signals and sum of instantaneous values, as displayed on a CRO screen.

— Fig. 4.7(*c*) shows the two signals on amplitude-frequency plane; these are portrayed on the CRO as two components of the composite signal, as the window of the spectrum analyser sweeps across the frequency range of the signal.

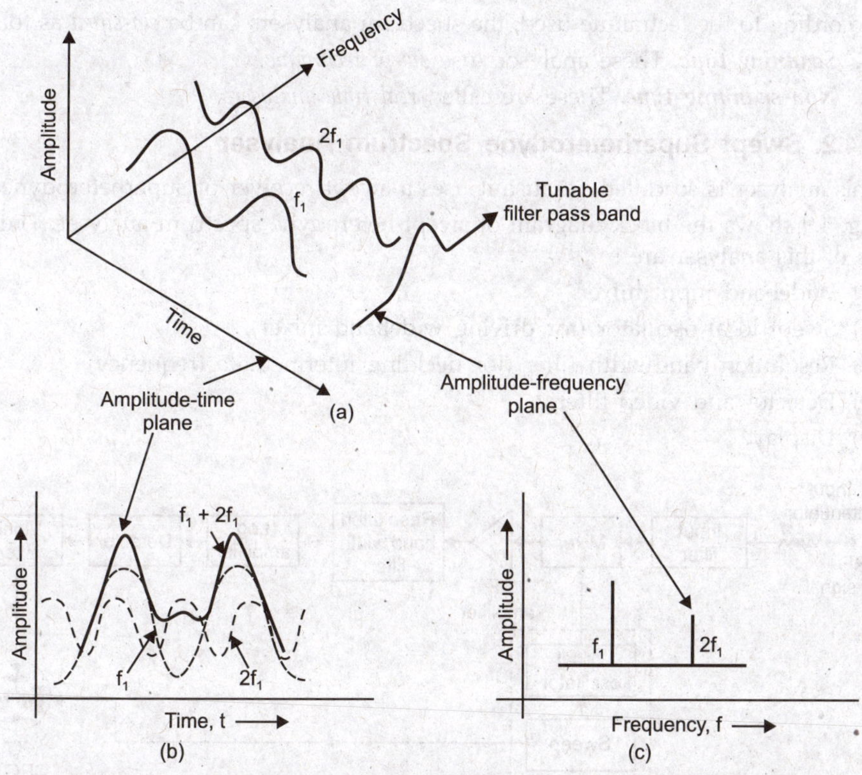


**Fig. 4.7.** A three-dimensional presentation of amplitude, frequency, and time.

- Spectrum analysers use either a parallel filter bank or swept frequency technique.

— In a *parallel filter bank analyser*, the frequency range is covered by a series of filters whose central frequencies and bandwidth (BW) are so selected that they overlap each other, as shown in Fig. 4.8.

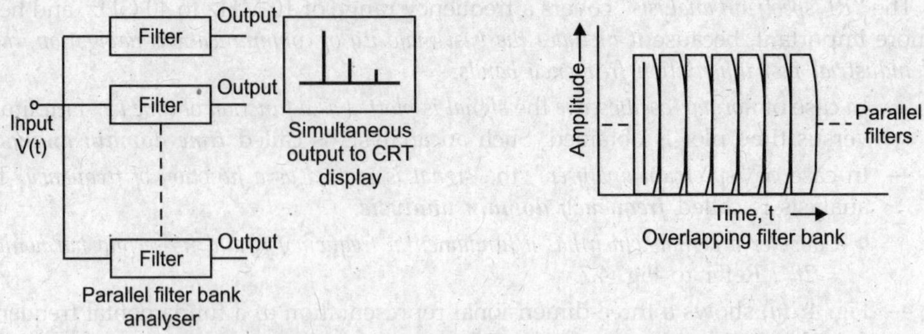


Fig. 4.8. Spectrum analyser (Parallel filter bank analyser).

— Typically, an audio analyser will have 32 of these filters, each covering one third of an octave.

For wide band narrow resolution analysis, particularly at RF or microwave signals, the *swept frequency technique* is preferred.

**Classification of spectrum analysers :**

According to the technique used, the spectrum analysers can be *classified* as follows :

1. ***Scanning type.*** These analysers use *swept technique.*
2. ***Non-scanning type.*** These are called *real time analysers.*

## 4.4.2. Swept Superheterodyne Spectrum Analyser

This analyser is so called because it uses a swept receiver of superheterodyne type.

Fig. 4.9 shows the block diagram of swept heterodyne spectrum analyser. The basic blocks of this analyser are :

(*i*) Wideband input mixer;

(*ii*) Swept local oscillator (for driving wideband mixer);

(*iii*) Resolution bandwidth filter (for deciding intermediate frequency);

(*iv*) Detector and video filter;

(*v*) Display.

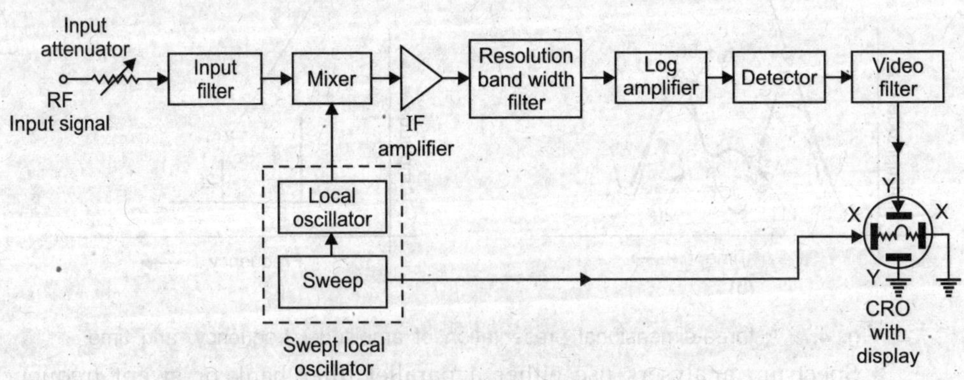


Fig. 4.9. Block diagram of swept heterodyne spectrum analyser.

— The RF input is applied to an input attenuator, which decides the level of input signal so as to keep it within the operating range of other blocks of the instrument. Normally spectrum analysers can handle 0 to 10 dBm (One dBm means decibel referred to 1 mW). Power handling range of input attenuator is 0.5 to 1 W and this limits the maximum input level, analyser can handle without any damage.

— The *"input filter"* (mostly a low pass filter) is used to reject unwanted signals; it suppresses the spurious signals.

— The "mixer" multiplies the input signal from filter and the local oscillator signal. It provides two signals at the output which are *proportional in amplitude to the input signal but having frequencies which are sum and difference of frequencies of input signal and local oscillators signal.*

— *"Intermediate frequency (IF) section"* is the most important stage in analyser where *real analysis takes place.* This section provides a wide selection of resolution width filters; these filters are described by their 3-dB bandwidth. These filters decide the resolving power of the analyser; the narrow the filter, the close two signals may be and still can be seen as separate responses. The filters are normally tuned to a narrow band of 2 GHz and thus the input signals which are separated from the local oscillator frequency by 2 GHz only are converted to the intermediate frequency band.

— The filter output is given to the *"log amplifier"*, which processes the incoming signal in a logarithmic fashion. The *logarithmic processing permits a large range of incoming signals to be measured and compared.*

— The signal from log amplifier (which is compressed one) is received by the *"detector"* and its output is given to the *"video filter"*. These filters are used *for post filtering or averaging the detector output.* The bandwidth setting of video filter is same or larger than resolution bandwidth filter. Averaging removes the pure signal remains and the random noise. The output of the video filter is given to the CRO for display purpose.

The *"swept local oscillator"* puts a limit on the stability and spectral purity in several performance areas.

**Characteristics of spectrum analyser :**

The characteristics of spectrum analyser are :

1. *"Frequency resolution"*. It is defined as the *ability of spectrum analyser to separate the signals which are closely spaced in the frequency.*
   - The following two factors determine the frequency resolution :
     (*i*) The bandwidth or selectivity of the IF amplifier.
     (*ii*) The frequency stability of the spectrum analyser as determined by the drift, the residual FM and phase noise of the local oscillators.
2. *"Dynamic range"*. It is ability of a spectrum analyser *to display the true spectra of large and small signals simultaneously.*
   - When the signal levels lie within the dynamic range, the spurious signals causing the distortion do not appear on the display.
   - It can be determined from the third order intermodulation distortion.
3. *"Sensitivity"*. It is the ability of a spectrum analyser *to measure the small signals as determined by its own internally generated noise.*
   - The sensitivity can be improved by the use of broadband preamplifiers with low noise figures, lower than that of spectrum analyser.
4. *"Sweep desensitization"*. It is an effect caused by scanning a spectrum analyser too

fast; it results in loss of amplitude, calibration, sensitivity and resolution.

- To avoid the sweep desensitization, the scan velocity in Hz/second must not exceed the square of 3-dB bandwidth of the IF filter in Hz.

### 4.4.3. Real Time Spectrum Analyser

A real time spectrum analyser is the one *which presents the effect of changes in all the input frequencies, on its spectrum display.* It is *used at audio frequencies.*

Fig. 4.10 shows the block diagram of a real time spectrum analyser :

— It consists of 24 channels and covers a range from 50 Hz to 10 kHz.

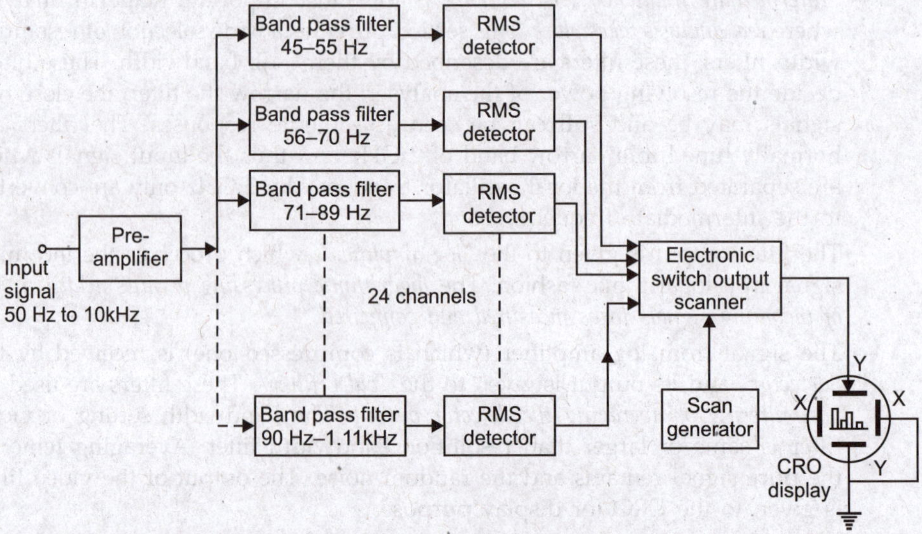


**Fig. 4.10.** Block diagram of real time spectrum analyser.

— Each channel consists of band pass filter and RMS detector. Since a number of band pass filters are used, it is called as a *multifilter real time spectrum analyser.*

— The electronic switch is used to scan the output of each filter sequentially.

— The output from electronic switch output scanner is given to the vertical deflecting plates of the CRO.

— The scan generator provides the horizontal deflection on CRO. The output of a scan generator is a sawtooth output which is synchronised with the electronic switch.

- In this analyser the composite amplitude of signal within the bandwidth of each filter is displayed as *a function of the frequency range of the filter.*
- These analysers are *only used in audio frequency applications* and can use as many as 32 filters.

**Advantages :**

1. Good resolution can be obtained by reducing the bandwidth of each filter.
2. The display rate can be increased at least upto 3 Hz.

**Disadvantages :**

1. Poor resolution.
2. Resolution cannot be adjusted. Hence, it is not possible to obtain the magnified portion of the display.

3. The total system is complex.
4. The frequency axis of the display is inflexible.

### 4.4.4. Fourier Analyser

**Fourier analyser** *uses digital signal processing techniques to provide measurement that is beyond the capabilities of spectrum analyser.* Some of the other applications are measurement of :

— very low-frequency or closely-spaced signals;

— random signals observed by noise;

— shared properties or relationships of two or more signals. There analysers are *based on the calculation of the discrete Fourier transformer using an algorithm called the "fast Fourier transform". This algorithm calculates the amplitude and phase of each signal component from a set of time domain samples of the input signal.*

Fig. 4.10. shows the block diagram of Fourier analyser.

— The signal applied to the analyser is *filtered* to remove out-of-band components.

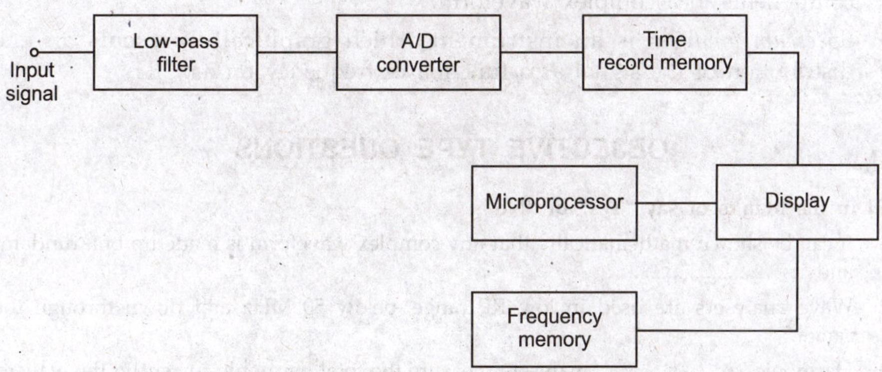


**Fig. 4.11.** Block diagram of Fourier analyser.

— The signal out of the *'low-pass filter'* is applied to the *'A/D converter'*, which samples and digitizes it at a regular time intervals until a full set of samples called a *time record* has been collected.

— The *'microprocessor'* then *performs the desired series of computations on the time data to obtain the frequency domain results.* These results, which are stored in the memory, *can be displayed on a CRT or recorded permanently with a recorder or plotter.*

- Since Fourier analysers are primarily digital instruments, interfacing to a computer or other digital system *is relatively simple.* Remote programming and transforming data via a bus or other interfacing systems can considerably *expand the range* of possible applications for the analyser. *By virtue of the digital nature of the analyser, it will provide a high degree of accuracy, stability, and repeatability in spectrum analysis.*

### 4.4.5. Applications of Spectrum Analyser

The applications of spectrum analyser are :

1. Amplitude modulation.
2. Frequency modulation.
3. Pulse modulation.
4. Measurement of harmonic distortion.

5. Noise measurement.
6. Determination of the frequency drift of a signal.
7. Analysis of air and water pollution.
8. Measurement of antenna pattern.
9. Tuning a parametric amplifier.
10. In the field of electronic testing related to troubleshooting and quality control.
11. In the fields of *biomedical*, *radars*, and *oceanography*.

## HIGHLIGHTS

1. Following instruments are used for signal analysis :

   Wave analysers; Distortion analysers; Spectrum analysers; Audio analysers; Modulation analysers.
2. A *wave analyser* is an instrument to measure relative amplitudes of single frequency components in a complex waveform.
3. *Spectrum analyser* is an instrument which graphically presents an energy distribution of the signal as a function of frequency on its CRT.

## OBJECTIVE TYPE QUESTIONS

**Fill in the blanks or say "Yes" or "No" :**

1. It can be shown mathematically that any complex waveform is made up of a fundamental and its .......................... .
2. Wave analysers are used in low RF range, below 50 MHz and down through the AF range.
3. Harmonic .......................... analysers measure the total harmonic current in the waveforms.
4. The wave analyser must have extremely high input distortion.
5. .......................... analyser is an instrument to measure relative amplitudes of single frequency components in a complex waveform.
6. Frequency selective wave analyser is used for measurements in the .......................... frequency range.
7. Heterodyne harmonic analyser makes use of a highly selective, fixed frequency filter.
8. .......................... analysis is defined as the study of energy distribution across the frequency spectrum of a given electrical signal.
9. .......................... analyser is an instrument which graphically presents an energy distribution of the signal as a function of frequency on its CRT.
10. Spectrum analysers use either a parallel filter bank or .......................... frequency technique.

## ANSWERS

| | | | |
|---|---|---|---|
| 1. harmonics | 2. Yes | 3. distortion | 4. No |
| 5. Wave | 6. audio | 7. Yes | 8. Spectrum |
| 9. Spectrum | 10. swept. | | |

## THEORETICAL QUESTIONS

1. Name the instruments which are used for signal analysis.
2. What is a "wave analyser"? Explain.
3. Discuss briefly a resonant or basic wave analyser.
4. Explain briefly, with a neat block diagram, the working of a frequency selective wave analyser.
5. Describe with the help of a block diagram the working of a heterodyne wave analyser.
6. What are the applications of wave analyser?
7. What do you mean by the term 'harmonic distortion'?
8. What is total harmonic distortion?
9. Explain briefly "Fundamental-suppression distortion analyser"? State its advantages also.
10. With the help of a block diagram, give the description of "Heterodyne harmonic analyser or wavemeter".
11. What is a "Spectrum analyser"?
12. What do you mean by "Spectrum analysis"?
13. Which information is provided by the spectrum analysis of a signal?
14. How are spectrum analysers classified?
15. Explain the working of a swept superheterodyne spectrum analyser, with a neat block diagram.
16. What are the characteristics of spectrum analyser?
17. Describe briefly "Real time spectrum analyser". State its advantages and disadvantages.
18. List the applications of spectrum analyser.
19. Write short notes on the following :
    (*i*) Wave analyser;
    (*ii*) Spectrum analyser;
    (*iii*) Harmonic distortion analyser.

CHAPTER

# 5 Digital Instruments

> **5.1. General aspects**; **5.2. Analog and digital systems**; **5.3. Basic concepts of digital instruments** – Introduction – Binary counters – Decimal counting units (DCUs) – Display devices – Frequency counters – Period counters – Characteristics of digital meters – Analog-to-digital (A/D) and digital-to-analog conversion **5.4. Digital voltmeters (DVMs)** – Introduction – Advantages of DVMs – Characteristic features of DVMs – Applications of DVMs – Classification of digital voltmeters – Ramp-type DVM – Dual-slope integrating type DVM – Integrating type DVM (voltage-to-frequency conversion) – Successive-approximation DVM – General specifications of a DVM; **5.5. Digital LCR meter**; **5.6. Digital multimeter (DMM)**; **5.7. Microprocessor-based instruments-intelligent instruments**; **5.8. Microprocessor-based RAM type DVM**; **5.9. IEEE-488 general purpose interface bus (GPIB) instruments** – *Highlights – Objective Type questions – Theoretical Questions.*

## 5.1 GENERAL ASPECTS

- *A **digital instrument** can be considered as a counter which counts the pulses in a predetermined time.* Digital transducers whose output is in the form of pulses are used to monitor the desired parameter.

*"Accuracy" of a digital instrument is dependent on the number of pulses generated by the transducer because the fraction of pulse cannot be generated and in counting there can be ambiguity of only one pulse at start/stop.* Hence *more are pulses* corresponding to a measure less the possibility of error corresponding to one pulse and *more the accuracy.*

— The digital instruments use *logic circuits and techniques* for carrying out measurements of quantities.

— The information in the electronic digital read-out (display) devices is *presented as a series of digits on the tubes, screen, or printed on a piece of paper.* The relevant characters (letters of alphabet from A to Z, numerals from 0 to 9, punctuation mark and other symbols in common use) can be generated by :

(*i*) Semiconductor light emitting diodes (LED).

(*ii*) Liquid crystal displays (LCD).

(*iii*) Numerical indicators tubes (NIT).

(*iv*) Hot filament or bar tubes.

— Digital instruments provide better resolution than analog instruments. A typical value of *resolution* of a digital instrument may be one part in $10^6$ *i.e.,* 1 μV can be read on the 1-V input range.

— *The **digital meters** work on the principle of "quantization".*

The analog quantity to be measured is first subdivided or quantized into a number of small intervals upto many decimal places. The objective of the digital instrument is then to determine in which portion of the subdivision the measurand can thus be identified as an integral multiple of the smallest unit called the *quantum,* chosen for subdivision.

The measuring procedure thus *reduces to one of counting the number of quanta present in the measurand.*

— Fig. 5.1. shows the building block of a digital instrument. Any digital instrument invariably consists of a an analog-to-digital converter (ADC) in its input stage. The display block may be analog or digital in nature. If an analog read out is desired, it becomes necessary to include a stage involving digital-to-analog (D/A) conversion.

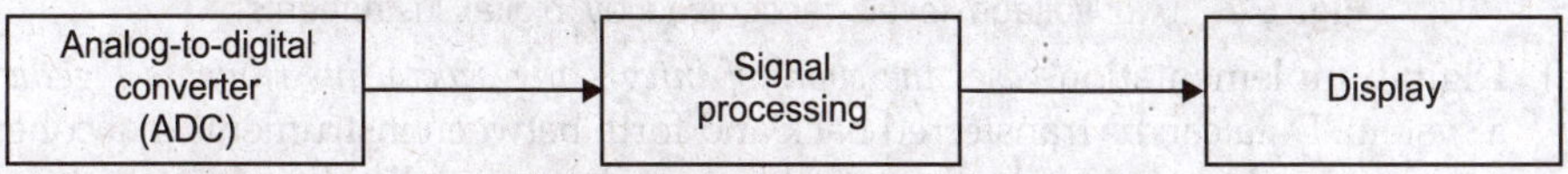


**Fig. 5.1.** Building block of a digital instrument.

## 5.2 ANALOG AND DIGITAL SYSTEMS

An ***analog system*** comprises devices that manipulate the physical quantities represented in *analog form.* Examples of analog systems are : Automobile speedometer, Magnetic tape recording and playback equipment etc.

A ***digital system*** is a combination of devices designed for manipulating physical quantities or information represented in *digital form.* Examples of digital systems are : Digital computers, typewriters, digital watches etc.

**Advantages of digital systems/techniques :**

1. Easier to design.
2. Easy storage of information.
3. Greater accuracy and precision.
4. Operation is programmable.
5. Digital circuits are less affected by the noise.

- Digital system may consist of the following *components* :

(*i*) Resistors; (*ii*) Capacitors;
(*iii*) Transistors; (*iv*) Linear ICs;
(*v*) Digital ICs; (*vi*) Digital devices;
(*vii*) Analog-to-digital converters (ADC) (*viii*) Digital-to-analog converters (DAC).

— *Data is digital form facilitates various operations that are normally required in signal processing.*

— An increase in the availability and type of computer facilities and a decrease in the cost of various methods required for digital systems is accelerating the development of digital instrumentation for measurement and signal processing.

## 5.3 BASIC CONCEPTS OF DIGITAL INSTRUMENTS

### 5.3.1. Introduction

A truly digital instrument uses *binary* (*i.e.,* base 2) *logic elements* such as *gates, flip-flops,* and so on, to perform the measurement function. These circuits operate with only two voltage levels (Fig. 5.2), and these represent the binary digits 0 and 1.

***Digital implementation*** entails the following *"advantages"* :

(*i*) The instrument is more *easily modified*, especially where the complexity is in embedded software.

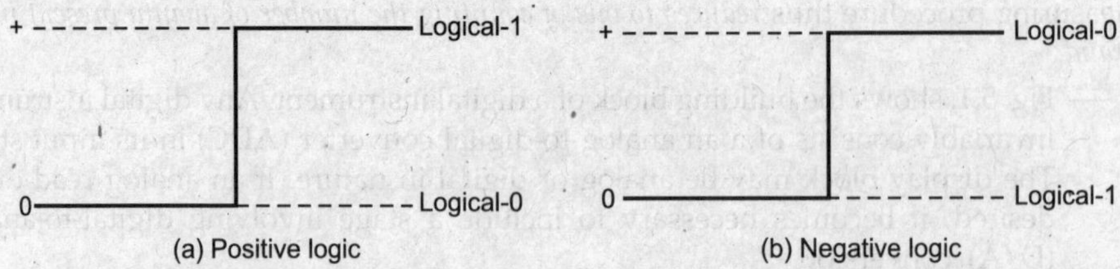


**Fig. 5.2.** Two voltage levels-recognized by digital instruments.

(*ii*) Digital implementation *eases the chore of integrating several instruments together* in a system. Data can be transferred back and forth between instruments in a coherent manner much more easily than analog signals, especially over long distances.

— For larger instrumentation systems, the use of general purpose interface bus (IEEE-488 GPIB) makes the chore of integrating instruments easy.

(*iii*) Even though software digital solutions are also very complex as in the case of analog solutions, yet the *hardware implications are minimal.* One can add RAM or ROM or use a hard drive to store a larger, more complex program that would be impossibly difficult to implement in analog hardware.

### 5.3.2. Binary Counters

The binary counter forms the heart of many "discrete logic" digital implementations.

- *A* ***binary counter*** *consists of a chain of locked flip-flops, each state is designed to count by 2.*
- *A* ***decoder*** *is used to indicate the state of the count* by examining the unique *Q* and not-*Q* states at the output of each stage.
- A 4-bit binary counter will count from $0000_2$ to $1111_2$ (*i.e.,* $0_{10}$ to $15_{10}$) while a *"binary coded decimal"* **(BCD)** counter counts only from $0000_2$ to $1001_2$ ($0_{10}$ to $9_{10}$). Table 5.1 shows both binary and BCD coding, the latter of which is a subset of 4-bit binary coding.

**Table 5.1. Binary and BCD Codes**

| Decimal | Binary | BCD | |
|---|---|---|---|
| 0 | 0000 | 0000 | |
| 1 | 0001 | 0001 | |
| 2 | 0010 | 0010 | |
| 3 | 0011 | 0011 | |
| 4 | 0100 | 0100 | |
| 5 | 0101 | 0101 | |
| 6 | 0110 | 0110 | |
| 7 | 0111 | 0111 | |
| 8 | 1000 | 1000 | |
| 9 | 1001 | 1001 | |
| 10 | 1010 | 0001 0000 | |
| 11 | 1011 | 0001 0001 | |
| 12 | 1100 | 0001 0010 | |
| 13 | 1101 | 0001 0011 | Two 4-bit |
| 14 | 1110 | 0001 0100 | counters |
| 15 | 1111 | 0001 0101 | required |

### 5.3.3. Decimal Counting Unit (DCUs)

**Decimal counting unit** *is the circuit of any electronic counter.* It consists of the following components :

(*i*) Decade counter;

(*ii*) Decoder/display-driver;

(*iii*) Data latch.

Fig. 5.3. shows the block diagram of an example of a decimal counting unit (DCU) :

— This circuit is constructed of TTL integrated circuit logic elements. The decade counter uses a chain of flip-flops, but it also contains some internal gates that sense the tenth state (*i.e.,* "9") and reset the counter to 0000 if another input pulse is received. In fact, now, there are several MSI and LSI integrated circuits available that contain all circuitry needed to form a complete electronic counter of several decades.

— The four output lines from the *"decade counter"* are coded in 8421 BCD format and they change state with every pulse.

— *"Input"* and *"reset"* are other two lines, in which the input line accepts the pulses being counted, while the reset line will reset the counter to 0000 when brought to ***High.*** For normal counting the line is kept ***Low***, and to clear the counter it is brought momentarily ***High.***

— The *"decoder"* is used to convert the 4-bit BCD code to the code needed to correctly drive the read-out device.

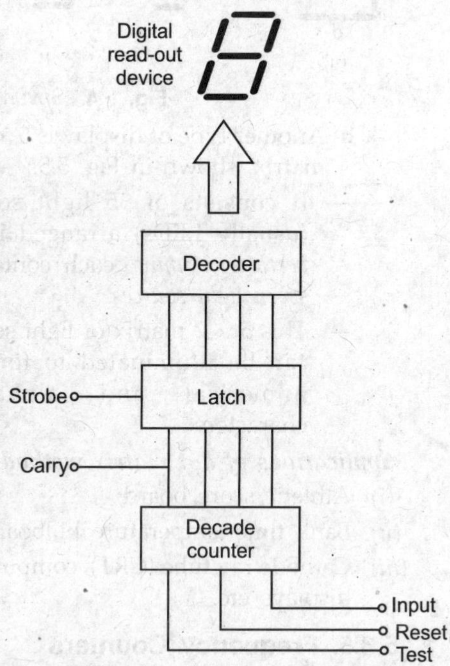


**Fig. 5.3.** Decimal decode counter.

— A quad *"latch"* circuit is used to eliminate the rolling effect (this effect is created because the display changes with the count).

— The *"strobe"* line, in practice, is held low while the count is being accumulated, and at the end of the count period the strobe line is brought high momentarily, just long enough to transfer the count data from the input lines to the output lines.

- The individual DCU can count only from 0 to 9, but if two DCUs are used in cascade, the assembly can count from 0 to 99, and three DCUs can count from 0 to 999 etc.; the assembly is known as ***decimal counting assembly (DCA).***

### 5.3.4. Display Devices

- In modern equipment the most common display is the *seven-segment* readout shown in Fig. 5.4(*a*). This type of display consists of seven illuminated bars arranged in a *figure-eight pattern.* It is possible to *form all ten decimal digits by lighting appropriate bars* [Fig. 5.4(*b*)]. A BCD-to-seven-segment decoder will ground those terminals *a* through *g* required to form the decimal digit represented in BCD form at the decoder input. Probably the most popular types as of this writing are the light-emitting diode (LED) and liquid crystal diode (LCD) displays.

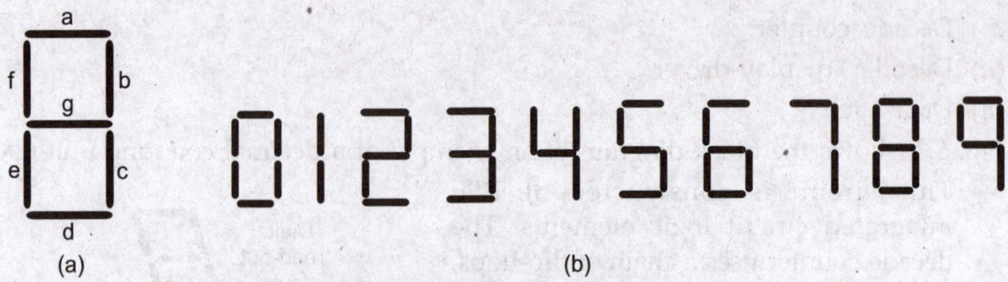


Fig. 5.4. Seven-segment decimal readout.

- Another type of display is 5 × 7 dot matrix, shown in Fig. 5.5.:
  — It consists of 35 light sources (usually LEDs) arranged in *five vertical columns,* each containing *seven light sources.*
  — This 5 × 7 matrix of light sources can be illuminated to form all numerical and alphabetic characters.

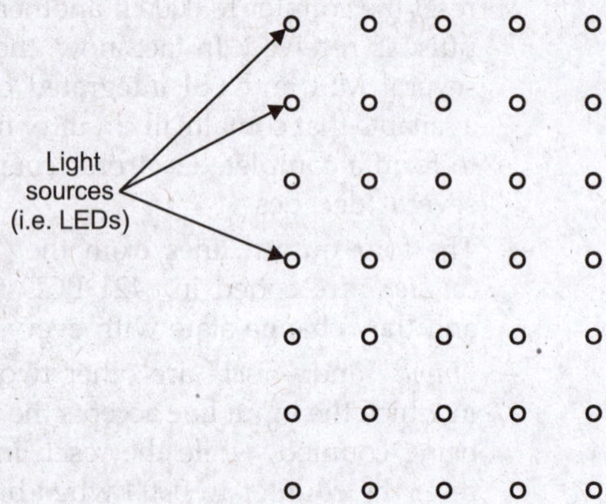


Fig. 5.5. 5 × 7 dot matrix.

***Applications of dot matrix method :***

(*i*) Athletic score board;
(*ii*) Bank time-temperature billboards;
(*iii*) Cathode ray tube (CRT), computer TV displays etc.

### 5.3.5. Frequency Counters

Refer to Fig. 5.6 :

- **Fig. 5.6(*a*)** shows the basic block diagram of a **"frequency counter"**. It consists of the following *sections :*

(*i*) Decade counter assembly (DCA); (*ii*) Main gate;
(*iii*) Trigger; (*iv*) Input amplifier;
(*v*) Main gate flip-flop; (*vi*) Time base;
(*vii*) Display clock.

— The frequency counter measures *"events per unit of time (EPUT)"* (*i.e.*, cycles per second), so the DCA must be tuned on only for a given period of time (*e.g.*, 0.1, 1, or 10 s).

— The main gate, main gate flip-flop, and time base sections are used *to allow input pulses into the DCA* for the designed period of time.

— The time base section consists of a crystal *oscillator* that produces pulses at precise rate such as 100 kHz, 1 MHz, 4 MHz, 10 MHz, and so on. A chain of *decade dividers* is used to reduce the crystal oscillator frequency *to a lower frequency.* The time base output frequency will be 10 Hz for 0.1 s, 1 Hz for 1 s and 0.1 Hz for 10-s measuring periods.

- **Fig. 5.6(*b*)** shows the timing diagram for one complete interval of EPUT counter.

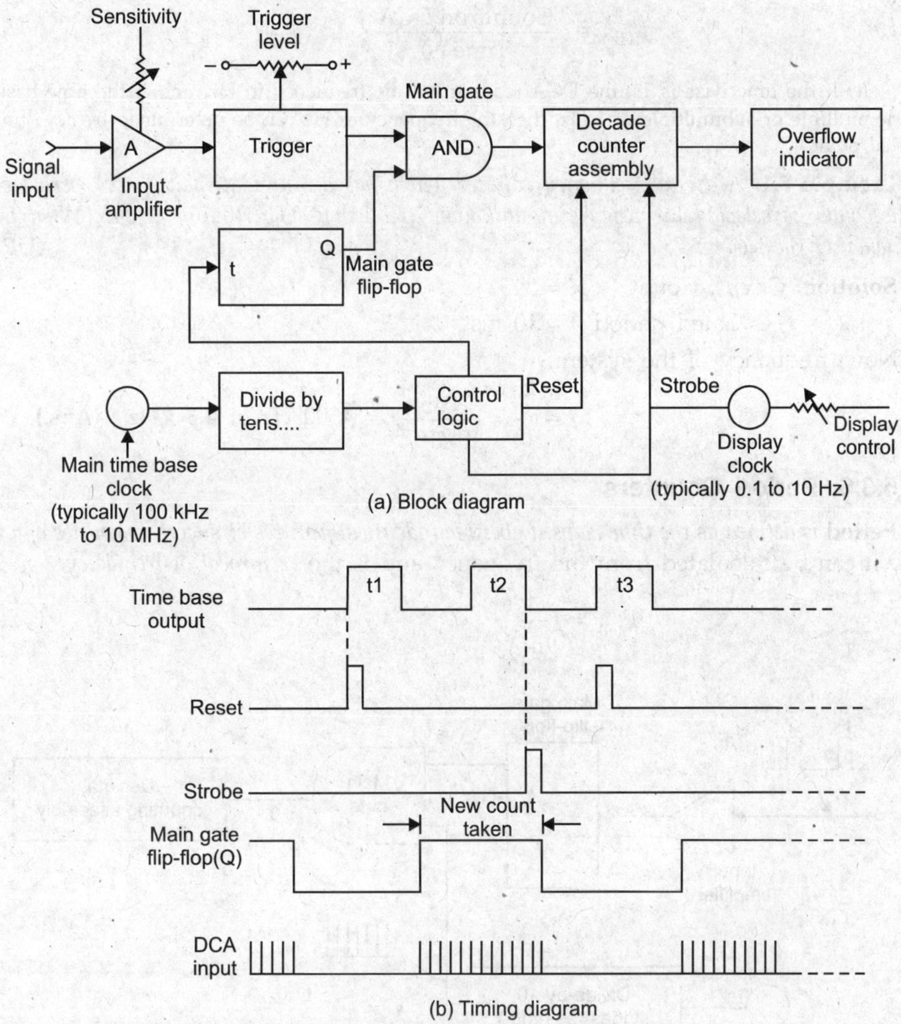


**Fig. 5.6.** Frequency counter.

— Pulses *t1*, *t2* and *t3* are output from the time base section. When pulse *t1* goes high, the control logic section generates a short pulse to reset the DCA to zero. When *t1* goes low again, the *Q* output of the J-K main gate flip-flop will go high. The main (AND) gate has one input tied to the *Q* output of the flip-flop and the other input is tied to the signal being counted. As a result, the main gate passes input pulses to the DCA *only* when the *Q* terminal of the flip-flop is high.

— Until the negative transition of *t2* occurs, the flip-flop remains set. At that time the *Q* output of the flip-flop drops low, turning off the flow of pulses into the DCA and causing the control logic section to generate a strobe pulse. This pulse tells the DCU latches to transfer data from the counter to the decoders. The display, then, shows *only completed count cycles and will hold the previous count until the end of the next interval.*

*The frequency of the input signal is the number of counts accommodated on the DCA divided by the time base period in seconds.*

*i. e.,*

$$f_{(Hz)} = \frac{\text{Counts on DCA}}{\text{Time (s)}} \quad ...(5.1)$$

***Note:*** If the time base is 1s, the DCA reading will be frequency in Hz, and if the time base is a decade multiple or submultiple of 1 Hz, then the frequency in Hz can be determined by adjusting the decimal point.

**Example 5.1.** *A digital frequency counter which has a 3-digit display, gated period of 10 milliseconds, is selected to measure an unknown frequency. The reading is 045. What is the frequency of the system ?* **(UPTU)**

**Solution.** *Given* : Count, $N = 045$

Gated period, t = 10 ms

Now, frequency of the system,

$$f = \frac{N}{t} = \frac{045}{10 \times 10^{-3}} = 4500 \text{ Hz or } \mathbf{4.5\ kHz} \quad \textbf{(Ans.)}$$

## 5.3.6. Period Counters

**Period** *is defined as the time elapsing between identical features or successive cycles of a wave form.* It can be calculated from the frequency and is the *reciprocal of frequency.*

$$P_{(s)} = \frac{1}{f_{(Hz)}} \quad ...(5.2)$$

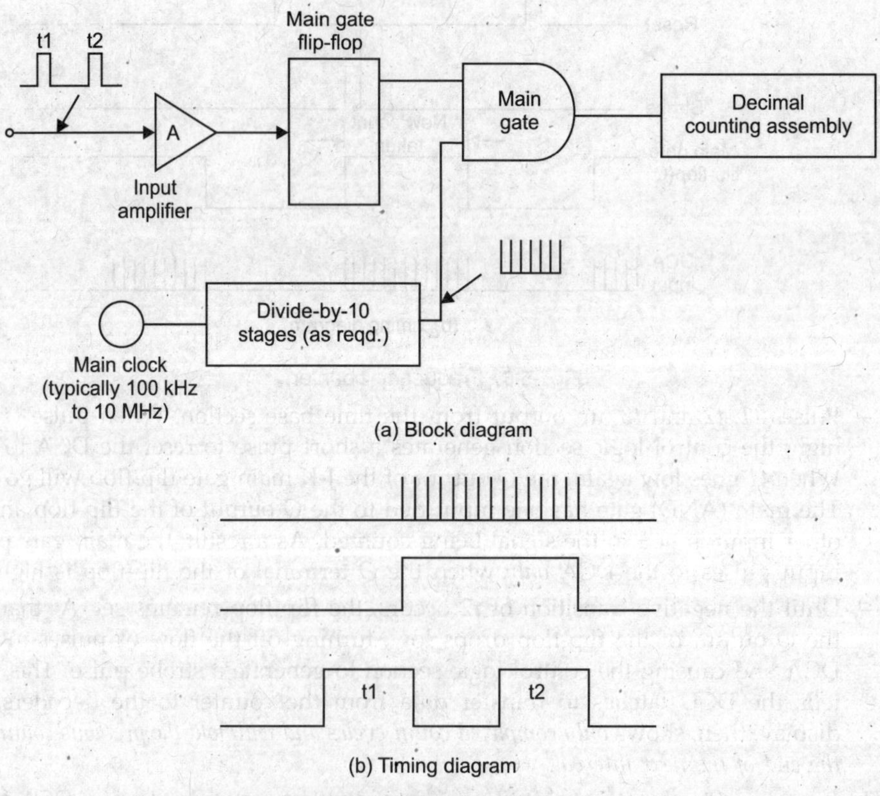


**Fig. 5.7.** Period counter.

*Refer to Fig. 5.7.*

- A period counter can be made by *reversing* the roles of the input stage and time base; the input amplifier is connected to the main gate flip-flop, and the time base is connected to the main gate as shown in **Fig. 5.7(*a*)**.
- **Fig. 5.7(*b*)** shows the timing diagram for a period counter. The flip-flop is set (*i.e.*, *Q* goes high) on the negative-going transitions of *t*1, and this allows the gate to pass pulses from the time base to DCA.

The time base frequency determines the time interval represented by each pulse. The period is given by the count accumulated by the DCA and the time base frequency :

$$P_{(s)} = \frac{\text{DCA count}}{f_{(\text{Hz})}} \qquad ...(5.3)$$

***Note:*** A time base frequency of 1000 Hz yields a period counter resolution of $\frac{1}{1000}$ s or 1 ms; similarly, one 10 kHz time base yields 0.1 ms resolution.

### 5.3.7. Characteristics of Digital Meters

Following are the few specifications which characterise digital meters :

1. Resolution; 2. Sensitivity; 3. Accuracy.

**1. Resolution.** *It is defined as the number of digit positions or simply the number of digits used in a meter.*

If number of full digits is *n*, then resolution,

$$R = \frac{1}{10^n} \qquad ...(5.4)$$

For $n = 4$, $R = \frac{1}{10^4} = 0.0001$ or 0.01%.

- A three-digit display on the digital meter for 0-1 V range will be able to indicate from 000 to 999 mV, with *smallest increment (resolution)* of 1 mV.

**2. Sensitivity.** *It is the smallest change in input which a digital meter is able to detect.* Thus, it is the full-scale value of the lowest voltage range multiplied by the resolution of the meter. In other words,

$$\text{Sensitivity, } S = (fs)_{min} \times R \qquad ...(5.5)$$

where, $(fs)_{min}$ = Lowest full-scale value of digital meter, and

$R$ = Resolution is decimal.

**3. Accuracy.** In the accuracy specifications of digital meters, the following two quantities are included :

(*i*) A percentage of range.

(*ii*) A percentage of reading.

Following solved examples, illustrate these characteristics :

**Example 5.2.** *What is the resolution of $3\frac{1}{2}$ digit display on 1 V and 10 V ranges? A $3\frac{1}{2}$ digital voltmeter has an accuracy specification of ± 0.5% of reading ± 1 digit. What is possible error in volt when the instrument is reading 2.00 V on 5-V scale.*

**Solution.** • Number of full digits on $3\frac{1}{2}$ digit display, $n = 3$

$\therefore$ Resolution, $R = \dfrac{1}{10^n} = \dfrac{1}{10^3} = 0.001$

It means, the meter cannot distinguish between values that differ from each other by less than 0.001 of full scale (*fs*).

*For full-scale range of 1 V:*

The *resolution* = (*fs*) × R = 1 × 0.001 = **0.001 V** **(Ans.)**

*For full-scale range of 10 V :*

The *resolution* = 10 × 0.001 = **0.01 V** **(Ans.)**

- The display for 2 V reading on 10 V scale of $3\frac{1}{2}$ digital meter would be 0.0200.

The digit in the least significant digit (LSD) has a value of :

$$(fs) \times R = 5 \times \frac{1}{10^3} = 0.005 \text{ V}$$

$$\textit{Total possible error} = \frac{0.5}{100} \times \text{reading} + \text{value of digit in LSD}$$

$$= \frac{0.5}{100} \times 2 + 0.005 = 0.01 + 0.005 = \mathbf{0.015\ V} \quad \textbf{(Ans.)}$$

**Example 5.3.** *What is the resolution of a $4\frac{1}{2}$ digit display? How would 16.95 V be displayed on a 10 V range and 0.6564 V on a 1 V and 10 V ranges ?*

**Solution.** • Number of full digits on $4\frac{1}{2}$ digit display, $n = 4$

$\therefore$ *Resolution,* $R = \dfrac{1}{10^n} = \dfrac{1}{10^4} = \mathbf{0.0001}$ **(Ans.)**

- There are 5 digit faces in $4\frac{1}{2}$ digit display, so *16.95 would be displayed as* **16.950** **(Ans.)**

Resolution on 1 V range is 0.000 × 1 = 0.0001 V.

Any reading upto the *4th decimal can be displayed.*

Hence, *0.6564 will be displayed as* **0.6564** **(Ans.)**

Resolution on 10 V range is 0.0001 × 10 = 0.001 V.

Hence, decimals upto *third decimal place can be displayed.*

Hence, *0.6564 will be displayed as* **0.656** **(Ans.)**

**Example 5.4.** *Explain briefly the $3\frac{1}{2}$-digit display in a digital meter.*

**Solution.** The number of digit positions in a digital meter determines the resolution. Hence a 3-digit display on digital voltmeter (DVM) for 0–1 V range will indicate values from 0–999 mV with a smallest increment of 1 mV.

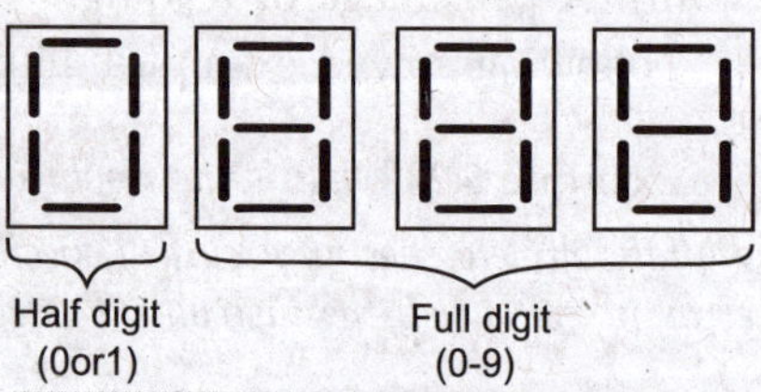

**Fig. 5.8.** 3½-digit display.

Normally, a *fourth* digit capable of indicating 0 or 1 (hence called a *half digit* is placed to the left. This permits the digital meter to read values above

999 upto 1999, to give overlap between ranges for convenience, a process called *"over-ranging"*. This type of display is called a $3\frac{1}{2}$ *digit display*, shown in Fig. 5.8.

### 5.3.8 Analog-to-Digital (A/D) and Digital-to-Analog Conversion

Refer to Article 9.4

## 5.4 DIGITAL VOLTMETERS (DVM)

### 5.4.1. Introduction

*A* **digital voltmeter** *(DVM) displays measurements of D.C. or A.C. voltage as discrete numerals* instead of pointer deflections on a continuous scale as in analog instruments.

- It is a *versatile and accurate instrument* that is employed in many laboratory measurement applications.
- Owing to the development and perfection of IC modules, the size, power requirements and cost of the digital voltmeter have been drastically reduced and, therefore, DVMs can actively compete with conventional analog instruments, both in price and portability.
- The various varieties of DVM differ in the following ways :

  (*i*) Number of digits; (*ii*) Number of measurements;
  (*iii*) Accuracy; (*iv*) Speed of reading;
  (*v*) Digital output of several types.

### 5.4.2. Advantages of DVMs

Following are the *advantages of DVM over other voltmeters* :

1. Higher accuracy and resolution.
2. Greater speed.
3. No parallax.
4. Reduced human error.
5. Compatibility with other digital equipment for further processing and recording.

### 5.4.3. Characteristic Features of DVMs

The characteristic features of digital voltmeters (DVMs) are :

1. *Input range* : From ± 1.000 V to ± 1000 V with automatic range selection and overload indication.
2. *Absolute accuracy* : As high as ± 0.005% of the reading.
3. *Resolution* : 1 part in million (1 μV reading can be read or measured on 1 V range)
4. *Stability* : Short-term, 0.002% of the reading for a 24-hr period; long term, 0.008% of the reading for a 6-month period.
5. *Input resistance* : Typically 10 MΩ;
   *Input capacitance* : 40 pF
6. *Calibration* : Internally from stabilised reference sources, independent of measuring circuit.
7. *Output signals* : In BCD (Binary-coded-decimal) form, for print output and further digital processing.
   - *Optional features* may include additional circuitry to measure current, ohms and voltage ratio.

### 5.4.4. Application of DVMs

DVMs are often used in *"data processing systems"* or *"data logging systems"*.

— In such systems, a number of analog input signals are scanned sequentially by an electronic system and then each signal is *converted to an equivalent digital value by the A/D (analog-to-digital) converter in the DVM.*

— The digital value is then transmitted to a printer along with the information about the input line from which the signal has been derived.

— The whole data is then printed out.

In this way, *a large number of input signals can be automatically scanned or processed and their values either printed or logged.*

### 5.4.5. Classification of Digital Voltmeters (DVMs)

The digital voltmeters can broadly be *classified* as follows :

1. Ramp-type DVM.
2. Dual-slope integrating type DVM.
3. Integrating type DVM (Voltage-to-frequency conversion).
4. Successive–approximation DVM.

### 5.4.6. Ramp-type DVM

The *principle of operation* of the ramp-type DVM is based *on the measurement of the time it takes for a linear ramp voltage to rise from 0 V to the level of input voltage, or decrease from the level of the input voltage to zero.* This interval of time is measured with an *electronic time-interval counter*, and the count is displayed as a *number of digits* on electronic indicating tubes.

- *Fig. 5.9* shows a ramp-type DVM.

**Fig. 5.9(*a*)** shows the *"voltage-to-time conversion"* using gated clock pulses.

— At the start of the measuring cycle, a ramp voltage is initiated; this voltage can be positive-going or negative going. The negative-going ramp, shown in Fig. 5.9(*a*) is continuously compared with the unknown input-voltage.

— At the instant that the ramp voltage equals the unknown voltage, a coincidence circuit, or comparator, generates a pulse which opens a gate **[See Fig. 5.9(*b*)].** The ramp voltage continues to decrease with time until it finally reaches 0 V (or ground potential) and a second comparator generates an output pulse which closes the gate.

— An oscillator generates clock pulses which are allowed to pass through the gate to a number of decade counting units (DCUs) which totalise the number of pulses passed through the gate.

— The *decimal number*, displayed by the indicator tubes associated with the DCUs, *is a measure of the magnitude of the input voltage.*

— The sample-rate multivibrator (MV) determines the rate at which the measurement cycles are initiated. The sample-rate circuit provides an initiating pulse for the ramp generator to start its next ramp voltage. At the same time, a reset pulse is generated which returns all the DCUs to their *zero state*, removing the display momentarily from the indicator tubes.

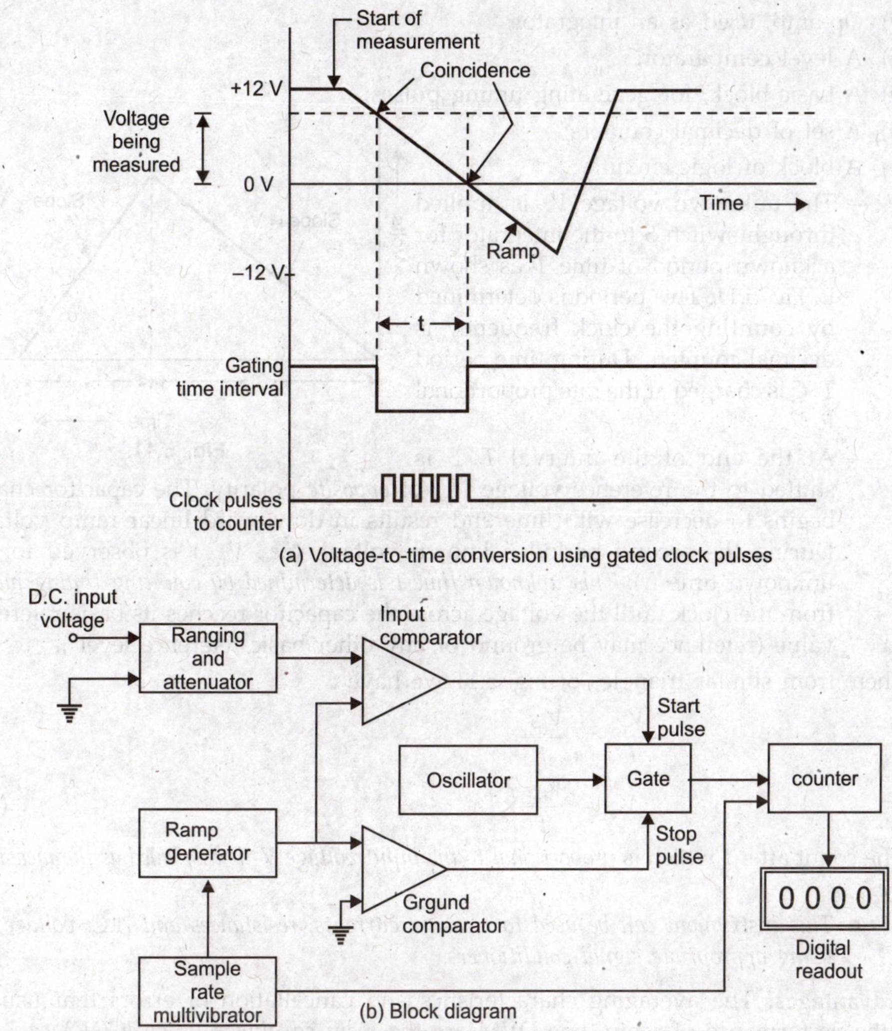


(a) Voltage-to-time conversion using gated clock pulses

(b) Block diagram

**Fig. 5.9.** Ramp-type DVM.

## 5.4.7. Dual-slope Integrating Type DVM

Fig. 5.10 shows the block diagram of a dual-slope integrating type DVM. It consists of the following *five* basic blocks:

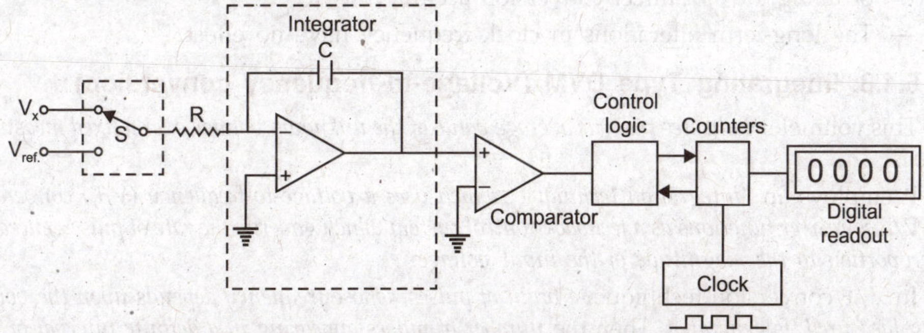


**Fig. 5.10.** Block diagram of a dual-slope integrating type DVM.

(*i*) Op-amp, used as an integrator;

(*ii*) A level comparator;

(*iii*) A basic block, for generating timing pulses;

(*iv*) A set of decimal counters;

(*v*) A block of logic circuitry.

— The unknown voltage $V_x$ is applied through switch $S$ to the integrator for a known period of time $T$ as shown in Fig. 5.11. This period is determined by counting the clock frequency is decimal counters. During time period $T$, $C$ is charged at the rate proportional to $V_x$.

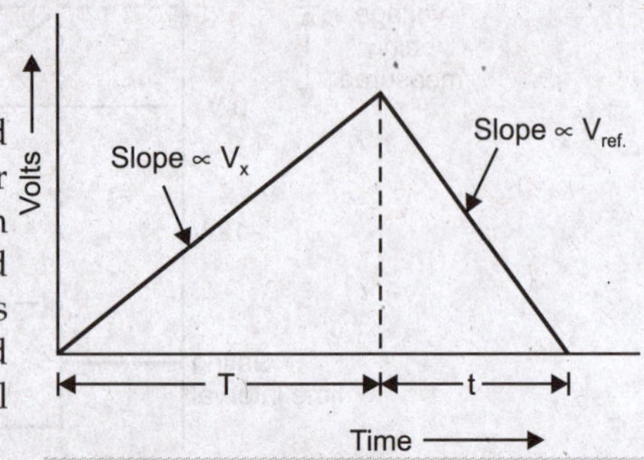


**Fig. 5.11.**

— At the end of the interval $T$, $S$ is shifted to the reference voltage $V_{ref.}$ of *opposite* polarity. The capacitor charge begins to decrease with time and results in downward linear ramp voltage. During the second period a known voltage (*i.e.*, $V_{ref.}$) is observed for an unknown time ($t$). *This unknown time t is determined by counting timing pulses* from the clock until the voltage across the capacitor reaches its basic reference value (reference may be ground or any other basic reference level ).

Then, from similar triangles of Fig. 5.11 we have :

$$\frac{V_x}{T} = \frac{V_{ref.}}{t}$$

or,
$$V_x = \frac{V_{ref.}}{t} \times T \quad ...(5.5)$$

The count after $t$ which is *proportional to the input voltage $V_x$ is displayed as the measured voltage.*

- *This instrument can be used to measure currents, resistances and A.C. voltages by using appropriate signal conditioners.*

**Advantages.** The averaging characteristics and cancellation of errors that usually limit the performance of ramp-type DVM are the *main advantages of such DVMs:*

— The integration characteristics provide the average value of the input signal during the period of first integration. Consequently, disturbance, such as spurious noise pulses, are *minimised.*

— Long-term drifts in the time constant, as may result from temperature variations or aging, do not affect conversion accuracy.

— The long-term alterations in clock frequency have no effect.

### 5.4.8. Integrating Type DVM (Voltage-to-frequency conversion)

This voltmeter *indicates the true average value of the unknown voltage over a fixed measuring period.*

It employs an *"integration technique" which uses a voltage-to-frequency (V/F) conversion. The V/F converter functions as a feedback control system which governs the rate of pulse generation in proportion to the magnitude of the input voltage.*

In V/F conversion technique, *a brain of pulses, whose frequency depends upon the voltage being measured, is generated.* Then the *number of pulses appearing in a definite interval of time is measured.*

The basic requirement of this technique is *"operational integrator"*, the circuit diagram and symbolic representation of which is shown in [Fig. 5.12(*a*, *b*)], it is a modification of operational amplifier. The operational integrator consists of an amplifier (having high gain) with a capacitor C connected between output and input. If $V_i$ is the input voltage, then $V_o$, of the operational integration is given by :

$$V_o = \frac{1}{RC}\int V_i \cdot dt \qquad ...(5.6)$$

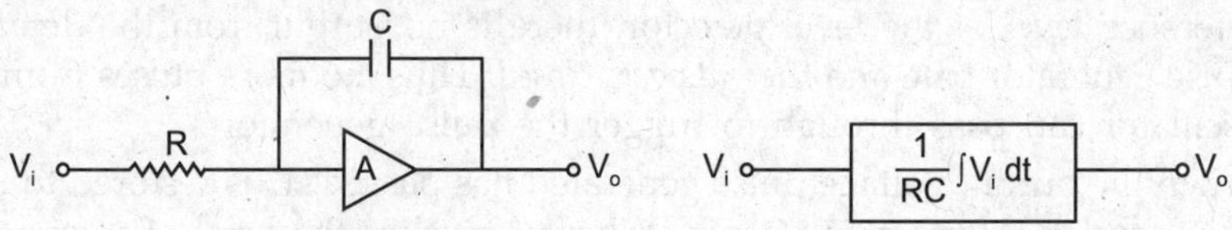


(a) Circuit diagram (b) Symbolic representation

**Fig. 5.12.** Operational integrator.

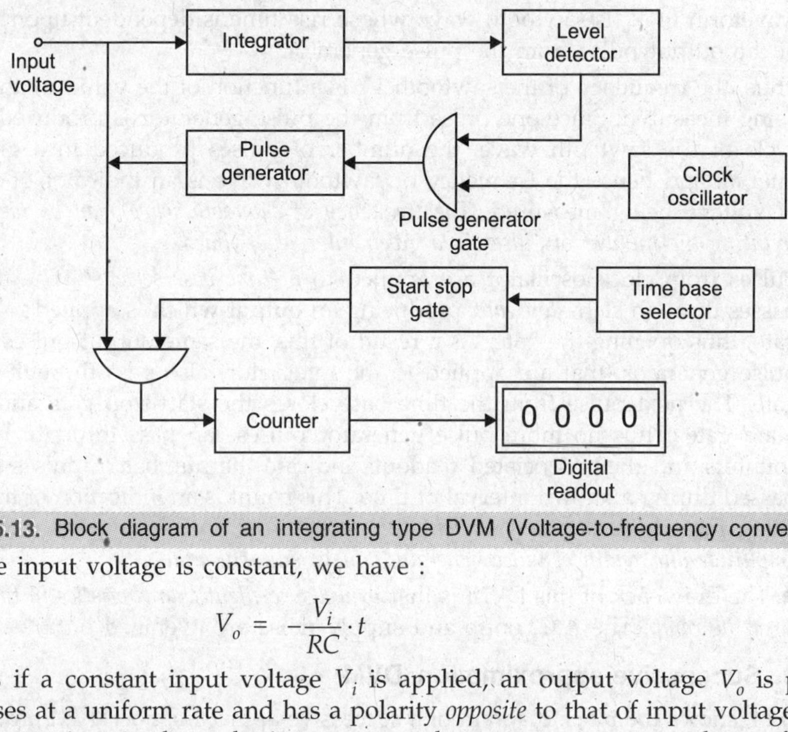


**Fig. 5.13.** Block diagram of an integrating type DVM (Voltage-to-frequency conversion).

If the input voltage is constant, we have :

$$V_o = -\frac{V_i}{RC}\cdot t \qquad ...(5.7)$$

Thus if a constant input voltage $V_i$ is applied, an output voltage $V_o$ is produced which rises at a uniform rate and has a polarity *opposite* to that of input voltage. That is, for a constant input voltage the integrator produces a ramp output voltage of opposite polarity such that the greater the input voltage, the sharper is the rate of rise, or step of the output voltage.

**Fig. 5.13.** shows the *block diagram* of this voltmeter :

— The unknown voltage, $V_i$ is applied to the input of the integrator and the output voltage $V_o$ starts to rise. *The slope of $V_o$ is determined by the value of input voltage, $V_i$.* This voltage is fed to a level detector, and, when $V_o$ reaches a certain reference level, the *detector* sends a pulse to *pulse generator gate.* The level detector is a device similar to a voltage comparator.

— The output voltage from integrator, $V_o$ is compared with the fixed voltage of an internal reference source and when $V_o$ reaches that level, the level detector produces an *output pulse.* This output pulse opens the pulse level gate, permitting pulse from a fixed frequency clock oscillator to pass through the *pulse generator*. This generator is similar to schmitt trigger that *produces an output pulse of fixed amplitude and width for every pulse it receives.*

— The output pulse whose polarity is *opposite* to that of $V_i$ and has greater amplitude is fed back to the input of the integrator. Hence, the net output to the integrator is now of reversed polarity (opposite to that of $V_i$). As a result of this reversed output, the output $V_o$ drops back to its original level. Since $V_o$ is now below the reference level of the level detector, there is no output from the detector to the pulse generator gate and the *gate gets closed*. Thus, no more pulses from the clock oscillator can pass through to trigger the pulse generator.

— When the output voltage pulse generated has passed, $V_i$ is restored to its original value and $V_o$ starts rising again. When it reaches the level of reference voltage again, the pulse generator gate is opened. The pulse generator is triggered by a pulse from the clock generator and the entire cycle is repeated again. Thus, the waveform of $V_o$ is sawtooth wave whose rise time is dependent upon the value of the output pulse from the pulse generator.

— Thus, the frequency of the sawtooth $V_n$ is a function of the value $V_i$, the voltage being measured. Since one pulse from the pulse generator is produced for each cycle of the sawtooth wave, the number of pulses produced in a given time interval and hence the frequency of sawtooth wave is an indicator of the value of voltage being measured. *The frequency of sawtooth wave may be measured by counting the number of pulses in a given interval of time.*

— Pulses from clock oscillator are applied to a *'time base selector'*. The first pulse passes through *start-stop gate*, producing an output which is applied to the main gate, thus opening the gate. As a result of this, the same output pulses from the pulse generator (that are applied to the integrator) also pass through the main gate. The next pulse from the time base closes the start-stop gate and also the main gate. Thus no more pulse generator pulses can pass through. Hence the counters and their associated readouts indicate the number of pulses that have passed during a known interval of time. This count is an indication of the voltage being measured. *In order to make the counter read directly in terms of voltage the amplitude and width of pulse generator can be suitably adjusted.*

- The drawback of this DVM is that it *needs excellent characteristics in the linearity of the ramp*. The A.C. noise and supply noise are averaged out.

### 5.4.9. Successive-approximation DVM

Fig. 5.14, shows the block diagram of a successive-approximation DVM, in which the electronic implementation of the successive-approximation technique is illustrated :

— A *"digital-to-analog (D/A) converter"* is used to provide the estimates. The "equal to or greater than" or "less than" decision is made by the *"comparator"*. The D/A converter provides the estimate and is compared to the input signal.

— A special shift register called a *"successive-approximation register (SAR)"* is used to control the D/A converter and consequently the estimates. At the *beginning* of the conversion all the outputs from the SAR are at *logic zero*. If the estimate is *greater than the input*, the comparator output is *high* and the first SAR output reverses state and the second output changes to a logic "one". If the comparator output is *low*, indicating that the estimate is *lower than the input signal* the first-output remains in the logic one state and the second output assumes the logic one state. *This continues to all the states until the conversion is complete.*

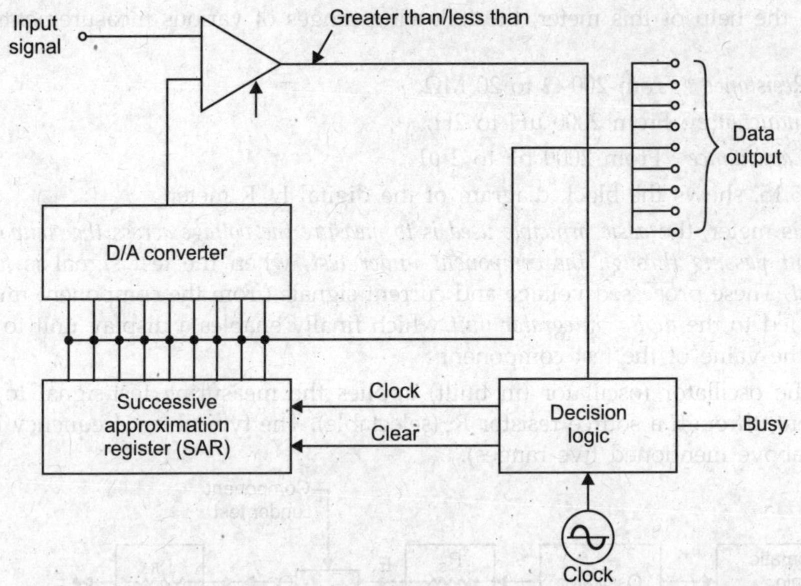


Fig. 5.14. Block diagram of a successive-approximation DVM.

— The above sequence of events are performed electronically. *For an N-bit conversion after N clocks, the actual value of the input is known*. The least significant bit is the state of the comparator. In some systems an additional clock is employed to store the last bit in the SAR and thus N + 1 clocks are required for a conversion.

## 5.4.10. General Specifications of a DVM

Following are the general specifications of a DVM :

(*i*) *Display* : 3 ½ digits, LCD; (*ii*) *Unit Annuciation*: mV, V, mA, Ω, kΩ, MΩ, buzzer, B (low battery); MANU (Manual), ac and ⏄ (diode test); (*iii*) *Max. Indication*: 1999 or – 1999; (*iv*) *Over-range indication*: only (1) or (–1) displayed at the MSB position; (*v*) *Polarity*: AUTO negative polarity indication; (*vi*) *Zero adjustment* : Automatic; (*vii*) *Functions*: DC volts, AC volts, DC amps, AC amps, Ohms, continuity test, diode test; (*viii*) *Ranging*: Sepectable automatic or mannual; (*ix*) *Automatic*: Instrument automatically selects maximum range for measurement and display. Auto ranging operates on all functions except for dc or ac current; (*x*) Manual switch selection as desired; (*xi*) *Sampling rate* : 2 samples/s, nominal; (*xii*); Low battery: β mark on LCD readout; (*xiii*) *Temperature*: Operating 0°C – 40°C, <80% RH(Relative humidity); Storage – 20°C – 60°C, < 70% RH; (*xiv*) *Power* : Two AR size 1.5 V batteries, Life 2000 hours typically with zinc-carbon; (*xv*) *Standard accessories*; Probe red-black; safety fuse 250 – 0.2A; (*xvi*) Size: 160(L) × 80(B) × 30(H); (*xvii*) *Weight*: 250 g without batteries; (*xviii*) *Input impedance*: 11 MΩ – 1000 MΩ; (*xix*) Accuracy : ± 0.5% – 0.7% or ± 5 digits for dc; 1.0% reading or ± 5 digits for ac at 40 – 500 kHz.

## 5.5 DIGITAL LCR METER

This type of meter is used *to measure the resistance, inductance, capacitance and dissipation factor*. The desired function can be selected by using a *rotary switch*. The various ranges available are :

(*i*) 200 μH/pF/Ω; (*ii*) 2000 μH/pF/Ω; (*iii*) 20 mH/nF/kΩ; (*iv*) 200 mH/nF/kΩ; (*v*) 2H/μF/MΩ.

With the help of this meter, the following ranges of various measurements can be made:

- *Resistance* : From 200 Ω to 20 MΩ;
- *Inductance* : From 2000 μH to 2H;
- *Capacitance* : From 2000 pF to 2 μF.

Fig. 5.15. shows the block diagram of the digital LCR meter :

In this meter, the *basic principle used is to measure the voltage across the component and the current passing through the component under test*, when the test signal is *fed to the component.* These processed voltage and current signals (from the component under test) are then fed to the *digital integrator unit* which finally enables a display unit to provide directly the value of the test component.

— The oscillator (oscillator (in built) applies the measuring test-signal to the test component, through a source resistor $R_s$ (selectable). The typical test frequency is 1 kHz (for the above mentioned five ranges).

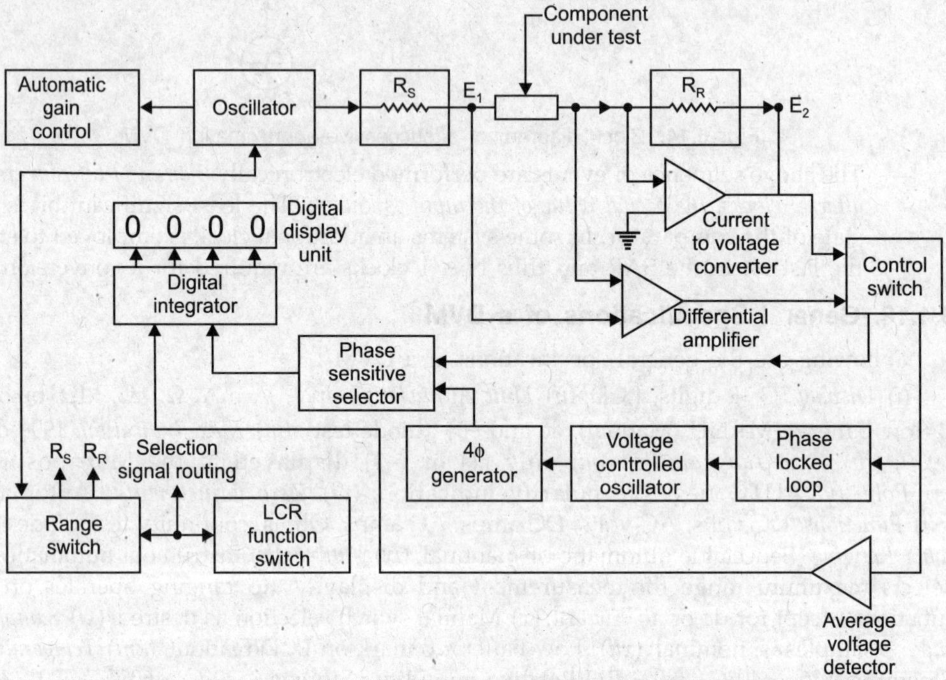


**Fig. 5.15.** Block diagram of a digitial LCR meter.

— The signal current then flows to current-to-voltage converter, which is essentially an op-amp with range resistor $R_R$, connected in the feedback path.

— The op-amp drives the junction of component and $R_R$ to a virtual ground and hence $R_R$ does not change the current through the component. Thus *signal current develops a voltage $E_2$ proportional to the current through the component.*

*The voltage $E_1$ and $E_2$ are the vector quantities and as such they define the characteristics of the component at a particular test frequency and signal level.*

Mathematically : $E_1 \propto V$ and $E_2 \propto I$

Thus, *Capacitance,* $C \propto \frac{I}{V} \propto \frac{E_2}{E_1}$; *Inductance,* $L \propto \frac{V}{I} \propto \frac{E_1}{E_2}$.

The above ratios are adopted in measurement modes and are obtained by dual-slope integration method. For *inductance* measurements, a *series* equivalent circuit of an inductor is assumed while for capacitance measurements, a parallel equivalent circuit of capacitor is assumed for five ranges.

— The values of $R_S$ and $R_R$ are selected, based on the impedance of the unknown component. For the measurement of *inductance,* the component impedance is usually low and hence $R_S$ is chosen much higher. This achieves a *constant current drive* to the component. The $R_S$ decides the value of current. For *capacitance* measurement, the component impedance is high and $R_S$ is chosen to be much low. Thus *constant voltage drive* is provided to the test component.

— The voltage $E_1$ is applied as one input to *'differential amplifier'*; $E_1$ is then fed to a *control switch,* along with $E_2$. The greater of the two is fed to *'average voltage detector (AVD)'* and lesser to the *'phase sensitive detector (PSD)'.*

— The signal given to AVD is also given to *'phase locked loop (PLL)'* and voltage controlled oscillator (VCD). These two produce the clock signals, which are locked in phase with the reference signal. The clock signal is then divided into phase shifted 90° and 270° signals. These are given to PSD, which detector the phase angle.

— The D.C. voltage outputs, from AVD and PSD are then given to *'digital integrator unit'* and final to *'digital display unit'* where the component value is displayed.

- For wide range of values, the basic accuracy of this meter is ± 0.15 percent of the reading.

— The range selection is fully automatic.

— The display used is normally 4 digit LEDs with automatic decimal point.

— The inductors and capacitors are automatically differentiated.

**Commercial digital LCR meter :**

Important features available in commercial digital LCR meter are :

(*i*) Auto-ranging and auto-computing facility.

(*ii*) High basic accuracy, upto 0.2 percent.

(*iii*) High frequency range, from 0.1 to 1 kHz.

(*iv*) Hold function.

(*v*) Multipoint measurement techniques.

(*vi*) Series or parallel modes of operation.

- Most of these meters are *"microprocessor controlled"* and multipoint measurement technique ensures better accuracies in measurement. Apart from the basic measurement of LCR and $Q$ they can also be used for series and parallel measurement *i.e.,* they can measure series and parallel components of $L$ and $C$.

— The microprocessor configuration provides operation flexibility *i.e.,* comparison with previous recorded values, storage of these values for future references (memory function).

— The user software provides user prompt facility by means of flashing LED for respective functions and to select appropriate, test frequencies to achieve optimum results.

— The *'Hold function'* is useful for selecting the components within a specified range.

## 5.6 DIGITAL MULTIMETER (DMM)

Digital multimeter offers *increased versatility due to its additional capability to measure A.C. voltage and current, D.C. current, and resistance.*

Fig. 5.16, shows the block diagram of a digital multimeter (DMM) :

- In the *"A.C. voltage mode"*, the applied input is fed through a calibrated, compensated attenuator, to a precision full-wave rectifier circuit followed by a ripple reduction filter. The resulting D.C. is fed to ADC and the subsequent display system.
- For *current measurements,* the drop across an internal calibrated shunt is measured, directly by the ADC in the *"D.C. current mode"*, and after A.C. to D.C. conversion in the *"A.C. current mode"*. This drop is often in the range of 200 mV (corresponding to full scale).
  - — Due to the lack of precision in the A.C.–D.C. conversions, the accuracy in the A.C. range is in general of the order of 0.2 to 0.5%. In addition, the measurement range is often limited to about 50 Hz at the lower frequency end due to the ripple in the rectified signal becoming a non-negligible percentage of the display and hence results in fluctuation of the displayed number. At the higher frequency end, deterioration of the performance of the AC/DC converter limits the accuracy.
  - — The A.C. measurement, is *often average reading, r.m.s. calibrated.*

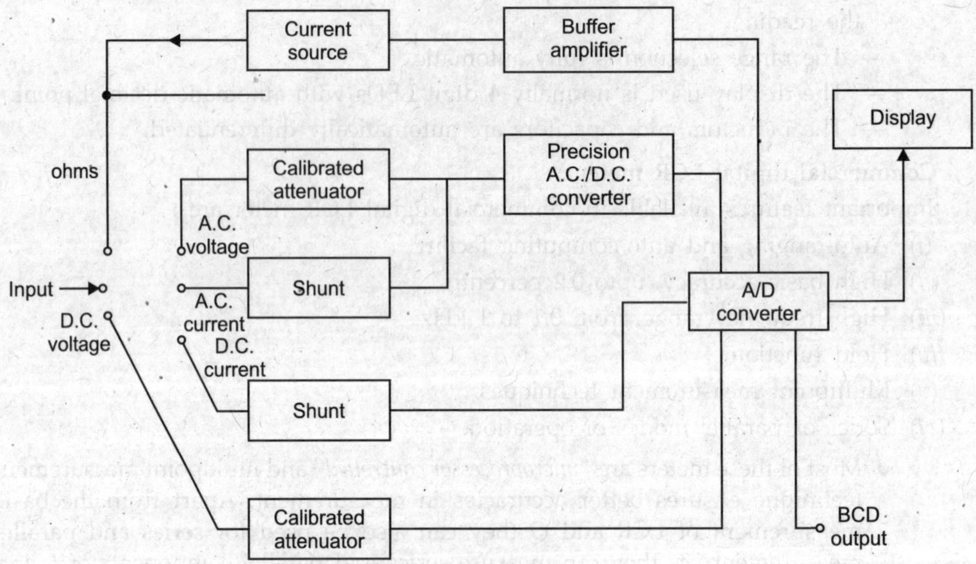


Fig. 5.16. Block diagram of a digital multimeter.

- In the *"resistance range"* the digital multimeter operates by measuring the voltage across the externally connected resistance, resulting from a current forced through it from a calibrated internal current source.
  - — The accuracy of resistance measurement is of the order of 0.1 to 0.5% depending on the accuracy and stability of the internal current sources. The accuracy may be *proper* in the *highest range* which is often about 10 to 20 MΩ. In the *lowest range,* the full scale may be $\simeq$ 200 Ω with a resolution of about 0.01 Ω for a $4\frac{1}{2}$ digit digital multimeter. In this range of resistance

measurement, the effect of the lead resistance will have to be carefully considered.

### Comparison between "Analog" and "Digital" Multimeters

| S. No. | Aspects | Analog multimeters | Digital multimeters |
|---|---|---|---|
| 1. | *Power supply* | Not required | Required |
| 2. | *Visual indication of changes in the reading* | Better | Not that much better |
| 3. | *Effect of electric noise* | Less | More |
| 4. | *Isolation problems* | Less | More |
| 5. | *Accuracy* | Less | High |
| 6. | *Interface of the output with external equipment* | Not possible | Possible |
| 7. | *Construction* | Simple | Complicated |
| 8. | *Size* | Bigger | Smaller |
| 9. | *Cost* | Less | More |
| 10. | *Nature of output* | Ambiguous–many times | Unambiguous reading |

## 5.7 MICROPROCESSOR-BASED INSTRUMENTS–INTELLIGENT INSTRUMENTS

*"Intelligent instruments"*, a new class of instruments, involve the *use of microprocessors (μP) as an integral part of the measuring instruments.*

Fig. 5.17. shows a block diagram of *microprocessor based impedance measuring instrument :*

— The operation makes interface with the instrument via the IEEE 488 bus to allow control by, or to make the measurement available to a large external computer system.

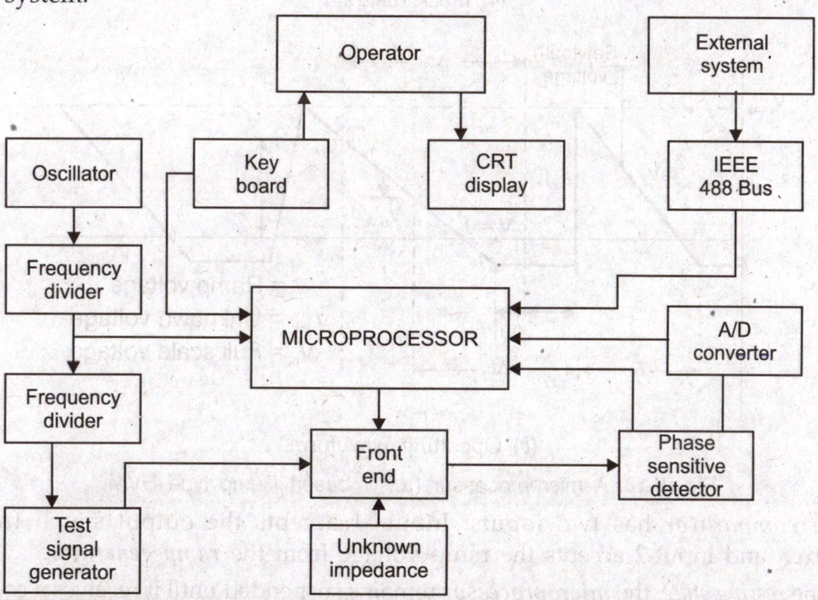


Fig. 5.17. Block diagram of microprocessor based impedance measuring instrument.

— The *frequency division of the oscillator signal* provides the timing clock signal and the A.C. test signal.

— The *front end circuit* applies the test signal to the unknown impedance and a standard impedance provides an output signal, proportional to the voltage across each, to the *phase sensitive detector* (PSD).

— PSD, which is also controlled by the microprocessor, converts the A.C. inputs of the impedance in vector form to a D.C. output.

— The A/D converter provides the digital data, which is used by the microprocessor to compute the value of the unknown impedance. This is then displayed on the CRT or sent as output to the IEEE 488 bus.

## 5.8 MICROPROCESSOR-BASED RAM TYPE DVM

Fig. 5.18 (a) and (b) shows the basic *block diagram* and *operating waveform* of a micro processor-based Ramp type DVM.

***Description***: The multiplexer has the following inputs:

(*i*) Input-1. Connected to the ground potential;

(*ii*) Input-2. The unknown input;

(*iii*) Input-3. Reference voltage input

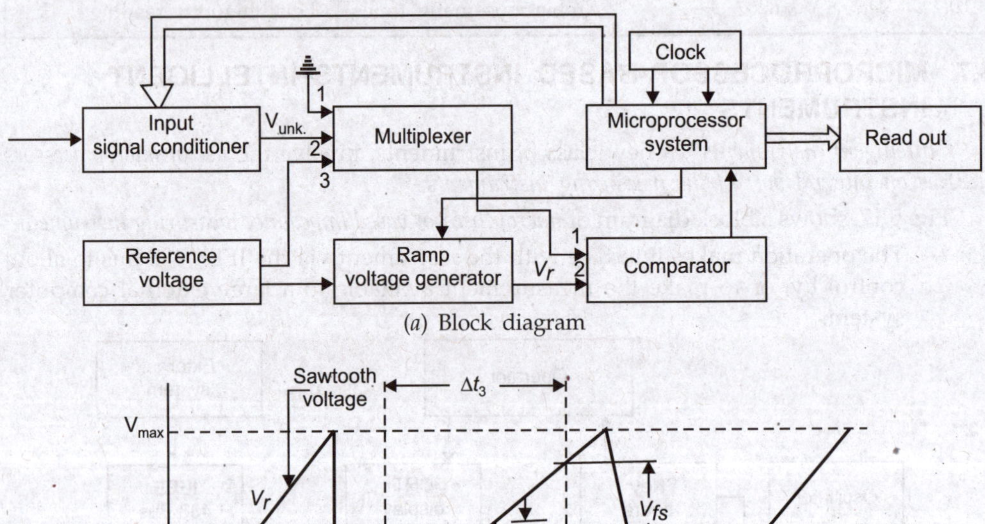


(*a*) Block diagram

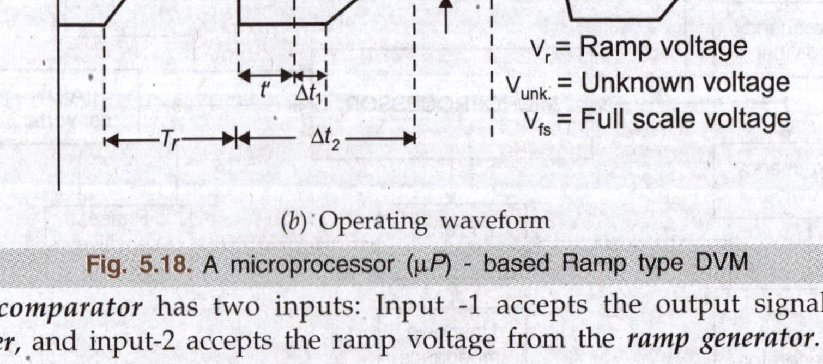


(*b*) Operating waveform

**Fig. 5.18.** A microprocessor ($\mu P$) - based Ramp type DVM

The ***comparator*** has two inputs: Input -1 accepts the output signal from the ***multiplexer***, and input-2 accepts the ramp voltage from the ***ramp generator***.

In the *resting state*, the ***microprocessor*** remains suspended until it receives a command to start conversion. During the resting period, the $\mu P$ regularly sends reset signals to the *ramp generator*. Whenever the ramp generator is reset, its capacitor discharges. It produces a ramp (i.e. a saw tooth voltage) whose duration $T_r$, and amplitude $V_{max}$ remain constant.

- As and when a conversion command arrives at the *microprocessor* at a time $t_1$, the multiplexer first connects Input-1 of the comparator to its Input-1 (*i.e.* ground potential) and brings the former to the ground potential.

– After a pause when Input-2 voltage arriving from the ramp generator becomes equal to Input-1 of the comparator, the comparator sends a signal to the microprocessor, that ramp voltage is zero. The microprocessor measures this time interval $\Delta t_1$ [Fig. 518(*b*)] by counting the number of clock pulses supplied by the clock generator during this interval. Let the count be $N_1$, which is there stored by the microprocessor.

– Then microprocessor sends a command due to which the comparator Input-1 is connected to Input-2 of the multiplexer; this connects the unknown voltage, $V_{unk.}$ to the Input-1 of the comparator. When the ramp voltage equals the unknown voltage, the comparator sends a signal to the microprocessor to measure the time internal $\Delta t_2$ . The count $N_2$, during this interval is also stored.

– Similarly, the next command from the microprocessor causes the comparator Input-1 to be connected to the Input-3 of the multiplexer, which is full scale reference voltage ($V_{fs}$). At the instant, when the ramp voltage equals $V_{fs}$, a pulse is sent to the microprocessor from the comparator output to measure this time internal, $\Delta t_3$ [Fig. 5.18 (*b*)]. The count $N_3$, during this interval is also stored.

The unknown voltge, $V_{unk.}$ is determined from the following equation:

$$V_{unk.} = C.\frac{(N_2 - N_1)}{(N_3 - N_1)}$$

where, C = Coefficient, it depends on the characteristics of the instrument and the units selected, and

$N_1, N_2, N_3$ = The count representing the zero drift, the unknown voltage and full scale voltage respectively.

In this method, the zero drift has practically no effect on the result, on account of the slope variation of the ramp.

**Advantages :**

1. The instrument's *accuracy is not affected* by the time and temperature instabilities of the element values of the circuit.

2. Good repeatability in switching instants in the presence of noise and interference.

3. The size of scale remains constant due to zero drift connection and maximum count.

**Disadvantage :** The suppression of noise and interference cannot be achieved.

## 5.9 IEEE-488 GENERAL PURPOSE INTERFACE BUS (GPIB) INSTRUMENTS

Now-a-days automatic test equipment is one of the leading methods for testing electronic equipment in factory production and troubleshooting situations. *The basic method is to use a programmable digital computer to control a bank of test instruments.* The bank of instruments can be configured for a special purpose or for general use.

The IEEE-488 bus or General purpose interface bus (GDIB) is a tool that is based on the IEEE (Institution of Electrical and Electronics Engineers) specifications.

The digital signals on IEEE-488 bus are generally similar to TTL (transistor-transistor logic), *i.e.,* a logic LOW is less than 0.8 V and a logic HIGH is greater than 2.0 V.

The digital signals can be connected to the instruments through a multiconductor cable upto 20 metres is length provided that an instrument load is placed every 2 metres.

Most IEEE-488 bus/GPIB systems operate unrestricted to 250 bilobytes per second or faster with some restrictions.

There are *two basic configurations* for the IEEE-488/GPIB system :

1. *Linear.* In this type of configuration a tap-off to the next instrument is taken from the previous one in series.
2. *Star.* In this case, the instruments are connected from a central point.

Fig. 5.19. shows the basic structure of IEEE-488/GPIB system. The figure indicates the following four different devices :

(*i*) Computer;
(*ii*) Frequency counter;
(*iii*) Signal generator;
(*iv*) Digital multimeter.

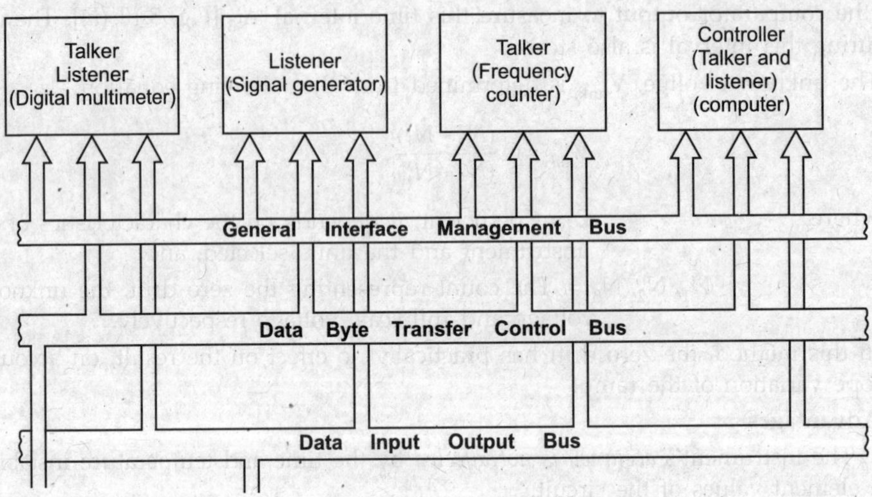


**Fig. 5.19.** Basic structure of IEEE-488/GPIB system.

The IEEE-488/GPIB system itself consists of *three major buses* :

(*i*) *General Interface Management (GIM) bus.* This bus co-ordinates the whole system and ensures an orderly flow of data over the data input/output (DIO) bus.

(*ii*) *Data I/O (DIO) bus.* This bus is a bi-directional 8-bit data bus that carries data, interface messages and device-dependent messages between the *controllers, talkers* and *listeners.* This bus sends asynchronously in byte-serial format.

(*iii*) *Data Byte Transfer (DBT) bus.* This bus controls the sending of data along the DIO bus.

- The signals defined for the above three buses are implemented as conductors in a system interface cable.

- Each IEEE-488/GPIB system compatible instrument will have a female 36-pin Amphenol-style connector on the rear panel.
- The devices connected to IEEE-488 GPIB system (*i.e.*, computer, frequency counter, signal generator and digital multimeter) are categorised as : *Controller; Listener and/or Talker*.

**(i) *Controller :***

— Its function is to communicate device addresses and other inface buses to instrument in the system.

**(ii) *Listener :***

— Its function is to receive commands from other instruments (usually the controller) when the correct address is placed on the base.

— The listener acts on the message received but does not send back any data to the controller.

**(iii) *Talker :***

— Its function is to respond to the message sent to it by the controller.

- ***V × I bus*** is another widely growing platform for instrumentation systems. It uses a mainframe chasis with a maximum of 13 slots to hold *modular instruments on plug-in boards*.

— The *V × I* backplane includes the 32-bit VME data bus, high performance instrumentation, buses for precision timing and synchronisation between instrument components, standardised initialisation and resource management to ease configuration.

## HIGHLIGHTS

1. A *digital instrument* can be considered as a counter which counts the pulses in a predetermined time.
2. The digital meters work on the principle of *"quantization"*,
3. Digital voltmeters (DVM) are broadly classified as :
   (*i*) Ramp-type DVM;
   (*ii*) Dual-slope integrating type DVM;
   (*iii*) Integrating type DVM (Voltage-to-frequency);
   (*iv*) Successive-approximation DVM.
4. Intelligent instruments involve the use of microprocessor as an integral part of the measuring instruments.

## OBJECTIVE TYPE QUESTIONS

1. The D' Arsonval meter movement can be converted into an audio-frequency A.C. ammeter by adding a .......................... to it.

   (*a*) thermocouple (*b*) rectifier
   (*c*) chopper (*d*) transducer.

2. In a linear meter, half-scale deflection occurs when there is .......................... per cent of the rated current through its coil

   (*a*) 100 (*b*) 25 (*c*) 50 (*d*) 75.

3. It is desired to convert a 0–1000 μA meter movement, with an internal resistance of 100 Ω into a 0–100 mA meter. The required value of shunt resistant is about
   (a) 1 Ω (b) 10 Ω
   (c) 99 Ω (d) 100 Ω.
4. Loading effect is principally caused by ......................... instruments.
   (a) high resistance (b) low-sensitivity
   (c) high-sensitivity (d) high-range.
5. A multimeter is used to measure
   (a) resistance (b) current
   (c) voltage (d) all of the above.
6. Digital instruments are those which
   (a) have numerical readout (b) use LED or LCD displays
   (c) have a circuitry of digital design (d) use deflection type meter movement.
7. The main difference between the electronic and electrical instruments is that an electronic instrument contains.
   (a) an electronic device (b) a transducer
   (c) a digital readout (d) electrons.
8. A sinusoidal voltage of rms value 10 V is applied to a D'Arsonval movement connected in series with a half-wave rectifier. It will show a reading of ......................... volt.
   (a) 9 (b) 4.5
   (c) 10 (d) 7.7
9. A VTVM produces negligible loading effect on a circuit under test primarily because
   (a) it virtually draws no current from the circuit
   (b) of its very high internal resistance
   (c) it uses high vacuum tubes
   (d) it is a null deflection instrument.
10. In a 3½ digit voltmeter, the largest number that can be read is
    (a) 0999 (b) 1999
    (c) 4999 (d) 9999.
11. A 3½ digit voltmeter having a resolution of 100 mV can be used to measure maximum voltage of
    (a) 100 V (b) 200 V
    (c) 1000 V (d) 5000 V.
12. The essential elements of an electronic instrument are
    (a) transducer (b) signal conditioner
    (c) indicating device (d) all of the above.
13. The current sensitivity of a meter is expressed in
    (a) ampere (b) ohm/ampere
    (c) ohm/volt (d) ampere/division.
14. The basic meter movement can be converted into an ohmmeter by connecting a ......................... with it.
    (a) high resistance in series (b) low resistance in parallel
    (c) battery in series (d) battery and a variable resistance in series.
15. A 0–1 mA meter has a sensitivity of
    (a) 1 kΩ/V (b) 1 mA
    (c) 1 kΩ (d) 1000 A.

**16.** A moving coil instrument has a resistance of 10 Ω and takes 40 mA to produce full-scale deflection. The shunt resistance required to convert this instrument for use as an ammeter of range 0 to 2A is

(*a*) 0.1021 Ω (*b*) 0.2041 Ω
(*c*) 0.2561 Ω (*d*) 0.4210 Ω.

**17.** A moving coil ammeter has a fixed shunt of 0.02 ohm resistance. If the coil resistance of the meter is 1000 Ω, a potential difference of 500 mV is required across it for full-scale deflection. Under this condition, the current in the shunt would be

(*a*) 2.5 A (*b*) 2.5 A
(*c*) 0.25 A (*d*) 0.25 A.

## ANSWERS

| | | | | | | |
|---|---|---|---|---|---|---|
| 1. (*b*) | 2. (*c*) | 3. (*a*) | 4. (*b*) | 5. (*d*) | 6. (*c*) | 7. (*a*) |
| 8. (*b*) | 9. (*b*) | 10. (*b*) | 11. (*c*) | 12. (*d*) | 13. (*c*) | 14. (*d*) |
| 15. (*a*) | 16. (*b*) | 17. (*b*) | | | | |

## THEORETICAL QUESTIONS

**1.** What is a digital instrument? Explain.
**2.** Explain briefly the principle on which the digital meters work.
**3.** What are the building blocks of a digital instrument? Explain briefly.
**4.** How does a digital system differ from an analog system? Explain.
**5.** What are the advantages of digital systems/techniques?
**6.** Enumerate the components of a digital system.
**7.** What are the advantages of digital implementation?
**8.** Explain briefly the following :
(*i*) Binary counters;
(*ii*) Decimal counting units (DCUs);
(*iii*) Display devices.
**9.** Explain with a block diagram the working of a 'frequency counter'.
**10.** What do you mean by the term 'Period'. How is it measured by a 'Period counter'?
**11.** Explain briefly the following characteristics of digital meters :
(*i*) Resolution;
(*ii*) Sensitivity;
(*iii*) Accuracy.
**12.** What is the function of a digital voltmeter?
**13.** State the advantages of digital voltmeters.
**14.** State the characteristic features of DVMs.
**15.** List the applications of DVMs.
**16.** Give the classification of digital voltmeters.
**17.** Explain briefly with neat diagrams the working of any two of the following instruments :
(*i*) Ramp-type DVM;
(*ii*) Dual-slope integrating type DVM;
(*iii*) Integrating type DVM (Voltage-to-frequency conversion)
**18.** With the help of a block diagram, describe briefly the working of a successive-approximation digital voltmeter.
**19.** What is the function of a digital LCR meter? Explain its working with the help of a neat block diagram.
**20.** State the important features available in a commercial digital LCR meter.
**21.** Describe briefly a "digital multimeter" with the help of a block diagram.
**22.** Give the comparison between analog and digital multimeters.
**23.** Give the description of a microprocessor based impedance measuring instruments.
**24.** Discuss briefly the working of IEEE-488 bus system.

CHAPTER

# 6 Instrument Transformers

**6.1. Introduction; 6.2. Terms relating to instrument transformers; 6.3. Current transformer (CT)** – General aspects – Relationships in a current transformer – Errors in current transformer – Characteristics of current transformers – Clip-on type current transformer; **6.4. Potential transformer (PT)** – General aspects – Comparison of current and potential transformers – Relationships in a potential transformer – Characteristics of potential transformers – Reduction of errors in instrument transformers – *Highlights – Objective Type Questions – Theoretical Questions – Unsolved Examples.*

## 6.1 INTRODUCTION

For the measurement of *large currents in D.C. circuits,* it is usual to use low-range ammeters with suitable *shunts*. For measuring *high voltages*, low range voltmeters are used with *high resistance connected in series with them*. However, this method is neither convenient nor practical for alternating current and voltage instruments. For this purpose, *specially constructed accurate-ratio* ***"instrument transformers"*** *are employed in conjunction with standard low-range A.C. instruments. Their purpose is to reduce the line current or supply voltage to a value small enough to be easily measured with meters of moderate size and capacity.* In other words, they are used for *extending the range of A.C. ammeters and voltmeters.*

The following are the *two basic advantages* inherent in the method using instrument transformers for measurement purposes :

(*i*) Standard rated instruments may be used.

(*ii*) Operating personnel coming in contact with the instruments are not subjected to high voltage and current of the lines, and so there is less danger to them. Even, with a low-voltage system, instrument transformers are used for measuring large currents, so that heavy loads to the instrument panel and to the ammeter and other current terminals are avoided.

*The principle of the instrument transformer is fundamentally the same as that of the power transformer.*

The instrument transformers are *classified* as follows :

1. *Current transformers (CT)* : These are used for the measurement of "current".
2. *Potential transformers (PT)* : These are used for the measurement of "voltage".

These transformers extensively used for *precise measurements*.

- The instrument transformers are widely *used* in protection circuits of power systems *for the operation of over-current,* under-voltage, earth-fault and various other relays.

Fig. 6.1 shows the wiring diagrams for current and potential instrument transformers.

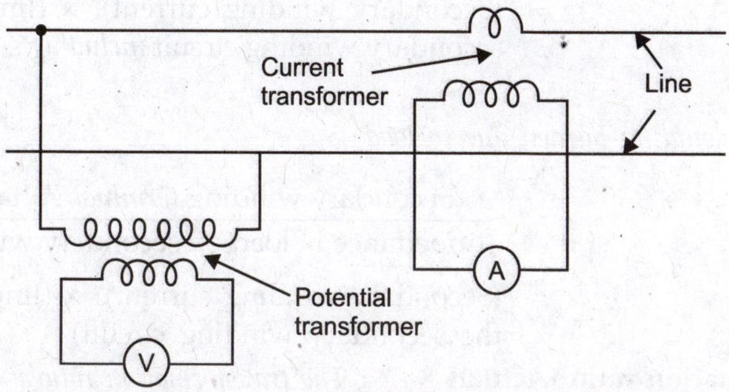


**Fig. 6.1.** Wiring diagrams for current and potential instrument transformers.

**Advantages and disadvantages of instrument transformers :**

The advantages and disadvantages of instrument transformers as *compared with ammeter shunts and voltmeter multipliers* are as follows :

***Advantages* :**

1. The instrument transformer enables the single range instrument to cover a *large* current or voltage range (normal ranges : 0-5A; 0-110V)
2. It is possible to locate the measuring instruments far away from the high voltage circuit by using long leads connecting the instrument transformer to the instrument.
3. When instruments are employed in conjunction with instrument transformers, their readings *do not depend* upon their constants viz., R, L and C as in the case of shunts and multipliers.
4. Many instruments can be operated from a single instrument transformer.
5. Small power loss.
6. By employing current transformer with a suitable split and hinged core, upon which the secondary winding is wound the current in a heavy current bus-bar can be measured without breaking the current circuit.

***Disadvantages* :** The instrument transformers *can not* be used for D.C. measurements.

- In using instrument transformers *for current (or voltage) measurements*, we must know the *ratio of primary current (or voltage) to the secondary current (or voltage)*. These ratios give the *multiplying factor* for finding the primary values from the instrument readings on the secondary side.

However, for *energy or power measurements*, it is essential to know not only the transformation ratio *but also the phase angle between primary and secondary currents (or voltages) because it necessitates further correction to the meter reading.*

## 6.2 TERMS RELATING TO INSTRUMENT TRANSFORMERS

The various terms relating to instrument transformers are discussed below :

1. **Burden of an instrument transformer :** The load across the secondary winding can be conveniently expressed as the output in volt-ampere at the rated secondary voltage. *The rated burden is the volt-ampere loading which is permissible without errors exceeding the limits for the particular class of accuracy.*

$$\text{Total secondary burden} = \frac{(\text{Secondary winding } \textit{induced} \text{ voltage})^2}{(\text{Impedance of the secondary winding circuit } \textit{including} \text{ impedance of secondary winding})}$$

= (Secondary winding current)$^2$ × (Impedance of secondary winding circuit *including* secondary winding) ...(6.1)

*Secondary winding burden due to load*

$$= \frac{\text{(Secondary winding } \textit{terminal} \text{ voltage)}^2}{\text{(Impedance of load on secondary winding)}}$$

= (Secondary winding current)$^2$ × (Impedance of *load* in the secondary winding circuit) ...(6.2)

2. **Transformation ratio (Actual) $K_{act.}$ :** *The transformation ratio is defined as the ratio of the primary phasor to the secondary phasor.*
   Transformer ratio (Actual) $K_{act.}$

$$= \frac{|\text{Primary phasor}|}{|\text{Secondary phasor}|} \quad ...(6.3)$$

$$= \frac{\text{Primary winding current (Actual)}}{\text{Secondary winding current (Actual)}} \text{ for a } CT \quad ...(6.4)$$

$$= \frac{\text{Primary winding voltage (Actual)}}{\text{Secondary winding voltage (Actual)}} \text{ for a } PT \quad ...(6.5)$$

3. **Nominal transformation ratio ($K_{nom.}$) :** *It is the ratio of rated primary winding current (or voltage) to the rated secondary winding current (or voltage).*

*i.e.,*
$$K_{nom.} = \frac{\text{Rated primary winding current}}{\text{Rated secondary winding current}} \text{ ... for a } CT \quad ...(6.6)$$

$$= \frac{\text{Rated primary winding voltage}}{\text{Rated secondary winding voltage}} \text{ ... for a } PT \quad ...(6.7)$$

4. **Turns ratio ($K_t$) :** This is defined as under :

$$\text{Turns ratio, } K_t = \frac{\text{Number of turns of secondary winding}}{\text{Number of turns of primary winding}} \text{ ...for a } CT \quad ...(6.8)$$

$$= \frac{\text{Number of turns of primary winding}}{\text{Number of turns of secondary winding}} \text{ ...for a } PT \quad ...(6.9)$$

- *Normally,* $K_{nom.} = K_t$

5. **Ratio correction factor (*RCF*) :** The ratio correction factor is *the transformation (actual) ratio divided by nominal ratio.*
   In other words,

$$RCF = \frac{K_{act.}}{K_{nom.}}$$

or,
$$K_{act.} = RCF \times K_{nom.} \quad ...(6.10)$$

- The ratio marked on transformers is their *nominal ratio.*

## 6.3 CURRENT TRANSFORMER (CT)

### 6.3.1. General Aspects

- Just as a shunt extends the range of a D.C. ammeter, so does the current transformer perform the same function in A.C. circuits. Thus a *high magnitude alternating current can be easily measured by a combination of a current transformer and a low range ammeter.*

- The primary of a current transformer (C.T.) consists of a few turns of thick cross-section connected in *series* with the high current line. Very often the primary is just one turn formed by taking the line conductor through the secondary winding (Fig. 6.2). The secondary winding consists of a large number of turns of fine wire designed for either 5A or 1A rating. Thus a current transformer is *step-up* transformer. The current transformer has the secondary effectively short-circuited through the low impedance of the ammeter. Fig. 6.3 shows the *current transformer connections*.

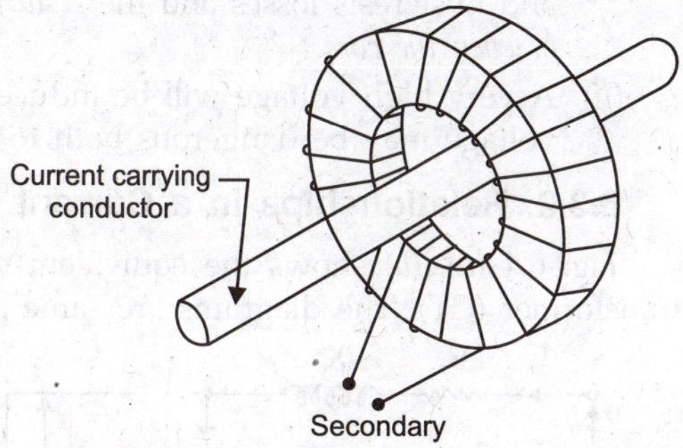


**Fig. 6.2.** Line conductor acting as primary.

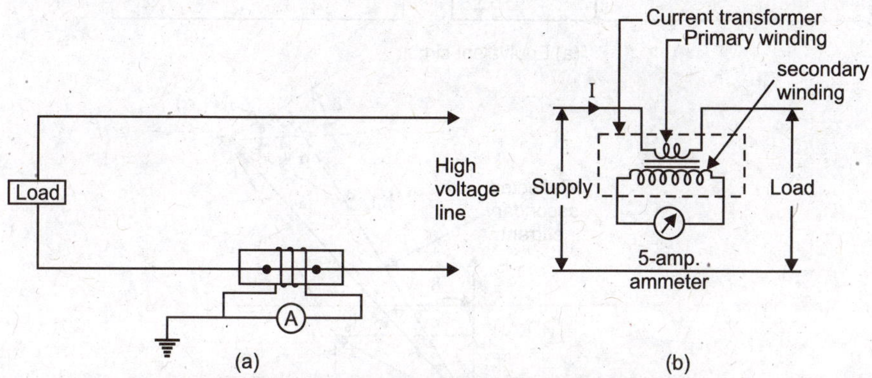


**Fig. 6.3.** Current transformer connections.

— *Core material* for CT is *usually low-loss silicon steel, mumetal.*

— Core should have *high permeability* and *low loss.*

— To keep losses low, flux density is kept low (≈ 0.1 tesla).

— Core lamination can be *rectangular or ring shaped.* Ring type core is jointless but difficult to wind.

— Windings must be rigid to withstand extremely large mechanical forces produced by short circuit currents.

— For low and medium voltages, tape and varnish insulations are used while for high voltage working CT may be compound filled or oil immersed.

- The *current transformer ratio is not equal to the ratio of secondary to primary turns, mainly because of the effect of the magnetising current.* The primary current can be thought of as the sum of two currents, the first to balance secondary current so that primary and secondary m.m.fs may balance and the second being the no-load current $I_0$. The component $I_0$ besides being responsible for a slight error in the current ratio, is also responsible for a phase angle error. The transformer *must be carefully designed to minimise the ratio and phase angle error.*

☞ - It may be noted that *current transformer* ***must never be operated on open-circuit*** *for the* following two *reasons* :

(*i*) There will be no secondary m.m.f. and since the primary current (and m.m.f.) is fixed, the core flux will increase enormously. This will cause large eddy current and hysteresis losses and the *resulting high temperature may damage the insulation or even the core.*

(*ii*) A very high voltage will be induced in the multi-turn secondary and this high voltage may be dangerous both to life and to the insulation.

### 6.3.2. Relationships in a Current Transformer

Fig. 6.4 (*a*), (*b*) shows the equivalent circuit and the phasor diagram of a current transformer (CT). The diagrams are same as for any other transformer. Here,

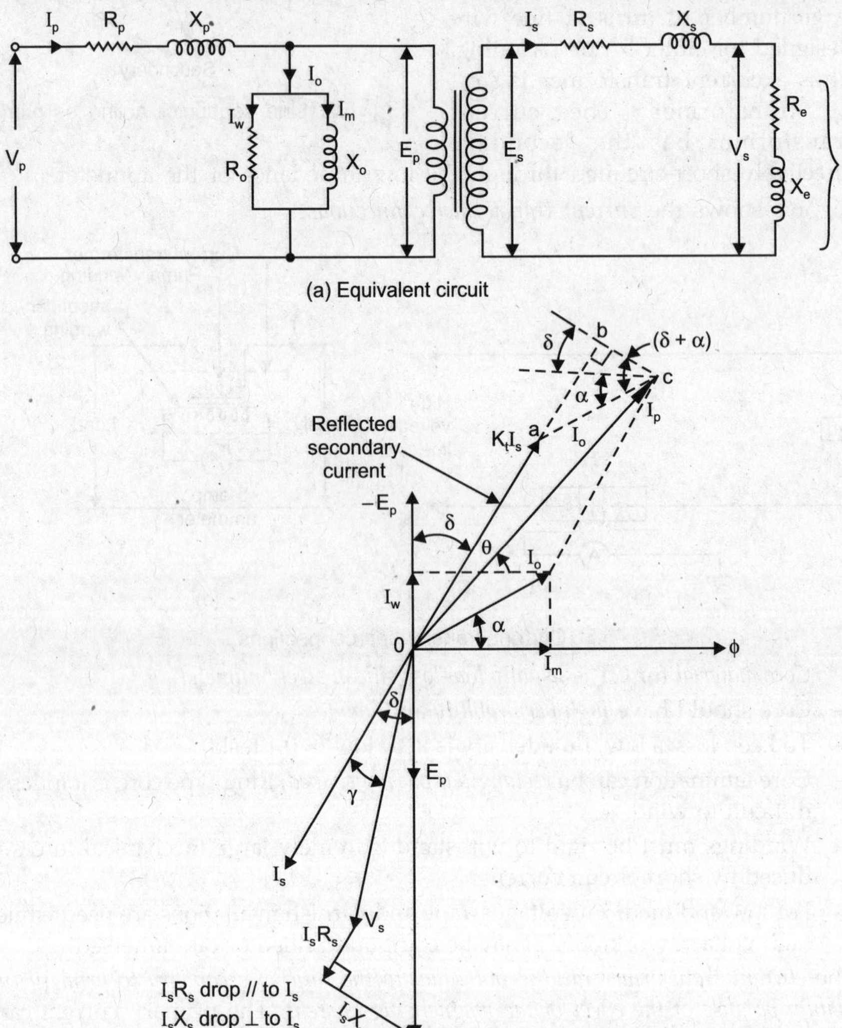


(a) Equivalent circuit

(b) Phase diagram (Not to scale)

**Fig. 6.4.** Current transformer (CT).

$\phi$ = Working flux,

$$K_t = \text{Turns ratio} = \frac{\text{Number of secondary winding turns}}{\text{Number of primary winding turns}},$$

$R_p$ = Resistance of primary winding,
$X_p$ = Reactance of primary winding,
$R_s$ = Resistance of secondary winding,
$X_s$ = Reactance of secondary winding,
$R_e$ = Resistance of external burden *i.e.*, load on secondary,
$X_e$ = Reactance of external burden *i.e.*, load on secondary,
$E_p$ = Primary induced voltage,
$E_s$ = Secondary induced voltage,
$V_p$ = Primary supply voltage,
$V_s$ = Secondary terminal voltage,
$I_0$ = Exciting current,
$I_p$ = Primary winding current,
$I_s$ = Secondary winding current,
$I_m$ = Magnetising component of $I_0$
$I_w$ = Loss component of $I_0$.
$\alpha$ = Angle between exciting current $I_0$ and working flux $\phi$,
$\theta$ = Phase angle of the transformer,
$\delta$ = Angle between secondary winding induced voltage ($E_s$) and secondary winding current ($I_s$)
= Phase angle of total impedance of secondary burden.

$$= \tan^{-1}\left(\frac{X_s + X_e}{R_s + R_e}\right), \text{ and}$$

$\gamma$ = Phase angle of load or burden *i.e.*, $R_e + jX_e$

$$= \tan^{-1}\frac{X_e}{R_e}$$

**Actual transformer ratio (current) ($K_{act.}$) :**

Consider the phasor diagram of CT with a lagging power factor, as shown in Fig. 6.4 (*b*).

Consider the right-angled triangle *obc* :

$$(oc)^2 = (ob)^2 + (bc)^2$$
$$= (oa + ab)^2 + (bc)^2$$

or,
$$I_p^2 = \left[K_t I_s + I_o \sin(\delta+\alpha)\right]^2 + \left[I_o \cos(\delta+\alpha)\right]^2$$
$$= K_t^2 I_s^2 + 2K_t I_s I_o \sin(\delta+\alpha) + I_o^2 \sin^2(\delta+\alpha) + I_o^2\cos^2(\delta+\alpha)$$
$$= K_t^2 I_s^2 + 2K_t I_s I_o \sin(\delta+\alpha) + I_o^2[\sin^2(\delta+\alpha) + \cos^2(\delta+\alpha)]$$
$$= K_t^2 I_s^2 + 2K_t I_s I_o \sin(\delta+\alpha) + I_o^2$$

or,
$$I_p = \sqrt{K_t^2 I_s^2 + 2K_t I_s I_o \sin(\delta+\alpha) + I_o^2} \qquad ...(6.11)$$

$\therefore$ Actual transformation (current) ratio,

$$K_{act.} = \frac{I_p}{I_s} = \frac{\sqrt{K_t^2 I_s^2 + 2K_t I_s I_o \sin(\delta+\alpha) + I_o^2}}{I_s} \qquad ...(6.12)$$

In a well designed transformer $I_o << K_t I_s$. Usually $I_o$ is less than 1% of $I_p$ and $I_p$ is, therefore, very nearly equal to $K_t I_s$.

∴ Eqn. (6.12) can be written as :

$$K_{act.} = \frac{\sqrt{K_t^2 I_s^2 + 2K_t I_s I_o \sin(\delta+\alpha) + I_o^2 \sin^2(\delta+\alpha)}}{I_s}$$

... Adjusting $I_o^2$ as $I_o^2 \sin^2(\delta+\alpha)$

or, $$K_{act.} = \frac{K_t I_s + I_o \sin(\delta+\alpha)}{I_s} = K_t + \frac{I_o}{I_s}\sin(\delta+\alpha) \quad ...(6.13)$$

Although (6.13) is approximate, yet it is *sufficiently accurate for all practical purposes.* The above *theory is applicable when the burden is "inductive" which is normally the case.* Eqn. (6.13) can be further expanded as,

$$K_{act.} = K_t + \frac{I_o}{I_s}(\sin\delta\cos\alpha + \cos\delta\sin\alpha)$$

$$= K_t + \frac{I_o \cos\alpha \,.\, \sin\delta + I_o \sin\alpha \,.\cos\delta}{I_s}$$

or, $$K_{act.} = K_t + \frac{I_m \sin\delta + I_w \cos\delta}{I_s} \quad ...(6.14)$$

$(\because I_o \cos\alpha = I_m; I_o \sin\alpha = I_w)$

Note that δ is *positive* for *lagging* p.f. load while *negative* for *leading* p.f. load.

**Phase angle (θ) :**

*The angle by which the secondary current phasor, when reversed, differs in phase from the primary current, is known as the* ***phase angle*** *of the transformer.*

This angle is taken to be +ve if the secondary current reversed leads the primary current. The angle is taken as –ve if the secondary current reversed lags behind the primary current.

The angle between $I_s$ reversed and $I_p$ is θ. Therefore, the phase angle is θ.

From the phasor diagram [Fig. 6.4 (*b*)], we have :

$$\tan\theta = \frac{bc}{ob} = \frac{bc}{oa+ab} = \frac{I_o\cos(\delta+\alpha)}{K_t I_s + I_o \sin(\delta+\alpha)}$$

As θ is very small, we can write,

$$\theta = \frac{I_o\cos(\delta+\alpha)}{K_t I_s + I_o \sin(\delta+\alpha)} \text{ rad.} \quad ...(6.15)$$

Since $I_o << K_t I_s$, therefore, we can neglect the term $I_o \sin(\delta+\alpha)$

$$\theta \simeq \frac{I_o\cos(\delta+\alpha)}{K_t I_s} \text{ rad.} \quad ...(6.16)$$

$$\simeq \frac{I_o\cos\delta\cos\alpha - I_o\sin\delta\sin\alpha}{K_t I_s}$$

$$\simeq \frac{I_m\cos\delta - I_w\sin\delta}{K_t I_s} \text{ rad.} \quad ...(6.17)$$

$$\simeq \frac{180}{\pi}\left(\frac{I_m\cos\delta - I_w\sin\delta}{K_t I_s}\right) \text{ degrees} \quad ...(6.18)$$

### 6.3.3. Errors in Current Transformer

It is evident from the eqn. 6.14 that the value of transformation ratio (actual ratio) is *not* equal to the turns ratio ($K_t$). Also the value is not constant and depends upon magnetising and loss components of the exciting current, the secondary winding load current and its power factor. This means the secondary winding current is not a constant fraction of the primary winding current but depends upon the above mentioned factors. This introduces *considerable errors* into current measurements.

In power measurements, due to the use of CT transformers, the following two types of errors are introduced :

1. *Ratio error* : Due to actual transformation ratio being *different* from the turns ratio ($K_t$).
2. *Phase angle error* : Due to secondary winding current *not* being exactly 180° out of phase with the primary winding current (difference being angle θ).

1. **Ratio error.** It is defined as :

$$\text{Percentage ratio error} = \frac{\text{Nominal ratio} - \text{Actual ratio}}{\text{Actual ratio}} \times 100$$

$$= \frac{K_{nom.} - K_{act.}}{K_{act.}} \times 100 \quad \text{...(6.19)}$$

2. **Phase angle error.** This error is denoted by angle θ, by which the phase difference between primary and secondary is different from 180°. It is given by :

$$\theta = \frac{180}{\pi}\left[\frac{I_m \cos\delta - I_w \sin\delta}{K_t I_s}\right] \text{degrees}$$

- ***Approximate formulae for errors :***

Usually, the *instrument burden* is largely resistive with some inductance and, therefore, *δ is +ve and is generally small.*

Hence, sin δ ≃ 0 and cos δ ≃ 1

Thus, from eqns. (6.14) and (6.18), we have :

$$K_{act.} = K_t + \frac{I_w}{I_s} \quad \text{...(6.20)}$$

and,

$$\theta = \left(\frac{180}{\pi}\right)\left(\frac{I_m}{K_t I_s}\right) \quad \text{...(6.21)}$$

But $I_p \simeq K_t I_s$, and, therefore, eqns. (6.20) and (6.21) can be rewritten as :

$$K_{act.} \simeq K_t + \frac{K_t I_w}{I_p} \simeq K_t\left(1 + \frac{I_w}{I_p}\right) \quad \text{...(6.22)}$$

and;

$$\theta = \left(\frac{180}{\pi}\right)\left(\frac{I_m}{I_p}\right) \quad \text{...(6.23)}$$

### 6.3.4. Characteristics of Current Transformers

The performance of current transformers under different operating conditions may be estimated by considering the phasor diagram. Brief statements of the performance variation with load conditions are as follows :

1. **Effect of change in power factor of secondary burden :**
The power factor of the secondary circuit depends on the power factor of the burden on secondary, this directly affects the two errors of the transformer as under :

- *Ratio error* : As δ is positive for all *inductive* loads, therefore, sin (δ + ∝) is positive and hence actual ratio $K_{act.}$ is always *greater* than the turns ratio $K_t$ [refer to eqn. (6.13)]. δ is negative for *capacitive* burdens and, consequently $K_{act.}$ is *less* than $K_t$.
- *Phase angle error* : When load is *inductive* and δ is small positive then θ is positive. But as δ approaches 90° as load becomes highly inductive then θ becomes negative. In case of *capacitive* load, δ is negative and θ is always positive.

2. **Effect of change in burden on secondary :**
— An increase in secondary winding circuit burden impedance means an increase in volt ampere rating. This necessitates an increase in the secondary winding induced voltage which can be generated by an increased flux and flux density. Therefore, both magnetising component $I_m$ and loss component $I_w$ are increased; subsequently *errors also increase.*
— A great burden impedance, in general, not only increases the transformation ratio, but also *shifts* the phase angle between primary winding current and secondary winding current reversed to *more positive values.*

3. **Effect of change in primary winding current ($I_p$) :**
— When the primary current $I_p$ changes, the secondary current $I_s$ changes proportionately.
— At *low values* of $I_p$ (or $I_s$), the exciting current component $I_m$ and $I_w$ are a greater proportion of $I_p$ and, therefore, the errors are *greater*.
— When $I_p$ increases, $I_s$ also increases and there is a *decrease in ratio error and phase angle* [refer to eqns. (6.22) and (6.23)].

4. **Effect of change in frequency :** The effect of increase in frequency is similar to that produced by a decrease in secondary burden, as increase in frequency will produce a corresponding *decrease in flux for the given e.m.f.* Consequently, there is reduction in $I_m$ and $I_w$ and hence *errors also get reduced.*

### 6.3.5. Clip-on Type Current Transformer

The main feature of this type of transformer is that *it can be used instantaneously for measurement of current in a current carrying conductor* (cable, insulated busbar or fuse link) *without interrupting the circuit for connection of ammeter in series with the circuit.*

Fig. 6.5, shows the schematic diagram of a clip-on type current transformer.

— It consists of a laminated core which is so arranged that it can be opened out at a hinged section by merely pressing a trigger-like projection. When the core is thus opened, it permits the admission of very heavy current-carrying busbars or feeders where upon the trigger is released and the core is tightly closed by a spring.

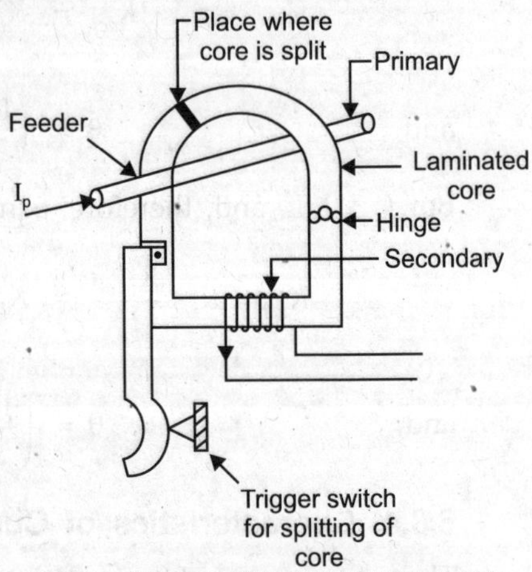


**Fig. 6.5.** Schematic diagram of a clip-on type current transformer.

— The current-carrying conductor of feeder acts as a single-turn primary whereas the secondary is connected across the standard ammeter conveniently mounted in the handle itself; the ammeter gives the value of current flowing through the conductor.

- The instrument can be used in A.C. circuits only.
- If can be operated by one hand.
- The accuracy claimed by the manufacturers is within 3% of full-scale with the conductor in a central position in the core.
- There transformers are designed for a voltage of 5 kV between the primary and secondary windings.

**Example 6.1.** *The following data relate to a current transformer :*

| | | |
|---|---|---|
| *Number of primary turns* | = | *single turn* |
| *Number of secondary turns* | = | *240* |
| *Secondary winding current* | = | *5 A* |
| *External burden* | = | *1.2 Ω (non-inductive)* |
| *Magnetomotive force (mmf) for setting up flux in the core* | = | *96 AT* |
| *Cross-section of the core* | = | *1200 mm²* |
| *Supply frequency* | = | *50 Hz* |

*Neglecting the effects of magnetic leakage, iron losses and $I^2R$ losses, calculate :*

*(i) Actual transformation ratio.*

*(ii) Phase angle.*

*(iii) Maximum flux density in the core.*

**Solution.** *Given* : $N_p = 1$; $N_s = 240$; $I_s = 5$ A; $R_e = 1.2\ \Omega$; mmf = 96 AT; $A_c = 1200\ \text{mm}^2$; $f = 50$ Hz.

***(i)* Actual transformation ratio ($K_{act.}$) :**

$$\text{Turns ratio, } K_t = \frac{N_s}{N_p} = \frac{240}{1} = 240$$

Voltage induced in the secondary winding,

$E_s$ = Current in the secondary winding ($I_s$) × impedance of secondary winding ($R_e$)

$= 5 \times 1.2 = 6.0$ V

As the secondary burden is purely resistive the $I_s$ is in phase with $E_s$, and the secondary power factor is unity or $\delta = 0$ (Refer to Figs. 6.5 and 6.4).

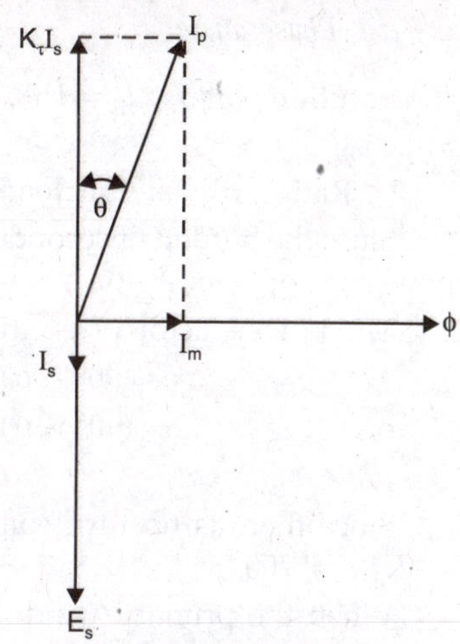


**Fig. 6.6.** Phasor diagram.

The loss component ($I_w$) of no-load current ($I_o$) is to be neglected and, therefore, $I_w = 0$

$$\therefore \quad I_o = \sqrt{I_m^2 + I_w^2} = I_m = \text{magnetising component}$$

and, $\alpha = 0$

$$\text{Also,} \quad I_m = \frac{mmf}{N_p} = \frac{96}{1} = 96\,A$$

Now, reflected secondary winding current = $K_tI_s$ = 240 × 5 = 1200 A

Referring to phasor diagram (Fig. 6.6), we have :

$$\text{Primary current, } I_p = \sqrt{(K_t I_s)^2 + (I_m)^2}$$

$$= \sqrt{(1200)^2 + (96)^2} = 1203.8\,A$$

$\therefore$ *Actual transformation ratio,* $K_{act.} = \frac{I_p}{I_s} = \frac{1203.8}{5} = \mathbf{240.76}$ **(Ans.)**

(*ii*) **Phase angle, θ :**

From Fig. 6.6, we get :

$$\tan\theta = \frac{I_m}{K_T I_s} = \frac{96}{1200} = 0.08$$

$$\therefore \quad \theta = \tan^{-1}(0.08) = \mathbf{4° \; 34'} \quad \textbf{(Ans.)}$$

(*iii*) **Maximum flux density in the core, $B_{max}$ :**

We know that : $E_s = 4.44\, f\, \phi_{max}\, N_s$

$$\therefore \quad \phi_{max} = \frac{E_s}{4.44\, fN_s} = \frac{6.0}{4.44 \times 50 \times 240} = 0.1126 \times 10^{-3}\,Wb$$

Hence, $$B_{max} = \frac{\phi_{max}}{A_c} = \frac{0.1126 \times 10^{-3}}{1200 \times 10^{-6}} = \mathbf{0.0938\ Wb/m^2} \quad \textbf{(Ans.)}$$

**Example 6.2.** *A ring core current transformer of ratio 1000/5 A is operating at full primary current with a secondary burden of non-inductive resistance of 1.1 Ω. Its exciting current is 1 A at a power factor of 0.45. Calculate :*

*(i) The ratio error at full load, assuming that there has been no compensation.*

*(ii) Phase angle.*

**Solution.** *Given* : $I_o' = 1$ A; Nominal ratio, $K_{nom.} = \frac{1000}{5} = 200; R_e = 1.1\Omega$, Power factor = 0.45.

(*i*) **Ratio error at full load :**

Since the burden of secondary winding is purely resistive therefore, secondary winding power factor is unity or δ = 0.

The power factor of exciting current is 0.45,

$$\therefore \quad \cos(90° - \alpha) = 0.45$$

$$\text{or} \quad (90° - \alpha) = \cos^{-1}(0.45) = 63.26°$$

$$\text{or} \quad \alpha = 90° - 63.26° = 26.74°$$

Since there is no turn compensation, the turns ratio is equal to nominal ratio or $K_t = K_{nom.} = 200$.

When the primary winding carries rated current of 1000 A, the secondary winding carries a current of 5 A.

Rated secondary winding current, $I_s = 5$ A

$$\therefore \quad K_t I_s = 200 \times 5 = 1000 \text{ A.}$$

Actual transformation ratio,

$$K_{act.} = K_t + \frac{I_o}{I_s}\sin(\delta + \alpha) \qquad \text{...[Eqn. (6.13)]}$$

$$= 200 + \frac{1}{5}\sin(0 + 26.74°) = 200.09$$

$$\therefore \quad \text{Ratio error} = \frac{\text{Nominal ratio} - \text{Actual ratio}}{\text{Actual ratio}} \times 100 = \frac{K_n - K_T}{R} \times 100$$

$$= \frac{200 - 200.09}{200.09} \times 100 = -\mathbf{0.045\%} \quad \textbf{(Ans.)}$$

(*ii*) **Phase angle, θ :** Refer to Fig. 6.6.

$$\theta = \frac{180}{\pi} \frac{I_o \cos(\delta + \alpha)}{K_t I_s} \text{ degrees} \qquad \text{...[Eqn. (6.16)]}$$

$$= \frac{180}{\pi} \frac{1 \times \cos(0 + 26.74°)}{200 \times 5} = \mathbf{3'4''} \quad \textbf{(Ans.)}$$

**Example 6.3.** *The primary winding of a 1000/5A, 50 Hz current transformer has a single turn. Its secondary burden consists of a non-inductive impedance of 1.4 Ω. If the iron loss in the core is 1.4 W at full load and magnetising mmf is 80 AT, calculate :*

*(i) Flux in the core.*

*(ii) Ratio error at full load.*

*Neglect leakage reactance.*

**Solution.** *Given* : Nominal ratio, $K_{nom.} = \frac{1000}{5} = 200$; $N_p = 1$; Secondary impedance = 1.4Ω; mmf = 80 AT; Iron loss in the core = 1.4 W.

(*i*) **Flux in the core, $\phi_{max}$ :**

Let us *assume* turns ratio to be equal to nominal, in the absence of any other data.

Then, Turns ratio, $K_t = K_{nom.} = 200$.

$$\therefore \quad N_s = K_t N_p = 200 \times 1 = 200 \qquad (\because N_p = 1)$$

Now, secondary induced voltage, $E_s = 5 \times 1.4 = 7$ V

Also, $E_s = 4.44 f \phi_{max} N_s$

$$\therefore \quad \phi_{max} = \frac{E_s}{4.44 \times f \times N_s} = \frac{7}{4.44 \times 50 \times 200} = \mathbf{0.1576 \times 10^{-3}\ Wb} \ \textbf{(Ans.)}$$

(*ii*) **Ratio error at full load :**

$$\text{Primary induced voltage, } E_p = \frac{E_s}{K_t} = \frac{7}{200} \text{ V}$$

$$\text{Loss component of exciting current, } I_w = \frac{\text{Iron loss}}{E_p} = \frac{1.4}{(7/200)} = 40 \text{ A}$$

$$\text{Magnetising component, } I_m = \frac{mmf}{N_p} = \frac{80}{1} = 80 \text{ A}$$

As the burden is purely resistive, therefore, the secondary load angle is zero or δ = 0.

We know that,

$$\text{Actual ratio, } K_{act.} = K_t + \frac{I_m \sin\delta + I_w \cos\delta}{I_s} \qquad \text{...[Eqn. (6.14)]}$$

$$= 200 + \frac{40}{5} = 208 \text{ V}$$

$$\therefore \quad \text{Ratio error} = \frac{K_{nom.} - K_{act.}}{K_{act.}} \times 100$$

$$= \frac{200 - 208}{208} \times 100 = -\mathbf{3.846\%} \quad \textbf{(Ans.)}$$

**Example 6.4.** *A current transformer with a bar primary has 250 turns in its secondary winding. The resistance and reactance of the secondary circuit are 1.4 Ω and 1.1 Ω respectively including the transformer winding. When 5A currents flows in the secondary winding, the magnetising mmf is 80 AT and the iron loss is 1.1 W. Determine the following :*

*(i) Ratio error.*

*(ii) Phase angle error.*

**Solution.** *Given* : $N_p = 1$; $N_s = 250$; $R_p = 1.4\ \Omega$; $X_s = 1.1\ \Omega$; $I_s = 5A$; Magnetising mmf = 80 AT; Iron loss = 1.1 W.

(*i*) **Ratio error :**

$$\text{Turns ratio, } K_t = \frac{N_s}{N_p} = \frac{250}{1} = 250$$

$$\text{Secondary circuit impedance} = \sqrt{(1.4)^2 + (1.1)^2} = 1.78\,\Omega$$

For secondary winding circuit :

$$\cos\delta = \frac{1.4}{1.78} = 0.7865; \ \sin\delta = \frac{1.1}{1.78} = 0.618$$

$$\text{Secondary induced voltage, } E_s = 5 \times 1.78 = 8.9 \text{ V}$$

$$\text{Primary induced voltage, } E_p = \frac{E_s}{250} = \frac{8.9}{250} = 0.0356\,\text{V}$$

Loss of component referred to primary winding,

$$I_w = \frac{\text{Iron loss}}{E_p} = \frac{1.1}{0.0356} = 30.9\,\text{A}$$

$$\text{Magnetising current, } I_m = \frac{mmf}{N_p} = \frac{80}{1} = 80\text{A}$$

$$\text{Actual transformation ratio, } K_{act.} = K_t + \frac{I_m \sin\delta + I_w \cos\delta}{I_s} \quad \text{...[Eqn. (6.14)]}$$

$$= 250 + \frac{80 \times 0.618 + 30.9 \times 0.7865}{5} = 264.75$$

Assuming, $K_{nom.} = K_t = 250$, since no other contrary information is given, we have :

$$\text{Ratio error} = \frac{K_{nom.} - K_{act.}}{K_{act.}} \times 100$$

$$= \frac{250 - 264.75}{264.75} \times 100 = -\mathbf{5.57\%} \quad \textbf{(Ans.)}$$

(*ii*) **Phase angle error, θ :**

We know that,

$$\theta = \frac{180}{\pi}\left[\frac{I_m \cos\delta - I_w \sin\delta}{K_t I_s}\right] \text{degrees} \quad \text{...[Eqn. (6.18)]}$$

$$= \frac{180}{\pi}\left[\frac{80 \times 0.7865 - 30.9 \times 0.618}{250 \times 5}\right] = \mathbf{2.01°} \quad \textbf{(Ans.)}$$

**Example 6.5.** *The secondary winding burden of current transformer, having a bar primary and 300 secondary turns, is an ammeter of resistance 1.0 Ω and reactance 0.55 Ω. The resistance and reactance of secondary winding are 0.3 Ω and 0.25 Ω respectively. The core requires the equivalent of an mmf of 90 AT for magnetisation and 45 AT for core losses. Determine the following :*

*(i) The primary winding current and actual transformation ratio when the ammeter in the secondary winding circuit indicates 5 A.*

*(ii) The number of turns which could be reduced in the secondary winding so that the ratio error be zero for this condition.*

**Solution.** *Given* : $N_p = 1$; $N_s = 300$; $R_e = 1.0\ \Omega$; $X_e = 0.55\ \Omega$; $R_s = 0.3\ \Omega$; $X_s = 0.25\ \Omega$; mmf for magnetisation = 90 AT; mmf for core loss = 45 AT; $I_s = 5$ A.

*(i)* $\boldsymbol{I_p\ ;\ K_{act.}}$ **:**

Total secondary circuit resistance $= R_s + R_e = 0.3 + 1.0 = 1.3\ \Omega$

Total secondary circuit reactance $= X_s + X_e = 0.25 + 0.55 = 0.8\ \Omega$

$$\text{Secondary circuit phase angle, } \delta = \tan^{-1}\left(\frac{0.8}{1.3}\right) = 31.6°$$

$$\cos\delta = \cos 31.6° = 0.8517, \text{ and}$$

or,
$$\sin\delta = \sin 31.6° = 0.524$$

$$\text{Turn ratio, } K_t = \frac{N_s}{N_p} = \frac{300}{1} = 300$$

$$\text{Magnetising current, } I_m = \frac{\text{Magnetising } mmf}{N_p} = \frac{90}{1} = 90\text{ A}$$

$$\text{Loss component, } I_w = \frac{mmf \text{ equivalent to iron loss}}{N_p} = \frac{45}{1} = 45\text{A}$$

$$\therefore \quad \textit{Actual ratio, } K_{act} = K_t + \frac{I_m \sin\delta + I_w \cos\delta}{I_s} \quad \text{...[Eqn. (6.14)]}$$

$$= 300 + \frac{90 \times 0.524 + 45 \times 0.8517}{5} = \mathbf{317.1} \quad \textbf{(Ans.)}$$

$$\text{Primary current, } I_p = K_{act.} \times I_s = 317.1 \times 5 = \mathbf{1585.5\ A} \quad \textbf{(Ans.)}$$

*(ii)* **Number of turns to be reduced :**

For eliminating the ratio error, we must reduce the secondary winding turns or in other words we must *reduce the turn ratio.*

*For zero ratio errors :*

$$K_{nom.} = K_{act.} \quad (\text{where, } K_{nom.} = K_t = 300)$$

or,
$$300 = K_t' + \frac{I_m \sin\delta + I_w \cos\delta}{I_s}$$

or,
$$300 = K_t' + \frac{90 \times 0.524 + 45 \times 0.8517}{5} = K_t' + 17.1$$

or
$$K_t' = 282.9$$

Hence secondary winding turns, $N_s = K_t' \times N_p = 282.9 \times 1 = 282.9$.

$\therefore$ *Reduction in secondary winding turns* = 300 – 282.9 $\simeq$ **17** **(Ans.)**

**Example 6.6.** *A current transformer has a bar primary, 100 secondary winding turns, and 100/1 nominal ratio. It is operating on an external burden of 1.45 Ω non-inductive, the secondary winding resistance being 0.25 Ω. The primary winding exciting current is 1.8 A, lagging 38.4° to the secondary voltage reversed. With 1A current flowing in the secondary winding, calculate :*

*(i) The actual ratio of primary winding current to secondary winding current;*

*(ii) The phase angle between these currents.*

**Solution.** *Given* : $N_p = 1$; $N_s = 100$; $K_{nom.} = 100/1$; $R_e = 1.45\ \Omega$; $R_s = 0.25\Omega$; $I_o = 1.8\text{A}$; Lagging angle, with secondary voltage reversed = 38.4°; $I_s = 1\text{A}$.

**(*i*) The actual ratio, $K_{act.}$ :**

$$\text{Turn ratio, } K_t = \frac{N_s}{N_p} = \frac{100}{1} = 100$$

Total secondary circuit resistance = 0.25 + 1.45 = 1.7 Ω

Secondary circuit reactance = 0

$\therefore$ The secondary p.f. is unity and its phase angle, $\delta = 0$

Now, $\alpha$ = Angle between $I_o$ and $\phi$

= 90° – (angle between $I_o$ and $E_s$ reversed)

= 90° – 38.4° = 51.6°

Also,
$$K_{act.} = K_t + \frac{I_o}{I_s}\sin(\delta+\alpha) \quad \text{...[Eqn. (6.13)]}$$

$$= 100 + \frac{1.8}{1.0}\sin(0+51.6°) = \mathbf{101.41} \quad \textbf{(Ans.)}$$

**(*ii*) Phase angle, θ :**

$$\theta = \frac{180}{\pi} \times \frac{I_o\cos(\delta+\alpha)}{K_t I_s}\text{ degrees} \quad \text{...[Eqn. (6.16)]}$$

$$= \frac{180}{\pi} \times \frac{1.8\cos(0+51.6°)}{100\times 1} = \mathbf{0.641°} \quad \textbf{(Ans.)}$$

**Example 6.7.** *A current transformer rated as 1000/5A, 25 VA has a turn ratio of 1 : 200. The core loss and magnetising components of the primary current are 5A and 8A under rated conditions. Determine the ratio and phase angle errors for the rated burden and rated secondary current at 0.8 p.f. lagging and 0.8 p.f. leading.*

*Neglect the resistance and leakage reactance of secondary winding.*

**Solution.** *Given* : $N_p = 1$, $N_s = 200$ *i.e.*, $K_t = \frac{200}{1} = 200$; Magnetising current, $I_m = 8\text{A}$; Loss component, $I_w = 5\text{A}$; $\cos\phi = 0.8$ lagging and leading.

**Case-1 :** *0.8 p.f. lagging* :

For lagging p.f. the secondary phase angle δ is + ve.

$$\cos\delta = \text{secondary p.f.} = 0.8, \quad \sin\delta = \sqrt{1^2 - (0.8)^2} = 0.6$$

$$\text{Actual ratio, } K_{act.} = K_t + \frac{I_m\sin\delta + I_w\cos\delta}{I_s} \quad \text{...[Eqn. (6.14)]}$$

$$= 200 + \frac{8 \times 0.6 + 5 \times 0.8}{5} = 201.76$$

$$\therefore \quad \textit{Ratio error} = \frac{K_{nom.} - K_{act.}}{K_{act.}} \times 100$$

$$= \frac{200 - 201.76}{201.76} = \mathbf{-0.872\%} \quad \textbf{(Ans.)} \quad (\because K_{nom.} = K_t = 200)$$

$$\textit{Phase angle, } \theta = \frac{180}{\pi}\left[\frac{I_m \cos\delta - I_w \sin\delta}{K_t I_s}\right] \text{degrees} \qquad \text{...[Eqn. (6.18)]}$$

$$= \frac{180}{\pi}\left[\frac{8 \times 0.8 - 5 \times 0.6}{200 \times 5}\right] = \mathbf{0.1948^\circ} \quad \textbf{(Ans.)}$$

**Case-2 :** *0.8 p.f. leading* :
For leading p.f. δ is –ve. cos δ = 0.8 and sin δ = – 0.6.

$$\text{Now,} \quad \text{Actual ratio, } K_{act.} = K_t + \frac{I_m \sin\delta + I_w \cos\delta}{I_s}$$

$$= 200 + \frac{8 \times (-0.6) + 5 \times 0.8}{5} = 199.84$$

$$\therefore \quad \textit{Ratio error} = \frac{K_{nom.} - K_{act.}}{K_{act.}} \times 100$$

$$= \frac{200 - 199.84}{199.84} \times 100 = \mathbf{+0.08\%} \quad \textbf{(Ans.)}$$

$$\textit{Phase angle, } \theta = \frac{180}{\pi}\left[\frac{I_m \cos\delta - I_w \sin\delta}{K_t I_s}\right] \text{degrees}$$

$$= \frac{180}{\pi}\left[\frac{8 \times 0.8 - 5(-0.6)}{200 \times 5}\right] = \mathbf{0.538^\circ} \quad \textbf{(Ans.)}$$

**Example 6.8.** *A 500/5A, 50 Hz current transformer has a one turn primary and an output of 20 VA. The number of secondary turns is 99, and the resistance and leakage reactance of the secondary winding are 0.4 Ω and 0.35 Ω respectively. At rated load with non-inductance burden, the inphase and quadrature components (referred to the flux) of the exciting mmf are 6AT and 8AT respectively.*

*Calculate the current and phase angle errors.*

**Solution.** *Given* : $N_p = 1$; $N_s = 99$; $\therefore$ $K_t = 99$; $K_{nom.} = \frac{500}{5} = 100$; $R_s = 0.4\ \Omega$; $X_s = 0.35\ \Omega$; Magnetising mmf = 6AT; Loss mmf = 8AT.

**(*i*) Ratio (current) error :**

$$\text{Magnetising current, } I_m = \frac{\text{Magnetising mmf}}{\text{Primary winding turns}} = \frac{6}{1} = 6\text{A}$$

$$\text{Loss component, } I_w = \frac{\text{Loss mmf}}{\text{Primary winding turns}} = \frac{8}{1} = 8\text{A}$$

Resistance of external burden (purely resistive – given), $R_e$

$$= \frac{VA}{I_s^2} = \frac{20}{5^2} = 0.8\,\Omega$$

Reactance of external burden, $X_e = 0\ \Omega$

Resistance of total secondary circuit burden

$$= R_s + R_e = 0.4 + 0.8 = 1.2\ \Omega$$

Reactance of total secondary circuit burden

$$= X_s + X_e = 0.35 + 0 = 0.35\ \Omega$$

$$\text{Secondary phase angle, } \delta = \tan^{-1}\frac{0.35}{1.2} = 16.26°$$

$\therefore$ $\cos\delta = \cos 16.26° = 0.96$ and $\sin\delta = \sin 16.26° = 0.28$

$$\text{Actual ratio, } K_{act.} = K_t + \frac{I_m \sin\delta + I_w \cos\delta}{I_s}$$

$$= 99 + \frac{6\times0.28 + 8\times0.96}{5} = 100.87$$

$$\textit{Ratio (current) error} = \frac{K_{nom.} - K_{act.}}{K_{act.}} \times 100$$

$$= \frac{100 - 100.87}{100.87} \times 100 = -\mathbf{0.86\%} \quad \textbf{(Ans.)}$$

(*ii*) **Phase angle error, θ :**

$$\theta = \frac{180}{\pi}\left[\frac{I_m \cos\delta - I_w \sin\delta}{K_t I_s}\right] \text{degrees}$$

$$= \frac{180}{\pi}\left[\frac{6\times0.96 - 8\times0.28}{99\times5}\right] = \mathbf{0.407°} \quad \textbf{(Ans.)}$$

**Example 6.9.** *A 100/5A current transformer, at its rated load of 20VA, has an iron loss of 0.18 W and a magnetising current of 1.4 A. It is supplying rated output to a meter having a ratio of resistance to reactance of 4. Calculate :*

*(i) Ratio error.*

*(ii) Phase angle.*

**Solution.** *Given* : $K_{nom.} = \frac{100}{5} = 20$; Burden of secondary circuit = 20 VA; Iron loss = 0.18 W; $I_m = 1.4$ A; $\frac{X}{R} = 4$.

(*i*) **Ratio error :**

Assuming $K_{nom.} = K_t = 20$ ... Any other data being absent.

Neglecting the secondary winding burden,

the total burden of the secondary circuit = Burden of the meter.

$\therefore$ Burden of secondary circuit = 20 VA

$$\text{Voltage across primary winding, } E_p = \frac{VA}{I_p} = \frac{20}{100} = 0.2V$$

$$\text{Loss component of exciting current, } I_w = \frac{\text{Iron loss}}{E_p} = \frac{0.18}{0.2} = 0.9\,\text{A}$$

$$\text{Secondary circuit phase angle, } \delta = \tan^{-1}\left(\frac{X}{R}\right)$$

or,
$$\delta = \tan^{-1}\left(\frac{1}{4}\right) = 14.04°$$

$\therefore$ $\cos\delta = 0.97$ and $\sin\delta = 0.2426$

$$\text{Actual ratio, } K_{act.} = K_t + \frac{I_m \sin\delta + I_w \cos\delta}{I_s}$$

$$= 20 + \frac{1.4 \times 0.2426 + 0.9 \times 0.97}{5} = 20.24$$

$$\therefore \quad \textit{Ratio error} = \frac{K_{nom.} - K_{act.}}{K_{act.}} \times 100$$

$$= \frac{20 - 20.24}{20.24} \times 100 = -\mathbf{1.186\%} \quad \textbf{(Ans.)}$$

(*ii*) **Phase angle, θ :**

$$\theta = \frac{180}{\pi}\left[\frac{I_m \cos\delta - I_w \sin\delta}{K_t I_s}\right] \text{degree}$$

or,
$$\theta = \frac{180}{\pi}\left[\frac{1.4 \times 0.97 - 0.9 \times 0.2426}{20 \times 5}\right] = \mathbf{0.653°} \quad \textbf{(Ans.)}$$

## 6.4 POTENTIAL TRANSFORMER (PT)

### 6.4.1. General Aspects

A **potential transformer** *is a step down transformer used along with a low range voltmeter for measuring a high voltage.*

- The primary is connected across the high voltage supply and the secondary to the voltmeter or potential coil of the wattmeter.
- Since the voltmeter (or potential coil) impedance is very high; the secondary current is very small and the potential transformer behaves as an ordinary two winding transformer operating on no-load.
- Fig. 6.7. shows a potential transformer used to measure the voltage of a circuit. It may be noted that the secondary is *grounded. This is done so that if the insulation breaks down, the high voltage does not endanger personnel who may be reading the meters.*
- These transformers are *made with high quality iron core operating at very low flux densities so that the magnetising current may be very small.*
- Careful design ensures minimum variation of voltage ratio with load and minimum phase shift between input and output voltages.

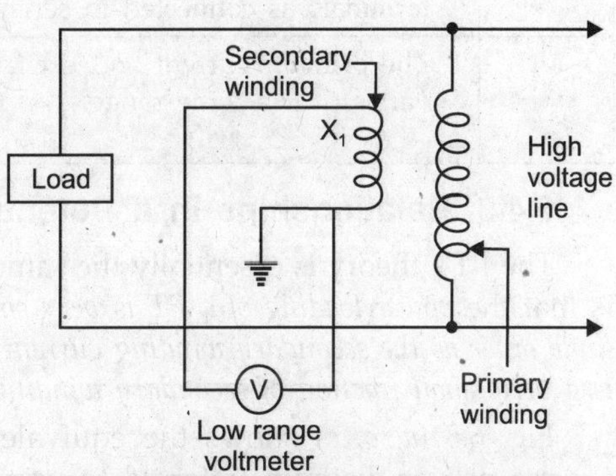


**Fig. 6.7.** Potential transformer connections.

- Potential transformer secondary is commonly designed for an output of **110 V.**
- PTs are usually of the following types :
  - — Upto voltage of 5000 V ..... Usually of *dry type.*
  - — Between 5000 V and 1,38000 V .... *Either dry type or oil immersed.*
  - — Voltages exceeding 1,38000V .... *Always immersed type.*
- As a rule, the PTs are of *shell type* because such a construction develops a *high degree of accuracy.*
- Potential transformer (PT) is different from power transformer. In power transformer current is 5% of full-load current, while in *case of PT, secondary current is of the same order as magnetising current.* Thus *PT has a small volt-ampere ratings (say about 15 VA).*
- The primary and secondary winding of the PT are *coaxial* in order to reduce the *leakage reactance to a minimum.* The low voltage winding (secondary winding) is placed next to the core.
- Oil filled bushings are used for oil filled PTs to minimise the overall size of the transformer. When neither side of the line is at ground potential, two bushings are used.

### 6.4.2. Comparison of Current and Potential Transformers

The comparison between current and potential transformers is given in the table below :

| S.No. | Current transformer (CT) | Potential transformer (PT) |
|---|---|---|
| 1. | Secondary must *always be shorted.* | Secondary is *nearly under open circuit conditions.* |
| 2. | The winding carries *full line current.* | The winding is impressed with *full-line voltage.* |
| 3. | The primary current is *independent* of the secondary circuit conditions. | The primary current depends on the secondary circuit conditions. |
| 4. | It can be treated as series transformer under short circuit conditions. | It can be treated as parallel transformer under open circuit secondary. |
| 5. | A small voltage exists across its terminals as connected in series. | Full line voltage appears across its terminals. |
| 6. | The primary current and excitation varies over a wide range. | The line voltage is almost constant hence exciting current and flux density varies over a limited range. |

### 6.4.3. Relationships in a Potential Transformer (PT)

The PT's theory is essentially the same as that of a power transformer. The difference is that the *power loading of a PT is very small and consequently the exciting current is of the same order as the secondary winding current, while in a power transformer the exciting current is a very small fraction of secondary winding load current.*

Fig. 6.8 (*a, b, c*) shows the equivalent circuit, phasor diagram and enlarged and concise phasor diagram (referred to primary) of a potential transformer. Here,

$\phi$ = Working flux,

$I_m$ = Magnetising current,

$I_w$ = Iron loss component of no load (exciting) current,
$I_o$ = No load (exciting) current,
$E_s$ = Secondary winding induced voltage,

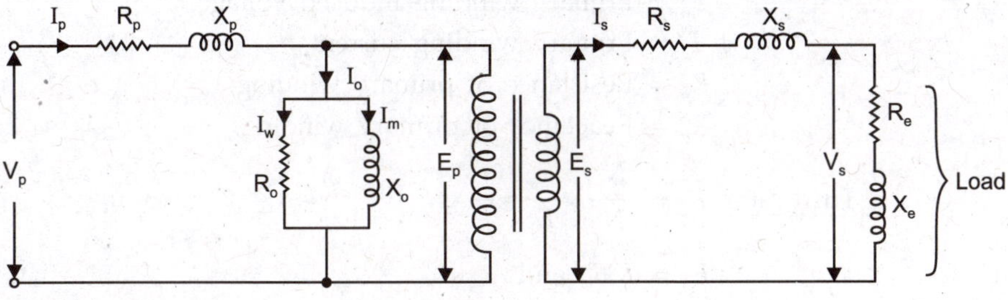

(a) Equivalent circuit

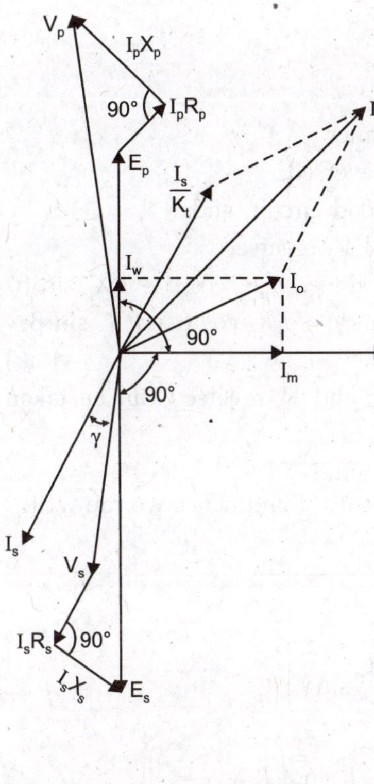

(b) Phasor diagram

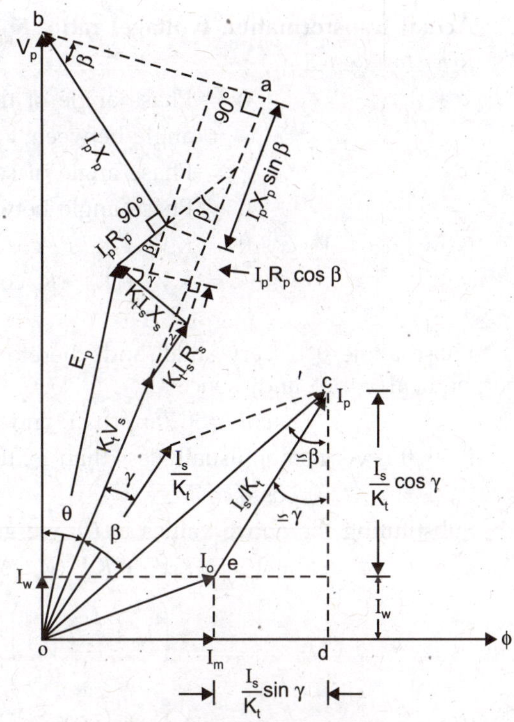

(c) Enlarged and concise phasor diagram of a PT (referred to primary)

**Fig. 6.8.** Potential transformer (PT).

$V_s$ = Secondary winding terminal voltage,
$I_s$ = Secondary winding current,
$R_s$ = Resistance of secondary winding,
$X_s$ = Reactance of secondary winding,
$R_e$ = Resistance of secondary load circuit,
$X_e$ = Reactance of secondary load circuit,
$\gamma$ = Phasor angle of secondary load circuit,

$$= \tan^{-1}\left(\frac{X_e}{R_e}\right)$$

$E_p$ = Primary winding induced voltage,
$I_p$ = Primary winding current,
$R_p$ = Resistance of primary winding,
$X_p$ = Reactance of primary winding.

$$\text{Turn ratio, } K_t = \frac{N_p}{N_s} = \frac{E_p}{E_s} = \frac{I_s}{I_p}$$

or, $E_p = K_t E_s$, and

$$I_p = \frac{I_s}{K_t}$$

**Actual transformation (voltage) ratio, $K_{act.}$ :**

*Refer to Fig. 6.8. (c) :*

$\theta$ = Phase angle of the transformer
= angle between $V_p$ and $V_s$ reversed,
$\gamma$ = Phase angle of secondary load circuit, and
$\beta$ = Phase angle between $I_p$ and $V_s$ reversed,

Now, $oa = V_p \cos\theta = K_t V_s + K_t I_s R_s \cos\gamma + K_t I_s X_s \sin\gamma + I_p R_p \cos\beta + I_p X_p \sin\beta$
$= K_t V_s + K_t I_s (R_s \cos\gamma + X_s \sin\gamma) + I_p R_p \cos\beta + I_p X_p \sin\beta$ ...(*i*)

Phase angle $\theta$ is very small and, therefore, both $V_p$ and $V_s$ reversed can be taken perpendicular to $\phi$ and, hence

$$\angle ocd = \beta \text{ (approx.), and } \angle ecd = \gamma \text{ (approx.)}$$

Now $\theta$ is very small usually less than 1°, therefore $\cos\theta = 1$ and hence we can write :

$$V_p \cos\theta = V_p$$

Substituting the above values in (*i*), we get :

$$V_p = K_t V_s + K_t I_s (R_s \cos\gamma + X_s \sin\gamma) + \left(I_w + \frac{I_s}{K_t}\cos\gamma\right)R_p + \left(I_m + \frac{I_s}{K_t}\sin\gamma\right)X_p$$

or,

$$V_p = K_t V_s + I_s \cos\gamma\left(K_t R_s + \frac{R_p}{K_t}\right) + I_s \sin\gamma\left(K_t X_s + \frac{X_p}{K_t}\right) + I_w R_p + I_m X_p \quad ...(6.24)$$

$$= K_t V_s + \frac{I_s}{K_t}\cos\gamma\left(K_t^2 R_s + R_p\right) + \frac{I_s}{K_t}\sin\gamma\left(K_t^2 X_s + X_p\right) + I_w R_p + I_m X_p$$

$$= K_t V_s + \frac{I_s}{K_t}\cos\gamma\,(R_p)_{eq} + \frac{I_s}{K_t}\sin\gamma\,(X_p)_{eq} + I_w R_p + I_m X_p$$

[where, $K_t^2 R_s + R_p = (R_p)_{eq}$ = equivalent resistance of transformer referred to 'primary'
and, $K_t^2 X_s + X_p = (X_p)_{eq}$ = equivalent reactance of transformer referred to 'primary']

$$\therefore \quad V_p = K_tV_s + \frac{I_s}{K_t}\left[(R_p)_{eq}\cos\gamma + (X_p)_{eq}\sin\gamma\right] + I_wR_p + I_mX_p \quad \text{...(6.25)}$$

∴ *Actual transformation (voltage) ratio,*

$$K_{act.} = \frac{V_p}{V_s} = K_t + \frac{\frac{I_s}{K_t}\left[(R_p)_{eq}\cos\gamma + (X_p)_{eq}\sin\gamma\right] + I_wR_p + I_mX_p}{V_s} \quad \text{...(6.26)}$$

Eqn. (6.24) can be rewritten as :

$$V_p = K_tV_s + K_tI_s\cos\gamma\left(R_s + \frac{R_p}{K_t^2}\right) + K_tI_s\sin\gamma\left(X_s + \frac{X_p}{K_t^2}\right) + I_wR_p + I_mX_p$$

$$= K_tV_s + K_tI_s\cos\gamma\,(R_s)_{eq} + K_tI_s\sin\gamma\,(X_s)_{eq} + I_wR_p + I_mX_p$$

[where, $R_s + \frac{R_p}{K_t^2} = (R_s)_{eq}$ = equivalent resistance of transformer *referred to 'secondary'*

and, $X_s + \frac{X_p}{K_t^2} = (X_s)_{eq}$ = equivalent reactance of transformer *referred to 'secondary'*]

$$\therefore \quad V_p = K_tV_s + K_tI_s\,[(R_s)_{eq}\cos\gamma + (X_s)_{eq}\sin\gamma) + I_wR_p + I_mX_p \quad \text{...(6.27)}$$

∴ *Actual transformation (voltage) ratio,*

$$\mathbf{K}_{act.} = \frac{V_p}{V_s} = K_t + \frac{K_tI_s\left[(R_s)_{eq}\cos\gamma + (X_s)_{eq}\sin\gamma\right] + I_wR_p + I_mX_p}{V_s} \quad \text{...(6.28)}$$

Using eqns. 6.26 and 6.28, the difference between $K_{act.}$ and $K_t$ is :

$$K_{act.} - K_t = \frac{\frac{I_s}{K_t}\left[(R_p)_{eq}\cos\gamma + (X_p)_{eq}\sin\gamma\right] + I_wR_p + I_mX_p}{V_s} \quad \text{...(6.29)}$$

$$= \frac{K_tI_s\left[(R_s)_{eq}\cos\gamma + (X_s)_{eq}\sin\gamma\right] + I_wR_p + I_mX_p}{V_s} \quad \text{...(6.30)}$$

**Phase angle, θ :**

From the phasor diagram [Fig. 6.8 (*c*)], we have :

$$\tan\theta = \frac{ab}{oa} = \frac{I_pX_p\cos\beta - I_pR_p\sin\beta + K_tI_sX_s\cos\gamma - K_tI_sR_s\sin\gamma}{K_tV_s + K_tI_sR_s\cos\gamma + K_tI_sX_s\sin\gamma + I_pR_p\cos\beta + I_pX_p\sin\beta}$$

The terms in denominator involving $I_p$ and $I_s$ are small and, therefore, they can be neglected as compared with $K_tV_s$

$$\therefore \quad \tan\theta = \frac{I_pX_p\cos\beta - I_pR_p\sin\beta + K_tI_sX_s\cos\gamma - K_tI_sR_s\sin\gamma}{K_tV_s}$$

$$= \frac{X_p\left(I_w + \frac{I_s}{K_t}\cos\gamma\right) - R_p\left(I_m + \frac{I_s}{K_t}\sin\gamma\right) + K_tI_sX_s\cos\gamma - K_tI_sR_s\sin\gamma}{K_tV_s}$$

$$= \frac{I_s \cos\gamma\left(\frac{X_p}{K_t} + K_t X_s\right) - I_s \sin\gamma\left(\frac{R_p}{K_t} + K_t R_s\right) + I_w X_p - I_m R_p}{K_t V_s}$$

$$= \frac{\frac{I_s \cos\gamma}{K_t}(X_p + K_t^2 X_s) - \frac{I_s \sin\gamma}{K_t}(R_p + K_t^2 R_s) + I_w X_p - I_m R_p}{K_t V_s}$$

$$= \frac{\frac{I_s \cos\gamma}{K_t}(X_p)_{eq} - \frac{I_s \sin\gamma}{K_t}(R_p)_{eq} + I_w X_p - I_m R_p}{K_t V_s}$$

$$= \frac{\frac{I_s}{K_t}\left[(X_p)_{eq} \cos\gamma - (R_p)_{eq} \sin\gamma\right] + I_w X_p - I_m R_p}{K_t V_s}$$

Since θ is small, therefore, θ = tan θ

$$\therefore \qquad \theta = \frac{\frac{I_s}{K_t}\left[(X_p)_{eq} \cos\gamma - (R_p)_{eq} \sin\gamma\right] + I_w X_p - I_m R_p}{K_t V_s} \text{ rad.} \qquad \text{...(6.31)}$$

In terms of quantities *referred to secondary,*

$$(R_s)_{eq} = \frac{R_p}{K_t^2} \text{ and } (X_s)_{eq} = \frac{X_p}{K_t^2}$$

$$\therefore \qquad \theta = \frac{K_t I_s\left[(X_s)_{eq} \cos\gamma - (R_s)_{eq} \sin\gamma\right] + I_w X_p - I_m R_p}{K_t V_s}$$

or,
$$\theta = \frac{I_s}{V_s}\left[(X_s)_{eq} \cos\gamma - (R_s)_{eq} \sin\gamma\right] + \frac{I_w X_p - I_m R_p}{K_t V_s} \text{ rad.} \qquad \text{...(6.32)}$$

**Ratio (voltage) error.** The actual ratio of transformation varies with operating conditions and the error in secondary voltage may be defined as :

$$\textit{Percentage ratio error} = \frac{K_{nom.} - K_{act.}}{K_{act.}} \times 100 \qquad \text{...(6.33)}$$

**Phase angle error.** In an ideal transformer there should not be any phase difference between primary winding voltage and the secondary winding voltage *reversed*. However, in an actual transformer there exists a phase difference between $V_p$ and $V_s$ reversed.

Phase angle error = θ, is given by : same as given in eqns. 6.31 and 6.32. The phase angle is taken as *+ve* when secondary winding voltage reversed *leads* the primary winding voltage. The angle is *–ve* when secondary winding voltage reversed *lags* the primary winding voltage.

- It may be remembered that *while measuring 'voltage', ratio error is only important while for 'power measurements' both ratio and phase angle errors are involved.*

### 6.4.4. Characteristics of Potential Transformers

1. **Effect of change in Secondary Burden :**
   - With the increase in secondary burden, the secondary current is increased and consequently the primary current is increased. Both the primary and secondary voltage drops increase and thus for a given value of $V_p$, there is a decrease in the value of $V_s$ and hence the *actual ratio increases as the burden increases.* This variation of error is *almost linear* with change in burden.
   - Further, owing to increased voltage drop, $V_p$ advances while $V_s$ retards in phase. As a result, the *phase angle* between $V_p$ and $V_s$ reversed *increases* and becomes *more negative as the current increases.*
2. **Effect of power factor of secondary burden :**
   - When the power factor of a secondary circuit burden is reduced, angle $\gamma$ increases. As a consequence, current $I_p$ shifts towards $I_o$ [see Fig. 6.8 (*b*)]. The voltages $V_p$ and $V_s$ come more nearly into phase with $E_p$ and $E_s$ respectively since the voltage drops are almost constant. The results in an increase in $V_p$ relative to $E_p$. But, since $V_p$ is constant, therefore, $E_p$ reduces relative to $V_p$; $V_s$ also reduces relative to $E_s$. Thus the *transformation ratio increases as the p.f. of secondary burden reduces.*
   - While $V_s$ advances in phase and $V_p$ retards hence *negative phase angle reduces as lagging p.f. of the secondary decreases.*
3. **Effect of change in frequency :**
   - The flux is inversely proportional to frequency, for a constant voltage. Thus, with increase in frequency, flux is reduced and, therefore $I_m$ and $I_w$ are decreased leading to *decrease in voltage ratio*. The decrease is not so much, as with increase in frequency the leakage reactance increases and, therefore, leakage reactance drops are increased giving an *increase in the ratio*. Thus changes in voltage ratio (due to change in frequency) are dependent upon relative values of $I_o$ and leakage reactances (as they produce opposite effects).
   - With the *increase in frequency*, the *phase angle is increased* since both the effects advance $V_s$ and the increase in secondary reactance retards $V_s$.
4. **Effect of primary voltage.** The variations of errors due to change in primary voltage are of no importance, as supply-system voltages do not vary widely.

### 6.4.5. Reduction of Errors in Instrument Transformers

The errors (ratio and phase angle) in instrument transformer may be reduced by using the following methods :

1. ***Reduction of magnetising and loss components (of exciting current $I_o$).*** From the equations (derived earlier) of ratio and phase angle errors it is evident that these errors depend on $I_m$ and $I_w$ (components of $I_o$). These components can be *reduced* by :

   (*i*) Using materials of high permeability; (*ii*) Keeping the flux density in the core to low value; (*iii*) Choosing low reactance core; (*iv*) Using a core of large cross-section; (*v*) Providing smaller magnetic paths to the flux; (*vi*) Designing the assembly and interleaving the core with proper precautions.
2. ***Reduction of resistance and leakage reactance.*** The errors depend on the voltage drops which in turn depend on the values of resistance and leakage reactance.
   - The *resistance* can be reduced by using thick conductors (*i.e.*, of large cross-section) and using the smallest length of mean turn.

- The *leakage reactance* of windings depends upon the leakage fluxes of primary and secondary windings and, therefore, windings *must be kept as close as possible.*
— Further, a high flux density means a high flux in the core and so the windings have lesser number of turns, which obviously results in smaller leakage reactance of the windings.

3. ***Turns compensation* :**

- At no load (when $I_s = 0$) the actual ratio ($K_{act.}$), in case of PT, *exceeds* the turns ratio ($K_t$) by an amount $\dfrac{(I_w R_p + I_m X_p)}{V_s}$ (Refer to eqn. 6.26). The ratio further increases with an inductive or resistive load because of voltage drops in resistance and leakage reactance of these windings. If $K_t = K_{nom.}$, the $K_{act.}$ differs from the nominal ratio and *errors creep* in.

  The remedy lies in making the $K_t$ less $K_{nom.}$ This can be accomplished by either *reducing* the number of turns of primary winding or *increasing* the number of secondary winding turns. By doing so, it becomes possible to make the actual ratio $K_{act.}$ equal to $K_{nom.}$ for one particular value of burden. At the same time, there is a general reduction in the ratio error over the entire working range of the burden. This is called *"turns compensation"*.
- *The phase angle θ, practically, remains unaffected due to a small change in $K_t$.*

**Example 6.10.** *The following data relate to a 1000/100 volts potential transformer :*

*Primary resistance* = 86.4 Ω

*Primary reactance* = 62.5 Ω

*Secondary resistance* = 0.78 Ω

*Total equivalent reactance* = 102 Ω

*No-load current* = 0.03 A at 0.42 *power factor*

*Calculate :*

*(i) Phase angle error at no load.*

*(ii) Burden in VA at unity power factor at which the phase angle will be zero.*

**Solution.** *Given :* $K_t = \dfrac{1000}{100} = 10$; $R_p = 86.4\ \Omega$; $X_p = 62.5\ \Omega$; $R_s = 0.78\ \Omega$; $X_e = 102\ \Omega$; $I_o = 0.03$ A at 0.42 p.f.

(*i*) **Phase angle error at no load, θ :**

No load p.f. $\cos\phi_o = 0.42$;

$$\therefore \quad \sin\phi_o = \sqrt{1-(0.42)^2} = 0.9075$$

Now,
$$I_w = I_o \cos\phi_o = 0.03 \times 0.42 = 0.0126 \text{ A}$$
$$I_m = I_o \sin\phi_o = 0.03 \times 0.9075 = 0.0272 \text{ A}$$

$$\text{Phase angle, } \theta = \frac{\dfrac{I_s}{K_t}\left[(X_p)_{eq}\cos\gamma - (R_p)_{eq}\sin\gamma\right] + I_w X_p - I_m R_p}{K_t V_s} \text{ rad.} \quad ...[\text{Eqn. (6.31)}]$$

At no load, $I_s = 0$

$$\therefore \qquad \theta = \frac{I_w X_p - I_m R_p}{K_t V_s} \text{rad.}$$

$$\text{or} \qquad \theta = \frac{0.0126 \times 62.5 - 0.0272 \times 86.4}{10 \times 100} = -\mathbf{5.37} \quad \textbf{(Ans.)}$$

(*ii*) **Burden in VA :**

$$\theta = 0 \qquad \text{...(Given)}$$

At unity p.f., $\cos \gamma = 1$, $\sin \gamma = 0$

$$\theta = \frac{\frac{I_s}{K_t}\ (X_p)_{eq} \cos \gamma - (R_p)_{eq} \sin \gamma \ + I_w X_p - I_m R_p}{K_t V_s} \text{rad}$$

Substituting the values, we have :

$$\text{or} \qquad \theta = \frac{\frac{I_s}{10}(102 \times 1 - 0) + 0.0126 \times 62.5 - 0.0272 \times 86.4}{10 \times 100}$$

$$\text{or,} \qquad \theta = 10.2\, I_s - 1.563$$

$$\therefore \qquad I_s = \frac{1.563}{10.2} = 0.1532\, A$$

$\therefore$ Burden load in $VA = V_s I_s = 100 \times 0.1532 =$ **15.32 VA** **(Ans.)**

**Example 6.11.** *A single-phase potential transformer has a turns ratio of 3810/63. The nominal secondary voltage is 63 V and the total equivalent resistance and leakage reactance referred to secondary are 2 Ω and 1 Ω respectively. Calculate the ratio and phase angle errors when the transformer is supplying a burden of 100 + j 200 Ω.* **(PTU – Dec. 2005)**

**Solution.** *Given* : $K_t (= K_{nom.}) = \frac{3810}{63} = 60.476$; $V_s = 63$ V; $(R_s)_{eq} = 2\Omega$; $(X_s)_{eq} = 1\ \Omega$; Burden = $100 + j\ 200\ \Omega$.

**Ratio error :**

*Neglecting no load components of exciting current,*

Transformation (actual) ratio,

$$K_{act.} = K_t + \frac{K_t I_s\ (R_s)_{eq} \cos \gamma + (X_s)_{eq} \sin \gamma}{V_s}$$

$$\text{or,} \qquad K_{act.} = K_t + \frac{K_t\ (R_s)_{eq} \cos \gamma + (X_s)_{eq} \sin \gamma}{(V_s / I_s)} \qquad \text{...[Eqn. (6.28)]}$$

$$\text{Burden} = R_e + jX_e = 100 + j\ 200\ \Omega \qquad \text{...}(i)$$

$$\therefore \qquad \gamma = \tan^{-1} \frac{X_e}{R_e} = \tan^{-1}\left(\frac{200}{100}\right) = 63.43°$$

$$\text{Also,} \qquad \frac{V_s}{I_s} = Z_s = \sqrt{100^2 + 200^2} = 223.6\,\Omega$$

Substituting the various values in eqn. (*i*), we get :

$$K_{act.} = 60.476 + \frac{60.476[2 \times \cos 63.43° + \sin 63.43°]}{223.6} = 60.9598$$

$$\therefore \quad \% \text{ Ratio error} = \frac{K_{nom.} - K_{act.}}{K_{act.}} \times 100$$

$$= \frac{60.476 - 60.9598}{60.9598} \times 100 = -\mathbf{0.7936\%} \quad \textbf{(Ans.)}$$

**Phase angle error, θ :**

$$\theta = \frac{I_s}{V_s}\left[(X_s)_{eq} \cos\gamma - (R_s)_{eq} \sin\gamma\right] \text{ rad.} \qquad \text{...[Eqn. (6.32)]}$$

$$= \frac{1}{223.6}[1 \times \cos 63.43° - 2 \sin 63.43°] \text{ rad.}$$

$$= -5.999 \times 10^{-3} \text{ rad.} = -5.999 \times 10^{-3} \times \frac{180}{\pi} \text{ degrees}$$

$$= -\ \mathbf{0.3437°} \quad \textbf{(Ans.)}$$

## HIGHLIGHTS

1. The principle of the instrument transformer is fundamentally the same as that of the power transformer.
2. Current transformers (CT) are used for the measurement of 'current'.
3. Potential transformers (PT) are used for the measurements of 'voltage'.
4. CT and PT are extensively used for precise measurements.
5. Ratio Correction Factor (RCF) is the ratio of actual transformation ratio to nominal ratio (*i.e.*, RCF = $\frac{K_{act.}}{K_{nom.}}$).
6. The *CT must never be operated on open circuit* because : (*i*) There will be no secondary mmf and since the primary current (and mmf) is fixed, the core flux will increase enormously. This will cause large eddy current and hysteresis losses and the resulting high temperature may damage the insulation or even the core; (*ii*) A very high voltage will be induced in the multi-turn secondary and this high voltage may be dangerous both to life and to the insulation.
7. A potential transformer (PT) is a *step down transformer* used along with a low range voltmeter for measuring a high voltage.
8. PT has a small volt-ampere ratings (say 15 VA).

## OBJECTIVE TYPE QUESTIONS

1. The ratio of transformation in the case of potential transformers
   (*a*) increases with increase in power factor of secondary burden
   (*b*) remains constant irrespective of the power factor of secondary burden
   (*c*) decreases with increase in power factor of secondary burden
   (*d*) none of the above.
2. In case of potential transformers :
   (*a*) the phase angle error is always positive

(*b*) the phase angle error is always negative

(*c*) the phase angle error is usually zero

(*d*) the phase angle error is positive when the secondary winding voltage reversed leads the primary winding voltage and is negative when the secondary winding voltage reversed lags behind the primary winding voltage.

**3.** The disadvantages of using multipliers with voltmeters at high voltages are :

(*a*) The power consumption of multipliers becomes large at large voltages.

(*b*) The multipliers at high voltages have to be shielded in order to prevent capacitive currents.

(*c*) The metering circuit is not electrically isolated from the power circuit.

(*d*) all of the above.

**4.** The nominal ratio of a current transformer is

(*a*) primary winding current/secondary winding current

(*b*) rated primary winding current/rated secondary winding current

(*c*) number of secondary winding turns/number of primary winding turns

(*d*) all of the above.

**5.** The errors in current transformers can be reduced by designing them with

(*a*) high permeability and low loss core materials, avoiding any joints in the core and also keeping the flux density to a low value

(*b*) using primary and secondary windings as close to each other as possible

(*c*) using large cross-sections for both primary and secondary winding conductors

(*d*) all of the above.

**6.** Capacitive potential transformers are used

(*a*) for primary winding phase voltages above 100 kV

(*b*) for keeping the value of transformation ratio constant irrespective of the burden by making certain adjustment

(*c*) because they are cheaper than the electromagnetic transformers above a certain voltage range

(*d*) all of the above.

**7.** The size of potential transformers

(*a*) is the same as that of power transformers of the same VA rating

(*b*) is much greater than that of power transformers of the same VA rating because they are designed for low ratio and phase angle errors which require large sized cores and winding conductors

(*c*) is smaller than that of power transformers of the same VA rating

(*d*) none of the above.

**8.** The transformation ratio in the case of a potential transformer is defined as ratio of

(*a*) primary winding voltage/secondary winding voltage

(*b*) rated primary winding voltage/rated secondary winding voltage

(*c*) number of turns of primary winding/number of turns of secondary winding

(*d*) all of the above.

**9.** The burden of current transformers is expressed in terms of

(*a*) secondary winding current

(*b*) VA rating of transformer

(*c*) voltage, current and power factor of secondary winding circuit

(*d*) none of the above.

**10.** The current in the primary winding of a current transformer depends upon

(*a*) burden of the secondary winding of transformer

(*b*) load connected to the system in which the C.T. is installed

(*c*) both burden of the transformer secondary winding and load connected to the system

(*d*) none of the above.

**11.** Turns compensation is used in current transformers primarily for reduction of

(*a*) phase angle error

(*b*) both ratio and phase angle errors

(*c*) ratio error, reduction in phase angle error is incidental

(*d*) none of the above.

**12.** The advantages of instrument transformers are :

(*a*) The readings of instruments used in conjunction with them do not depend upon their resistance, inductance etc.

(*b*) The ratings of instrument transformers have been standardized and the ratings of instruments used in conjunction with them also get standardized. Therefore, there is reduction of cost and ease in replacement.

(*c*) The metering circuit is electrically isolated from the power circuit thereby providing safety to operating personnel.

(*d*) all of the above.

**13.** The disadvantages of shunts for use at high currents are :

(*a*) It is difficult to achieve good accuracy with shunts.

(*b*) Power consumption of the shunts is large.

(*c*) The metering circuit is not electrically isolated from the power circuit

(*d*) All of the above.

**14.** When the secondary winding of a current transformer is open-circuited with the primary winding energized

(*a*) the whole of the primary current produces large value of flux in the core (limited to only saturation) thereby producing a large voltage in the secondary winding.

(*b*) the large voltage may act as safety hazard for the operators and may even rupture the insulation.

(*c*) when the large magnetizing force is taken off it leaves a large value of residual magnetism.

(*d*) all of the above.

**15.** Clamp on meters are used because

(*a*) with their use it is possible to measure current flowing in a line without breaking the circuit.

(*b*) their accuracy is high.

(*c*) it is not possible to measure voltage without them.

(*d*) all of the above.

**16.** A short-circuiting link is provided on the secondary side of a current transformer because.

(*a*) when the secondary winding of the C.T. is short-circuited by the link with the primary winding energized a very high current flows on the primary side.

(*b*) when the secondary winding of the C.T. is short circuited by the link with the primary winding energized, it is possible to make any adjustments in the secondary winding circuit like replacing a faulty ammeter.

(*c*) when the short-circuiting link is opened with the secondary winding open-circuited, the current on the primary side falls to almost zero.

(*d*) all of the above.

**17.** When the secondary winding of a potential transformer is suddenly open circuited with primary winding excited

(*a*) large voltages are produced in the secondary winding which may be a safety hazard to operating personnel.

(*b*) the large voltages so produced may rupture the insulation.

(*c*) the primary winding draws only the no load current.

(*d*) none of the above.

**18.** The ratio and phase angle errors in potential transformers may be reduced by

(*a*) increasing the exciting current

(*b*) increasing the resistance and leakage reactance in the transformer

(*c*) not employing turns compensation

(*d*) none of the above.

## ANSWERS

| | | | | | | |
|---|---|---|---|---|---|---|
| 1. (*c*) | 2. (*d*) | 3. (*d*) | 4. (*b*) | 5. (*a*) | 6. (*d*) | 7. (*b*) |
| 8. (*a*) | 9. (*b*) | 10. (*b*) | 11. (*c*) | 12. (*d*) | 13. (*d*) | 14. (*d*) |
| 15. (*a*) | 16. (*b*) | 17. (*c*) | 18. (*d*). | | | |

## THEORETICAL QUESTIONS

**1.** Explain the use of instrument transformers.

**2.** What are the basic advantages inherent in the method of using instrument transformers for measurement purposes ?

**3.** List the advantages and disadvantages of instrument transformers.

**4.** Draw the wiring diagrams for current and potential transformers ?

**5.** What are the advantages of instrument transformers over ammeter shunts and voltmeter multipliers ?

**6.** Explain the meaning of the term "burden" in the instrument transformer.

**7.** What is meant by "ratio error" and "phase angle error" in CTs ?

**8.** What is meant by turns compensation and why is it done ?

**9.** Derive expressions for the ratio and phase angle errors of a current transformer.

**10.** Draw the phasor diagram of a CT and discuss the effect of variation of power factor of the secondary burden upon the performance.

**11.** What is phase angle error and on which factors does it depend ?

**12.** Define the following terms as applied to instrument transformers :

(*i*) Burden of an instrument transformer. (*ii*) Actual transformation ratio.

(*iii*) Nominal transformation ratio. (*iv*) Turn ratio.

**13.** What is "Ratio correction factor"; ? Explain briefly.

**14.** State the reasons why current transformer must never be operated on open circuit.

**15.** Write a short note on 'clip-on type current transformer'.

**16.** How does a PT differ from a power transformer ?

**17.** Explain the use of PT with the help of a connection diagram.

**18.** Give the comparison between CT and PT.

**19.** Derive the expressions for 'actual transformation (voltage), ratio and phase angle' in case of a potential transformer.

**20.** Explain the following in relation to a PT :

(*i*) Effect of change in secondary burden.

(*ii*) Effect of power factor of secondary burden.

(*iii*) Effect of change in frequency.

**21.** How errors can be reduced in instrument transformers ?

## UNSOLVED EXAMPLES

### Current transformer (CT)

**1.** The turns of primary and secondary windings of a current transformer are 1 and 200 respectively. The secondary winding supplies a current of 5A to a non-inductive burden of 1 Ω resistance. The requisite flux is set up in the core by an mmf of 80 AT. The frequency is 50 Hz and net cross-section of the core is 1000 mm$^2$. Neglecting the effects of magnetic leakage, iron losses and $I^2R$ losses, calculate :

(*i*) The actual transformation ratio. (*ii*) Phase angle.

(*iii*) Maximum flux density in the core. [**Ans.** (*i*) 200.64; (*ii*) 4°34′; (*iii*) 0.1126 Wb/m$^2$]

**2.** A ring core current transformer of ratio 1000/5A is operating at full primary current, with a secondary burden of non-inductive resistance of 1 Ω. Its exciting current is 1 A at a p.f. of 0.4.

Calculate : (*i*) The ratio error at full load, assuming there has been no compensation; (*ii*) Phase angle. [**Ans.** (*i*) – 0.04%; (*ii*) 3′9″.]

**3.** The primary winding of a 1000/5A, 50 Hz current transformer has a single turn. Its secondary burden consists of a non-inductive impedance of 1.6 Ω. If the iron loss in the core is 1.5 W at full load and magnetising mmf is 100 AT, calculate : (*i*) Flux in the core; (*ii*) Ratio error at full load. Neglect leakage reactance.

[**Ans.** (*i*) 0.18 × 10$^{-3}$ Wb; (*ii*) – 3.61%]

**4.** A current transformer with a bar primary has 300 turns in its secondary winding. The resistance and reactance of the secondary circuit are 1.5 Ω and 1.0 Ω respectively including the transformer winding. With 5A flowing in the secondary winding, the magnetising mmf is 100 AT and the iron loss is 1.2 W. Calculate :

(*i*) Ratio error; (*ii*) Phase angle error. [**Ans.** (*i*) – 5.54%; (*ii*) 2.34°]

**5.** The secondary winding burden of a current transformer, having a bar primary and 200 secondary winding turns, is an ammeter of resistance 1.2 Ω and reactance 0.5 Ω. The resistance and reactance of secondary winding are 0.2 Ω and 0.3 Ω respectively. The core requires the equivalent of an mmf of 100 AT for magnetisation and 50 AT for core losses. Determine the following :

(*i*) The primary winding current and actual transformation ratio when the ammeter in the secondary winding indicates 5 A.

(*ii*) The number of turns which could be reduced in the secondary winding so that the ratio error be zero for this condition.

**6.** A current transformer has a bar primary, 100 secondary winding turns and 100/1 nominal ratio. It is operating on an external burden of 1.6 Ω non-inductive, the secondary winding resistance being 0.2 Ω. The primary winding current is 1.9 A, lagging 40.6° to the secondary voltage reversed. With 1 A current flowing in the secondary winding, calculate :

(*i*) The actual ratio of primary winding current to secondary winding current;

(*ii*) The phase angle between these currents. [**Ans.** (*i*) 101.44; (*ii*) 0.708°]

**7.** The current transformer rated as 1000/5A, 25 VA has a turn ratio of 1 : 200. The core loss and magnetising components of the primary current are 4A and 7A under rated conditions. Determine the ratio and phase angle errors for the rated burden and the rated secondary current at 0.8 p.f. lagging and 0.8 p.f. leading.

Neglect the resistance and leakage reactance of secondary winding.

[**Ans.** – 0.24%, + 0.185°; + 0.603%, 0.46°]

8. A 500/5A, 50 Hz current transformer has a one turn primary and an output of 15 VA. The number of secondary turns is 98 and the resistance and leakage reactance of the secondary winding are 0.35 Ω and 0.3 Ω respectively. At rated load with non-inductive burden, the inphase and quadrature components (referred to the flux) of the exciting mmf are 8AT and 10 AT respectively. Calculate the current and phase angle errors.

[**Ans.** – 0.38%; 0.537°]

9. A 100/5A current transformer, at its rated load of 25 VA, has an iron loss of 0.2 W and a magnetising current of 1.5 A. It is supplying rated output to a meter having a ratio of resistance to reactance of 5. Calculate it ratio error and phase angle.

[**Ans.** – 1.075%; 0.75°]

**Potential transformer (PT)**

10. The following data relate to a 1000/100 volts potential transformer :
Primary resistance = 94.5 Ω; Primary reactance = 66.2 Ω; Secondary resistance = 0.86 Ω; Total equivalent reactance = 110 Ω. Calculate :
(*i*) Phase angle error at no load;
(*ii*) Burden in VA at unity p.f. at which the phase angle will be zero. [**Ans.** –4.1′; 10.9 VA]

# PART II : *Electronic Instruments*

CHAPTER

# 7 Sensors And Transducers



## 7.1 INTRODUCTION

The *primary sensing element* (*sensor*) is the first and foremost requirement for measurement and automatic controls. The ***sensors*** *sense the condition, state or value of the process variable and produce an output which reflects this condition, state or value. The* **transducers** *transform the energy of the process variable to an output of some other type of energy which is able to operate some control device.* Sometimes a *secondary transducer* may be employed to transform the output of the primary sensor to still another type of energy.

**Examples :**

- In the ordinary *dial indicator* the indicating spindle acts as a sensor/detector for displacement. It simply performs the function of sensor/detector and nothing else.
- The function of a *Bourdon tube* of a pressure gauge is *twofold*: Firstly to sense the pressure and secondly to give the resulting effect or output in the form of displacement. Here the tube acts a sensor/detector transducer.
- In a *compressive load cell,* the platform detects the force and gives an output in the form of deflection. This deflection may be further converted into an electrical output by strain gauges (called *secondary transducer*).

For the measurement of particular quantity, different types of sensors and transducers are available and the choice of a suitable unit depends upon the static and dynamic performance characteristics.

## 7.2 DEFINITION OF TRANSDUCER

A broad definition of a transducer is as follows:

*"A* **transducer** *is a device which converts the energy from one form to another".*

Most of the transducers either convert electrical energy into mechanical displacement and/or convert some non-electrical physical quantity (*e.g.,* force, sound, temperature etc.) to an electrical signal.

A transducer performs the following *functions* in an electronic instrumentation system :

1. *Detects or senses the presence, magnitude and changes in physical quantity being measured.*
2. *Provides a proportional electrical output signal* (see Fig. 3.1.)

- A **transducer** can be broadly defined as a *device which converts a non-electrical quantity into an electrical quantity.*
- An **electrical transducer** *is a sensing device by which a physical, mechanical or optical quantity to be measured is transformed directly, with a suitable mechanism, into an electrical voltage or current proportional to the input measurand.*

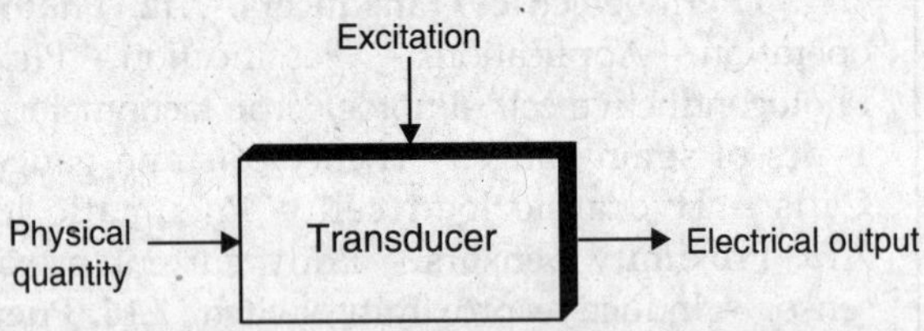


**Fig. 7.1.** Transducer.

- An **inverse transducer** *is defined as a device which converts an electrical quantity into a non-electrical quantity.* It is a precision actuator which has an electrical input and a low power non-electrical output. A *piezoelectric* crystal acts as an inverse transducer because when a voltage is applied across its surfaces, it changes its dimensions causing a mechanical displacement.
- The significant parameters which *dictate the transducer capability* are :

  (*i*) Linearity; (*ii*) Repeatability;
  (*iii*) Resolution; (*iv*) Reliability.

## 7.3 CLASSIFICATION OF TRANSDUCERS

**A.** Transducers are broadly *classified* into two groups as follows:

**1.** ***Active transducers.*** They are also known as *self-generating type transducers.* These transducers *develop their own voltage or current. The energy required for production of an output signal is obtained from the physical phenomenon being measured.*

**Examples:** *Thermocouples and thermopiles, piezoelectric pick-up, photovoltaic cell.*

2. ***Passive transduces.*** *They are known as externally-powered transducer.* These transducers *derive the power required for the energy conversion from an external power source.* However, they *may absorb some energy from the physical phenomenon under study.*

   **Examples:** *Resistance thermometers and thermistors, potentiometric devices, differential transformer, photoemission cell etc.*

**B. Classification based on the type of output :**

1. ***Analogue transducers.*** These transducers convert the input physical phenomenon into an analogous output which is a continuous function of time.

   **Examples:** *Strain gauge, a thermocouple, a thermistor or an LVDT* (linear voltage differential transformer).

2. ***Digital transducers.*** These transducers convert the input physical phenomenon into an electrical output which may be in form of pulse.

**C. Classification based on electrical principle involved :**

1. ***Variable-resistance type :***

   (*i*) Strain and pressure gauges. (*ii*) Thermistors, resistance thermometers.
   (*iii*) Photoconductive cell etc.

2. ***Variable-inductance type :***

   (*i*) Linear voltage differential transformer (LVDT).
   (*ii*) Reluctance pick-up. (*iii*) Eddy current gauge.

3. ***Variable-capacitance type :***

   (*i*) Capacitor microphone. (*ii*) Pressure gauge.
   (*iii*) Dielectric gauge.

4. ***Voltage-generating type :***

   (*i*) Thermocouple. (*ii*) Photovoltaic cell.
   (*iii*) Rotational motion tachometer. (*iv*) Piezoelectric pick-up.

5. ***Voltage-divider type :***

   (*i*) Potentiometer position censor. (*ii*) Pressure-actuated voltage divider.

Table 7.1. shows the *measurements versus transduction methods*.

- While *describing a particular transducer* the information must be available about the following aspects :

  (*i*) The measurand.
  (*ii*) The sensing element which responds directly to the measurand.
  (*iii*) The principle of operation of the transducer and where the output of the transducer originates.
  (*iv*) The useful range.

- While *selecting a dector-transducer element*, the following major consideration need to be looked into:

  (*i*) Mechanical suitability in terms of
  — Physical size, weight and shape;
  — Mounting arrangement;
  — Ruggedness.

  (*ii*) Electrical suitability in terms of:
  — Sensitivity;
  — Frequency response;
  — Ease of signal transmission.

(*iii*) Environmental suitability in terms of
— Sensitivity to temperature and self-heating effects;
— Magnetic fields;
— Vibration; dust and humidity;
— Supply frequency etc.

(*iv*) Transducer performance in terms of calibration accuracy.

(*v*) Desired measurement accuracy and range, power requirements, overload protection and vulnerability to sudden failure.

(*vi*) Purchase aspects.

**Table 7.1: Measurements versus Transduction Methods**

| S. No. | Quantity to be measured | Type of transducer |
|---|---|---|
| 1. | *Displacement* | — Resistive<br>— Inductive<br>— Capacitive<br>— Piezoelectric<br>— Magnetoelectric<br>— Radioactive<br>— Electron tube. |
| 2. | *Thickness* | — Inductive<br>— Capacitive<br>— Piezoelectric<br>— Photoelectric<br>— Radioactive.<br>— Magnetoelectric<br>— Electron tube. |
| 3. | *Velocity* | — Resistive<br>— Inductive<br>— Capacitive<br>— Piezoelectric<br>— Photoelectric<br>— Magnetoelectric<br>— Radioactive<br>— Electron tube. |
| 4. | *Acceleration* | — Resistive<br>— Inductive<br>— Capacitive<br>— Piezoelectric<br>— Magnetoelectric<br>— Radioactive. |
| 5. | *Mass* | — Inductive<br>— Piezoelectric<br>— Magnetoelectric<br>— Radioactive. |
| 6. | *Force* | — Resistive<br>— Inductive<br>— Piezoelectric<br>— Radioactive. |
| 7. | *Pressure* | — Resistive<br>— Inductive<br>— Capacitive<br>— Piezoelectric<br>— Thermoelectric<br>— Magnetoelectric<br>— Radioactive<br>— Electron tube. |
| 8. | *Flow* | — Resistive<br>— Inductive<br>— Capacitive<br>— Piezoelectric |
| 9. | *Level* | — Resistive<br>— Capacitive<br>— Piezoelectric<br>— Photoelectric<br>— Radioactive. |
| 10. | *Temperature* | — Resistive<br>— Photoelectric<br>— Thermoelectric<br>— Radioactive. |
| 11. | *Humidity* | — Resistive<br>— Capacitive. |
| 12. | *Viscosity* | — Resistive<br>— Capacitive<br>— Piezoelectric<br>— Magnetostrictive. |

### 7.3.1. Basic Requirements of a Transducer

Transducers should meet the following *basic requirements* :

1. Ruggedness;
2. Linearity;
3. Repeatability;
4. High output signal quality;
5. High reliability and stability;
6. Good dynamic response;
7. No residual deformation;
8. No hysteresis.

### 7.3.2. Transducer Sensitivity

*The relationship between the measurand and the transducer output signal is referred to as* **transducer sensitivity**.

$$i.e., \text{ Transducer sensitivity} = \frac{\text{Output signal increment}}{\text{Measuand increment}}$$

Sensitivity of a transducer should be *usually as high as possible* since then it becomes easier to take the measurements.

### 7.3.3. Specifications for Transducers

While selecting the proper transducer for any applications, or ordering the transducers, the following *specifications* should be thoroughly considered:

(*i*) Ranges available.
(*ii*) Squaring system.
(*iii*) Sensitivity.
(*iv*) Maximum working temperature.
(*v*) Method of cooling employed.
(*vi*) Mounting details.
(*vii*) Maximum depth.
(*viii*) Linearity and hysteresis.
(*ix*) Output for zero input.
(*x*) Temperature coefficient of zero drift.
(*xi*) Natural frequency.

## 7.4 ADVANTAGES AND DISADVANTAGES OF ELECTRICAL TRANSDUCERS

The advantages and disadvantages of electrical transducers are as follows :

**Advantages :**

1. Very small power is required for controlling the electrical or electronic system.
2. The electrical output can be amplified to any desired level.
3. The effect of friction is reduced to the minimum possible.
4. Mass-inertia effects are reduced to minimum possible (In fact when dealing with electrical or electronic signals the inertia effects are because of electrons, which have negligible mass).
5. The size and shape of the transducers can be suitably designed to achieve the optimum weight and volume.
6. The output can be indicated and recorded remotely at a distance from the sensing medium. Also, more than one indicator can be actuated simultaneously.
7. The output can be modified to meet the requirements of the indicating or controlling equipment.
8. The signals can be conditioned or mixed to obtain any combination with outputs of similar transducers or control signals as in air data computer or adaptive control systems.

*Example.* The measurement of 'Mach number' with two measurands.

9. The contour design and dimensions can be so chosen as not to disturb the measurand phenomena as in the case of turbulence measurements.

Modern digital computers have made the use of electrical transducers absolutely essential.

**Disadvantages :**

1. Low reliability in comparison to mechanical types, due to ageing and drift of the active components.
2. Sensing elements and associated signal conditioners are comparatively expensive.
3. In some cases, the accuracy and resolution attainable are not as high as in mechanical devices. The best accuracy available is of the order of 0.01 percent.

## 7.5 TRANSDUCERS ACTUATING MECHANISMS

Transducers are also known as *gauges, pick-ups and signal generators.* Most of the pick-ups have following *two basic elements :*

(*i*) Activating device.

(*ii*) Transducing element.

Fig. 7.2. shows some typical actuating mechanisms.

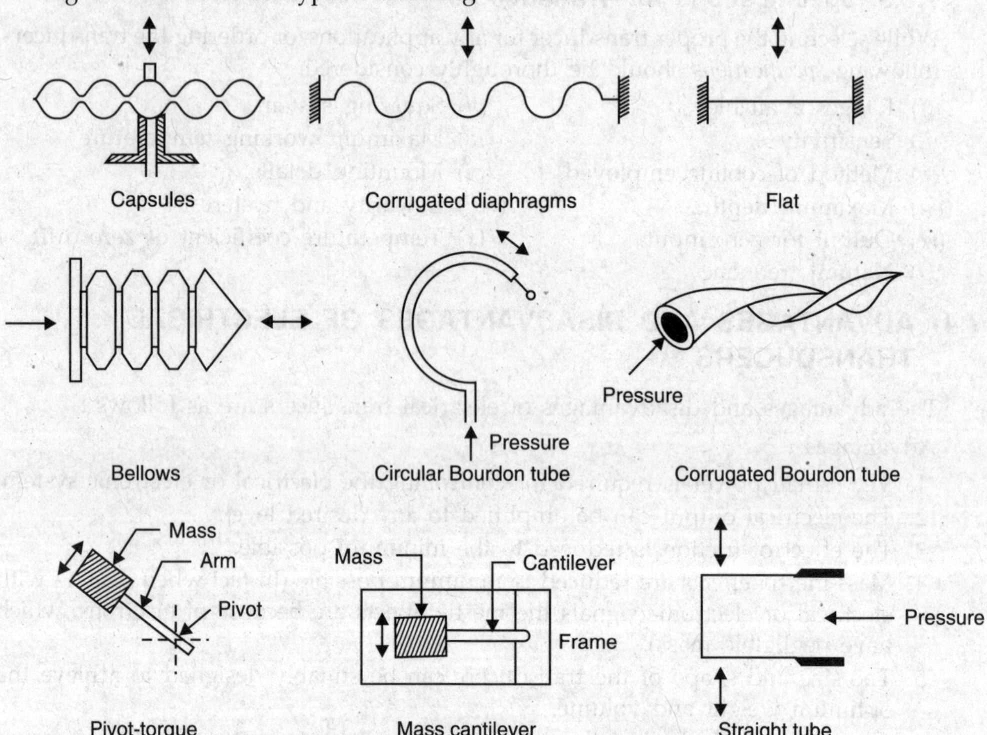


**Fig. 7.2.** Transducer actuating mechanisms.

## 7.6 RESISTANCE TRANSDUCERS

The resistance of a metal conductor is expressed by a simple equation that involves a few physical quantities. The relationship is given by $R = \frac{\rho L}{A}$;

where, $R$ = Resistance, Ω,

$\rho$ = Resistivity of conductor materials, Ω-*m*,

$L$ = Length of conductor, *m*, and

$A$ = Cross-sectional area of the conductor, $m^2$.

Any method of varying one of the quantities involved in the above relationship can be the designed basis of an *electrical resistance transducer.* There are a number of ways in which resistance can be changed by a physical phenomenon.

The *translational and rotational "potentiometers"* which work on the basis of change in the value of resistance with change in length of the conductor can be used for measurement of translational or rotary displacements.

*"**Strain gauges**" work on the principle that the resistance of a conductor or a semiconductor changes when strained.* This property can be used for measurement of displacement, force and pressure.

The resistivity of materials changes with the change of temperature thus causing a change of resistance. This property may be used for measurement of *"temperature"*.

In a *resistance transducer* an indication of measured physical quantity is given by *a change in the resistance.* It may be *classified* (as discussed above) as follows :

1. Mechanically varied resistance — *Potentiometer*
2. Thermal resistance change — *Resistance thermometers*
3. Resistivity change — *Resistance strain gauge.*

### 7.6.1. Linear and Angular Motion Potentiometers

Such potentiometers *convert the linear motion or the angular motion of a rotating shaft into changes in resistance.* The device is a variable resistor whose resistance is varied by the movement of a slider over a resistance element.

— Translatory devices have strokes from 2.5 mm to 5 mm.

— Rotational devices have full scale from 10° to 60° full turn.

The potentiometer shown in Figs. 7.3 and 7.4 form a part of the bridge circuit whose output voltage is changed by the slider position.

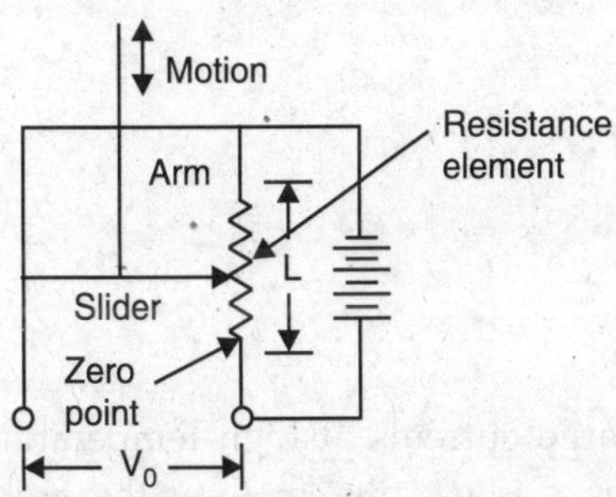


**Fig. 7.3.** Linear motion potentiometer.

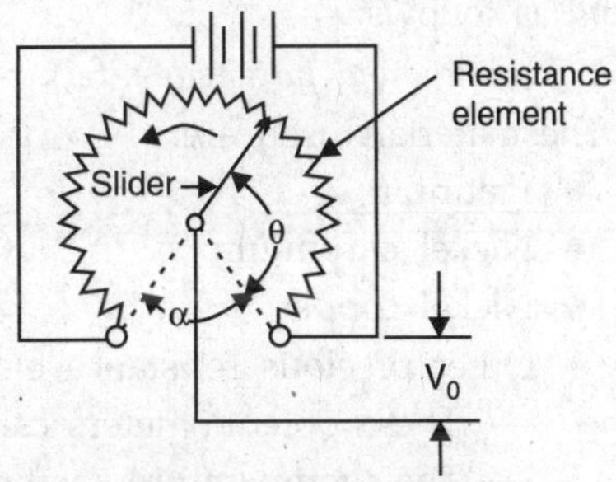


**Fig. 7.4.** Rotary motion potentiometer.

— The slider is powered by the mechanical part on which the linear displacement or angular measurement are to be made.

— Due to arm movement, the slider moves over the resistance element and thus shorts out a portion of the resistance. *The change in resistance in the potentiometer is then an indication of the amount of motion and the direction of movement is indicated by whether the resistance is increasing or decreasing.* The unbalanced voltage is measured directly or fed into an amplifier and recorded.

The potentiometers are used in many transducers designed to measure :

(*i*) Pressure (*ii*) Force

(*iii*) Acceleration (*iv*) Liquid level.

The potentiometers have the folllowing *advantages* and *disadvantages* :

**Advantages :**

(*i*) High output.
(*ii*) Less expensive.
(*iii*) Available in different sizes, shapes and ranges.
(*iv*) Simple to operate.
(*v*) Their electrical efficiency is very high.
(*vi*) Rugged construction.
(*vii*) Insensitivity towards vibration and temperature.

**Disadvantages :**

(*i*) Limited life due to early wear of the sliding ram.
(*ii*) The output tends to be noisy and erratic in high speed operation or when in high vibration environment.
(*iii*) In wire wound potentiometers the resolution is limited while in cermet and metal film potentiometers, the resolution is infinite.

**Power rating of potentiometers :**

The potentiometers are designed with a definite power rating which is related directly to their heat dissipating capacity. The manufacturer normally designs a series of potentiometers of single turn with a diameter of 50 mm with a wide range of ohmic values ranging from *100 Ω to 10 kΩ* in *steps of 100 Ω.* These potentiometers are essentially of the *same size* and of the *same mechanical configuration.* They have the *same heat transfer capabilities.* Their rating is typically *5 W at an ambient temperature of 21°C.* This limits their input excitation voltage.

**Materials used for potentiometers :**

The materials used for potentiometers may be classified as *wire wound* and *non-wire wound* as follows:

***1. Wire wound potentiometers :***

The materials used are:

- Platinum;
- Nickel chromium;
- Nickel copper;
- Other precious resistance elements.
  - — These potentiometers carry relatively large currents at high temperatures.
  - — Their temperature coefficients of resistance is usually small, of the order of $20 \times 10^{-6}/°C$ or less.
  - — Their resolution is about 0.025 – 0.05 mm and is limited by the number of turns that can be accommodated on the body.
  - — The response is limited to about 5 Hz.
  - — The maximum speed with which a wire wound potentiometer may be turned is about 300 r.p.m.

***2. Non-wire wound potentiometers :***

These are also called *continuous potentiometers.*

The *materials* used are :

- Cermet;
- Hot moulded carbon;

- Carbon film;
- Thin metal film.
  - — These potentiometers provide *improved resolution and life* (since resolution is no longer limited by the number of turns that can be wrapped onto a body)
  - — A continuous potentiometer may be turned at a speed of 2000 r.p.m.
  - — These are *more sensitive to temperature changes* and have a *higher wiper contact resistance*, which is variable and can *carry only moderate currents*.

**Example 7.1.** *Explain briefly the types of errors encountered in a transducer.*

**Solution.** The types of errors encountered in a transducer are briefly discussed below:

1. Scale errors.
2. Dynamic errors.
3. Noise and drift errors.

1. ***Scale errors.*** *Calibration* in general may be defined as the *process for determination, by measurement or comparison with a standard, of the correct value of each scale reading on the measuring instrument.* It is the determination of the settings of a control device that correspond to particular values of voltage, current, frequency, pressure or some other output.
2. ***Dynamic errors.*** *Fidelity* of an instrument system is defined as the *degree of closeness with which the system records the signal which is impressed upon it.* It refers to the ability of the system to reproduce the output in the same form as the input. "*Dynamic error*" is the *difference between the true value of a quantity changing with time and the value indicated by the instrument if no static error is assumed.*
3. ***Noise and drift errors.*** *Drift* is undesired change or a gradual variation in output over a period of time that is unrelated to changes in output, operating conditions or load. Drift for a measuring device can either be systematic, random, or some combination of the two. For most devices, drift is measured and specified as a percentage of output span.

**Example 7.2.** *A linear resistance potentiometer is 50 mm long and is uniformly wound with wire having a resistance of 10000 Ω. Under normal conditions, the slider is at the centre of the potentiometer. Find the linear displacement when the resistances of the potentiometer as measured by a Wheatstone bridge for two cases are (i) 3850 Ω (ii) 7560 Ω.*

*Are the two displacements in the same direction? If it is possible to measure a minimum value of 10 Ω resistance with the above arrangements, find the resolution of the potentiometer in mm.* **(Anna University)**

**Solution.** Under normal condition, it is given that the slider is at the centre of the potentiometer, hence under normal position the resistance of the potentiometer

$$= \frac{10000}{2} = 5000\ \Omega$$

The resistance of the potentiometer wire per unit length

$$= \frac{10000}{50} = 200\ \Omega/\text{mm}$$

(*i*) Change in resistance of potentiometer from its normal position

$$= 5000 - 3850 = 1150\ \Omega$$

$$\therefore \quad \textbf{Displacement} = \frac{1150}{200} = \textbf{5.75 mm (Ans.)}$$

(*ii*) Change in resistance of potentiometer from its normal position

$$= 7560 - 5000 = 2650\ \Omega \text{ (in the } \textit{opposite direction})$$

$$\therefore \quad \textbf{Displacement} = \frac{2560}{200} = \textbf{12.8 mm. (Ans.)}$$

The two displacements are in the *opposite direction*.

**Resolution of the potentiometer**

$$= \frac{\text{Min. measurable resistance}}{\Omega/\text{mm}} = \frac{10}{200} = \textbf{0.05 mm. (Ans.)}$$

### 7.6.2. Thermistors and Resistance Thermometers

These transducers are thermally sensitive variable resistors made of certain conducting and ceramic-like semiconducting materials. They are used *as temperature detecting elements used to sense temperature for the purpose of measurements and control.*

**Thermistors** are essentially semiconductors which behave as resistors with a *high negative temperature coefficient of resistance*. The high sensitivity to temperature changes makes the thermistors extremely useful for precision temperature (–60°C to + 15°C) measurements, control and compensation. Their resistance ranges from 0.5 Ω to 0.75 MΩ.

Thermistors are composed of sintered mixture of metallic oxides such as manganese, nickel, cobalt, copper, iron and uranium.

Fig. 7.5. shows the commercial forms of thermistors.

**Applications of thermistors :**

1. *Measurements of temperature* .................... (major application).
2. Temperature compensation in complex electronic equipment, magnetic amplifiers and instrumentation equipment.
3. Measurement of power at high frequencies.
4. Vacuum measurements.
5. Measurements of level, flow and pressure of liquid.
6. Measurement of thermal conductivity.

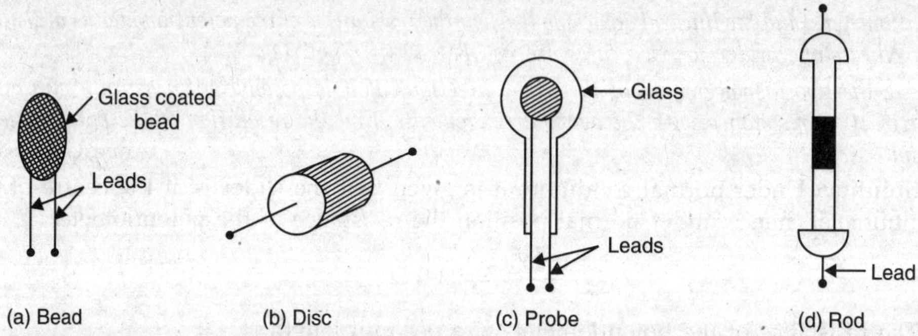


**Fig. 7.5.** Commercial forms of thermistors.

### 7.6.3. Wire Resistance Strain Gauges

Refer to article 7.14.

**Salient features of thermistors :**

1. The thermistors are compact, rugged and inexpensive.
2. They have good stability, when properly aged.

3. Their response time can vary from a fraction of a second to minutes depending on the size of the detecting mass and thermal capacity of the thermistors. It varies *inversely* with dissipation factor.
4. The upper limit of temperature for thermistors is dependent on physical changes in the material or solder used in attaching the electrical connections and is usually 400°C or less.
5. These can be installed at a distance from their associated measuring circuits if elements of high resistances are used such that the resistance of leads is negligible.
6. The measuring current should be maintained to as low a value as possible so that self-heating of thermistors is avoided otherwise errors are introduced on account of change of resistance caused by self-heating.

- *For more details refer to* **Art. 12.2.**

**Example 7.3.** *(a) A thermistor has a resistance temperature coefficient of –5% over a temperature range of 25°C to 50°C. If the resistance of the thermistor is 100 W at 25° C, what is the resistance at 35°C?*

*(b) Suggest a complete instrumentation scheme in block diagram form to measure the temperature in a closed oven with the help of thermistor.*

**Solution.** (*a*) $R_{35} = R_{25}[1 + \alpha(35 - 25) = 100[1 - 0.05(35 - 25)] = \mathbf{50\ \Omega}$ **(Ans.)**

(*b*) Fig. 7.6. shows the complete instrumentation scheme for the measurement of temperature with the help of a thermistor. Thermistor is mounted in the oven at a place where temperature is to be sensed. With the increase in temperature, resistance of the thermistor decreases causing imbalance in Wheatstone bridge circuit whose output balance voltage is amplified by signal conditioning device, the amplified output when connected to a suitable output device gives the value of the temperature of the oven.

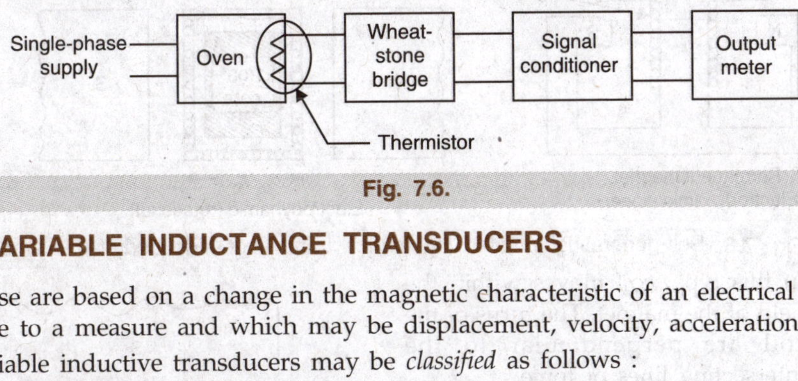


**Fig. 7.6.**

## 7.7 VARIABLE INDUCTANCE TRANSDUCERS

These are based on a change in the magnetic characteristic of an electrical circuit in response to a measure and which may be displacement, velocity, acceleration etc.

Variable inductive transducers may be *classified* as follows :

**1. Self-generating type.** *In this type voltage is generated because of the relative motion between a conductor and a magnetic field.*

These may be further *classified* as follows:

(*i*) Electromagnetic type.
(*ii*) Electrodynamic type.
(*iii*) Eddy current type.

**2. Passive type.** In this type the *motion of an object results in changes in the inductance of the coils of the transducer.*

These may be further *classified* as follows :

(*i*) Variable reluctance.

(*ii*) Mutual inductance.

(*iii*) Differential transfer type.

### 7.7.1. Self-generating Type

#### 7.7.1.1. Electromagnetic type

Fig. 7.7 shows an electromagnetic type of self-generating variable inductance transducer :

— It consists of a permanent magnet core on which a coil is directly wound.

— When a plate of iron or other ferromagnetic material is moved with respect to the magnet, the flux field expands or collapses and a voltage is induced in the coil.

— This device is used for *indication of angular speed*. The measurements of speed can be made with great accuracy when the pick-up is placed near the teeth of a rotating gear.

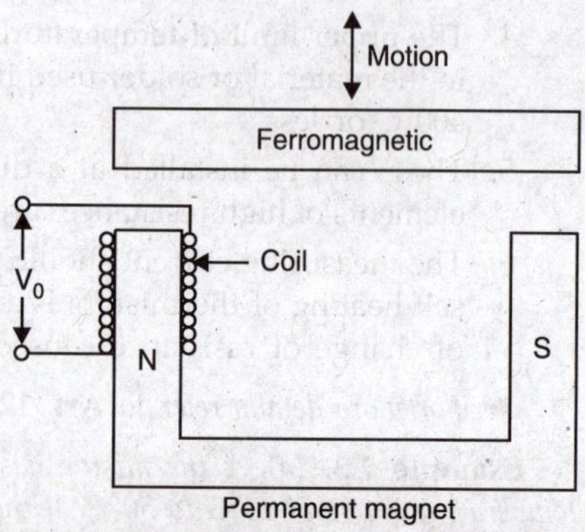


Fig. 7.7. Self-generating variable inductance transducer—Electromagnetic type.

#### 7.7.1.2. Electrodynamic type

This type of transducer, linear and rotational is shown in Fig. 7.8.

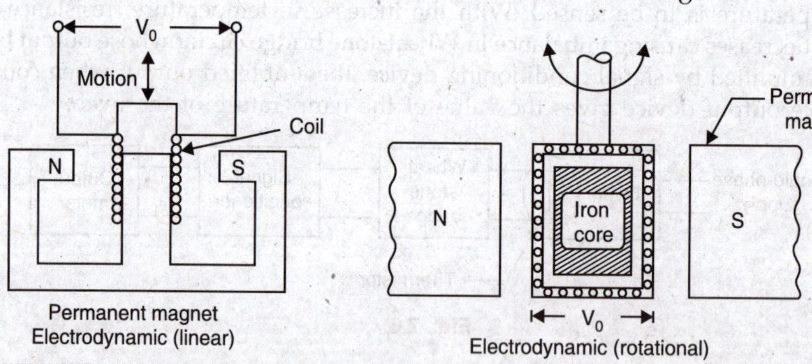


Fig. 7.8. Self-generating variable inductance transducer—Electrodynamic type.

— In this type, coil moves within the field of the magnet. The turns of the coil are perpendicular to the intersecting lines of force.

— When the coil moves it induces a voltage which at any moment is *proportional to the velocity of the coil.*

- The principle of these transducers is used in the *magnetic flow meters.*

#### 7.7.1.3. Eddy current type

Fig. 7.9 shows an eddy current type of self-generating variable inductance transducer.

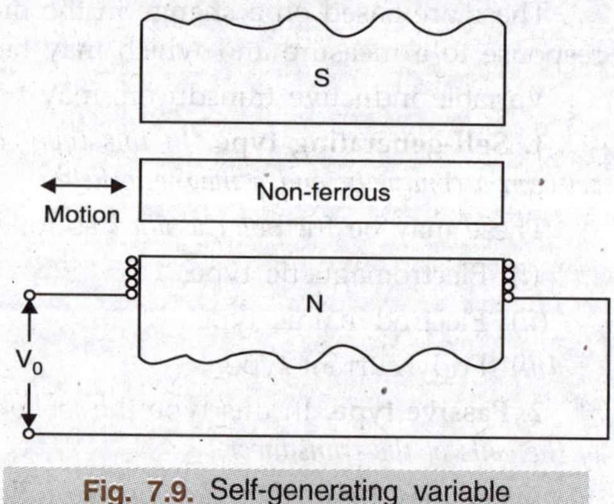


Fig. 7.9. Self-generating variable inductance transducer—Eddy current type.

**Eddy current or drag cup tachometer :**

In this type of tachometer (Fig. 7.10) the test-shaft rotates a permanent magnet and this induces eddy currents in a drag cup of disc held close to the magnet. The eddy currents produce a torque which rotates the cup against the torque of a spring. The disc turns in the direction of rotating magnetic field until the torque developed equals that of the spring. A pointer attached to the cup indicates the rotational speed on a calibrated scale.

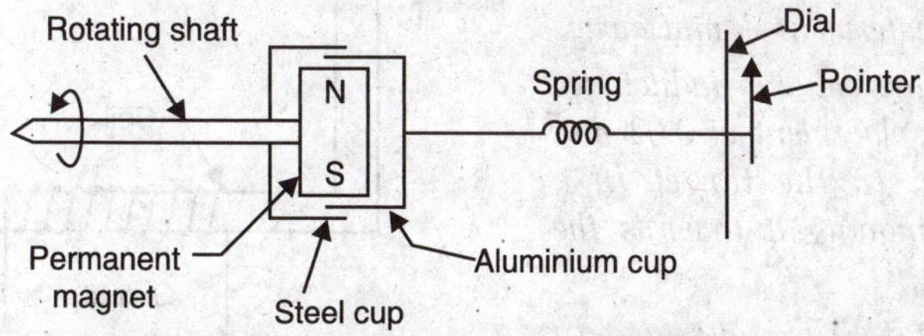


Fig. 7.10. Eddy current or drag type tachometer.

- The *automobile speedometers* operate on this principle.
- These tachometers are used for measuring rotational speeds upto 12000 r.p.m. with an accuracy of ± 3 per cent.

## 7.7.2. Passive Type

### 7.7.2.1. Variable reluctance transducer

In these transducers (comprising of a magnetic field and core with a gap between the core and the fixed coils) a change in the reluctance of the magnetic circuit by a mechanical input results in a similar change both in the inductance and inductive reactance of the coils. The change in inductance is then measured by suitable circuitry and related to the value of mechanical input.

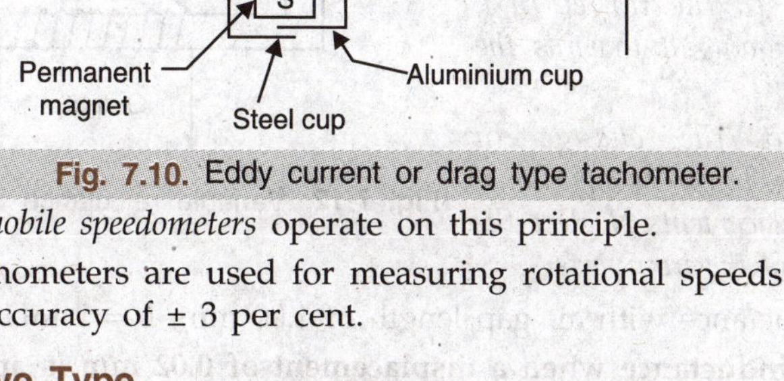


Fig. 7.11. Variable reluctance transducer.

The magnetic circuit reactance may be changed by affecting a change :

(*i*) in the *air gap* or

(*ii*) in the *amount/type of core material*.

— Transducers which make use of *air gap change* are referred as *reluctance type.*

— Transducers which utilize a *variable core* are referred as *permeance type.*

A variable *reluctance transducer* is shown in Fig. 7.11. Here the inductance of a single coil is changed through the variable air gap. The change in inductance may be calibrated in terms of movement of the armature.

This principle of variable reluctance is used for the measurement of dynamic quantities such as :

(*i*) Pressure (*ii*) Force
(*iii*) Displacement (*iv*) Acceleration
(*v*) Angular position etc.

**Example 7.4.** *Fig. 7.12 shows a variable reluctance type proximity inductive transducer in which the coil has inductance of 2 mH when the target made of ferromagnetic material is 1 mm away.*

(*i*) *Calculate the value of inductance when a displacement of 0.02 mm is applied to the target in a direction moving it towards the core.*

(*ii*) *Show that the change in inductance is linearly proportional to the displacement. Neglect the reluctance of the iron parts.*

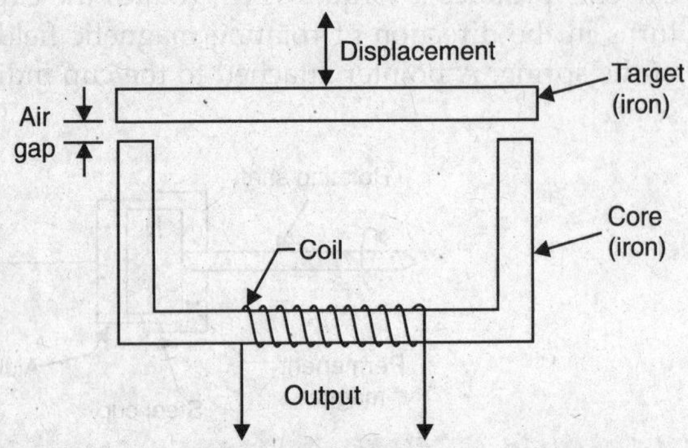


**Fig. 7.12.** Variable reluctance inductive transducer.

**Solution.** Inductance with air gap length of 1.00 mm, $L = 2$ mH

(*i*) **Value of inductance when a displacement of 0.02 mm is applied :**

Length of air gap when a displacement of 0.02 mm is applied towards the core

$$= 1.00 - 0.02 = 0.98 \text{ mm}$$

Now, the inductance is inversely proportional to the length of air gap as the reluctance of flux paths through iron are neglected. *Since the gap length decreases the inductance increases by ΔL.*

$$\therefore \quad L + \Delta L = 2 \times \frac{1}{0.98} = \mathbf{2.04\ mH\ (Ans.)}$$

or, $$\Delta L = 2.04 - 2 = 0.04 \text{ mH}$$

(*ii*) **$\Delta L$ α displacement :**

The ratio of change in inductance to the original inductance

$$= \frac{\Delta L}{L} = \frac{0.04}{2} = 0.02$$

Also, the ratio of displacement to original gap length

$$= \frac{0.02}{1} = 0.02$$

Hence the $\Delta L$ α **displacement** .... *Proved.*

- This relationship, however, is *true of only very small values of displacement.*

**Variable permeance transducer :**

Fig. 7.13 shows a *variable permeance transducer* in which the inductance of coil is changed by *varying the core material.*

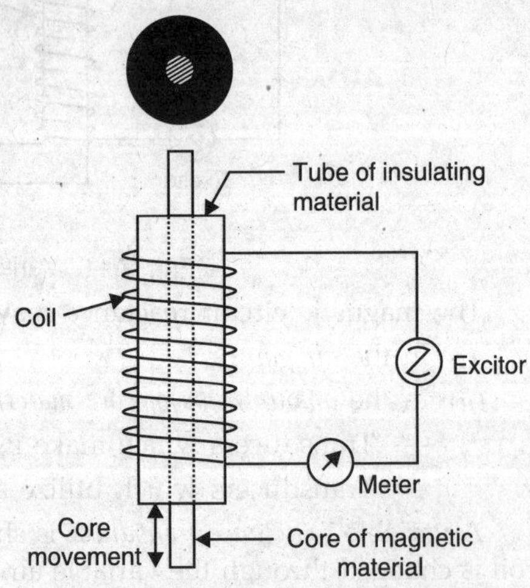


**Fig. 7.13.** Variable permeance transducer (self inductance arrangements).

— The transducer consists of a coil of many turns of wire wound on a tube or insulating material with a moveable core of magnetic material.

— When the coil is energized and the core enters the solenoid cell, the *inductance of the coil increases in proportion to the amount of metal within the coil.*

It is primarily used for the measurement of :

(*i*) Displacement; (*ii*) Strain; (*iii*) Force.

**7.7.2.2. Mutual inductance transducer**

A two-coil mutual inductance transducer is illustrated in Fig. 7.14. It consists of an energising coil *X* and a pick-up coil *Y*. A change in the position of the armature by a mechanical input changes the air gap. This causes a change in the output from coil *Y*, which may be used as measure of the displacement of the armature *i.e.*, the mechanical input.

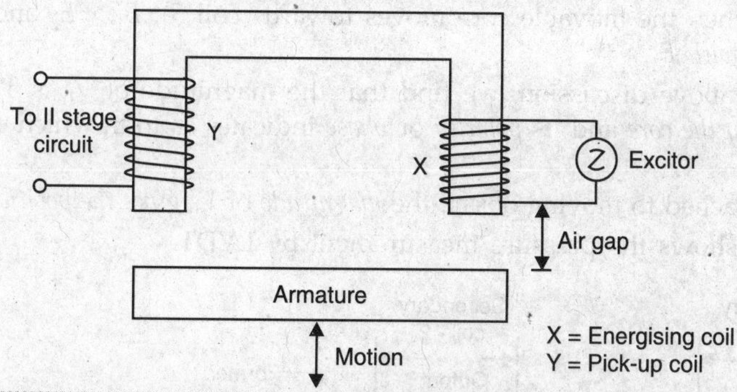


**Fig. 7.14.** Mutual inductance transducer.

**7.7.2.3. Linear-variable-differential transformer (LVDT)**

LVDT *is a passive inductive transducer and is commonly employed to measure force (or weight, pressure and acceleration etc. which depend on force) in terms of the amount and direction of displacement of an object.*

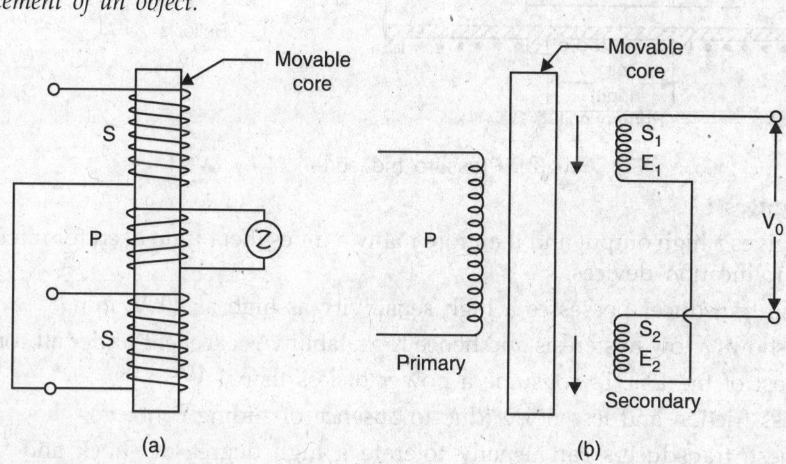


**Fig. 7.15.** Linear-variable-differential transducer (LVDT).

**Construction.** Refer to Fig. 7.15(*a*).

— It consists of one primary winding (*P*) and two secondary windings ($S_1$ and $S_2$) which are placed on either side of the primary mounted on the same magnetic core. The magnetic core is free to move axially inside the coil assembly and the motion being measured is mechanically coupled to it.

— The two secondaries $S_1$ and $S_2$ have equal number of turns but are connected in series opposition so that e.m.fs, ($E_1$ and $E_2$) induced in them are 180° out of phase with each other and hence, cancel each other out. [See Fig. 7.15 (*b*)]

— The primary is energised from a suitable A.C. source.

**Working :**

— When the core is in the centre (*called reference position*) the induced voltages $E_1$ and $E_2$ are *equal and opposite.* Hence they cancel out and the output voltages $V_0$ is zero.

— When the external applied force moves the core towards coil $S_2$, $E_2$ is increased but $E_1$ is *decreased* in magnitude though they are still antiphase with each other. The net voltage available is $(E_2 - E_1)$ and is *in phase with* $E_2$.

Similarly, when the movable core moves towards coil $S_1$, $E_1 > E_2$ and $V_0 = E_1 - E_2$ and is in *phase with* $E_1$.

Thus, from above discussion, we find that the magnitude of $V_0$ is a *function of the distance moved by the core* and its *polarity* or *phase* indicates as to in which direction it has moved.

If core is attached to moving object, the *magnitude* of $V_0$ *gives the position of that object.* Fig. 7.15(*c*) shows the pressure measurement by LVDT.

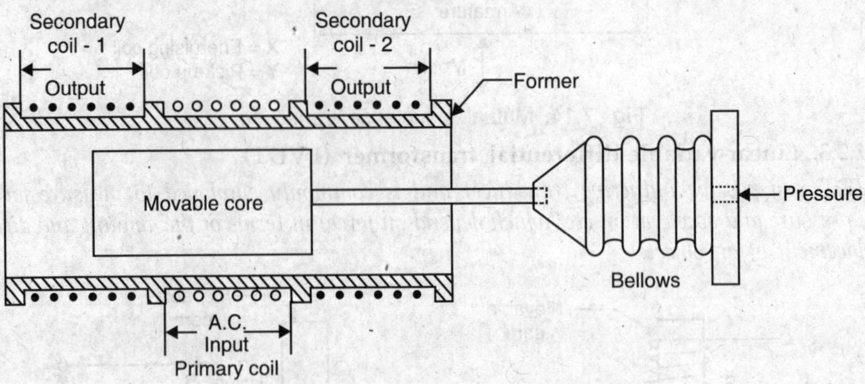


**Fig. 7.15 (*c*). Pressure measurement by LVDT.**

***Advantages :***

1. It gives a high output and therefore many a times there is no need for intermediate amplification devices.
2. The transducer possesses a high sensitivity as high as 40 V/mm.
3. It shows a low hysteresis and hence repeatability is excellent under all conditions.
4. Most of the LVDTs consume a power of less than 1 W.
5. Less friction and less noise (due to absence of sliding contacts).
6. These transducers can usually tolerate a high degree of shock and vibration without any adverse effects.
7. It can operate over a temperature range from –265°C to 600°C.
8. It is available in radiation-resistant design for operation in nuclear reactors.

***Disadvantages :***

1. These transducers are sensitive to stray magnetic fields but shielding is possible. This is done by providing magnetic shields with longitudinal slots.

2. Relatively large displacements are required for appreciable differential output.
3. The receiving instrument must be selected to operate on A.C. signals or demodulator network must be used if a D.C. output is required.
4. Several times, the transducer performance is affected by vibrations.
5. The dynamic response is limited mechanically by the mass of core and electrically by the frequency of applied voltage. The frequency of the carrier should be at least ten times the highest frequency component to be measured.

***Applications:***

1. Measurement of material thickness in hot strip or slab steel mills.
2. In accelerometers.
3. Jet engine controls in close proximity to exhaust gases.

**Note.** LVDT *is not suited for fast dynamic measurements on account of mass of the core.*

**Example 7.5.** *In a linear voltage differential transformer (LVDT) the output voltage is 1.8 V at maximum displacement. At a certain load the deviation from linearity is maximum and it is ± 0.0045 V from a straight line through the origin. Find the linearity at the given load.*

**Solution.** *Given :* The output voltage of LVDT at maximum displacement = 1.8 V

The deviation from a straight line through the origin = ± 0.0045 V

$$\therefore \quad \text{\%age linearity} = \pm\frac{0.0045}{1.8}\times 100 = \mathbf{\pm 0.25\%} \textbf{ (Ans.)}$$

**Example 7.6.** *The output of a LVDT is connected to a 4 V voltmeter through an amplifier whose amplification factor is 500. An output of 1.8 mV appears across the terminals of LVDT when the core moves through a distance of 0.6 mm. If the millivoltmeter scale has 100 divisions and the scale can be read to* $\frac{1}{4}$ *of a division, calculate:*

*(i) The sensitivity of LVDT.*

*(ii) The resolution of the instrument in mm.*

**Solution.** (*i*) **The sensitivity of LVDT :**

$$\text{The sensitivity of LVDT} = \frac{\text{Output voltage}}{\text{Displacement}} = \frac{1.8}{0.6} = \mathbf{3\ mV/mm} \textbf{ (Ans.)}$$

(*ii*) **The resolution of the instrument :**

Sensitivity of measurement = Amplification factor × sensitivity of LVDT

$$= 500 \times 3 = 1500 \text{ mV/mm}$$

$$1 \text{ scale division} = \frac{4}{100} \text{ V} = 40 \text{ mV}$$

Minimum voltage that can be read on the voltmeter

$$= \frac{1}{4}\times 40 = 10 \text{ mV}$$

∴ Resolution of the instrument

$$= 10\times\left(\frac{1}{1500}\right) = \mathbf{0.0067\ mm} \textbf{ (Ans.)}$$

## 7.8 CAPACITIVE TRANSDUCERS

The principle of operation of capacitive transducers is based upon the familiar equation for capacitance of a parallel plate capacitor :

$$\text{Capacitance, } C = \frac{\in A}{d} = \frac{\in_r \in_o A}{d} \qquad ...(7.1)$$

where, $\in = \in_r \in_o$ = Permittivity of medium, F/m,

$\in_r$ = Relative permittivity, (for air $\in_r = 1$),

$\in_o$ = Permittivity of free space = $8.85 \times 10^{-12}$ F/m,

$A$ = Overlapping area of plates, and

$d$ = Distance between the two plates.

Any physical quantity which can cause a change in $\in$, $A$ or $d$ can be measured by the *capacitance gauge.*

The *displacement is measured by measuring the change in capacitance brought about by :*

(*i*) Change in area, or

(*ii*) Change in distance between the plates.

*The change in capacitance on account of change in dielectric is used to measure change in liquid and gas levels.*

### 7.8.1. Capacitive Transducers—Using Change in Area of Plates

Figure 7.16(*a*), (*b*) shows the elementary diagrams of the arrangements of a capacitive transducer where capacitance change occurs because of change in the area of plates.

*Since capacitance is directly proportional to the effective area of the plates, response of such a system is linear.*

Fig. 7.16(*c*) shows variation of the capacitance.

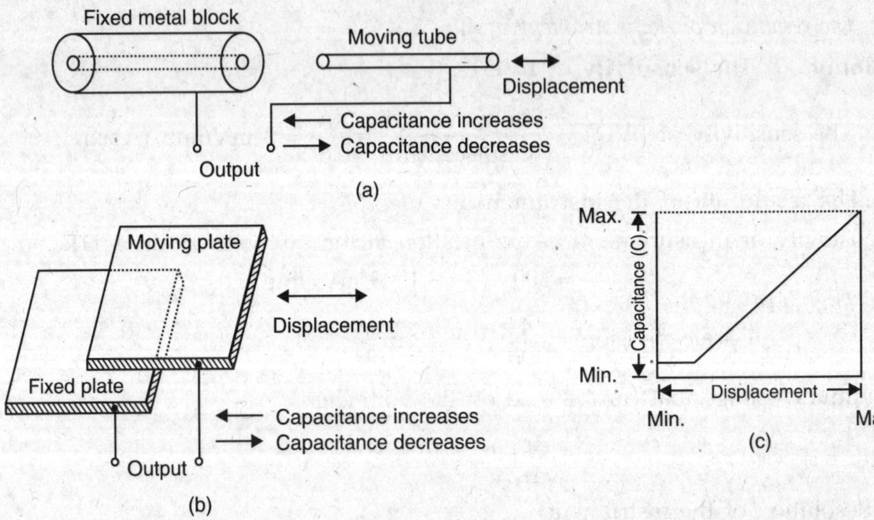


**Fig. 7.16.** Capacitive transducers working on the principle of change of capacitance with change of area.

### 7.8.2. Capacitive Transducer–Using Change in Distance Between the Plates

Fig. 7.17 shows the basic form of a capacitive transducer utilizing the effect of *change of capacitance with change in distance between the plates.*

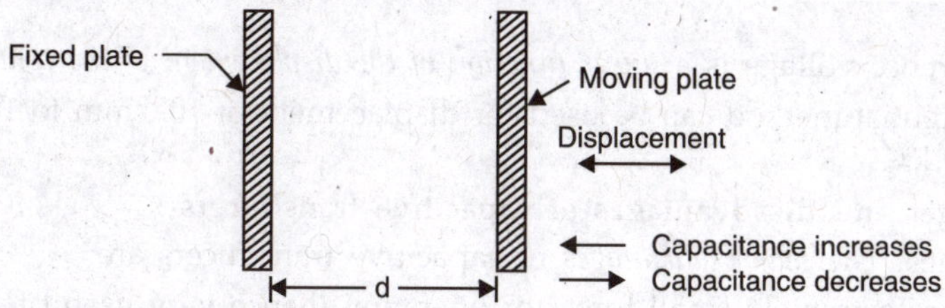


**Fig. 7.17.** Capacitive transducer.

One is a *fixed plate* and the displacement to be measured is applied to the other plate which is *movable.* Since, the capacitance, C varies inversely as the distance between the plates the response of this transducer is *not linear.*

**Differential capacitor system:**

In a differential capacitor system, let the normal position of the central plate be represented by a solid line as shown in Fig. 7.18. The capacitances $C_1$ and $C_2$ are then identical.

*i.e.,*
$$C_1 = C_2 = C = \frac{\in A}{d} \quad \text{...(7.2)}$$

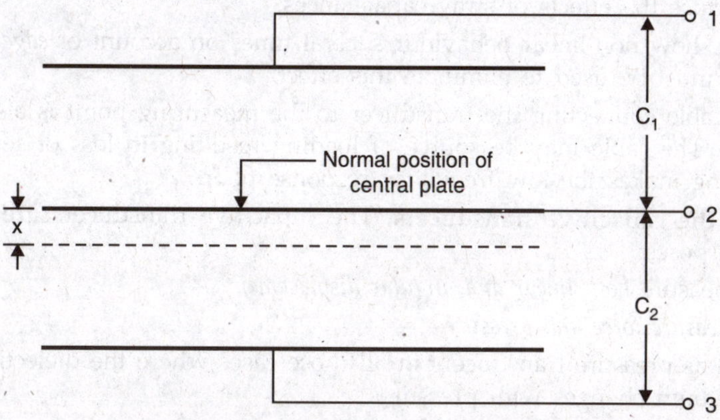


**Fig. 7.18.** Differential capacitor system.

When the central plate is displaced parallel to itself through a distance *x*, the capacitances are

$$C_1 = \frac{\in A}{d+x}, C_2 = \frac{\in A}{d-x} \quad \text{...(7.3)}$$

For an alternating voltage *E* applied between the terminals 1 and 2, the voltages across $C_1$ and $C_2$ are given by

$$E_1 = \frac{EC_2}{C_1+C_2} = E\frac{d+x}{2d}$$

and,
$$E_2 = \frac{EC_1}{C_1+C_2} = E\frac{d-x}{2d} \quad \text{...(7.4)}$$

When the differential measurement circuit is fed with output from the terminals pairs 1 and 3, and 2 and 3, the difference voltage would be recorded.

$$E_1 - E_2 = E\frac{x}{d} \quad \text{...(7.5)}$$

The difference voltage is a *linear function of the displacement of the central plate.*

The differential method can be used for displacement of $10^{-8}$ mm to 10 mm with an accuracy of 0.1%.

**Advantages and disadvantages of capacitive transducers :**

**Advantages.** The *major advantages* of capacitive transducers are:

1. Require extremely small force for operation (hence very useful for use in small systems).
2. Extremely sensitive.
3. Require small power for operation.
4. High input impedance; therefore, loading effects are minimum.
5. Frequency response is good.
6. A resolution of the order of $2.5 \times 10^{-3}$ mm can be obtained.
7. Can be used for applications where stray magnetic fields render the inductive transducers useless.

**Disadvantages.** The *principal disadvantages* of capacitive transducers are:

1. The metallic parts must be insulated from each other. The frames must be earthed to reduce the effects of stray capacitances.
2. They show non-linear behaviour several times on account of *edge effects* ; *guard rings* must be used to eliminate this effect.
3. The cable connecting the transducer to the measuring point is also a source of error. The cable may be source of loading resulting in loss of sensitivity. Also loading makes the low frequency response poor.

**Uses of the capacitive transducers.** The capacitive transducers are used for the following purposes:

1. To measure *both linear and angular displacements.*
2. To *measure force and pressure.*
3. Used as pressure transducers in all those cases where the dielectric constant of a medium changes with pressures.
4. To measure humidity in gases.
5. Used in conjunction with mechanical modifiers for measurement of *volume, density, weight, input level* etc.

**Example 7.7.** *A parallel plate capacitive transducer uses plates of area 300 mm² which are separated by a distance 0.2 mm.*

*(i) Determine the value of capacitance when the dielectric is air having a permittivity of* $8.85 \times 10^{-12}$ *F/m.*

*(ii) Determine the change in capacitance if a linear displacement reduces the distance between the plates to 0.18 mm. Also determine the ratio of per unit change of capacitance to per unit change of displacement.*

*(iii) If a mica sheet 0.01 mm thick is inserted in the gap, calculate the value of original capacitance and change in capacitance for the same displacement. Also calculate the ratio*

*of per unit change in capacitance to per unit change in displacement. The dielectric constant of mica is 8.*

**Solution.** *Given* : $A$ = 300 mm$^2$ = 300 × 10$^{-6}$ m$^2$; $d$ = 0.2 mm; $\epsilon_0$ = 8.85 × 10$^{-12}$ F/m; $\epsilon_r$ (mica) = 8

**(*i*) Value of capacitance:**

$$\textit{Value of capacitance, } C = \frac{\epsilon_o \epsilon_r A}{d} = \frac{\epsilon_o A}{d} \qquad (\because \epsilon_r = 1)$$

$$= \frac{8.85\times10^{-12}\times300\times10^{-6}}{0.2\times10^{-3}} \text{F} = \mathbf{13.275\ pF\ (Ans.)}$$

**(*ii*) Change in capacitance, $\Delta C$:**

Change is displacement Δd= 0.2 – 0.18 = 0.02 mm.

Capacitance after application of displacement,

$$C + \Delta C = \frac{8.85\times10^{-12}\times300\times10^{-6}}{0.18\times10^{-3}} \text{F} = 14.75 \text{ pF}$$

*Change in capacitance,* ΔC = 14.75 – 13.275 = **1.475 pF (Ans.)**

*Ratio of per unit change of capacitance to per unit change of displacement,*

$$\frac{\Delta C / C}{\Delta d / d} = \frac{(1.475/13.275)}{(0.02/0.2)} = \mathbf{1.111\ (Ans.)}$$

**(*iii*) *C*, Δ*C*, Ratio (Δ*C*/*C*)/(Δ*d*/*d*) when mica sheet is inserted:**

Initially, the displacement between the plates is 0.2 mm. Since the thickness of mica is 0.01 mm, the length of air between the plates = 0.2 – 0.01 = 0.19 mm.

Initial capacitance of transducer,

$$C = \frac{\epsilon_o A}{\dfrac{d_1}{\epsilon_{r_1}} + \dfrac{d_2}{\epsilon_{r_2}}}$$

or,

$$C = \frac{8.85\times10^{-12}\times300\times10^{-6}}{\left(\dfrac{0.19}{1}+\dfrac{0.01}{8}\right)\times10^{-3}} \text{F} = \mathbf{13.88\ pF\ (Ans.)}$$

When a displacement of 0.02 mm is applied, the length of air gap is reduced to 0.19 – 0.02 = 0.17 mm.

∴ Capacitance with displacement applied

$$= \frac{8.85\times10^{-12}\times300\times10^{-6}}{\left(\dfrac{0.17}{1}+\dfrac{0.01}{8}\right)\times10^{-3}} \text{F} = 15.5 \text{ pF}$$

*Change in capacitance,* ΔC = 15.5 – 13.88 = **1.62 pF (Ans.)**

$$\textit{Ratio } \frac{\Delta C / C}{\Delta d / d} = \frac{(1.62/13.88)}{(0.02/0.2)} = \mathbf{1.167\ (Ans.)}$$

### 7.8.3. Capacitive Tachometer

Refer to Fig. 7.19. A capacitive pick-up tachometer consists of a vane attached to one end of the rotating machine shaft. When the shaft rotates between the field capacitive plates, there occurs a change in the capacitance. The capacitor forms a part of an oscillator tank so that *number of frequency changes per unit of time is a measure of the shaft speed.* The pulses thus produced are *amplified* and *squared,* and may then be *fed to frequency measuring unit* or to a digital counter so as to provide a digital analog of the shaft rotation.

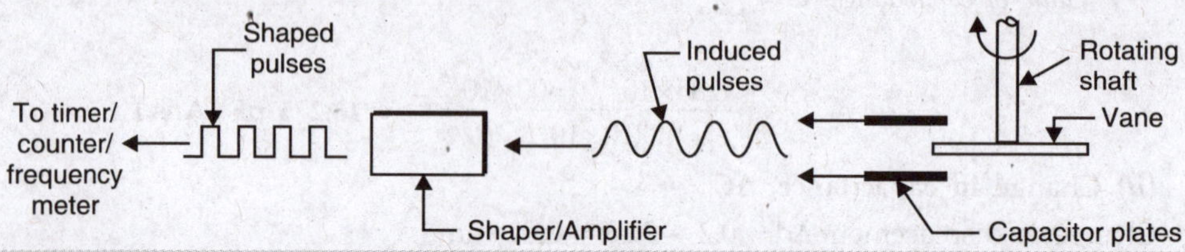


**Fig. 7.19.** Capacitive pick-up tachometer.

## 7.9 PIEZOELECTRIC TRANSDUCERS

### 7.9.1. Piezoelectric Materials

*A **"piezoelectric material"** is one in which an electric potential appears across certain surfaces of a crystal if the dimensions of the crystals are changed by the application of a mechanical force.* This potential is produced by the displacement of external charges. The effect is reversible, *i.e.,* conversely, if a varying potential is applied to the proper axis of the crystal, it will change the dimensions of the crystal thereby deforming it. This effect is known as **piezoelectric effect.**

Elements exhibiting piezoelectric qualities are sometimes known as *electro-resistive elements.* Common piezoelectric materials are : *Ammonium dihydrogen phosphate, Rochelle salt, lithium sulphate, dipotassium tartrate, potassium dihydrogen phosphate, quartz and ceramics A and B.*

There are two main groups of piezoelectric crystals:

1. ***Natural crystals*** ..... such as *quartz and tourmaline.*
2. ***Synthetic crystals*** ..... such as *Rochelle salt, lithium sulphate, dipotassium tartrate etc.*

### 7.9.2. Desirable Properties of Piezoelectric Materials

The *desirable properties* of piezoelectric materials are :

(*i*) Stability.
(*ii*) High output.
(*iii*) Insensitivity to temperature and humidity.
(*iv*) The ability to be formed into most desirable shape.

***Natural crystals*** entail the following *advantages* :

(*i*) Higher mechanical and thermal stability.
(*ii*) Ability to withstand higher stresses.
(*iii*) Low leakage.
(*iv*) Good frequency response.

***Synthetic materials,*** in general, have a *higher voltage sensitivity.*

- *"Quartz"* is the most stable piezoelectric material. However, its output is quite small.

- *"Rochelle"* salt provides the highest output but it can be worked over a limited humidity range and has to be protected against moisture. The highest temperature is limited to 45°C.
- *"Barium titanate"* has the advantage that it can be formed into a variety of shapes and sizes since it is polycrystalline. It has also a higher dielectric constant.

### 7.9.3. Working of a Piezoelectric Device

A typical mode of operation of a piezoelectric device employed for measuring varying force applied to a simple plate is shown in Fig. 7.20. *The magnitude and polarity of the induced charge on the crystal surface is proportional to the magnitude and direction of the applied force.* The charge at the electrode gives rise to voltage ($E$), given by,

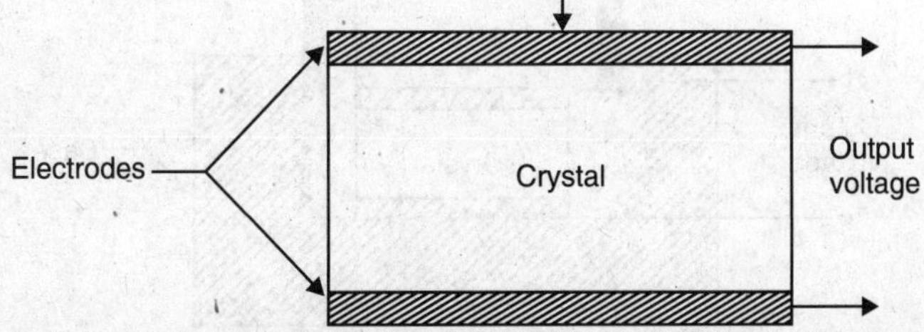


Fig. 7.20. Piezoelectric transducer.

$$E = \frac{gtF}{A} = gtp \qquad ...(7.6)$$

where, $g$ = Voltage sensitivity in Vm/N, $\left[ g = \frac{K}{t},\ k = \text{Piezoelectric constant},\ t = \text{Thickness of the crystal.} \right]$

$F$ = Force in $N$ (newton),

$A$ = Area of the crystal in $m^2$, and

$p$ = Pressure $\left(= \frac{F}{A}\right)$ in $N/m^2$.

### 7.9.4. Advantages and Disadvantages of Piezoelectric Transducers

**Advantages :**

1. High frequency response.
2. Small size.
3. High output.
4. Rugged construction.
5. Neglible phase shift.

**Disadvantages :**

1. Output affected by changes in temperature.
2. Cannot measure static conditions.

**Applications :**

These transducers find the following *fields of application:*

1. Accelerometers.

2. Pressure cells.
3. Force cells.
4. Ceramic microphones.
5. Phonograph pick-up.
6. Cartridges.
7. Industrial cleansing apparatus.
8. Under-water detection system.

### 7.9.5. Piezoelectric Accelerometer

Refer to Fig. 7.21. A piezoelectric accelerator is probably the *simplest* and *most commonly used transducer for measuring acceleration.*

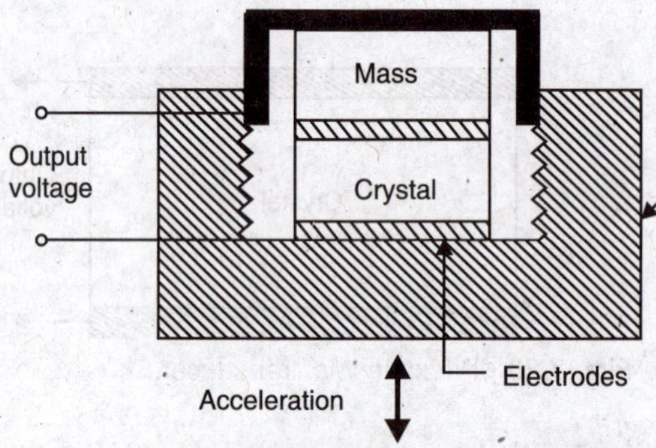


Fig. 7.21. Piezoelectric accelerometer.

**Construction.** It consists of a *piezoelectric crystal* sandwitched between two *electrodes* and has *mass*'placed on it. The unit is fastened to the base whose acceleration characteristics are to be obtained. The *can* threaded to the base acts as a spring and squeezes the mass against the crystal. *Mass exerts a force on the crystal and a certain voltage output is generated.*

**Working.** When the base is accelerated *downward* inertial reaction force on the base acts upward against the top of the can. This relieves stress on the crystal. According to Newton's second law of motion, *force* = *mass* × *acceleration,* since the mass is a fixed quantity, the *decrease in force is proportional to the acceleration.* Similarly, an acceleration in the upward direction would increase the force on the crystal in proportion to the acceleration. The resulting change in the output voltage is recorded and correlated to the acceleration imposed on the base.

*Advantages :*

1. Small size and a small weight.
2. High output impedance.
3. Can measure acceleration from a fraction of *g* to thousands of *g*.
4. High sensitivity.
5. High frequency response (10 Hz to 50 kHz).

*Disadvantages :*

1. Unsuitable for applications where the input frequency is lower than 10 Hz.
2. Subject to hysteresis errors.
3. Sensitive to temperature changes.

**Example 7.8.** *A 2.5 mm thick quartz piezoelectric crystal having a voltage intensity of 0.055 Vm/N is subjected to a pressure of 1.4 MN/m$^2$. If the permittivity of quartz is $40.6 \times 10^{-12}$ F/m, calculate:*

*(i) Voltage output.*

*(ii) Charge sensitivity of the crystal.*

**Solution.** *Given:* $t$ = 2.5 mm or $2.5 \times 10^{-3}$ m; $g$ = 0.055 Vm/N; $p$ = 1.4 MN/m$^2$; $\in (= \in_o \in_r) = 40.6 \times 10^{-12}$ F

**(*i*) Voltage output, *E*:**

$$E = gtp \qquad \text{...[Eqn. (7.6)]}$$
$$= 0.055 \times 2.5 \times 10^{-3} \times 1.4 \times 10^{6} = \mathbf{192.5\ V\ (Ans.)}$$

**(*ii*) Charge sensitivity of the crystal:**

$$\textit{Charge sensitivity} = \in_o \in_r g = \in . g$$
$$= 40.6 \times 10^{-12} \times 0.055 \text{ C/N}$$
$$= \mathbf{2.233\ pC/N\ (Ans.)}$$

**Example 7.9.** *A piezoelectric crystal measuring 6 mm × 6 mm × 1.8 mm is used to measure a force. Its voltage sensitivity is 0.055 Vm/N. Calculate the force if voltage developed is 120 V.*

**Solution.** *Given:* $A$ = 6 mm × 6 mm = $36 \times 10^{-6}$ m$^2$; $t$ = 1.8 mm or $1.8 \times 10^{-3}$ m; $g$ = 0.055 Vm/N; $E$ = 120 V

**Force *F*:**

We know that,
$$E = gtp \qquad \text{...[Eqn. (7.6)]}$$
$$120 = 0.055 \times (1.8 \times 10^{-3}) \times p$$

or
$$p = \frac{120}{0.055 \times 1.8 \times 10^{-3}} \text{ N/m}^2 = 1.212 \text{ MN/m}^2$$

∴ Force
$$F = p \times A = 1.212 \times 10^{6} \times 36 \times 10^{-6} = \mathbf{43.63\ N\ (Ans.)}$$

**Example 7.10.** *The following data relate to a barium titanate pick-up:*

| | |
|---|---|
| *Dimensions* ................. | *6 mm × 6 mm × 1.5 mm* |
| *Force acting on the pick-up* ................. | *6N* |
| *The charge sensitivity of the crystal* ................. | *150 pC/N* |
| *Permittivity* ................. | *$12.5 \times 10^{-9}$ F/m* |
| *Modulus of elasticity* ................. | *$12 \times 10^{6}$ N/m$^2$* |

*Calculate the following:*

*(i) The strain.*

*(ii) The charge and capacitance.*

**Solution.** *Given:* $A = 6 \times 6$ = 36 mm$^2$ or $36 \times 10^{-6}$ m$^2$; $t$ = 1.5 mm or $1.5 \times 10^{-3}$ m; $\in = 12.5 \times 10^{-9}$ F/m; $F$ = 6 N; $d$(charge sensitivity) = 150 pC/N; Modulus of elasticity, $E = 12 \times 10^{6}$ N/m$^2$;

**(*i*) The strain, *e*:**

$$\text{Pressure, } p = \frac{F}{A} = \frac{6}{36 \times 10^{-6}} \text{ N/m}^2 = 0.167 \text{ MN/m}^2$$

$$\textbf{Strain, } e = \frac{\text{Stress}}{\text{Young's modulus}} = \frac{0.167 \times 10^{6}}{12 \times 10^{6}} = \mathbf{0.0139\ (Ans.)}$$

**(*ii*) Charge and capacitance; Q, C:**

$$\text{Voltage sensitivity, } g = \frac{d}{\epsilon_0 \epsilon_r} = \frac{d}{\epsilon} = \frac{150\times10^{-12}}{12.5\times10^{-9}} = 12 \times 10^{-3}\ \text{Vm/N}$$

$$\therefore \quad \text{Voltage generated, } E = gtp$$

$$= 12 \times 10^{-3} \times 1.5 \times 10^{-3} \times 0.167 \times 10^{6} = 3\text{V}$$

Hence, *charge*, $Q = d \times F = 150 \times 10^{-12} \times 6$ C = **900 pC (Ans.)**

and, *Capacitance*, $C = \frac{900\times10^{-12}}{3}$ F = **300 pF (Ans.)**

## 7.10 HALL EFFECT TRANSDUCERS

### 7.10.1. Hall Effect

When a current carrying conductor is placed in a magnetic field, a *transverse effect* is noted. This effect is called Hall effect (discovered by Hall in 1879). Hall found that: *"When a magnetic field is applied at right angles to the direction of electric current an electric field is set up which is perpendicular to both the direction of electric current and the applied magnetic field".*

In other words:

*"When any specimen carrying a current I is placed in the transverse magnetic field B, then an electric field E is induced in the specimen in the direction perpendicular to both I and B. The phenomenon is known as* **Hall effect".**

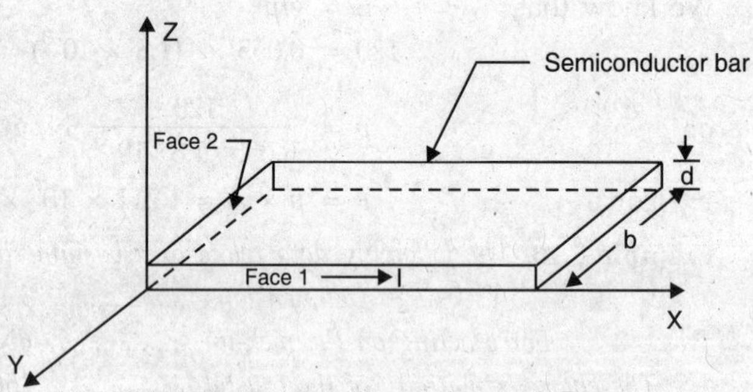


**Fig. 7.22.** Current carrying semiconductor bar subject to transverse magnetic field.

In the Fig. 7.22 is shown a specimen bar carrying a current *I* in the positive-*X* direction. Let a magnetic field *B*, be applied in the positive-*Z* direction. Then according to Hall effect, a force gets exerted on the charge carriers (whether electrons or holes) in the negative-*Y* direction. This current *I* may be due to holes moving in the positive-*X* or due to *free electrons* moving in the negative-*X* direction through the *semiconductor specimen.* Hence irrespective of the nature of the charge carriers, whether holes or electrons, these charge carriers get passed *downwards towards face* 1 of the specimen shown in Fig. 7.22.

The current, in an N-type specimen, is carried almost fully by electrons. These electrons, as a result of Hall effect, accumulate on side 1 which surface then gets negatively charged relative to side 2. Consequently, a potential difference develops between surfaces 1 and 2 and is called the **'*Hall voltage*'** ($V_H$). This Hall voltage in an N-type semiconductor is positive at surface 2. On the other hand, in a P-type specimen, the Hall voltage is positive at surface 1. These two results have been verified experimentally.

*The Polarity of Hall voltage enables us to determine experimentally whether the semiconductor specimen is of N-type or P-type.*

The magnitude of Hall voltage ($V_H$) is given by the expression :

$$V_H = \frac{R_H BI}{b} \quad ...(7.7)$$

where, $R_H$ = Hall coefficient,

$B$ = Magnetic field strength,

$I$ = Current carried by the specimen, and

$b$ = Width of the specimen along the magnetic field.

The *Hall effect* may be used for:

1. Determining whether a semiconductor is N-type or P-type.
2. Determining the carrier concentration.
3. Calculating the mobility having measured the conductivity.
4. *Magnetic field meter.* The Hall voltage $V_H$ for a given current is proportional to $B$. Hence measurement of $V_H$ measures the magnetic field $B$.
5. *Hall effect multiplier.* The instrument gives an output proportional to the product of two signals. Thus if current $I$ is made proportional to one input and if $B$ is proportional to the second input, then *Hall voltage $V_H$ is proportional to the product of the two inputs.*

## 7.10.2. Hall Effect Transducers

Hall effect transducers are the transducers in which Hall effect is utilised to measure various electrical or non-electrical quantities.

*Commercial Hall effect transducers are made from germanium or other semiconductor materials.*

The following are the *applications of Hall effect transducers :*

**1. Displacement measurement:**

Hall effect transducer may be used to measure a linear displacement or to locate a structural element is cases *where it is possible to change the magnetic field strength by variation in the geometry of a magnetic structure.*

Fig. 7.23 shows the arrangement of Hall-effect displacement (linear) transducer . The Hall effect element is located in the gap, adjacent to the permanent magnet. The field strength produced in the gap due to the permanent member is changed by changing the position of a ferromagnetic plate. The voltage output of the Hall effect element is proportional to the field strength in the gap which is function of the position (*i.e.,* displacement) of ferromagnetic plate with respect to the structure.

- With this method the *displacements as small as 0.025 mm can be measured.*

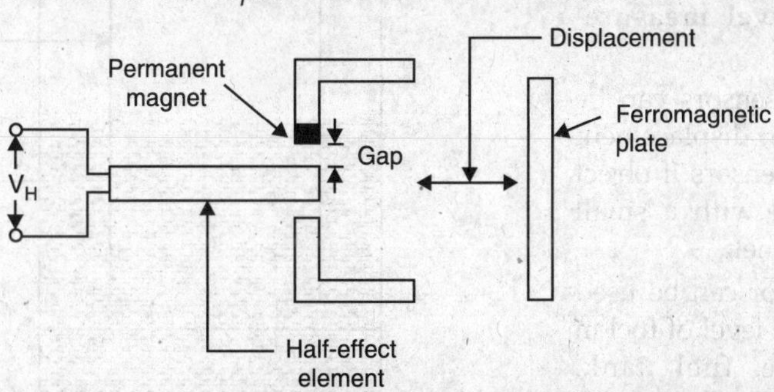


**Fig. 7.23.** Hall-effect displacement transducer.

**2. Current measurement:**

Hall effect transducer can be used to measure current in a conductor without interrupting the circuit and without making electrical connection between the conductor circuit and the meter.

When a D.C. or A.C. current flows through the conductor, it sets up a magnetic field around. This *magnetic field is proportional to the current.* A Hall effect transducer is inserted in a slotted ferromagnetic tube which acts as a magnetic concentrator. *The voltage produced at the output terminals is proportional to the magnetic field strength and hence is proportional to the current, flowing through the conductor.*

— The magnetic concentrator can be omitted at high current levels since the magnetic field is fairly strong in the vicinity of the Hall element and thus can cause output voltages which can be detected easily.

- This method can be used *to measure current from less than a mA to thousands of amperes.*

**3. Magnetic flux measurement:**

The magnetic flux can be measured by using Hall effect transducer. Here, a semiconductor plate is inserted into the magnetic field which is to be measured. The magnetic lines of force are perpendicular to the semiconductor. The transducer gives an output voltage which is *proportional to the magnetic field intensity (B).*

Following are the *advantages* and *disadvantages* of the system:

***Advantages:***

(*i*) The system requires a very small space in the direction of the magnetic field and thus the Hall effect element can be inserted in narrow gaps for magnetic measurements in air spaces.

(*ii*) The Hall effect element gives out a continuous electric signal in direct response to the magnetic field strength.

***Disadvantages:***

High sensitivity to temperature variations, and Hall coefficient may vary from plate to plate which may need individual calibration in each case.

**4. Fluid level measurement:**

Hall effect sensors can be used as position, displacement and proximity sensors if object is being sensed with a small permanent magnet.

Such a sensor can be used to determine the level of fuel in an automobile fuel tank. Fig. 7.24 shows a fluid level detector.

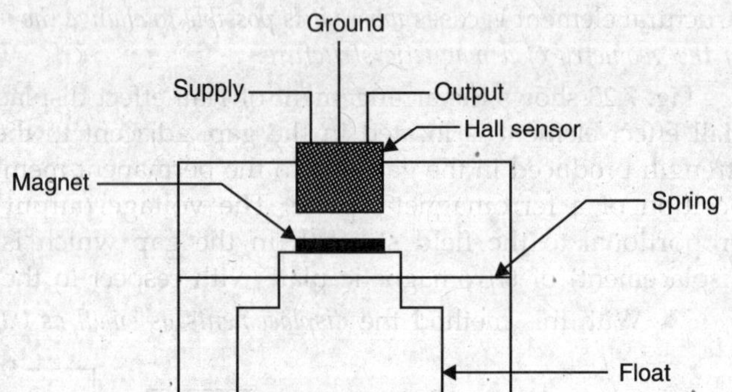


**Fig. 7.24.** Fluid level detector.

A magnet is attached to float and as the level of fuel changes and so the float distance from the Hall sensor changes. The result is a Hall voltage output which is a measure of the distance of the float from the sensor and hence the level of the fuel in the tank.

**Example 7.11.** *The resistivity of semiconductor material was known to be 0.00912 Ω m at room temperature. The flux density in the Hall model was 0.45 Wb/m$^2$.*

*Calculate the Hall angle for a Hall co-efficient of 3.55 × 10$^{-4}$ m$^3$/coloumb.*

**Solution.** Refer to Fig. 7.25.

Resistivity of the semiconductor material, $\rho = 0.00912\ \Omega\text{m}$

Flux density in the Hall model,

$$B = 0.48\ \text{Wb/m}^2$$

Hall co-efficient,

$$R_H = 3.55 \times 10^{-4}\ \text{m}^3/\text{C}$$

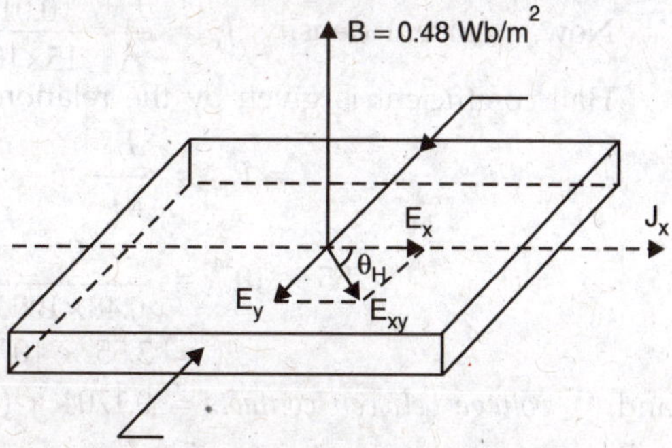


Fig. 7.25

**Hall angle, $\theta_H$:**

Resistivity, $$\rho = \frac{E_x}{J_x}$$

$$0.00912 = \frac{E_x}{J_x}$$

$$E_x = 0.00912\ J_x$$

Also, $$R_H = \frac{e_y}{B\ J_x}$$

$$3.55 \times 10^{-4} = \frac{e_y}{0.48 J_x}$$

$$E_y = 3.55 \times 10^{-4} \times 0.48\ J_x = 1.704 \times 10^{-4}\ J_x$$

Now, $$\tan\theta_H = \frac{E_y}{E_x}$$

$$\frac{1.704 \times 10^{-4} J_x}{0.00912\ J_x} = 0.01868$$

$$\theta_H = \mathbf{1°\ 4'}\ \textbf{(Ans.)}$$

**Example 7.12.** *Fig. 7.26 shows a specimen of silicon doped semiconductor having the Hall co-efficient of 3.55 × 10$^{-4}$ m$^3$/coloumb. Calculate the voltage between contacts when a current of 15 mA is flowing.*

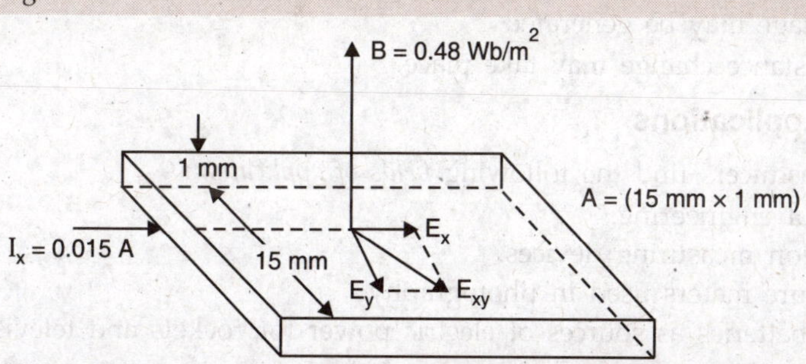


Fig. 7.26

**Solution.**

Hall co-efficient, $R_H = 3.55 \times 10^{-4}\ \text{m}^3/\text{C}$

Current, $I = 15\ \text{mA} = 0.015\ \text{A}$

Area, $A = 15\ \text{mm} \times 1\ \text{mm} = 15 \times 10^{-6}\ \text{m}^2$

Flux density, $B = 0.48\ \text{Wb/m}^2$.

**Voltage between contacts:**

Now, current density, $J_x = \dfrac{I_x}{A} = \dfrac{0.015}{15 \times 10^{-6}} = 1000\ \text{A/m}^2$

Hall co-efficient is given by the relation:

$$R_H = \frac{E_y}{BJ_x}$$

$$3.55 \times 10^{-4} = \frac{E_y}{0.48 \times 1000}$$

$$E_y = 3.55 \times 10^{-4} \times 0.48 \times 1000 = 0.1704\ \text{V/m}$$

and, *voltage between contacts* $= 0.1704 \times (15 \times 10^{-3}) =$ **0.002556 V. (Ans.)**

## 7.11 THERMOELECTRIC TRANSDUCERS

Two dissimilar metal conductors when joined at the ends and the two junctions kept at different temperatures, then a small e.m.f. is produced in the circuit. *The magnitude of this voltage depends upon the materials of conductors and the temperature difference between the two junctions.* This thermoelectric effect is used in **thermocouples** for the *measurement of temperature.*

Any number of combination of metals may be used. Two commonly employed combinations are:

1. *Iron and constantan* (an alloy of copper and nickel).
2. *Chromel* (an alloy of chromium and nickel) and *alumel* (an alloy of aluminium and nickel).

- For more details refer to **Article 12.25.**

## 7.12 PHOTOELECTRIC TRANSDUCERS

### 7.12.1. Principle of Operation

The photoelectric transducers *operate on the principle that when light strikes special combination of materials then following may result:*

(*i*) Electrons may flow.
(*ii*) A voltage may be generated.
(*iii*) A resistance change may take place.

### 7.12.2. Applications

These transducers find the following *fields of application:*

1. Control engineering.
2. Precision measuring devices.
3. Exposure meters used in photography.
4. Solar batteries as sources of electric power for rockets and television, counting machines etc.
5. Satellites used in space research.

### 7.12.3. Classification

Photoelectric transducers may be *grouped as follows:*

1. Photoemissive cell.
2. Photovoltaic cell.
3. Photoconductive cell.

### 7.12.4. Photoemissive Cell

This cell is also known as *photo tube.* It is based on the emission of electrons from a metal cathode (or photo-sensitive surface) when it is exposed to light or other radiation.

Refer to Fig. 7.27. It consists of two metallic electrodes (*i.e.,* a cathode and an anode) supported in an evacuated glass bulb fitted with a base like a thermionic valve. The cathode is either semi-cylindrical or V-shaped and is made of a metal coated with an emissive material. The anode is in the form of a thin wire facing the cathode.

When the light falls on the cathode photo-electrons are emitted which are attracted by the positive anode. Subsequently current is produced whose magnitude (for a given cathode) depends on (*i*) intensity of incident radiation and (*ii*) anode cathode voltage.

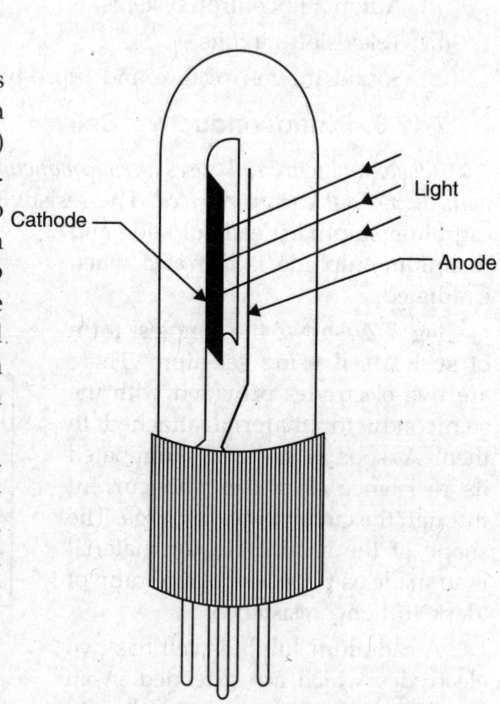

**Fig. 7.27.** Photoemissive cell.

Photo-emissive cell finds use in: (*i*) field of *photometry and calorimetry,* (*ii*) *sound reproduction* from a *motor-picture film,* (*iii*) *'on and off' circuits* and *other circuits* concerning the counting or sorting of objects on a conveyor belt, automatic opening of a door etc.

### 7.12.5. Photovoltaic Cell

In this cell sensitive element is a *semiconductor* (not metal) which generates voltage in proportion to the light or any radiant energy incident on it. The most commonly used photo-voltaic cells are barrier layer type like iron-selenium cells or Cu-$CuO_2$ cells.

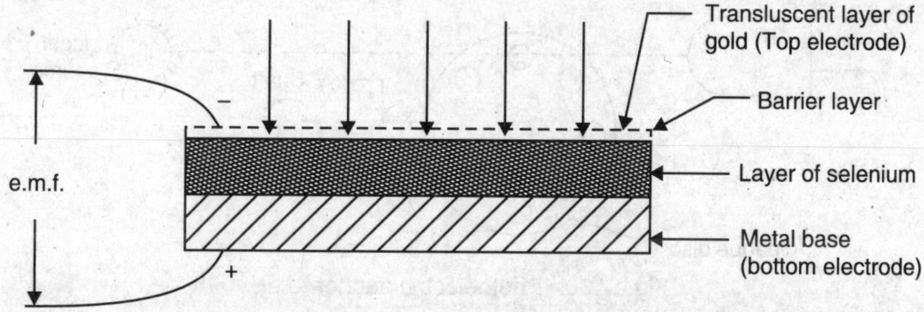

**Fig. 7.28.** Photovoltaic cell.

Fig. 7.28 shows a typical widely used photo-voltaic cell—"Selenium cell". It consists of a metal electrode on which a layer of selenium is deposited; on the top of this a barrier layer is formed which is coated with a very thin layer of gold. The latter serves as a transluscent electrode through which light can impinge on the layer below. Under the influence of this light, a negative charge will build up on the gold electrode and a positive charge on the bottom electrode.

Photo-voltaic cells are widely used in the following *fields:*

(*i*) Automatic control systems.

(*ii*) Television circuits.

(*iii*) Sound motion picture and reproducing equipment.

### 7.12.6. Photoconductive Cell

*"Photoconductive cell" uses a semiconductor material whose resistance changes in accordance with the radiant energy received.* The resistivity of semiconductor materials like selenium, cadmium sulphide, lead sulphide and thalmium sulphide is *decreased* when irradiated.

Fig. 7.29 shows the simplest form of such a cell using selenium. There are two electrodes provided with the semiconductor material attached to them. As soon as the cell is illuminated its resistance decreases and current through the circuit becomes large. The shape of the semiconductor material is so made as to obtain a large ratio of 'dark to light' resistance.

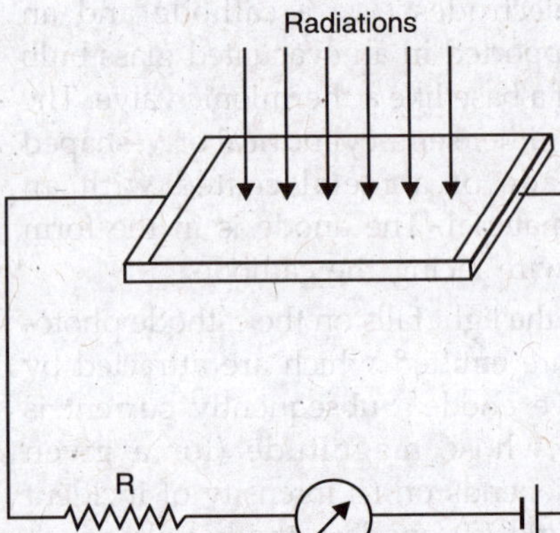


Fig. 7.29. Photoconductive cell.

A cadmium sulphide cell has two electrodes which are extended in an inter-digital pattern in order to increase the contact area with the sensitive material. It has high 'dark to light' ratio.

Photoconductive cells are generally used for *detecting ships and aircrafts by the radiations given out by their exhausts or (funnels) and for telephony by modulated infrared lights.*

### 7.12.7. Photoelectric Tachometer

Fig. 7.30 shows a photoelectric tachometer.

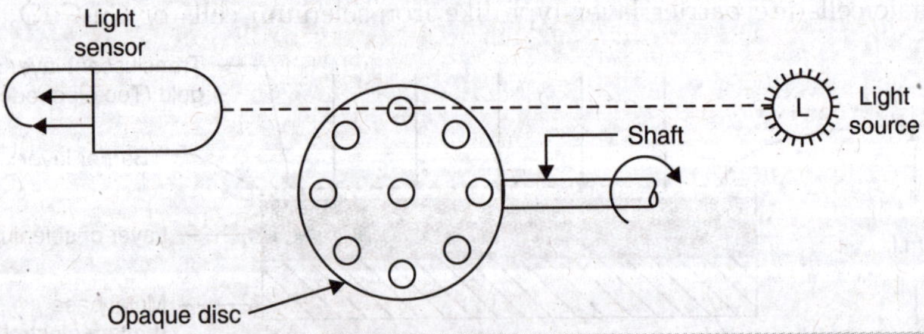


Fig. 7.30. Photoelectric tachometer.

- It consists of an *opaque disc* mounted on the shaft whose speed is to be measured. The disc has a number of equivalent holes around the periphery. On one side of the disc there is a source of light (*L*) while on the other side there is a light sensor (may be a photosensitive device or phototube) in line with it (light-source).

- On the rotation of the disc, holes and opaque portions of the disc come alternatory in between the light source and the light sensor. When a hole comes in between the *two*, light passes through the holes and falls on the light sensor, with the result that an *output pulse is generated.* But when the opaque portion of the disc comes in between, the light from the source is blocked and hence there is no pulse output. Thus *whenever a hole comes in line with the light source and sensor, a pulse is generated.* These pulses are counted/measured through an *electric counter.*

The number of pulses generated depends upon the following *factors* :

(*i*) The number of holes in the disc;

(*ii*) The shaft speed.

*Since the number of holes are fixed, therefore, the number of pulses generated depends on the speed of the shaft only.* The electronic counter may therefore be calibrated in terms of speed (r.p.m.)

***Advantages.*** It is a *digital instrument.*

***Disadvantages.*** It is required to replace the light source periodically and if the grating period is small then errors might creep in the output.

## 7.13 STRAIN GAUGES

### 7.13.1. Introduction

*When a metal conductor is stretched or compressed, its resistance changes on account of the fact that both length and diameter of conductor change. The value of resistivity of the conductor also changes. When it is strained its property is called* ***piezo-resistance****.* Therefore, resistance strain gauges are also known as ***piezo-resistive gauges.***

The **strain** *gauge is a measurement transducer for measuring strain and associated stress in experimental stress analysis.* Secondly many other detectors, and transducers, notably the load cells, torque meters, diaphragm type pressure gauges, temperature sensors, accelerometers and flow meters, employ *a strain gauge as a secondary transducer.*

### 7.13.2. Type of Strain Gauges

Four types of strain gauges are:

1. Wire-wound strain gauges.
2. Foil-type strain gauges.
3. Semiconductor strain gauges.
4. Capacitive strain gauges.

#### 7.13.2.1. Wire-wound strain gauges

There are two main classes of wire-wound strain gauges:

1. Bonded strain gauge.
2. Unbonded strain gauge.

***Bonded strain gauge:***

It is composed of fine wire, wound and cemented on a resilient insulating support, usually a wafer unit. Such units may be mounted upon or incorporated in mechanical elements or structures whose deformations under stress are to be determined. While there are no limits to the basic values which may be selected for strain-gauge resistances, a typical example may be taken as of the order or 100 to 500 Ω.

Fig. 7.31 shows the commonly used form of resistance wire strain gauges.

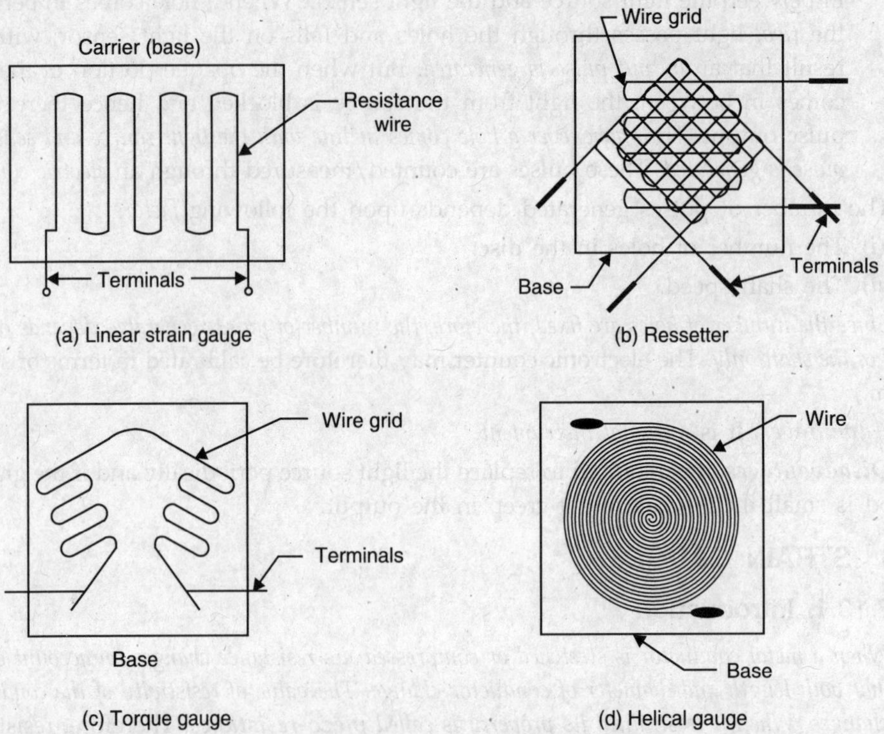


**Fig. 7.31.** Resistance wire strain gauges.

***Unbonded strain gauge:***

Fig. 7.32 shows an *unbonded strain gauge. M* and *N* are attached by rods *m* and *n*, respectively, to points between which *displacement is to be measured*. Pick-up and measuring networks are energized from similar but isolated source. Unbalance originating in pick-up is detected and balanced by servo-actuated measuring network, *providing a reading of strain on graduated scale.*

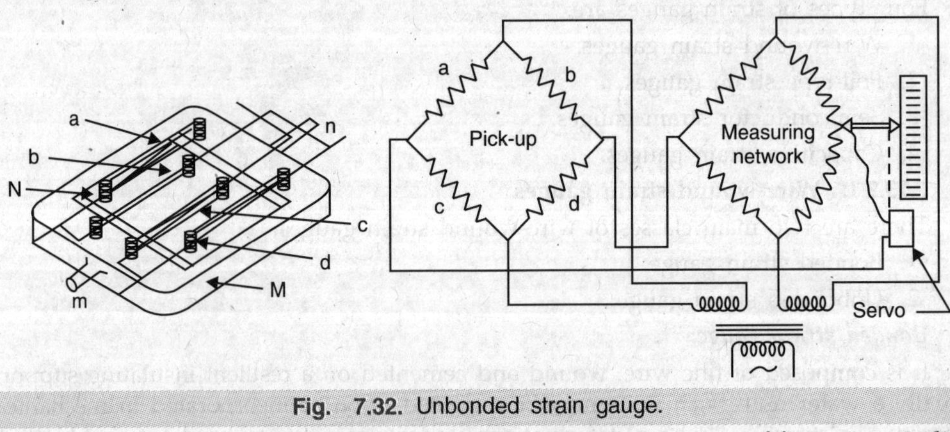


**Fig. 7.32.** Unbonded strain gauge.

In the unbonded strain gauge the resistance structure comprises of fine wire winding stretched between insulating supports mounted alternately on the two members between which displacement is to be measured (see Fig. 7.32). These wires comprise the four arms of a Wheatstone-bridge network of *which two opposite arms are tightened and the other two slackened by the displacement.*

— *While a bonded gauge tends to respond to the average strain in the surface to which it is cemented, the unbonded form measures displacement between the two points to which the respective supports are attached.*

— *Unbonded wire strain gauges* are usually operated on input potentials ranging upto 35 V direct or alternating current. Under conditions of extreme balance, corresponding to full operating range, the open-circuit e.m.f. may be of the order of 8 to 10 mV and closed circuit current upto 100 μA.

Strain gauges for use on A.C. circuits are supplied in both capacitive and inductive forms, wherein the corresponding characteristics of A.C. circuit components are varied by the displacement to be measured.

***Requirements/Characteristics of resistance wire strain gauges:***

*The resistance wire strain gauges should have the following characteristics to have excellent and reproducible results.*

1. The strain gauge should have a *high value of gauge factor.* A high value of gauge factor indicates a large change in resistance for a particular strain resulting in high sensitivity.
2. The *resistance of strain gauge should be as high as possible* since this minimizes the effects of undesirable variations of resistance in the measurement circuit.
3. The strain gauges should have a *low resistance temperature co-efficient.* This is essential to minimize errors on account of temperature variations which affect the accuracy of measurements.
4. The strain gauge *should not have any hysteresis effect in its response.*
5. In order to maintain constancy of calibration over the entire range of strain gauge, it should have *linear characteristics i.e.,* the variations in resistance should be a linear function of the strain.
6. The strain gauges are frequently used for dynamic measurements and hence their frequency response should be good. The linearity should be maintained within accuracy limits over the entire frequency range.

**7.13.2.2. Foil strain gauges**

In these gauges *the strain is sensed with the help of metal foil.* Foil gauges have a *much greater dissipation capacity as compared with wire wound gauges on account of their greater surface area for the same volume.* Due to this reason they can be employed for *higher operating temperature range.*

In these gauges, the bonding is better due to large surface area of the foil. The *bonded foil* gauges *find* a wider field of action.

Fig. 7.33 shows a typical foil gauge.

- The characteristics of foil type strain gauges are similar to those of wire wound strain gauges and their gauge factors are typically the same as that of wire wound strain gauges.
- The resistance value of foil gauges which are commercially available is between 50 and 1000 Ω.
- The maximum gauge current is about 30 mA, maximum strain for ± 1 percent, linearity is about ± 0.3 percent and strain variation upto at least 50 kHz can be measured.

**Advantages :**

1. Excellent strain reproducibility.
2. Low hysteresis and creep effects.

3. Superior mechanical stability under prolonged strain and higher temperature conditions (owing to larger ratio of surface area to cross-sectional area), so these gauges can be used for higher currents.
4. Longer life (since terminals have no stress concentration due to absence of joints).
5. Can be fabricated economically on a mass scale.

**Uses.** The foil strain gauges are employed for both *stress analysis as well as for construction of transducers.*

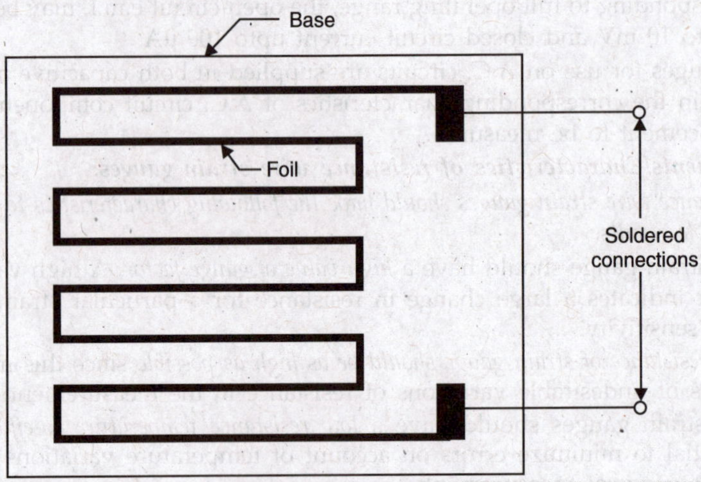


**Fig. 7.33.** Foil gauge.

### 7.13.2.3. Semiconductor strain gauges

- *Semiconductor strain gauges depend for their action upon piezo-resistive effect i.e., the change in value of the resistance due to change in resistivity.*
- These gauges are used where a *very high gauge factor and small envelope are required.*

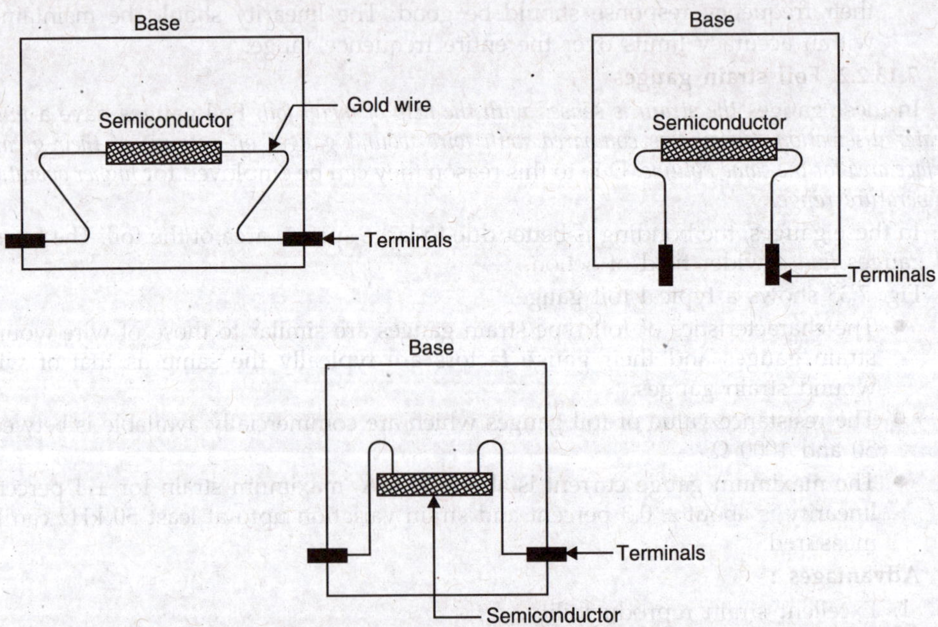


**Fig. 7.34.** Semiconductor strain gauges.

- For semiconductor strain gauges semiconducting materials such as *silicon* and *germanium* are used.
- A typical strain gauge consists of a *strain sensitive crystal material and leads that are sandwiched in a protective matrix.* The production of these gauges employs conventional semiconductor technology using semiconducting wafers or filaments which have a thickness of 0.05 mm and bonding them on suitable insulating substances, such as teflon. Gold, lead are generally applied for making the contacts.

Fig. 7.34 shows some typical semiconductor strain gauges.

***Advantages:***

1. These gauges have high gauge factor.
2. They are chemically inert and have low cross-sensitivity.
3. They are almost free from hysteresis and creep effects because of integrated format of the actual gauge itself.
4. Fatigue life is in excess of $10 \times 10^5$ operations and the frequency response is upto $10^{12}$ Hz.
5. These gauges can be very small ranging in length from 0.7 to 7 mm. *They are very useful for measurement of local strains.*

***Disadvantages:***

1. The *major and serious disadvantage is that these gauges are very sensitive to change in temperature.*
2. Linearity of these gauges is *poor.*

#### 7.13.2.4. Capacitive strain gauges

Fig. 7.35 shows a capacitive strain gauge. It uses the *principle of variation of capacitance with variation of distance between electrodes.* The electrodes are flexible metal strips of about 0.1 mm thickness. The strain to be measured is applied to the top plate. This changes the distance between the curved electrodes resulting in change of capacitance.

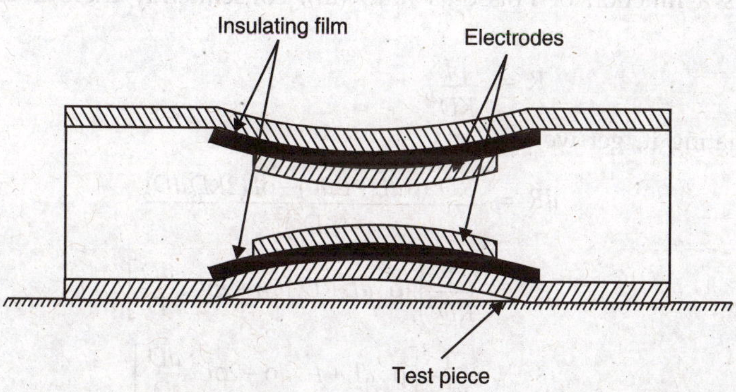


**Fig. 7.35.** Capacitive strain gauge.

The strain-capacitance relationship, in general, is *not linear* but variations in dimensions and shape allow gauge characteristics to be chosen so as to match the range of capacitance to be measured with a good degree of accuracy.

- A capacitance strain gauge has a capacitance of about 0.5 pF.
- Its overall size is 5 mm × 17 mm × 1 mm.
- It uses a polyamide film of insulating material.
- It can be used upto a temperature of 300°C.

### 7.13.3. Theory of Strain Gauges

When a strain gauge is subjected to tension (*i.e.*, positive strain) its length *increases* while its cross-sectional area *decreases*. Since the resistance of a conductor is proportional to its length and inversely proportional to its area of cross-section, the resistance of the gauge increases with positive strain. The change in the value of resistance of strained conductor is more than what can be accounted for an increase in resistance due to dimensional changes. The extra change in the value of resistance is attributed to a *change in the value of resistivity of a conductor when strained; this property is known as* ***piezo-resistive effect.***

Strain gauges are most commonly used in Wheatstone bridge circuits to measure the change of resistance of grid of wire for calibration proposes; the "**gauge factor**" *is defined as the ratio of per unit change in resistance to per unit change in length.*

*i.e.*,
$$\text{Gauge factor } (G_f) = \frac{\Delta R/R}{\Delta L/L} \quad \text{...(7.8)}$$

where, $\Delta R$ = Corresponding change in resistance $R$, and

$\Delta L$ = Change in length per unit length $L$.

The resistance of the wire of strain gauge $R$ is given by :

$$R = \frac{\rho L}{A}$$

where, $\rho$ = Resistivity of the material of wire (of strain gauge),

$L$ = Length of the wire, and

$A$ = Cross-sectional area of the wire,

$= KD^2$, $K$ and $D$ being a constant and diameter of the wire respectively.

As earlier stated, when the wire is strained its length increases and lateral dimension is reduced as a function of Poisson's ratio (m); consequently there is an increase in resistance.

Now,
$$R = \frac{\rho L}{KD^2}$$

Differentiating it, get we

$$dR = \frac{KD^2(\rho.dL + L.d\rho) - \rho L(2KD.dD)}{(KD^2)^2}$$

$$= \frac{1}{KD^2}\left[(\rho\,.\,dL + L\,.\,d\rho) - 2\rho L\,.\,\frac{dD}{D}\right]$$

or,
$$\frac{dR}{R} = \frac{\frac{1}{KD^2}\left[\rho\,.\,dL + L\,.\,d\rho - 2\rho L\,.\,\frac{dD}{D}\right]}{\frac{\rho L}{KD^2}}$$

$$= \frac{dL}{L} + \frac{d\rho}{\rho} - 2\frac{dD}{D}$$

Now, Poisson's ratio,
$$\mu = \frac{\text{Lateral strain}}{\text{Longitudinal strain}} = \frac{-dD/D}{dL/L}$$

or,
$$\frac{dD}{D} = -\mu \times \frac{dL}{L}$$

For small variations, the above relationship can be written as:

$$\frac{\Delta R}{R} = \frac{\Delta L}{L} + 2\mu \frac{\Delta L}{L} + \frac{\Delta \rho}{\rho} \quad ...(7.9)$$

Gauge factor, $G_f = \dfrac{\Delta R / R}{\Delta L / L}$

or, $$\frac{\Delta R}{R} = G_f \cdot \frac{\Delta L}{L} = G_f \times e \quad ...(7.10)$$

where, $e$ = strain = $\dfrac{\Delta L}{L}$

The *gauge factor* can be written as:

$$G_f = 1 + 2\mu + \frac{\Delta \rho / \rho}{e} \quad ...(7.11)$$

$$= 1 \quad + \quad 2\mu \quad + \quad \frac{\Delta \rho / \rho}{e}$$

= Resistance change due to change of length + Resistance change due to change in area + Resistance change due to piezo-resistive effect

or, $$G_f = 1 + 2\mu + \frac{\Delta \rho / \rho}{\Delta L / L} \quad ...(7.12)$$

The strain is usually expressed in terms of *microstrain;* 1 micro strain = 1 μm/m

If the change in the value of resistisity of a material when strained is *neglected*, the gauge factor can be rewritten as:

$$G_f = 1 + 2\mu \quad ...(7.13)$$

Eqn. (7.13) is valid only when *piezo-resistive effect* (*i.e.*, change in resistisity due to strain) is *almost negligible.*

- The Poission's ratio for all metals lies between 0 and 0.5. This give $G_f$ as 2 approximately. In case of wire wound strain gauges where the common value for Poisson's ratio is 0.3, the value of $G_f$ amounts to 1.6.
- The value of the gauge factor varies from material to material but it is generally assumed that it remains constant in the working range of strain required. *Its value is determined experimentally.*
- *Knowing the gauge factor ($G_f$), the strain in the member can be directly found out by the change of resistance.*

**Properties of gauge materials:**

The grid material for its proper functioning must possess the following desirable *properties:*

1. High resistivity.
2. High gauge factor.
3. High mechanical strength.
4. High electrical stability.
5. Low temperature sensitivity.
6. Low hysteresis.
7. Low thermal e.m.f. when joined with other materials.

8. Good corrosion resistance.
9. Good weldability.

**Adhesing techniques:**

For proper mounting of the strain gauges, the following *steps* should be strictly followed:

1. Before mounting the strain gauge on the surface, the surface must be preferably cleaned by emery cloth and base material exposed.
2. Remove the various traces of grease or oil etc. by using a solvent like acetone.
3. Swab the back of the strain gauge by cotton dipped in acetone once, to ensure that the back is free from grease etc.
4. Apply a generous quantity of cement to the cleaned resistance and then simply place the cleaned gauge on it and excess cement worked out. Make sure that there is no bubble between the surface and the gauge, if any one is there, that should be removed. Avoid using heavy pressure, otherwise cement may puncture the paper and short the grid.
5. Allow the gauge to sit for at least eight or ten hours before using it. If possible a slight weight might be placed by keeping a strong rubber on the gauge.
6. After cement has been fully cured, check the continuity of wire by an ohmmeter and weld the electric leads.

**Example 7.13.** *The gauge factor of a resistance wire strain gauge using a soft iron wire of small diameter is 4.2. Neglecting the piezo-resistive effect, calculate the Poisson's ratio.*

**Solution.** *Given:* $G_f = 4.2$

When the *piezo-resistive effect* is neglected, the gauge factor is given by:

$$G_f = 1 + 2\mu \qquad \text{...[Eqn. (7.13)]}$$

or,

$$4.2 = 1 + 2\mu$$

$$\therefore \quad \mu = \frac{4.2-1}{2} = \mathbf{1.6\ (Ans.)}$$

**Example 7.14.** *A simple electrical strain gauge of resistance 120 Ω and having a gauge factor of 2 is bonded to steel having an elastic limit stress of 400 MN/m² and modulus of elasticity is 200 GN/m². Calculate the change in resistance,*

*(i) due to a change in stress equal to* $\frac{1}{10}$ *of the elastic range;*

*(ii) due to change of temperature of 20°C if the material is advance alloy. The resistance temperature coefficient of advance alloy is 20 × $10^{-6}$/ °C.*

**Solution.** *Given:* $R = 120\ \Omega$; $G_f = 2$; Elastic limit stress = 400 MN/m²; Modulus of elasticity = 200 GN/m²; Resistance temperature coefficient, $\alpha_0 = 20 \times 10^{-6}$/°C.

**Change in resistance:**

(*i*) $$\text{Change in stress} = \frac{1}{10} \times 400\ \text{MN/m}^2 = 40 \times 10^6\ \text{N/m}^2$$

$$\text{Modulus of elasticity} = 200\ \text{GN/m}^2 = 200 \times 10^9\ \text{N/m}^2$$

$$\text{Strain, } e = \frac{\text{Stress}}{\text{Modulus of elasticity}} = \frac{40\times10^6}{200\times10^9} = \frac{1}{5}\times10^{-3}$$

$$\text{Gauge factor } G_f = \frac{\text{Per unit change in resistance}}{\text{Per unit change in length}}$$

or, $$G_f = \frac{\Delta R / R}{e} \quad \text{or} \quad \Delta R = R\, G_f\, e$$

or, $$\Delta R = 120 \times 2 \times \frac{1}{5} \times 10^{-3} = 48 \times 10^{-3}\ \Omega = \mathbf{45\ m\Omega}\ \textbf{(Ans.)}$$

(ii) $$R_{t2} = R_{t1}\,[1 + \alpha_0(t_2 - t_1)]$$

∴ Change in resistance $R_{t2} - R_{t1} = R_{t1}\ \alpha_0(t_2 - t_1)$

or, $$\Delta R = R_{t2} - R_{t1} = 120 \times 20 \times 10^{-6} \times (20)$$
$$= 48 \times 10^{-3}\ \Omega = \mathbf{48\ m\Omega}\ \textbf{(Ans.)}$$

**Example 7.15.** *A strain gauge is bonded to a beam which is 12 cm long and has a cross-sectional area of 3.8 cm$^2$. The unstrained resistance and gauge factor of the strain gauge are 220 Ω and 2.2 respectively. On the application of load the resistance of the gauge changes by 0.015 Ω. If the modulus of elasticity for steel is 207 GN/m$^2$, calculate:*

*(i) The change in length of the steel beam.*

*(ii) The amount of force applied to the beam.*

**Solution.** *Given:* $L$ = 12 cm = 0.12 m; $A$ = 3.8 cm$^2$ = 3.8 × 10$^{-4}$ m$^2$; $R$ = 220 Ω; $G_f$ = 2.2; $\Delta R$ = 0.015 Ω; $E$ = 207 GN/m$^2$.

**(i) The Change in length of steel beam. ΔL:**

Gauge factor, $$G_f = \frac{\Delta R / R}{\Delta L / L}$$

∴ $$\Delta L = \frac{(\Delta R/R).L}{G_f} = \frac{(0.015/220) \times 0.12}{2.2} = \mathbf{3.72 \times 10^{-6}\ m}\ \textbf{(Ans.)}$$

**(ii) The amount of force applied to the beam, F:**

$$E = \frac{\text{Stress}}{\text{Strain}} = \frac{\sigma}{e}$$

∴ $$\sigma = E \times e = E \times \frac{\Delta L}{L}$$

$$= (207 \times 10^9) \times \frac{3.72 \times 10^{-6}}{0.12} = 6.417 \times 10^6\ \text{N/m}^2$$

∴ $$\text{Force, } F = \sigma \cdot A = 6.417 \times 10^6 \times 3.8 \times 10^{-4}\ \text{N} = \mathbf{2.438\ kN}\ \textbf{(Ans.)}$$

### 7.13.4. Strain-gauge Circuits

The following strain gauge circuits will be discussed :

1. Ballast circuit.
2. Wheatstone bridge circuit.
   (i) Balanced (null) condition
   (ii) Unbalanced (deflection) condition
   — Quarter bridge
   — Half bridge
   — Full bridge.

#### 7.13.4.1. Ballast-circuit (Voltage-sensitive potentiometric circuit)

Fig. 7.36 shows a ballast circuit-voltage-sensitive potentiometric circuit. Here,

$v_i$ = Input supply voltage,

$v_0$ = Output voltage,

$R_b$ = Ballast resistance, and

$R_g$ = Resistance of the unstrained resistance gauge.

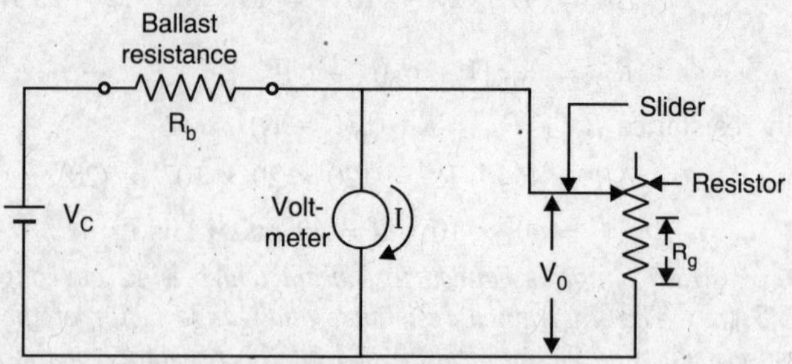


**Fig. 7.36.** Ballast circuit-voltage-sensitive potentiometric circuit.

The output voltage, when *no stress* is applied to the strain gauge, is given by

$$v_0 = \left(\frac{R_g}{R_g + R_b}\right) v_i \quad \text{...(7.14)}$$

When the gauge is strained, the gauge resistance changes to $(R_g + dR_g)$ and the output voltage becomes,

$$v_0 + dv_0 = \left[\frac{(R_g + dR_g)}{(R_g + dR_g) + R_b}\right] v_i \quad \text{...(7.15)}$$

∴ The *change in the output voltage,*

$$dv_0 = \left[\frac{(R_g + dR_g)}{(R_g + dR_g) + R_b} - \frac{R_g}{R_g + R_b}\right] v_i$$

$$= \left[\frac{(R_g + dR_g)(R_g + R_b) - R_g\{(R_g + dR_g) + R_b\}}{\{(R_g + dR_g) + R_b\}(R_g + R_b)}\right] v_i$$

$$= \left[\frac{\left(R_g^2 + R_g R_b + dR_g R_g + dR_g R_b - R_g^2 - dR_g R_g - R_g R_b\right)}{(R_g + R_b)^2}\right] v_i$$

$$\text{...}(\because dR_g > R_g)$$

$$= \left[\frac{dR_g \cdot R_b}{(R_g + R_b)^2}\right] v_i = \frac{R_g \cdot R_b}{(R_g + R_b)^2} \cdot \frac{dR_g}{R_g} v_i \quad \text{...(7.16)}$$

... Multiplying numerator and denominator by $R_g$.

Also, *condition of maximum sensitivity* is given by : $R_g = R_b$

Hence, $$dv_0 = \frac{v_i}{4} \cdot \frac{dR_g}{R_g}$$

Also, $$\frac{dR_g}{R_g} = G_f \cdot e$$

$$\therefore \qquad dv_0 = \left(\frac{G_f}{4}\right) e\, v_i \quad \text{...where, } G_f \text{ denotes the gauge factor ...(7.16a)}$$

From eqn. (7.16*a*) it is evident that *change in output-voltage when gauge is strained is directly proportional to strain.*

- The ballast circuit is used for *dynamic strain measurements where static strain components are ignored.*

***Limitations of potentiometric circuits:***

(*i*) No possibility of compensation for temperature variations.

(*ii*) High sensitivity to A.C. interference giving hum due to ground loops, induction from high current lines and poor connections.

**7.13.4.2. Wheatstone bridge circuit**

The wheatstone bridge technique can be used in the following *two ways*: (*i*) Null mode; (*ii*) Deflection mode.

**1. Null mode:**

Refer to Fig. 7.37. The resistances, with *no straining*, are so arranged that $v_B = v_D$ and the galvanometer gives zero deflection.

Then, $$\frac{R_1}{R_3} = \frac{R_2}{R_4} \qquad \text{...(7.17)}$$

where, $R_1 = R_g$ = Unstrained resistance of the gauge.

In measurement of strains, generally $R_1$ is the strain gauge, $R_2$ and $R_4$ are the fixed resistances and $R_3$ is a variable resistor.

When the gauge is strained, its resistance $R_1$ changes by an amount $dR_1$. This change unbalances the bridge resulting into the deflection of the galvanometer. The balance is then regained by adjusting $R_3$ by an amount $dR_3$. The rebalanced conditions gives:

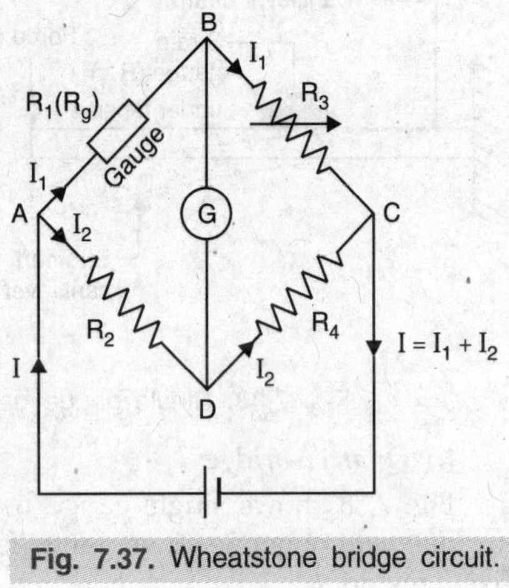

Fig. 7.37. Wheatstone bridge circuit.

$$\frac{R_1 + dR_1}{R_3 + dR_3} = \frac{R_2}{R_4}$$

or, $$R_1 + dR_1 = (R_3 + dR_3)\frac{R_2}{R_4}$$

or, $$R_1 + dR_1 = R_3 \times \frac{R_2}{R_4} + dR_3 \times \frac{R_2}{R_4}$$

or, $$R_1 + dR_1 = R_1 + dR_3 \times \frac{R_2}{R_4} \qquad \left[\because R_1 = \frac{R_3}{R_4} \times R_2 \text{ from eqn. (7.17)}\right]$$

or, $$dR_1 = dR_3 \times \left(\frac{R_2}{R_4}\right) \qquad \text{...(7.18)}$$

If the resistances, of all the limbs of the wheatstone bridge are equal, then

$$R_1 = R_2 = R_3 = R_4 = R_g$$

and, $$dR_1 = dR_3 \quad ...(7.19)$$

The change in resistance $dR_1$, in terms of strain, is given by

$$dR_1 = G_f\, e\, R_g,$$ where $G_f$ is the gauge factor and $e$ is the strain

$\therefore$ $$dR_3 = G_f\, e\, R_g \quad ...(7.20)$$

Eqn. (7.20) indicates that the *change in the value of resistance $R_3$ is direct measurement of strain.*

**2. Deflection mode:**

Initially the bridge resistances are so adjusted that the bridge is in balanced. The equilibrium gets disturbed when the gauges are strained. Then, the voltage $v_0$ is measured under this unbalanced condition.

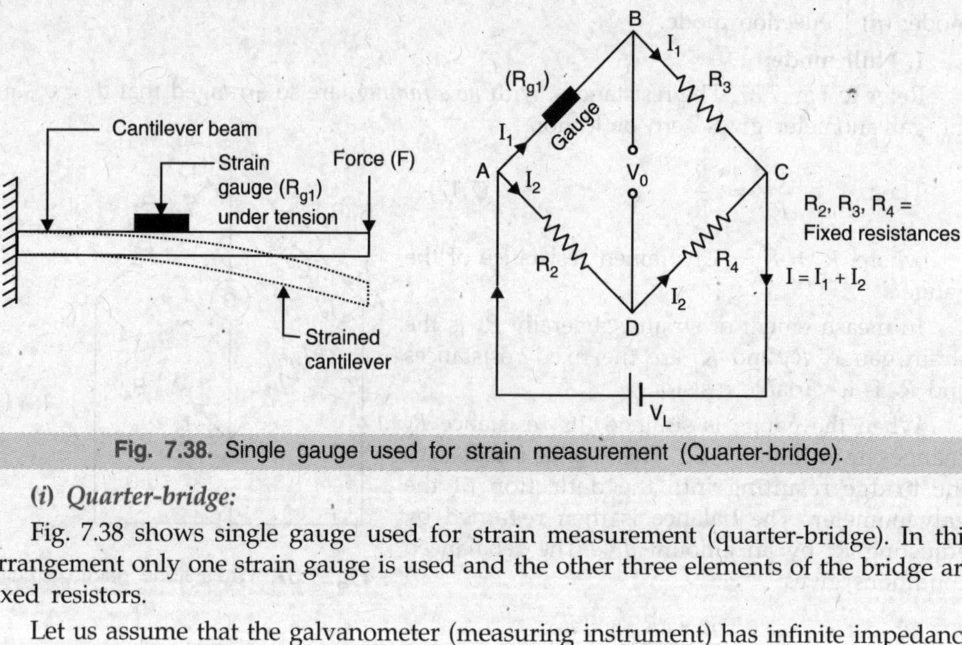


**Fig. 7.38.** Single gauge used for strain measurement (Quarter-bridge).

***(i) Quarter-bridge:***

Fig. 7.38 shows single gauge used for strain measurement (quarter-bridge). In this arrangement only one strain gauge is used and the other three elements of the bridge are fixed resistors.

Let us assume that the galvanometer (measuring instrument) has infinite impedance and therefore no current flows through it. Then,

Current flowing through the limbs $AB$ and $BC$, $I_1 = \dfrac{v_i}{R_{g1} + R_3}$ ...(7.21)

Voltage drop in limb $AB$ (or voltage at terminal $B$),

$$v_{AB} = I_1 R_{g1} = \frac{R_{g1}}{R_{g1} + R_3} \cdot v_i \quad ...(7.22)$$

Similarly, $$I_2 = \frac{v_i}{R_2 + R_4} \quad ...(7.23)$$

and, $$v_{AD} = \frac{R_2}{R_2 + R_4} \cdot v_i \quad ...(7.24)$$

*Initially,* $R_{g1} = R_2 = R_3 = R_4 = R$

$$\therefore \quad v_{AB} = v_{AD} = \frac{v_i}{2} \qquad \text{...(7.25)}$$

and, $v_0$ = Voltage across the terminals $B$ and $D$

$$= v_{AB} - v_{AD} = 0$$

Obviously, the bridge is balanced under *unstrained conditions.*

When the gauge is strained (see Fig. 7.38), the resistance $R_{g1}$ changes by an amount $dR_{g1}$. Then,

$$v_{AB} = \left[\frac{R_{g1} + dR_{g1}}{(R_{g1} + dR_{g1}) + R_3}\right] v_i = \left(\frac{R + dR}{2R + dR}\right) v_i$$

$[\because Rg_1 = R_3 = R \text{ and } dR_{g1} = dR]$

$$v_{AD} = \left(\frac{R_2}{R_2 + R_4}\right) v_i = \frac{v_i}{2} \qquad (\because R_2 = R_4 = R)$$

The changed output voltage,

$$v_0 + dv_0 = \left(\frac{R + dR}{2R + dR} - \frac{1}{2}\right) v_i$$

$$= \left(\frac{2R + 2dR - 2R - dR}{2(2R + dR)}\right) v_i = \left(\frac{dR}{4R + 2dR}\right) v_i$$

Since $dR << R$ and $v_i = 0$ (under unstrained conditions), therefore

$$dv_0 = \frac{v_i}{4} \cdot \frac{dR}{R} \qquad \text{...(7.26)}$$

or,

$$dv_0 = \left(\frac{G_f}{4}\right) e \cdot v_i \qquad \text{...(7.27)}$$

...in terms of gauge factor $G_f$ and applied strain $e$

From eqn. (7.27) it is obvious that the *output voltage is directly proportional to the applied strain.*

***(ii) Half-bridge:***

Fig. 7.39, shows two gauges used for strain measurement (Half-bridge). In this arrangement two of the bridge elements are strain gauges and the other two are fixed resistors. The strain gauge-1 is bonded to the upper surface of the cantilever beam and a second strain gauge-3 is bonded to the lower surface and located precisely underneath the gauge-1. These gauges are connected electrically to form adjacent limbs of the Wheatstone bridge circuit.

The temperature effects are cancelled out by having $R_2 = R_4$ and using *two identical gauges* in the opposite arms of the bridge.

Suppose, $R_{g1} = R_{g3} = R_2 = R_4 = R$

Under *no strain conditions:*

$$v_{AB} = v_{AD} = \frac{v_i}{2}; \; v_B = v_D \text{ and } v_0 = 0.$$

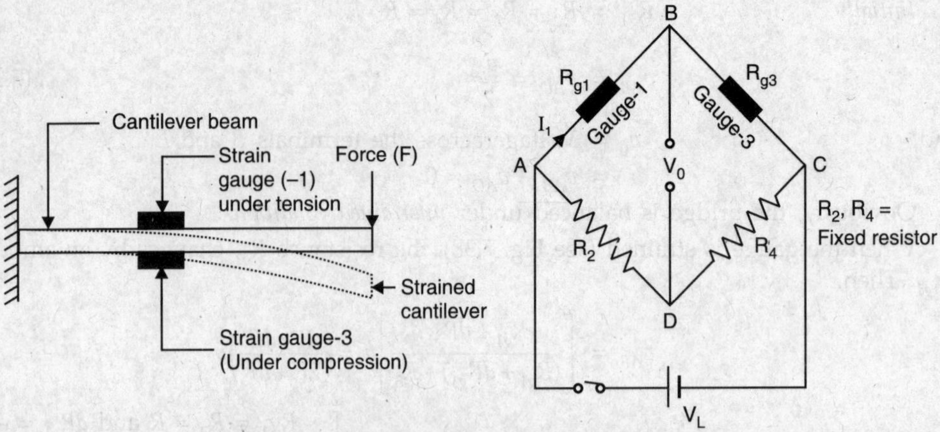


**Fig. 7.39.** Two gauges used for strain measurement (Half-bridge).

On the application of load to the cantilever beam, the resistance of the gauge $R_{g1}$ *increases* due to *tensile load* whilst $R_{g3}$ *decreases* due to equal *compressive strain* so that,

Resistance of gauge–1 = $R_{g1} + dR_{g1}$

and, Resistance of gauge–3 = $R_{g3} - dR_{g3}$

Now,

$$v_{AB} = \frac{R_{g1}}{R_{g1}+R_{g3}} \cdot v_i$$

$$= \frac{R+dR}{(R+dR)+(R-dR)} v_i = \frac{R+dR}{2R} \cdot v_i \quad \text{...(7.28)}$$

and,

$$v_{AD} = \frac{R_2}{R_2+R_4} v_i = \frac{v_i}{2} \quad \text{...(7.29)} \quad (\because R_2 = R_4)$$

The changed output voltage,

$$v_0 + dv_0 = \frac{R+dR}{2R} \cdot v_i - \frac{v_i}{2}$$

$$= v_i\left(\frac{R+dR}{2R} - \frac{1}{2}\right) = v_i\left(\frac{2R+2dR-2R}{4R}\right)$$

or,

$$v_0 + dv_0 = \frac{v_i}{2} \cdot \frac{dR}{R} \quad \text{...(7.30)}$$

Since under unstrained conditions $v_0 = 0$, therefore, change in output voltage due to applied strain becomes,

$$dv_0 = \frac{v_i}{2} \cdot \frac{dR}{R}$$

or,

$$dv_0 = \left(\frac{G_f}{2}\right) e \cdot v_i \quad \text{...(7.31)}$$

which is *twice* the output of Wheatstone bridge using one gauge only.

The eqn. (7.31) can be rewritten as:

$$dv_0 = \frac{v_i}{4}\left[\frac{dR}{R} - \left(\frac{-dR}{R}\right)\right]$$

or, $$dv_0 = \frac{v_i}{4}\left[\begin{Bmatrix}\text{Fractional change in}\\ \text{resistance of gauge in limb } AB\end{Bmatrix} - \begin{Bmatrix}\text{Fractional change in}\\ \text{resistance of gauge in limb } BC\end{Bmatrix}\right] \quad ...(7.32)$$

The –ve sign with fractional change in resistance of the gauge in limb $BC$ is due to the fact that compressive and tensile strains are of *opposite signs.*

*In general,* for the *two gauges connected in the adjacent limbs of a bridge circuit,* we have:

$$dv_0 = \frac{v_i}{4}\left(\frac{dR_{g1}}{R} - \frac{dR_{g3}}{R}\right) \quad ...(7.33)$$

Thus, *when both the gauges are mounted on the top of the cantilever beam, the two effects cancel each other and the output voltage is zero.*

***(iii) Full-bridge:***

Fig. 7.40, shows four gauges used for stain measurement (Full-bridge). In this arrangement all the four elements of the bridge are strain gauges.

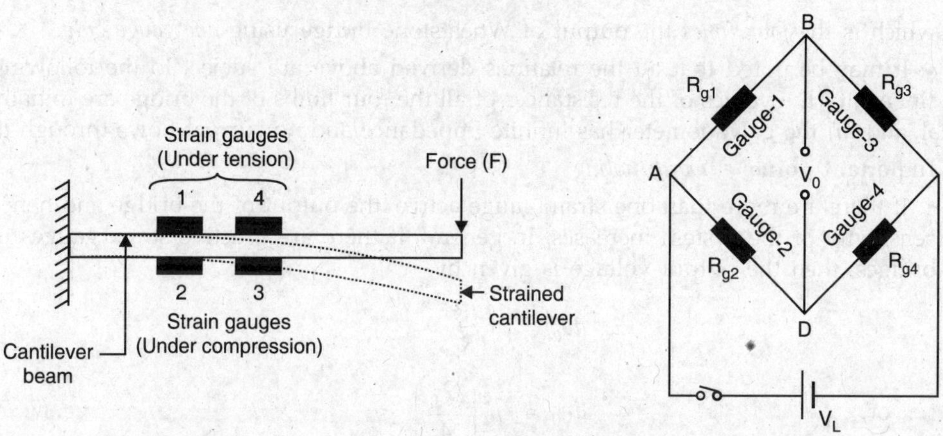


**Fig. 7.40.** Four gauges used for strain measurement (Full-bridge).

All the four gauges are similar and have equal resistances when unstained, *i.e.,*

$$R_{g1} = R_{g2} = R_{g3} = R_{g4} = R.$$

Under no-strain condition : $v_{AB} = v_{AD} = \frac{v_i}{2}$; $v_B = v_D$ and $v_0 = 0$.

When the load is applied to the cantilever beam, the resistance $R_{g1}$ and $R_{g4}$ *increase* due to *tensile load* whilstresistances $R_{g2}$ and $R_{g3}$ *decrease* due to equal *compressive strain.* When strained the resistances of the various gauges are

$$R_{g1} = R_{g4} = R + dR \text{ (tension)}$$

$$R_{g2} = R_{g3} = R - dR \text{ (compression)}$$

and, $$v_{AB} = \frac{R_{g1}}{R_{g1} + R_{g3}} \cdot v_i$$

or, $$v_{AB} = \frac{R + dR}{(R + dR) + (R - dR)} v_i = \frac{R + dR}{2R} \cdot v_i \quad ...(7.34)$$

and, $$v_{AD} = \frac{R_{g2}}{R_{g2}+R_{g4}} v_i$$

or, $$v_{AD} = \frac{R-dR}{(R-dR)+(R+dR)} \cdot v_i = \frac{R-dR}{2R} \cdot v_i \quad ...(7.35)$$

The changed output voltage,

$$v_0 + dv_0 = \frac{R+dR}{2R} \cdot v_i - \frac{R-dR}{2R} \cdot v_i$$

$$= v_i \left( \frac{R+dR}{2R} - \frac{R-dR}{2R} \right) = v_i \frac{dR}{R}$$

Since the output voltage under unstrained conditions, $v_0 = 0$, therefore, change in output voltage due to applied strain becomes,

$$dv_0 = v_i \frac{dR}{R}$$

*i.e.*, $$dv_0 = G_f \, e.v_i \quad ...(7.36)$$

which is the *four times* the output of Wheatstone bridge using *one gauge only*.

• It may be noted that all the relations derived above are subject to the following conditions : (*i*) the values of the resistances of all the four limbs of the bridge are initially equal, and (*ii*) the galvanometer has infinite impedance and no current flows through it.

**Important points** – *worth noting* :

1. If there are more than one strain gauge active, the output of the bridge and hence the sensitivity of the system increases. In general, if there are $n$ active strain gauges in the bridges, then the output voltage is given by

$$dv_0 = n \frac{dR}{R} \cdot v_i$$

or, $$dv_0 = n \cdot \left( \frac{G_f}{4} \right) e \cdot v_i \quad ...(7.37)$$

($G_f$ and $e$ are gauge factor and strain respectively).

The increased bridge output is expressed in terms of *"bridge constant"* (it represents the ratio of the actual bridge output to that if only one gauge were effective). The bridge constants for the three arrangements discussed above are 1, 2 and 4 respectively.

2. *High gauge sensitivity* can be obtained with :

(*i*) *High gauge factor*: It depends upon :

— The gauge material;

— The configuration of the gauge wire;

— The mechanical loading.

- In general the *foil* and *wire gauges* have gauge factor of about 2 and semiconductor gauges have typical values of about – 100 to + 200 (approx).

(*ii*) *Large excitation voltage*. It depends upon current or power rating of the gauge; typical values being 15 mA and 15 mW respectively.

#### 7.13.4.3. Temperature compensation

In order to measure static strains accurately, it is absolutely necessary that adequate temperature compensation may be provided in the Wheatstone bridge. The errors due to change of temperature arise due to the following two *reasons* :

(*i*) Resistance change of the wire in the strain gauge with change in temperature.

(*ii*) Differential expansion existing between the gauges and the metal to which they are bonded, if the temperature coefficients of the two are not same.

Temperature effects may be dealt with as follows :

(*i*) *By compensation or cancellation.* This method is extensively used for both metallic as well as semiconductor gauge.

(*ii*) *By evaluation as a part of the reduction problem.* This method is employed only for semiconductor gauges.

**Temperature compensation or cancellation techniques :**

Temperature compensation may be provided by :

- Adjacent arm balancing or compensating gauge.
- Self-temperature compensation.
- Use of special external control circuitry.

**1. Adjacent arm compensating gauge :**

(*i*) ***Use of dummy gauge.*** The arrangement is shown in Fig. 7.41. The active strain gauge is installed on the test specimen (cantilever beam in this case) while the dummy or compensating gauge is installed on a like piece of material and is *not subjected to strain.*

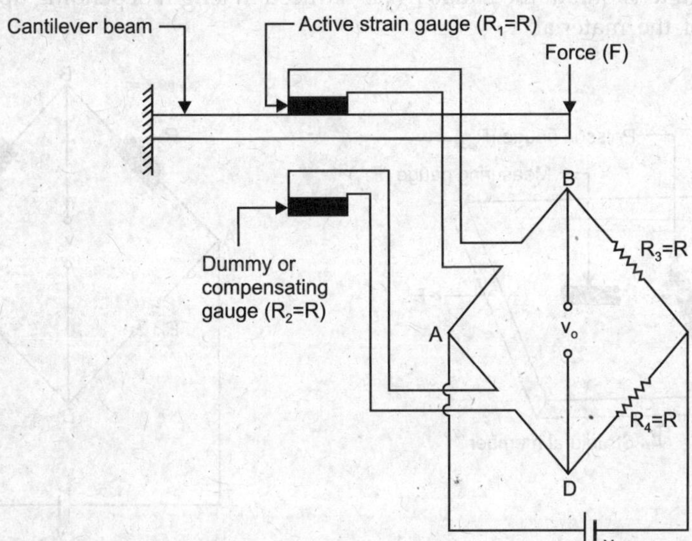


**Fig. 7.41.** Use of dummy gauge for temperature compensation.

The bridge is initially balanced and therefore $\frac{R_1}{R_2} = \frac{R_3}{R_4}$. Suppose a change in temperature occurs, and the resistances of active and dummy gauges change by $dR_1$ and $dR_2$ respectively. In order that there should be no output the bridge should be balanced and this requires,

$$\frac{R_1 + dR_1}{R_3} = \frac{R_2 + dR_2}{R_4}$$

or,
$$\frac{dR_1}{R_3} = \frac{dR_2}{R_4}$$

Let $R_3 = R_4 = R$ and therefore for bridge balance $dR_1$ should be equal $dR_2$. Now if active and dummy gauges are identical $R_1 = R_2 = R$ and if dummy gauge is placed in the same environments as the active strain gauge $dR_1 = dR_2$, therefore there will be no output due to changes in ambient temperature.

**(ii)** ***Use of two active gauges in adjacent arms.*** In some applications, where equal and opposite strains are known to exist, it is possible to attach similar gauges in such a way that one gauge experiences a positive strain and the other a negative strain (see Fig. 7.41).

- The arrangement shown in Fig. 7.41, *not only gives increased sensitivity but also eliminates the effects caused by temperature variations.*

**(iii)** ***Use of four active gauges.*** This arrangement is illustrated in Fig. 7.42.

- Four active-arm bridges are extensively used *when strain gauges are used as secondary transducers to give maximum sensitivity combined with full temperature compensation.* The effect of increasing the number of active gauges is the same as if a low impedance detector is used.

**(iv)** ***Poisson's method.*** In this method of temperature compensation two active gauges $R_{g1}$ and $R_{g3}$ bonded at right angles are used as shown in Fig. 7.42. Here gauge $Rg_1$ is in tension but the gauge $Rg_3$ is reduced in length depending upon Poisson's ratio of the material.

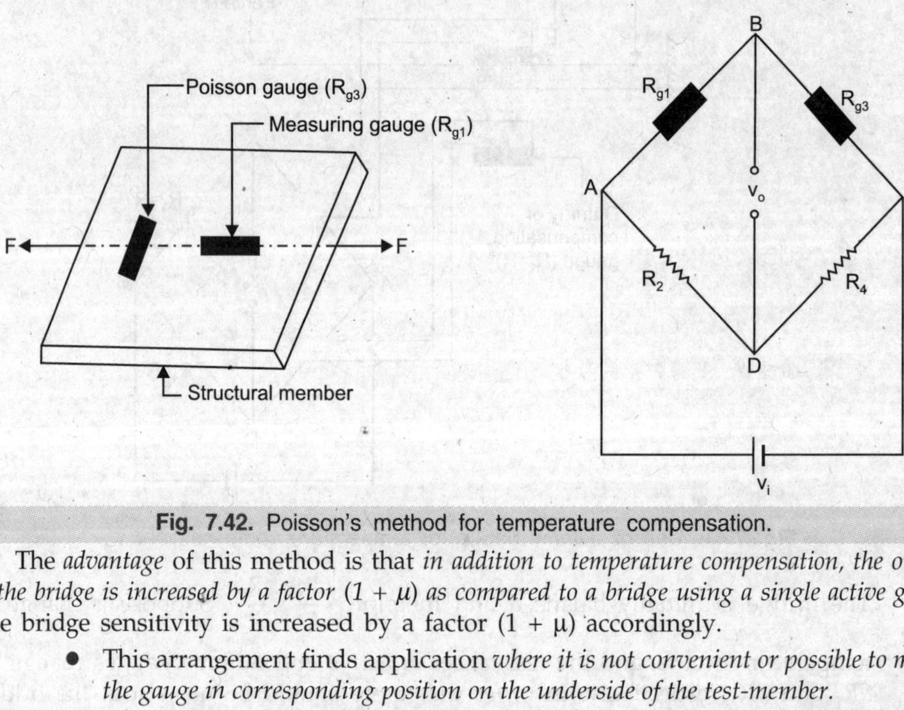


**Fig. 7.42.** Poisson's method for temperature compensation.

The *advantage* of this method is that *in addition to temperature compensation, the output of the bridge is increased by a factor* $(1 + \mu)$ *as compared to a bridge using a single active gauge.* The bridge sensitivity is increased by a factor $(1 + \mu)$ accordingly.

- This arrangement finds application *where it is not convenient or possible to mount the gauge in corresponding position on the underside of the test-member.*

**2. Self temperature compensation :**

In several applications it may be difficult or impossible to obtain temperature compensation by using adjacent arm compensating gauge acting either as a dummy gauge or a compensating gauge. Under these circumstances, it is desirable to use *self temperature compensation.*

Following two types of self-compensated gauges are available :

***(i) Selected melt gauge :***

- This method of self-compensation is based upon the fact that through proper manipulation of alloy and processing, usually through cold working, some control over the temperature sensitivity of the grid material may be obtained.
- Both wire and foil grid materials may be prepared which show very low apparent strain variation with temperature over a certain temperature range.

***(ii) Dual element-gauge :***

- In this case, two wire elements are connected in series to make one strain gauge assembly.
- *The two elements have different temperature characteristics and are so chosen that the net strain produced on account of temperature changes is minimised when the gauge is mounted on the test surface.*

### 7.13.4.4. Strain-gauge calibration

In most applications, a theoretical objection to the use of strain gauges is that it is impossible to check the accuracy of the readings obtained from them. Once the strain gauge is bonded in its measuring position, it cannot be removed, transferred or subjected to a known strain or calibration.

The calibration, in most of the cases, is done by introducing an accurately known sample of the variable and then observing response of the system. But this practice cannot be extended to bonded resistance strain-gauge because of the nature of transducer. In the case of strain gauges, the manufacturer has to maintain strict quality control and ensure that all the strain gauges produced by him are identical. Generally the gauge factors are usually quoted with a tolerance limit of about ± 0.2 percent or less. Thus if gauge factor and gauge resistance are known, any resistance strain-gauge system can be easily calibrated.

The method of strain gauge calibration consists of determining the response of the system to the introduction of a known small resistance change at the gauge and of calculating an equivalent strain therefrom. The small resistance change is introduced by shunting a relatively high value precision resistance $R_{sh}$ across the gauge (gauge resistance $R_g$) and to introduce this resistance $R_{sh}$ in parallel to $R_g$, a switch $S$ is used, as shown in Fig. 7.43.

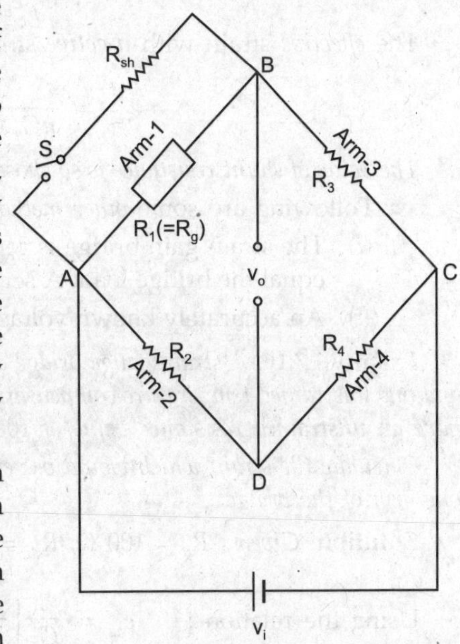


**Fig. 7.43.** Calibration of strain gauges.

Resistance of arm-1 of the bridge *before* switch $S$ is closed $= R_g$

Resistance of arm-1 of the bridge *after* switch is closed

$$= \text{Resistance of parallel combination of } R_g \text{ and } R_{sh}$$

$$= \frac{R_g \, . \, R_{sh}}{R_g + R_{sh}}$$

∴ The change in resistance of arm-1 resulting from introduction of $R_{sh}$,

$$dR = R_g - \frac{R_g \, . \, R_{sh}}{R_g + R_{sh}} = \frac{R_g^2}{R_g + R_{sh}} \quad \text{...(7.38)}$$

Also, strain $e = \dfrac{1}{G_f} \cdot \dfrac{dR}{R}$, where $G_f$ is the gauge factor.

Substituting $dR_g$ for $dR$ in eqn. (7.38), we get the *equivalent strain* as,

$$e_{eq} = \frac{1}{G_f} \cdot \frac{dR_g}{R_g} = \frac{1}{G_f} \cdot \frac{1}{R_g} \cdot \frac{R_g^2}{(R_g + R_{sh})} = \frac{1}{G_f}\left(\frac{R_g}{R_g + R_{sh}}\right) \quad \text{...(7.39)}$$

∴ Shunt resistance,

$$R_{sh} = \frac{R_g \, (1 - G_f \, . \, e_{eq})}{G_f \, . \, e_{eq}} \quad \text{...(7.40)}$$

In practice $G_f \, . \, e_{eq} << 1$ and therefore the term $G_f \, . \, e_{eq}$ in the numerator of eqn. 7.40 can be *neglected without-introducing any significant error.*

Then,

$$R_{sh} = \frac{R_g}{G_f \, e_{eq}} \quad \text{...(7.41)}$$

The *effective* strain with *n-active strain gauges* in the bridge is given as :

$$e_{eff.} = \frac{R_g}{nG_f \, (R_g + R_{sh})} \quad \text{...(7.42)}$$

*The value of shunt resistance is so chosen that the read-out system gives a full scale deflection.*

- Following are some *other methods of electrical calibration :*
  - (*i*) The strain gain bridge is replaced with a substitute load, initially adjusted to equal the bridge load. A series resistance is then used for calibration.
  - (*ii*) An accurately known voltage is injected into the bridge circuit.

**Example 7.16.** *A strain gauge bridge consists of two fixed 100 Ω resistors; one active gauge and one unstrained temperature compensating gauge connected in adjacent arms. The two gauges have an unstrained resistance value of 100 Ω each. The gauge factor is 2.1.*

*Calculate the strain which would be represented by 80000 Ω calibration resistance shunting one arm of the bridge.*

**Solution.** *Given* : $R_g = 100\ \Omega$; $R_{sh} = 80000\ \Omega$; $G_f = 2.1$.

Using the relation :

$$e_{eq} = \frac{1}{G_f}\left(\frac{R_g}{R_g + R_{sh}}\right) \quad \text{...(7.39)}$$

or,

$$e_{eq} = \frac{1}{2.1}\left(\frac{100}{100 + 80000}\right)$$

$$= 5.945 \times 10^{-4} \text{ or } \mathbf{594.5 \ microstrain} \ \textbf{(Ans.)}$$

**Example 7.17.** *In a four arm bridge circuit four strain gauges are mounted on a simple flat tensile specimen arranged for complete temperature compensation and maximum sensitivity. A calibration resistor of $100 \times 10^3$ Ω is connected across one of the strain gauges. Gauge resistances are each 200 Ω and each gauge has a gauge factor of 2.0.*

*Calculate the axial strain in the specimen, if the strain gauge read-out has 140 divisions when calibration switch is closed and 220 divisions when the load is applied.*

**Solution.** *Given* : $n = 4$; $R_g = 200\ \Omega$; $R_{sh} = 100 \times 10^3\ \Omega$;

$$\text{Effective strain, } e_{eff.} = \frac{R_g}{n\,G_f\,(R_g + R_{sh})} \qquad ...(7.42)$$

$$= \frac{200}{4 \times 2.0(200 + 100 \times 10^3)} = 249.5 \times 10^{-6}$$

*When the calibration switch is closed,* the read out gives a reading of 140 divisions.

$$\therefore \qquad 1 \text{ divisions} = \frac{249.5 \times 10^{-6}}{140} = 1.782 \times 10^{-6} = 1.782 \text{ microstrain.}$$

When the *strain gauge is loaded, the strain* = 1.782 × 220 = **392.04 microstrain (Ans.)**

**7.13.4.5. Stress-strain relationships and strain gauge rosettes**

Here, we shall discuss cases of uniaxial and biaxial strain measurements :

**1. Simple uni-axial stresses.** In this case, the axial stress (σ) can be calculated from the observed strain (*e*) using the modulus of elasticity, *E*, of the material used (within the elastic limits) as follows :

$$\sigma = e\,.\,E \qquad ...(7.43)$$

**2. Biaxial stresses.** Fig. 7.44 shows a cylindrical pressure vessel (closed at both ends) subjected to biaxial stresses. Consider an element subjected to stresses $\sigma_x$ and $\sigma_y$ applied simultaneously.

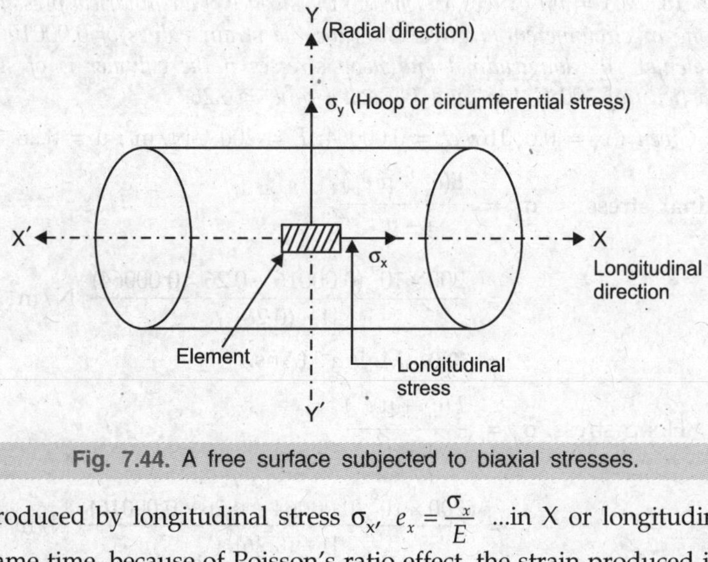


Fig. 7.44. A free surface subjected to biaxial stresses.

Strain produced by longitudinal stress $\sigma_x$, $e_x = \dfrac{\sigma_x}{E}$ ...in X or longitudinal direction.

At the same time, because of Poisson's ratio effect, the strain produced in *Y* or radial direction $= -\dfrac{\mu\,\sigma_x}{E}$, where μ is the Poisson's ratio.

Now, the stress $\sigma_y$ applied in $Y$ or radial direction will produce a strain equal to $\frac{\sigma_y}{E}$ in $Y$-direction and $-\frac{\mu\,\sigma_y}{E}$ in X or longitudinal direction.

The net strains in the two directions are :

$$e_x = \frac{\sigma_x}{E} - \frac{\mu\,\sigma_y}{E} = \frac{1}{E}(\sigma_x - \mu\,\sigma_y) \quad ...(7.44)$$

and,

$$e_y = \frac{\sigma_y}{E} - \frac{\mu\,\sigma_x}{E} = \frac{1}{E}(\sigma_y - \mu\,\sigma_x) \quad ...(7.45)$$

∴ We get expressions for the stresses as :

$$\sigma_x = \frac{E(e_x + \mu\, e_y)}{1-\mu^2} \quad ...(7.46)$$

and.

$$\sigma_y = \frac{E(e_y + \mu\, e_x)}{1-\mu^2} \quad ...(7.47)$$

If a third stress $\sigma_3$ acts in the third orthogonal direction, then the stress-strain relationships are given as :

$$e_x = \frac{1}{E}[\sigma_x - \mu(\sigma_y + \sigma_z)] \quad ...(7.48)$$

$$e_y = \frac{1}{E}[\sigma_y - \mu(\sigma_z + \sigma_x)] \quad ...(7.49)$$

$$e_z = \frac{1}{E}[\sigma_z - \mu(\sigma_x + \sigma_y)] \quad ...(7.50)$$

**Example 7.18.** *Two strain gauges attached to the surface of a cylindrical pressure vessel, one in axial and one in circumferential direction, gave the strain values of 0.00016 and 0.00064 respectively, calculate the longitudinal and hoop stresses if the cylinder is of steel having a modulus of elasticity of 200 GN/m$^2$ and Poisson's ratio = 0.26.*

**Solution.** *Given* : $e_x$ = 0.00016; $e_y$ = 0.00064; $E$ = 200 GN/m$^2$; $\mu$ = 0.26.

**Longitudinal stress,**

$$\sigma_x = \frac{E(e_x + \mu\, e_y)}{1-\mu^2} \quad ...(7.46)$$

$$= \frac{200\times10^9\,(0.00016 + 0.26\times0.00064)}{[1-(0.26)^2]}\ \text{N/m}^2$$

$$= \mathbf{70.01\ MN/m^2\ (Ans.)}$$

Hoop stress,

$$\sigma_y = \frac{E(e_y + \mu\, e_x)}{1-\mu^2}$$

$$= \frac{200\times10^9\,(0.00064 + 0.26\times0.00016)}{[1-(0.26)^2]}\ \text{N/m}^2$$

$$= \mathbf{146.2\ MN/m^2\ (Ans.)}$$

**Example 7.19.** *A simple tension member having an area of 110 mm$^2$ is subjected to 25 kN. Strains of 1540 and – 420 micro-strain are measured in axial and transverse directions respectively. Calculate :*

*(i) Modulus of elasticity.*

*(ii) Poisson's ratio.*

**Solution.** Given : Area, $A = 110 \text{ mm}^2 = 110 \times 10^{-6} \text{ m}^2$; Load, $P = 25$ kN;

$e_x = 1540$; $e_y = -420$ micro-strain.

**(i) Modulus of elasticity, E :**

$$\text{Axial stress, } \sigma_x = \frac{P}{A} = \frac{25 \times 10^3}{110 \times 10^{-6}} = 227.27 \times 10^6 \text{ N/m}^2$$

Transverse stress, $\sigma_y = 0$

$$\text{Axial stress, } e_x = \frac{1}{E}(\sigma_x - \mu\,\sigma_y) = \frac{\sigma_x}{E}$$

$$\therefore \quad E = \frac{\sigma_x}{e_x} = \frac{227.27 \times 10^6}{1540 \times 10^{-6}} \text{ N/m}^2 = 147.6 \times 10^9 \text{ N/m}^2$$

$$= \mathbf{147.6 \ GN/m^2 \ (Ans.)}$$

***(ii)* Poisson's ratio, μ :**

Strain in the transverse direction,

$$e_y = \frac{1}{E}(\sigma_y - \mu\,\sigma_x) \qquad \text{...(7.45)}$$

$$-420 \times 10^{-6} = \frac{1}{147.6 \times 10^9}(0 - \mu \times 227.27 \times 10^6)$$

$$\text{or,} \quad \mu = \frac{420 \times 10^{-6} \times 147.6 \times 10^9}{227.27 \times 10^6} = \mathbf{0.2728 \ (Ans.)}$$

## Strain gauge rosettes :

*"Rosette gauges" are used for measuring strain in complex parts.* These gauges have *three or four separate grids with various angular orientations* and they can be cemented to the part with no particular attention being paid to the overall gauge orientation. The resultant strain on each of the grids can then be recorded and the true magnitudes and directions of the significant surface strains in the part can be calculated from these data.

Depending on the arrangement of the grids, we have the following types of rosettes :

- Rectangular rosettes.
- Delta rosettes.
- T-delta rosettes.

*The formulae for calculation of principal strains and stresses for various arrangements are given below :*

**1. Two element rosette.** Refer to Fig. 7.45.

- *Principal strains :* $e_1$ ; $e_2$
- *Principal stresses :*

$$\sigma_1 = \frac{E}{1 - \mu^2}(e_1 + \mu\, e_2) \qquad \text{...(7.51)}$$

$$\sigma_2 = \frac{E}{1-\mu^2}(e_2 + \mu e_1) \qquad ...(7.52)$$

- *Maximum shear stress* :

$$\tau_{max} = \frac{E}{2(1-\mu)}(e_1 - e_2) \qquad ...(7.53)$$

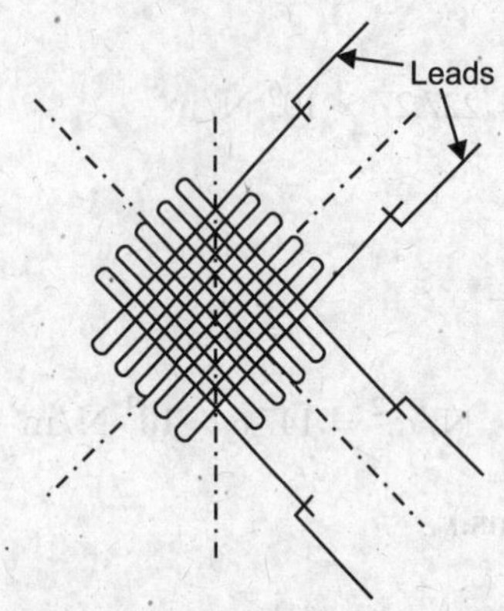


**Fig. 7.45.** Two element rosette. Two grid elements are mounted in between two thin papers.

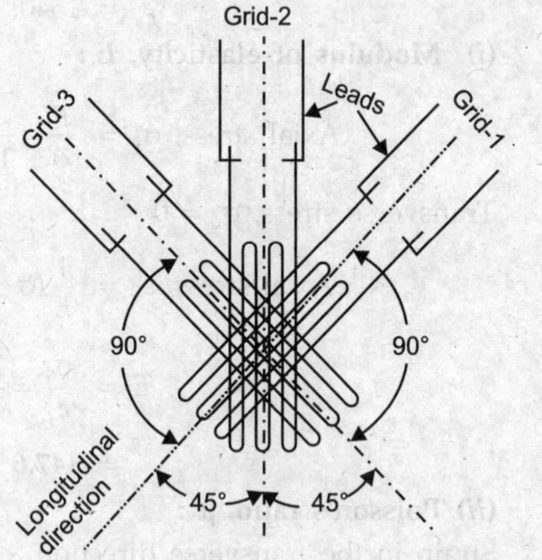


**Fig. 7.46.** Rectangular rosette. Grids do not make electrical contact. Grids are kept in-between two thin papers.

2. **Three-element rectangular rosette.** Refer to Fig. 7.46.
   - *Principal strains* :

$$e_{max}, e_{min} = \frac{e_1 + e_3}{2} \pm \frac{1}{\sqrt{2}}\left[(e_1 - e_2)^2 + (e_2 - e_3)^2\right]^{1/2} \qquad ...(7.54)$$

   - *Principal stresses* :

$$\sigma_{max}, \sigma_{min} = \frac{E(e_1 + e_3)}{2(1-\mu)} \pm \frac{E}{\sqrt{2}(1+\mu)}\left[(e_1 - e_2)^2 + (e_2 - e_3)^2\right]^{1/2} \qquad ...(7.55)$$

   - *Maximum shear stress* :

$$\tau_{max} = \frac{E}{\sqrt{2}(1+\mu)}\left[(e_1 - e_2)^2 + (e_2 - e_3)^2\right]^{1/2} \qquad ...(7.56)$$

The orientation of principal stress is given by

$$\tan 2\theta = \frac{2e_2 - e_1 - e_3}{e_1 - e_3} \qquad ...(7.57)$$

This is the axis at which $\tau_{max}$ occurs.

3. **Three-element delta rosette.** Refer to Fig. 7.47.
   - *Principal strains* :

$$e_{max}, e_{min} = \frac{e_1 + e_2 + e_3}{3} \pm \frac{\sqrt{2}}{3}\left[(e_1 - e_2)^2 + (e_2 - e_3)^2 + (e_3 - e_1)^2\right]^{1/2} \qquad ...(7.58)$$

- *Principal stresses :*

$$\sigma_{max}, \sigma_{min} = \frac{E(e_1 + e_2 + e_3)}{3(1-\mu)} \pm \frac{\sqrt{2}E}{3(1+\mu)}\left[(e_1 - e_2)^2 + (e_2 - e_3)^2 + (e_3 - e_1)^2\right]^{1/2} \quad ...(7.59)$$

- *Maximum shear stress :*

$$\tau_{max} = \frac{\sqrt{2}E}{3(1+\mu)}\left[(e_1 - e_2)^2 + (e_2 - e_3)^2 + (e_3 - e_1)^2\right]^{1/2} \quad ...(7.60)$$

The orientation of principal stress is given by

$$\tan 2\theta = \frac{\sqrt{3}(e_3 - e_2)}{2e_1 - e_2 - e_3} \quad ...(7.61)$$

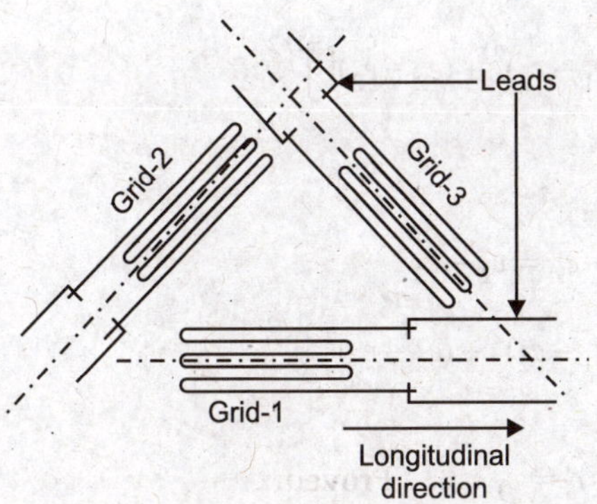


Fig. 7.47. Delta rosette.

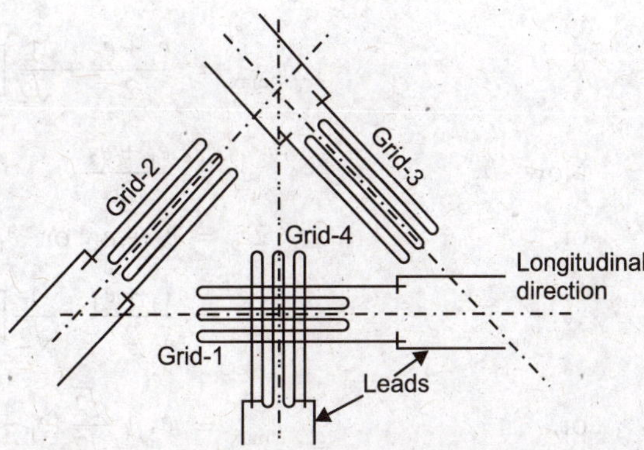


Fig. 7.48. T-delta rosette. Girds do not make electrical contacts.

**4. T-delta rosette.** Refer to Fig. 7.48.

- *Principal strains :*

$$e_{max}, e_{min} = \frac{1}{2}\left[e_1 + e_4 \pm (e_1 - e_4)^2 + \frac{4}{3}(e_2 - e_3)^2\right]^{1/2} \quad ...(7.62)$$

- *Principal stresses :*

$$\sigma_{max}, \sigma_{min} = \frac{E}{2}\left[\frac{e_1 + e_4}{1-\mu} \pm \frac{1}{1+\mu}\sqrt{(e_1 - e_4)^2 + \frac{4}{3}(e_2 - e_3)^2}\right] \quad ...(7.63)$$

- *Maximum shear stress :*

$$\tau_{max} = \frac{E}{2(1+\mu)}\sqrt{(e_1 - e_4)^2 + \frac{4}{3}(e_2 - e_3)^2} \quad ...(7.64)$$

The principal axis is located at $\theta$ according to,

$$\tan 2\theta = \frac{2}{\sqrt{3}}\left(\frac{e_3 - e_2}{e_1 - e_4}\right) \quad ...(7.65)$$

The T-delta rosette has the following *advantages* over others due to the fourth gauge element (three strain measurements are essential to completely define any stress situation) :

(*i*) The fourth gauge gives a *check on the results* obtained from other three elements.

(*ii*) Due to four readings, averaging effect results in *better accuracy*.

**Example 7.20.** *In a rectangular rosette, the maximum and minimum strains are given by*

$$e_{max}, e_{min} = \frac{e_1 + e_3}{2} \pm \frac{1}{\sqrt{2}}\left[(e_1 - e_2)^2 + (e_2 - e_3)^2\right]^{1/2}$$

*and then are located at an angle,* $\tan 2\theta = \dfrac{2e_2 - e_1 - e_3}{e_1 - e_3}$

*Show that :* $e_{max} = e_1$ *and* $e_{min} = e_3$ *when* $\theta = 0$.

**Solution.** When $\theta = 0$, $\tan 2\theta = 0$ and hence,

$$\frac{2e_2 - e_1 - e_3}{e_1 - e_3} = 0 \quad \text{or} \quad e_2 = \frac{e_1 + e_3}{2}$$

$$\therefore \quad e_{max} = \frac{e_1 + e_3}{2} + \frac{1}{\sqrt{2}}\left[(e_1 - e_2)^2 + (e_2 - e_3)^2\right]^{1/2}$$

Now, $$e_2 = \frac{e_1 + e_3}{2}$$

or, $$2e_2 = e_1 + e_3 \quad \text{or} \quad e_1 - e_2 = e_2 - e_3.$$

$$\therefore \quad e_{max} = \frac{e_1 + e_3}{2} + \frac{1}{\sqrt{2}}\left[(e_1 - e_2)^2 + (e_2 - e_3)^2\right]^{1/2}$$

or, $$e_{max} = e_2 + \frac{\sqrt{2}}{\sqrt{2}}(e_1 - e_2) = e_1 \; \text{.......} \; \textbf{Proved.}$$

Similarly, we can get, $e_{min} = e_3$.

**Example 7.21.** *A rectangular rosette is mounted on a steel plate having a modulus of elasticity 200 GN/m² and Poisson's ratio 0.3. The three strains measured are :*

$e_1 = 60$ *microstrain;* $e_2 = 48$ *microstrain;* $e_3 = -12$ *microstrain;*

*Calculate the following :*

(*i*) *The principal strains.*

(*ii*) *The principal stresses.*

(*iii*) *The maximum shear stress*

(*iv*) *Location of the principal planes.*

**Solution.** *Given :* $e_1 = 60$ microstrain; $e_2 = 48$ microstrain;

$e_3 = -12$ microstrain; $E = 200 \text{ GN/m}^2$; $\mu = 0.3$.

**(*i*) The Principal strains, $e_{max}$ ; $e_{min}$ :**

$$e_{max} ; e_{min} = \frac{e_1 + e_3}{2} \pm \frac{1}{\sqrt{2}}\left[(e_1 - e_2)^2 + (e_2 - e_3)^2\right]^{1/2} \quad \text{...(7.54)}$$

$$= \frac{60 - 12}{2} \pm \frac{1}{\sqrt{2}}\left[(60 - 48)^2 + (48 + 12)^2\right]^{1/2}$$

or, $$e_{max}, e_{min} = 24 \pm \frac{61.19}{\sqrt{2}} = \mathbf{67.27 \times 10^{-6}, - 19.27 \times 10^{-6}} \; \textbf{(Ans.)}$$

***(ii)* The principal stresses, $\sigma_{max}$ ; $\sigma_{min}$ :**

$$\sigma_{max}, \sigma_{min} = \frac{E(e_1 + e_3)}{2(1-\mu)} \pm \frac{E}{\sqrt{2}\,(1+\mu)}\left[(e_1 - e_2)^2 + (e_2 - e_3)^2\right]^{1/2} \quad ...(7.55)$$

$$= \frac{200 \times 10^9\,(60 \times 10^{-6} - 12 \times 10^{-6})}{2(1-0.3)} \pm \frac{200 \times 10^9}{\sqrt{2}(1+0.3)}$$

$$[(60 \times 10^{-6} - 48 \times 10^{-6})^2 + (48 \times 10^{-6} + 12 \times 10^{-6})^2]^{1/2}$$

or,
$$\sigma_{max}, \sigma_{min} = (6.857 \pm 6.656) \times 10^6 \text{ N/m}^2$$

$$= \mathbf{13.513 \text{ MN/m}^2, 0.201 \text{ MN/m}^2 \text{ (Ans.)}}$$

***(iii)* The maximum shear stress, $\tau_{max}$ :**

$$\tau_{max} = \frac{E}{\sqrt{2}\,(1+\mu)}\left[(e_1 - e_2)^2 + (e_2 - e_3)^2\right]^{1/2} \quad ...(7.56)$$

$$= \frac{200 \times 10^9}{\sqrt{2}(1+0.3)}[(60 \times 10^{-6} - 48 \times 10^{-6})^2 + (48 \times 10^{-6} + 12 \times 10^{-6})^{1/2}$$

or,
$$\tau_{max} = 6.656 \times 10^6 \text{ N/m}^2 \text{ or } \mathbf{6.656 \text{ MN/m}^2 \text{ (Ans.)}}$$

***(iv)* Location of the principal planes, $\theta$ :**

$$\tan 2\theta = \frac{2\,e_2 - e_1 - e_3}{e_1 - e_3} \quad ...(7.57)$$

$$= \frac{2 \times 48 - 60 + 12}{60 + 12} = 0.667$$

$$\therefore \quad 2\theta = \tan^{-1}(0.667) = 33.7° \text{ or } 213.7°$$

or,
$$\theta = \mathbf{16.85° \text{ or } 106.85° \text{ (Ans.)}}$$

## 7.14 LOAD CELLS

**Load cells** *are elastic devices that can be used for measurement of force through indirect methods i.e., through use of secondary transducers.*

Load cells utilize an elastic member as the primary transducer and strain gauges as secondary transducer. When the combination of the *strain gauge–elastic member* is used for weighing, it is called a *"load cell"*.

While designing load cells using strain gauges the following *factors* should be considered :

(*i*) Stiffness of the elastic element.

(*ii*) Optimum positioning of gauges on the element.

(*iii*) Provision for compensation of the temperature.

When large loads are to be measured, the direct tensile-compressive member may be used, whereas, in case of small loads, strain amplification provided by bending may be used with advantage.

### 7.14.1. Hydraulic Load Cell

Fig. 7.49 shows a *hydraulic load cell.*

Here the force variable is impressed upon a diaphragm which deflects and thereby transmits the force to a liquid. The liquid medium contained in a confined space, has a preload pressure of 2 bar. On the application of the force the liquid pressure increases and *equals the force magnitude divided by the effective area of the diaphragm*. The pressure is transmitted to and read on an accurate pressure gauge calibrated directly on force units.

- These cells have been used to measure loads upto about 25 MN (with an *accuracy of 0.1% of full scale); resolution is about 0.02 per cent.*

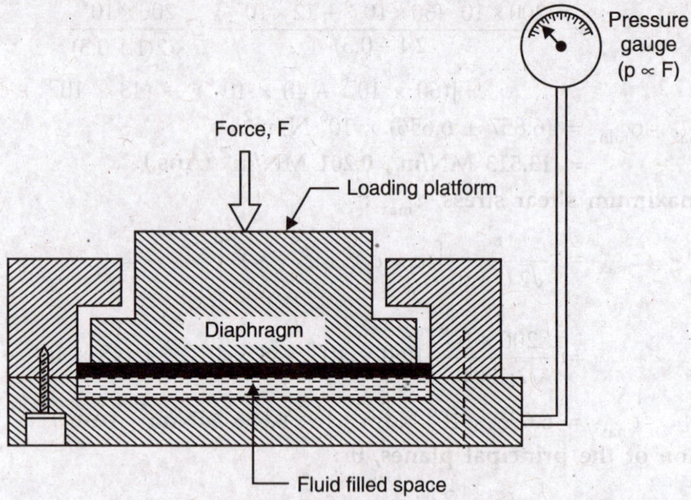


**Fig. 7.49.** Hydraulic load cell.

### 7.14.2. Pneumatic Load Cell

This cell operates on the *force-balance principle.* It employs a nozzle-flapper transducer similar to the conventional relay system. For any constant applied force, the system attains equilibrium at a specific nozzle opening and *corresponding pressure is indicated by the height of mercury column in a manometer.*

- The commercially available load cells (operating on this principle) can measure loads upto 25 kN with an accuracy of 0.5% of full scale.

### 7.14.3. Strain-Gauge Load Cells

These cells convert weight or force into electrical outputs which are provided by the strain gauges; these outputs can be connected to various measuring instruments for indicating, recording and controlling the weight or force.

Usually the *strain gauges are directly applied to the force-developing device and the device is calibrated against strain-gauge output.*

- These are excellent force-measuring devices, particularly for *transient and non-steady forces.*
- These are used in conjunction with CRO (for display purposes) for measurement of *rapidly changing loads..*

**Construction and working of the load cell :**

Fig. 7.50 shows a simple strain gauge load cell. It consists of a steel cylinder, on which are mounted four identical strain gauges. The gauges $R_{g1}$ and $R_{g4}$ are along the direction of applied load and the gauges $R_{g2}$ and $R_{g3}$ are attached circumferentially to gauges $R_{g1}$ and $R_{g4}$. All the four gauges are connected electrically to the four limbs of a Wheatstone bridge circuit.

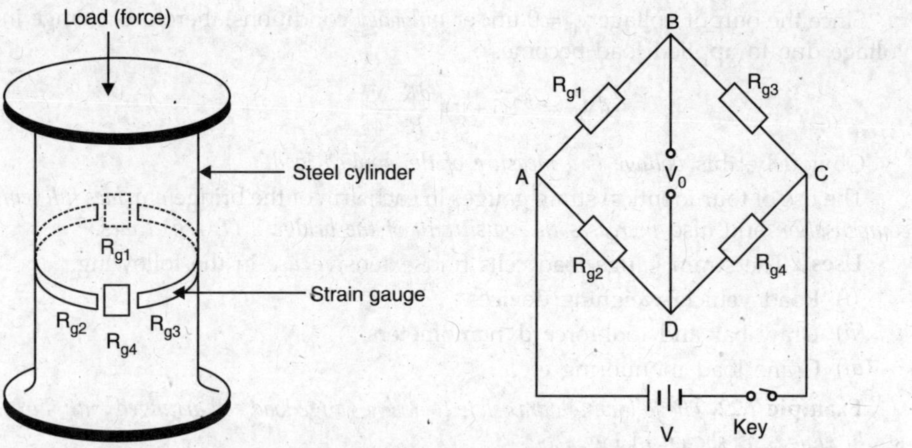


(a) Load cell (b) Wheatstone bridge circuit

**Fig. 7.50.** Strain gauge load cell.

When there is *no load on the cell*, all the four gauges have the same resistance (*i.e.*, $R_{g1} = R_{g2} = R_{g3} = R_{g4}$). Obviously the terminals *B* and *D* are at the same potential, the bridge is balanced and the output voltage is zero.

*i.e.*, $$v_{AB} = v_{AD} = \frac{v}{2} \quad ...(7.66)$$

and, $$v_{AB} - v_{AD} = v_0 = 0 \quad ...(7.67)$$

On the application of a compressive load to the unit, the *vertical gauges* ($R_{g1}$ and $R_{g4}$) undergo compression (*i.e.*, negative strain) and, therefore, there is *decrease in resistance*. The *circumferential gauges* $R_{g2}$ and $R_{g3}$, simultaneously, undergo tension (*i.e.*, positive strain) leading to *increase in resistance*. The two strains are not equal; these are related to each other by a factor, μ, the Poisson's ratio. Thus when strained, the resistances of various gauges are :

$$R_{g1} = R_{g4} = R - dR \quad ...(\text{compression})$$
$$R_{g2} = R_{g3} = R + dR \quad ...(\text{tension})$$

Potential at terminal *B*, $$v_{AB} = \frac{R_{g1}}{R_{g1}+R_{g3}} v = \frac{R-dR}{(R-dR)+(R+\mu.dR)} \times v$$

$$= \frac{R-dR}{2R-dR(1-\mu)} \times v \quad ...(i)$$

Potential at terminal *D*, $$v_{AD} = \frac{R_{g2}}{R_{g2}+R_{g4}} v = \frac{R+\mu.dR}{(R+\mu.dR)+(R-dR)} \times v$$

$$= \frac{R+\mu.dR}{2R-dR(1-\mu)} \times v \quad ...(ii)$$

The changed output voltage,

$$v_0 + dv_0 = \frac{R-dR}{2R-dR(1-\mu)} \times v - \frac{R+\mu.dR}{2R-dR(1-\mu)} \times v$$

...[Using (*i*) and (*ii*)]

$$= \frac{dR(1+\mu)}{2R} = 2(1+\mu)\left(\frac{dR}{R}\cdot\frac{v}{4}\right) \quad ...(7.68)$$

...*in magnitude*

Since the output voltage $v_0 = 0$ under *unloaded* conditions, therefore, change in output voltage due to applied load becomes :

$$dv_0 = 2(1+\mu)\left(\frac{dR}{R}\cdot\frac{v}{4}\right) \quad ...(7.69)$$

Obviously, this *voltage is a measure of the applied load.*

The use of four identical strain gauges in each arm of the bridge *provides full temperature compensation* and also *increases the sensitivity of the bridge 2 (1 + μ) times.*

**Uses :** The strain gauge load cells find extensive *use* in the following :

(*i*) Road vehicle weighing devices.

(*ii*) Draw bar and tool-force dynamometers.

(*iii*) Crane load monitoring etc.

**Example 7.22.** *The following data relate to strain gauge load cell arranged with four identical strain gauges as shown in Fig. 7.50.*

*Diameter of the steel cylinder = 60 mm; Nominal resistance of each gauge = 120Ω; Gauge factor = 2.0; Supply voltage (v) = 6V; Modulus of elasticity for steel = 200 GN/m²; Poisson's ratio = 0.3.*

*Calculate the sensitivity of the load cell.*

**Solution.** *Given* : $d$ = 60 mm = 0.06 m; $R_g$ (each gauge) = 120 Ω; $G_f$ = 2.0, v = 6V; $E$ = 200 GN/m²; μ = 0.3.

**Sensitivity of the load cell :**

Consider a load of 1 kN applied to the load cell.

$$\text{Stress } (\sigma) = \frac{\text{Load}}{\text{Cross-sectional area}} = \frac{1\times10^3}{\frac{\pi}{4}\times(0.06)^2} = 0.3537\times10^6\,\text{N/m}^2$$

$$\text{Strain, } e = \frac{\text{Stress } (\sigma)}{\text{Modulus of elasticity (E)}} = \frac{0.3537\times10^6}{200\times10^9} = 1.7685\times10^{-6}$$

Fraction change in resistance,

$$\frac{dR}{R} = G_f \times e = 2.0 \times 1.7685 \times 10^{-6} = 3.537 \times 10^{-6}$$

$$\text{Output voltage, } dv_0 = 2(1+\mu)\left(\frac{dR}{R}\cdot\frac{v}{4}\right) \quad ...[\text{Eqn. } (7.68)]$$

$$= 2(1+0.3)\left(3.537\times10^{-6}\times\frac{6}{4}\right) = 13.79\times10^{-6}\,\text{V} = 13.79\,\mu\text{V}$$

Hence, the *sensitivity of the load cell* = **13.79 μV/kN (Ans.)**

## 7.15 PROXIMITY SENSORS

A ***proximity sensor*** consists of an *element that changes either its state or an analog signal when it is close to, but often not actually touching, an object.*

Magnetic, electrical capacitance, inductance, and eddy current methods are particularly suited to the design of a proximity sensor.

- A ***photoemitter-detector pairs*** represents another approach, where *interruption or reflection of a beam of light is used to detect an object in a non-contact manner.* The emitter and detecter are usually a phototransistor and a photodiode.

*Common* ***applications*** for proximity sensors and limit switches include :

(*i*) Counting moving objects;

(*ii*) Limiting the traverse of a mechanism.

### 7.15.1. Eddy Current Proximity Sensors

**Working principle :**

When a coil is supplied with an alternating current an alternating magnetic field is produced. If there is a metal object in close proximity to this attending magnetic field, then eddy currents are induced in it. The eddy currents themselves produce a magnetic field which *distorts* the magnetic field responsible for their production. Consequently, the *impedance of the coil changes and so the amplitude of the alternating current. This change, at some preset level, can be used to trigger a switch.*

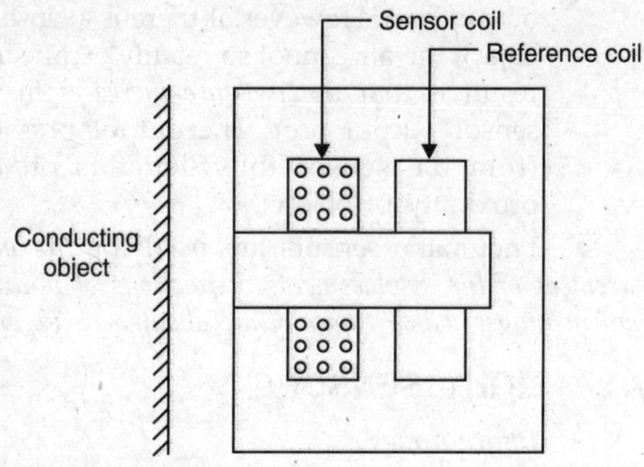


**Fig. 7.51.** Eddy current proximity sensor.

Fig. 7.51. shows the basic form of an eddy current proximity sensor. It is *used for the detection of non-magnetic but conductive materials.*

***Advantages :***

(*i*) Small in size.
(*ii*) Relatively inexpensive.
(*iii*) High flexibility.
(*iv*) High sensitivity to small displacements.

### 7.15.2. Capacitance Proximity Sensor

Fig. 7.52. shows a schematic diagram of a capacitance proximity sensor.

— It consists of a simple plate (one of the forms), with the object (*earthed*) acting as the other plate.

— As the object approaches the sensor, separation between the plate of the capacitor and object changes which becomes significant as the object is close to the sensor.

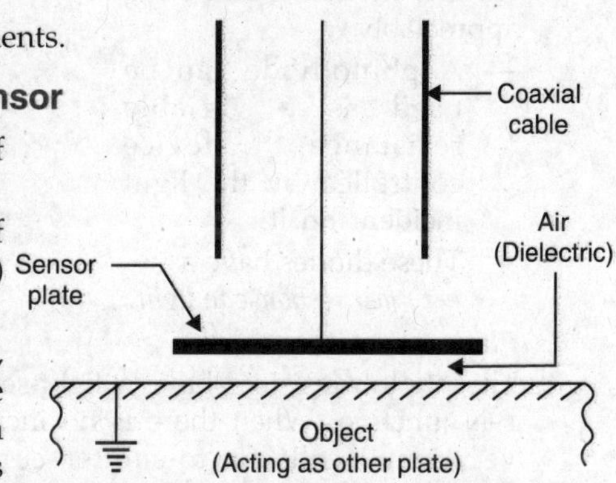


**Fig. 7.52.** Capacitance proximity sensor.

### 7.15.3. Inductive Proximity Switch

— An inductive proximity switch consists of a coil wound round a core.

— When the end of the coil is close to a metal object its *inductance changes*. This change can be monitored by its effect on a resonant circuit and the change used to trigger a switch.

- It can only be used *for the detection of metal objects* and is best with ferrous metals.

## 7.16 PNEUMATIC SENSORS

These sensors involve the use of compressed air, displacement or proximity of an object being transformed into a *change in air pressure.*

Fig. 7.53. shown the basic form of a pxeumatic sensor.

— Low pressure air is allowed to escape through a port in front of the sensor.

— This escaping air, in the absence of any close by object, escapes and in doing so also *reduces the pressure* in the nearby sensor output port. However, if there is a close by object, the air cannot so readily escape and result is that the *pressure increases* in the sensor output port. The output pressure from the sensor thus depends on the proximity of objects.

- Pneumatic sensors are used for the *measurement of the displacements of fractions of millimeters in ranges which typically are about 3 to 12 mm.*

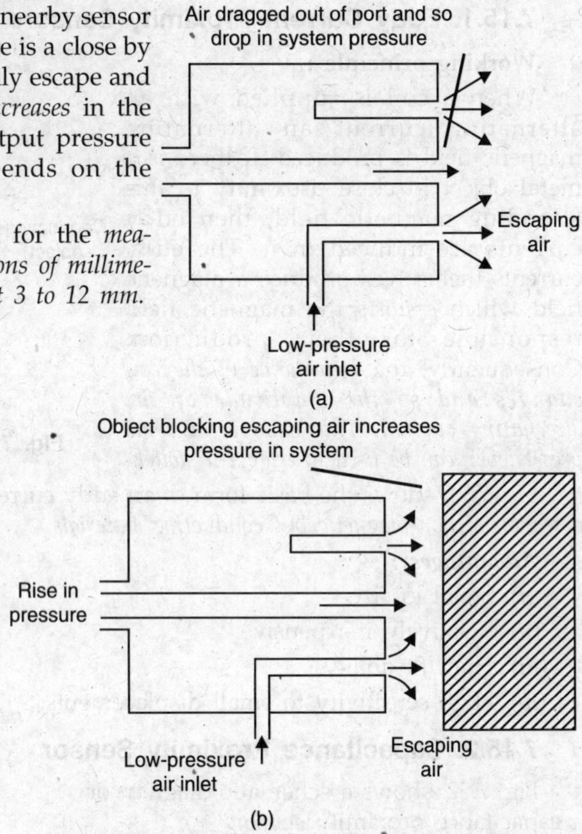


Fig. 7.53. Pneumatic proximity sensor.

## 7.17 LIGHT SENSORS

1. ***Photodiodes* :**

   "*Photodiodes*" are semiconductor junction diodes which are connected into a circuit in reverse bias, so giving a very high resistance, so that when light, falls on the junction the diode resistance drops and the current in the circuit rises appreciably

   — A photodiode can be used as a variable resistance device controlled by the light incident on it.

   — These diodes have a *very fast response to light*.

2. ***Phototransistors* :**

   The phototransistors have a light-sensitive collector-base P-N junction. When there is no incident light there is a very small collector-to-emitter current. When light is incident, a base current is produced that is *directly proportional to the light intensity*. This leads to the production of a collector current which is then a *measure of the light intensity*.

   — Phototransistors are often available as integrated packages with the phototransistor connected in a Darlington arrangement with a conventional transistor (Fig. 7.54). Since this arrangement gives a higher current gain, the device gives a *much greater collector current for a given light intensity*.

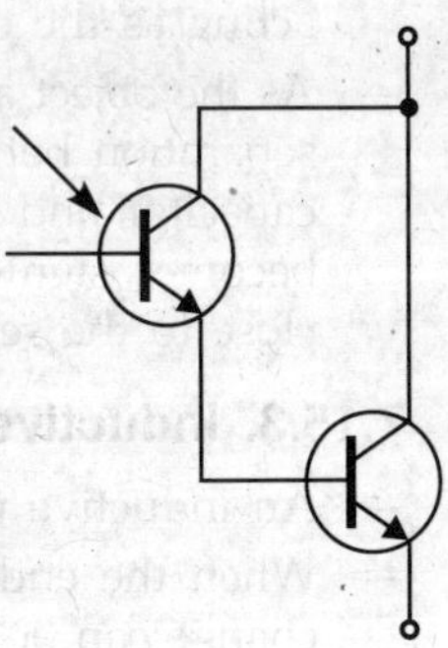

Fig. 7.54. Photo Darlington.

3. ***Photoresistor* :**

   It has a *resistance* which depends on the intensity of the light falling on it, *decreasing linearly as the intensity increases*.

   — The cadmium sulphide photoresistor is most responsive to light having wavelengths shorter than about 515 nm and the cadmium selinide photoresistor for wavelengths less than about 700 nm.

- An array of light sensors is often required in a small space in order to determine the variations of light intensity across that space.

## 7.18 TACTILE SENSOR

Human sense of touch is a marvelous phenomenon and scientists have a real challenge in developing robot fingers that can actually "feel" the difference between various textures and surface shapes.

One of the forms of *tactile sensor uses polyvinylidene fluoride ($PVF_2$) sheet which has piezo-electric qualities*. A soft film is sand wiched between two layers of $PVF_2$. The soft film transmits vibrations. The principle used in this case is both piezo-electric and its reverse effect.

A.C. voltage is supplied to the lower layer (i.e., which is a piezo-electric material) and as a result generates vibrations (Reverse of piezo-electric effect). These vibrations are picked up by the soft film and transmitted to the upper layer which converts these vibrations to alternate voltage (piezo-electric effect).

## 7.19 SMART SENSORS

*The sensors having decision-making and communication logic added to the basic sensors, are called* **smart or intelligent sensors.**

In such sensors, *other features* included are :

(*i*) Compensation for interferring inputs ;

(*ii*) Linearisation;

(*iii*) Self test;

(*iv*) Calibration facility.

— Whenever possible, there are usually microcomputers and other elements on the same chip.

- Fig. 7.55 shows the functional diagram for *data acquisition system using the smart pressure sensor*:

— When a piezoresistive sensor is used, as primary sensor for pressure, *the main interferring input is temperature and so a secondary temperature sensor is used to compensate for the effect of temperature on resistance change.*

— The memory of microcontroller has calibration data stored in it. The microcontroller is essentially a microprocessor with input/output (I/O) facility.

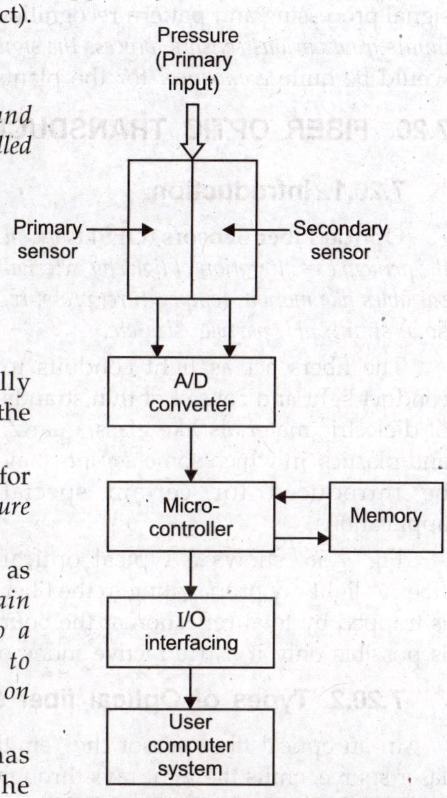


**Fig. 7.55.** Data acquisition system using smart pressure sensor.

— The IEEE 1451.2 standard for smart sensors includes the following *modules* :

(*i*) STIM (Smart Transducer Interface Module)

(*ii*) NCAP (Network Application Processor)

(*iii*) TEDS (Transducer Electronic Data Sheet)

(*iv*) TII (Transducer Independent Interface)

— The TEDS contain information of the transducer's identification (viz. make, model, specification, calibration, location etc.)

— Any transducer can be interfaced through STIM, performing the functions of signal conditioning, linearisation, signal processing, patterns recognition of monitored signals etc.

— The TII interfaces STIM with NCAP. It is a 10-wire serial I/O bus that defines data transport from STIM to NCAP, address and data transport from NCAP to STIM, latching data, performing triggering functions, data transport acknowledge, interrupt used by STIM to request services of NCAP etc. NCAP is a device used for supporting a network interface, and may provide local intelligence including the possible use of a microcontroller. It is the smart sensor's window to the outside world. *Any control network may be connected to a group of sensors, with an appropriate NCAP.*

A useful application of the above unit is for a condition monitoring system for a plant, to monitor the health of machines and process, to detect possible damage. Instead of using a dedicated computer with spectrum analyses or data acquisition system or signal processing and pattern recognition, *a smart unit on the lines of IEEE 1451.2 can accept inputs from various sensors, process the signals and diagonise fault if any and communicate* ; this would be quite *economical* for the plants.

## 7.20. FIBER OPTIC TRANSDUCERS

### 7.20.1. Introduction

**Optical Fiber Sensors (OFS)** *work on the principle of alteration of light by external variables like motion, temperature, pressure, flow, strain, pH, particle size etc.*

The fibers act as light conduits to conduct light and consist of thin strands of dielectric materials like glass, quartz, and plastics, in which some *dopants* may be introduced for certain special applications.

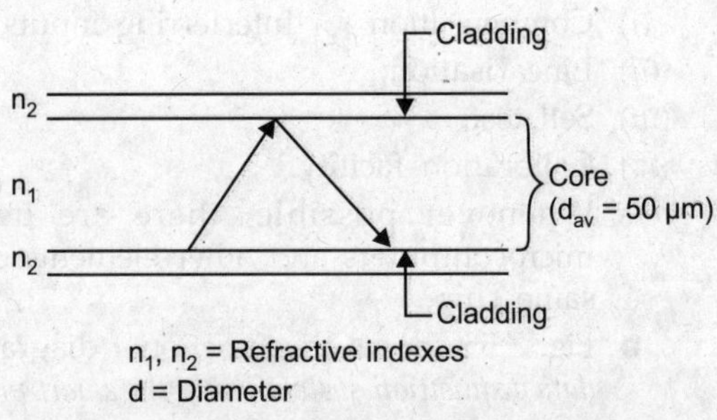


**Fig. 7.56.** A typical optical fiber.

Fig. 7.56, shows a typical optical fiber. A light ray propagating in the fiber is trapped by total reflection at the boundary of the core and the cladding material. This is possible only if $n_2$ (refractive index of the material) < $n_1$ (refractive index of core)

### 7.20.2. Types of Optical fiber sensor configurations

In an optical fiber sensor the "emitters", which may be a light emitting diode or a laser source, emits the light rays through the fiber, which get modulated due to the outer signal which is to be measured. The output fiber is connected to the "detector", which converts the optical energy into electrical energy. These detectors work on the principle of creation of an electron-hole pair in semiconductors or the release of the electrons from the cathode of the photomultiplier tube.

Basically, following are the *two types of optical fiber sensor configurations* :

1. Extrinsic (or incoherent) sensors :

   - In this type of sensor (Fig.7.57) the *intensity indulation of light takes place outside the fiber.*

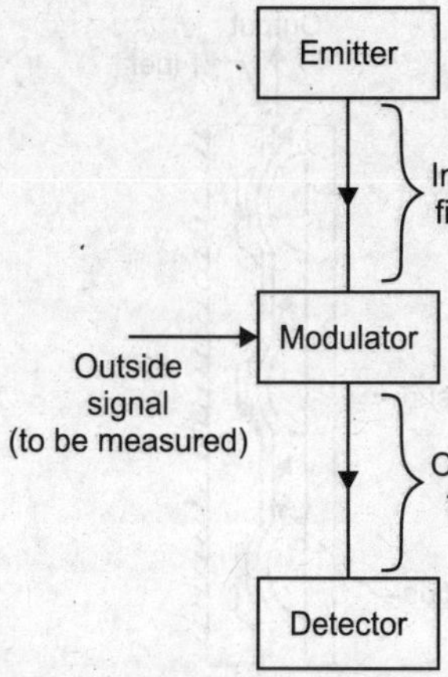


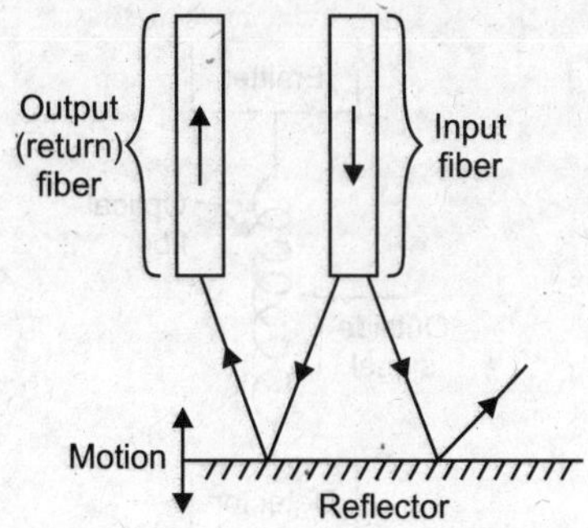


**Fig. 7.57.** Extrinsic sensor configuration. **Fig. 7.58.** Example of an extrinsic type sensor.

- Fig. 7.58, shows an extrinsic (or external) intensity modulator type of sensor in which the position of the reflector due to motion to be measured may change, thus *changing the light intensity in the output fiber*, which is detected by a detector.
- Fig. 7.59, shows a *typical fiber optic flow sensor* – an extrinsic type sensor.

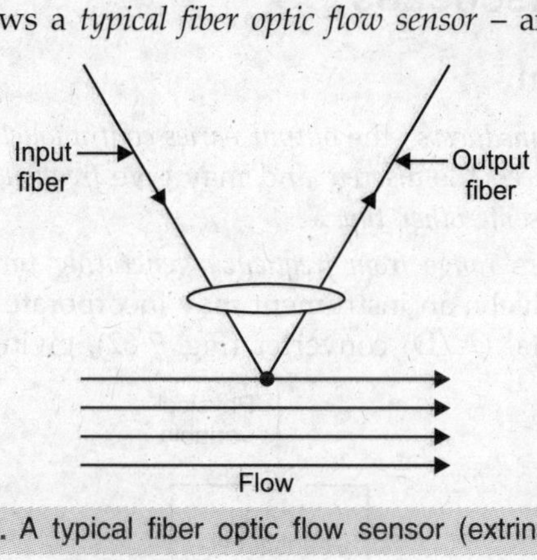


**Fig. 7.59.** A typical fiber optic flow sensor (extrinsic type).

— In this type of flow sensor, the frequency of light waves scattered from the particles in the moving fluid is Doppler shifted, the *frequency shift being proportional to velocity*.

— The input optical fiber carries light, which is focused on the fluid flow.

— The output fiber carries the light scattered by the particles in the flowing liquid and particle velocity is found from the modulated frequency spectrum.

2. Intrinsic (or coherent) sensor : Refer to Fig. 7.60
   - In this case, the *intensity modulation of light takes place within the fiber*.
   - Fig. 7.61 shows an example of an intrinsic type sensor :

     — Here, the fiber bending due to pressure, which is the input variable (to be measured), *induces radiation losses, changing the intensity at the output*. This happens within the fiber itself and hence the configuration is called as intrinsic type. This causes radiation of light even at small deformation and is also called *micro-bend sensor*.

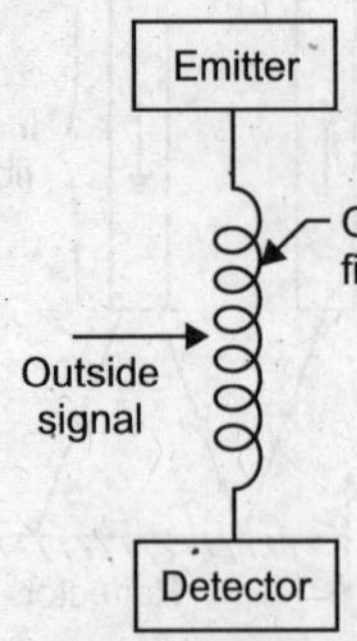


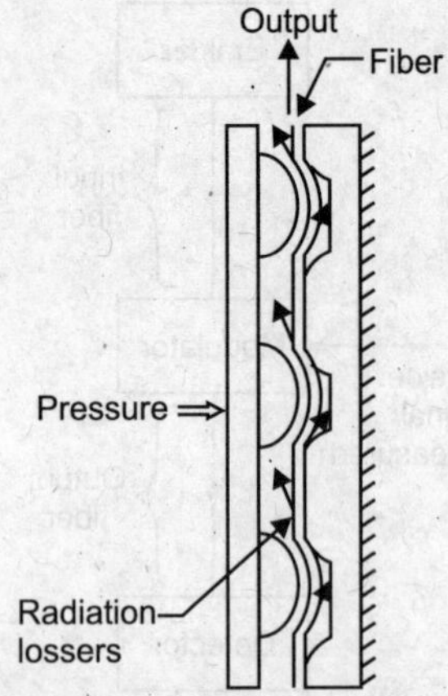


Fig. 7.60. Intrinsic sensor configuration. Fig. 7.61. Example of an intrinsic type sensor.

— In this type of sensor configuration, apart from the intensity modulation, *there may be phase or frequency modulation*, which after detection may be employed for many applications.

## 7.21 DIGITAL TRANSDUCERS

### 7.21.1. Introduction

Whereas in "*analog transducers*" the *output varies continuously according to the input*, the output in "**digital** *transducers*" is *discrete* and may give *frequency type output or a digitally coded output, of binary or some other type.*

- Digital transducers *range from frequency generating types of transducers to digital encoders.* Alternatively, an instrument may incorporate an analog transducer and an analog-to-digital (A/D) converter (Fig. 7.62), giving a digital output.

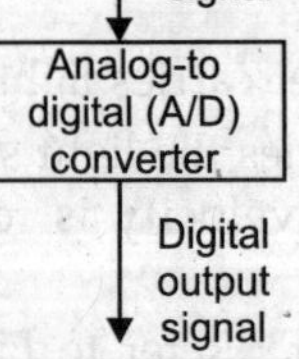


Fig. 7.62. Digital instrument incorporating A/D converter.

**Advantages :**

The main advantages of *digital transducers* are :

1. It is possible to obtain increased accuracy in pulse count.
2. It becomes easy to use digital computers along with the transducers, for data manipulation.
3. There are ergonomic advantages in presenting digital data.

4. It is easy to transmit digital signals without distortion and external noise (since digital signals are not dependent on signal amplitudes).

### 7.21.2. Frequency Domain Transducers

The *output* in these transducers is in the form of *pulses or sinusoidal wave forms, the frequency of which is a measure of the magnitude of the physical variable.*

The following three types of frequency domain transducer are described :

1. Electromagnetic frequency domain transducer.
2. Opto-electrical frequency domain transducer.
3. Vibrating string transducer.

1. **Electromagnetic frequency domain transducer :**

Fig. 7.63 shows an electromagnetic frequency domain transducer which can employed for the *measurement of speed.*

— It consists of a permanent magnent on a solenoid.

— A gear of ferromagnetic material is mounted on the rotating shaft whose speed is to be measured.

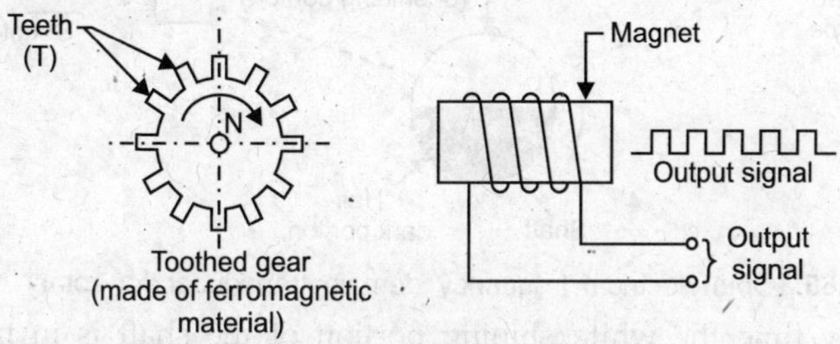


**Fig. 7.63.** Electromagnetic frequency domain transducer.

— As each gear tooth passes in front of the magnet, the gap length changes. Consequently there is a change in the flux density and a voltage pulse is induced in the coil. *Pulse frequency equal speed N times the number of teeth T.* The form of the output signal is shown in Fig. 7.63.

Thus, pulse frequency is a measure of rotational speed.

2. **Opto-electrical frequency domain transducer :**

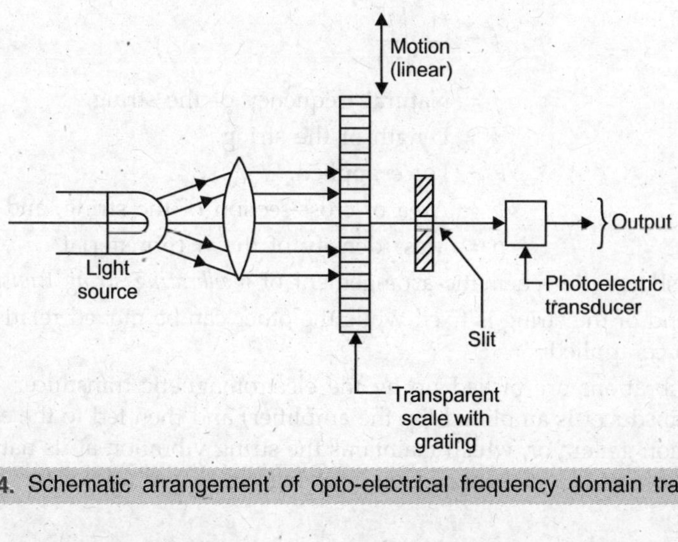


**Fig. 7.64.** Schematic arrangement of opto-electrical frequency domain transducer.

Fig. 7.64, shows schematic arrangement of this type of transducer for *linear motion* :

— The arrangement includes a transparent scale with a grating.

— The moving object is attached to the transparent scale.

— Light from a source passes through the scale and the slit and then falls on a photoelectric transducers. The slit width is such that a motion equal to the pitch of the grating produces one complete cycle of light and darkness at the photoelectric cell. Thus, a pulse output is obtained. From the number of pulses, the change in motion of the scale and the object attached to it can be determined.

- Fig. 7.65, shows opto-electrical frequency domain transducer for measurement of *rotational speed of a shaft* :

— The shaft has half dark and half white or shining portion.

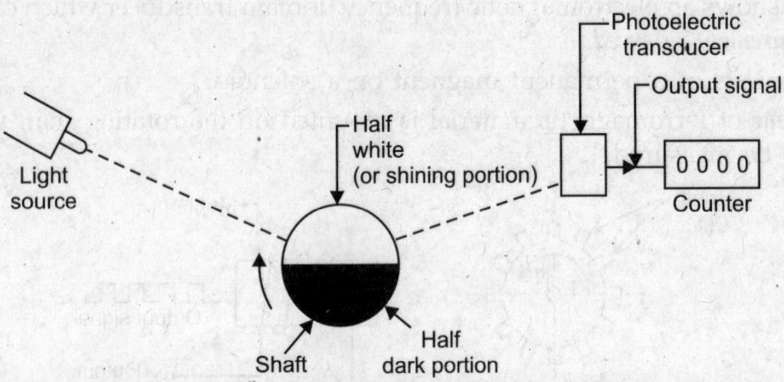


**Fig. 7.65.** Opto-electrical frequency domain transducer for rotary motion.

— Every time the white/shining portion of the shaft is in front of the light source, the reflected light, falling on the photoelectric transducer, gives an electrical pulse output. The frequency of the pulses is thus a measure of speed of rotation of the shaft.

3. **Vibrating string transducer** :

This type of transducer is essentially employed to measure the *force* applied to a metal string, which is kept-vibrating, the *frequency* of which is dependent on the force applied. The natural frequency of the string is given by :

$$f = \frac{1}{2l}\sqrt{\frac{P}{a\rho}}$$

where, $f$ = Natural frequency of the string,

$l$ = Length of the string,

$P$ = Force applied,

$a$ = Area of cross-section of the string, and

$\rho$ = Mass density of the wire material.

Fig. 7.66, shows the schematic arrangement of a *vibrating string transducer* :

— One end of the string is fixed while the other can be moved relative to it, due to the force applied.

— The vibrations are picked up by the electromagnetic transducer. The output of the transducer is amplified (by the amplifier) and then fed to the electromagnetic vibration generator, which maintains the string vibration at its natural frequency $f$. Frequency $f$ gets changed due to change in magnitude of force $P$.

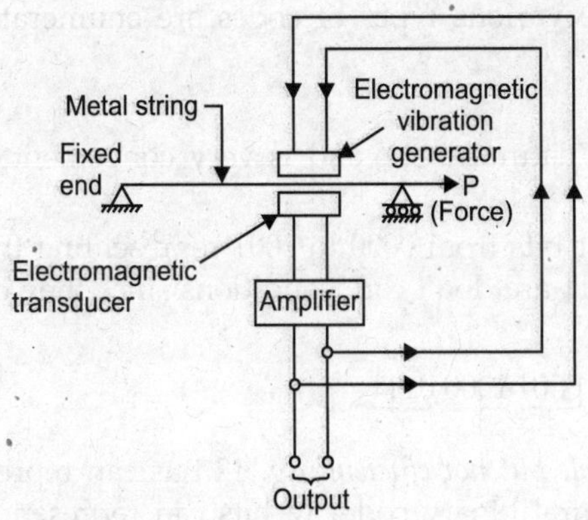


**Fig. 7.66.** Vibrating string transducer.

— The frequency is measured by a *frequency counter and is a measure of the force applied to the string.*

- This transducer can be used for *measurement of force and displacement*.
- Vibrating diaphragm and cylinder types of transducers have been developed to measure *pressure*.

### 7.21.3. Digital Encoders

*The* ***digital encoders*** *convert analog motion (rotary or linear) to a digital output form.*

**7.21.3.1. Introduction to digital coding**

In digital circuits, each number of piece of information is defined by an equivalent combination of binary digits. *A complete group of these combinations which represent numbers, letters or symbols is called a* ***digital code.***

The group of 0s and 1s in the binary number can be thought of as a code representing the decimal numbers. When a decimal number is represented by its equivalent binary number, it is called *a straight binary coding*.

In modern digital equipment, *codes are used to represent and process numerical information.*

The digital output is usually required to be in *binary form*, especially for use with a *digital computer*. Unlike the decimal system where there are ten possible states for a digit (0-9), the binary system has two states, viz. ON or OFF or 1 and 0, respectively. To form a number, these are arranged with least significant bit (LSB) on the right and most significant bit (MSB) on the left. In natural or straight binary code, the arrangement is as shown in Fig. 7.67.

Thus, 11101 in binary system would correspond to the number 29 in the decimal system, as shown in Fig. 7.67. This arrangement uses 5 bits and this can specify any number upto *31*, in steps of 1 or the resolution is $\frac{1}{31}$. More bits would be needed for smaller resolution. *Each bit may be considered to be replaced by a switch, whose state ON would correspond to the state 1 and OFF to state 0.* Thus, a number of bits or switches can specify any desired number.

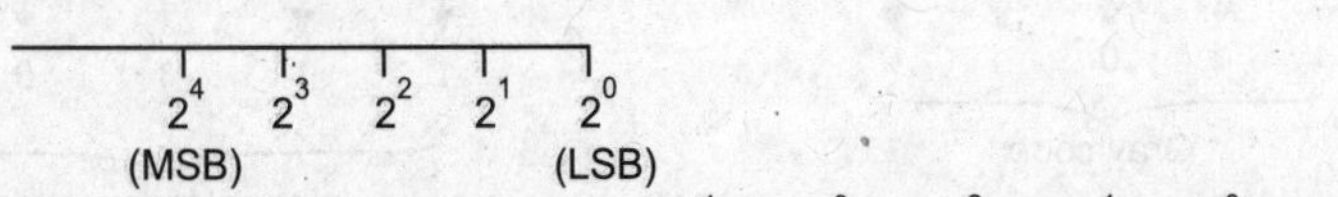


**Fig. 7.67.** Binary code arrangement.

**Types of codes.** The various types of codes are enumerated and briefly discussed below :

**1. BCB Code**

It is also known as 'natural BCD' and is very convenient for representing decimal digits in digital circuits.

- It consists of four bits from 0000 to 1001 representing the decimal numbers from 0 to 9.1010 to 1111 are don't care conditions since they do not have any meaning in BCD.

  *Example* : 187 = $\underbrace{0001}_{1}\underbrace{1000}_{8}\underbrace{0111}_{7}$

- This code is *simple but not efficient, e.g.* 12 bits can represent a number upto 999, while in the natural binary code, 12 bits can represent number upto 4095.

**2. Excess-3 Code**

- The code can be derived from BCD by *adding 3 to each coded number.*

  It is useful when it is desired to obtain the 9's complement of a decimal digit represented by this code. The 9's complement is obtained simply by complementing each bit.

- This code can be *conveniently used for performing subtracting operations in digital computers.*

**3. Gray Code**

- In this code any *one bit changes between any two successive numbers.*
- This code is used to overcome the inherent disadvantage of the straight binary code. In a straight binary code, a change of number 1 might result in a change of number of bits. Thus a change of number 7 to 8 would need a change in all the four bits as 7 is represented by 0111 and 8 by 1000. Thus an error of one bit in a large digital number, in practice, would result in a large error in decimal reconversion. In the case of a *digital transducers*, an increment in the measured variable would thus have to change all the bits, and any error in a bit, would change the number considerably. This is *overcome in the Gray code*, in which with a change from one value to the next, only one bit is changed at a time. The *conversion from natural binary to Gray code is done as follows, keeping the same number of bits* :

  (*i*) MSB is unchanged during conversion ; (*ii*) Gray code bit for other bits in the natural binary is *same* if the digit to the *left* in the natural binary is zero ; (*iii*) The bit is changed, if the bit to the left is 1. *Example* : Fig. 7.68 shows the application of these rules for changing the number 10011 in natural binary code by 11010 in Gray Code.

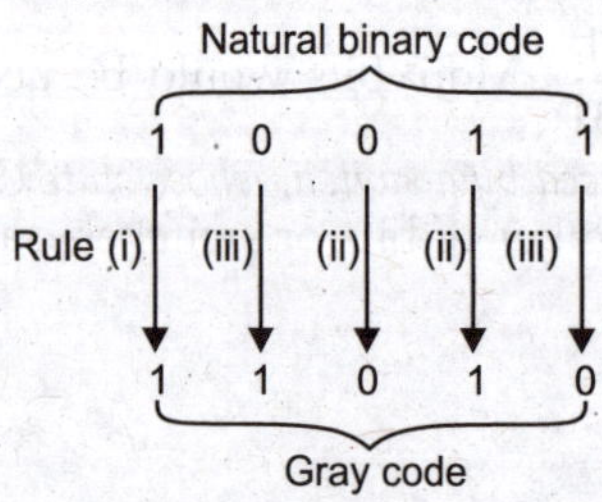


**Fig. 7.68.** Natural binary to gray code conversion.

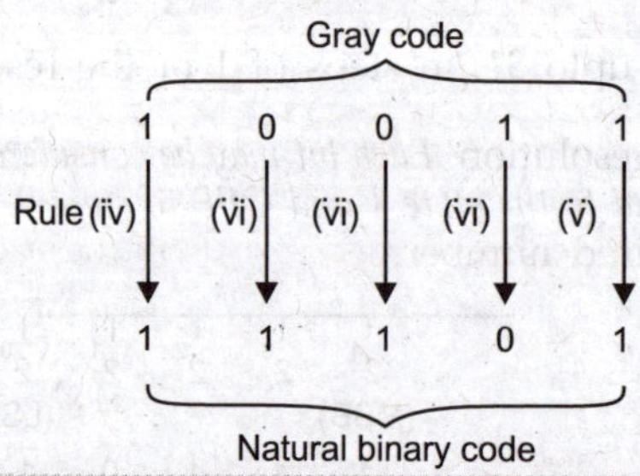


**Fig. 7.69.** Gray Code to natural binary conversion.

While changing from Gray to natural code, the following rules apply :

(*iv*) MSB is unchanged during conversion ; (*v*) for every other bit in Gray code, natural binary is *same*, if the number of 1s to the left is even ; (*vi*) The bit is *changed* if the number of 1s to the left is *odd*. *Example* : 10011 in Gray code would be changed to 11101 in natural binary code, as shown in Fig. 7.69.

- Gray code is mainly used in the *location of angular positions of a rotating shaft*.

**4. Octal Code**

- The octal system is a 8 base system.

  It uses 3 bits to represent one octal digit.

**5. Hexadecimal Code**

- The hexadecimal system is a base 16 system.
- It uses four bits to represent one hexadecimal digit.
- The hexadecimal digits are represented as 0 to 9 continued by alphabetical characters from *A* to *F*.

### 7.21.3.2. Description of ditigal encoders

In the *contact type* arrangement there is direct contact between the brushes and encoder disc. The disadvantage of such a system is the wear of brushes and disc (due to friction between the brushes and disc)

**1. Optical type encoder :**

Fig. 7.70. illustrates the arrangement of an *optical type encoder* :

— It is a non-contact encoder.

— The disc has transparent and opaque areas, corresponding to the conducting and non-conducting one respectively, in the contacting type encoder disc.

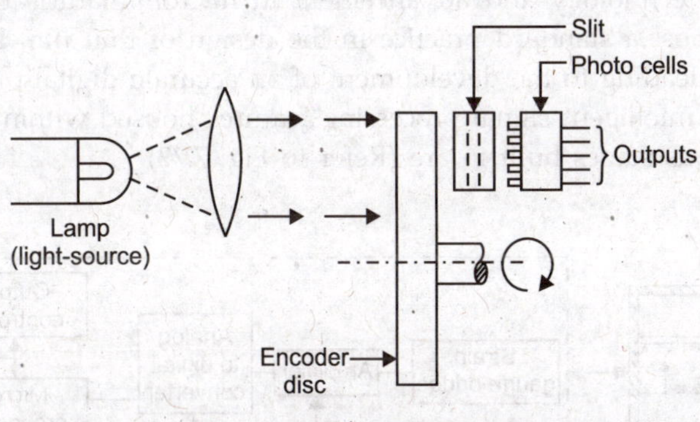


**Fig. 7.70.** Optical type encoder.

— A light source is used along with slits and photo cells. The photo cell, corresponding to a particular track, would produce an electrical output if the transparent portion is in front of the slit and light source, giving state ON (or 1) while no electrical output from a cell would correspond to OFF (or 0) state.

**2. Magnetic type encoder :**

Fig. 7.71 shows a magnetic encoder circuit :

— It is also a non-contact type encoder.

— This encoder uses a number of small toroidal magnets with coils around them.

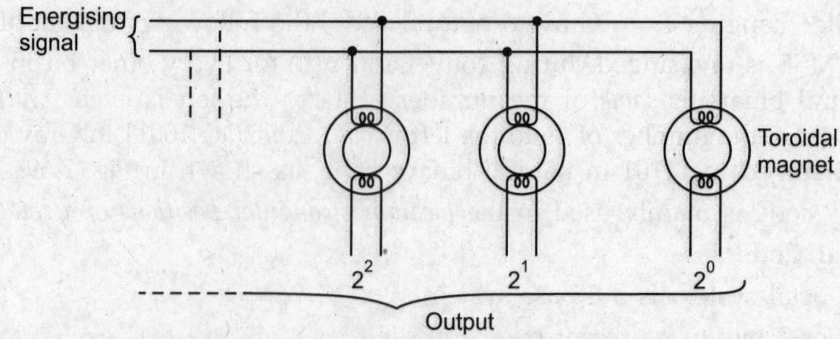


**Fig. 7.71.** Magnetic encoder circuit.

— The area corresponding to the conducting and non-conducting ones in the contact type encoder disc form the non-magnetic and magnetic areas on the disc. The presence or absence of such areas is detected by coils which are in close proximity to each track on the disc.

— One of the coils in each toroidal magnet is energised with high frequency A.C. carrier signal. If the toroid is over a non-magnetic area of the encoder disc, a voltage due to transformer action is induced in the output coils (state 1) while a magnetic area would saturate the magnetic circuit and output is very small (state 0). The signals on the output windings are *demodulated* to remove the high frequency carrier signal.

### 7.21.4. Digital Pressure Transducer

These days it has become feasible to incorporate the signal conditioner and associated electronic circuitry within the body of the transducer itself, owing to the advent of integrated circuit technology and advancement in microminiaturisation innovation. It has now been almost a standard practice in the design of thin film based and surface mounted devices, leading to the development of an accurate digital pressure transducer with a number of intelligent signal processing features housed within a relatively small volume. The various stages built in are (Refer to Fig. 7.72) :

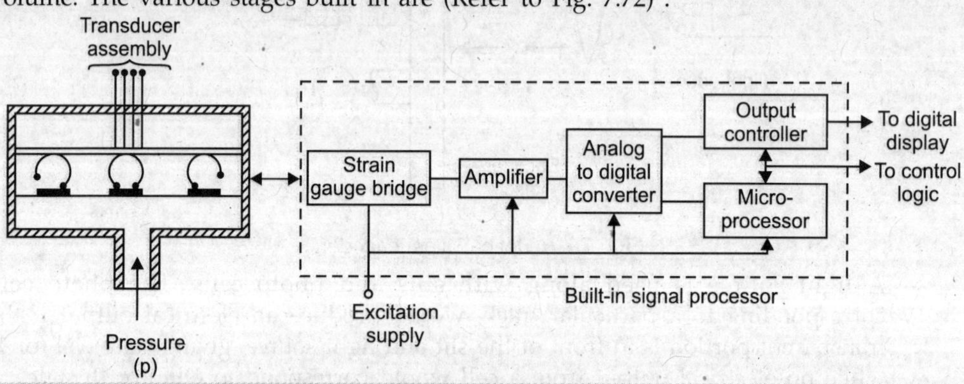


**Fig. 7.72.** Digital pressure transducer.

- Preamplifier ;
- Analog to digital (A/D) converter ;
- Output processor, controlled with a microprocessor (μP)

  The final electrical output can be directly connected to a digital display or modified to give a direct output to 0 to +/– 5V or 4 to 20 mA.

## 7.22 RECENT TRENDS–SMART PRESSURE TRANSMITTERS

The microprocessors are now being used in *transmitters* also; as a consequence of the availability of computing power the transmitters have become more intelligent.

The output in case of smart transmitters is 4–20 mA on 2-wire but with the added capability of digital communication from a hand-held interface connected anywhere on 4–20 mA signal, the remote adjustment of the transmitter data base and acquisition diagnostic information to minimise loop downtime is possible. It has *high rangeability and much better performance.*

The transmitter senses all the *three parameters, 'differential pressure', 'static pressure'* and *'temperature'*. The meter body is pre-programmed in manufacturing to characterise the unit for linearity, static pressure and temperature effects, and it *computes a highly repeatable and accurate pressure measurement*. These characteristics are held in PROM memory and being specific to one meter are kept with the meter body. The combination of characterised meter body and digital electronics has enabled a quantum leap forward in performance.

- The *rangeability* to smart transmitters is very high (400 : 1). Thus only 3 sensors would be required to cover the entire range of 2.5 millibar to 700 bar.
- The *reliability* is very high due to use of minimum number of components and protection against all foreseeable damping like radio frequency, reverse polarity, overpressure, surge voltage and lightning.

**Advantages of digital transmitters :**

The major advantages of digital (so called *"smart"*) transmitters over their conventional analog counterparts are :

(*i*) Increased rangeability (400 : 1 against 6 : 1 of analog transmitters).
(*ii*) Higher accuracy.
(*iii*) Self-diagnostic facilities.
(*iv*) Almost no drift with time.
(*v*) Reduced cabling cost due to the use of a field bus cuts.
(*vi*) Better noise immunity.
(*vii*) Economical, because of improved overall performance.
(*viii*) Ambient temperature compensation.
(*ix*) Remote adjustability of range, damping, polarity etc. (This makes the commissioning of the entire system simpler).

## 7.23 SELECTION OF SENSORS

A number of factors need to be considered for selecting a sensor for a particular application are :

1. The nature of the measurement required *e.g.,*
   — The variable to be measured, its nomial value, the range of values;
   — The accuracy required;
   — The required speed of measurement;
   — The reliability required;
   — The environmental conditions under which the measurement is to be made.
2. The nature of the output required from the sensor, thus determining the signal conditioning requirements in order to give suitable output signals from the measurement.

Then possible sensors can be identified taking into account such factors as their range, accuracy, linearity, speed of response, reliability, maintainability, life, power supply requirements, ruggedness, cost, availability.

## 7.24 ROTARY VARIABLE DIFFERENTIAL TRANSFORMER (RVDT)

RVDT is used for the measurements of *angular displacements*. It *converts the angular displacements into electrical signals*. Its operating principle is similar to that of LVDT.

Fig. 7.73, shows the constructional details of a RVDT :

— The construction of RVDT is similar to that of LVDT (*i.e.* one primary winding *P* and two secondary windings $S_1$ and $S_2$), the only dfference is that the core is cam-shaped and rotated inside the windings through a shaft.

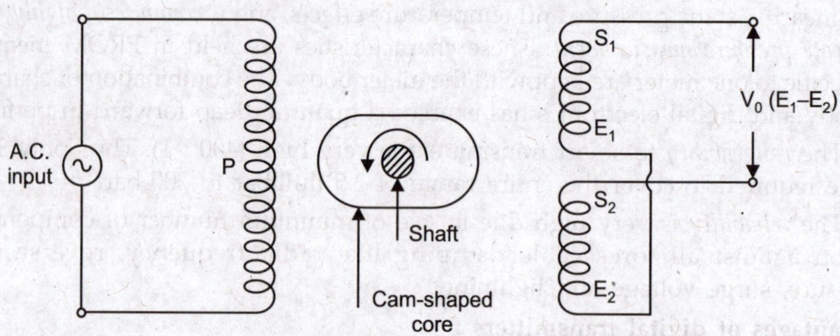


P = Primary winding ; $S_1$, $S_2$ = Secondary windings;
$E_1$, $E_2$ = Induced e.m.fs ; $V_0$ = Output voltage.

**Fig. 7.73.** Rotary variable differential transformer–Constructional details

— The core is rotated in clockwise and anticlockwise directions with respect to *"Null"* position. When the core is at *"Null"* position the voltage is *zero* because the voltages induced in both the secondary windings $S_1$ and $S_2$ are *equal and in opposite directions*.

— When any angular displacement to be measured (w.r.t. Null position) is applied to the shaft, differential voltage gets developed across the secondary windings at the output. When the core is rotated in clockwise direction, the voltage gets increased. If the core is rotated in anticlockwise direction, the voltage increases. The output voltage is increasing in both the directions, but there is opposition with respect to each other according to the position of shaft (clockwise and anticlockwise).

## 7.25 SYNCHROS AND RESOLVERS

Basically, *"synchros and resolvers" are a class of A.C. electromechanical variable-coupling transformer devices primarily employed for angular data transmission and computing systems.*

**Synchro :**

— The basic electromagnetic structure of a synchro consists of a *wound rotor* and a *wound stator*, concentrically arranged to give adjustable mutual coupling between the windings of the two members.

— The rotor is wound with a *single-phase* concentrated winding and the stator is wound with *three-phase distributed windings connected in a Y-configuration*.

— The rotor is excited with an A.C. signal, and by the transformer action between the rotor and stator windings, voltages are induced in the stator coils. *The magnitudes of these voltages depend upon the angular position of the rotor with respect to the stator*.

Fig. 7.74, shows a typical practical set-up in which a pair of synchros are used – one acts as a transmitter and the other as a receiver.

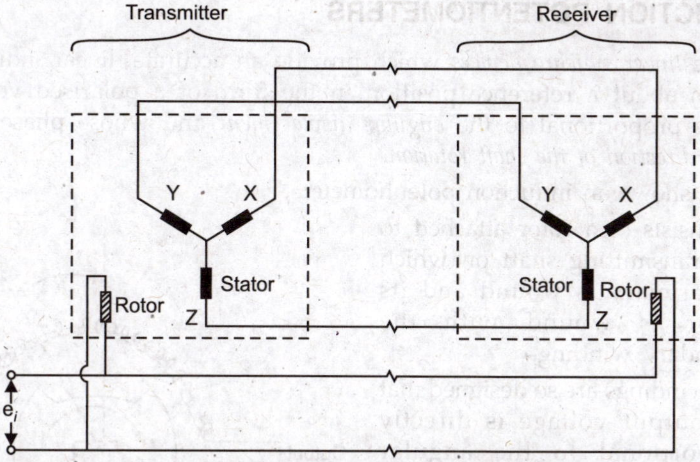


**Fig. 7.74.** Practical set-up of a pair of synchros.

**Resolver :**

— An inductive resolver is essentially a variable transformer with a *rotary electromagnetic coupling between the primary and secondary windings.*

— In a typical angle-positioning resolver, two windings are provided at *right angles to one another* for both stator and rotor. The windings are located in such a manner that output voltage has an amplitude *proportional to the sine or cosine of the input shaft position of the rotor with respect to the stator.* Depending upon the designer's particular application, either the rotor or stator may be used as the primary.

Fig. 7.75, shows the electrical connections of a resolver.

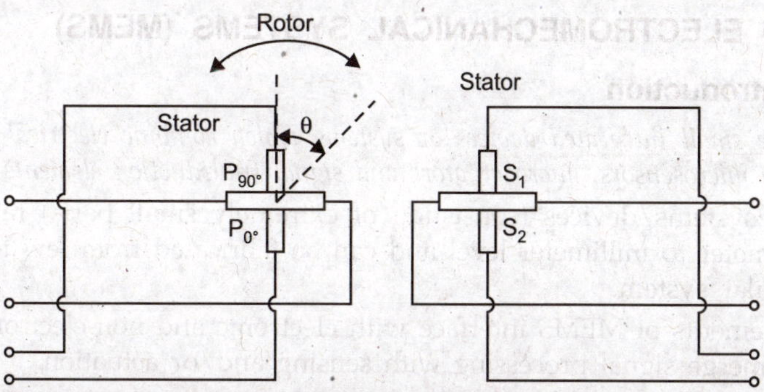


**Fig. 7.75.** Electrical connections of a resolver.

**Advantages of synchros and resolvers :**

The main advantages of synchros and resolvers are :

1. High reliability.
2. High accuracy (0.01 percent feasible).
3. Infinite resolution.
4. Useful operating angle of 360° and capable of continuous rotations.
5. No rotational wear, except at slip rings.
6. Relatively insensitive characteristics to stray cable capacitances.
7. The operational speeds of the system are quite high.

## 7.26 INDUCTION POTENTIOMETERS

These are *linear synchro devices* which provide an accurate, linear indication of the shaft rotation about a reference position in the form of a polarised voltage whose magnitude is proportional to the *angular displacement* and whose phase relationship indicates the *direction of the shaft rotation.*

Fig. 7.76, shows an induction potentiometer :

— It consists of a rotor attached to the transmitting shaft on which the primary is wound and its stator is wound with the secondary winding.

— The windings are so designed that the output voltage is directly proportional to the angular position of the rotor.

— Good linearity is feasible for angular displacements in the range of ± 45° only, even though the shaft rotation is continuous.

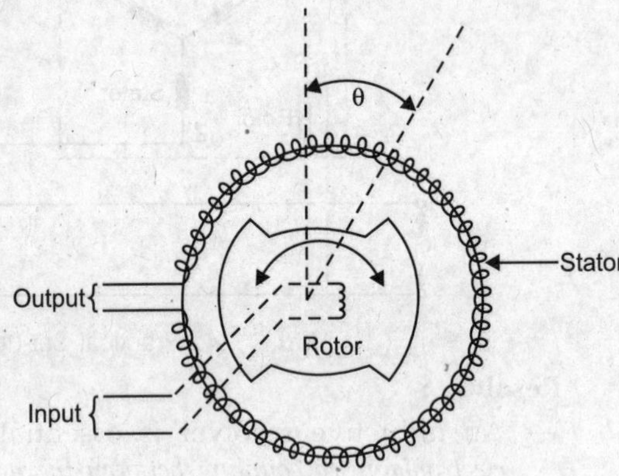


**Fig. 7.76.** Induction potentiometer.

Some of the *errors* that can arise are due to : (*i*) Mechanical asymmetry, (*ii*) Variable output impedance, and (*iii*) Thermal effects.

These devices are analogous to resistance potentiometers but since they are induction-type components, they have less restraining forces acting upon the rotors and hence are *capable of providing higher resolution.*

## 7.27 MICRO ELECTROMECHANICAL SYSTEMS (MEMS)

### 7.27.1. Introduction

**MEMS** *are small integrated devices or systems which combine electrical and mechanical components like microsensors, microactuators and signal transduction elements.*

— These systems/devices (consisting of extremely small parts) range from sub-micrometer to millimeter level and can be fabricated from few to millions in a particular system.

— The elements of MEMS interface with electronic and non-electronic signals and often merge signal processing with sensing and/or actuation.

- *"MEMS" are used to sense, control and activate mechanical processes on the microscale and function individually or in arrays.*

### 7.27.2. Advantages of MEMS

MEMS claim the following advantages over conventional electromechanical systems :

1. Better stability and higher accuracy in the performance.
2. Can be produced on mass scale (once the fabrication design and process is finalised).
3. As these devices are available in very small sizes, they can be easily inserted with other applications (*Miniaturization*).
4. Compact integration of comprehensive functionality.

5. Depending upon the requirements, the MEMS can be designed for various signals (*e.g.*, mechanical, optical, electrical, biological etc.)

### 7.27.3. Applications of MEMS

The various fields of application of MEMS are :

1. Consumer products.
2. Process and manufacturing industries.
3. Aerospace industry.
4. Automobile engineering.
5. Telecommunication engineering.
6. Medical field.

### 7.27.4. Description of MEMS

Fig. 7.77 shows the general block diagram of MEMS :

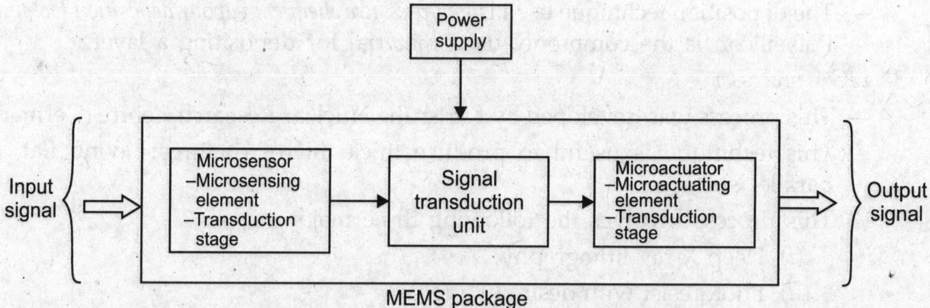


**Fig. 7.77.** General block diagram of MEMS.

1. ***Input signals :***

- These signals are received from various sources/systems such as mechanical, electrical, optical, chemical etc.
- They are admitted/supplied to the MEMS package.

2. ***MEMS package :***

- It is semiconductor chip (IC), which can be fabricated by suitable micromachining technique.
- It consists of the following components :

  (*i*) *Microsensors :*

  — Microsensing element;

  — Signal processing hardware.

  These sensors are used to *sense* different physical parameters such as pressure, temperature, force, humidity, sound etc.

  (*ii*) *Microactuators :*

  — Microactuating unit;

  — Transduction stage.

  The microactuator is an element that *converts electrical signals into physical signals. Example:* Microswitches, microvalves, micropumps, microgrippers, etc.

3. ***Output signals :***

- These are the signals interfaced with the system to be controlled or monitored.

*Owing to monolithic integration of sensors, actuators and signal transduction unit, closed loop control system can be developed.*

### 7.27.5. Manufacturing of MEMS

*The technique employed for manufacturing MEMS is called micromachining or micromanufacturing.*

The various techniques of micromachining are :

1. *Bulk micromachining :*
   — It is used for production of microsensors and accelerometers.
   — In this method, isotropic and anisotropic etching techniques are used.
2. *Surface micromachining :*
   — In this method, MEMS are produced by depositing, patterning and etching a sequence of thin films, typical 1 to 100 μm thick.
   — The deposition technique used is *low-pressure chemical vapour deposition technique.* Polysilicon is the commonly used material for depositing a layer.
3. *LIGA process :*
   — This process was developed by Karlsruhe Nuclear Research Centre (Germany).
   — This technique is useful to produce thick microstructures having flat and parallel surfaces.
   This process involves the following three major steps :
   (*i*) Deep X-ray lithography;
   (*ii*) Photoresist with desired pattern;
   (*iii*) Electroplating.
   — It is widely used to produce *microgears* and *microturbines.*

### 7.27.6. MEMS Accelerometer—Vibration Microsensor

The most common method of observing the conditions of a rotating machine is the vibration analysis. *"Vibration" is measured by its frequency and amplitude.* The most common type of vibration sensor is **accelerometer**. The accelerometers are designed for the purpose of measurement of shock and vibrations.

- *"Piezoelectric-type accelerometer"* has the highest frequency response and high output voltage, making it convenient for measurement of shock and vibration.
  — The most suitable piezoelectric material for fabrication of MEMS accelerometer is PZT (lead zirconate titanate oxide).
  — The piezoelectric sensor is a self-generating device which is ideal for sensing dynamic events such as shocks and vibrations.
  Following are the advantages of PZT sensor :
  (*i*) Fast response;
  (*ii*) Better stability;
  (*iii*) High frequency bandwidth;
  (*iv*) High charge sensitivity (upto 110 pF/N)
  — In its *construction* (Fig. 7.78), a membrane on silicon wafer with $SO_2$ layer is fabricated by bulk

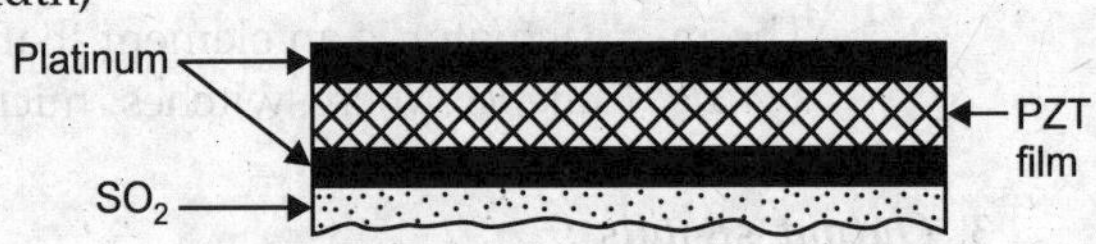

Fig. 7.78. Construction of MEMS accelerometer.

micromachining. The film of PZT is sandwiched between platinum electrodes by deposition technique.

- When the accelerometer is connected to the vibrating surface, according to the working principle, the *PZT film experiences a force and an electric charge is developed across it. This charge variation can be converted into voltage variation at the output by charge amplifier.*

**Applications.** The MEMS accelerators is suitable for Z measuring and analysing the vibration levels at rotating machines, pumps, compressors and turbines; these measurements may be useful for controlling the vibrations to check noise and damage of rotating machines.

### 7.27.7. MEMS Humidity Microsensor

Fig. 7.79 shows the construction of a MEMS humidity microsensor. For such a sensor barium titanate ($BaTiO_3$) is used because it undergoes strong changes in its dielectric constant when exposed to humid atmosphere.

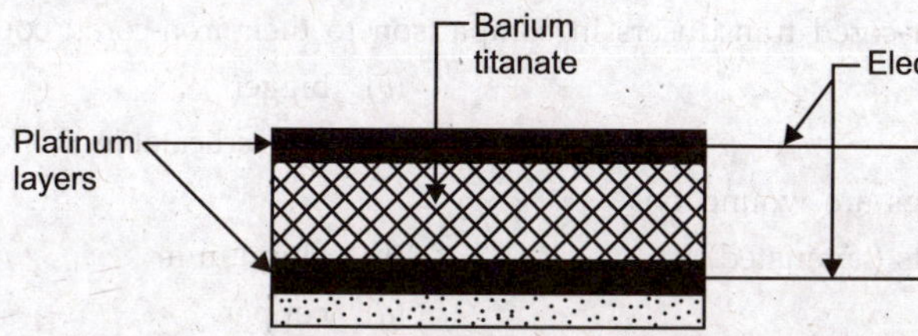


**Fig. 7.79.** Construction of a MEMS humidity microsensor.

$BaTiO_3$ (a synthetically produced piezoelectric material-ferroelectric ceramic) has high permittivity and with induced polarization it is conveniently used in sensors. It advantages are : Good mechanical and thermal stability, low leakage and good frequency response, large voltage sensitivity and large dielectric constant (2000–5000 appr.). A nanometric $BaTiO_3$ film can be fabricated on a silicon ($SiO_2/Si$) substrate.

The capacitance of the capacitor formed by placing $BaTiO_3$ between silver-palladium and nickel electrodes layer is given by :

$$C = \frac{K \epsilon_0 A}{d} \quad \text{...(7.70)}$$

where, $K$ = Dielectric constant,

$\epsilon_0$ = Permittivity of free space = $8.85 \times 10^{-12}$ F/m,

$A$ = Surface area, and

$d$ = Distance between the electrode layers.

When assembly is exposed to *humidity, the capacitance C changes due to change in the value of K. This change in C is calibrated into voltage using charge amplifiers.*

## HIGHLIGHTS

1. The technology of using instruments to measure and control the physical and chemical properties of materials is called *instrumentation*.
2. Modes of measurements are :

   (*i*) Primary measurements (*ii*) Secondary measurements

   (*iii*) Tertiary measurements.

3. A transducer is a device which converts the energy from one from to another.
4. Transducers may be classified as follows :
   A. (*i*) Active transducers (*ii*) Passive transducers
   B. (*i*) Variable-resistance type (*ii*) Digital transducers.
   C. (*i*) Variable-resistance type (*ii*) Variable-inductance type
   (*iii*) Variable-capacitance type (*iv*) Voltage-generating type
   (*v*) Voltage-divider type.

## OBJECTIVE TYPE QUESTIONS

1. LVDT is a
   (*a*) capacitive transducer (*b*) resistive transducer
   (*c*) inductive transducer (*d*) none of them.
2. The size of air-cored transducers in comparison to their iron-cored counter parts is
   (*a*) smaller (*b*) bigger
   (*c*) same (*d*) unpredicatable.
3. LVDT windings are wound on
   (*a*) steel sheets (laminated) (*b*) aluminium
   (*c*) ferrite (*d*) copper.
4. Piezoelectric crystals are used for measurement of ........ changes.
   (*a*) static (*b*) dynamic
   (*c*) static and dynamic (*d*) any of these.
5. Piezoelectric crystals produce an e.m.f.
   (*a*) when external mechanical force is applied
   (*b*) when external magnetic field is applied
   (*c*) when radiant energy stimulates the crystal
   (*d*) when the junction of two such crystals is heated.
6. Piezoelectric crystals are used for the measurement of
   (*a*) temperature (*b*) velocity
   (*c*) sound (*d*) none of the above.
7. In the given circuit, how much the voltmeter will read ?
   (*a*) 0 V (*b*) 10 V
   (*c*) 4 V (*d*) 5 V
   (*e*) 3.33 V.

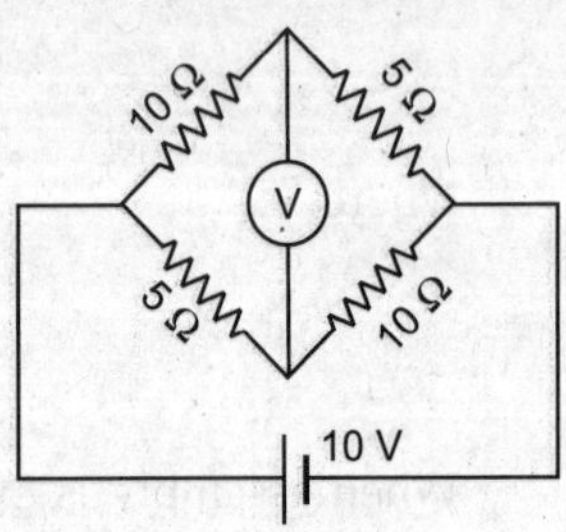

Fig. 7.80

8. Capacitive transducers operate upon the principle (*s*) of
   (*a*) variation of over-lapping area of plates
   (*b*) variation of separation of plates
   (*c*) variation of relative permittivity of dielectric material between two plates
   (*d*) all of the above.
9. Capacitive transducers are normally employed for ........ measurements.
   (*a*) static (*b*) dynamic
   (*c*) both static and dynamic (*d*) transient.

**10.** The thermo-electric effect was first observed by

(*a*) Seebeck (*b*) Thomas Young
(*c*) Pirani (*d*) Thermus.

**11.** Thermocouples are ............. transducers.

(*a*) active (*b*) passive
(*c*) adhesive (*d*) both (*a*) and (*c*).

**12.** Nitro-cellulose cement is used in strain gauges as

(*a*) carrier (*b*) base
(*c*) adhesive (*d*) lead.

**13.** A resistance thermometer is basically a/an

(*a*) active transducer (*b*) passive transducer
(*c*) potentiometer (*d*) none of these.

**14.** Platinum resistance thermometer can be used upto

(*a*) 200°C (*b*) 850°C
(*c*) 1200°C (*d*) 1500°C.

**15.** Which of the following should be incorporated in the RTD to make a temperature sensitive bridge most sensitive to temperature ?

(*a*) Platinum (*b*) Nickel
(*c*) Thermistor (*d*) Copper.

**16.** Bourdon tubes have the advantages of

(*a*) high accuracy and good dynamic response
(*b*) high sensitivity and good repeatability
(*c*) not being prone to shock vibrations
(*d*) not being susceptible to hysteresis.

**17.** A transducer is basically a device which converts

(*a*) mechanical energy into electrical
(*b*) energy or information from one form to another
(*c*) mechanical displacement into electrical
(*d*) none of these.

**18.** The gauge factor of a strain gauge is given as

(*a*) $G = \dfrac{\Delta R/R}{\Delta l/l}$ (*b*) $G = \dfrac{\Delta l/l}{\Delta R/R}$

(*c*) $G = \dfrac{\Delta R/R}{\Delta D/D}$ (*d*) none of these.

**19.** The gauge factor *G* and the Poisson's ratio μ are related as

(*a*) $g = 1 + \mu$ (*b*) $G = \mu$

(*c*) $G = 2 + \mu$ (*d*) $G = \dfrac{1+\mu}{2}$.

**20.** Thermocouples are generally used for accurate temperature measurement upto

(*a*) 350°C (*b*) 550°C
(*c*) 1400°C (*d*) 3500°C.

**21.** For surface temperature measurement one can use

(*a*) strain gauges (*b*) diaphragm
(*c*) RTD (*d*) thermocouple.

**22.** LVDT can be used for

(*a*) vibration measurement (*b*) angular velocity measurement
(*c*) force measurement in beam (*d*) load measurement on a column.

23. The principles of operation of LVDT is based on variation of
(a) self inductance (b) mutual inductance
(c) reluctance (d) permeance.

24. An LVDT has an output in the form of
(a) linear displacement of core (b) pulse
(c) rotary movement of core (d) none of the above.

25. Hall effect transducers have the drawbacks of
(a) high-sensitivity to temperature variation
(b) variation of Hall's coefficient from plate to plate
(c) poor resolution
(d) both (a) and (b).

26. A Hall's effect pick-up can be used for measuring
(a) pressure (b) magnetic flux
(c) relative humidity (d) current.

27. Self generating transducers are ........ transducers.
(a) active (b) passive
(c) secondary (d) inverse.

28. The transducer that converts the input signal into the output signal, which is a continuous function of time, is known as ....... transducer.
(a) active (b) passive
(c) analog (d) digital.

29. A transducer that converts measurand into the form of pulse is called the ........ transducer.
(a) active (b) analog
(c) digital (d) pulse.

30. Certain types of materials generate an electrostatic charge or voltage when mechanical force is applied across them. Such materials are called the
(a) Piezo-electric (b) photo-electric
(c) thermo-electric (d) none of these.

31. Piezoelectric transdcuers are .......... transducers.
(a) active (b) passive
(c) inverse (d) both (a) and (c).

32. Piezoelectric transducers work when we apply to it
(a) heat (b) mechanical force
(c) vibrations (d) illumination.

33. In semiconductor strain gauges, the change in resistance on application of strain is mainly on account of change in
(a) length of wire (b) diameter of wire
(c) resistivity (d) both (a) and (b).

34. The drawbacks of semiconductor strain gauges are
(a) low fatigue life (b) poor linearity
(c) that they are expensive and brittle (d) none of these.

35. In a resistance potentiometer, the non-linearity ............. the ratio of potentiometer to load resistance
(a) decreases with the increase in (b) increases with the increase in
(c) is independent of (d) none of the above.

36. High value pot resistance leads to
(a) low sensitivity (b) high sensitivity
(c) low non-linearity (d) less error.

37. A strain gauge is a passive transducer and is employed for converting
   (*a*) mechanical displacement into a change of resistance
   (*b*) pressure into a change of resistance
   (*c*) force into displacement
   (*d*) none of the above.
38. Semiconductor strain gauges have
   (*a*) high gauge factor (–100 to 150) (*b*) very small size (0.7 to 7 mm)
   (*c*) higher fatigue value (*d*) all of these.
39. The strain gauges should have low
   (*a*) gauge factor (*b*) resistance
   (*c*) resistance temperature coefficient (*d*) all of the above.
40. The carrier material employed with strain gauges at room temperature is
   (*a*) impregnated paper (*b*) bakelite
   (*c*) aluminium foil (*d*) epoxy.
41. Seismic transducer is used for measurement of
   (*a*) linear velocity (*b*) angular velocity
   (*c*) acceleration (*d*) pressure.
42. The accelerometer using LVDT has the advantage of
   (*a*) high natural frequency (*b*) being contactless device
   (*c*) better resolution (*d*) all of these.
43. Bonded strain gauges are
   (*a*) exclusively used for construction of transducers
   (*b*) exclusively used for stress analysis
   (*c*) used for both stress analysis and for construction of transducers
   (*d*) none of the above.
44. A load cell is essentially a
   (*a*) strain gauge (*b*) thermistor
   (*c*) resistive potentiometer (*d*) inductive transducer.
45. A load cell is an electro-mechanical device and is widely used for measurement of
   (*a*) static forces (*b*) dynamic forces
   (*c*) temperature (*d*) both (*a*) and (*b*).
46. Which of the following can be used for pressure measurement ?
   (*a*) Thermometer (*b*) Pyrometer
   (*c*) Bolometer (*d*) Piezoelectric crystal.
47. Radiation pyrometers are used for measuring temperature in the range of
   (*a*) 500 – 1000°C (*b*) 1000–2000°C
   (*c*) 1200 – 3500°C (*d*) above 4000°C.
48. The best method of measuring the temperature of hot bodies radiating energy in the visible spectrum is
   (*a*) thermocouple (*b*) optical pyrometer
   (*c*) RTD (*d*) thermistor.
49. Pirani gauge is used for measuring ........ pressure.
   (*a*) very high (*b*) high
   (*c*) very low (*d*) atmospheric.

**50.** Pirani gauge are used for measurement of pressure ranging from

(a) $10^{-4}$ to 1 torr (b) 1 to 10 torrs

(c) 10 to 100 torrs (d) above 100 torrs.

**51.** The ionization vacuum gauge, in construction, is similar to a

(a) vacuum diode (b) vacuum triode

(c) thyratron (d) none of these.

**52.** The device used for measuring temperatures exceeding 1500°C is

(a) radiation pyrometer (b) RTD

(c) thermocouple (d) bimetallic thermometer.

**53.** The most suitable device for measuring temperature of a furnace is

(a) RTD (b) thermistor

(c) optical pyrometer (d) bimetallic thermometer.

**54.** Which of the following devices cannot be used for measurement of temperature ?

(a) RTD (b) Thermocouple

(c) LVDT (d) Pyrometer.

**55.** Which of the following is not the drawback of radiation pyrometers ?

(a) Their initial as well as installation costs are high

(b) Poor precision and slow response

(c) They need maintenance

(d) Each pyrometer needs individual calibration.

**56.** Pyrometer is used to measure

(a) strain (b) pressure

(c) displacement (d) temperature.

**57.** The device used for measuring low pressure, of the order of $10^{-2}$ torr, is

(a) strain gauge (b) Pirani gauge

(c) ionization gauge (d) any of these.

**58.** Moving-coil pick-up is used for measuring

(a) linear velocity (b) vibrations

(c) displacement (d) pressure.

**59.** Electronic counters are used for measuring

(a) linear velocity (b) angular velocity

(c) acceleration (d) pressure.

**60.** Angular velocity is measured by

(a) strain gauge (b) solar cell

(c) A.C. tacho-generator (d) none of the above.

**61.** A Wheatstone bridge circuit using strain gauges can be used for measuring

(a) static strains (b) dynamic strains

(c) both (a) and (b) (d) none of these.

**62.** Dummy strain gauges are used for

(a) calibration of strain gauges (b) compensation of temperature variations

(c) increasing bridge sensitivity (d) all of the above.

**63.** In measurements using two strain gauges, the second strain gauge is provided for

(a) temperature compensation (b) increasing sensitivity

(c) stability (d) linearity

(e) both (a) and (b).

**64.** Which of the following additional devices is required for measuring pressure with an LVDT ?

(*a*) Bellows (*b*) Bourdon tubes
(*c*) Bolometer (*d*) Rotameter
(*e*) either (*a*) or (*b*).

**65.** Which of the following devices cannot be used for measurement of pressure ?

(*a*) LVDT (*b*) RTD
(*c*) Piezoelectric transducers (*d*) Piezoresistive transducers.

**66.** The transducers used for measurement of linear displacement are

(*a*) strain gauges and resistive potentiometers
(*b*) LVDTs, capacitive transducers and Hall effect transducers
(*c*) thermocouples, thermistors and RTDs
(*d*) both (*a*) and (*b*).

**67.** Rotational displacement can be measured by

(*a*) strain gauges (*b*) resistive potentiometers
(*c*) reluctance transducers (*d*) both (*b*) and (*c*).

**68.** Temperature compensation, in bridge circuit arrangement, is affected by

(*a*) using dummy strain gauges
(*b*) using strain gauges of smaller gauge factor
(*c*) reversing strain gauges
(*d*) any of these.

**69.** Which one of the following devices cannot be used to measure pressure ?

(*a*) Strain gauge (*b*) LVDT
(*c*) Piezoelectric crystal (*d*) Pyrometer.

**70.** Which of the following additional is required for measuring pressure with piezoelectric crystal ?

(*a*) Bellows (*b*) Strain gauges
(*c*) Rotameter (*d*) RTD.

## ANSWERS

| | | | | | | |
|---|---|---|---|---|---|---|
| 1. (*c*) | 2. (*b*) | 3. (*c*) | 4. (*b*) | 5. (*a*) | 6. (*a*) | 7. (*e*) |
| 8. (*d*) | 9. (*c*) | 10. (*a*) | 11. (*a*) | 12. (*e*) | 13. (*b*) | 14. (*b*) |
| 15. (*c*) | 16. (*b*) | 17. (*b*) | 18. (*a*) | 19. (*c*) | 20. (*c*) | 21. (*d*) |
| 22. (*c*) | 23. (*b*) | 24. (*a*) | 25. (*d*) | 26. (*d*) | 27. (*a*) | 28. (*c*) |
| 29. (*c*) | 30. (*a*) | 31. (*d*) | 32. (*b*) | 33. (*c*) | 34. (*c*) | 35. (*b*) |
| 36. (*b*) | 37. (*a*) | 38. (*d*) | 39. (*c*) | 40. (*a*) | 41. (*c*) | 42. (*d*) |
| 43. (*c*) | 44. (*a*) | 45. (*d*) | 46. (*d*) | 47. (*c*) | 48. (*b*) | 49. (*c*) |
| 50. (*a*) | 51. (*b*) | 52. (*a*) | 53. (*c*) | 54. (*c*) | 55. (*b*) | 56. (*d*) |
| 57. (*b*) | 58. (*a*) | 59. (*b*) | 60. (*c*) | 61. (*c*) | 62. (*b*) | 63. (*e*) |
| 64. (*e*) | 65. (*b*) | 66. (*d*) | 67. (*d*) | 68. (*a*) | 69. (*d*) | 70. (*a*). |

## THEORETICAL QUESTIONS

1. Define the term "instrumentation".
2. List the various modes of measurement.
3. Enumerate the elements of a measurement system.
4. What is transducer ?
5. What are the functions of a transducer in an electronic instrumentation system ?
6. How are transducers classified ?
7. What are the advantages of electromechanical transducers ?
8. Explain briefly with diagrams important transducer actuating mechanisms.
9. Describe briefly the following :
   (*i*) Thermistors and resistance thermometers.
   (*ii*) Wire resistance strain gauges.
10. Give the classification of variable inductance transducers.
11. Explain briefly any two of the following transducers :
    (*i*) Self-generating variable inductance transducer – Electromagnetic type
    (*ii*) Variable reluctance transducer
    (*iii*) Mutual inductance transducer
    (*iv*) Linear-variable-differential transformer (LVDT).
12. What is the principle on which a capacitive transducer works ?
13. What are the advantages and disadvantages of capacitive transducers ?
14. Give the applications of capacitive transducers.
15. What is a piezoelectric transducer ? List the advantages and disadvantages of piezo-electric transducers.
16. How are photoelectric transducers classified ?
17. Explain briefly the following :
    (*i*) Photoemissive cell (*ii*) Photoconductive cell
    (*iii*) Photovoltaic cell.
18. What is a strain gauge ?
19. Explain briefly with neat diagrams, any two of the following :
    (*i*) Wire-wound strain gauges (*ii*) Foil-type strain gauges
    (*iii*) Semiconductor strain gauges (*iv*) Capacitive strain gauges.
20. What is a load cell ?
21. Explain briefly any two of the following :
    (*i*) Hydraulic load cell.
    (*ii*) Pneumatic load cell.
    (*iii*) Strain-gauge load cell.
22. Describe briefly the following sensors :
    (*i*) Eddy current proximity sensor.
    (*ii*) Capacitance proximity sensor.
23. Write a short note on "Pneumatic sensor".

## UNSOLVED EXAMPLES

1. A linear resistance potentiometer is 50 mm long and is uniformly wound with a wire having a resistance of 10000 Ω. Under normal conditions, the slider is at the centre of the potentiometer.

(i) Find the linear displacements when the resistances of the potentiometer are measured by a Wheatstone bridge for two cases are : (a) 3800 ohms and (b) 7500 ohms.

(ii) If it is possible to measure a minimum value of 12 ohms resistance with the above arrangement, find the resolution of the potentiometer in mm.

[**Ans.** (i) 6 mm, 12.5 mm; (ii) 0.06 mm]

2. In a linear voltage differential transformer the output voltage is 2.0 V at maximum displacement. At a certain load, the deviation from linearity is maximum and it is ± 0.004 V from a straight line through origin, find the linearity at the given load.

[**Ans.** ±2%]

3. The output of a LVDT is connected to a 5V voltmeter through an amplifier whose amplification factor is 250. An output of 2 mV appears across the terminals of LVDT when the core moves through a distance of 0.5 mm. If the multimeter has 100 divisions and the scale can be read to a $\frac{1}{5}$ of a division. Calculate :

(i) The sensitivity of LVDT, and

(ii) The resolution of the instrument in mm, [**Ans.** (i) 4 m V/mm, (ii) 0.01 mm]

4. A parallel plate capacitive transducer uses plates of area 250 $mm^2$ which are separated by a distance 0.2 mm.

(i) Calculate the value of capacitance when the dielectric is air having a permittivity of $8.85 \times 10^{-12}$ F/m.

(ii) Calculate the change in capacitance if a linear displacement reduces the distance between the plates to 0.18 mm. Also calculate the ratio of per unit change of capacitance to per unit change of displacement.

(iii) If a mica sheet 0.01 mm thick is inserted in the gap, calculate the value of original capacitance and change in capacitance for the same displacement. Also calculate the ratio of per unit change of capacitance to per unit change in displacement. The dielectric constant of mica is 8.

[**Ans.** (i) 11.06 pF; (ii) 1.23 pF, 1.11; (iii) 11.57 pF, 1.35 pF, 1.167]

5. A capacitive transducer uses two quartz diaphragms of area 675 $mm^2$ separated by a distance of 3.8 mm. A pressure of 850 kN/$m^2$ when applied to the top diaphragm produces a deflection of 0.55 mm. The capacitance is 330 pF when no pressure is applied to the diaphragms. Determine the value of capacitance after the application of a pressure of 850 kN/$m^2$. [**Ans.** 385.8 kN/$m^2$]

6. A capacitive transducer, used in pressure measuring instrument has a spacing of 4.2 mm between its diaphragms. A pressure of 600 kN/$m^2$ produces an average deflection of 0.28 mm of the diaphragm of the transducer. A transducer which has a capacitance of 250 pF before the application of pressure is connected in an oscillation circuit having a frequency of 120 kHz. Determine the change in frequency of oscillator after the application of pressure to the transducer. [**Ans.** 4.1 kHz approx]

7. A 2 mm thick quartz piezoelectric crystal having a voltage intensity of 0.055 Vm/N is subjected to a pressure of 1.8 MN/$m^2$. Calculate the voltage output and charge density of the crystal. Take the permittivity of quartz as $40.6 \times 10^{-12}$ F/m.

[**Ans.** 198 V, 2.23 pC/N]

8. A piezoelectric material measuring 5 mm × 5 mm × 1.5 mm is used to measure a force. Its voltage sensitivity is 0.055 Vm/N. Calculate the force if voltage developed is 110V.

[**Ans.** 33N]

9. The following data relate to a barium titrate pick-up :

Dimensions : 5 mm × 5 mm × 1.25 mm; Force acting on the pick-up = 5N. The charge sensitivity of the crystal = 150 pC/N; Permittivity = $12.5 \times 10^{-9}$ F/m; Modulus of elasticity = $12 \times 10^6$ N/$m^2$. Calculate strain, charge and capacitance.

[**Ans.** 0.0167 : 750 pC; 250 pF]

**10.** A strain of 5 micro-strain is caused in a structural member when subjected to a compressive force. Two separate strain gauges are attached to the structural member, one is nickel wire strain gauge (gauge factor = –12.1) and other is nichrome wire strain gauge (gauge factor = 2). If the resistance of strain gauges before being strained is 130 Ω, calculate the change in the value of resistance of the gauges after they are strained.

[**Ans.** 7.865 mΩ (increase); 1.3 mΩ (decrease)]

**11.** A strain gauge is bonded to a beam which is 10 cm long and has a cross-sectional area of 4 $cm^2$. The unstrained resistance and gauge factor of the strain gauge are 220Ω and 2.2 respectively. On the application of load the resistance of the gauge changes by 0.013Ω. If the modulus of elasticity for steel is 207 $GN/m^2$, calculate : (*i*) the change in length of the steel beam, and (*ii*) the amount of force applied to the beam.

[**Ans.** (*i*) $2.68 \times 10^{-6}$ m; (*ii*) 2.219 kN]

**12.** A single strain gauge having resistance 144 Ω is mounted on a steel cantilever beam at a distance of 0.15 m from the free end. The beam dimensions are 25 cm (length) × 2.0 cm (width) × 0.3 cm (depth). An unknown force applied at the free end produces a deflection of 12.7 mm of the end. If the change in gauge resistance is found to be 0.1824Ω, calculate the gauge factor. Take Young's modulus for steel as 200 $GN/m^2$. [**Ans.** 2.3]

CHAPTER

# 8 Signal Conditioning

**8.1 Introduction**-General measurement system components-signal conditioning and its necessity-Processes adopted in signal conditioning-mechanical amplification and electrical signal conditioning **8.2 Functions of signal conditioning equipment; 8.3 Amplification 8.4 Types of amplifiers; 8.5 Mechanical amplifiers; 8.6 Fluid amplifiers; 8.7 Optical amplifiers; 8.8 Electrical and electronic amplifiers-** Desirable characteristics of electronic amplifiers-electronic amplification of gain-A.C. and D.C. amplifiers-modulated and unmodulated signals-integrated circuit (ICs)-Operational amplifiers (Op-amp)-attenuators-filters-input circuitry; *Highlights–Objective Type Questions – Theoretical Questions*

## 8.1 INTRODUCTION

### 8.1.1. General Measurement System Components

Fig. 8.1 shows the general measurement system components :

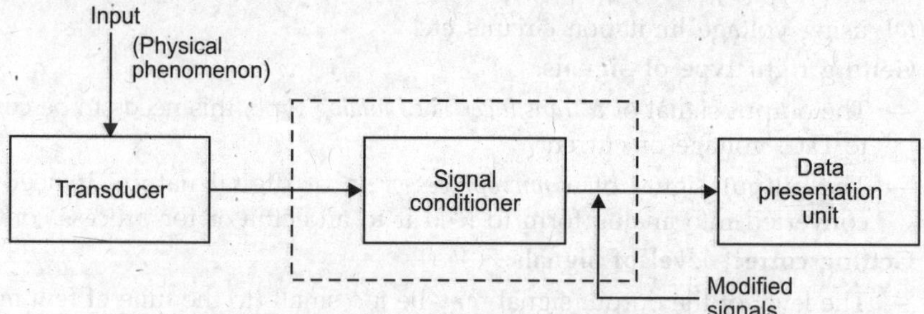


**Fig. 8.1.** Components of a general measurement system.

— The *"first stage" of the instrumentation or measurement system which detects the measurand (which is basically a physical quantity) is termed as* **detector-transducer stage.** In this stage, in most of the cases, the quantity is detected and is transduced into an electrical form.

— The output from the first stage *needs certain modifications* before it becomes compatible with the data presentation stage. The necessary modification is carried out in the *"intermediate stage"*, more commonly referred to as the **signal conditioning stage**.

— The *"last stage"* of the measurement system may consist of indicating, recording, displaying, data processing elements or may consist of control elements.

- *Measurement of dynamic mechanical quantities* places special requirements on the elements in the signal conditioning stage.

*Large amplifications, as well good transient response, are often desired,* both of which are difficult to obtain by mechanical hydraulic, or pneumatic methods. Consequently, *electrical or electronic elements are usually required.*

### 8.1.2. Signal Conditioning and its Necessity

**Signal conditioning** *may be defined as the process of modifying the output signals from the transducer into usable and satisfactory signal using amplification, attenuation, non-linearisation, linearisation or multiplication by another function.*

The necessity of signal conditioning may be due to following *reasons* :

1. Signals may be too noisy due to electromagnetic interference.
2. Signals may be too small, usually is mV range.
3. Signals may be non-linear and require to be converted into digital form.
4. Signals may be analog one and require to be converted into digital form.
5. Signals may be digital one and need to be converted into analog signals.
6. It may be required to improve the quality of digital signals.

### 8.1.3. Processes Adopted in Signal Conditioning

Following processes are usually adopted in signal conditioning :

1. **Protection.** The range of the output signals from the transducer may be so high that it may damage the next unit or element which needs to be protected.

   *Example:* If a high voltage/current signals are fed to the microprocessor, it will get damaged. The microprocessor can be protected by :

   (*i*) employing a series of current limiting resistors, fuses to break if current is too high;

   (*ii*) using a step down transformer if the voltage is too high;

   (*iii*) employing polarity protection;

   (*iv*) using voltage limitation circuits etc.

2. **Getting right type of signals:**

   — The output signal of a *transducer* is of *analog* type, this needs to be converted to D.C. voltage or current.

   — The output signal of a *microprocessor* is of digital nature, it needs to be converted into analog form to feed it to an actuator for process controlling.

3. **Getting correct level of signals:**

   — The level of the output signal may be too small (to the tune of few mV), this needs to be amplified for feeding it into an analog-to-digital converter. It may also be difficult to measure such low level signals.

   — For amplification operational amplifiers (Op-amp) are widely used.

4. **Elimination of interferences :**

   — Some undesired signals or disturbances (say noise disturbance due to electromagnetic interference) may be associated with the output signals, these need to be eliminated.

   — Such interferences are eliminated by the use of ***filters***.

5. **Manipulation of signals:** The output signals may be non-linear in nature, these need to be linearised and vice versa.

### 8.1.4. Mechanical Amplification and Electrical Signal Conditioning

**Limitations/disadvantages of mechanical amplification:**

In the field of dynamic measurements, strictly, mechanical systems are much more uncommon than they were in the years past, largely because of several inherent *disadvantages.* Mechanical amplification by the elements such as linkages, gearing, or cams (these elements present design problems of immense magnitudes particularly if dynamic inputs are to be handled) is quite limited because of the following *reasons* :

(*i*) When amplification is required *frictional forces are also amplified, resulting in considerable undesirable signal loading.* These effects, coupled with backlash and elastic deformations, result in *poor response.*

(*ii*) Initial loading results in reduced frequency response and in certain cases, depending on the particular configuration of the system, *phase response is also a problem.*

**Advantages of electrical signal conditioning:**

In several detector-transducer combinations which provide an output in electrical form, it is convenient to perform further signal conditioning electrically.

- Such conditioning may typically *include:*
  - — Converting resistance changes to voltage changes;
  - — Subtracting offset voltages;
  - — Increasing signal voltages;
  - — Removing unwanted frequency components.
- Electrical methods are also preferred for their ease of *power amplification.*
  - — *Additional power may be fed into the system to provide a greater output power than input by the users of "power amplifiers"*, which have no important mechanical counterpart in most instrumentation. (It is true that hydraulic and pneumatic systems may be set up to increase signal power; however, their use is limited to relatively slow-acting control applications, primarily in the fields of chemical processing and electric power generation). This technology is of a particular value when recording procedures employ stylus-type recorders, mirror galvanometers, or magnetic-disc methods.

## 8.2 FUNCTIONS OF SIGNAL CONDITIONING EQUIPMENT

The signal conditioning equipment may be required to perform the following *functions on the transduced signal:*

1. Amplification
2. Modification or modulation
3. Impedance matching
4. Data processing
5. Data transmission.

1. **Amplification.** It means *enhancement of the signal level* which is often in the low level range. The amplification system must bring the level of transducer signal to a value adequate enough to make it useful for *conversion, processing, indicating* and *recording.*
2. **Modification or modulation.** It means to *change the form of signal.* The signal may be *smoothened, linearised, filtered or converted into digital form.*
3. **Impedance matching.** The signal conditioning equipment arranges the input, and output impedances of the matching device so as to prevent loading of the transducer and to maintain a high signal level at the recorder.
4. **Data processing.** To carry out mathematical operations (*e.g.*, addition, subtraction, differentiation, integration etc.) before indication or recording of data.
5. **Data transmission.** To transmit signal from one location to another without changing the contents of the information.

The whole task of signal conditioning requires the following:

(i) Ingenuity;

(ii) Proper selection of components;

(iii) Faithful reproduction of signal..

- The elements of signal conditioning are designed in such a fashion as to be *insensitive to all extraneous inputs. The accuracy, range and dynamic response are all designed to be compatible with the detector transducer.*
- In several situations the *"signal conditioning"* or *"data acquisition equipment"* is an *excitation* and *amplification system for passive transducers.* It may be an *amplification system for active transducers.* In both the applications, the transducer output is brought upto adequate level to make it useful for *conversion, processing, indicating* and *recording.*
- In case of *"passive transducers"* (*e.g.*, strain gauges, potentiometer resistance thermometers, inductive and capacitive transducers) excitation is needed because these transducers do not generate their own voltage or current; the excitation is provided from external sources.
- The *"active transducers"* (*e.g.*, technogenerators, thermocouples, inductive pick-ups and piezoelectric crystals) do not require excitation from an external source since they produce their own electrical output. However, these signals have a low voltage level and as such they need to be amplified.

The excitation sources may be:

- D.C. voltage source.
- A.C. voltage source.

Figures 8.2 and 8.3 show D.C. and A.C. signal conditioning systems respectively.

**D.C. signal conditioning system:**

Refer to Fig. 8.2. The resistance transducers like strain gauges constitute one or more than one arm of a Wheatstone bridge which is excited by an isolated D.C. source. The bridge can be balanced by a potentiometer and can also be calibrated for unbalanced conditions.

***Characteristics of a D.C. amplifier:***

(i) It should have extremely good thermal and long term stability.

(ii) It may require balanced differential inputs giving high mode rejection ratio (CMRR); CMRR is a *measure of ratio of desired signal to undesired signal, this value should be as high as possible.*

***Advantages:***

(i) D.C. amplifier is easy to calibrate at low frequencies.

(ii) It is able to recover from an overload condition unlike its A.C. counterpart.

***Disadvantages:***

- The major disadvantage of a D.C. amplifier is that it suffers from the *problem of drift.* As a result, the *low frequency spurious signals come out as data information.* This problem is overcome by the *use of the drift amplifiers.*
- The D.C. amplifier is followed by a lowpass filter which eliminates high frequency components or noise from the data signal.

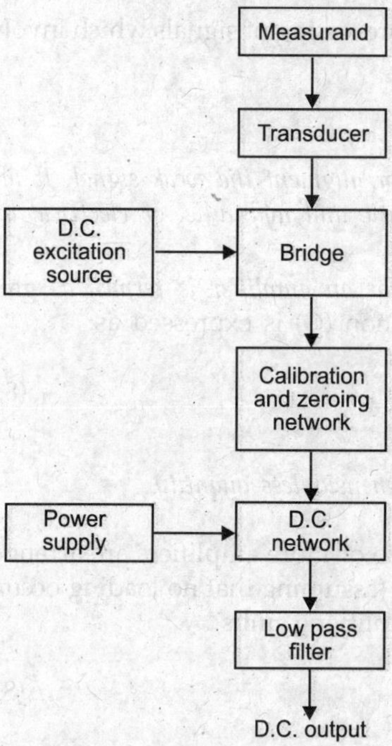


**Fig. 8.2.** D.C. signal conditioning system.

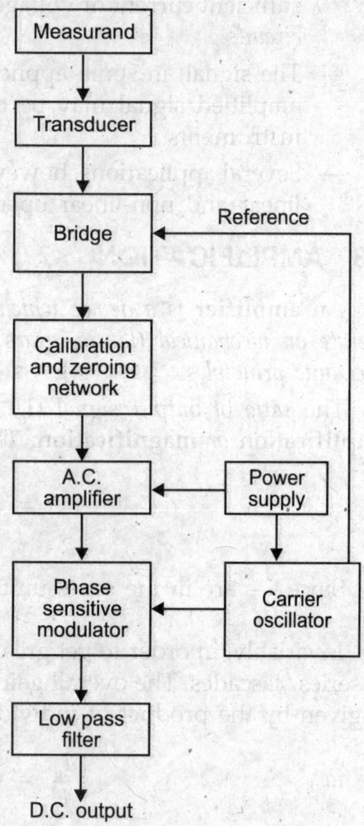


**Fig. 8.3.** A.C. signal conditioning system.

***Uses.*** D.C. systems are generally used for common resistance transducers such as *potentiometers and resistance strain gauges.*

**A.C. signal conditioning system:**

Refer to Fig. 8.3. The problems which are encountered in D.C. systems are overcome through *carrier type A.C. signal conditioning system.*

The transducer parameter variations amplitude modulate the carrier frequencies at the bridge output and the waveform is *amplified and demodulated.* The demodulation is phase sensitive so that polarity of D.C. output indicates the direction of the parameter change in the bridge output.

In carrier systems, it is *very easy to obtain very high rejection of mains frequency pick-up.*

— Active filters can be used to reject this frequency and prevent overloading of A.C. amplifier.

— The carrier frequency components of the data signal are filtered out by the phase-sensitive demodulators.

***Uses.*** A.C. systems are used for variable reactance transducers and for systems where signals have to be transmitted long via cables to connect the transducers to the signal conditioning equipment.

— The physical quantities like pressure, temperature, acceleration, strain etc. after having being *transduced* into their analogous electrical form and *amplified* to sufficient current or voltage levels (say 1 to 10 V) are further *processed by electronic circuits.*

— The signal, in some applications, does not need any further processing and the amplified signal may be directly applied to indicating or recording or control instruments.

— Several applications, however, involve further processing of signals which involve linear and non-linear operations.

## 8.3 AMPLIFICATION

An **amplifier** *is a device which is used to increase or augment the weak signal. It may operate on mechanical (levers, gears etc.) optical, pneumatic and hydraulic, or electrical and electronic principles.*

The *ratio of output signal* ($I_o$) *to input signal* ($I_i$) *for an amplifier is termed as* **gain, amplification** *or* **magnification**. The gain of amplification (G) is expressed as:

$$G = \frac{I_o}{I_i} \qquad \text{...(8.1)}$$

Since $\frac{I_o}{I_i}$ are in the same units, the gain G is a *dimensionless quantity.*

Invariably, in order to get greater magnification, two or more amplifiers are arranged in series/cascades. The overall gain of the arrangement (assuming that no loading occurs) is given by the product of individual gains of the amplifying units,

*i.e.,*
$$\frac{I_o}{I_i} = G_1 \cdot G_2 \cdot G_3 \cdots\cdots \qquad \text{...(8.2)}$$

## 8.4 TYPES OF AMPLIFIERS

The amplifiers, on the basis of principle of working, may be *categorised* as follows:

1. Mechanical amplifiers.
2. Fluid amplifiers.
3. Optical amplifiers.
4. Electrical and electronic amplifiers.

## 8.5 MECHANICAL AMPLIFIERS

The mechanical amplifiers may be further *classified* as follows:

**(*i*)** ***Simple and compound levers :*** The compound lever has two or more levers linked together so that output from one lever provides the input to the other.

**Example.** The *Huggenberger extensometer* is one of the most popular and *accurate mechanical amplifier*. It uses a system of compound levers to give *very high magnification* to the order of *2000 or even more.*

**(*ii*)** ***Simple and compound gears :*** The simple and compound gear trains are used quite frequently *to provide mechanical amplification of either angular displacement or rotary speed.*

A *"compound gear train"* gives *greater modification with the additional advantage of no change in the direction of input signal.*

**Examples.** The gear trains are used for the magnification of displacement in the *Bourdon tube pressure gauge* and in the *dial-test indicator* where linear movement is translated into rotation by means of rack and pinion.

***Limitations of mechanical amplification:***

The mechanical amplification, as earlier stated, usually suffers from the errors caused by the following *factors*.

(*i*) Internal loading;

(*ii*) Friction at the mating parts;

(*iii*) Elastic deformation;

(*iv*) Backlash.

## 8.6 FLUID AMPLIFIERS

Fluid amplifiers may be *classified* as follows:

(*i*) ***Hydraulic amplifier*** **:** When a small displacement is applied to a piston operating inside a cylinder containing some liquid, there occurs a large displacement of the liquid in the output tube which has a small diameter.

**Example.** This principle is employed in the *mercury-in-glass thermometer* and the *single-column manometers.*

(*ii*) ***Pneumatic amplifier*** **:** Pneumatic methods are extensively used and can be applied to any type of measurement.

## 8.7 OPTICAL AMPLIFIERS

In optical amplification, a ray of light strikes a mirror with an angle of incidence $i$ and gets reflected with angle of reflection equal to the angle of incidence. When the mirror rotates through an angle $\theta$, the angle of incidence change to $(i + \theta)$. Before rotation of the mirror, the angle between the incident ray and reflected ray is $2i$ and after rotation it is $2(i + \theta)$. Obviously there is angular magnification of $2\theta$ between the incident and reflected rays. In order to get a greater magnification, more number of mirrors surfaces may be used.

**Examples.** This principle to amplify the input signals is used in the following cases.

— Optical levers;

— U.V. galvanometers;

— Mechanical-pointer galvanometers.

## 8.8 ELECTRICAL AND ELECTRONIC AMPLIFIERS

The electrical amplifiers are used to *increase the magnitude of weak voltage or current signals resulting from electromechanical transducers.*

### 8.8.1. Desirable Characteristics of Electronic Amplifiers

The following are the desirable characteristics of electronic amplifiers:

(*i*) *High input impedance* so that its loading effect on the transducer in minimum.

(*ii*) *Low output impedance* so that the amplifier is *not unduly loaded by the display or recording device.*

(*iii*) *Frequency response* should be as *good as that of the transducer.*

### 8.8.2. Electronic Amplification of Gain

● The following are the several *generalities* that can be listed for the ideal (but non-existent) electronic amplifier:

(*i*) Infinite gain (lower gain can be obtained by adding attenuation circuits).

(*ii*) Infinite input impedance; no input current, hence no load on the previous stage or device.

(*iii*) Zero output impedance (low noise).

(*iv*) Instant response (wide frequency bandwidth).

(*v*) Zero output for zero input.

(*vi*) Ability to ignore or reject, extraneous inputs.

Of course, none of these aims can be completely realized, it is often possible to approach them, and their assumption simplifies circuit analysis.

● In an electronic amplifier, separate power is provided so that the output power may exceed the input if that is required.

Here, if $v_i$ = Input voltage,

$i_i$ = Input current,

$v_o$ = Output voltage, and

$i_o$ = Output current,

Then :

$$\text{Gain} = \frac{\text{Power output}}{\text{Power input}} = \frac{v_o i_o}{v_i i_i} \quad \text{...(8.3)}$$

$$\text{Voltage amplification} = \frac{\text{Voltage output}}{\text{Voltage input}} = \frac{v_o}{v_i} \quad \text{...(8.4)}$$

$$\text{Current amplification} = \frac{\text{Current output}}{\text{Current input}} = \frac{i_o}{i_i} \quad \text{...(8.5)}$$

● Another way of expressing *power gain* is through the use of *decibel.*

The common logarithm (log to the base 10) of power gain is known as *bel power gain.*

$$\text{Power gain} = \log_{10}\left(\frac{P_o}{P_i}\right) \text{bel}$$

$$1 \text{ bel} = 10 \, dB$$

$$\therefore \quad \text{Power gain} = 10 \log_{10}\left(\frac{P_o}{P_i}\right) \text{dB} \quad \text{...(8.6)}$$

If the two powers are developed in the same resistance or equal resistance, then

$$P_i = \frac{V_i^2}{R} = I_i^2 R$$

$$P_o = \frac{V_0^2}{R} = I_o^2 R$$

$$\therefore \quad \textit{Voltage gain} = 10 \log_{10} \frac{V_0^2/R}{V_i^2/R} = 20 \log_{10} \frac{V_o}{V_i} \text{ dB} \quad \text{...(8.7)}$$

$$\text{Current gain} = 10 \log_{10} \frac{I_o^2 R}{I_i^2 R} = 20 \log_{10} \frac{I_o}{I_i} dB \qquad ...(8.8)$$

**Example 8.1.** *A three-stage amplifier has a first voltage gain of 100, second stage voltage gain of 200 and third stage voltage gain of 400. Find the total voltage gain in dB.*

**Solution.** First-stage voltage gain in *dB*

$$= 20 \log_{10} 100 = 20 \times 2 = 40$$

Second-stage voltage gain in *dB*

$$= 20 \log_{10} 200 = 20 \times 2.3 = 46$$

Third-stage voltage gain in *dB*

$$= 20 \log_{10} 400 = 20 \times 2.6 = 52$$

$$\text{Total voltage gain} = 40 + 46 + 52 = \mathbf{138\ dB\ (Ans.)}$$

**Example 8.2.** *(i) A multistage amplifier employs five stages each of which has a power gain of 30. What is the total gain of the amplifier in dB?*

*(ii) If a negative feedback of 10 dB is employed, find the resultant gain.*

**Solution.** Absolute gain of each stage = 30

Number of stages = 5

(*i*) *Power gain of one stage* = $10 \log_{10} 30$ dB = 14.77

∴ Total *power gain* = 5 × 14.77 = **73.85 dB (Ans.)**

(*ii*) *Resultant power gain with negative feedback*

= 73.85 – 10 = **63.85 dB (Ans.)**

### 8.8.3. A.C. and D.C. Amplifiers

The instrumentation systems usually employ the following two types of electronic amplifiers.

(*i*) A.C. coupled amplifiers.

(*ii*) D.C. coupled amplifiers.

- For an *"A.C. amplifiers"* bandwidth is the range of frequencies between which gain or amplitude ratio is constant to within – 3dB (3dB down points). This corresponds to the frequencies at which the voltage output amplitude falls by 29.3% to 70.7% of the maximum value.
  — The *"A.C. amplifiers"* are *only capable of dealing with rapid, repetitive signals but are usually simpler and cheaper* when compared with their D.C. counterparts.
  — In an "A.C. amplifier system" the amplifier drift and spurious noise are *not significant*; the mains frequency pick-up *rejection is very high.*
- The *"D.C. amplifiers"* are capable of *amplifying static, slowly changing or rapid-repetitive input signals.*
  — The "D.C. amplifier systems" are *easy to calibrate at low frequencies, and have the ability to recover rapidly from overload conditions.*

### 8.8.4. Modulated and Unmodulated Signals

The measurands may be *"pure"* in the sense that analog electrical signal contains *nothing more than the real time variation of the measurand information itself.*

On the other hand, the signal may be *"mixed"* with a **carrier** which consists of a *voltage oscillation at some frequency higher than that of the signal.* A common rule of thumb is that the *frequency ratio should be at least 10:1.*

The measurand affects the *carrier* by varying either its *amplitude* or its *frequency:*

— In the *former* case the *carrier frequency is held constant and its amplitude is varied by the measurand. This process is known as* **Amplitude modulation** (or **AM**).

— In the *latter* case the *carrier amplitude is held constant and its frequency is varied by the measurand.* This is known as **Frequency modulation** (or **FM**).

The most familiar use of AM and FM transfer of signals is in AM and FM *radio broadcasting.*

When *"modulation"* is used in instrumentation *"amplitude modulation"* (AM) is the *more common form.*

— Nearly any mechanical signal from a passive pick-up can be transduced into an analogous AM form. Sensors based on either *inductance* (*e.g.* differential transformer) or *capacitance* (*e.g.* capacitance pick-up for liquid level) require an A.C. excitation.

In addition, however, resistance-type sensors may also use an A.C. excitation, as with some strain gauge circuits.

It is required to extract signal information from the modulated carrier.

- This operation, *when AM is used,* may take several forms:
  - — The simplest is merely to display the entire signal using an *oscilloscope or oscillograph,* and then to *"read"* the result from the envelope of the carrier.
  - — More commonly, the *mixed signal and carrier are* ***"demodulated"*** by *"rectification* and *filtering".*
- FM demodulation is more complex operation and may be accomplished through the use of
  - — Frequency discrimination,
  - — Ratio detection, or
  - — IC phase-locked loops.

### 8.8.5. Integrated Circuits (ICs)

The integrated circuits (ICs), as the name implies, are groups of circuit elements combined to perform specific purposes. For the most part the elements consist of *transistors, diodes, resistors,* and, to lesser extent, *capacitors,* all connected and packaged in convenient plug-in units.

ICs from the building blocks are used to construct more complex circuits such as :

— Differential amplifiers;

— Mixers (for combining signals);

— Timers;

— Filters;

— Audio preamps;

— Auto-power amplifiers;

— Voltage references;

— Regulators and comparators;

— Several digital devices.

### 8.8.6. Operational Amplifiers (Op-amp)

An ***operational amplifier (Op-amp)*** *is a linear integrated circuit* (IC) *that has a very high voltage gain, a high input impedance and a low output impedance.*

It is so called because it can be employed to carry out many different mathematical operations like *"addition"*. *"subtraction"*, *"multiplication"*, *"division"*, *"integration"*, *"differentiation"* etc.

- Operational amplifiers are linear integrated circuits that work on *relatively low supply voltage.*
- They are *reliable and inexpensive.*
- An *deal operational amplifier* is device of infinite voltage gain, infinite bandwidth, infinite input impedance (open) and zero output impedance.
- An Op-amp may contain two dozen transistors, a dozen resistors and one or two capacitors.

  **Examples :** μA 709, LM 108-LM 208, CA 741 CT and CA741T.

### 8.8.6.1. Specifications/Characteristics of an Op-amp

While selecting an Op-amp, the following characteristics need to be considered:

1. *Input offset voltage.* It is the voltage that must be applied at the input terminals to make the output voltage zero (This is about 2 mV for a 741 amplifier). The offset voltage changes with temperature.
2. *Input offset current.* It is defined as the net difference in current that must be applied at the input terminals to make the output voltage zero (This is 20 nA for a 741 amplifier).
3. *Input check current.* It is the mean of the two input currents to make the voltage zero.
4. *Slew rate.* It is the maximum rate at which the output can change. It is expressed as volts/microseconds.
5. *Unity gain frequency.* This is the frequency at which the open loop gain of the amplifier becomes unity.
6. *Common mode rejection ratio (CMRR).* It is the ratio of desirable signals to undesirable signals.

- An Op-amp is the *basic building block* for:

— Amplifiers
— Integrators
— Summers
— Differentiators
— Comparators
— A/D and D/A converters
— Active filters
— Sample and hold amplifiers.

### 8.8.6.2. Op-amp description

Fig. 8.4(*a*) shows a standard symbol (a triangle having two input labelled differently and a single output) for an Op-amp, the one shown in Fig. 8.4(*b*) is also oftenly used.

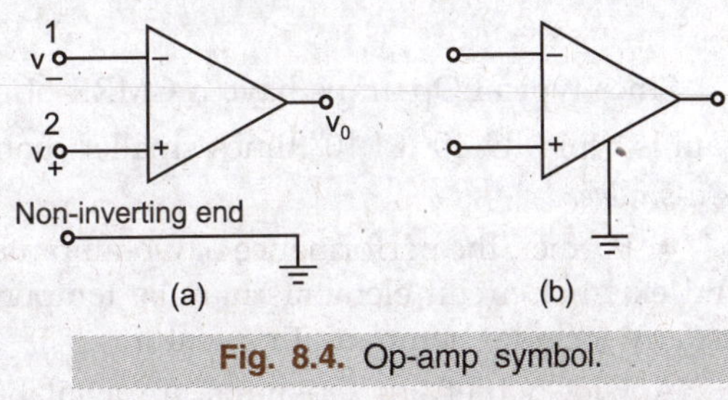


**Fig. 8.4.** Op-amp symbol.

One input terminal is designated by *–ve* sign, it is called *inverting end* while other input terminal is designated by a *+ve* sign, it is called *non-inverting end.*

These plus (+) and minus (–) polarities indicate *phase reversal* only. It does not mean that $v_1$ and $v_2$ [Fig. 8.4(*a*)] are negative and positive respectively. Additionally, it also *does not imply* that a positive input voltage has to be connected to the plus-marked non-inverting terminal 2 and negative input voltage to negative-marked inverting terminal 1. *In fact amplifier can be used either way up so to speak.* It may also be noted that all input and output voltages are referred to a common reference usually the ground shown in Fig. 8.4(*a*).

*Operational amplifier operating with –ve feedback possesses stable closed loop gain and also insensitivity to variation of supply voltage and ambient temperature.*

- The voltage at the output terminal $v_o$, is the product of the amplifier gain $G$, and the voltage difference :

$$v_o = G(v_+ - v_-) \qquad ...(8.9)$$

The output voltage is roughly limited to the power supply voltages $V_{cc}$ and $V_{ee}$, as the voltage difference increases; if the voltage difference becomes too large, the output *saturates* near one of these values and remains constant.

- *The differential characteristic of op-amp has great importance in instrumentation because it eliminates offset voltages and noise signals common to both input terminals.* For example, nearby power lines may induce 50-cycle noise in the exterior circuitry leading to the amplifier. Such line noise is often present in identical form at both input terminals. This behaviour is known as *common-mode rejection*. If, instead, an op-amp receives the output of a voltage-sensitive Wheatstone bridge, the common offset voltages of the two voltage dividers are cancelled, and only the desired difference voltage is applied.

**Limitations of Op-amp :** Most of the Op-amps have a nonideal characteristic according to which they *do not completely satisfy the differential amplifying property*. With *both inputs grounded residual output voltage remains.*

— The multitude of transistors, resistors, and other elements within the Op-amp are *never perfectly matched*, so the amp output actually reaches zero at some small *non-zero* input voltage. To accommodate this input offset voltage, the common Op-amp is provided with pins marked "offset null" or "balance" which provides a means for adjusting the unwanted offset voltage towards zero.

Another limitation is that the actual common-mode rejection is finite. If the two input signals each include a common-mode voltage $v_{cm}$, the Op-amp's actual voltage will be,

$$v_o = G(v_+ - v_-) + G_{cm}\, v_{cm} \qquad ...(8.10)$$

The finite common-mode rejection is characterised by the "*common-mode rejection ratio* (CMRR)" in decibels:

$$\text{CMRR} = 20 \log_{10}\left(\frac{G}{G_{cm}}\right) \text{dB} \qquad ...(8.11)$$

Since typical Op-amps have a CMRR of 60 to 120 dB, therefore, the common-mode gain is typically $10^3$ to $10^6$ times smaller than the differential gain, hence a *high CMRR is desirable.*

- Further, the performance of Op-amp may be limited by thermal drift. Both internal and external circuit elements may be temperature sensitive, and design of each circuit usually includes compensating features.

A wide variety of Op-amps are available, and their differences largely represent attempts to *improve*:

— Thermal stability;
— CMRR;
— Offset voltages;
— Frequency response .... These refinements, however, increase the cost.

**8.8.6.3. Applications of Op-amp**

Operational amplifiers may be used as the *basic components of* :

- Linear voltage amplifiers;
- Integrators and differentiators;
- Function generators;
- Impedance transformers;
- Differential amplifiers;
- Voltage comparators;
- Filters;
- Many other devices.

**8.8.6.4. Op-amp circuits used in Instrumentation**

Some of the commonly used Op-amp circuits are described below:

1. Inverter;
2. Adder;
3. Subtracter;
4. Multiplier and divider;
5. Integrator;
6. Differentiator;
7. Buffer amplifier;
8. Differential amplifier.

**1. Inverter.** Fig. 8.5, shows the circuit of an Op-amp used as *inverter.* The feedback resistance $R_f$ is made equal to the resistance $R_1$, connected to the inverting end of the amplifier,

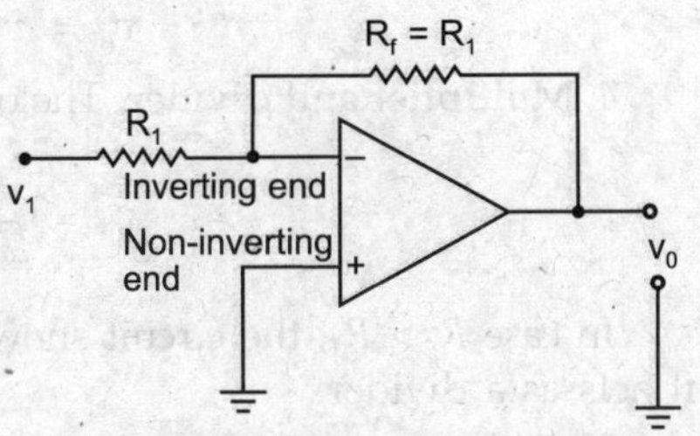


**Fig. 8.5.** Op-amp as an inverter.

$$\text{Output voltage, } v_o = -\frac{R_f}{R_1} v_1 = -v_1 \quad ...(8.12)$$

$$(\because R_f = R_1)$$

Obviously, the output voltage is 180° out of phase with the input voltage.

**2. Adder.** Fig. 8.6 shows an Op-amp circuit that performs the signals with amplification (if desired); using superposition theorem, we get

Output voltage,

$$v_o = -\left(\frac{R_f}{R_1} v_1 + \frac{R_f}{R_2} v_2 + \frac{R_f}{R_3} v_3\right) \quad ...(8.13)$$

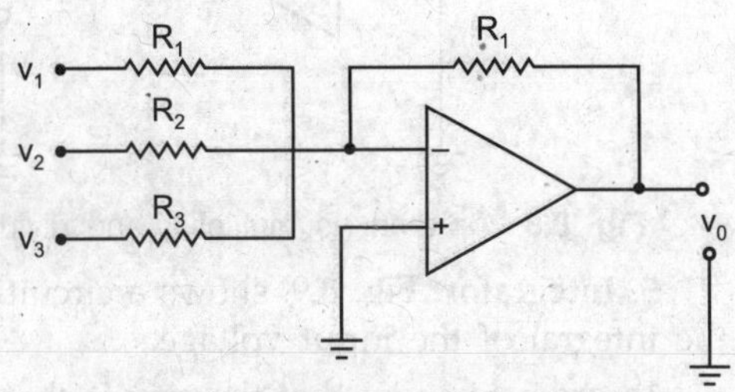


**Fig. 8.6.** Op-amp as an adder.

If $R_1 = R_2 = R_3 = R_f$, then

$$v_o = -(v_1 + v_2 + v_3) \quad ...(8.14)$$

*i.e.*, sum of the individual input voltages. The inversion that occurs cannot be avoided.

**3. Subtracter.** The Op-amp circuit used for subtraction of two input signals is shown in Fig. 8.7. The output of the 2nd Op-amp is given by :

$$v_o = -\left(-v_1 \frac{R_f}{R_1} \cdot \frac{R_f}{R_3} + v_2 \frac{R_f}{R_2}\right) \qquad ...(8.15)$$

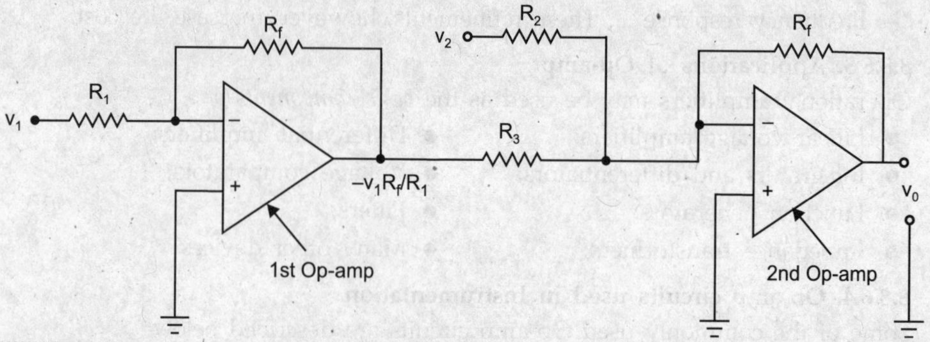


Fig. 8.7. Op-amp as a subtractor.

If $R_f = R_1 = R_2 = R_3$, the circuit acts as a *pure subtractor* and the output, in this case, is given by

$$v_o = v_1 - v_2 \qquad ...(8.16)$$

**4. Multiplier and divider.** The output of an Op-amp in the inverting mode is given by

$$v_o = -\left(\frac{R_f}{R_1}\right) \times v_1$$

In case $R_f > R_1$, the circuit shown in Fig. 8.8 acts as a multiplier and in case $R_f < R_1$ it acts as a divider.

Thus, by choosing the values of $R_f$ and $R_1$, the multiplier and the divider circuits can be designed for multiplication and division by any number.

Fig. 8.8. Op-amp as multiplier and divider.

Fig. 8.9. Op-amp as an integrator.

**5. Integrator.** Fig. 8.9. shows a circuit in which the output voltage is proportional to the integral of the input voltage.

In order to show that the circuit shown in Fig. 8.9. acts as an integrator using KCL (Kirchoff's Current Law) at node $v_-$, we have,

$$i_n = i_C.$$

or,

$$\frac{v_- - v_1}{R} = C\frac{d}{dt}(v_o - v_-)$$

For *infinite differential gains,* $v_- = 0$

$$\therefore \qquad -\frac{v_1}{R} = C\frac{d}{dt}v_o$$

By integration, we have

$$v_o = -\frac{1}{RC}\int v_1\, dt \qquad ...(8.17)$$

The convenient values of $R$ and $C$ are MΩ and μ$F$ range respectively.

**6. Differentiator.** The differential amplifier circuit is obtained by *interchanging* the positions of resistance $R$ and capacitor $C$ as shown in Fig. 8.10.

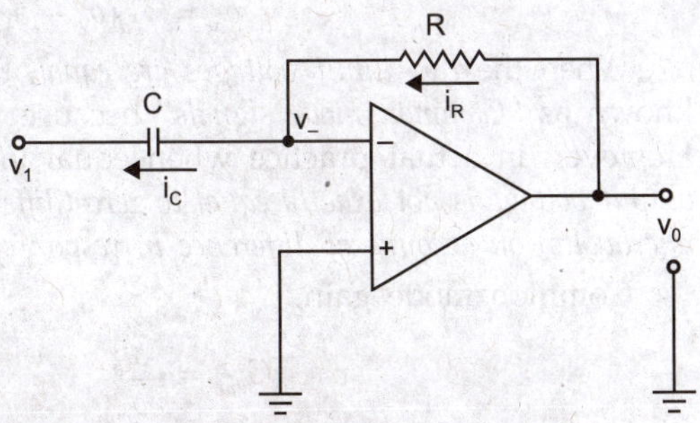


**Fig. 8.10.** Op-amp as differentiator.

At node $v_-$, we have :

$$i_C = i_R$$

$$C\frac{d}{dt}(v_- - v_1) = \frac{v_o - v_-}{R}$$

Now, $\qquad v_- = 0$

$$\therefore \qquad -C\frac{d}{dt}(v_1) = \frac{v_o}{R}$$

or, $\qquad$ Output voltage, $v_o = -RC\frac{d}{dt}(v_1) \qquad ...(8.18)$

Thus, the output voltage is equal to the differentiated input voltage.

- The *Op-amps are normally used as differentiators as they tend to decrease the signal to noise (S/N) ratio.*

**7. Buffer amplifier.** The buffer amplifier is essentially an impedance transformer which converts a voltage at high impedance to the same voltage at low impedance. The circuit of a unity gain buffer amplifier also called a *"voltage follower"* is shown in Fig. 8.11.

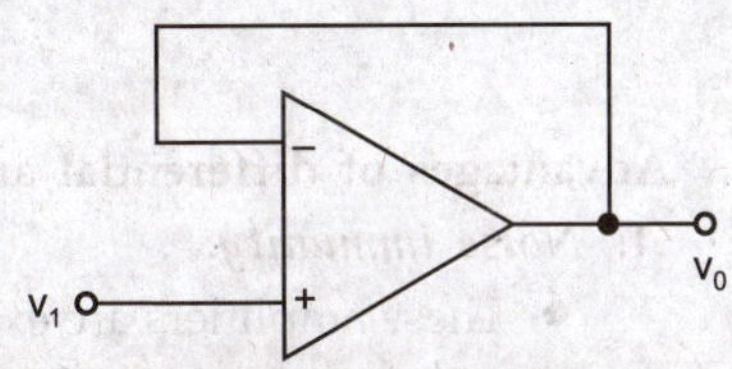


**Fig. 8.11.** Unity gain buffer amplifier voltage follower.

- *The use of unity gain buffer amplifiers greatly reduces the loading effects in measurements systems.*

**8. Differential amplifier.** A differential amplifier (an Op-amp) is of significant importance in an instrumentation system.

In its basic form it has two *inputs* and *outputs*. The signals available to the two outputs are identical except that the two are 180° out-of-phase with each other.

*The output voltage of the amplifier is proportional to the difference between the two input voltages.*

Fig. 8.12, shows an Op-amp used as a differential amplifier.

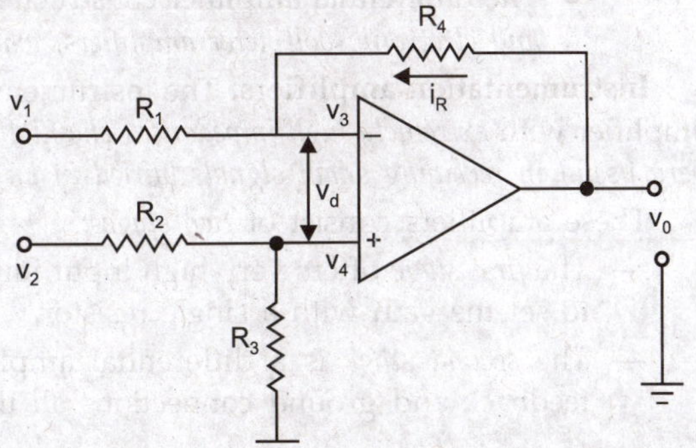


**Fig. 8.12.** Op-amp used as a differential amplifier.

Here, $v_o = G_d(v_+ - v_-)$

where, $G_d$ = Differential gain.

The signal $v_d = (v_+ - v_-)$ is called *"Difference Mode Signal"* or simply *"Difference Signal"*

$$G_d = \frac{v_o}{v_+ - v_-} = \frac{v_o}{v_d} \quad ...(8.19)$$

It can be proved that,

$$v_o = G_d(v_2 - v_1) \quad ...(8.20)$$

When the *two input voltages are equal,* the output voltage is zero. Equal inputs are known as *"Common mode signals"* because the input signal is common to both inputs. However, in actual practice when equal input voltages are applied to the inputs, the *output voltage is not exactly equal to zero (difference is typically of the order of several hundred microvolts) on account of difference in response of the two inputs to common mode signals.*

Common mode gain,

$$G_{cm} = \frac{v_o}{v_{cm}} \quad ...(8.21)$$

where, $G_{cm}$ = Common mode gain, and

$v_{cm}$ = Common mode input signal.

The *"Common mode rejection ratio (CMRR)"* is defined as

$$CMRR = \frac{G_d}{G_{cm}} \quad ...(8.22)$$

Also,

$$CMRR = 20 \log_{10}\left(\frac{G_d}{G_{cm}}\right) \text{ dB} \quad ...(8.22a)$$

... when expressed in *dB*.

**Advantages of differential amplifiers :**

1. ***Noise immunity:***
   - These amplifiers are extensively used in equipment such as *electronic voltmeters and oscilloscopes.*
2. ***Drift immunity :***
   - The differential amplifier *has inherent capabilities of eliminating problem of drift.*
   - The differential amplifier construction is used for the *early stages of oscilloscope and electronic voltmeter amplifiers, where low drift is extremely important.*

**Instrumentation amplifiers.** The instrumentation amplifier is a dedicated differential amplifier with *extremely high impedance. The high common mode rejection makes this amplifier very useful in receiving small signals buried in large common-mode offsets and noise.*

These amplifiers consist of *two stages:*

— The *first stage* offers very high input impedance to both input signals and allows to set the gain with a single resistor.

— The *second stage* is a differential amplifier (unity gain) with output, negative feedback and ground connections all throughout.

### 8.8.7. Attenuators

*An* **attenuator** *is a two-port resistive network and is used to reduce the signal level by a given amount.*

In a number of applications, it is necessary to introduce a specified loss between the source and a matched load without altering the impedance relationship. Attenuators may be used for this purpose.

Attenuator may be *symmetrical* or *asymmetrical*, and can be either *fixed or variable. A fixed attenuator with constant attenuation* is called a **pad**.

- Variable attenuators are used as control volumes in radio broadcasting sections.
- Attenuators are also used in *laboratory to obtain small value of voltage or current for testing circuits.*

The attenuation is expressed in **decibels (dB)** or, in *naper*. The attenuation offered by a network in decibels is given by

$$\text{Attenuation in } dB = 10\log_{10}\left(\frac{P_i}{P_o}\right) \quad ...(8.23)$$

where, $P_i$ is the input power and $P_o$ is the output power.

The attenuators may be of the following types:

1. Resistance attenuator.
2. Symmetrical T-attenuator.
3. L-type attenuator.
4. $\pi$-type attenuator.

### 8.8.8. Filters

**Filtering** *is the process of attenuating unwanted components of a measurement while permitting the desired component to pass.*

The **filter** *is an electronic circuit which can pass or stop a particular band of frequencies through it.* The filters was first designed by G.A. Campbell and D.Z. Zobel at Bell laboratories.

*The band of frequencies which will pass through filter is called the* ***pass band*** and the band of all remaining frequencies is called ***attenuation band****.* In case of **ideal filter**, all frequencies of pass band will pass without suffering from any attenuation while the band of all remaining frequencies of attenuation band will be suppressed completely.

**Classification of filters:**

The filters may be *classified* as follows:

***A. On the basis of passing and attenuating of frequencies:***

**1. Low pass filters :**

- These are those filters which pass only low frequencies through them and which reject all high frequencies above the cut-off frequencies.
- A low pass filter is also called *"lag network"* because it causes a phase lag in the output signal.
- This type of filter is also called *"integrating network"*.

**2. High pass filters :**

- These are those filters which pass only high frequencies through them and which reject all low frequencies below the cut-off frequency.
- The high pass filter is a differentiating network and is also called as *"lead network"* because it will cause a phase lead in the output signal.

**3. Band pass filters :**

- These are those filters which pass a band of frequencies through them and which reject all other frequencies to pass through them.

4. **Band stop filters :**
   - These filters, which are also known as *"band elimination filters"*, are those which reject a band of frequencies to pass through them and which allow the other frequencies to pass though them.

***B. On the basis of relation between series and shunt impedances :***

1. **Constant filters** (*or prototype filters*). In this filter the series impedance $z_1$ and stunt impedance $z_2$ are interrelated by the relation:
$$z_1 z_2 = k^2, \qquad ...(8.24)$$
where, $k$ is a constant independent of frequency.
2. ***m*-derived filters.** These filters do not have the product of series and shunt impedances equal to $k^2$, but have the same characteristic impedance as the corresponding $k$ section, with sharper attenuation characteristic.
   - Fig. 8.13(*a*) shows some terminology as applied to a low pass filter (Similar terms are applicable to the high pass and notch or band reject filters, respectively) while in the Fig. 8.13(*b*) are shown the band-pass filter characteristics.

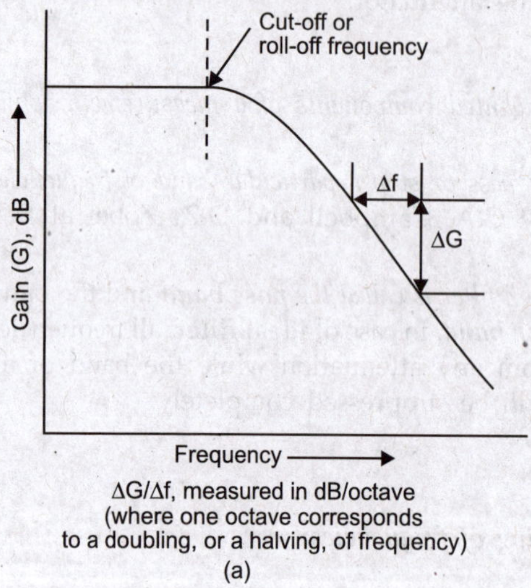


(a)

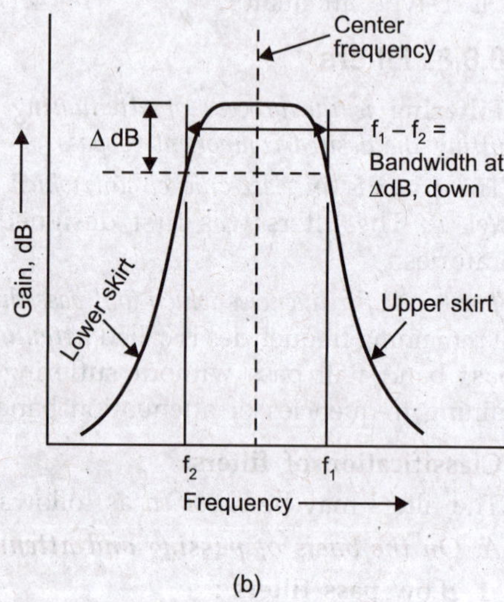


(b)

**Fig. 8.13**

- Fig. 8.14 shows the ideal characteristics of filters.

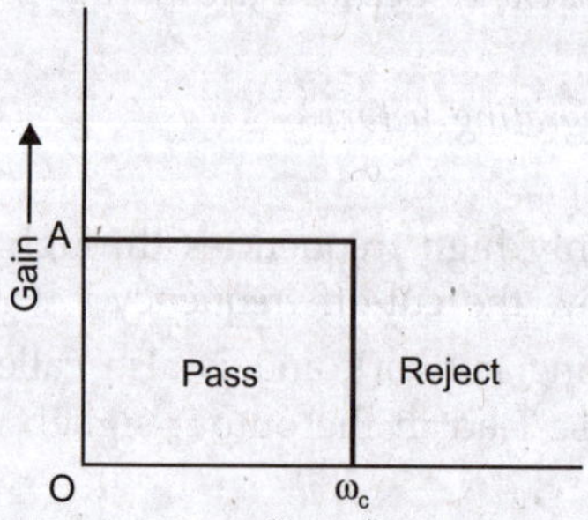


(i) Low pass filter

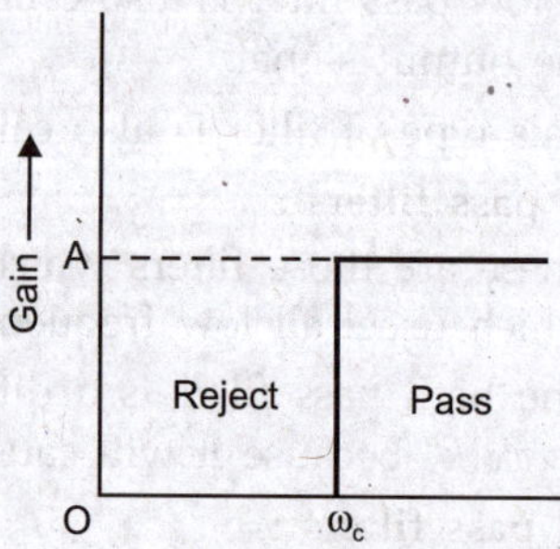


(ii) High pass filter

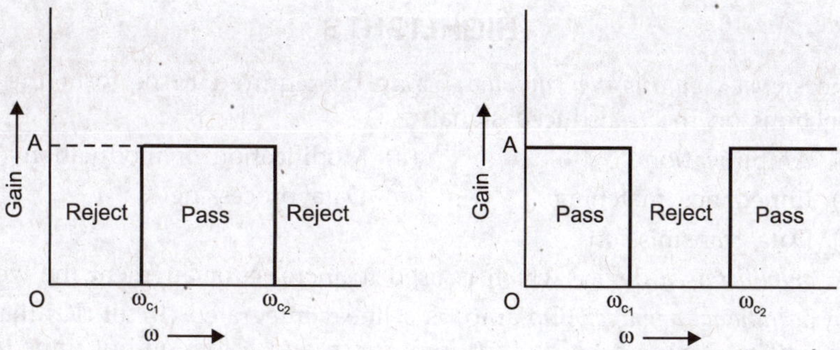


(iii) Band pass filter (iv) Band stop filter

**Fig. 8.14.** Ideal characteristics of filters.

## 8.8.9. Input Circuitry

Although not all-inclusive, the following types of input circuits are used for signal conditioning of electrical transducers :

1. Simple current-sensitive circuits.
2. Ballast circuits.
3. Voltage-dividing circuits.
4. Bridge circuits.
5. Resonant circuits.
6. Amplifier input circuits.

| | | | |
|---|---|---|---|
| 1. 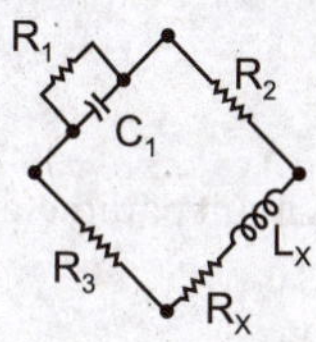 Maxwell circuit | *Measures L (best for $\omega L_x/R_x < 10$)*<br>*Balance equations :*<br>$L_x = R_2R_3C_1$<br>$R_x = \dfrac{R_2R_3}{R_1}$ | 2. 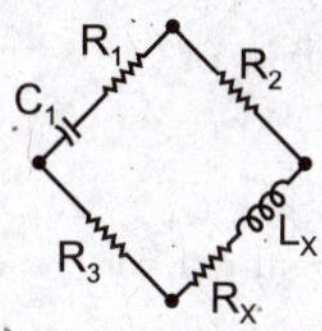 Hay circuit | *Measures L (best of $\omega L_x/R_x > 10$)*<br>*Balance equations :*<br>$L_x = \dfrac{R_2R_3C_1}{1+\omega^2C_1^2R_1^2}$<br>$R_x = \dfrac{\omega^2C_1^2R_1R_2R_3}{1+\omega^2C_1^2R_1^2}$ |
| 3. 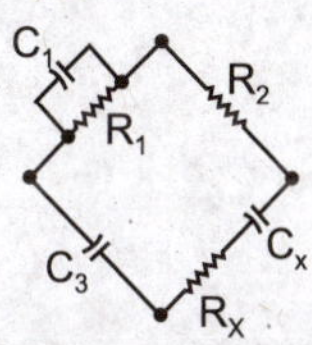 Schering circuit | *Measures C*<br>*Balance equations :*<br>$C_x = C_3\dfrac{R_1}{R_2}$<br>$R_x = R_2\dfrac{C_1}{C_3}$ | 4. 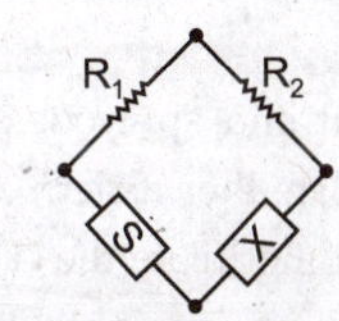 Comparison with series constants | *Meausres L or C*<br>*Balance equations :*<br>$R_x = R_s\dfrac{R_1}{R_2}$<br>If inductive, $L_x = L_s\dfrac{R_2}{R_1}$<br>If capacitive, $C_x = C_s\dfrac{R_1}{R_2}$ |
| 5. 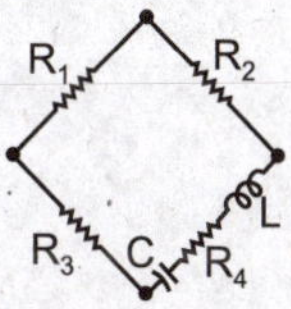 Resonant circuit | *Measures L or C (f known), f (L and C known)*<br>*Balance equations :*<br>$X_L = X_C$ or $L.\ C = \dfrac{1}{\omega^2}$<br>$f = \dfrac{1}{2\pi\sqrt{LC}}$ | 6. 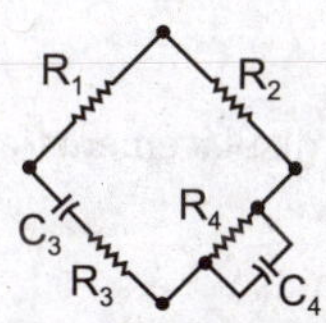 Wien or RC frequency bridge | *Measures f*<br>*Balance equations :*<br>$f = \dfrac{1}{2\pi\sqrt{R_3R_4C_3C_4}}$<br>$\dfrac{R_1}{R_2} = \dfrac{R_3}{R_4} + \dfrac{C_4}{C_3}$ |

**Fig. 8.15.** Impedance bridge arrangements.

## HIGHLIGHTS

1. The *signal conditioning equipment* may be required to perform the following functions on the transduced signal :
   (*i*) Amplification (*ii*) Modification or modulation
   (*iii*) Impedance matching (*iv*) Data processing
   (*v*) Data transmission.
2. An *amplifier* is a device which is used to increase or augment the weak signal.
3. An *operational amplifier* (Op-amp) is a linear integrated circuit (IC) that has very high voltage gain, a high input impedance and a low output impedance.
4. *Filtering* is process of attenuating unwanted components of a measurement while permitting the desired component to pass.

## OBJECTIVE TYPE QUESTIONS

1. The closed loop gain of an Op-amp is dependent upon whether the Op-amp is used
   (*a*) in inverting mode (*b*) in non-inverting mode
   (*c*) is independent of the fact whether the input is connected to inverting or non-inverting terminal.
   (*d*) is dependent upon the fact whether the input is connected to inverting or the non-inverting terminal.
2. A buffer amplifier has gain of
   (*a*) infinity (*b*) zero
   (*c*) unity (*d*) dependent upon the circuit parameters.
3. A.C. amplifiers are best suited for
   (*a*) steady-state signals (*b*) low frequency signals
   (*c*) rapidly varying signals (*d*) none of these.
4. The amplifier drift and spurious noise signals are not significant in
   (*a*) a.c. amplifiers (*b*) d.c. amplifiers
   (*c*) charge amplifiers (*d*) none of these.
5. In a carrier system, drift and spurious signals are important
   (*a*) because they modulate the carrier
   (*b*) because they do not modulate the carrier
   (*c*) because it is easier to achieve a stable carrier than a stabilized d.c. source.
   (*d*) none of the above.
6. When using d.c. signal conditioning system, with a carrier of 3 kHz, the data frequency should be limited to :
   (*a*) 1 kHz (*b*) 5 Hz
   (*c*) 600 Hz (*d*) 2 MHz.
7. The input and output displacements are of opposite phase in
   (*a*) simple lever (*b*) compound lever
   (*c*) compound gear trains (*d*) none of these.
8. What is the desirable feature in an electronic amplifier?
   (*a*) High output impedance (*b*) Low input impedance
   (*c*) good frequency response (*d*) All of these.

9. Charge amplifiers are used in order to amplify the output signals of
(a) inductive (b) capacitive
(c) resistive
(d) piezoelectric and capacitive transducers.

10. Filters that transmit all frequencies below a defined cut-off frequency are known as
(a) low-pass filters (b) high-pass filters
(c) band-pass filters (d) any of these.

11. Excitation and amplification systems are needed
(a) for active transducers only (b) for passive transducers only
(c) for both active and passive transducers
(d) for both passive and output transducers.

12. A d.c. amplifier
(a) needs to have a balanced differential inputs with a high common mode rejection ratio (CMRR) to give very good thermal and long term stability.
(b) easy to calibrate at low frequencies and has ability to recover from overload conditions.
(c) is immune to drift and low frequency spurious signals come out as data information.
(d) is followed by a low pass filter to eliminate high frequency components including noise from the data signal.
(e) all of the above.

13. The output from frequency-modulation system is
(a) a.c. voltage (b) d.c. voltage
(c) a.c. and d.c. voltage (d) any of these.

14. The data transmission with synchro systems employs .............. telemetering to convey the requisite information.
(a) frequency (b) position
(c) impulses (d) voltage.

15. When using a.c. signal conditioning system for capacitive transducers, the carrier frequencies
(a) range between 50 Hz and 20 kHz (b) should be of the order of 0.5 MHz
(c) should be of the order to 20 MHz (d) none of the above.

16. An a.c. signal conditioning system is normally used for
(a) resistive transducers like strain gauges (b) inductive and capacitive transducers
(c) piezoelectric transducers (d) all of the above.

17. The overall gain or amplification of a system of two amplifiers arranged in series is
(a) $G_1 + G_2$ (b) $G_1 - G_2$
(c) $G_1 \times G_2$
(d) $\frac{G_1}{G_2}$ where $G_1$ and $G_2$ are the two gains expressed as pure numbers.

18. The properties of an ideal Op-amp are :
(a) It should have zero input impedance (b) It should have high input impedance
(c) It should have a zero open loop gain (d) None of the above.

19. The moving iron voltmeters indicate :
(a) the same value of d.c. and a.c. voltages.
(b) lower values for a.c. voltages than the corresponding values of d.c. voltages.
(c) higher value for a.c. voltages than the corresponding values of d.c. voltages.
(d) none of the above.

## ANSWERS

| | | | | | | |
|---|---|---|---|---|---|---|
| 1. (*a*) | 2. (*c*) | 3. (*c*) | 4. (*a*) | 5. (*a*) | 6. (*c*) | 7. (*a*) |
| 8. (*c*) | 9. (*d*) | 10. (*a*) | 11. (*d*) | 12. (*e*) | 13. (*b*) | 14. (*b*) |
| 15. (*b*) | 16. (*b*) | 17. (*c*) | 18. (*b*) | 19. (*b*). | | |

## THEORETICAL QUESTIONS

1. What do you mean by the following terms as applied to instrumentation or measurement system?
   (*i*) Detector-transducer stage.
   (*ii*) Signal conditioning stage.
2. State the limitations of mechanical amplification.
3. What are the advantages of electrical signal conditioning?
4. Explain briefly the following functions of signal conditioning equipment:
   (*i*) Amplification (*ii*) Modification or modulation
   (*iii*) Impedance matching (*iv*) Data processing
   (*v*) Data transmission.
5. Explain briefly the following:
   (*i*) D.C. signal conditioning system.
   (*ii*) A.C. signal conditioning system.
6. Describe briefly the term "Amplification".
7. Explain briefly any two of the following amplifiers:
   (*i*) Mechanical amplifiers (*ii*) Fluid amplifiers
   (*iii*) Electrical and electronic amplifiers.
8. State the generalities that can be listed for an ideal electronic amplifier.
9. What are A.C. and D.C. amplifiers? Explain briefly.
10. What do you mean by "Modulated and unmodulated signals"?
11. What is an Op-amp? State its limitations as well.
12. Explain briefly the term "Common-mode rejection ratio (CMRR).
13. State the applications of Op-amp.
14. Enumerate some of the commonly used Op-amp circuits.
15. Explain briefly the following:
    (*i*) Buffer amplifier. (*ii*) Differential amplifier.
16. State the advantages of differential amplifiers.
17. What is an attenuator? How are the attenuators classified?
18. What do you mean by the terms "Filtering" and "Filter"?
19. How are filters classified?

CHAPTER

# 9 Data Acquisition Systems and Conversion

**9.1 Introduction; 9.2 Objectives** and configuration of data acquisition system **9.3 Data acquisition systems**-Analog and automated data acquisition systems-single channel data acquisition system-Multi-channel data acquisition system-applications of data acquisition systems; **9.4 Data conversion**-Analog-to-digital conversion (ADC)-Digital-to-analog (D/A) conversion. *Highlights – Objective Type Questions – Theoretical Questions.*

## 9.1 INTRODUCTION

**Data acquisition system** (DAS) *may be defined as a system used for data processing, data conversion, data transmission and data storage.*

A typical data acquisition consists of individual sensors with necessary signal conditioning, multiplexing, data conversion, data processing, data handling and associated transmission, storage and display systems.

— *The analog data is usually acquired and converted into digital form for the purpose of processing, transmission, display, and storage.* Data may be transmitted over long distance (from one location to another one) or short distances (from a test centre to a nearby computer).

— The data my be displayed on a digital panel meter or as a part of CRT presentation. The same may be stored temporarily (in either raw or processed form) for immediate use or permanently for ready reference later.

Fig. 9.1 shows a generalised data acquisition system.

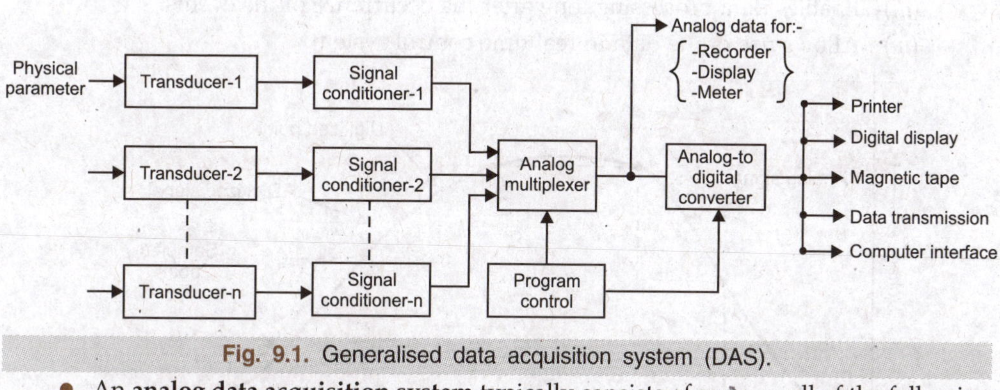


**Fig. 9.1.** Generalised data acquisition system (DAS).

- An **analog data acquisition system** typically consists of same or all of the following elements.

(*i*) Transducers

(*ii*) Signal conditioners

(*iii*) Calibrating equipment
(*iv*) Integrating equipment
(*v*) Visual devices
(*vi*) Graphic recording instruments
(*vii*) Magnetic tape instrumentation
(*viii*) Analog computers
(*ix*) High speed cameras and TV equipment.

- A ***digital data acquisition system*** may include some or all of the following elements :
  (*i*) Transducer.
  (*ii*) Signal conditioner.
  (*iii*) Scanner or multiplexer.
  (*iv*) Signal converter.
  (*v*) Analog-to-digital converter (ADC).
  (*vi*) Auxiliary equipment and system programmer.
  (*vii*) Digital recorder.
  (*viii*) Digital printer.

Now-a-days, in mechatronic and measurement systems, *microprocessors, microcontrollers, single-board computers, and personal computers* are widely used. As such, it is increasingly important for engineers to understand how to directly access information and analog data from the surrounding environment with these devices.

Consider a signal from a sensor as illustrated by the analog signal in Fig. 9.2. In this case there are *two options* :

- Firstly, one could record the signal with an analog device such as chart recorder (which physically plots the signal on the paper) or display *it with an oscilloscope.*
- Secondly, *the data may be stored by using a microprocessor or computer. This process is called computer* ***"data acquisition"*** and entails the following *merits*:
  (*i*) Can result in greater data accuracy.
  (*ii*) Provides more compact storage of the data.
  (*iii*) Enables data processing long after the occurrence of the events.
  (*iv*) Allows use of the data in real time control system.

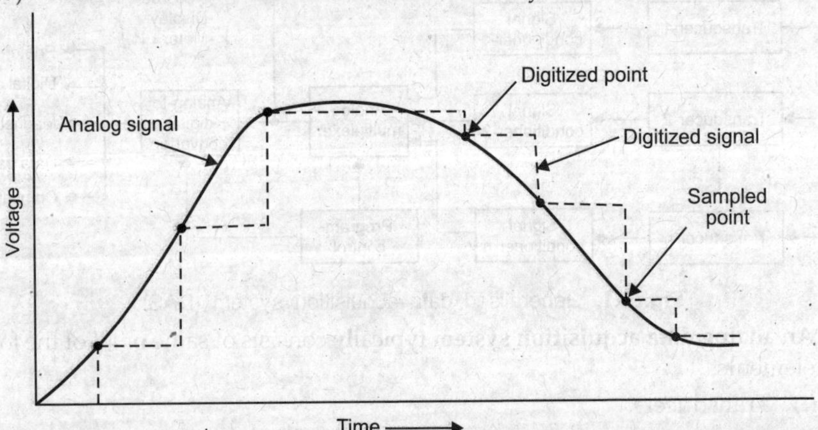


**Fig. 9.2.** Analog signal and sampled equivalent.

In order to input analog data to a digital circuit or microprocessor, the analog data *must be transformed into digital values.* The first step is to *numerically evaluate the signal at discrete instants in time. This process is called* ***"sampling"***, and the result is *"digitized signal"* composed of discrete values corresponding to each sample (See Fig. 9.2).

## 9.2 OBJECTIVES AND CONFIGURATION OF DATA ACQUISITION SYSTEM (DAS)

**Objectives of DAS :**

The objectives of a data acquisition system may be summed up as follows :

1. To be reliable, flexible and capable of being expanded for future requirements.
2. To acquire the necesary data, at correct speed and at correct time.
3. Down time not be more than 0.1 percent.
4. To be able to compute unit performance indices using on-line, real-time data.
5. To maintain on-line optimum and safe operations, it must monitor the complete plant operation.
6. To make use of all data efficiently to inform the operator about the state of the plant.
7. To be able to collect, summarise and store data for diagnosis of operation and record purpose.
8. To provide an effective human communication system and be able to identify problem areas, thereby minimising unit availability and maximising unit through point at minimum cost.

**Configuration of DAS :**

The important factors that decide the configuration and the sub-systems of a data acquisition system are as follows :

(*i*) The number of channels to the monitored.
(*ii*) Sampling rate per channel.
(*iii*) Signal conditioning requirement of each channel.
(*iv*) Resolution and accuracy.
(*v*) Cost.

The various general configurations include :

1. *Signal channel possibilities :*
   (*i*) Direct conversion.
   (*ii*) Pre-amplification and direct conversion.
   (*iii*) Sample and hold, and conversion.
   (*iv*) Pre-amplification, sample and hold, and conversion.
   (*v*) Pre-amplification, signal conditioning, and any of the above.
2. *Multi-channel possibilities :*
   (*i*) Multiplexing the outputs of single channel converters.
   (*ii*) Multiplexing the outputs of sample-hold circuits.
   (*iii*) Multiplexing the inputs of the sample-hold circuits.
   (*iv*) Multiplexing low level data.
3. *Noise-reduction options :*
   (*i*) Filtering.
   (*ii*) Integrating converters and digital processing.

## 9.3 DATA ACQUISITION SYSTEMS

**Data acquisition** *is the process of using output signals and inputting that into a computer.* The output signal may be one that originates from direct measurement of electrical quantities such as voltage, frequency, resistance etc. or that originates from sensors.

The description of various types of data acquisition systems is given in the following articles.

### 9.3.1. Analog and Automated Data Acquisition Systems

Fig. 9.3, shows the block diagram of elements of ***analog data acquisition system***:

— This system consists of a sensor-transducer the output of which is connected to DAC board (this is a PCB) through a signal conditioning unit. The DAC board is plugged to a computer. The DAC board consists of a *multiplexer, amplifier, ADC, register* and *control circuitry*, the output of control circuitry connected to a computer system.

— A software is employed to control the acquisition of data through DAC. When the program requires input from a particular sensor, it activates the DAC board by sending control word to the control and status register. The control word indicates what type of operation the board has to carry out.

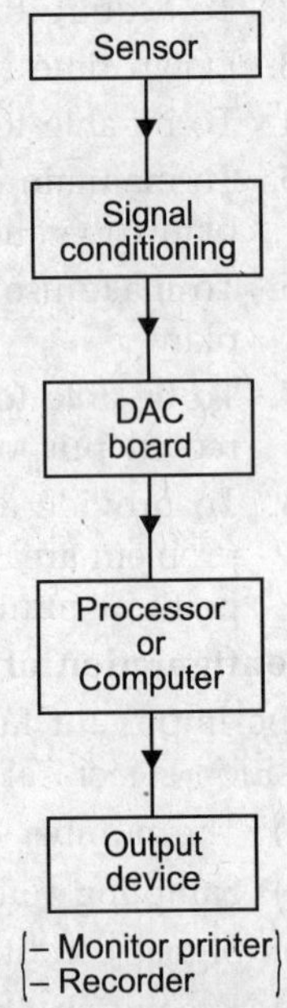


Fig. 9.3. Block diagram of analog data acquisition system.

**Automated data acquisition systems** may take the following forms:

1. Data loggers;
2. Computer with plug-in boards.

**1. Data loggers :**

— A *data logger* can monitor the inputs from a large number of sensors.

— Inputs from individual sensors, after suitable signal conditioning, are fed into the *multiplexer*. The multiplexer is used *to select one signal* which is then fed, after amplification, to the analog-to-digital converter.

— The digital signal is then processed by a microprocessor. The microprocessor is able to carry out simple arithmetic operations, perhaps taking the average of a number of measurements.

— The output of the system might be displayed on a digital meter that indicates the output and channel number, used to give a permanent record with a printer, stored on a floppy disc or transferred to perhaps a computer for analysis.

- As data loggers are often used with *thermocouples*, there are often special inputs for thermocouples, these providing cold junction compensation and linearisation. The multiplexer can be switched to each sensor in turn and so the output consists of a sequence of samples. Scanning of the inputs can be selected by programming the microprocessor to switch the multiplexer to just sample a single channel, carry out a single scan of all channels, a continuous scan of all channels, or perhaps carry out a periodic span of all channels.

**2. Computer with plug-in boards :**

Fig. 9.4, shows the basic elements of a data acquisition system using plug-in boards with a computer.

The signal conditioning prior to the inputs to the board depends on the sensors concerned.

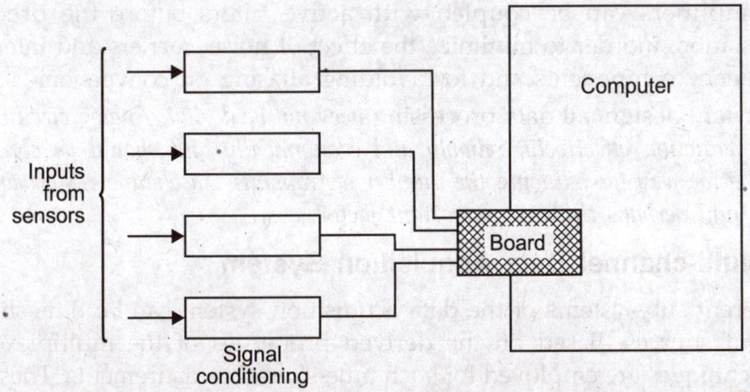


**Fig. 9.4.** Data acquisition system (Using plug-in-boards with a computer).

***Examples:***

(i) *Thermocouples* — Amplification, cold junction compensation and linearisation;

(ii) *Strain gauges* — Wheatstone bridge, voltage supply for bridge and linearisation;

(iii) *RTDs* — Current supply, circuitry and linearisation.

### 9.3.2. Single Channel Data Acquisition System

Fig. 9.5 shows the block diagram of a single channel data acquisition system.

— The system consists of a *"signal conditioner"* followed by an *"analog-to-digital (A/D) converter,"* performing repetitive conversions at a free running, internally determined rate.

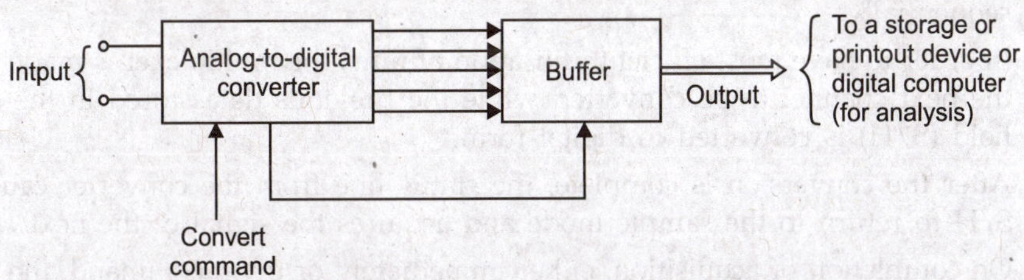


**Fig. 9.5.** Single channel data acquisition system.

— The digital outputs from the "buffer" are further fed to a storage or printout device or to a digital computer for analysis.

*Example* : Digital panel meter (DPM)

— Analog-to-digital converters used in this system applications are usually designed to receive external commands to *convert and hold*.

Although these converters based on dual-slope techniques are useful for conversion of low frequency data, such as from thermocouples, especially in the presence of noise, the most popular type of converter for data system applications is the "*successive approximation*" type since it is *capable of high resolution and high speed at moderate cost.*

- Invariably it is observed that the signal level is very low compared to input requirement. In such cases, the amplification of the input signal is done to bring its level to match the input requirements. This is called *pre-amplification.* Preamplifiers can be coupled with active filters before the process of data acquisition, inorder to minimize the effect of noise, carriers and interfering high-frequency components and for avoiding aliasing on conversion.
- In system design, all data processing *need not be digital. Analog circuits can perform data reduction effectively, reliably and economically and should be considered as an alternative way for reducing the number of transmission channels, software complexity, noise and perhaps the most important factor, cost.*

### 9.3.3. Multi-channel Data Acquisition System

The different sub-systems of the data acquisition system can be time-shared by two or more input sources. Based on the derived properties of the multiplexed system a number of techniques are employed for such time-shared measurements. These techniques are enumerated and discussed below :

1. Multi-channel analog multiplexed system.
2. Multiplexing the outputs of sample-holds.
3. Multiplexing after A/D conversion.
4. Multiplexing low-level data.

**1. Multi-channel analog multiplexed system :**

Fig. 9.6 shows a conventional multi-channel data system having a single analog-to-digital (A/D) converter preceded by a multiplexer :

— In this system the individual signals are applied directly, or after preamplification and/or signal conditioning wherever necessary, to the multiplexer.

— These analog signals are then converted to digital signals by the A/D converter, sequentially.

— In order to have most efficient utilisation of time, the multiplexer is made to seek the next channel to be converted while the previous data stored in the sample hold (S/H) is converted to digital form.

— After the conversion is complete, the status line from the converter causes the S/H to return to the sample mode and acquires the signal of the next channel.

— On completion of acquisition, either immediately or upon command, the S/H is switched to the hold mode; a conversion begins again and the multiplexer switches on to the subsequent channel.

In cases where the signal variations are extremely slow, sufficient accuracy in measurement can be achieved even without the S/H.

- The *advantage* of this system is that it has a lower cost due to the sharing of a majority of sub-system.
- This method is relatively *slower than systems where the S/H output or even A/D converter outputs are multiplexed.*

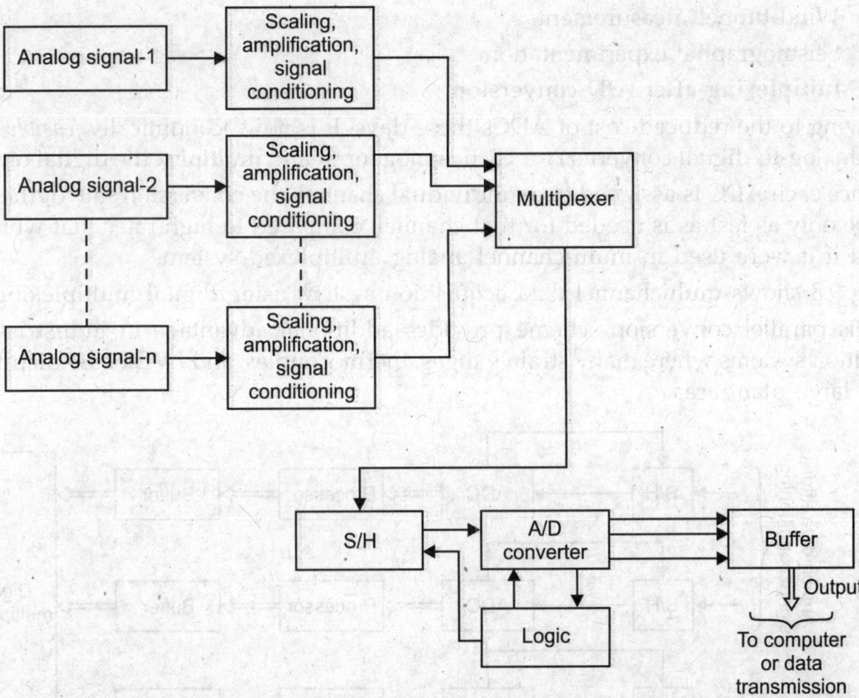


**Fig. 9.6.** Multichannel analog multiplexed system using a single A/D converter.

**2. Multiplexing the outputs of sample-holds :**

Fig. 9.7 shows a simultaneous sampled system multiplexer :

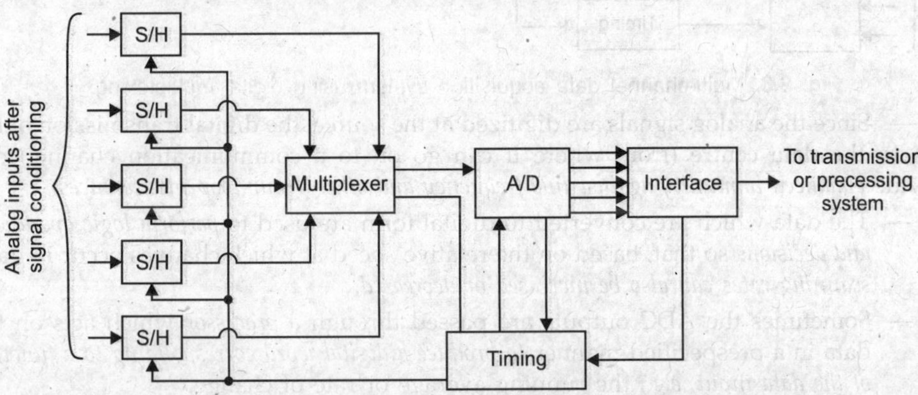


**Fig. 9.7.** Simultaneous sampled system multiplexer.

— The technique of multiplexing the outputs of the sample-hold (S/H) is particularly useful when a large number of channels are to be monitored at the same time (*i.e.*, synchronously) but at moderate speeds.

— An individual S/H is assigned to each channel, and they are updated synchronously by a timing circuit.

— The outputs of S/H outputs are connected to an A/D converter through a multiplexer resulting in sequential read-out of the outputs.

*This approach may be used in the following applications :*

- Radar and fire control systems.

- Wind-tunnel measurements.
- Seismographic experimentation.

**3. Multiplexing after A/D conversion :**

Owing to the reduced cost of ADCs these days, it is now economically *feasible to use ADC* (analog-to-digital converter) for each analog input and multiplex the digital outputs.

Since each ADC is assigned to an individual channel, the conversion rate of the ADC need be only as fast as is needed for that channel, compared to high rates that would be needed if it were used in multi-channel analog multiplexed system.

Fig 9.8 shows multichannel data acquisition system using digital multiplexing :

This parallel conversion scheme provides additional advantage in industrial data acquisition systems where many strain gauges, thermocouples, and LVDTs are distributed over a large plant area.

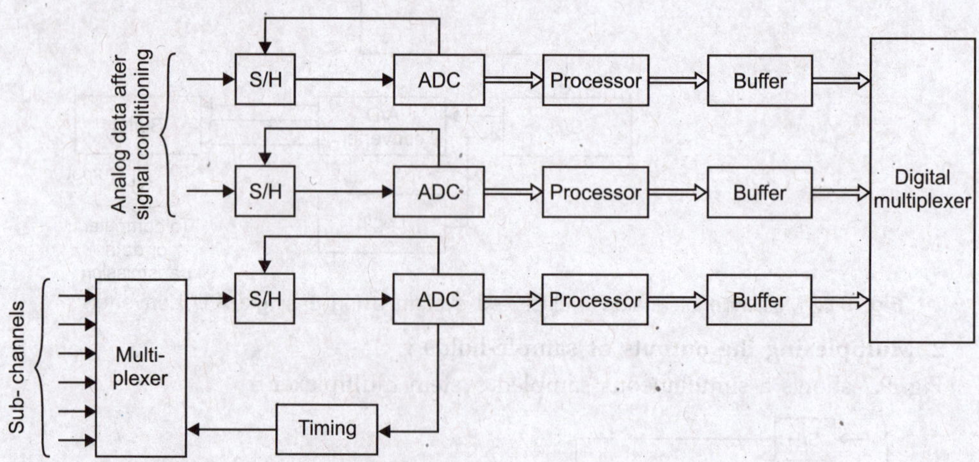


**Fig. 9.8.** Multi-channel data acquisition system using digital multiplexing.

— Since the analog signals are digitized at the source, the digital transmission of the data to the data centre (from where it can go on to a communication channel) can provide *enhanced immunity against line frequency and other group-loop interferences*.

— The data which are converted to digital form are used to *perform logic operations and decisions* so that, based on the relative speed at which changes occur in data, *scanning rates can also be increased or decreased.*

— Sometimes the ADC outputs are passed through a *processor* which acts on the data in a prespecified manner *to produce a digital word corresponding to a function of the data input, e.g.*, the running average or rate of change.

**4. Multiplexing low-level data :**

- A "low-level multiplexing system" enables the use of a *single high-quality data amplifier for handling multi-channel low-level inputs.*
- This technique can be attractive when a large number of channels (250 or more) all having low-level outputs need to be monitored at moderate speeds. *A typical application is a 200 channel stress measurement system in a transmission tower set-up.*

### 9.3.4. Applications of Data Acquisition Systems

Data acquisition systems find wide use in a variety of *"industrial and scientific areas"*, including the *aerospace, biomedical* and *telemetry industries.* The type of data acquisition system to be used depends upon the application and the intended use of recorded input data.

- The *"analog data acquisition systems"* are generally used when wide bandwidth is required or when lower accuracy can be tolerated.
- *"Digital data acquisition systems"* are used when the physical quantity being monitored has a *narrow bandwidth* (i.e., when the quantity varies slowly). These systems are also used when *high accuracy and low per channel cost is required.* These systems are in general, *more complex than analog systems, both in terms of instrumentation involved and the volume and complexity of input data they can handle.*

## 9.4 DATA CONVERSION

### 9.4.1. Analog-to-Digital Conversion (ADC)

#### 9.4.1.1. Digital signals

The majority of sensors supply the output which tends to be in analog form. Where a *microprocessor* is used as part of the measurement or control system, the analog output from the sensor *has to be converted into a digital form* before it can be used as an input to the microprocessor. Similarly, most actuators operate with analog inputs and so the digital output from a microprocessor has to be converted into an analog form before it can be used as input by the actuator.

#### 9.4.1.2. ADC process

The *"analog-to-digital conversion" process* changes a sampled analog voltage into digital form. This process, conceptually involves the following *two steps:*

(*i*) **Quantizing.** It is defined as the *transformation of a continuous analog input into a set of discrete output states.*

(*ii*) **Coding.** It is *assignment of a digital code word or number to each output state.*

**Procedure of conversion:**

*Analog-to-digital conversion* involves converting analog signals into binary words. Fig. 9.9 shows the basic elements of analog-to-digital conversion. The procedure of A/D conversion is that a *clock supplies regular time signal pulses to ADC (analog-to-digital converter) and every time it receives a pulse it samples the analog signal.* The types of signals involved at various stages of analog-to-digital conversion are shown in Fig. 9.10:

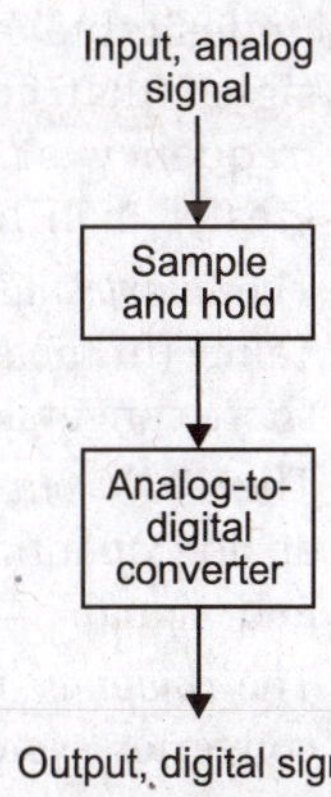


**Fig. 9.9.** Basic elements of A/D conversion.

- Fig. 9.10(*a*), shows the analog signal;
- Fig. 9.10(*b*), shows the clock signals which supply the time signals at which the sampling occurs;
- Fig. 9.10(*c*), shows a series of various pulses which is the result of sampling (sampled signal);
- Fig. 9.10(*d*), shows the sampled and held signal which is obtained by using a *sample and hold unit.* (This unit is necessary because A/D converter requires a finite amount of time, termed the '*conversion time*', to convert analog signal into a digital one) which holds each sampled value until the next pulse occurs.

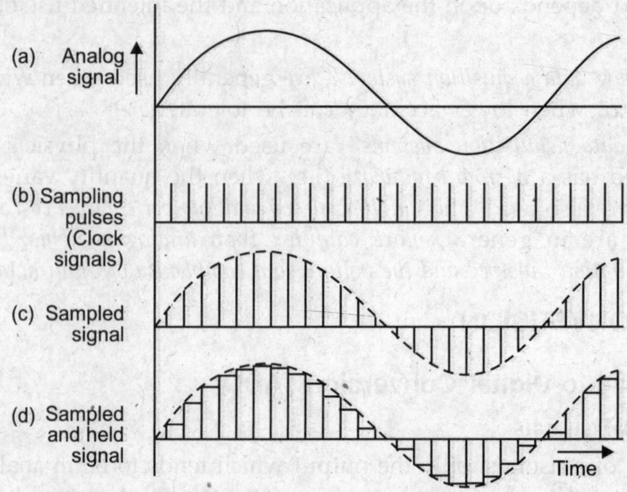


**Fig. 9.10.** Signals : (*a*) Analog; (*b*) Clock; (*c*) Sampled; (*d*) Sampled and held.

### 9.4.1.3. Components used in A/D Conversion

In order to acquire an analog voltage for digital processing, it is imperative to properly select the following components and apply them this sequence :

(*i*) Buffer amplifier; (*ii*) Low-pass filter;
(*iii*) Sample and hold amplifier; (*iv*) Analog-to-digital (A/D) converter;
(*v*) Computer.

Figure 9.11, shows the components used in A/D conversion:

- The *buffer amplifier* provides a signal in a range close to but not exceeding the full input voltage range of the A/D converter.
- The *low-pass filter* is necessary to remove any undesirable high-frequency components in the signal that could produce aliasing. The cut-off frequency of the low-pass filter should not be greater than half the sampling rate.
- The *sample and hold amplifier* maintains a fixed input value (from an instantaneous sample) during the short conversion time of the A/D converter.
- The *A/D converter* should have a resolution and analog quantization size appropriate to the system and signal.
- The *computer* must be properly interfaced to A/D converter system to store and process the data.
- The analog-to-digital conversion process requires a small but finite interval of time that must be taken into consideration when assessing the accuracy of the results.
- The *conversion time depends on the design of the converter, the method used for conversion, and the speed of the components used in the electronic design.*

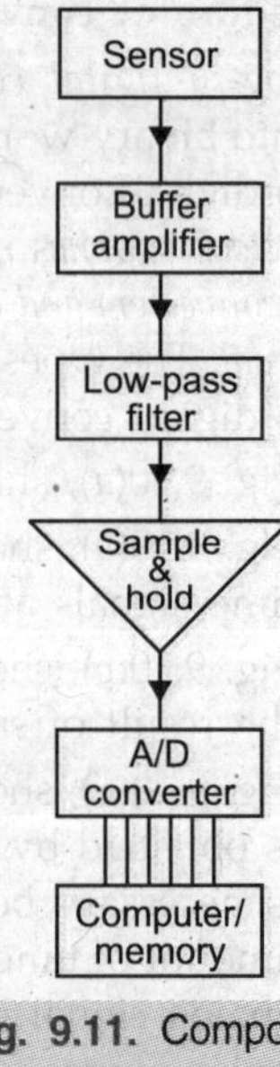


**Fig. 9.11.** Components used in A/D conversion.

— Owing to the continuous change in the analog signals, the uncertainty about when in the sample time window the conversion occurs causes corresponding uncertainty in the digital value. This is of significant importance if there is no sample and hold amplifier on the A/D input. The term *"aperture time"* refers to the duration of the time window and is associated with any error in the digital output due to changes in the input during this time.

— In a signal, sampling at or about Nyquist frequency will yield the correct frequency components. In order to obtain accurate amplitude resolution, we must have an A/D converter with an adequately small aperture time.

### 9.4.1.4. Analog-to-Digital (A/D) Converter

An ***analog-to-digital (A/D)*** *converter is an electronic device that converts an analog voltage to a digital code.* The output of the A/D converter can be directly interfaced to digital devices such as microcontroller and computers.

The ***"resolution"*** *of an A/D converter is the number of bits used to digitally approximate the analog value of the input.* The number of possible states $N$ is equal to the number of bit combinations that can be output from the converter : $N = 2^n$

where, $n$ = The number of bits.

The number of analog *"decision points"* that occur in the process of quantizing is $(N - 1)$. The *"analog quantization size"* $Q$ is defined as the *full scale range of the A/D converter by the number of output states.*

**Design Principles:**

***Analog-to-Digital (A/D) converters*** *are designed based on a number of different principles; these are:*

(*i*) Successive approximations.
(*ii*) Flash or parallel encoding.
(*iii*) Single-slope and dual-slope integration.
(*iv*) Switched capacitor.
(*v*) Delta sigma.

Some of these are discussed below:

**(*i*) *Successive approximation A/D converter :***

The successive approximation A/D converter is very widely used because :

- It is fast in operation;
- It has high resolution;
- It is less expensive.

The various subsystems involved in this type of converter are shown in Fig. 9.12.

— The *"clock"* generates a voltage, emitting a regular sequence of pulses which are counted, in a *binary manner*, and the resulting binary word is converted into an analog voltage by a *"DAC"* (*digital-to-analog converter*). This voltage rises in steps and is *compared with the 'analog input voltage'* from the sensor.

— When the clock-generated voltage passes the input analog voltage the pulses from the clock are stopped from being counted by a *"gate"* being closed. The output from the counter at that time is then a digital representation of the analog voltage.

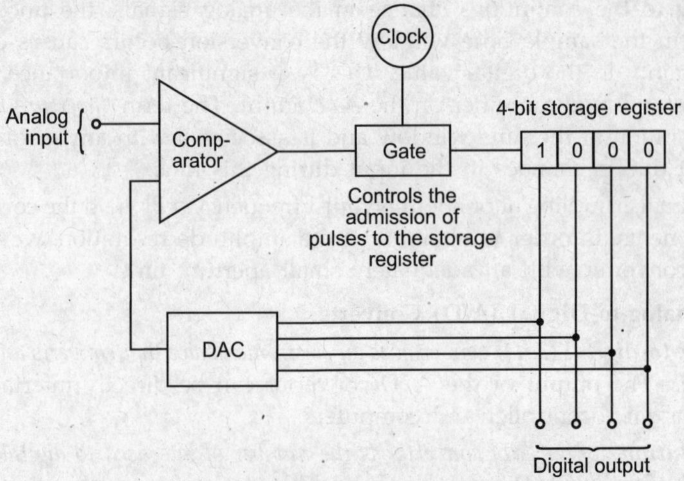


**Fig. 9.12.** Successive approximation analog-to-digital converter (ADC).

**Note:** When frequency of the clock is $f$, the time taken between the pulses is $\frac{1}{f}$; hence the time taken to generate the word, *i.e.*, the conversion time is $\frac{n}{f}$.

***(ii) Flash A/D converter :***

*The fastest type of A/D converter is known as a* ***flash converter.***

Fig. 9.13, shows a flash ADC:

— For an $n$-bit converter, $2^n-1$ separate voltage comparators are used in parallel, with each having the analog input voltage as one input.

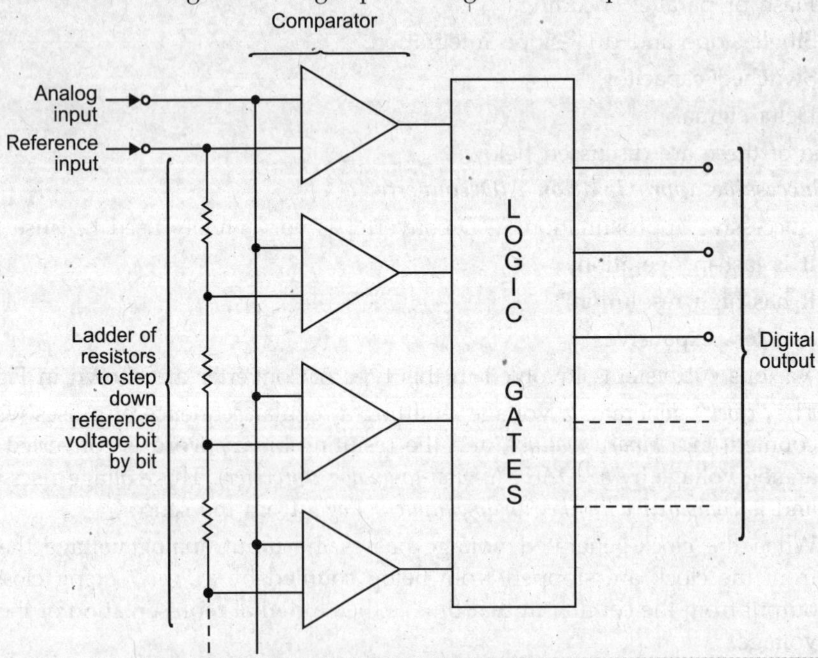


**Fig. 9.13.** Flash analog-to-digital converter (ADC).

— A *reference voltage* is applied to a ladder of resistors so that the voltage applied as the other input to each comparator is one bit *larger* in size than the voltage applied to the previous comparator in the ladder. Thus when the analog voltage is applied to the ADC, all these comparators for which the analog voltage is greater than the reference voltage of a comparator will give a *high* output and those for which it is less will be *low.*

— The resulting outputs are fed in parallel to a *logic system* which translates into a digital word.

***(iii) Single-slope and dual-slope integration :***

***(a) Single-slope or ramp or voltage-to-time A/D converter :***

Fig. 9.14 shows the schematic of a ramp ADC :

— A ramp analog-to-digital converter (ADC) involves an analog voltage which is increased at constant rate (and hence called *ramp* voltage) and is applied to a comparator where it is compared with the analog voltage from the sensor. The time consumed by the ramp voltage to increase to the value of the sensor voltage will depend on the size of the sampled analog voltage.

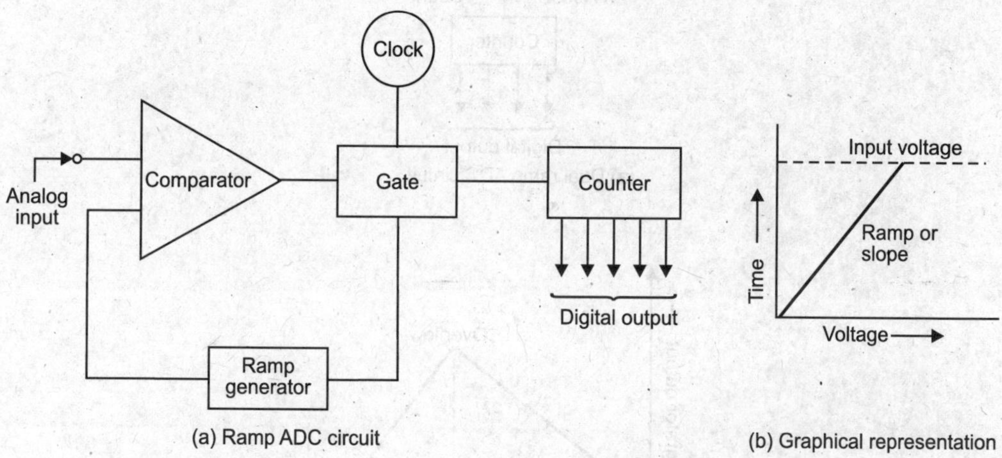


**Fig. 9.14.** Single slope or ramp ADC.

— When the ramp voltage starts, a gate is opened which starts a binary counter counting the regular pulses from a clock. When the two voltages are equal, the gate closes and the word indicated by the counter is the digital representation of the sampled analog voltage.

***(b) Dual-slope integration A/D converter or dual ramp converter :***

This type of converter, as compared to single ramp converter, is more commonly used.

Fig. 9.15, shows the dual slope/dual ramp *ADC.* The analog input voltage and the reference input voltage are successively connected to the integrator with the help of a switch. These two voltages (analog input and reference input) *must be of opposite polarities.*

— The fixed voltage is integrated for a fixed sample time.

— The integrated value is then discharged at a fixed rate and the time to do this is measured by a counter. The count is then a measure of the analog input voltage.

- The *"advantage"* of these converters is that they have *excellent noise rejection* because the integral action averages out random negative and positive contributions over the sampling period.
- Their *"limitation"* is that they are *very slow* in operation.

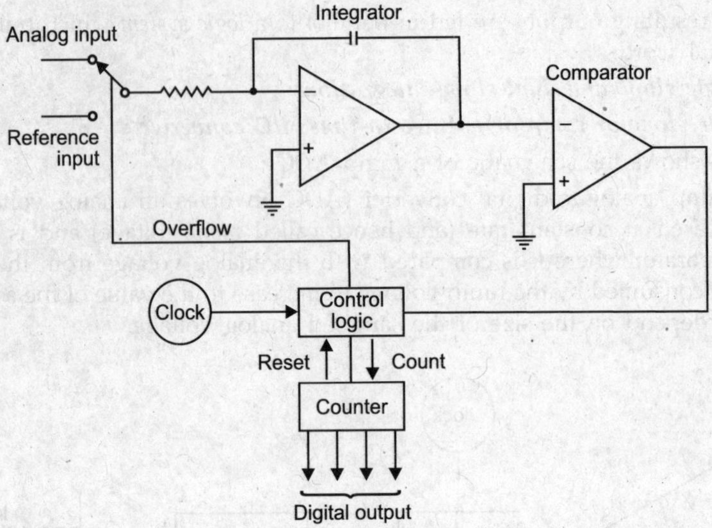


(a) Dual ramp ADC circuit

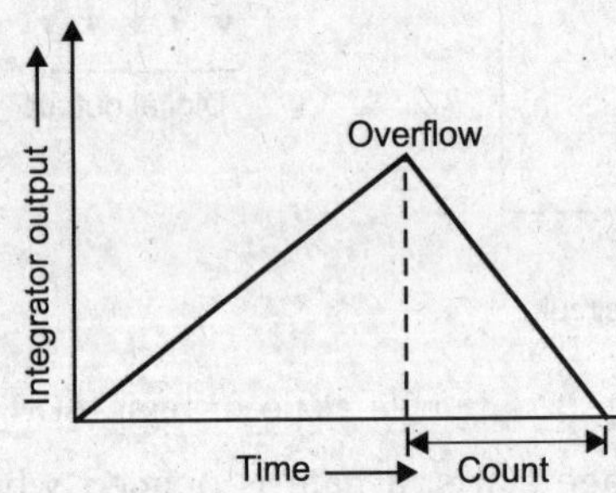


(b) Graphical representation

**Fig. 9.15.** Dual slope/dual ramp A/D converter.

**Multiplexers :**

*The* ***"multiplexer"*** *is essentially an electronic switching device which enables each of the inputs to be sampled in turn.*

A *"multiplexer"* is a circuit that is able to have inputs of data from a number of sources and then, by selecting an input channel, give an output from just one of them.

- In *applications* where there is a need for measurement to be made at a number of different locations, rather than use a separate ADC and microprocessor of each measurement, a *"multiplexer"* can be used *to select each input in turn and switch it through a single ADC and microprocessor* (Fig. 9.16).

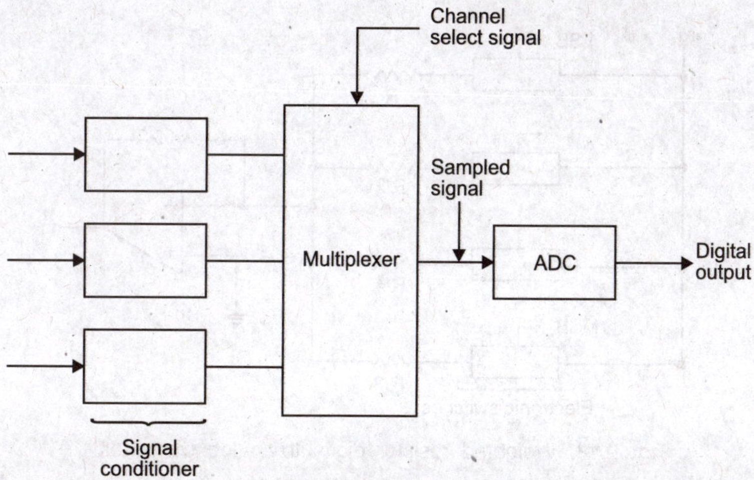


Fig. 9.16. Multiplexer.

***Digital multiplexer :***

Fig. 9.17 shows a two channel multiplexer. The logic level applied to the select input determines which AND gate is enabled so that its data input passes through the *OR* gate to the output.

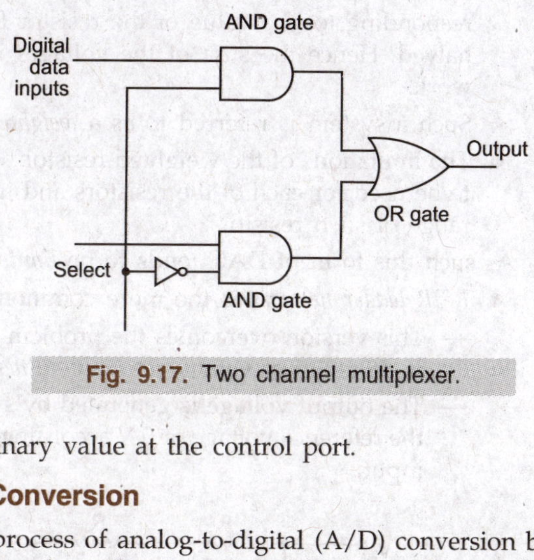


Fig. 9.17. Two channel multiplexer.

A number of forms of multiplexers are available in *IC* packages.

- ***"Demultiplexer"*** is similar to multiplexer but with *reversed action.* It accepts a digital signal through its one input and then channelises it to a particular output selected by binary value at the control port.

## 9.4.2. Digital-to-Analog (D/A) Conversion

Invariably we have to reverse the process of analog-to-digital (A/D) conversion by changing a digital value to an analog value. This is called ***digital-to-analog (D/A)*** *conversion.*

A D/A converter (DAC) *allows a computer or other digital device to interface with external analog circuits and devices.*

- The input to a digital-to-analog converter (DAC) is a *binary word;* the output is an analog signal that represents the weighted sum of the non-zero bits represented by the word.

  ***Example:*** An output of 0010 must give an analog output which is twice that given by an input of 0001.

**Digital-to-Analog Converters :**

— Fig. 9.18, shows a simple form of DAC using a summing amplifier to form the weighted sum of all the non-zero bits in the input word.

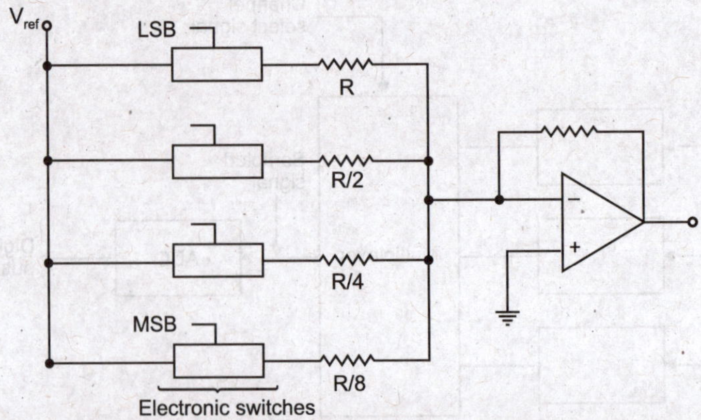


Fig. 9.18. Weighted resistor digital-to-analog converter.

— The reference voltage ($V_{ref}$) is connected to the resistors by means of electronic switches which respond to binary 1.

— The values of input resistances depend on which bit in the word a switch is responding to, the value of the resistor for successive bits from the LSB being halved. Hence the sum of the voltages is a weighted sum of the digits in the work.

- Such a system is referred to as a *weighted-resistor network.*
- The limitations of the weighted-resistor network is that accurate resistances have to be used for each of the resistors and it is difficult to obtain the required wide range of such resistors.

As such this form of DAC tends to be *limited to 4-bit-conversions.*

- *R-2R ladder network* is the more commonly used version (Fig. 9.19).
  - — This version overcomes the problem of obtaining accurate resistances over a wide range of values, *only two values being required.*
  - — The output voltage is generated by switching sections of the ladder to either the reference voltage or 0 V according to whether there is a 1 or 0 in the digital input.

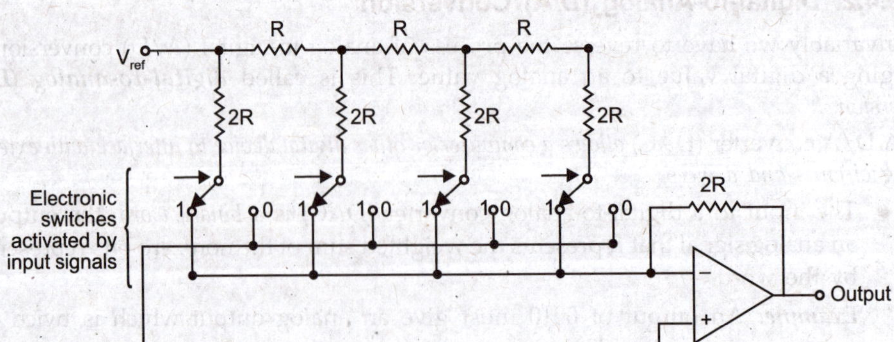


Fig. 9.19. R-2R ladder digital-to-analog converter.

- Fig. 9.20 shows computer control hardware, illustrating the *roles that A/D and D/A converters play in a* ***mechatronic control system.***
  - — An analog voltage signal from a *sensor* (*e.g.*, a thermocouple) is converted to a digital value.
  - — The *computer* uses this value in a control algorithm, and the computer outputs an analog signal to an *actuator* (*e.g.*, an electric motor) to cause some change in the system being controlled.

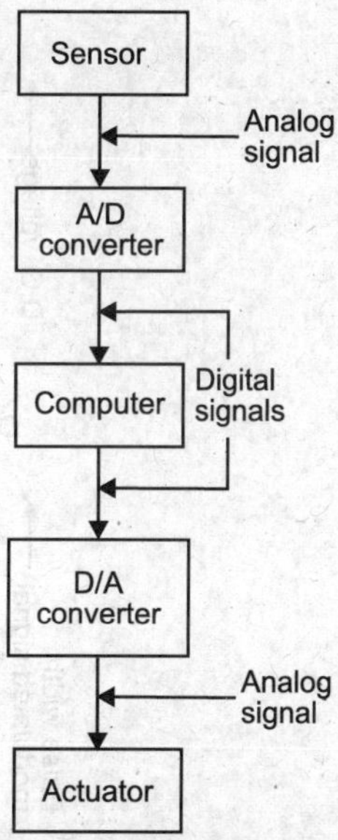


**Fig. 9.20.** Computer control hardware.

**Pulse-Modulation:**

While dealing with the transmission of low-level D.C. signals from sensors, a problem that is encountered is that the gain of Op-amp (operational amplifier) used to amplify them may drift causing a drift in the output. This problem can be solved if the signal is a *sequence of pulses rather than a continuous-time signal.* This can be achieved in the following *two* ways:

1. Pulse amplitude modulation (PAM).
2. Pulse width modulation (PWM).

**1. Pulse amplitude modulation:**

— In this method of conversion, D.C. signal (Fig. 9.21(*a*)] is chopped in the way as shown in Fig. 9.21(*b*). The output from the chopper is a chain of pulses, the heights of which depends on the D.C. level of the input signal. This process is called *"pulse amplitude modulation"*.

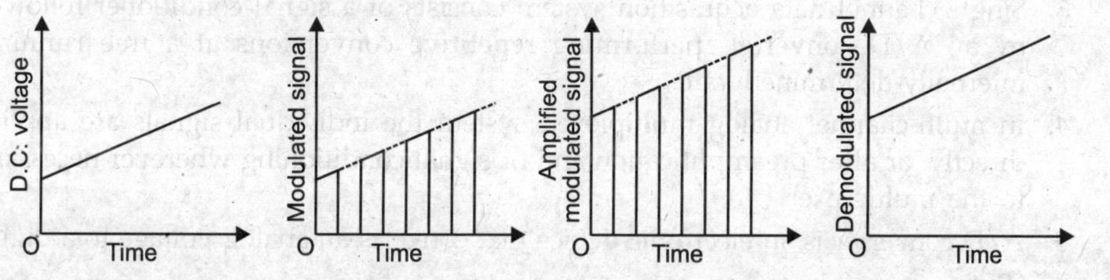


**Fig. 9.21.** Pulse amplitude modulation.

— After amplification and any other signal conditioning, the modulated signal [Fig. 9.21(*c*)] can be demodulated [Fig. 9.21(*d*)] to give a D.C. output.

**2. Pulse width modulation (PWM) :**

This type of modulation is used where the width, *i.e., duration of a pulse rather than its amplitude* depends on the size of the voltage, as shown in Fig. 9.22.

— PWM is widely used with control systems as a *means of controlling the average value of a D.C. voltage.*

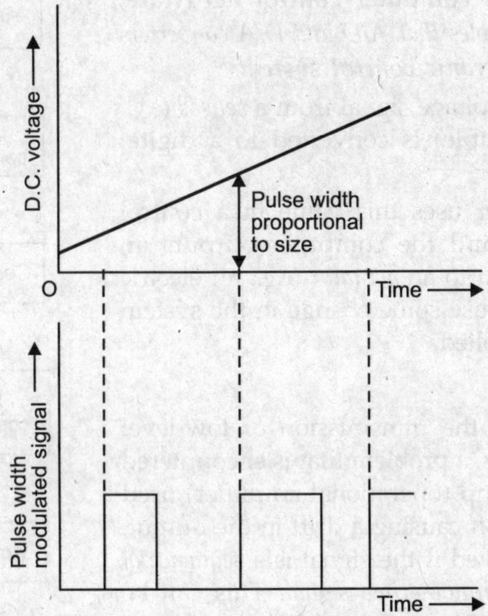


Fig. 9.22. Pulse width modulation (PWM).

## HIGHLIGHTS

1. The system used for data processing, data conversion, data transmission and data storage is called data *acquisition system.*
2. Automated data acquisition system may take the following forms : (*i*) Data loggers; (*ii*) Computer with plug-in-boards.
3. Single channel data acquisition system consists of a signal conditioner followed by an A/D converter, performing repetitive conversions at a free running, internally-determined rate.
4. In multi-channel analog multiplexed system the individual signals are applied directly, or after preamplification and or signal conditioning wherever necessary, to the multiplexer.
5. A/D converter is an electronic device that converts an analog voltage to a digital code.
6. A D/A converter allows a computer or other digital device to interface with external analog circuits and devices.

## OBJECTIVE TYPE QUESTIONS

**Fill in the Blanks or Say "Yes" or "No"**

1. .............. is the process of using output signal and inputing that into a computer.
2. A data logger cannot monitor the inputs from a large number of sensors.
3. Digital panel meter (DPM) is an example of .............. channel data acquisition system.
4. A low level multiplexing system enables the use of a single high-quality data amplifier for handling multi-channel low level inputs.
5. The analog data acquisition systems are generally used when .............. bandwidth is required or when .............. accuracy can be tolerated.

6. Digital data acquisition systems are less complex than analog systems.
7. ADC process changes a sampled analog voltage into digital form.
8. ............. is defined as the transformation of a continuous analog input into a set of discrete output states.
9. ............. is the assignment of a digital code word or number to each output state.
10. The faster type of A/D converter is known as ............. converter.
11. A ............. is essentially an electronic switching device which enables each of the inputs to be sampled in turn.

## ANSWERS

| | | | |
|---|---|---|---|
| 1. Data acquisition | 2. No | 3. Single | 4. Yes |
| 5. Wide, lower | 6. No | 7. Yes | 8. Quantizing |
| 9. Coding | 10. flash | 11. multiplexer | |

## THEORETICAL QUESTIONS

1. What is meant by data acquisition ?
2. What is a data acquisition system ?
3. Draw the block diagram of a generalised data acquisition system and explain it briefly.
4. Explain the different components of a data acquisition system.
5. What are the objectives of a data acquisition system ?
6. What are the factors that decide the configuration of a data acquisition system ?
7. With the help of suitable block diagram, explain analog data acquisition system.
8. Write a note on "Automated data acquisition systems."
9. Explain briefly a digital acquisition system.
10. Explain with a neat diagram the working of a single-channel data acquisition system.
11. Explain the working of a multi-channel analog multiplexed data acquisition system.
12. How is low-level multiplexing achieved ?
13. Give the applications of data acquisition systems.
14. Explain briefly the procedure of analog-to-digital (A/D) conversion.
15. Define the following terms :
    (*i*) Quantizing. (*ii*) Coding.
16. Describe briefly the components use in A/D conversion.
17. Explain briefly the following
    (*i*) Successive approximation A/D converter
    (*ii*) Flash A/D converter.
18. What is a muliplexer ? Explain briefly.
19. Explain briefly the following :
    (*i*) Weighted resistor D/A converter.
    (*ii*) R-2R ladder D/A converter.
20. Discuss briefly the following :
    (*i*) Pulse amplitude modulation (PAM)
    (*ii*) Pulse width modulation (PWM).

CHAPTER

# 10 Data Transmission and Telemetry

> **10.1. Data/Signal transmission;** Mechanical transmission-Hydraulic transmission-Pneumatic transmission-Magnetic transmission-Electric type of transmitters-Converters; **10.2. Telemetry**-Definition-General telemetering system; Types of telemetry systems-Voltage telemetering system-Current telemetering system-Position (or ratio) telemetering system-RF telemetry system – Introduction to types of signals and transmission paths-Modulation and demodulation-FM telemetry system-PAM telemetry system; *Highlights; Objective Type Questions; Theoretical Questions.*

## 10.1 DATA/SIGNAL TRANSMISSION

*The terms* **"Data transmission"** *and* **"Telemetry"** *refer to the process by which the measurand is transferred to a remote location for the purpose of being processed, recorded and displayed.*

The terms "*measuring devices*" and "*transmitters*" generally go by side by side and it is very difficult to make any distinction between them.

A measuring device converts a primary indication into some form of energy that can easily be displayed on a scale; some transmitters also do the same things. In the *stricter sense* **"transmitters"** *could be considered as devices which transmit the value of the primary variable at a considerable distance from the primary element.* If transmission is to be carried over *very long distances,* then devices are known as **"telemeters"**.

For transmission purposes, the measured variable is converted into a transmittable signal (either pneumatic or electrical), so that it can be received by a remote indicating, recording, or controlling device. The *selection of transmission device* depends upon the *nature of the variable and the distance the signal is required to be sent.*

For data transmission various methods have been developed; the choice of a particular method depends upon :

(*i*) The physical variable; (*ii*) The distance involved.

- The **hydraulic and pneumatic methods** are employed for transmission over a *short distance.*
  - — The pneumatic type transmission devices are generally suitable for transmission upto maximum distance of 200 m.
- The **electrical/electronic methods** are suitable *equally for short as well as long distance transmission.*
  - — Generally short transmission is carried out on own communication connections between sending and receiving devices.
  - — The *telemeters* which are designed for *long-distance transmission* may be designed to transmission over their own wire or over phone wires or by microwave.

### 10.1.1. Mechanical Transmission

The *"rack and pinion arrangement"* and the *"gear trains"* as used in *Bourdon-tube pressure* gauge and *dial indicator gauge* constitute mechanical transmission. They *amplify the displacement* and also *transmit the signal to a pointer which moves across a calibrated dial.*

### 10.1.2. Hydraulic Transmission

Fig. 10.1. shows the hydraulic method of transmission, which is commonly used. Here four bellows are employed, two at the transmission end and two at the receiving end. The four bellows are connected by an impulse pipeline and the whole system is filled with liquid. When the actuating link, on the transmission end, is operated by the measurand, then one bellow is expanded and other is contracted. This expansion and contraction is communicated to receiving end, which moves the receiving pointer an equal amount. The purpose of using two bellows on either side is to compensate for changes in ambient temperature.

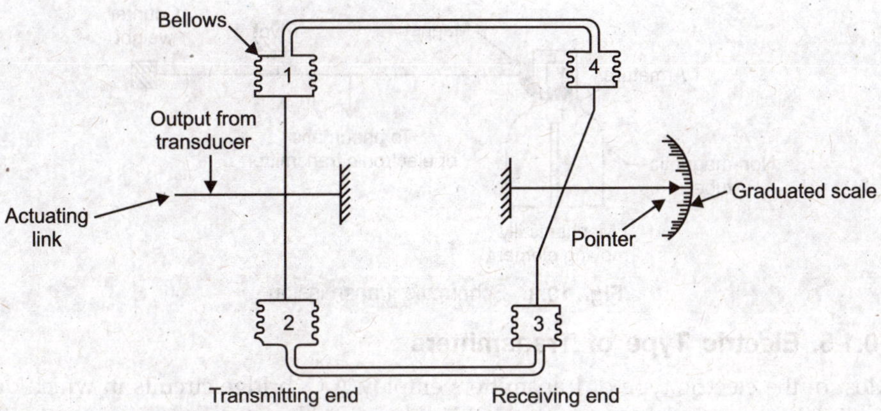


**Fig. 10.1.** Hydraulic method of transmission.

### 10.1.3. Pneumatic Transmission

Fig. 10.2, shows the one of the pneumatic methods of transmission **(Flapper nozzle mechanism).**

It consists of an open nozzle which is supplied with air through a restriction/orifice (its diameter being *smaller* than nozzle diameter for proper functioning). In front of the nozzle there is a flapper which is positioned by the measuring element. The force on the flapper is produced by a transducer which converts the measurand into linear displacement. The flapper is pivoted about a point and the other end, it contains some balancing counter weight.

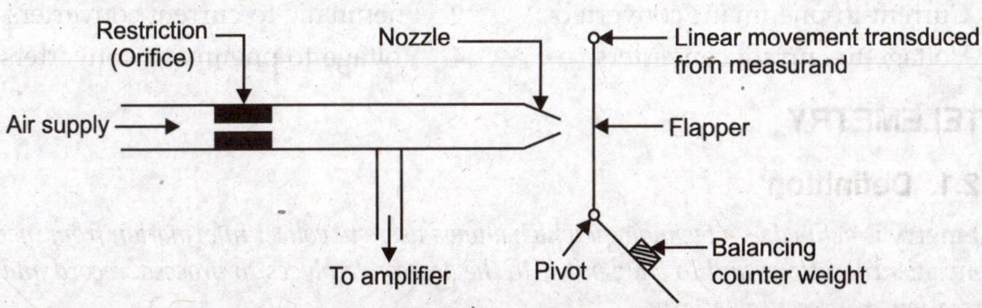


**Fig. 10.2:** Schematics of pneumatic transmission–Flapper nozzle mechanism.

When the flapper is moved against, the nozzle, the air cannot escape and maximum air passes to the *amplifier*, and when flapper is moved away from the nozzle, minimum air passes to the amplifier as most of the air escapes to atmosphere. Thus, the movement of flapper from one extreme position to another serves to control the amplifier, which produces an air pressure proportional to the measurand of adequate strength for transmission over the required distance.

### 10.1.4. Magnetic Transmission

Fig. 10.3, shows the schematics of magnetic transmission. In this arrangement/device, an armature is attached at the end of the mechanical moving part whose movement is to be transmitted outside the armature moving inside a non-magnetic tube. A magnet is placed around the armature outside the tube. The magnet follows the movement of the armature and repositions a pneumatic transmitter. The magnet movement could also be utilised to operate an electronic transmitter.

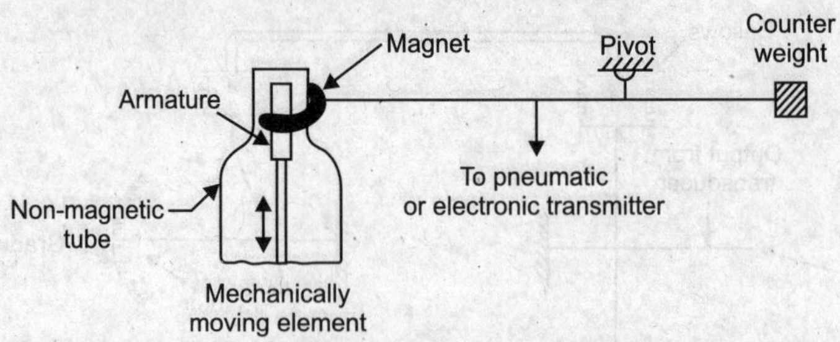


**Fig. 10.3.** Schematic transmission.

### 10.1.5. Electric Type of Transmitters

Most of the electric type of transmitters employ A.C. bridge circuits in which degree of coupling between inductances is varied by changing the amount of iron core within a coil. The common *examples* are:

1. Wheatstone bridge transmitter.
2. Inductance bridge.
3. Impedance bridge.
4. Differential transformer.
5. Self synchronous motor (Selsyn)
6. Resistance manometers.

### 10.1.6. Converters

The converters are series of transducers which play an important role in the modern instrumentation, linking electrical (voltage and current based) and pneumatic systems together.

Following are the most commonly used converters :

1. Current-to-pneumatic converters.
2. Pneumatic-to-current converters.
3. Voltage-to-current converters.
4. Voltage-to pneumatic converters.

## 10.2 TELEMETRY

### 10.2.1. Definition

**Telemetry** *is defined as a technology which allows users to collect information from inaccessible and inconvenient locations and to transmit it to the accessible places to process, record and display the information in presentable form.*

Or

**Telemetry** *(as the name implies) means measuring at a distance.*

### 10.2.2. General Telemetring System

Fig. 10.4, shows the block diagram of a general telemetring.

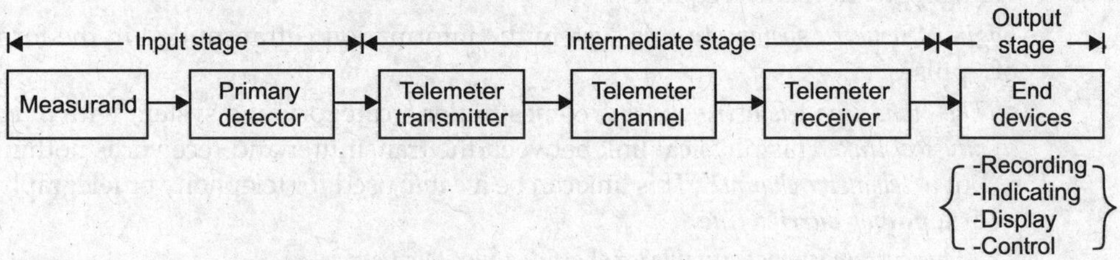


**Fig. 10.4.** Block diagram of general telemetring system.

- In the 'primary stage', the "measurand" is detected by the "*primary detector*" and sent to the channel transmitter.
- The 'intermediate stage' consists of the following three elements :
  (*i*) Telemeter transmitter;
  (*ii*) Telemeter channel;
  (*iii*) Telemeter receiver.

The "*telemeter transmitter*" converts the output of the primary detector into an electrical signal and transmits it to the "*telemetring channel*". This signal (in electrical format) is received by a "*telemeter receiver*" placed at a remote location. This receiver converts the signal into a usable form and is indicated or recorded or displayed by the "*end device*" (output stage) which is calibrated in terms of the measurand. The control device is provided to take corrective action on measurand through feed back loop to control the output.

For telemetry, some basic instruments are required for :

— translating the measurand into a suitable electrical signal;

— processing and transmission of signal over the channel;

— receiving the signal and converting it into a readily comprehensible reading.

### 10.2.3. Types of Telemetry Systems

The telemetry systems can be *classified* as follows :

**I. According to the link between the telemeter transmitter and the telemeter receiver :**

1. *Land line (**D.C.**) telemetry system :*
   (*i*) Voltage telemetering system.
   (*ii*) Current telemetering system.
   (*iii*) Position (or ratio) telemetering system.
2. *Ratio frequency (RF) telemetry system :*
   — Frequency telemetering system
   — Pulse telemetering system.
   - **A.C.** telemetering system is employed *both for land line and radio-frequency airborne telemetry techniques.* A.C. telemetry is employed for sensors, which provide an A.C. output or voltage to frequency converters. The data is available in the form of current or voltage, which is usually very weak for such transmission. Hence *it is modulated by an A.C. carrier waves generated by an oscillator.*

**II. According to the form in which the information is transmitted :**

1. *Analog telemetry system.* Here information is transmitted in the form of current voltage, position and frequency.
2. *Digital telemetry system.* In this system the information is transmitted in the form of "pulses".
   - *The "land line telemetry system"* comes under a category of a system with *direct physical link.* This physical link between the transmitter and receiver is nothing but a *'telemeter channel'*. This link can be a ***cable*** used for telephony or telegraphy or a ***power carrier line.***
     - *This system is most suitable for short distance telemetry.*
   - In *"radio frequency (RF) telemetry system"* there is *no physical link* between the telemeter transmitter and the telemeter receiver. The link between the transmitting end and the receiving end is established through the ***radio links.***

**Advantages and disadvantages of land line telemetry system :**

***Advantages :***

(*i*) Very effective system for short-distance transmission (as the telemeter or communication channel can be established with the help of a simple cable or transmission line).

(*ii*) Very simple circuitry is required for the measurement set-up.

(*iii*) The information can be transmitted (in the form of voltage, current and position) easily by the use of modern electronic circuitry available now-a-days.

(*iv*) A wide variety of detectors (or the primary sensing elements) are available which produce electrical signal (voltage or current) in relation with the variable being measured at the system's input stage.

***Disadvantages :***

(*i*) Limited frequency response.

(*ii*) Signal multiplexing is very difficult.

(*iii*) The distortions in the transmission links are directly incorporated in the main system.

(*iv*) The information transmitted through the link is affected by the EMI effects of main frequency from nearby cable.

(*v*) The effects of thermoelectric emfs are significant in case of telemetry systems transmitting D.C. signals.

### 10.2.4. Voltage Telemetering System

- In these systems the *measurand is converted to A.C. or D.C. voltage.*
- For such systems, the *self-balancing potentiometers are the usual receivers.*

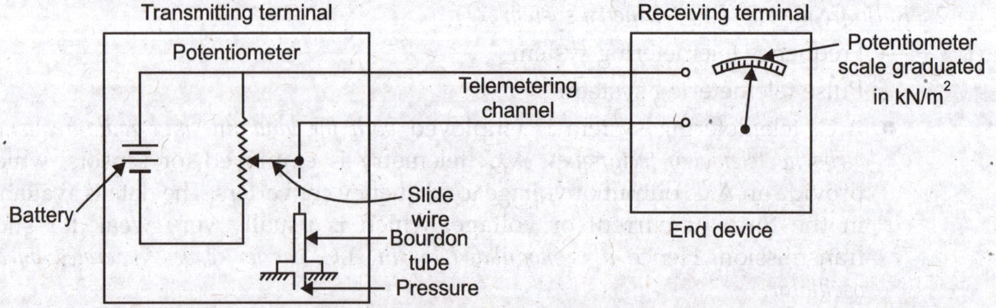


**Fig. 10.5.** Voltage telemetering system.

Fig. 10.5, shows a voltage telemetering system :

***Construction :***

— It consists of a slide-wire potentiometer connected in series with a battery at the transmitting terminal. The sliding contact is connected to the bourdon tube used for pressure measurement.

— At the receiving terminal a null balance D.C. potentiometer or a recorder is used.

— In between the transmitting terminal and the receiving terminal, a pair of wires form a telemeter channel.

***Working/Operation :***

— With the change of pressure in the system, the bourdon tube actuates the slider of the potentiometer. Consequently, the voltage changes at the transmitting end.

— This voltage is carried by the telemetering channel to the receiving terminal, where it is measured with the help of null balanced D.C. potentiometer indicator calibrated in terms of pressure scale ($kN/m^2$) or recorded if required. Some systems make use of the deflection type indicators to measure voltage; these indicators are calibrated for the line resistance. Basically, a null balance D.C. potentiometer reduces the current carried by the telemetering channel with negligible resistance.

Such systems use primary sensing elements (detectors) such as *microphones, tachometers, bourdon tubes or differential transformers as all produce voltage signal.*

***Disadvantage :*** These systems are affected by line resistance, leakage, interfering sources nearly, noise and require higher-quality circuits than current systems, especially for low voltages.

- The voltage telemetering system is *limited for transmission upto 300 metres distances.*
- A voltage telemetering system is *very much suited for adding several input voltages in series* provided the measurement is *linear.* However, this system needs a relatively *more expensive receiving terminal.* This system is normally *not suitable* to the use of many receivers at the same time.

## 10.2.5. Current Telemetering System

Fig. 10.6 shows a basic current telemetering system :

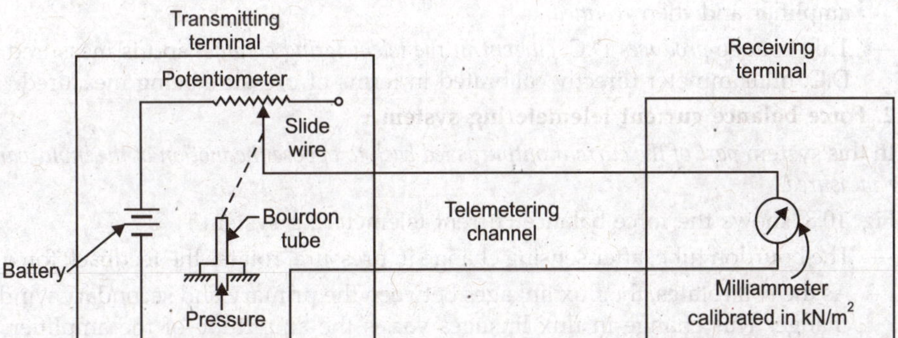


**Fig. 10.6.** Basic current telemetering system.

***Construction :***

— It consists of a slide-wire potentiometer in series with a battery.

— The slider is connected to the bourdon tube which measures pressure.

— At the receiving terminal, a multimeter is connected in series, which is calibrated in terms of pressure scale ($kN/m^2$).

***Working/Operation :***

— When the pressure in the system changes, the bourdon tube moves the sliding contact thereby changing the current at the transmitting terminal.

— This current passes to the receiving terminal through the pairs of wires (telemetering channel) and at the receiving terminal it is measured by the milliammeter.

These days, the following *improved versions* of this system are used :

1. Motion balance system; 2. Force balance system

**1. Motion balance current telemetering system :**

In this system slide-wire potentiometer is replaced by a position detector like inductor transducer or capacitor transducer.

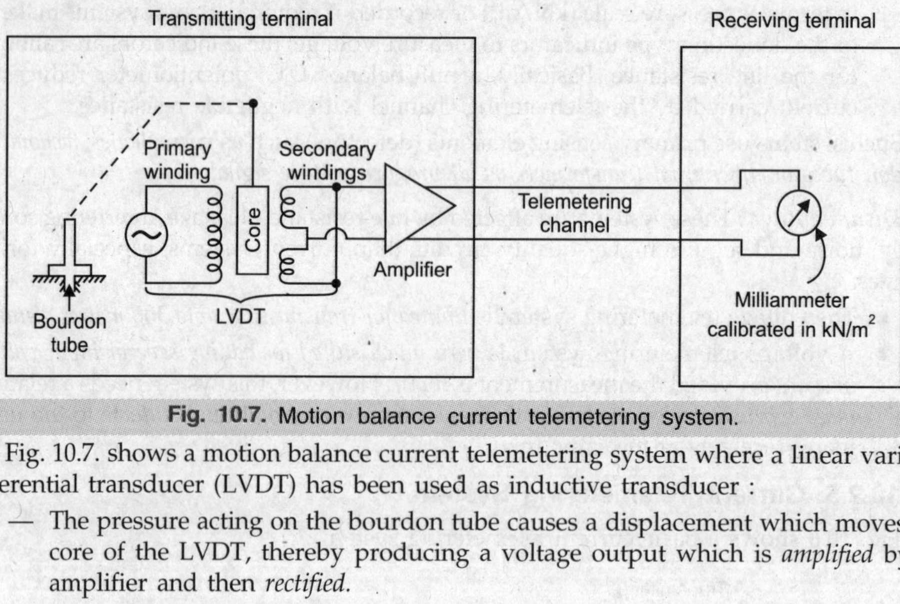


**Fig. 10.7.** Motion balance current telemetering system.

Fig. 10.7. shows a motion balance current telemetering system where a linear variable differential transducer (LVDT) has been used as inductive transducer :

— The pressure acting on the bourdon tube causes a displacement which moves the core of the LVDT, thereby producing a voltage output which is *amplified* by an amplifier and then *rectified.*

— This voltage *produces D.C. current in the telemetering channel* and is measured by a D.C. milliammeter directly calibrated in terms of pressure being measured.

**2. Force balance current telemetering system :**

In this system *part of the current output is fed back to oppose the motion of the input variable being measured.*

Fig. 10.8. shows the force balance current telemetering system :

— The bourdon tube, after sensing change in pressure, rotates the feedback force coil.

— As the coil rotates, the flux linkages between the primary and secondary windings change. This change in flux linkages varies the amplitude of the amplifier. The output is connected to the feedback force coil which produces a force opposing the bourdon tube input.

This system of current telemetering *increases the accuracy as small variations are required which result in better linearity.*

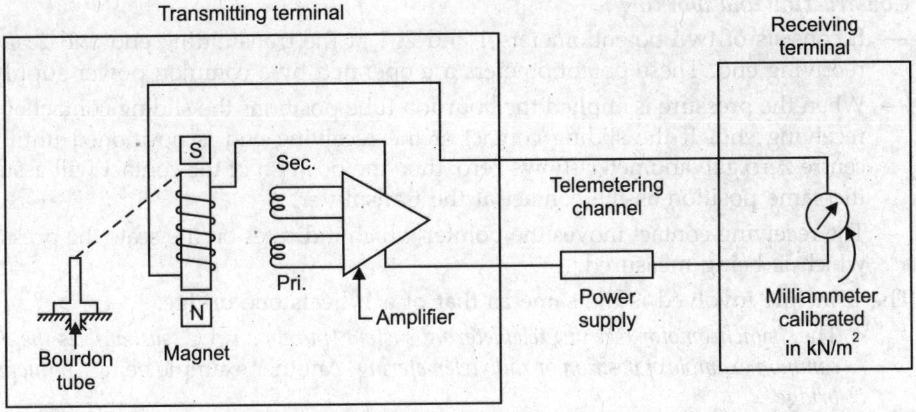


**Fig. 10.8.** Force balance current telemetering system.

***Advantages of current telemetering system :***

(*i*) The current systems can develop higher voltages than most voltage systems and, consequently, it can be made more immune to the effect of thermal and inductance voltages in the interconnecting leads as well as line resistance.

(*ii*) Simple D.C. milliammeters can be used with special calibration for line resistance.

(*iii*) Several receivers can be operated simultaneously.

(*iv*) The received signals can be added or subtracted directly.

(*v*) Changes in line resistance are compensated by basic feedback method.

(*vi*) The response of the system to an input change is almost instantaneous.

(*vii*) The energy level is adequately high to minimise the effects of extraneous voltages.

- This system is *also not suitable for long distance* since the current output is varied by means of an adjustable resistance in the line.

### 10.2.6. Position or (Ratio) Telemetering System

This system transmits and reproduces the measured variable by positioning variable resistors or other electrical components in a *bridge circuit* form so as to produce proportional changes at both the transmitter and the receiver ends.

Fig. 10.9 shows a position telemetering *bridge type* system :

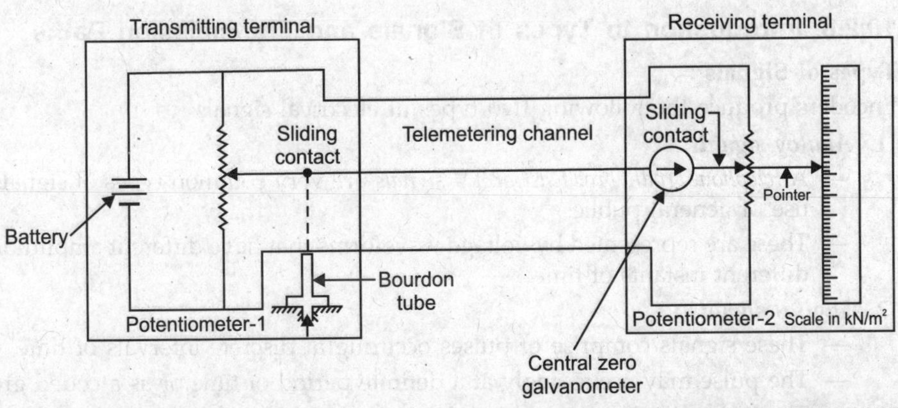


**Fig. 10.9.** Position telemetering bridge type system.

*Construction and working :*

— It consists of two potentiometers 1 and 2, 1 at the transmitting end and 2 at the receiving end. These potentiometers are operated by a common power supply.

— When the pressure is applied the bourdon tube positions the sliding contact at the receiving end. If the sliding contact at the receiving end is positioned until the centre zero galvanometer shows zero, then the position of the contact will assume the same position as the contact at the transmitter.

— The receiving contact moves the pointer which indicates on the scale the pressure which is being measured.

The principle involved is the same as that of a Wheatstone bridge.

- The *"synchromotor (selsyn) telemetering system (purely an A.C. system)" is the most common example of position or ratio telemetering*. Another example being *"inductance bridge"*.

— In this system *angular input displacement is converted into relative magnitude of three phase A.C. voltages.*

***Advantages :***

(*i*) Requires no intermediate amplifiers or conversions.

(*ii*) Relatively inexpensive.

(*iii*) Minimum moving parts, so the maintenance is low.

(*iv*) Instantaneous response.

(*v*) Power taken for their operation directly from the line.

***Limitation :*** These systems are *affected by excessive line resistance.*

## 10.2.7. Radio Frequency (RF) Telemetry System

In this telemetry system the *link between the transmitting end and the receiving end is established through the "radio links"* (there being no physical link).

*Example :* The controlling of aircraft on test flights, rockets and spacecrafts.

- The RF telemetry systems are *more suitable for transmission of data over distances more than 1 km.*
- In this system, along with certain band of the radio frequency spectrum, a microwave link above 4 MHz is also allocated. This is because the radio waves at these frequencies travel in straight lines with some repeaters, located every 30 km to 60 km, on high buildings or towers.

## 10.2.8. Introduction to Types of Signals and Transmission Paths

**Types of Signals :**

Encoders produce the following two types of electrical signals :

**1. *Analog signals :***

— A *telephone, radio broadcast or TV signals* are very common types of signals for use of general public.

— These are represented by voltage waveforms that have different amplitudes at different instants of time.

**2. *Digital signals :***

— These signals comprise of pulses occuring at discrete intervals of time.

— The pulse may occur singly at a definite period of time or as a coded group.

— These signals play a very important role in the transmission and reception of coded messages.

— Simplest types of digital signals are the Telegraph and Teleprinter signals but sometimes analog messages are also converted into digital form before being transmitted for certain reasons.

**Transmission Paths :**

The transmission of message can be carried out by means of the following :

1. **Transmission lines** ... termed as *Line Communications.*
2. **Radio waves** ... termed as *Radio Communications.*

**1. Line communications :**

- The most common form of line communication is *telephone* and *telex networks.*
- These are carried out using overhead lines on poles and also by use of buried cables.
- In UHF range and above line communications are carried out by the use of **wave guides.** The following points are worth noting :
  (*i*) In *wave guides* the signal propagation takes place by means of *radio waves* provided the cross-sectional dimensions of the proper wave guide rectangular hollow pipe maintain a proper relation with a wavelength of the wave.
  (*ii*) *Wave guides* are employed as transmission lines for microwave communication in the frequency range 2 GHz to 11 GHz.
  (*iii*) *Wave guides* do not transmit signals *below cut-off frequency* and can have band-width in excess of 20% of their operating frequencies.
  (*iv*) *Optical wave guides* have been developed in the form of glass fibres to carry light modulated signals at frequencies of the order of 100 THz (1 Tera hertz = $10^{12}$ hertz) from *lasers.* This will have a very vast capacity to carry signals.

**2. Radio communications :**

- In this type of communication propagation of signals through atmosphere is used.
- Non-directional antennas are employed for broadcast transmission and reception while directional antennas are employed in radio telephony and other point-to-point links.
- Radio waves are commonly termed as unbounded waves but generally they are bounded by the surface of earth and various ionospheric conducting layers that lie between a height of 50 km and 400 km above ground.

**Major Radio Frequency Bands**

| | *Name* | *Frequencies* | | |
|---|---|---|---|---|
| 1. | Very low frequency (VLF) | Below 30 kHz | ... | Radio location equipment |
| 2. | Low frequency (LF) | 30 kHz to 300 kHz | ... | Wartime radio navigation |
| 3. | Medium frequency (MF) | 300 kHz to 3 MHz | ... | Includes AM radio broadcast band |
| 4. | High frequency (HF) | 3 MHz to 30 MHz | ... | Radio |
| 5. | Very high frequency (VHF) | 30 MHz to 300 MHz | ... | Includes FM broadcast band and television VHF channels |

6. Ultra high frequency (UHF) 300 MHz to 3 GHz ... Includes television UHF channels
7. Super high frequency (SHF) 3 GHz to 30 GHz ... Satellite communications
8. Extremely high frequency (EHF) 30 GHz to 300 GHz ... Satellite communications

- In a *radio receiver* by providing an *RF amplifier* the following *advantages* are obtained :

(*i*) Signal-to-noise ratio is improved.
(*ii*) Availability of greater gain and better sensitivity.
(*iii*) Image frequency rejection is improved.
(*iv*) Spurious frequencies are prevented.
(*v*) Better selectivity.
(*vi*) Improved coupling of the receiver to the antenna.

## 10.2.9. Modulation and Demodulation

### 10.2.9.1. Introduction

Several signals out of different types of signals that are generally encountered in communication systems have frequency spectra that is not suitable for direct transmission especially when atmosphere is used as the transmission channel. In such a case, the frequency spectra of the signal may be translated by *modulating high frequency carrier wave with signal.*

In order to transmit and receive the intelligence (code, voice, music etc.) successfully, the following two processes are essential :

(*i*) Modulation; (*ii*) Demodulation.

### 10.2.9.2. Modulation :

**Definition :**

Modulation may be defined as follows :

- ***Modulation*** is *the process of combining the low-frequency signal with a very high-frequency radio wave called 'carrier wave (CW)'*. The resultant wave is called *modulated carrier wave.* This job is done at the transmitting station.

*Or*

***Modulation*** is *a process in electronic circuits by which the characteristics of one wave form (carrier) is modified by the variations in another wave (audio signal).*

*Or*

***Modulation*** is *the process of combining an audio-frequency (AF) signal with a radio frequency (RF) carrier wave (Fig. 10.10).* AF signal is called a *modulating wave* and the resulting wave produced is called *modulated wave.* During modulation, some characteristic of the carrier wave is varied in time with the modulating signal and is accomplished by combining the two.

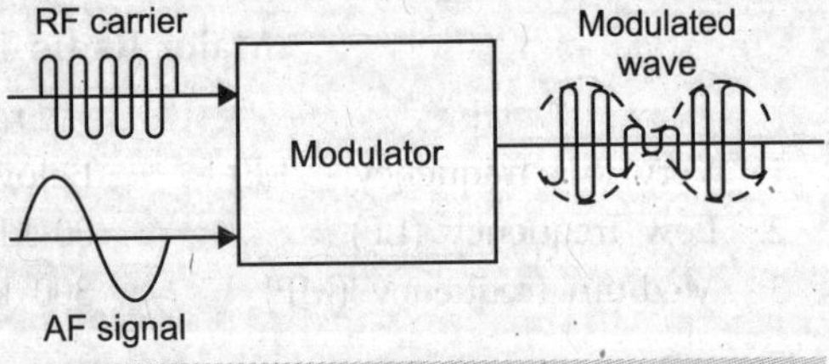

**Fig. 10.10**

### 10.2.9.3. Need of modulation

In electronic communication, the modulation is necessary for the following *reasons :*

1. Modulation increases operating range.

2. It reduces the size of transmitting and receiving antennas.
3. It permits transmission without wire.
4. It is extremely difficult to radiate low frequency signals through earth's atmosphere in the form of electromagnetic energy.

### 10.2.9.4. Methods of modulation

For a sinusoidal carrier wave, the mathematical expression is given as

$$e = E_c \sin(\omega_c t + \phi) = E_c \sin(2\pi f_c t + \phi) \quad ...(1)$$

Thus, the waveform can be varied by any of its following three factors or parameters :

(*i*) $E_c$ – The *amplitude* ;
(*ii*) $f_c$ – The *frequency* ;
(*iii*) $\phi$ – The *phase.*

Accordingly, there are three types of sine-wave modulations known as :

1. Amplitude Modulation (AM)
2. Frequency Modulation (FM)
3. Phase Modulation (FM).

**1. Amplitude modulation (AM ) :**

*The process by which the amplitude of a carrier wave is varied in accordance with the modulating signal is called* **amplitude modulation.**

The process of amplitude modulation is shown graphically in Fig. 10.11.

— For the sake of simplicity, the AF/message signal has been assumed sinusoidal [Fig. 10.11(*a*)].

— The carrier wave by which it is desired to transmit the AF signal is shown in [Fig. 10.11(*b*)].

— The resultant wave called *modulated wave* is shown in [Fig. 10.11(*c*)].

The function of the modulation is to mix these two waves in such a way that (*a*) is transmitted along with (*b*).

— The fluctuations in the amplitude of the carrier wave depend on the *signal amplitude* and the rate at which these fluctuations take place depends on the *frequency* of the audio signal.

— *All stations are broadcasting on the standard broadcast band (550 – 1550 kHz) use AM modulation.*

— **Percent modulation** *indicates the degree to which the AF signal modulates the carrier wave.*

— *Methods of amplitude modulation are :*

(*i*) Amplifier modulation.
(*ii*) Oscillator modulation.

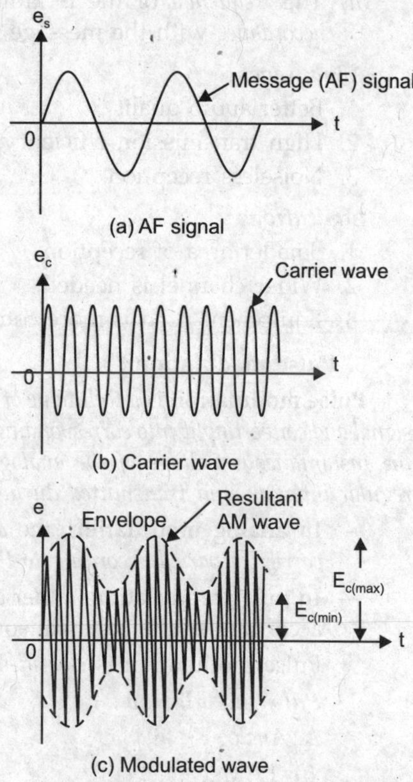


**Fig. 10.11.** Amplitude modulation.

**Limitations of amplitude modulation :**

Following are the *limitations* of amplitude modulation :

1. Smaller operating range.
2. Poor efficiency.
3. Poor audio quality.
4. Noisy reception.

**2. Frequency modulation (FM) :**

*The process by which the frequency of a carrier wave is varied in accordance with the modulating signal is called* **frequency modulation.**

The process of frequency modulation is shown graphically in Fig. 10.12.

The three waves shown in the figure are : (*a*) Signal; (*b*) Carrier; (*c*) The resultant frequency modulated wave.

When a signal of frequency $f_s$ is modulated with carrier wave of frequency $f_c$, a resultant modulated wave is produced.

The following points are worth noting :

(*i*) The *amplitude* of the modulated wave is the *same* as that of the carrier wave.

(*ii*) The *frequency* of the modulated wave *varies in accordance* with the message signal.

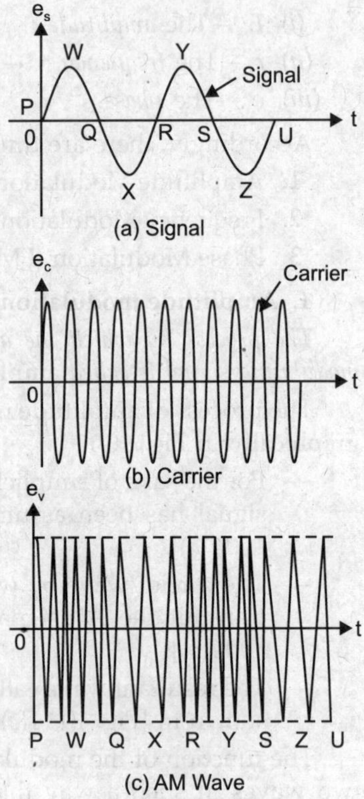


Fig. 10.12. Frequency modulation.

*Advantages :*

1. Better audio quality.
2. High transmission efficiency.
3. Noiseless reception.

*Limitations :*

1. Smaller area of reception.
2. Wider channel is needed.
3. Equipment used is more complex and costly.

**3. Pulse modulation :**

**Pulse modulation** is *a technique of modulating the analog signal and converting it into corresponding values. In this process the instantaneous voltage of the analog signal is sampled at regular intervals and transmitted during these sampling periods only.*

— **In analog modulation**, the *amplitude or frequency or phase (any one parameter) of carrier is varied according to the instantaneous voltage of signal.*

— **In pulse modulation**, either amplitude or width or position of the pulse is varied according to the instantaneous voltage of signal (except for PCM).

- Pulse modulation is *classified* as follows :

***Pulse modulation :***

A. *Analog :*

(*i*) PAM

(*ii*) PTM → PDM, PWM
→ PPM

B. *Digital* :

(*i*) PCM, (*ii*) Delta-modulation.

- **Sampling theorem** : This theorem was developed by *Nyquist*.

It states : *"If sampling rate is any pulse modulating system exceeds twice the maximum signal frequency, the original signal can be constructed at the receiver with minimum distortion."*

— If $f_m$ is the maximum signal frequency then sampling rate is greater than $2f_m$.

— The minimum samplings rate is called the *Nyquist rate.*

— *The reciprocal of the sampling rate* (or sampling frequency) is called ***sampling line*** or ***Nyquist interval.***

- **Pulse Amplitude Modulation (PAM) :**

— PAM can be *generated* by using an AND gate.

— PAM can be *demodulated* by passing through low pass filter with cut-off frequency as the highest signal frequency.

— Generally FM is used to modulate the pulses so that it can be transmitted; such a system is called PAM-FM.

- **Pulse Width Modulation (PWM) :**

— It is also called Pulse Duration Modulation (PDM) or Pulse Length Modulation (PLM).

— It can be generated using a monostable multivibrator.

— It can be demodulated by feeding the PWM signal to an integrating circuit.

— In PWM the width of the pulse is changed in accordance with the instantaneous value of the modulating signal with amplitude remaining constant.

— The most oftenly used integrating circuit is the loudspeaker itself.

- **Pulse Position Modulation (PPM) :**

— It can be generated in the similar way as PWM but the pulse width is kept *constant* from the starting point of occurrence of pulse.

— It can be *demodulated* by converting into PWM using a flip-flop.

— In PPM the position of the pulse or the time of occurrence of the pulses is changed in accordance with the instantaneous magnitude of the modulating signal.

- **Pulse Code Modulation (PCM) :**

— In this type of modulation, digital equivalents of the instantaneous voltage levels of the signals are transmitted in the form of pulses.

— The total amplitude range which the signal may occupy is divided into a number of standard levels; this process is called *Quantizing*. The level actually set is the nearest standard level. The *quantization levels depend upon number of bits per sample.*

— PCM encoder functions similar to A/D converter; PCM decoder functions similar to D/A converter.

#### 10.2.9.5. Digital modulation techniques

In digital communications, the modulating signal consists of binary data or an *M*-ary encoded version of it. This data is used to modulate a carrier wave (usually sinusoidal) with fixed frequency. In fact the input data may represent the digital computer outputs or PCM waves, generated by digitizing voice or video signals. The channel may be a telephone channel, microwave radio link, satellite channel or an optical fiber.

In ***digital communication***, *the modulation process involves switching or keying the amplitude, frequency or phase of the carrier in accordance with the input data.* Thus, there are three basic modulation techniques for the transmission of digital data; they are known as :

(*i*) Amplitude-shift keying (ASK).

(*ii*) Frequency-shift keying (FSK).

(*iii*) Phase-shift keying (PSK).

The above techniques can be viewed as special cases of amplitude modulation, frequency modulation and phase modulation respectively.

***Types of digital modulation techniques :***

Various types of digital modulation techniques are :

I. *Coherent digital modulation techniques :*

1. Coherent binary modulation techniques.
2. Coherent binary amplitude shift keying or on-off keying.
3. Coherent demodulation of binary ASK.
4. Binary phase shift keying (BPSK).
5. Coherent binary frequency shift keying (BFSK).

II. *Non-coherent binary modulation techniques :*

1. Non-coherent binary ASK.
2. Non-coherent detection of FSK.
3. Differential phase shift keying (DPSK).
4. Quadrature phase shift keying (QPSK).

**10.2.9.6. Demodulation or detection**

*The process of extracting the low frequency modulating signal from the modulated wave is known as* ***demodulation*** *or* ***detection.***

- The demodulation of an *'AM wave'* involves *two operations :*
  (*i*) Rectification of the modulated wave ;
  (*ii*) Elimination of the RF component of the modulated wave.
- The demodulation of an *'FM wave'* involves *three operations :*
  (*i*) Conversion of frequency changes produced by modulating signal into corresponding amplitude changes ;
  (*ii*) Rectification of the modulating signal ;
  (*iii*) Elimination of RF component of the modulated wave.

**10.2.9.7. Noise**

- **Noise** may be defined, in an electrical sense, *as any extraneous form of energy tending to interfere with the proper and easy reception and reproduction of those signals which it is desirable to receive.*
- **Noise Power :**
  — Noise power,
    (*i*) is *directly proportional to temperature ;*
    (*ii*) is *directly proportional to the bandwidth of system ;*
    (*iii*) is *independent of resistance value.*
  — Unit of *noise power is watts.*
  — Units of *noise power density are watts/hertz.*

— The *average thermal noise power is constant as long as the bandwidth is fixed.*

— ***Thermal noise voltage*** *is proportional to temperature, bandwidth and resistance.*

- **Noise figure :** *It is the ratio of S/N ratio at input to S/N ratio at output.* Ideal noise figures is '1' or 0 dB. Noise figure is *always greater than 1. Higher the noise figure of the amplifier, noiser the amplifier.*
- **Classification of noise :**

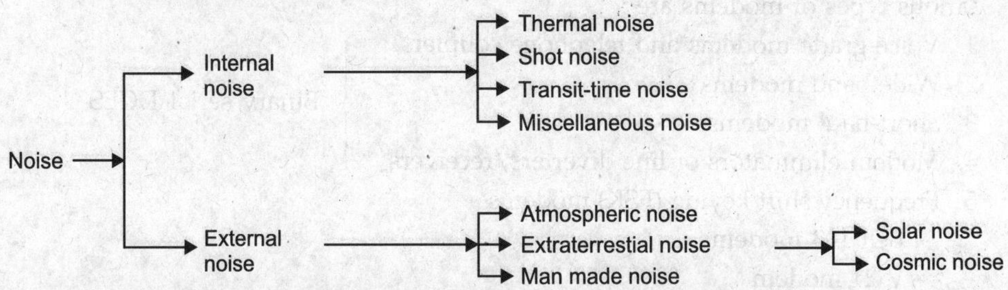


All receivers have their basic components consisting of tubes or transistors. The different types of noise present in these devices are thermal noise, shot noise, partition noise etc.

— In superheterodyne receivers, mixers are much noiser than amplifiers. It is so, because, the plate current of the partition noise is greatly increased in mixers.

— At short wave frequencies, the image frequency rejection is very poor and this introduces noise.

— A low value of transductance also helps in increasing the noise.

Thus, the effect of noise is maximum in mixers of all the receivers.

### 10.2.9.8. Modems

- Modem is an *acronym for MODulator DEModulator.*
- *A* **modem** *is a device that converts data from digital computer signals to analog signals that can be sent over a phone line.* This is called *"modulation"*. The *analog signals are then converted back into digital data by the receiving modem.* This is called *"demodulation"*.
- Fig. 10.13 shows a modem interfacing block diagram :

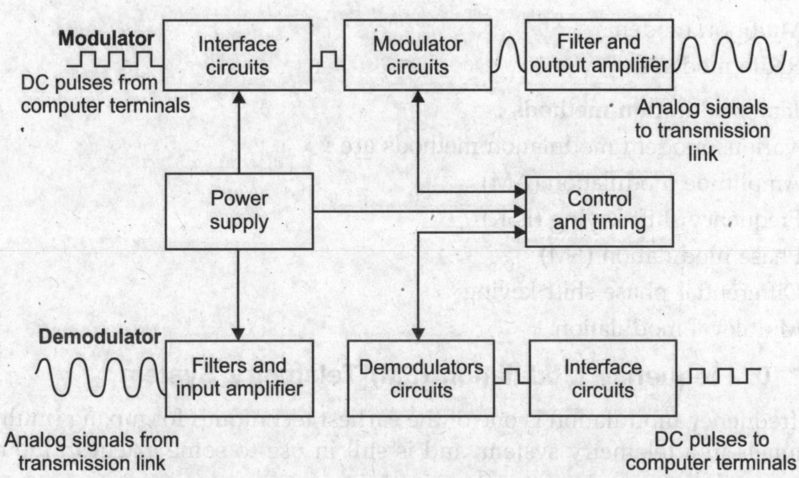


**Fig. 10.13.** Modem interfacing block diagram.

— A modem is fed with digital information in the form of ones and zeros, from the CPU (central processing unit). The modem then analyzes this information and converts it into analog signals, that can be sent over a telephone line.

— Another modem then receives these signals, converts them back into digital data, and sends the data to the receiving CPU.

**Types of Modems :**

Various types of modems are :

1. Voice-grade modems and telephone couplers
2. Wide-band modems
3. Short-haul modems
4. Modem eliminators or line diverters/receivers

(Items 1–4: Binary serial DCES)

5. Frequency shift keying (FSK) modems
   — Bell 103 modem
   — V 21 modem
   — Bell 202 modem
6. Phase shift keying (PSK) modems
   — 201 modem
   — V.26 modem
   — Bell 212 modem
   — V.22 modem
   — V.22 bis modem
   — V.27 modem
   — V.29 modem
   — V.32 modem
   — V.32 bis modem
   — V.FC modem
   — V.34 modem
7. Data modems
8. Smart modems
9. Multiport modems
10. Radio modems.

**Modem modulation methods :**

The various modem modulation methods are :

1. Amplitude modulation (AM)
2. Frequency shift keying (FSK)
3. Phase modulation (FM)
4. Differential phase shift keying
5. Multilevel modulation.

### 10.2.10. Frequency Modulation (FM) Telemetry System

The frequency modulation is one of the earliest techniques for mixing (multiplexing) data channels in a telemetry system, and is still in use to some extent in the telemetry field.

Fig. 10.14 shows a basic frequency modulation (FM) telemetry system :

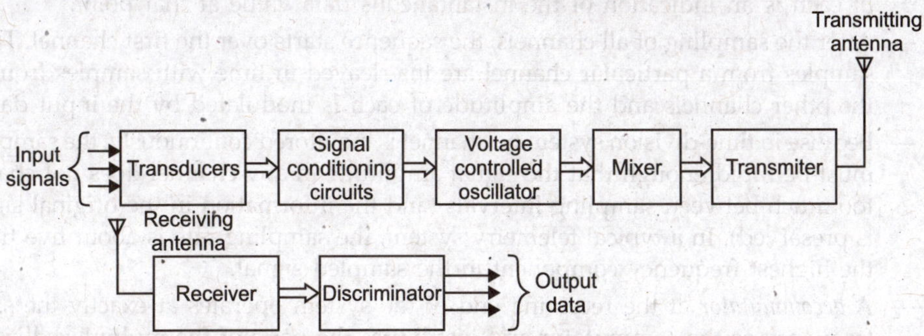


**Fig. 10.14.** Basic FM telemetry system.

— The output of each transducer/signal conditioner modulates the frequency of a voltage-controlled subcarrier oscillator. Many oscillators, each operating in a dedicated part of the frequency spectrum, are mixed for radio transmission. As each voltage controlled oscillator is assigned with a separate frequency of the frequency spectrum, so each signal can be modulated without interfering with other signals.

— At the receiving station, a FM demodulator (usually called *discriminator*) is tuned to the frequency of each subcarrier. Thus, when the measurement value changes at the source, the output voltages of the discriminator change accordingly.

### 10.2.11. Pulse Amplitude Modulation (PAM) Telemetry System

The primary limitation of the FM telemetry system is that the *capacity of channels offered is less.* Thus, to have still higher capacity of channel within a linked bandwidth, the PAM telemetry system is widely used which makes use of the *"time division multiplexing"* technique.

Fig. 10.15 shows a basic time division multiplexing system :

— All channels use same frequency band of the allotted frequency spectrum but *not at the same time.*

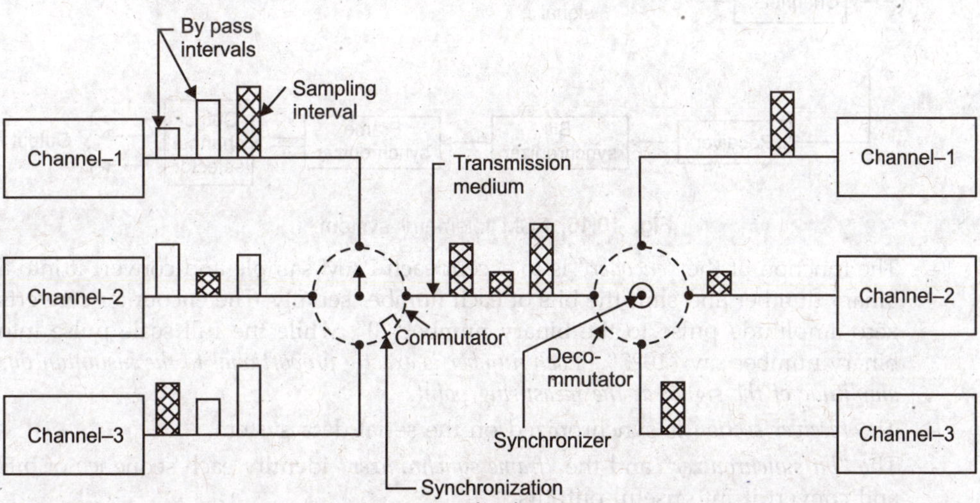


**Fig. 10.15.** Basic time division multiplexing system.

- The *commutator* samples the signal in each channel in sequence, and the amplitude of each is an indication of the instantaneous data value at that point.
- After the sampling of all channels, the sequence starts over the first channel. Thus, samples from a particular channel are interleaved in time with samples from all the other channels and the amplitude of each is modulated by the input data.
- Because in time-division system no channel is monitored continuously, the sampling must be rapid enough that the signal amplitude in any channel does not change too much between sampling intervals (and the information in the original signal is preserved). In a typical telemetry system, the sampling rate is about five times the highest frequency component in the sampled signal.
- A *decommutator* at the receiving end of the system operates at exactly the same frequency as the commutator and distributes the parts of the multiplexed signal to the proper output channels.

Because a time-division system is *based on precise timing*, it is of significant importance that the *demodulator be synchronized exactly with the commutator.*

### 10.2.12. Pulse Code Modulation (PCM) Telemetry System

This system is so named, since the measured data is represented in the form of *binary weighted codes.*

*This system also makes use of time division multiplexing system* (as used in PAM); however, in this system an *encoder is used after the commutator stage.*

Fig. 10.16 shows the block diagram of the PCM telemetry system :

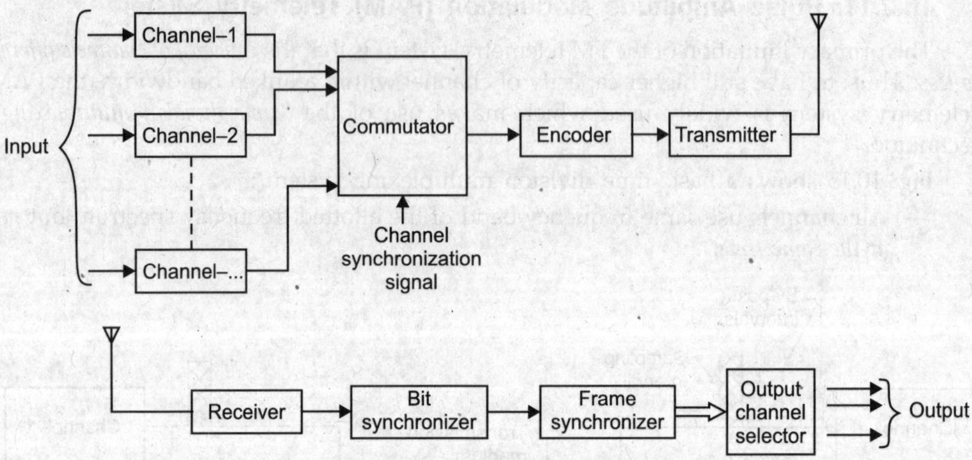


**Fig. 10.16.** PCM telemetry system.

- The function of the *"encoder"* is to accept each PAM sample and convert it into a binary number and shift the bits of each number serially. The encounter converts zero amplitude pulse to the binary number "0", while the full scale pulse into binary number say "1023". *Each number is exactly proportional to the instantaneous amplitude of the signal at the measuring point.*
- The *receiver section* is synchronized on the serial data system.
- The *"bit synchronizer"* and the *"frame synchronizer"* identify each sequence of bits and convert it into useful outputs.

## Comparison of FM, PAM and PCM

| S. No. | Aspects | FM | PAM | PCM |
|---|---|---|---|---|
| 1. | *Accuracy* | Poor | Poor | Excellent |
| 2. | *Efficiency in use* | Medium | Best | Worst |
| 3. | *Percentage of use in practical systems* | 20% | 15% | 65% |

## HIGHLIGHTS

1. *Telemetry* is defined as a technology which allows users to collect information from inaccessible and inconvenient locations and to transmit it to the accessible places to process, record and display the information in presentable form.
2. The RF telemetry systems are more suitable for transmission of data over distances *more* than 1 km.
3. In pulse code modulation (PCM), digital equivalents of the instantaneous voltage levels of the signals are transmitted in the form of pulses.
4. A *'modem'* is a device that converts data from digital computer signals to analog signals that can be sent over a phone line. This is called modulation. The signals are then converted back into digital one by receiving modem. This is called 'demodulation'.

## OBJECTIVE TYPE QUESTIONS

1. The FM telemetry as compared with AM telemetry requires a channel that is
   (*a*) smaller than what is required for AM telemetry
   (*b*) equal to that of AM telemetry
   (*c*) 10 times of that required for AM telemetry
   (*d*) 100 times of that required for AM telemetry.
2. FM systems operate at
   (*a*) VHF and UHF (*b*) only VHF
   (*c*) only MF (*d*) MF and HF.
3. FM systems as compared to AM systems are
   (*a*) less affected by noise
   (*b*) equally affected by noise
   (*c*) more affected by noise
   (*d*) highly affected by noise since they operate at VHF and UHF.
4. Time division multiplexing requires
   (*a*) constant data transmission
   (*b*) transmission of data samples
   (*c*) transmission of data at random
   (*d*) transmission of data of only one measurand.
5. Modem is an acronym of
   (*a*) modulation (*b*) demodulation
   (*c*) modulation and demodulation (*d*) all of the above

6. Time division multiplexing is used when the data to be transmitted
   (a) is slow changing
   (b) has a small bandwidth
   (c) is slow changing and has a low bandwidth
   (d) is fast changing and has a high bandwidth.

## ANSWERS

1. (c) 2. (a) 3. (a) 4. (b) 5. (c) 6. (c)

## THEORETICAL QUESTIONS

1. Define the terms "Data transmission" and "Telemetry'.
2. What is the difference between a 'transmitter' and a 'telemeter' ?
3. Explain briefly any two of the following :
   (i) Mechanical transmission.
   (ii) Hydraulic transmission.
   (iii) Pneumatic transmission.
   (iv) Magnetic transmission.
4. Discuss briefly the working of a general telemetering system, with the help of a block diagram.
5. How are telemetry systems classified ?
6. List the advantages and disadvantages of land line telemetry system.
7. Describe briefly the following :
   (i) Voltage telemetering system.
   (ii) Current telemetering system.
8. Explain briefly 'Force balance current telemetering system'.
9. State the advantages of current telemetering system.
10. Discuss the working of a 'position telemetering system' with the help of a neat diagram.
11. Write a short note on 'RF telemetry system'.
12. What do you mean by the terms 'Modulation' and 'Demodulation' ?
13. Explain briefly the following :
    (i) Amplitude modulation.
    (ii) Frequency modulation.
14. What do you mean by the following :
    PAM; PWM; PPM; PCM.
15. What is a modem? Explain.
16. Explain briefly the following telemetry systems :
    (i) Frequency modulation (FM) telemetry system.
    (ii) Pulse amplitude modulation (PAM) telemetry system.

CHAPTER

# 11 Display Devices and Recorders



## 11.1 INTRODUCTION

The main purpose of any measurement system is to provide information concerning the state and condition of the physical phenomenon being investigated. The measuring systems may be activated either directly from the measuring means (*e.g.*, bellows, pressure spiral etc., to which indicating pointer is attached directly through lever and leverage system) or by means of servo-operated system (null-balance system which incorporates a feedback circuit in a closed loop). The last stage of a measurement system is the *data presentation stage;* if the results of the system are meaningful they must be displayed for *instant observation by a display device* or *for storage for observation at a later stage by a recorder* (The data presentation devices may be called as *"output devices"*).

The following factors decide about the *choice between the display devices and recorders* :

(*i*) The information content of the output.

(*ii*) The expected use of the output.

The output devices may be *categorized* as follows :

1. Single number output devices.
2. Time domain output devices.

**1. Single number output devices :**

Such devices indicate the value of some particular quantity under condition such that the value to be measured can be regarded as time variant over the timer interval during which measurements are made; thus a *single number will represent measurement.*

*"Indicating instruments"* and *"Digital display units"* belong to this class.

**2. Time domain output devices :**

The indicating instruments or the digital display (suitable only when the *output varies at a very slow rate*) do not serve the purpose *when the values of the quantity are to be taken as a function of time.*

*For fast changing outputs* (where signal waveform or shape is the desired information) ... *"Cathode ray oscilloscope (CRO)* is used.

For keeping a *permanent record of the variation of the output with time ..... "Cathode ray tube photographs", direct writing recorders" strips chart recorders," magnetic tape-recorders* etc. are used.

The machine interpretable outputs can be had from :

— Magnetic tapes;

— Punched paper tapes;

— Punched cards;

— Pulsed signals.

- The information available from an instrument may take the following forms :

(*i*) *Quantitative information* (*e.g.*, angular spread in r.p.m; force in newtons).

(*ii*) *Qualitative information* (*e.g.*, the approximate value or direction of change of some variable, a check reading).

(*iii*) Status information (*e.g.*, On/Off, in/out).

(*iv*) Alphanumeric and symbolic information (*e.g.*, the labels and instructions; letters A to Z, the numerals 0 to 9 punctuation marks and various other simple symbols can be generated and intepreted)

- *A good display, functionally, is one which permits the best combination of speed accuracy and sensitivity when transferring the necessary information from the instrument to the operator.*

## 11.2 ELECTRICAL INDICATING INSTRUMENTS

Qualitatively the electrical indicating instruments are widely used for measurements of current, voltage, resistance and power. These instruments can be *classified* as follows :

1. Analog instruments
2. Digital instruments.

**Analog** type of meters use scale and needle (pointer) type arrangement to display the value of the measured parameter. *The measured parameter is converted into electrical signal which further actuates some electromechanical device to move the pointer continuously as the parameter changes.*

**Digital** type meters *indicate the reading in exact numerals.*

### 11.2.1. Analog Instruments

**Analog indicators** *are basically current-operated devices, working on the "d'Arsonval type galvanometer principle"*. The device consists essentially of a circular or rectangular coil of a large number of turns of fine wire, suspended between the poles of a permanent magnet. The coil is mounted in position with a jewel bearing and is coupled to a pointer-indicator moving over a graduated scale. The restoring force for equilibrium is provided through a control spring. The sensitivity and the torque produced depend upon the coil dimensions, number of turns, flux density, and the control spring constant. The resulting deflection (δ) is given as :

$$\delta = KIB \cos \theta \qquad \text{...(11.1)}$$

where, $K$ = Constant of proportionality,

$I$ = Current through the coil,

$B$ = Flux density, and

$\theta$ = Angle between the plane of the coil and magnetic field when no current is flowing.

*The basic movement can be adapted to measure D.C. voltages with a series resistor and to measure A.C. voltages with a rectifying diode and resistor.*

**Pointer-scale analog indicators :**

In analog instruments the value of the measured parameter is indicated by positioning of the indicating pointer against a calibrated scale. The purpose can be achieved either by moving the pointer with relation to a stationary scale (fixed-scale moving-pointer indicators) or the scale may be moved with relation to a fixed reference (moving-scale fixed-pointer indicators).

**1. *Single-point indicators :***

The *fixed-scale and movable-pointer indicators,* available in a variety of forms, are shown in 11.1. Fig. 11.2 shows the fixed-pointer and movable-scale indicators.

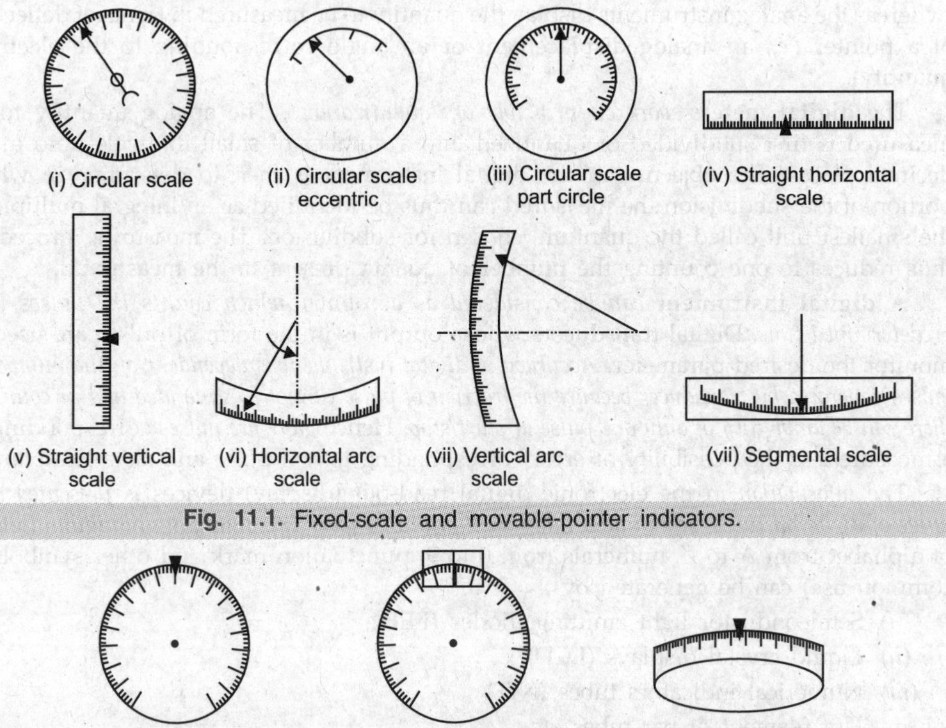


**Fig. 11.1.** Fixed-scale and movable-pointer indicators.

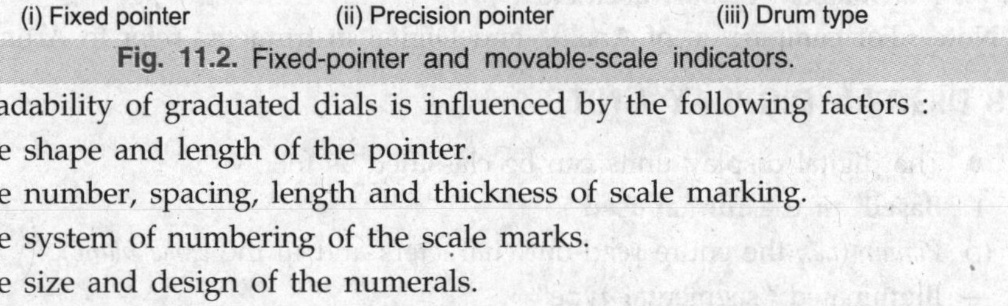


**Fig. 11.2.** Fixed-pointer and movable-scale indicators.

The readability of graduated dials is influenced by the following factors :

- The shape and length of the pointer.
- The number, spacing, length and thickness of scale marking.
- The system of numbering of the scale marks.
- The size and design of the numerals.

**2. *Multi-point, multi-pointer and multi–range indicators :***

***Mult-point indicator.*** In this system the indicator pointer can be connected to a number of inputs, one at a time with the help of a selector switch.

The selector switch may be operated either manually or automatically after a pre-determined time. The observed reading is multiplied by a factor corresponding to the particular measurand.

Generally such systems are confined where measurable variables are of electrical signals as the selection is accomplished by switching electrical circuits. However, gas selector switches also exist which connect one gas pipe at a time to the measuring instrument and are well designed to avoid leakage of gas.

***Multi-pointer indicator.*** This type of indicator contains more than one number of pointers and above each point the identification number of the medium being measured is marked. *Usually this arrangement is used in recorders and not in indicators.*

***Multi-range indicators.*** An instrument with multi-range indicators has *different scales for different ranges;* the choice of a particular scale is made by a selector switch.

### 11.2.2. Digital Instruments

The **digital instruments** *indicate the value of the measurand in the form of decimal number"* (whereas the analog instruments display the quantity to be measured in terms of deflection of a pointer *i.e.,* an analog displacement or an angle corresponding to the electrical quantity).

The **digital meters** *work on principle of "quantization".* The analog quantity to be measured is first subdivided or quantized into a number of small intervals upto many decimal places. The objective of the digital instrument is then to determine in which portion of the subdivision the measured can thus be identified as an integral multiple of the smallest unit called the quantum, chosen for subdivision. The measuring procedure thus reduces to one counting the number of quanta present in the measurand.

A **digital instrument** *can be considered as a counter which counts the pulses in a predetermined time.* Digital transducers whose output is in the form of pulses are used to monitor the desired parameter. *Accuracy of digital instrument is dependent on the number of pulses generated by transducer because the fraction of pulse cannot be generated and in counting there can be ambiguity of only one pulse at start/stop.* Hence *more are pulses* corresponding to a measure less the possibility of error corresponding to one pulse and *more the accuracy.*

The *information* in the electronic digital read-out (display) devices is *presented as a series of digits on tubes, screen or printed on a piece of paper.* The relevant characters (letters of alphabet from A to Z, numerals from 0 to 9, punctuation mark and other symbols in common use) can be generated by :

(*i*) Semiconductor light emitting diodes (LED)

(*ii*) Liquid crystal displays (LCD)

(*iii*) Numerical indicators tubes (NIT)

(*iv*) Hot filament or bar tubes etc.

**Note :** For comparison of Analog and Digital instruments refer to Article 1.13.

## 11.3 DIGITAL DISPLAY UNITS

- The digital display units can be classified as follows :

**1. Based on the format used :**

(*i*) *Planar* (*i.e.,* the entire read-out characters and in the *same plane*)

— Illuminated "segmental type"

— Illuminated "dot matrix type"

— Displays using *rear projection* and *gaseous discharge type* segmental displays.

(*ii*) *Non-planar :* Here the characters are displayed in *different planes.*

— *Gaseous discharge tubes* like ***nixies.***

— *Displays using "illuminated lucite sheets."*

In general displays may be *classified* as follows :

**2. Based on methods of conversion of electrical data into visible light :**

(*i*) *Active displays* :

— CRTs

— Gas discharge plasma

— LEDs etc.

(*ii*) *Passive displays* :

— Light controllers

— LCDs

— EPIDs etc.

**3. Based on the display size and physical dimensions :**

(*i*) *Symbolic displays* :

— Alpha numeric

— Nixie tubes

— LEDs etc.

(*ii*) *Console displays* :

— CRTs

— LEDs etc.

(*iii*) *Large screen display* :

— Enlarged projection system

**4. Based on resolution and legibility of characters :**

(*i*) *Simple single element indicator.*

(*ii*) *Multi-element displays.*

**5. Based on the applications :**

(*i*) *Analog displays* :

— Bar graph displays (CRT)

(*ii*) *Digital displays* :

— Nixies

— Alphanumeric

— LEDs etc.

In digital electronic field, the *commonly used displays are* :

1. Cathode ray tube (CRT)
2. Light emitting diode (LED)
3. Liquid crystal display (LCD)
4. Gas discharge plasma displays (cold cathode display or *Nixies*)
5. Electro-luminescent (EL) displays
6. Incandescent display
7. Electrophoretic image displays (EPID)
8. Liquid vapour display (LVD).

### 11.3.1. Cathode Ray Tube (CRT)

**Cathode ray tube (CRT)** is the *usually commonly used 'read out system' for displaying large quantities of alphanumeric data.* It forms the basis of CROs and TV systems conventionally.

- The generation system of characters on CRTs require relatively simple electronic circuitry. A typical CRT display has *easy facilities for the control of digit size* by controlling the deflection sensitivity of the system. The number of characters displayed can be changed with the help of time shared deflection and modulator circuits.
- The intensity and brightness can be realized with different gray scales and the display can have different colours depending on the phosphor used in the screen; generally the phosphor is chosen to be green or white.
- CRTs of storage type facilitate showing a stationary pattern on the screen without flickering display. It is also possible to retain the pattern for a long time independent of the phosphor persistence.

### 11.3.2. Light Emitting Diode (LED)

A P-N junction can absorb light energy and produce electric current. The opposite process is also possible, that is a junction diode can emit light. The emission of light occurs under forward bias condition due to recombination of electrons and holes. The emitted light may be visible or invisible.

A P-N *junction diode, which emits light when forward biased is known as a* **light emitting diode (LED)**.

The amount of light output is *directly proportional to the forward current*. Thus, higher the forward current, higher is the light output.

Fig. 11.3 (*a*) shows the schematic symbol of a light emitting diode. The arrows pointing away from the diode symbol represent the *light*, which is being transmitted away from the junction.

Fig. 11.3 (*b*) shows the basic structure of a light emitting diode.

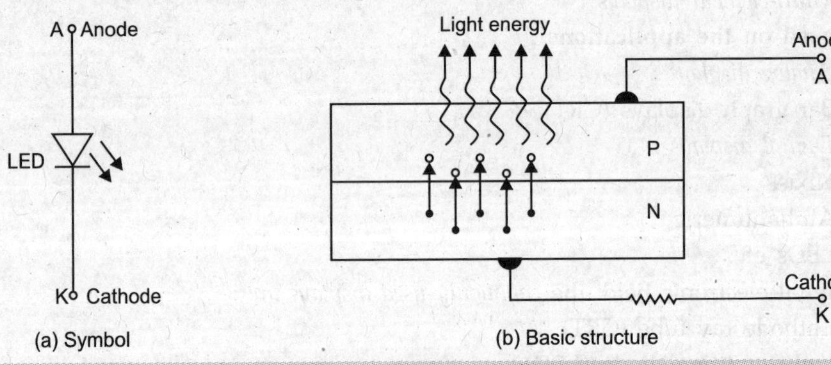


**Fig. 11.3.** Light emitting diode (LED).

In a forward based *P-N* junction, free electrons from *N*-type material diffuse into P-region. Once in *P*-region these free electrons encounter holes and eventually recombine. Thus the free conduction electron fills a vacancy in covalent structure and thus becomes a valence electron. In doing so the electron loses a certain amount of energy as it jumps form conduction band to the valence band. In Si or Ge *diode, the energy that recombining electrons lose is dissipated in the form of heat. But if other semiconductor material such as gallium arsenide and gallium phosphide are used to form P-N diode, the energy lost by recombining electrons is given off in the form of* **light energy**.

- The buffer requirements for CRT are complex and have to fulfil many functions such as :

(*i*) The buffer must translate data from digital to analog form, which determines the beam position and intensity;

(*ii*) The buffer must refresh the image on the tube face in excess of thirty times a second to avoid flicker;

(*iii*) The buffer must maintain sufficient accuracy to hit any one place out of about five million possible places on the tube face;

(*iv*) To reduce the amount of data needed from the computers to drive the roster and vector CRTs, the buffer must be able to fill in details of the intensity and movement needed for the continual control of the beam and to combine this with data needed to refresh the image.

— Diodes made from gallium arsenide (GaAs) emit infrared radiation invisible to eyes and such diodes are referred to as IRED—Infrared emitting diodes.

Fig. 11.4 shows two curves used to determine LED **operating characteristics**.

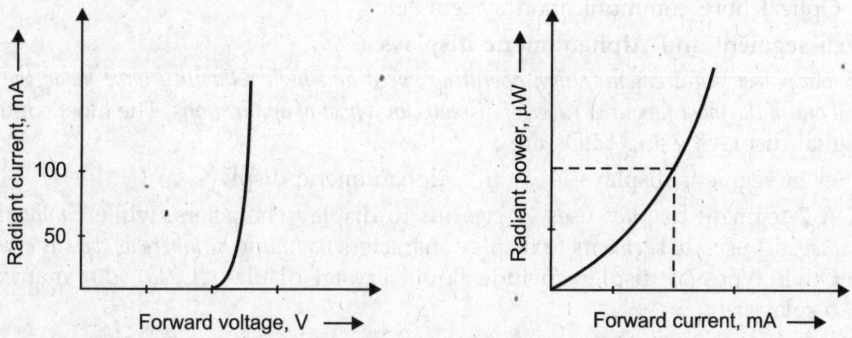


**Fig. 11.4.** Operating characteristics-LED.

— Fig. 11.4 (*a*) is forward bias V-I curve for a typical IRED, the type used in burglar alarms. Forward bias of around 1 V is required to produce significant forward current.

— Fig. 11.4 (*b*) is a plot of radiant output power *vs* forward current. The radiant output power is rather small (μW) and *indicates a very low efficiency of electrical to radiant energy conversion.*

- They have *fast switching characteristics.*

**Advantages :**

1. LEDs are *miniature in size* and they can be stacked together to form numeric and alphanumeric display in high density matrix.
2. The light output from an LED is a function of current flowing thought it. Therefore, intensity of light emitted from LEDs can be *smoothly controlled.*
3. LEDs have a *high efficiency* as emitted of electromagnetic radiation. They require moderate power for their operation. They are useful where miniaturization in the low D.C. power are important.
4. They are very useful where *dynamic operation* of large number of arrays is involved.
5. *Economical* and *high degree of reliability.*
6. Being rugged, they *can withstand shocks and vibrations.*
7. *Long life.*
8. They *occupy small area.*
9. *Available in different colours* like red, yellow,. green and amber.

**Disadvantage :**

LEDs are not suited for large areas display, primarily because of their high cost.

**Applications :**

Since LEDs operate at voltage levels 1.5 V to 3.3 V, they are highly compatible with solid state circuitry. Their *uses* include the following :

(*i*) Panel indicator.
(*ii*) Digital watches.
(*iii*) Multimeters.
(*iv*) Calculators.
(*v*) Itercoms.
(*vi*) Switch boards.
(*vii*) Burglar-alarm systems.
(*viii*) Optical fibre communication systems etc.

**Seven segment and Alphanumeric displays :**

*The low power requirements, high operating speed and high reliability have made the LEDs in general one of the most favoured types of displays for a host of applications.* The most commonly used digital displays using LEDs are :

(*i*) Seven segment displays; (*ii*) Alphanumeric displays.

- A 7-segment display uses 7 *segments* to display characters, while *"alphanumeric display"* uses 14 *segments* to display characters *including numbers as well as alphabets.* Other types of display include liquid crystal display (LCD), dot matrix type displays etc.

**1. Seven segment display :**

- In segmental displays, it is usual to employ a single LED for each segment [Fig. 11.5 (*a*)].
- For conventional seven segment LED displays [including the decimal point (DP) *i.e.,* the 8th segment], the wiring pattern is simplified by making one terminal common to all LEDs and other terminals corresponding to different segments. The terminals can be either of common anode (CA) form or common cathode (CC) form, shown in Fig. 11.5 (*b*) and (*c*).

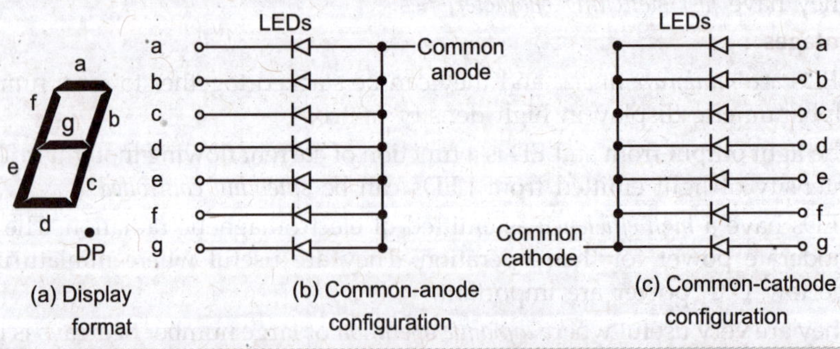


**Fig. 11.5.** LED seven segment display.

- The common cathode LEDs have to be driven by "active high" decoder ICs, and the common-anode LEDs have to be driven by "active-low" decoders. Fig. 11.6 shows this type of decoder connected to a seven-segment display digit. It may be noted that series resistors are included for each of the LEDs corresponding to the seven segments. Further, an *additional LED corresponding to the decimal point (DP) is also provided in the display,* which again has a current limiting series resistance. The typical value of the series resistance is 150 Ω for a segment current of about 20 mA.

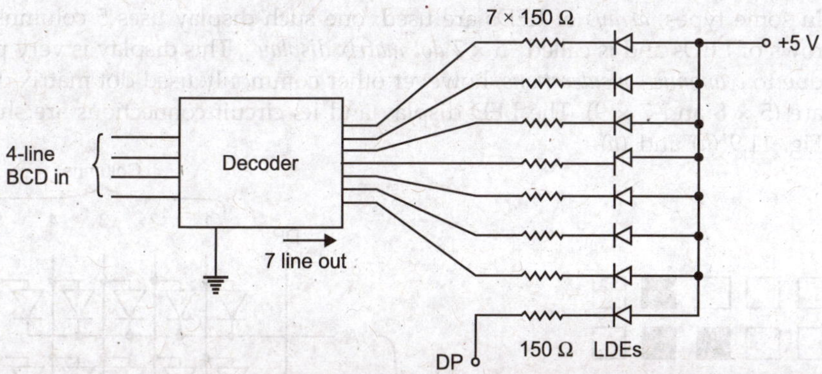


**Fig. 11.6.** LED seven segment display driver.

- By forward biasing different LEDs, we can display the digits 0 through 9. For *example* to display a zero, the LEDs a, b, c, d, e and f are forward biased. To light up 5, we need to forward bias segments a, f, g, c, d. Thus in a seven segment display, depending upon the digit to be displayed, the *particular set of LEDs is forward biased*. The various digits from 0 to 9 which can be displayed using seven segment display are shown in Fig. 11.7.

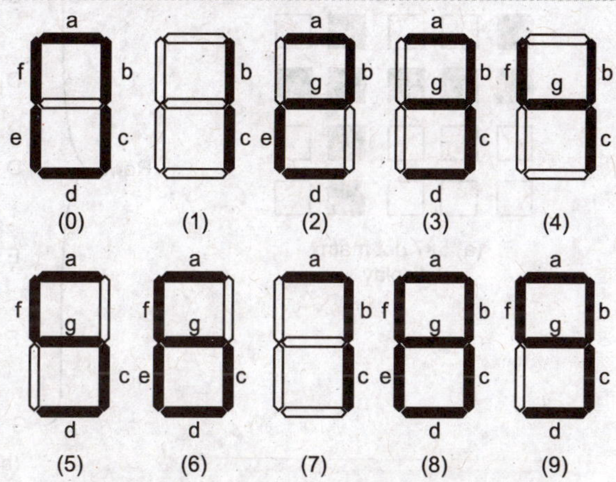


**Fig. 11.7.** Various digits displayed with seven-segment display.

**2. Alphanumeric displays :**

- The alphanumeric displays are used to display alphabets as well as numeric characters in response to the electrical inputs. Arrays of LEDs as well as CRTs are extremely useful in this application. "Read only memories" that are used to facilitate generation of alphanumeric characters, and long serial shift registers for memories are now available in MOS/LSI form.
- In such displays, the LEDs are arranged in various ways. One such alphanumeric display using 18 LEDs, called **eighteen-segment display** is shown in Fig. 11.8.

— Due to the diagonally arranged LEDs such as j, h, m, k the characters like Y, K, M etc. can be

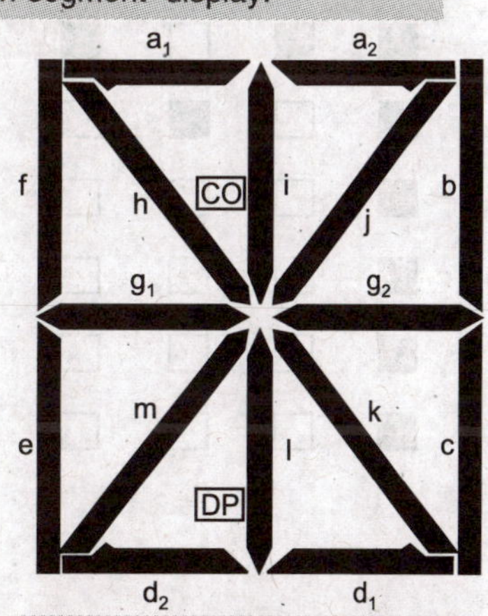


**Fig. 11.8.** Eighteen- segment display.

displayed with such a display. Separate LEDs are used for Decimal point (DP) and Colon Operator (CO).

— The various alphanumeric characters are displayed by lighting up the LEDs located in specific positions.

— In some types, *arrays* of LEDs are used; one such display uses 5 columns and 7 rows of LEDs and is called "*5 × 7 dot matrix display*". This display is very popular due to *economic considerations*, however other commonly used dot matrix displays are (5 × 8 and 7 × 9). The LED display and its circuit connections are shown in Fig. 11.9 (*a*) and (*b*).

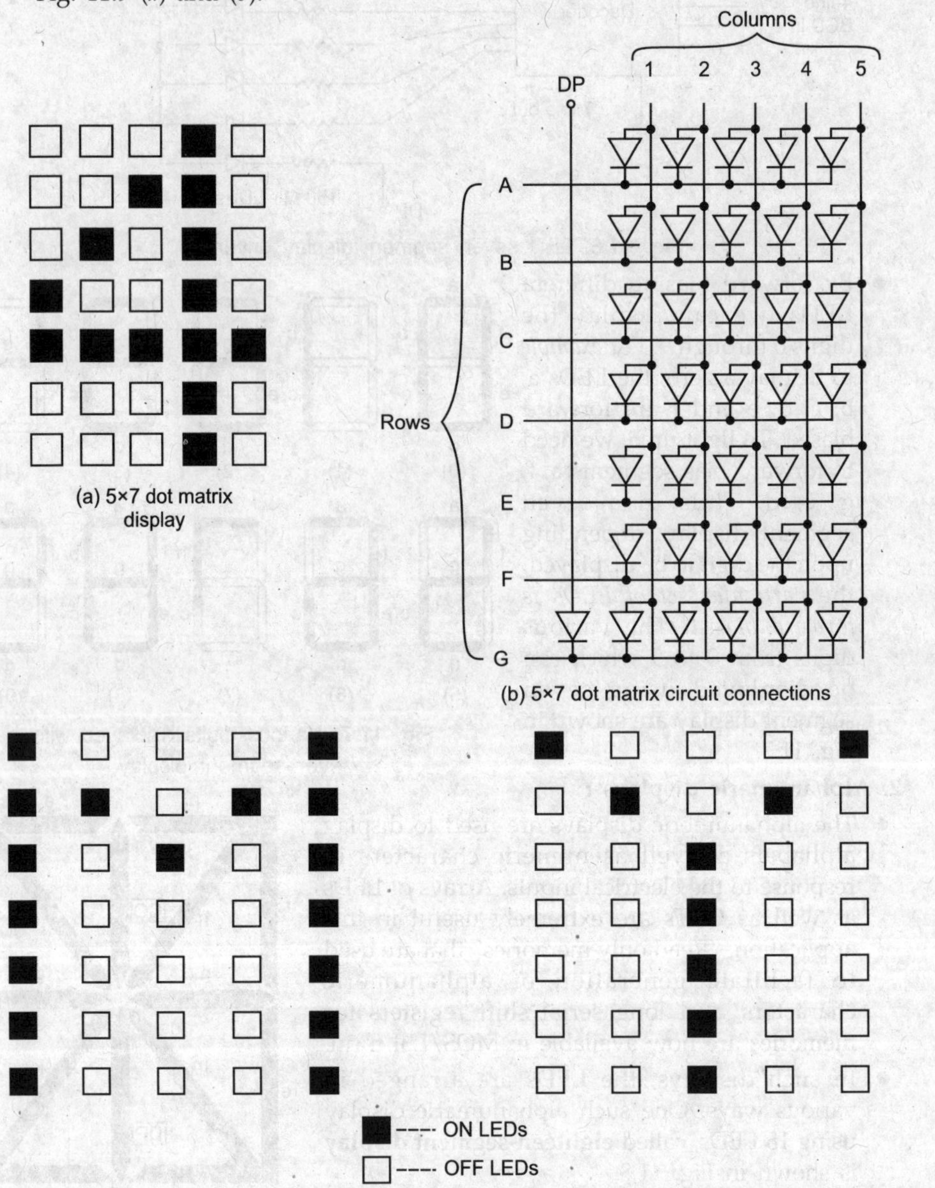


(c) 5×7 dot matrix display

**Fig. 11.9.** 5 × 7 dot matrix display.

- Depending upon the required character, the corresponding LEDs are switched on, in the display. Fig. 11.9(*c*) shows the display of characters M and Y with the help of 5 × 7 dot matrix display.
- In order to improve the display's brightness, a dynamic system is used. In this arrangement the LEDs are not lit continuously but are sequentially lit by scanning in a "vertical strobe" or horizontal strobe" mode (This is similar to running lights used in modern advertisements).
  — In the "*vertical strobe mode*", a *single row is selected at a time*, the appropriate LEDs are energised in that row, and then the signal is applied to next row.
  — In the "*horizontal strobe mode*", a *single column is selected at a time.*
- Alphanumeric displays using LEDs employ a number of square and oblong emitting areas, arranged either as dot matrix or segmented bar matrix. The alphanumeric LEDs are normally lard out on a single slice of semiconductor material, all the chips being enclosed in a package, similar to an IC, except that the packaging compound is transparent and not opaque.

**Bar-graph display :**

The **bar graph display** is basically an array of LEDs, LCDs, or gas-discharge panels, arranged in an *n × m matrix* form, by which the signal magnitude is indicated in a linear or angular format.

The light-emitting diode bar-graph consists of a number of LED arrays (row-wise) arranged in a single or multiple column. The packed density may be as much as 50 to 200 devices per $cm^2$, and the display signs can be as large as 5 cm × 10 cm.

The *main advantages* of this type of display are :

(*i*) Solid-state nature (without any moving parts)

(*ii*) Good speed of response.

- Such displays find wide applications in *process instrumentation systems.*

### 11.3.3. Liquid Crystal Displays (LCDs)

LCDs are passive displays characterised by *very low power consumption and good contrast ratio.*

- A liquid crystal is a material, usually an organic compound, which flows like a liquid at room temperature; its molecular structure has some properties normally associated with solids (*e.g.*, cholesterol nonanoate and p-azoxyanisole).
- When light is incident on an activated layer of a liquid crystal, it is *either absorbed or else is scattered by the disoriented molecules.*
- A liquid crystal 'cell' (Fig. 11.10) consists of a thin layer (about 10 μm) of a liquid crystal sandwiched between two glass sheets with transparent electrodes deposited on their inside faces.
  — When both glass sheets are transparent, the cell is known as *transmittive type cell.*
  — When one glass is transparent and the other has a reflective coating, the cell is called *reflective type.*

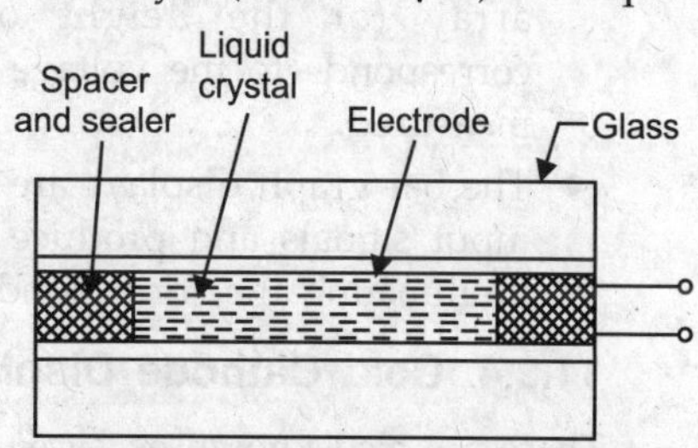


**Fig. 11.10.** A liquid crystal cell.

**Features of LCDs :**

(*i*) They are light scattering.

(ii) They can operate in a reflective or transmissive configuration.

(iii) They do not actively generate light and depend for their operation on *ambient or back lighting*.

(iv) The power required by an LCD to scatter or absorb light is extremely small of the order of a few μW/cm. They operate at low voltage ranging from 1 to 15 V.

(v) The response time of LCDs typically range between 10 and 100 ms.

**Advantages :**

1. Extremely low power requirement.
2. Low cost
3. Long life time–about 50,000 hours.

**Disadvantages :**

1. LCDs are *very slow devices*. They have a turn-on time of a few milli-seconds, and a turn-off time of tens of milliseconds.
2. They occupy a large area.
3. When used on D.C. their life span is quite small. Therefore, they are used with A.C. supplies having a frequency less than 500 Hz.
4. Poor visibility in low ambient temperature.
5. Limited temperature range of operation (0–60°C).
6. Low reliability.

LCDs are usually of the *'seven segment type' for numeric use* and have one common back electrode and seven transparent front electrodes characters.

**Uses :**

1. Cellular phone display.
2. Desk top LCD monitors.
3. Note book computer display.
4. Watches and portable instruments.
5. Pocket T.V. receiver.

**Bar graph displays :**

- These displays are *analog displays* which are an alternative to conventional D'Arsonval moving coil meters.
- They use a closely packed linear array or column of display elements, *i.e.*, "DOT-LEDs", which are independently driven so that the array (or the height of the column) corresponds to the voltage or current being measured.

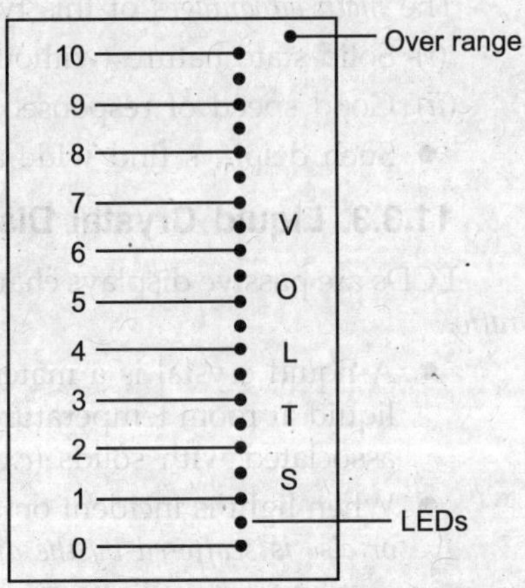

Fig. 11.11. Analog meter using bar graph of LEDs.

- The bar graph displays are generally used in the panel meters to accept analog input signals and produce an equivalent display of the input signal level by illuminating the corresponding LEDs as depicted in Fig. 11.11.

### 11.3.4. Cold Cathode Displays or Nixies

These are also known as *"neon or gas-discharge displays"*.

These are the most well known type of alphanumeric displays.

*Their operation is based on the emission of light in a cold cathode gas filled tube under breakdown condition*. The cold cathode indicators are called **Nixie tubes**.

Fig. 11.12 shows the construction of the nixie tube.

— It consists of 10 cathodes (each in the shape of a numeral from 0 to 9) and one anode, all are made of *thin wires*. The anode is in the form of *thin frame*.

— The electrodes are enclosed in a glass filled envelope with connecting pins at the bottom. Neon gas is usually emplayed and it gives an orange-red glow when activated. However, other colours are available when different gases are used.

— During normal operation, the anode is returned to *positive* supply through a suitable current limiting resistor, the value of supply being *greater than the worst-case breakdown voltage of the gas within the tube. The gas in the vicinity of the appropriate cathode glows when the cathode is switched to ground potential.*

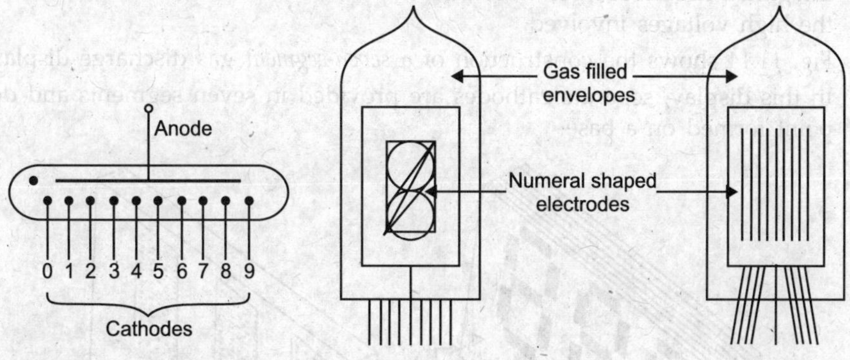


(a) Schematic symbol of Nixie (b) Front view (c) Side view

**Fig. 11.12.** Nixie tube.

— Because 10 cathodes have to be associated with a single anode inside the glass bulb, they have necessarily to be stacked in *different planes. This requires different voltages for different cathodes to enable the glow discharge.*

- Several nixie tubes possess *dot-cathodes* either on the left or right of the character to serve as decimal points.
- The standard nixie is not the only format used with cold cathode technology–both bar and dot matrix versions are available.

— The bar types have a cathode which forms the segment and operates in a fashion similar to the standard neon tube. *Identical supply voltage and drivers are required.* In the dot type display, each dot is in matrix fashion and operates as an individual glow discharge light source. The required dots are selected by an X-Y addressing array of thin film metal lines.

**Fig. 11.13.** Fifteen segment nixie.

Fig. 11.13 shows a special nixie having 15 cathodes constituting 15 segments that be used to *produce numeric as well as alphanumeric characters.*

***Characteristics of nixie tubes :***

1. The numerals are usually large, typically 15-30 mm high and appear in the same base line for in-line read-out.
2. These are single digit devices with or without a decimal point.
3. These can also display alphabetical symbols.

4. They are either side viewing or top viewing.
5. Most of the nixie tubes require D.C. supply of 150-220V, and the selected cathode carries current in the range of 1-5 mA.
6. A nixie tube can be pulse operated and hence can be used in multiplexed displays.

**Segmental gas discharge displays :**

- These displays work on the principle of gas discharge glow, similar to the case of nixie tubes.
- They are available in *seven-segment* or *fourteen-segment* format for display of numeric and alphanumeric characters.
- Integrated circuit (ICs) have been developed to drive these devices and to develop the high voltages involved.

Fig. 11.14 shows the construction of a *seven-segment* gas discharge display.

— In this display, separate cathodes are provided in seven segments and decimal point formed on a base.

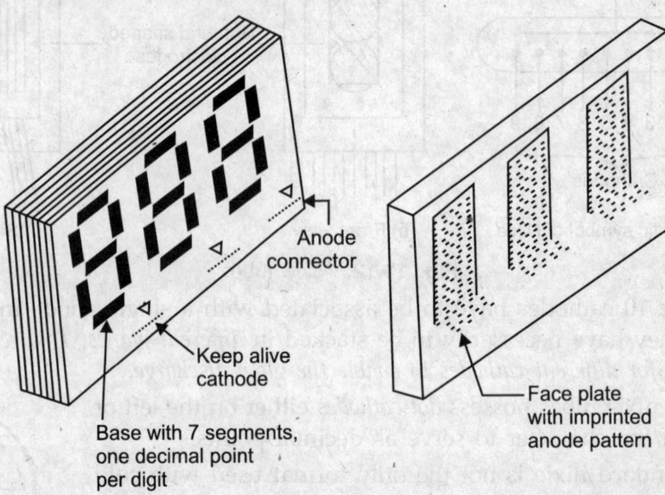


**Fig. 11.14.** Seven-segmented display using gaseous discharge.

— Each seven segment group has a single anode deposited on the covering face plate.

— The gas is contained in the space between anodes and cathodes.

— The connections are made with the help of pins for all electrodes at the rear. A keep alive cathode is also enclosed with each group of segments. A 50 μA current maintained through the keep alive cathode provides a source of ions; this improves the display's switching speed.

— All numeric characters can be displayed by activating the appropriate segment.

— Display panels of rows or columns of such characters can be easily constructed by extending a single character.

– *The power requirements of such devices are more or loss in the same range as these for nixie tubes.*

- A *major advantage* is that the power consumed is extremely small as we can have bright displays with the tubes currents as low as 200 μA.
- The *main disadvantage* of such displays is the *high voltage required to operate them;* this voltage is between 150-200 V. This high voltage transistors must be employed as switches for the cathodes.

### 11.3.5. Electro-luminescent (EL) Displays

These displays are important means of light generation.

They can be fabricated using poly crystalline semiconductors, which are essentially *phosphor powder or film type structures.*

- The **powder type** consists of powder phosphor with some binding material e.g., organic liquids deposited on a glass sheet.

The glass has transparent conductive segments (*e.g.*, seven-segment displays) or dots (dot matrix display) along with the required conductive leads on the side on which the phosphor is coated for electrical connections.

A metallic electrode, usually aluminium, is placed over the phosphor in a pressure cell by vacuum evaporation, so as to from an electrical connection on the other side of the phosphor. The resulting device is *capacitive*, because of poor conduction paths in phosphor. An A.C. field applied across the chosen segment (or dot) and aluminium electrode *excites the phosphor resulting in the emission of light.*

- In **film type** *structures* the electro-luminescent powder structure is replaced by a polycrystalline phosphor film which is deposited on a glass substrate using a vacuum or pressure cell.

These devices can be operated by A.C. as well as by D.C.

### 11.3.6. Incandescent Display

Incandescence has been a basic process of producing light for several decades.

- Such displays using *sixteen* segments as well as *5 × 7 dot matrix formats* fabricated using thin film micro electronics are now available for alphanumeric characters.
- These displays are characterised by *simple technology, bright output and compatibility with ICs but at very low operating speeds.*

### 11.3.7. Electrophoretic Image Display (EPID)

**Electrophoresis** is *the movement of charged pigment particles suspended in a liquid under the influence of an electric field.* This phenomenon has been utilised in EPIDs.

- These displays are characterised by *large character size, low power dissipation and internal memory.*

The relatively slow speed of these displays is a *major limitation*, particularly for use as a *dynamic display*.

- The life span of an EPID is a few thousand hours.
- The panel of EPID makes use of the electrophoretic migration of charged particles in suspension. The extremely thin suspension which largely contains the pigment particles and suspending liquid, is sandwiched between a pair of electrodes one of which is transparent. The application of a D.C. electric field, across the electrodes, moves the particles electrophoretically towards either electrode, the movement depending mainly on the polarity of the charge on the particles. The reflective colour of the suspension layer changes on account of this migration.

It is usual to have the transparent electrode as a common electrode. The back electrodes are generally segmented. During the normal operation of display the transparent electrode is maintained at ground potential and segmented electrodes at the back are given different potentials.

- EPID panels generally follow a segmented character format–*typically seven segments for numeric characters.*

EPID panel has a *memory* because the pigment particles deposited on an electrode surface remain there even after the applied voltage is removed.

- In order to achieve a *desired colour display*, the colour combinations of both the pigment particles and suspending liquid can be used.

### 11.3.8. Liquid Vapour Display (LVD)

*These displays* (latest in economical display technology) *employ a new reflective passive delay principle and depend on the presence of ambient lights for their operation.*

Fig. 11.15 shows the structure of a typical LVD cell.

— It consists of a transparent volatile liquid encased between two glass plates and side spacers.

— The rear glass plate has a black background and the front glass surface in contact with the liquid is roughened, so that the liquid wets it.

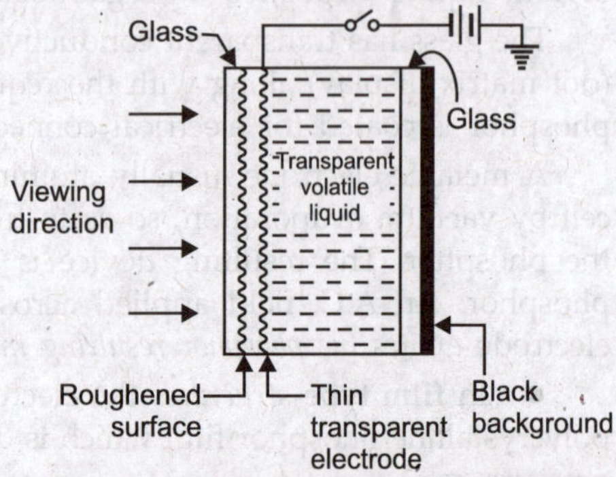


**Fig. 11.15.** Structure of LVD cell.

— *The transparent electrode is heated by using a voltage drive, which is the basis for the display function.*

— When the display is in 'OFF' condition with no voltage applied across the transparent electrode the viewer sees the background through the front transparent glass electrode and the liquid.

— In order to obtain an 'ON' condition of the display a voltage is applied to the transparent electrode. This causes sufficient heat in the electrode which evaporates the liquid in contact with it, and a combination of vapour film and vapour bubbles is formed around the roughened glass. As the evaporative index of vapour is approximately one, there is discontinuity established at the interface between the front glass plate and the liquid which *gives rise to light scattering*. This makes it a simple display device.

The organic liquid used in this type of display should have : (*i*) Refractive index close to that of the glass plate; (*ii*) Minimum energy for vaporising the liquid in contact with the roughened surface.

- LVD has a *better contrast ratio than LCD.*
- The *speed of operation of LVDs is low.*

### 11.3.9. Comparison of Various Display Devices

The comparison of various display devices is given in table 11.1 below :

**Table 11.1.** Comparison of display devices

| S. No. | Display devices | Advantages | Disadvantages | Applications |
|---|---|---|---|---|
| 1. | *CRTs* | — Bright<br>— Efficient | — Bulky<br>— High voltage | — Large display<br>— Small and large group viewing |

| | | | | |
|---|---|---|---|---|
| | | — Uniform<br>— Planar display -all colours<br>— High reliability. | — Non-digital address<br>— High initial cost. | — Console display. |
| 2. | *LEDs* | — Bright<br>— Efficient<br>— Red, yellow, amber, green colours<br>— Compatible with ICs<br>— Small size. | — High cost per element<br>— Limited reliability<br>— Low switching speed. | —Indicators and small displays<br>— Individual viewing<br>— Flat panel. |
| 3. | *LCDs* | — Good contrast in bright ambient light<br>— Low power<br>— Compatible with ICs<br>— Low cost element. | — Limited temperature range (0–60°C)<br>— Limited reliability<br>— A.C. operation necessary<br>— Low switching speed. | — do — |
| 4. | *NIXIEs* | — Bright<br>— Range of colours<br>— Low cost element<br>— Compatible with ICs. | — High drive power. | — Indicators<br>— Small, medium and large displays<br>— Small group viewing. |
| 5. | *ELs* | — Low cost element<br>— Many columns. | — Not compatible with ICs. | — Indicator and small display<br>— Flat panel. |

## 11.4 RECORDERS

### 11.4.1. Introduction

A **recorder** *records electrical and non-electrical quantities as a function of time.* The record may show how one variable varies with respect to another, or how the input signal varies with time.

The record serves the following *objectives* :

(*i*) It preserves the details of measurement at a particular time.

(*ii*) It provides at a glance the overall picture of the performance of unit.

(*iii*) It provides immediate reflection on the action taken by the operator.

### 11.4.2. Types of Recorders

In an instrumentation system, one of the important considerations is the method by which the data required is recorded. The recording method should be consistent with the type of system. If we are dealing with a wholly analog system, then *analog recording techniques* should be used. While, on the other hand, if the system has a digital output, *digital recording devices* are employed.

Two types of recorders are:

**1. *Analog recorders :***

(*i*) *Graphic recorders*

(*a*) Strip chart recorders
- Galvanometer type
- Null type
  — Potentiometric recorders
  — Bridge recorders
  — LVDT recorders

(*b*) Circular chart recorders

(*c*) X-Y recorders

A *"graphic recorder"* may be defined as an instrument, which draws a graph that relates two or more variables to time or to each other. A *"graph"* is an analog representation of discrete information.

(*ii*) *Oscillographic recorders.*

(*iii*) *Magnetic tape recorders.*

**2. *Digital recorders.***

The above recorders are discussed briefly below :

### 11.4.3. Strip Chart Recorders

Fig. 11.16 shows the basic constructional features of a strip chart recorder.

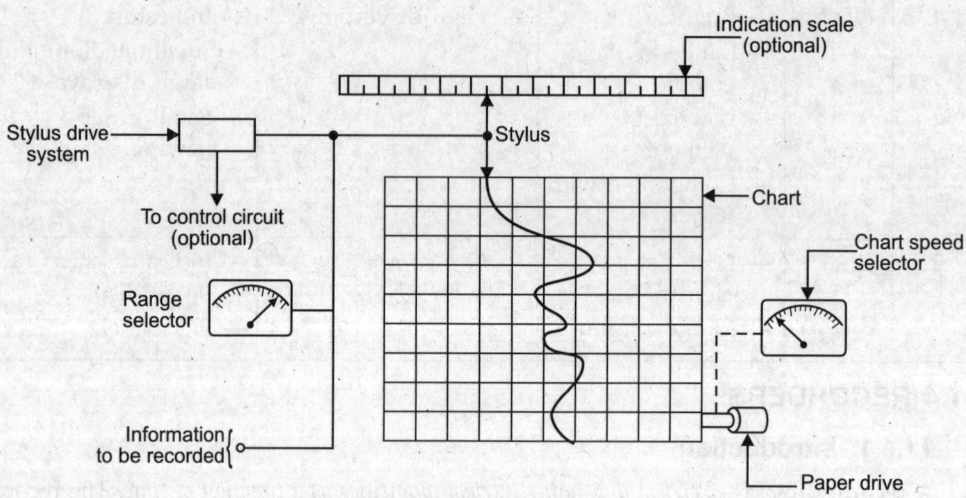


**Fig. 11.16.** Strip chart recorder.

A strip chart consists of the following:

- A long wall of graph paper moving *vertically*.
- A system for driving the paper at some selected speed.
- A stylus for marking paper on the moving graph paper (Most recorders use a pointer attached to the stylus, which (pointer) moves over a calibrated scale thus showing instantaneous value of the quantity being measured).
- A stylus driving system which moves the stylus in nearby exact replica or analog

of the quantity being measured (A spring wound mechanism may be used but in most of the recorders a *synchronous motor* is used for driving the paper).

**Marking mechanisms.** The most commonly used mechanisms employed for marking marks on the paper are :

(*i*) Marking with ink filled stylus.
(*ii*) Marking with heated stylus.
(*iii*) Chopper bar.
(*iv*) Electric stylus marking.
(*v*) Electrostatic stylus.
(*vi*) Optical marking method.

**Tracing systems.** For producing graphic representations, the following two types of tracing systems are used:

(*i*) Curvilinear system.
(*ii*) Rectilinear system.

**11.4.3.1. Galvanometer type strip chart recorders**

— This type of recorder operates on the *"deflection principle"*.

— The deflection is produced by a galvanometer, (d'Arsonval) which produces a torque on account of a current passing through its coil. This current is proportional to the quantity being measured. (See Fig. 11.17)

— These recorders can work on ranges for a few mA to several mA or from a few mV to several mV.

— The moving galvanometer type recorder is *comparatively inexpensive instrument*, having a narrow bandwidth of 0 to 10 Hz. It has a sensitivity of 0.4 mV/mm or from a chart of 100 mm width a full scale deflection of 40 mV is obtained. Linear amplifiers are used for measurement of smaller voltages.

— This type of recorder is not useful for recording fast variations in either current or voltage or power.

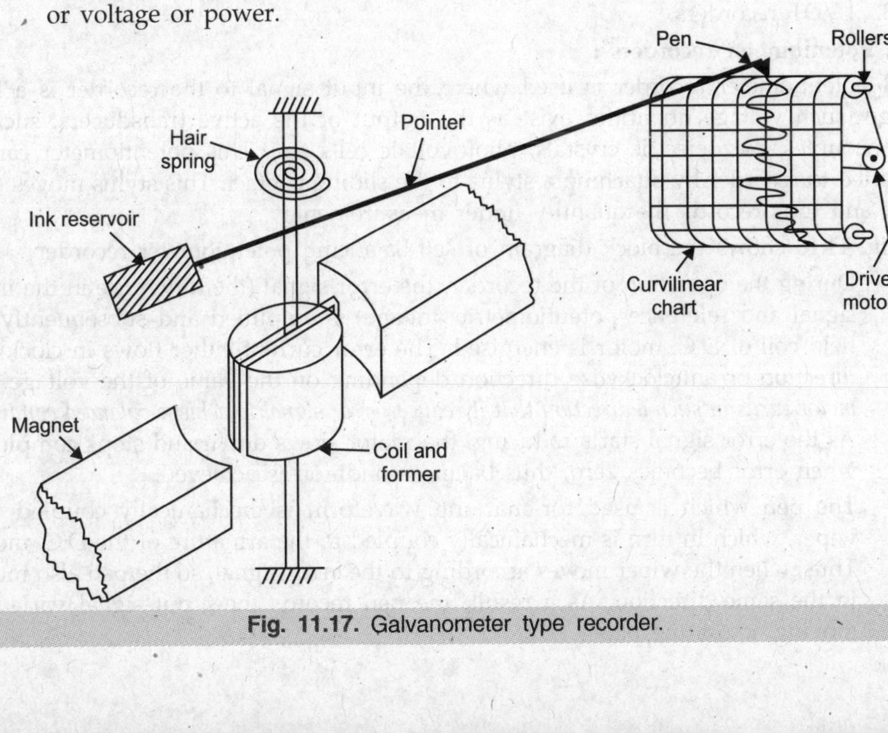


**Fig. 11.17.** Galvanometer type recorder.

- Some recorders contain a timing mechanism that prints a series of small dots along the edge of the paper chart, as the paper moves through the recorder. This time marker produces *one mark per second*. These types of recorders are mostly used as ***optical recorders***, and contain a light source provided by either an ultra violet or tungsten lamp. *A small mirror is connected to the galvanometer movement and this light beam is focussed on the mirror. The beam reflected from the mirror is focussed into a spot on a light sensitive paper. As the current passes through the coil the mirror deflects. The movement of the light beam is affected by the deflection of the small mirror, and the spot on the paper also varies for the same reason, thus tracing the waveform on the paper.*

**11.4.3.2. Null type strip chart recorders**

*The main disadvantage of a galvanometer type recorder* is that it has a *low input impedance and a limited sensitivity.* This disadvantage can be overcome by using an *amplifier* between the input terminals and the display or indicating instruments. This amplifier provides a high input impedance and improved sensitivity *at the cost of low accuracy.* For improving the accuracy of the instrument the input signal is *compared with a reference voltage using a potentiometer circuit.*

- These recorders *operate on the principle of "self balancing" or "null conditions".*

When the input signal given by the sensor or transducer is applied to the measuring unit of the recorder the balanced condition of the instrument gets disturbed. This unbalanced signal produces *error signal.* This operates on some device which converts this error signal to some action so that the balanced condition of the instrument is achieved. Now, this action is nothing but the magnitude of the error signal which is *directly proportional to the change in input signal.*

The null type strip chart recorders are of the following *types* :

1. Potentiometer recorders.
2. Bridge recorders.
3. LVDT recorders.

**1. Poteniometer recorders :**

A potentiometer recorder is used where the input signal to the recorder is a D.C. voltage (such voltage quantities exist as the output of the active transducers, such as thermocouples, piezoelectric crystals, photovoltaic cells etc). This potentiometer can be converted to recorder by attaching a stylus to the sliding contact. This stylus moves over paper and thus records the quantity under measurement.

Fig. 11.18 shows the block diagram of self-balancing potentiometer recorder :

— During the operation of the recorder, the error signal (the difference in the input signal and reference potentiometric voltage) is amplified and subsequently the field coil of D.C. motor is energised. The error current either flows in clockwise direction or anticlockwise direction depending on the value of the voltage. *The motor turns in such a direction that it reduces error signal to achieve balanced condition.* As the error signal starts reducing, the motor slows down and stops completely when error becomes zero; thus balance condition is achieved.

— The pen which is used for marking waveform is mechanically coupled to a wiper, which in turn is mechanically coupled to the armature of the D.C. motor. Thus, when the wiper moves according to the error signal, so the pen also moves in the same direction. As a result, the pen records the input signal variations moving across the paper.

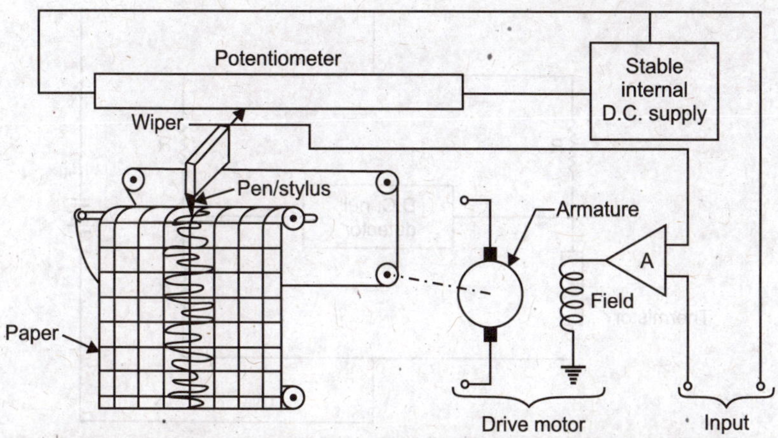


**Fig. 11.18.** Block diagram of self-balancing potentiometer recorders.

- Such recorders have a very high input impedance, infinity at balance conditions, and a high sensitivity of the order of 4V/mm with an error of less than ± 0.25% with a bandwidth of 0.8 Hz.
- The most common application of potentiometric recorder is for *recording and control of process temperatures.* Self balancing potentiometers are unduly used in industry because of the following *reasons:*

(*i*) Their action is automatic and thus eliminate the constant operation of an operator.

(*ii*) They draw a curve of the quantity of being measured with the help of recording mechanism.

(*iii*) They can be mounted on switchboard or panel and thus act as mounting devices for the quantity under measurement.

**Single-point and multi-point recorders:**

- *Instruments that record changes of only one measured variable are called* **single-point recorders.**
- A **multi-point recorder** may have as many as 24 inputs, with traces displaced in 6 colours.

**2. Bridge recorders :**

These recorders are used when the signals to be monitored are due to variation in some electrical parameter of a passive transducer such as *variation in resistance which occurs in strain gauges, thermistors and photoconductive cells.*

Temperature is one of the non-electrical quantities which can be measured by this type of recorder.

Fig. 11.19 shows a bridge recorder using a *thermistor.*

— The resistance changes in the thermistor cause corresponding changes is in the bridge output. These changes are applied to the detector.

— The bridge balance (null) can be restored by varying the resistance of another arm of the bridge, while recording in terms of current, voltage or temperature. Based on the kind of voltage supplied to the bridge, the output can be chosen to be D.C. or A.C.

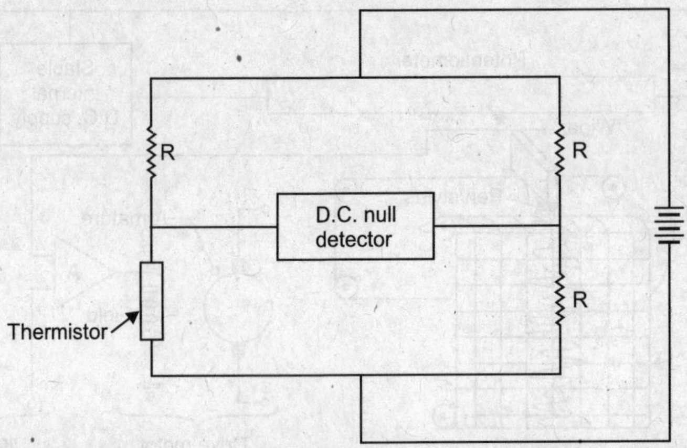


**Fig. 11.19.** Bridge recorder (using thermistor).

### 3. LVDT recorder

This recorder is used to record *force, pressure* etc. which can be translated into displacement with the help of a force assuming device.

Fig. 11.20 shows a LVDT recorder.

— It consists of linear voltage differential transformers (LVDT) $T_1$ and $T_2$.

— The displacement to be recorded is applied to the soft iron of differential transformer $T_1$. $T_2$ is associated with the recording circuit.

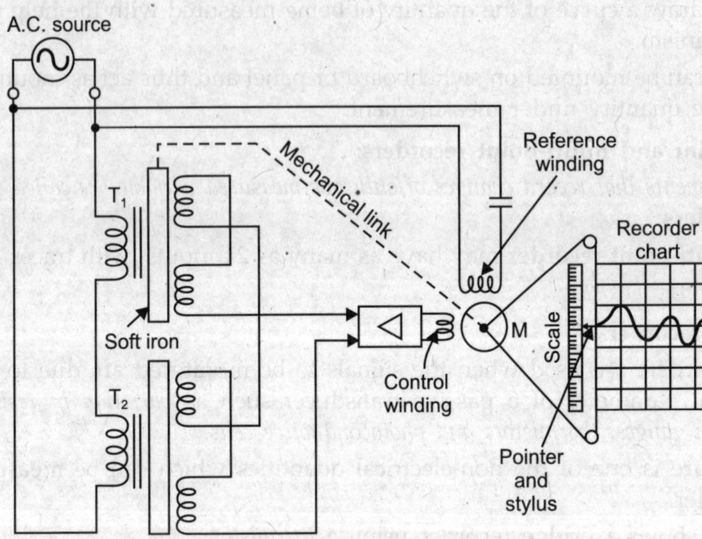


**Fig. 11.20.** LVDT recorder.

— The primary windings of $T_1$ and $T_2$ are connected in series to the same A.C. source. The secondary windings are connected in *series opposition*. Thus at the balance condition (*i.e.*, null condition), the output voltages of the transformers are equal and opposite and the net output (A.C. signal source) is *zero*.

— Now if a displacement is applied to the core of the transformer $T_1$ the output voltages of the two transformer will be different and a different output voltage is obtained. The differential output voltage upsets the balance and an A.C. error signal (with reference to the phase of the voltage of A.C. supply) depends upon the direction of the applied displacement.

— The amplified A.C. signal is then sent to the control winding of the servometer M causing it to rotate in a direction that moves the iron core of transformer $T_2$ to *rebalance the circuit*. As the motor rotates, it also moves the stylus across the chart of the recorder and the pointer across the scale. When the balance is restored the error signal vanishes and the motor stops. The position of the stylus reflects the new position of the movable core (*i.e.*, the displacement) and the trace produced by the stylus on the paper shows the change from the previous position. This pointer indicates the new position on the scale which is calibrated directly in terms of displacement.

#### 11.4.3.3. Advantages and disadvantages of strip chart recorders

**Advantages :**

(*i*) The rate of movement of the chart can easily be changed to spread out the trace of the variable being observed.

(*ii*) Data conversion is easier when rectangular coordinates are used.

(*iii*) These recorders require the use of servo-mechanisms to position the pointer or pen. Therefore more than adequate power is available, there being no real limitations on the weight of the pen, pressure between pen and paper or length of the pointer.

(*iv*) Relatively large amount of paper can be inserted at one time in the form of a nell.

(*v*) Many more separate variables can be recorded on a strip chart than on circular chart.

**Disadvantages :**

(*i*) The mechanism is considerably more complicated than is required to drive a circular chart.

(*ii*) Observing behaviour several hours or days back is not as easy as picking out one circular chart which covers the desired period of time.

#### 11.4.3.4. Applications of strip chart recorder

Following are some of the important applications of strip chart recorders :

**1. *Temperature recorder.*** The two primary methods used for recording temperature are :

(*i*) Thermocouple methods :

— This method utilises commercially available thermocouples that cover a wide range of temperatures.

— The thermocouples serve very well as temperature sensing elements and are readily compatible with strip-chart recorders.

(*ii*) The resistance method.

**2. *Sound level recording.*** Invariably it is required to obtain a record of sound level over a period of time near hospitals, schools or residences, airports etc. It can be done by with an ordinary microphone and a strip-chart recorder, provided the output signal from the microphone is of adequate amplitude to drive the recorder.

Fig. 11.21 shows a set up for obtaining sound level data using a sound level meter and a strip chart recorder.

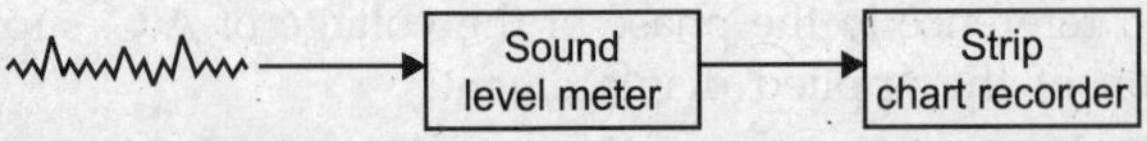


**Fig. 11.21.** Set-up for measuring sound level as a function of time.

*3. Recording amplifier drift.* Transistor amplifiers are sensitive to temperature changes. The changes in temperature *cause the bias voltage of the transistor to change*, thereby changing the operating or quiescent point, generally by a small amount. This change of the point is called *drift*. Because this is generally a slow process a strip chart recorder may be used to monitor and record the drift by connecting the recorder to the output of the amplifier.

**11.4.3.5. Comparison between potentiometric recorder and galvanometeric recorder**

The comparison between potentiometric and galvanometric recorders is given in table 11.2 below.

**Table 11.2. Comparison between Potentiometric recorder and Galvanometric recorder**

| S. No. | Aspects | Potentiometric recorder | Galvanometric recorder |
|---|---|---|---|
| 1. | *Working principle* | — *Self balancing or null conditions.* It compares the input signal with the reference and *produces error signal.* | — *D'arsonval movement.*<br>— It produced deflection when current passes through coil. This produces a *torque* which is used for driving a pointer and recording input signal. |
| 2. | *Recording mechanism* | — The *error signal* produced according to the difference between the input signal and reference is *amplified.*<br>— The current corresponding to the error signal drives the motor which moves the pen coupled with the wiper. | — The deflection of the galvanometer is proportional to the current through the coil (which is proportional to the input signal measured).<br>— Due to the deflection, torque is produced, which moves the pointer to which pen-ink system is fitted. |
| 3. | *Response* | Faster | Slower |
| 4. | *Bandwidth* | Higher–of the order of few hundreds of Hz. | Smaller of the order of 0-10 Hz. |
| 5. | *Input impedance* | Very high, hence loading problems are eliminated. | Very low, hence loading problems are significant. |
| 6. | *Sensitivity* | Higher–equal to 4V/mm for error less than ± 0.25% | Lower–equal to 0.4 V/mm. |
| 7. | *Applications* | Used for recording the output from the transducer in the form of voltage, current, change in resistance or inductance or capacitance, pressure temperature etc. | Commonly used as optical recorder which uses ultraviolet source. |

### 11.4.4. Circular Chart Recorders

As the name implies, the data in this type of recorder is *recorded on a flat circular chart*. These recorders were developed mainly to take advantage of the availability and convenience of spring-wound clock and synchronous motor movements to drive the chart in a circular direction.

— The circular chart used here has concentric circles ruled on it to form its scales. In addition, there are printed arcs extending from the centre of the chart to the paper's edge. As the pen of the recorder is moved it swings along these arcs; these arcs are called the *"time arcs"*.

— The speed of the rotation of chart is usually one revolution per 24 hours or per seven days or any other speed, which can be conveniently obtained by using synchronous motor with suitable gear assembly.

*The radial position of the pen at any time indicates the instantaneous value of the quantity under measurement.*

— Chart diameter is limited to a maximum of 0.3 m, speed of the chart is also limited, resolution along the scale length is usually non-uniform and the charts do not run for a long period.

— Magnitude of several variables can be recorded on a single chart which makes it easy and convenient to analyse the interrelationship of various measurements and also it saves the panel mounting space.

The various drives for circular charts are *classifieds* as follows :

(*i*) Mechanical (spring clock drive).

(*ii*) Pneumatic (air lock drive).

(*iii*) Electric (synchronous regulated D.C. motor or motor wound spring).

(*iv*) Dual powered drive (duplex) *i.e.*, a synchronous motor and spring clock mechanical drive.

(*v*) Externally controlled drives.

- *Circular chart recorders are particularly suitable for direct actuation by a number of mechanical sensors such as bellows, bourdon tubes etc.*

The comparison between circular chart and strip recorders is given in table 11.3.

**Table 11.3. Comparison between Circular chart and Strip chart recorders**

| S. No. | Aspects | Circular chart recorders | Strip chart recorders |
|---|---|---|---|
| 1. | *Handling and storing* | Easy | Very easy |
| 2. | *Shape and size of chart* | Circular, varying in size from 100 mm to 250 mm diameter. | Curvilinear type, available in the form of long strips usually rolled on to a drum. |
| 3. | *Usable recording area* | 40 to 50% area of chart is calibrated and rest is the space covered by mechanism involved. | 90% or more of the chart width is usable recording area, very small portion being taken up by punched holes for guide purposes. |
| 4. | *Amount of information that can be carried.* | Strictly limited amount. | It can be packed with information. |
| 5. | *Exhibition of information* | It shows all the information recorded at a glance. | It needs to be unrolled to see past records. |

| | | | |
|---|---|---|---|
| 6. | *Facility to record* | It is possible to simultaneously record on the full-chart range upto four separate variables. | It is possible to record upto 4 to 6 points simultaneously and thus afford saving a lot of panel space. |
| 7. | *Cost* | Low initial cost. | Cost though high is justified, considering its versatility, predictive diagnostic capability, invaluable tool for analysing the overall dynamic response. |
| 8. | *Range of chart speeds* | Usually the circular chart moves at one constant speed and high speed phenomenon cannot be recorded. | It is possible to have wide range of chart speeds and records fast changing phenomenon. |
| 9. | *Chart speed* | The chart speed is limited and as such recording cycle takes longer time for multiple points. | The availability of wide range of chart speeds enables the recording of greater number of points and at a much higher speed than is practical with circular chart instruments. |
| 10. | *Type of operators required* | Less skilled operators can do the job since it is easy to adjust and repair instruments. | Skilled operators are required. |

### 11.4.5. X-Y Recorders

A **X-Y recorder** is *an instrument which gives a graphic record of the relationship between two variables.*

This system has a pen which can be positioned along the two axes with the writing paper remaining stationary. There are two amplifier units, one amplifier actuates the pen in the Y-direction as the input signal is applied, while the second amplifier actuates the pen in X-direction. The movements of the pen in X-and Y-directions are automatically controlled by means of a motor, pulleys and a linear potentiometer. Obviously, trace of the marking pen will be due to the combined effects of two signals applied simultaneously. In these recorders, *an e.m.f is plotted as a function of another e.m.f.* There are many variations of X-Y recorders. With the help of these *recorders and appropriate tranducers a physical quantity may be plotted against another physical quantity.*

Fig. 11.22 shows a block diagram of a typical analog X-Y recorder :

— A signal enters in each of the two channels.

— The signals are *attenuated* to the inherent full scale range of the recorder (often 0.5 mV). The signal then passes to a *balance circuit* where it is compared with an internal reference voltage.

– The *error signal* (*i.e.* the difference between the input signal voltage and the reference voltage) is fed to a "***chopper***" which converts D.C. signal to an A.C. signal.

— The signal is then *amplified* in order to actuate a *servomotor* which is used to balance the system and hold it in balance as the value of the quantity being recorded changes.

The action described above takes place in both the axes simultaneously. Thus we get a record of one variable with respect of another.

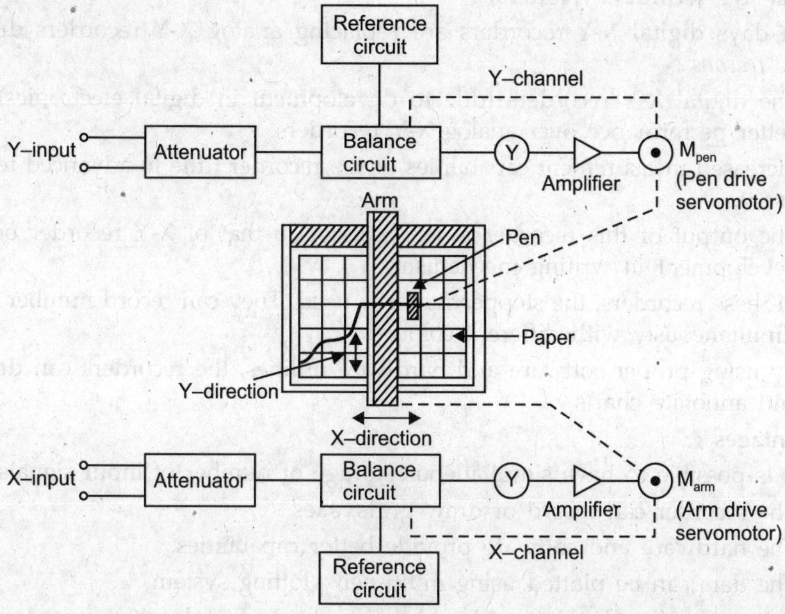


**Fig. 11.22.** X-Y recorder.

**Advantages :**

1. The instantaneous relationship between two physical quantities can be recorded.
2. The relationship between either electrical or nonelectrical quantities can be recorded.
3. In modern types of recorders, zero offset adjustments are available.

**Applications.** A few examples in which use of X-Y recorders is made are as under :

1. Plotting of stress-strain curves, hysteris curves and vibrations amplitude against swept frequency.
2. Pressure-volume diagrams for I.C. engines.
3. Pressure-flow studies for lungs.
4. Lift drag wind tunnel tests.
5. Electrical characteristics of materials such as resistance versus temperature and plotting the output from electronic calculators and computers.
6. Speed-torque characteristics of motor.
7. Regulation curves of power supplies.
8. Plotting of characteristics of vacuum tubes, zener diodes, rectifiers and transistors etc.

**Specifications :** Some important specifications of X-Y recorder are :

(*i*) Sensitivity ............. 10 μV/mm.

(*ii*) Slewing speed ............. 1.5 m/s

(*iii*) Frequency response ............. About 6 Hz both the axes

(*iv*) Chart size ............. 250 mm × 180 mm

(*v*) Accuracy ............. ± 0.3 per cent.

**Digital X-Y recorders (Plotters) :**

These days digital X-Y recorders are replacing analog X-Y recorders due to the following *reasons* :

- The digital X-Y recorders (due to development in digital electronics) provide better performance over analog X-Y recorders.
- Increased measurement capabilities of the recorder (due to advanced techniques used).
- The output of this recorder is far superior to that of X-Y recorder because of development in writing mechanism.
- In these recorders, the stepper motor is used. They can record number of inputs simultaneously with different colours.
- By using proper software and hardware utilities, the recorders can draw grids and annotate charts.

**Advantages :**

(*i*) It is possible to have simultaneous storage of number of input signals.

(*ii*) The recorder can record or draw grids, axes.

(*iii*) The hardware and software provide better capabilities.

(*iv*) The data can be plotted using multi-pen plotting system.

- The comparison between strip chart recorder and analog X-Y recorders is given in table 11.4 below :

**Table 11.4. Comparison between "Strip chart recorder"and "Analog X-Y recorder".**

| S.No. | Strip chart recorder | X-Y recorder (analog) |
|---|---|---|
| 1. | Also called as *X-t plotter*. | Also called X-Y plotter. |
| 2. | Input variable is plotted as a function of time. | One input variable is plotted as a function of the other. |
| 3. | The paper is kept *rotating*. | The paper is held *stationary*. |
| 4. | Generally a self-balancing potentiometer is used to plot the emf as a function of time. | One self-balancing potentiometer controls the position of roll while the other controls the position of the recording pen. |
| 5. | The zero offset adjustments are *not available*. | The zero adjustments *are available*. |

## 11.4.6. Oscillographic Recorders

*The* **oscillographic recorders** *are instruments which record information on paper or film with ink, powder, electrostatic charge, liquid, electric arc or light.*

Although galvanometer recorders, pen recorders, X-Y recorders, strip chart recorders and CRT recorders all may be considered oscillographic recorders, yet *here word 'oscillograph'* implies either a *galvanometer or a CRT recorder–the recorders having bandwidths more than 20 kHz.*

### 11.4.6.1. Ultraviolet (U.V.) recorders:

These recorders are basically *electro-mechanical oscillographic recorders* and modified version of Duddel's oscillographs.

An ultraviolet recorder consists of a number of galvanometer (moving coil) elements mounted in a single magnet block as shown in Fig. 11.23. A paper sensitive to ultraviolet light is used for producing a trace for the purpose of recording. The u.v. light is projected on the paper with the help of mirrors attached to the moving coils.

**Working.** When a current is passed through the moving (galvanometer) coil, it deflects under the influence of the magnetic field of the permanent magnet. The ultraviolet light falling on the mirrors is deflected and projected on to the u.v. light sensitive paper through a lens and mirror system. The paper is driven past the moving high spot and thus a trace of variation of current with respect to time is produced.

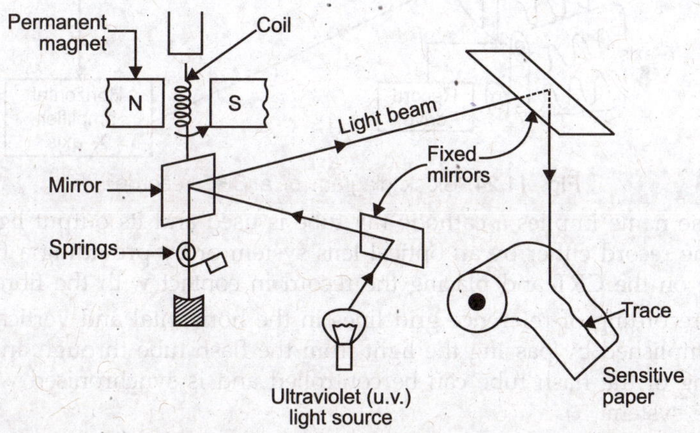


**Fig. 11.23.** Ultraviolet (u.v.) recorder.

The recorder, in addition to the input currents, may have the following additional traces:

(*i*) Grid lines.
(*ii*) Timing lines.
(*iii*) Trace identification.

- The ultraviolet (u.v.) recorders, compared to the mechanical and pen recorders, have *better frequency and response characteristics*; the typical values are :

  Frequency response = 0 to 300 Hz; 0 to –12 kHz (maximum)
  Response time = 16 μs

  — The maximum frequency that may be recorded depends upon the frequency response of the galvanometer used. When high frequency signals are to be recorded, the marking paper is moved with sufficient speed so as to spread out the trace along the time-axis. *The u.v. recorders have an additional advantage of multi-trace recording.*

- *"Typical applications"* of U.V. recorders are in recording :
  (*i*) Regulation transients of generators.
  (*ii*) Output of transducers.
  (*iii*) Control system performance.

- These recorders are also used for *recording the magnitude of low frequency signals which cannot be measured with analog (pointers) type instruments.*

#### 11.4.6.2. Cathode ray tube (CRT) recorders

These recorders are *four-dimensional display and recording devices.*

Fig. 11.24 shows the block diagram of a CRT recorder.

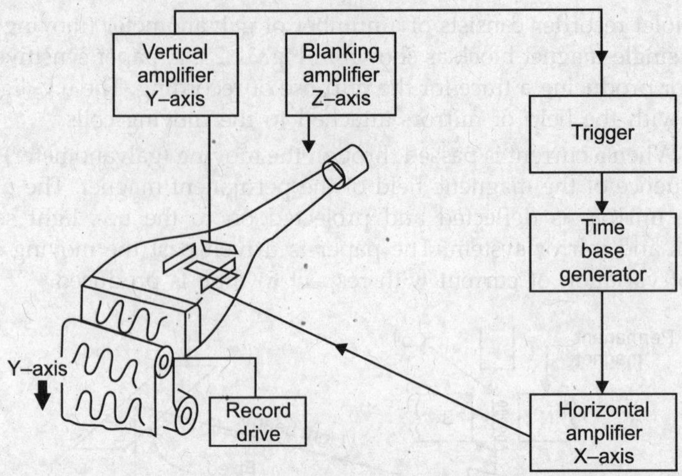


Fig. 11.24. Block diagram of a CRT recorder.

— As the name implies a cathode ray tube is used and its output light is focussed on the record either by an optical lens system or by providing a fibre optic face plate on the CRT and placing the record in contact with the fibre optics.

— The recording of reference grid lines in the horizontal and vertical directions is accomplished by passing the light from the flash tube through an aperture. The timing of the flash tube can be controlled and is synchronised with the record drive system.

- The CRT recorders can record data upto 1000 times faster than galvanometer recorders. Further because of four axes they have capability of recording continuously.

## 11.4.7. Magnetic Tape Recorders

These recorders have response characteristics which enable them to be used at *higher frequencies*; hence they find an *extensive use in Instrumentation systems.*

### 11.4.7.1. Construction and operating principle

**Construction :**

A magnetic tape recorder consists of the following *basic components*:

1. Recording head.
2. Magnetic tape.
3. Reproducing head.
4. Tape transport mechanism.
5. Conditioning devices.

**1. *Recording head*** :

Fig. 11.25 shows the construction of a magnetic tape recording head :

— Its construction is similar to that of transformer having a toroidal core with a coil.

— The fine air gap of lengths 5-15 μm is shunted by the passing magnetic tape.

— When the current used for recording is passed through the coil wound around magnetic core, it produces magnetic flux.

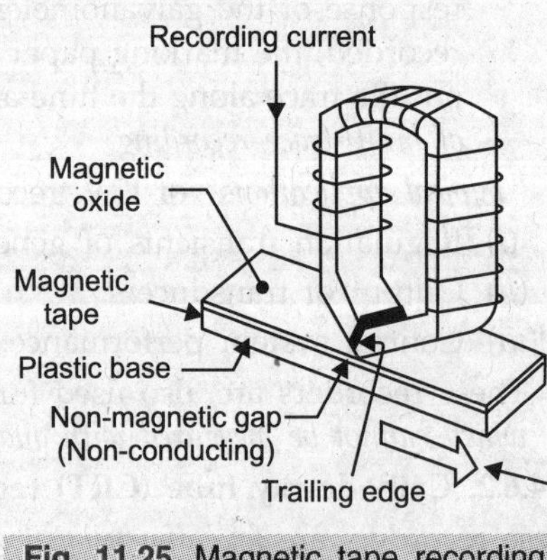


Fig. 11.25. Magnetic tape recording head.

— As the magnetic tape having iron oxide particles passes the head, the magnetic flux produced gets linked with the iron oxide particles and these particles get magnetised. The state of magnetization of the oxide as it leaves the gap is retained, thus the actual recording takes place at the trailing edge of the gap.

— Any signal recorded on the tape appears as a *magnetic pattern dispersed in space along the tape, similar to the original coil current variation with time.*

**2. *Magnetic tape :***

— It is made of a thin sheet of tough, dimensionally stable plastic, one side of which is coated with a magnetic material. Typically, the plastic base is polyvinyl chloride (PVC) or polyethylene terephthalate, for example *mylar*.

— The magnetic coating consists of a dispersion of very small particles of iron oxide ($Fe_2O_3$) in a plastic binder.

— A typical tape is 12.7 mm wide and 25.4 μm thick.

— The magnetic particles conform to the magnetic pattern and retain it.

**3. *Reproducing head :***

— It is similar to that of a recording head in appearance.

— Its *function is to detect the stored magnetic pattern and to convert it back to original electrical signal.*

**4. *Tape transport mechanism :***

Fig. 11.26 shows a simple tape transport mechanism. This mechanism moves the magnetic tape along the recording head or reproducing head with a constant speed.

— The magnetic tape is wound on two reels; one is called as supply reel and other is called as tape-up reel. Both the reels rotate in same direction. The fast winding of the tape or the reversing of the tape is done by using special arrangements.

— The rollers are used to drive and guide the tape.

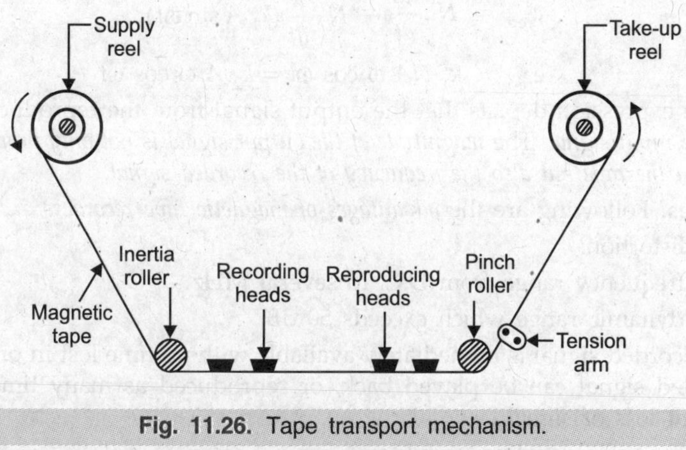


**Fig. 11.26.** Tape transport mechanism.

The tape transport mechanism must perform the following *tasks* :

(*i*) To handle the tape without straining and wearing it.

(*ii*) To guide the tape across magnetic heads with great precision.

(*iii*) To maintain uniform and sufficient gap between the tape and heads.

(*iv*) To maintain proper tension of magnetic tape.

***5. Conditioning devices :***

- These devices consists of *'amplifiers'* and *'filters' to modify signal to a format that can be properly recorded on a tape.*

— *Amplifiers* amplify the signal to be recorded while *filters* remove unwanted ripple quantities.

**Operating principle of tape recorders :**

The *operating principle* of tape recorders is as follows :

When a magnetic tape is passed through a recording head, the signal to be recorded appears as some magnetic pattern on the tape. *This magnetic pattern is in accordance with the variations of original recording current. The recorded signal can be reproduced back by passing the same tape through a reproducing head where the voltage in induced corresponding to the magnetic patten on the tape.* The induced voltage depends on the direction of magnetization and its magnitude on the tape. The e.m.f thus produced is proportional to the rate of change of magnitude of magnetisation

$$\therefore \quad i.e., \quad e \propto N.\frac{d\phi}{dt},$$

where, N is the number of turns of the winding put on reproducing head.

Since the voltage in the reproducing head is proportional to $\frac{d\phi}{dt}$, the reproducing head acts as a differentiator.

Let us assume that the orignal signal is A sin$\omega t$. The current in the recording head winding and the flux produced will be proportional to this voltage.

$$\therefore \quad \phi = k_1 \text{ A sin } \omega t, \quad \text{where } k_1 = \text{constant}$$

Assuming that the tape retains this flux pattern and regenerates it in the reproducing head, the voltage induced in the reproducing head winding,

$$e_{rep.} = N.\frac{d\phi}{dt} = N.\frac{d}{dt}(k_1 A \sin \omega t)$$

or, $$e_{rep.} = k_1 NA\, \omega \cos \omega t = k_2\, A\, \omega \cos \omega t \quad ...(11.1)$$

The above expression depicts that the output signal from the reproducing head is a *derivative of the input signal.* The *magnitude of the output signal is not only proportional to the flux recorded on the tape but also the frequency of the recorded signal.*

**Advantages.** Following are the *advantages of magnetic tape recorders* :

1. Low distortion.
2. Wide frequency range from D.C. to several MHz.
3. Wide dynamic range which exceeds 50 dB.
4. The recorded signal is immediately available with no time lost in processing. The recorded signal can be played back, or reproduced as many times as desired without loss of signal.
5. Multi-channel recording possible.
6. Exceedingly high density of data points giving simplified storage and handling.

**Applications.** These recorders are available in various sizes and in various forms (*e.g.*, portable, semi-portable and rack mounted); due to their unique features they have wide applications, some of these are :

1. Communication surveillance and spying.
2. Data recording and analysis on missiles, aircrafts and satellites.

3. Medical research and patient monitoring.
4. Industrial research and production monitoring and control.

### 11.4.7.2. Recording methods

Following are the three methods of magnetic tape recording which are used for instrumentation purposes :

1. Direct recording.
2. Frequency modulation (FM) recording.
3. Pulse Duration Modulation (PDM) recording.

**1. Direct reading :**

It is the simplest method of recording and *usually requires one tape track for each channel. The signal to be recorded is amplified, mixed with a high frequency bias and fed directly to the recording head as a varying electric current.*

***Advantages:***

(*i*) This requires simple electronic circuits.

(*ii*) This recording process requires only simple, moderately priced electronic circuitry.

(*iii*) It has a wide frequency response ranging from 50 Hz to about 2 MHz for a tape speed of 3.05m/s. It provides the *greatest bandwidth obtainable from a given recorder*.

(*iv*) It has a good dynamic response and takes overload without increase in distortion.

(*v*) It can be used for recording voice and multiplexing a number of channels of information into one channel of tape recording.

(*vi*) It is used to record signal where information is contained in the relation between frequency and amplitude, such as spectrum analysis of noise.

***Limitations :***

(*i*) This recorder is only used when maximum bandwidth is required.

(*ii*) This recording process is characterised by some amplitude instability caused by random surface inhomogeneities in the tape coating. Some portion may not be perfectly recorded owing to dirt or poor manufacture.

**2. Frequency modulation (FM) recording :**

This system is generally used when a more accurate response to D.C. voltage is required.

Fig. 11.27 shows a FM recording system :

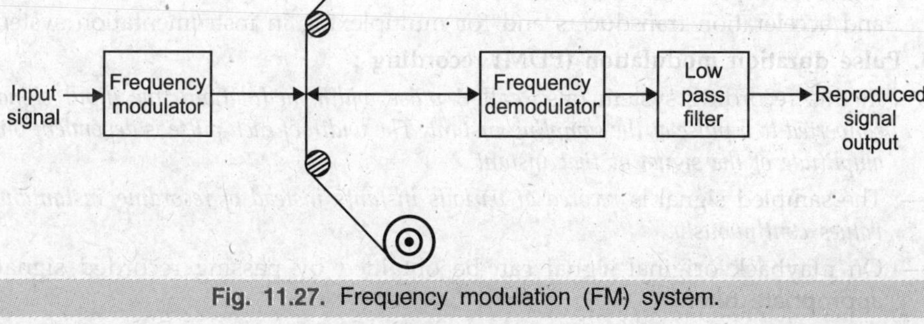


Fig. 11.27. Frequency modulation (FM) system.

- In this system a carrier oscillator frequency $f_c$ called the centre frequency, is modulated by the level of the input signal. The centre frequency is selected with respect to the tape speed and frequency deviation selected for the tape recorders is ± 40% about the carrier frequency.
- When there is no input signal, *i.e.*, zero input, the modulation is at centre frequency $f_c$. If a positive input signal is applied, the frequency deviates from the centre frequency by some amount in a certain direction, the application of a *negative* input voltage deviates the centre frequency in the *opposite direction.*
- The output of the modulation, which is fed to the tape, is a signal of constant frequency for D.C. inputs, and varying frequency for A.C. inputs. The variation of frequency is directly proportional to the amplitude of the input signal.
- On playback, the output of the reproducing head is demodulated and fed through a *"low pass filter" which removes the carrier and other unwanted frequencies* reproduced due to the modulation process.
- The *"frequency demodulator"* converts the difference between the centre frequency and the frequency on the tape, to a voltage proportional to the difference in the frequencies. This system can thus record frequencies from D.C. to several thousand hertz. Residual carrier signals and out of band noise are removed by a low pass filter.

***Advantages :***

1. Primarily useful when the D.C. component of the input signal is to be preserved.
2. No drop-out effect due to inhomogeneities of the tape material.
3. Wide frequency range, can record from D.C. voltages to several KHz.
4. Independent of amplitude variations and accurately reproduces the waveform of the input signal.

***Disadvantages :***

1. Limited frequency range.
2. High tape speed is required.
3. Extremely sensitive to tape speed fluctuations.
4. Recording circuitry is more complicated than that of direct recording system.
5. Requires a high quality tape transport and speed control mechanisms.
6. Comparatively expensive.

***Applications :***

- This type or recording system is primarily used when the D.C. component of the input signal is to be preserved or when the amplitude variations of the direct recording process are not tolerable.
- This system is widely used for recording the voltages from the pressure, force and acceleration transducers and for multiplexing in instrumentation systems.

**3. Pulse duration modulation (PDM) recording :**

- In this recording system (also called *pulse width modulation), the input signal is converted to a pulse at the sampling instant. The width of each pulse is dependent on the amplitude of the signal at that instant.*
- The sampled signal is *recored at various instants instead of recording instantaneous values continuously.*
- On playback original signal can be obtained by passing recorded signal to appropriate filter.

***Advantages :***

1. High accuracy.
2. High signal-to-noise ratio.
3. The system is capable to record information from a large number of channels simultaneously.

***Disadvantages :***

1. Limited frequency response.
2. Low reliability.
3. Complex circuitry.

***Applications :*** It is used in instrument systems for special applications such as for *simultaneous recording of large number of slowly changing variables.*

**11.4.7.3. Magnetic recording to store data on the floppy discs and hard discs of computers**

The use of *'magnetic recording' is restored to store data on the floppy disc and hard discs of computers.*

The basic principles are that *a recording head, which responds to the input signal, produces corresponding magnetic patterns on a thin layer of magnetic material and a read head gives an output by converting the magnetic patterns on the magnetic material to electrical signals.* Besides these heads the systems require a transport system which moves the magnetic material in a controlled way under the heads..

**1. Magnetic recording codes :**

- In digital recording, the signals are recorded as a *coded combination of bits.*
- A *bit cell* is the element of the magnetic coating where the magnetism is either completely saturated in one direction or completely saturated in the reverse direction. Saturation is when the magnetising field has been increased to such an extent that the magnetic material has reached its maximum amount of magnetic flux and further increases in magnetising current produce no further change.
- For getting proper flux reversals, some of the commonly used methods (involving encoding) are :
  - (*i*) Phase encoding (PE);
  - (*ii*) Non-return-to-zero (NRZ);
  - (*iii*) Frequency modulation (FM);
  - (*iv*) Modified frequency modulation (MFM);
  - (*v*) Run length limited (RLL).

*Optimum code is the one that allows the bits to be packed as close as possible and which can be read without error*. The read head can locate reversals quite easily but they must not be too close together.

— The RLL code has the advantage of being *more compact* than the other codes, PE and FM taking up the most space.

— MFM and NRZ take up the same amount of space. *NRZ* has the disadvantage of, unlike the other codes, *not being self locking.*

**2. Magnetic discs :**

*"Digital recording"* is commonly done *on a floppy or hard disc.*

The digital data is stored on the disc surface along concentric circles called *tracks*, a single disc having many such tracks. A single read-write head is used for each disc surface and heads are moved, by means of a mechanical actuator, backwards and forwards to access different tracks. The disc is spun by the drive and the read/write heads *read or write data into a track.*

The 3.5 *"floppy disc"* used in the personal computer can store 1.4 *Mbytes* of data.

***"Hard discs".*** These are sealed units with data stored on the disc surface along concentric circles.

— A hard disc assembly has more than one such disc and the data is stored on magnetic coatings on both sides of the discs.

— The discs are rotated at high speeds and tracks assessed by moving the read-write heads.

— Large amounts of data can be stored on such assemblies of discs; storage of the order of many *G bytes* are now common.

### 11.4.8. Printers

The *printers provide a record of data on paper.* Such printers are available in the following *versions* :

1. The dot matrix printer.
2. The ink jet printer.
3. The laser printer.

**1. The dot matrix printer :**

Fig. 11.28 shows the head mechanism of a dot matrix printer.

— It consists of either 9 or 24 pins in a vertical line.

— Each pin is controlled by an electromagnet which when turned on propels the pin onto the inking ribbon. This transfers a small bob of ink onto the paper behind the ribbon. A character is formed by moving the print head in horizontal lines back-and-forth across the *paper* and *firing* the appropriate pins.

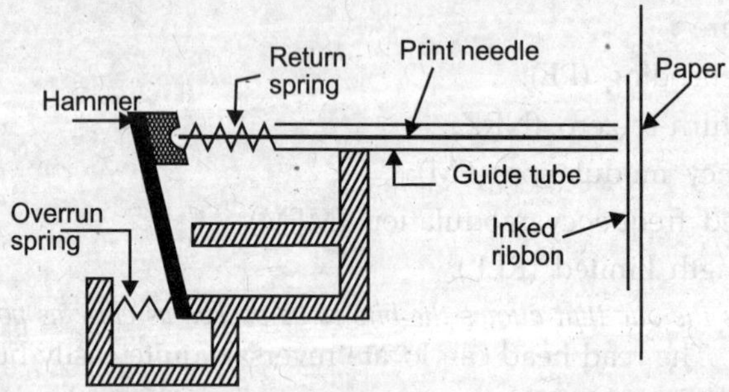


Fig. 11.28. Head mechanism of a dot matrix printer.

**2. The ink jet printer :**

This type of the printer uses a conductive ink which is forced through a small nozzle to produce a jet of very small drops of ink of constant diameter at a constant frequency.

— In one form a constant stream of ink passes along a tube and is pulsed to form fine drops by a piezoelectric crystal which vibrates at a frequency of about 100 kHz. (Fig. 11.29).

— In another form is used a small heater in the print head with vaporized ink in a capillary tube, so producing gas bubbles which push out 'ink drops'.

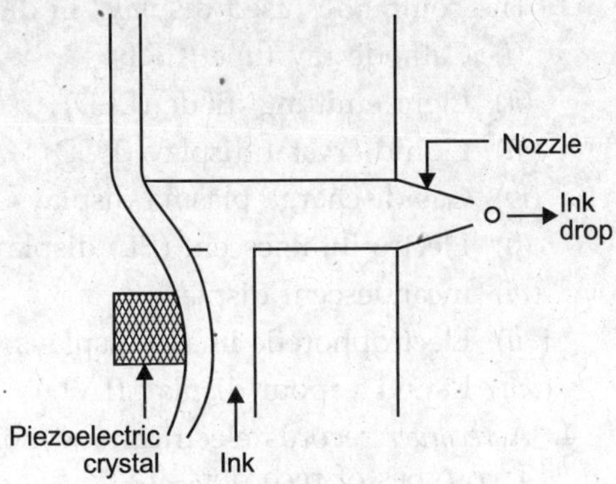


**Fig. 11.29.** Production of stream of drops.

**3. The laser printer :**

Fig. 11.30 shows the basic elements of a laser printer.

— It has a photosensitive drum which is coated with a selenium-based light sensitive material. The selenium, in the dark, has a high resistance and consequently becomes charged as it passes close to the charging wire; this is a wire at a high voltage and off which charge leaks.

— A light beam is made to scan along the length of the drum by a small rotating eight-sided mirror. When light strikes the selenium its resistance drops and it can no longer remain charged. By controlling the brightness of the beam of light the points on the drum can be discharged or left charged.

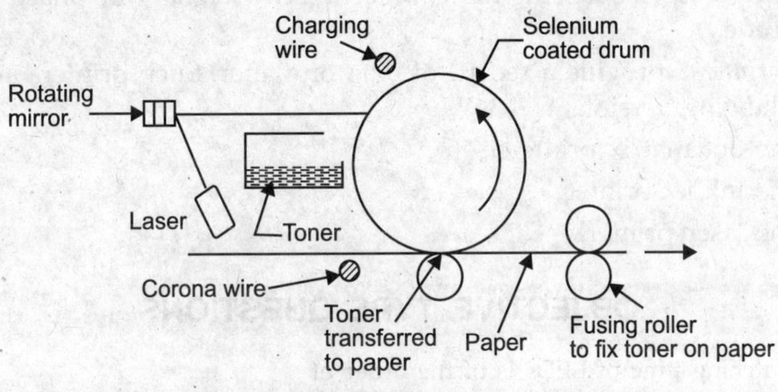


**Fig. 11.30.** Laser printer's basic elements.

— As the drum passes the toner reservoir, the charged areas attract particles of toner which thus stick to the areas that have not been exposed to light and do not stick on the areas that have been exposed to light.

— The paper is given a charge as it passes another charging wire, the so called corona wire, so that as it passes close to the drum it attracts the toner off the drum. A hot fusing roller is then used to melt the toner particles so that, after passing between rollers, they finely adhere to paper.

## HIGHLIGHTS

1. *"Analog indicators"* are basically current-operated devices working on the d'Arsonval type galvanometer principle.
2. *"Digital type meters"* indicate the reading in exact numerals.

3. The commonly used displays in digital electronic field are :
   (*i*) Cathode ray tube (CRT)
   (*ii*) Light emitting diode (LED)
   (*iii*) Liquid crystal display (LCD)
   (*iv*) Gas discharge plasma displays (cold cathode displays or Nixies)
   (*v*) Electro-luminescent (EL) displays
   (*vi*) Incandescent displays
   (*vii*) Electrophoretic image displays (EPID)
   (*viii*) Liquid vapour display (LVD)
4. A *recorder* records electrical and non-electrical quantities as a function of time. Two types of recorders are :
   (*i*) Analog recorders.
   (*ii*) Digital recorders.
5. Instruments that record change of only one measured variable are called *single point recorders.* A *multi-point recorder* may have as many as 24 inputs, with traces displaced in 6 colours.
6. A *X-Y recorder* is an instrument which gives a graphic record of the relationship between two variables.
7. *Cathode ray oscilloscope (CRO)* is an instrument which presents signal waveforms visually. It is also useful for comparing two signals in phase, frequency or amplitude.
8. The *"printers"* provide a record of data on paper. Such printers are available in the following versions :
   (*i*) The dot matrix printers.
   (*ii*) The ink jet printer.
   (*iii*) The laser printer.

## OBJECTIVE TYPE QUESTIONS

1. The switching time of LEDs is of the order of
   (*a*) 1 s (*b*) 1 ms
   (*c*) 1 μs (*d*) 1 ns.
2. LEDs emit light
   (*a*) only in red colour (*b*) only in yellow colour
   (*c*) only in green colour colour (*d*) in red, green, yellow and amber
3. An LCD requires a power of
   (*a*) 20 W (*b*) 20 mW
   (*c*) 20 μW (*d*) 20 nW.
   to light up one segment of a seven-segmented display system.
4. X-Y recorders
   (*a*) record one quantity with respect to another quantity
   (*b*) record one quantity on X axis with respect to time on Y axis
   (*c*) record one qunatity on Y axis with respect to time on X axis.
   (*d*) any of the above.

5. In a direct magnetic tape recording system, the reproducing head and amplifier characteristics are
   (a) – 6 db/octave. +6 db/octave (b) – 6 db/octave, + 20 db/decade
   (b) + 6 db/octave, + – 6 db/octave (d) – 20 db/decade, + 20 db/decade.
6. A Nixie tube requires
   (a) 10 cathodes (b) 14 cathodes
   (c) 15 cathodes (d) 20 cathodes.
7. Nixie tubes require
   (a) 150 V – 220 V (b) 100 V – 150 V
   (c) 150 mV – 220 mV (d) 100 mV – 150 mV
   to produce glow discharge.
8. If an information is required to be stored over a short interval of time
   (a) a single number devices should be used
   (b) a storage type oscilloscope should be used
   (c) a CRO with photographic equipment should be used
   (d) a direct writing recorder or a magnetic tape recorder should be used.
9. The machine interpretable outputs from an analog transducer can be had from
   (a) magnetic tapes (b) punched cards and tapes
   (c) teletypewriter (d) all of the above.
10. Digital instruments have input inpedance of the order of
   (a) Ω (b) kΩ
   (c) MΩ (d) mΩ.
11. The advantages of F.M. magnetic tape recording are :
   (a) It can record from d.c. to several kHz.
   (b) It is free from dropout effects.
   (c) It is independent of amplitude variations and accurately reproduces the waveform of input signal.
   (d) All of the above.
12. The turn on and turn off times of a LCD are of the order of
   (a) 1 s (b) 1 ms
   (c) 10 μs (d) 10 ns.
13. The power requirement of an LED is
   (a) 40 mW per numeral (b) 40 μW per numeral
   (b) 10 W per numeral (d) 10 μW per numeral.
14. If four flip-flops are connected in cascade, they serve as
   (a) 2 : 1 scalar (b) 4 : 1 scalar
   (c) 8 : 1 scalar (d) 16 : 1 scalar.
15. The LEDs, for their display require
   (a) a voltage of 1.2 V and a current of 20 mA
   (b) a voltage of 25 V and a current of 20 mA
   (c) a voltage of 1.2 V and a current of 100 mA
   (d) a voltage of 25 V and a current of 100 mA.
16. Which of the following is the visual display unit?
   (a) Cathode any oscilloscope (b) U.V. recorder
   (c) Storage oscilloscope (d) Moving coil oscillograph.
17. Which of the following units has a high frequency response but presents difficulty in getting a permanent record?

(a) Servo recorder (b) Moving coil oscillograph
(c) X-Y recorder (d) Cathode ray oscilloscope.

**18.** The advantages of F.M. magnetic tape recording are :
(a) It can record from d.c. to several kHz. (b) It is free from dropout effects.
(c) It is independent of amplitude and accurately reproduces the waveform of input signal.
(d) All of the above.

**19.** The source of emission of electrons in a CRT is :
(a) PN junction diode
(b) A barium and strontium oxide coated cathode
(c) Accelerating anodes (d) Post accelerating anodes.

**20.** The pointer-scale instruments have a
(a) very high (b) very low
(c) linear (d) stable frequencies response.

**21.** The operation of a moving-coil current recording instruments is based on
(a) photo-electric principle (b) D' Arsonval principle
(c) piezo-electric principle (d) thermo-electric principle.

**22.** The time bases of an oscilloscope are generated by
(a) horizontal amplifier (b) vertical amplifier
(c) sweep generators (d) storage oscilloscope.

**23.** Which of the following devices requires a matching network to avoid overloading of the signal source and prevent damage from excessive current?
(a) U.V. recorder (b) X-Y recorder
(c) Storage oscilloscope (d) Servo recorder.

**24.** X-Y recorders record a quantity
(a) with respect to another quantity
(b) on X-axis with respect to time on Y-axis.
(c) on Y-axis with respect to time on X-axis
(d) any of these.

**25.** In an electrodynamometer type of wattmeter
(a) the current coil is made fixed (b) the pressure coil is made fixed
(c) any of the two coils can be made fixed (d) both the coils should be movable.

**26.** An average reading VTVM uses one diode with an external series resistance. A high value of series resistance is used so that the instrument should have
(a) a high input impedance (b) a linear $v$-$i$ characteristics
(c) low power consumption (d) all of the above.

**27.** In a CRT the focusing anode is located
(a) between pre-accelerating and accelerating anode
(b) after accelerating anode
(c) before pre-accelerating anode (d) none of the above.

**28.** Post acceleration is needed in a CRO if the frequency of the signals is
(a) less than 1 MHz (b) more than 1 MHz
(c) more than 10 MHz (d) more than 10 Hz.

**29.** The slewing speed in an X-Y recorder refers to
(a) time base
(b) maximum constant velocity that the marking pen achieves
(c) frequency response

(*d*) relationship between inputs to *X* and *Y* channels.

**30.** Which of the following recorder/display units has the highest frequency response and least response time?

(*a*) X-Y plotters (*b*) U.V. recorders

(*c*) Pen recorders (*d*) CRO.

**31.** An LCD requires a power of approximately

(*a*) 20 W (*b*) 20 mW

(*c*) 20 μW (*d*) 20 nW.

## ANSWERS

| | | | | | | |
|---|---|---|---|---|---|---|
| 1. (*d*) | 2. (*d*) | 3. (*c*) | 4. (*a*) | 5. (*c*) | 6. (*c*) | 7. (*a*) |
| 8. (*b*) | 9. (*d*) | 10. (*c*) | 11. (*d*) | 12. (*c*) | 13. (*a*) | 14. (*d*) |
| 15. (*a*) | 16. (*a*) | 17. (*d*) | 18. (*d*) | 19. (*b*) | 20. (*b*) | 21. (*b*) |
| 22. (*c*) | 23. (*a*) | 24. (*a*) | 25. (*a*) | 26. (*d*) | 27. (*a*) | 28. (*c*) |
| 29. (*b*) | 30. (*d*) | 31. (*c*). | | | | |

## THEORETICAL QUESTIONS

**1.** What are the factors that decide about the choice between the display devices and recorders ?

**2.** How are output devices categorized ?

**3.** What are the differences between the analog type and digital type instruments.

**4.** How are digital display units classified ?

**5.** Name the displays which are commonly used in digital electronic field.

**6.** Explain briefly the following dispalys :

(*i*) Cathode ray tube (CRT).

(*ii*) Light emitting diode (LED).

**7.** State the advantages, disadvantges and applications of LEDs.

**8.** Discuss briefly the various types of displays.

**9.** Light emitting diodes, liquid crystal diodes play very important role in digital display devices. Describe them in detail. What are their advantages and disadvantages? Compare.

**10.** Briefly discuss the use of LED and LCD as display devices in instrumentation. Comment on their relative merits and demerits.

**11.** State the difference between an analog and a digital indicator.

**12.** Explain a seven-segment LED display.

**13.** Discuss with a neat diagram a method of realising a seven segment numeric display using LEDs.

**14.** What are the operating principles of LCD display ?

**15.** What are the advantages of LCD display over Nixie tube and LED display ?

**16.** Compare the relative performance of the following display devices in numeric display applications :

(*i*) Flat panel alphanumeric CRT.

(*ii*) Liquid vapour display.

(*iii*) Electrophoretic image display.

(*iv*) Nixie tubes.

**17.** What is the function of the display or recording element of a generalised measurement system?

18. How does a display unit differ from a recorder?
19. How are the output devices categorized? Explain briefly.
20. How can we get machine interpretable outputs?
21. List the different forms in which the display is available from an instrument.
22. How are electrical indicating instruments classified?
23. How analog display meters differ from digital type meters?
24. Give a comparison between analog type and digital type instruments.
25. Give four examples each of the analog type and digital type instrumentation.
26. Explain briefly the following:
    (*i*) Single-point indicators.
    (*ii*) Multi-point multi-pointer and multi-range indicators.
27. Explain briefly the essential features of indicating instruments.
28. Describe briefly any two of the following :
    (*i*) Moving-iron instruments; (*ii*) Moving-coil instruments;
    (*iii*) Rectifier instruments.
29. What are advantages of electronic voltmeters?
30. What are digital instruments? State their principle of operation.
31. Explain briefly any two of the following:
    (*i*) Semiconductor light emitting diodes (LED);
    (*ii*) Liquid crystal displays;
    (*iii*) Hot filament or bar tubes;
    (*iv*) Numerical indicator tubes (NIT).
32. What is a recorder?
33. Elaborate the difference between a display unit and a recorder.
34. What is meant by a direct reading instrument?
35. Explain the functioning of a basic type of strip chart recorder. Enumerate the different types of marking mechanisms used in it.
36. Distinguish between single point and multi-point recorders.
37. What is a X-Y recorder? State its applications.
38. Explain the working of an ultraviolet (U.V) recorder. State its applications.
39. What are the basic components of a magnetic tape recorder for instrumentation applications? List its advantages and disadvantages.
40. Explain with neat diagram the construction and working of a cathode ray oscilloscope (CRO).
41. Describe the graphic recorders.
42. What is a strip chart recorder? Why is it called X-t recorder too? Describe the working of a basic type of strip chart recorder.
43. Explain with the help of a diagram the CRT recorder.
44. Explain with the help of a diagram the working of a galvanometer oscillographic recorder.
45. Describe the FM magnetic tape recording. Give its merits, demerits and applications.
46. What is pulse duration modulation (PDM)? Give its merits and demerits.
47. Describe the digital tape recorders.
48. Write short notes on the following :
    (*i*) Potentiometer type recorder.
    (*ii*) Circular chart recorders.
    (*iii*) X-Y recorders.

CHAPTER

# 12 Measurement of Non-Electrical Quantities

> **12.1. Pressure measurement** – Definition of pressure – Units for pressure and pressure measuring instruments – Electrical pressure transducers – Measurement of high pressure – Measurement of low pressure – Calibration of pressure measuring equipment; **12.2. Temperature measurement** – Introduction – Temperature measuring instrument – Electrical resistance thermometer or resistance temperature detector (RTD) – Thermistors – Thermocouple thermometers – Radiation pyrometers – Optical pyrometers; **12.3. Flow measurement** – Introduction – Flow measurement methods/devices – Rotameter and elbow meter – Electromagnetic flow meters – Hot wire anemometers – Ultrasonic flow meters; **12.4. Measurement of displacement** – Measurement of linear displacement – Measurement of angular displacement; **12.5. Measurement of velocity/speed** – Measurement of linear velocity – Measurement of angular velocity – Electrical tachometers; **12.6. Measurement of acceleration** – Piezoelectric accelerometer – Seismic accelerometer; **12.7. Measurement of force** – Load cells; **12.8. Measurement of torque** – Introduction – Torque measurement methods – Electrical torsion meter – Strain gauge torsion meter; **12.9. Measurement of shaft power** – Eddy current dynamometers – D.C. dynamometer; **12.10. Measurement of liquid level** – Capacitance type level gauge – Gamma rays liquid level gauge – Ultrasonic liquid level gauge – Electrical level gauge – *Highlights* – *Objective Type questions* – *Theoretical Questions.*

## 12.1 PRESSURE MEASUREMENT

### 12.1.1. Definition of Pressure

**Pressure** *is defined as a force per unit area.*

Pressures are exerted by gases, vapours and liquids. The instruments that we generally use, however, record pressure as the difference between the two pressures. Thus, it is the *difference between the pressure exerted by a fluid of interest and the ambient atmospheric pressure.* Such devices indicate the pressure either above or below that of the atmosphere.

**Atmospheric pressure :**

*The atmospheric air exerts a normal pressure upon all surfaces with which it is in contact, and it is known as* **atmospheric pressure.** The atmospheric pressure is also known as *Barometric pressure.*

The atmospheric pressure at sea level (above absolute zero) is called *standard atmospheric pressure.*

***Note:*** The local atmospheric pressure may be a little lower than these values, if the place under question is higher than sea level, and higher values if the place is lower than sea level, due to the corresponding decrease of the column of air standing, respectively.

**Gauge pressure :**

*It is the pressure, measured with the help of pressure measuring instrument, in which atmospheric pressure is taken as datum.*

The *atmospheric pressure* on the scale is marked as *zero.*

***Gauges*** record the pressure above or below the local atmospheric pressure, since they measure the difference in pressure of the liquid to which they are connected and that of surrounding air.

If the pressure of the liquid is below the local atmospheric pressure, then the gauge is designated as '*vacuum gauge*' and the recorded value indicates the amount by which the pressure of the liquid is *below* local atmospheric pressure, *i.e., negative pressure.*

(*Vacuum pressure* is defined as the pressure *below the atmospheric pressure*).

**Absolute pressure :**

It is necessary to establish an absolute pressure scale which is independent of the changes in atmospheric pressure. *A pressure of absolute zero can exist only in complete vacuum.*

*Any pressure measured above the absolute zero of pressure is termed as an* **'absolute pressure'.**

A schematic diagram showing the gauge pressure, vacuum pressure and the absolute pressure is given in Fig. 12.1.

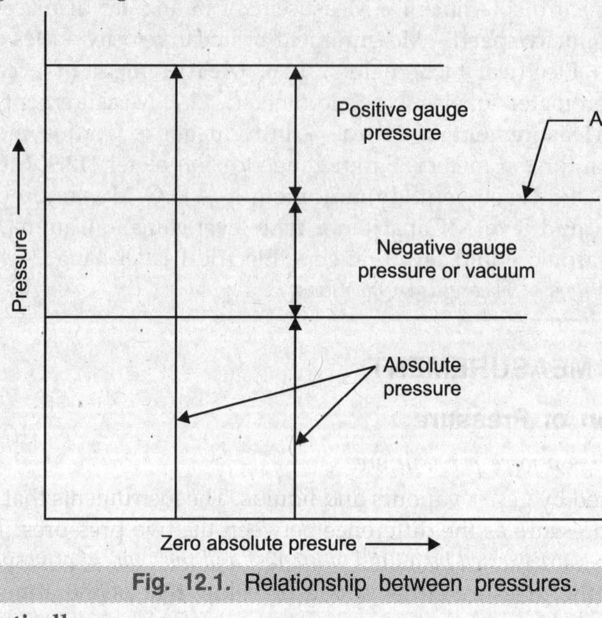


Fig. 12.1. Relationship between pressures.

**Mathematically :**

1. Absolute pressure = Atmospheric pressure + gauge pressure

   *i.e.,* $p_{abs.} = p_{atm.} + p_{gauge.}$

2. Vacuum pressure = Atmospheric pressure – absolute pressure.

**Static pressure ($p_s$) and total pressure ($p_t$) :**

***Static pressure*** *is defined as the force per unit area acting on the wall by a fluid at rest or flowing parallel to the wall in a pipeline.*

***Total or stagnation pressure*** *is defined as the pressure that would be obtained if the fluid stream were brought to rest isentropically.* For an incompressible fluid or gas flowing at low velocities, the *dynamic pressure* equals $\frac{V^2}{2g}$ where, $V$ is the velocity of the fluid flow.

Thus, Total pressure = Static pressure + dynamic (or velocity) pressure

$$\therefore \quad p_t = p_s + \frac{V^2}{2g} \qquad ...(12.1)$$

### 12.1.2. Units for Pressure and Pressure Measuring Instruments

The fundamental S.I. units of pressure is pressure is newton per square meter (N/m$^2$). This is also known as *Pascal* ($P_a$).

*"Low pressures"* are often expressed in terms of *mm of water* or *mm of mercury*. This is an abbreviated way of saying that the pressure is such that will support a liquid column of stated height.

- *Standard atmospheric pressure has the following equivalent values :*
  101.3 kN/m$^2$ or 101.3 kPa; 10.3 m of water; 760 mm of mercury; 1013 mb (millibar); $\simeq$ 1 bar = 100 kPa = $10^5$ N/m$^2$.

(When the local atmospheric pressure is not given in a problem, it is taken as 100 kN/m$^2$ or 10 m of water for simplicity of calculations.)

**Pressure measuring instruments :**

The pressure measurement instruments can be *categorized* as follows :

1. *Instruments for measuring low pressures (below 1 mm of Hg) :*
   - Manometers.
   - Low pressure gauges.
2. *Instrument for medium and high pressures (between 1 mm of Hg to 1000 atmospheres) :*
   - Bourdon tube.
   - Diaphragm gauges.
   - Bellow pressure gauge.
   - Dead-weight pressure gauge.
3. *Instrument for measuring low vacuum and ultra high vacuum (760 torr to $10^{-9}$ torr and beyond; 1 torr = 1 mm of Hg) :*
   - Mcleod.
   - Thermal conductivity.
   - Ionization gauges.
4. *Instruments for measuring very high pressures (1000 atmospheres and above) :*
   - Bourdon tube.
   - Diaphragm gauges.
   - Electrical resistance pressure gauges.
5. *Instruments for measuring varying pressure :*
   - Engine indicator.
   - Cathode ray oscilloscope (CRO).

Fig. 12.2. shows the summary of applicable range of pressure gauges.

Some of these instruments are discussed in the following Articles.

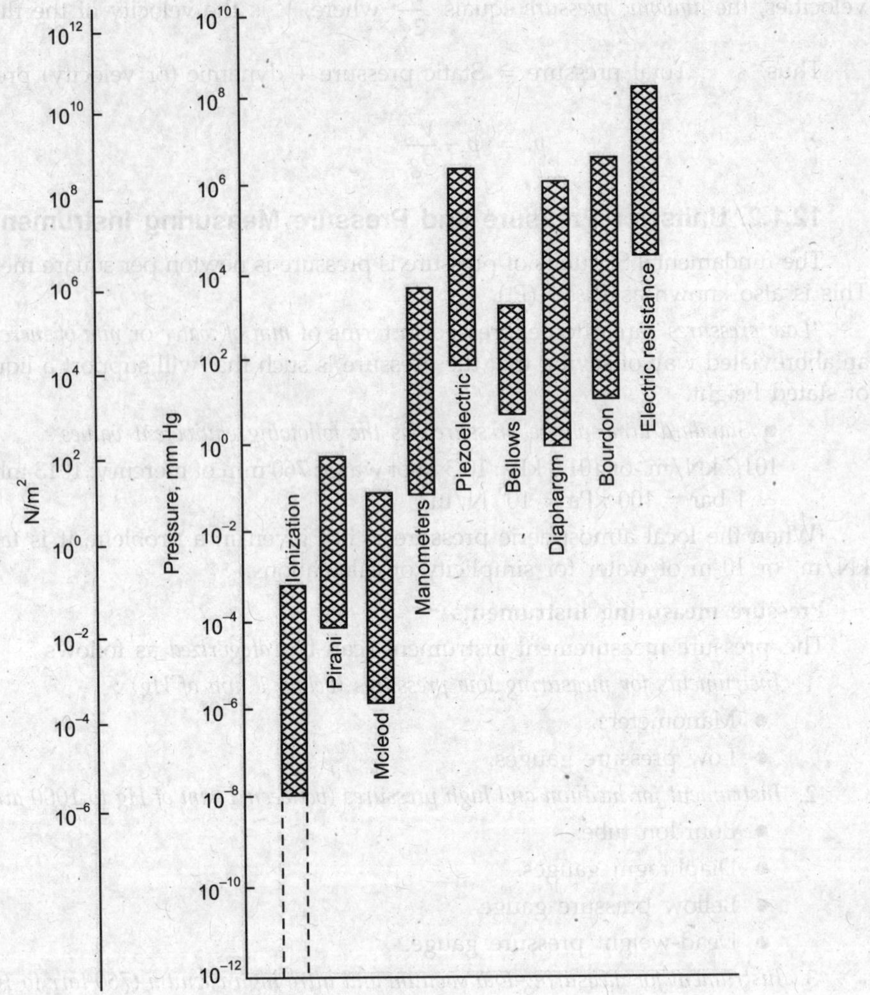


Fig. 12.2. Summary of applicable range of pressure gauges.

### 12.1.3. Electrical Pressure Transducers

A **transducer** *is a device for converting a signal of one sort into the other type.* In other words, a **pressure gauge** *is a transducer for converting the distortion of a Bourdon spring into an indication through use of a calibrated dial* and so on. In case of electrical transducers for measuring pressure, various transformations which are needed *to convert pressure to some sort of usable electrical signal* can be :

1. The change of resistance of an elastic material such as a wire.
2. The change in capacitance when the distance between capacitor plates is varied or the dielectric is changed.
3. The conversion of a changing pressure into electrical voltage signals.
4. Creation of a voltage by the movement of a coil within a magnetic field.
5. Generation of a voltage or current through the use of photoelectric effects.
6. Torque balance system (electromagnetic).

The following **pressure transducers** will be discussed :

1. Resistance type pressure transducers.
2. Pressure voltage type.
3. Inductive type.
4. Capacitive type.
5. Carbon pile type.
6. Piezoelectric type.
7. Photoelectric type.
8. Electromagnetic type.

**12.1.3.1. Resistance type pressure transducers**

Resistance type devices used in pressure transducers include;

1. Strain gauges.
2. Moving contacts.

**1. *Strain gauge* :**

A strain gauge is simply a fine wire in the form a grid (Fig. 12.3). When the grid is distorted, the resistance ($R$) of the wire changes according the formula :

$$R = k\frac{L}{A} \quad ...(12.2)$$

where $k$ = A constant for the particular kind of wire

$L$ = Length of the wire, and

$A$ = Cross-section area.

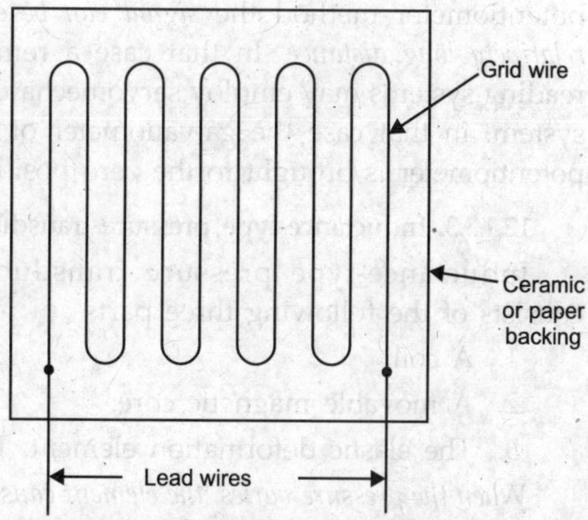


**Fig. 12.3.** Strain gauge.

As the strain gauge is distorted by the elastic deformation element, its *length is increased and its cross-section area is reduced*. Both of these changes *increase the resistance.*

- Since little distortion is required to change the resistance of a strain gauge through its total range, this type of transducer can be used to detect *very small movement* and, therefore, *very small pressure changes.*
- Since current flows in the strain gauge, some heat will be generated which must be compensated for. One method of compensation employs a second strain gauge placed close to the measuring unit. These two are used in a bridge circuit so that the *heating effect of the two resistive elements is cancelled out*, and the only resistance change in the bridge is due to the deformation.

Most often, strain gauge bridges are A.C. bridges.

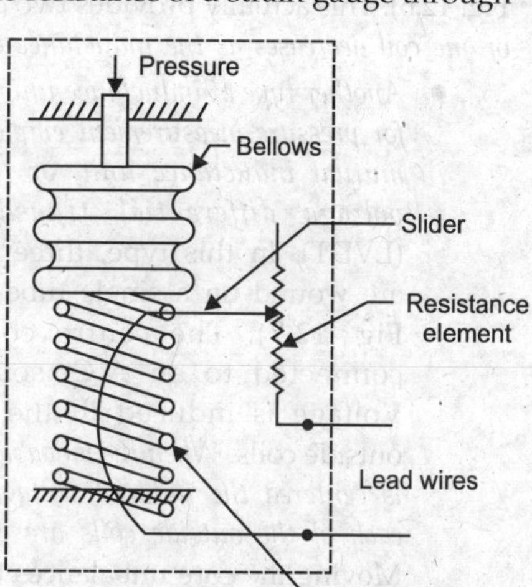


**Fig. 12.4.** Moving contact type resistance transducer.

**2. *Moving contact type resistance transducer* :** This type of transducer is most often used with bellows because of the force required. Fig. 12.4 shows a typical circuit for such a transducer.

**12.1.3.2. Pressure-voltage transducer (potentiometer devices)**

One of the simplest of pressure-voltage transducer uses a potentiometer. *The potentiometer is used to convert a pressure signal into an electrical voltage as shown* in Fig. 12.5.

The potentiometer should have *little friction.* Levers may be used to match the required movement of the potentiometer with that of the bellows, the signal applied to potentiometer may be relatively larger of the order of several volts. By the use of potentiometer method the *signal can be sent relatively long distance.* In that case a remote reading systems may employ servomechanism system. In that case the galvanometer of the potentiometer is brought to the zero position.

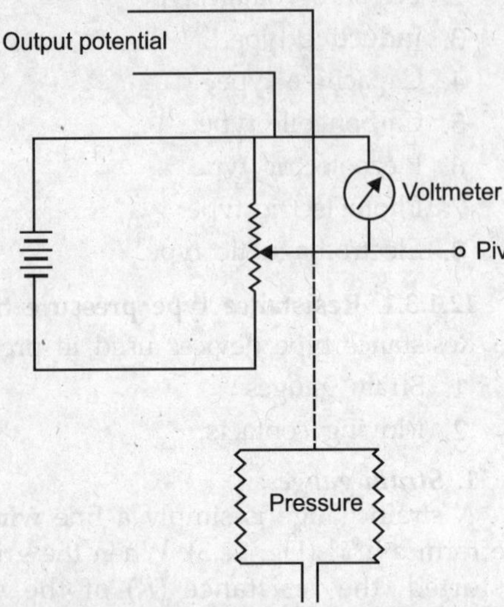


Fig. 12.5. Pressure-voltage transducer.

**12.1.3.3. Inductance-type pressure transducer**

Inductance-type pressure transducers consists of the following three parts :

1. A coil.
2. A movable magnetic core.
3. The elastic deformation element. The element is attached to the core.

*When the pressure varies, the element causes the core to move within the coil.* An alternating current is passed through the coil, and as the core moves the inductance of the coil changes. *The current passing through the coil increases as the inductance decreases.* This type of transducer is used in a *current-sensitive coil.*

To get increased sensitivity the coil may be divided into two, using a centre tap (See Fig. 12.6). This actually provides two coils. As the core moves inside the coils, the *inductance of one coil decreases as the inductance of the other increases.*

- *Another type of inductance unit used for pressure measurement employs a mutual inductance unit, or linear voltage differential transducer* (LVDT). In this type, three coils are wound on a single tube (See Fig. 12.7). The centre coil is connected to an A.C. source. Voltage is induced in the two outside coils. *When the movable core is centered the induced voltages in each of the outside coils are equal.* Moving the core unbalances these voltages, and the difference between them can be measured.

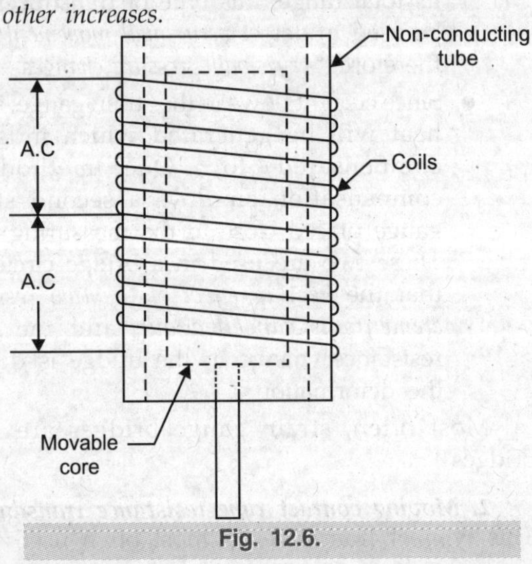


Fig. 12.6.

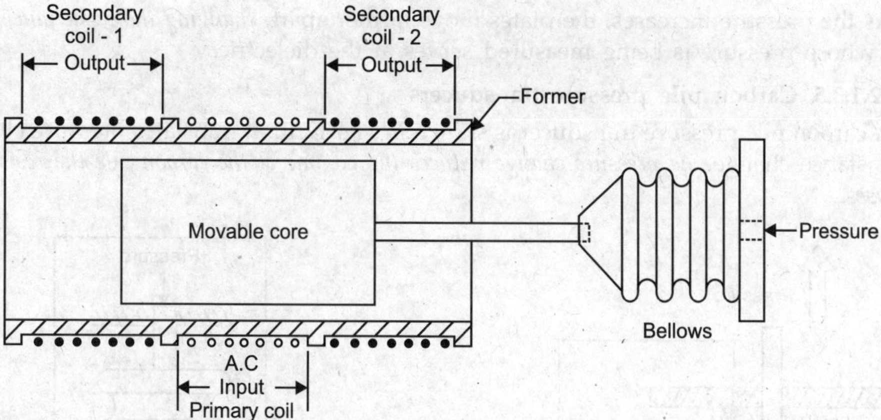


Fig. 12.7. Linear voltage differential transducer (LVDT).

**Reluctance-type pressure transducer :**

Similar to the inductance type pressure transducer is the reluctance-type. Fig. 12.8 shows a reluctance-type pressure trans-ducer. Here the electrical energy is provided by the field of a permanent magnet rather than A.C. source.

For pressure measurement, the elastic deformation element moves an armature closer to or farther from the permanent magnet. *The reluctance of a coil wrapped around the permanent magnet varies with the movement.*

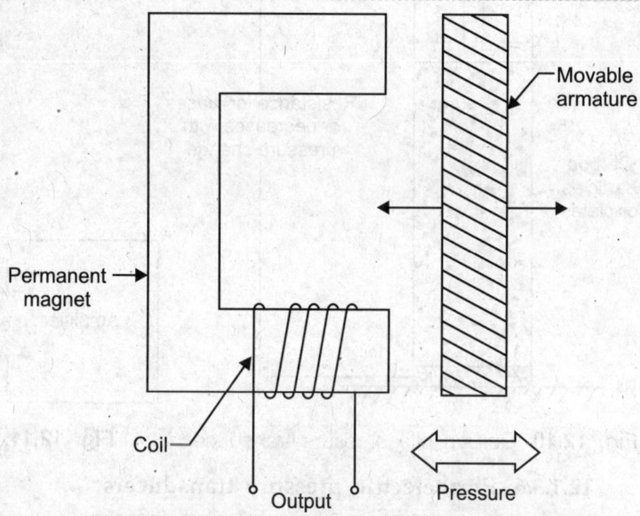


Fig. 12.8. Reluctance-type pressure transducer.

### 12.1.3.4. Capacitive-type pressure transducer

Capacitive pressure transducers consists of two conductive plates and a dielectric (See Fig. 12.9).

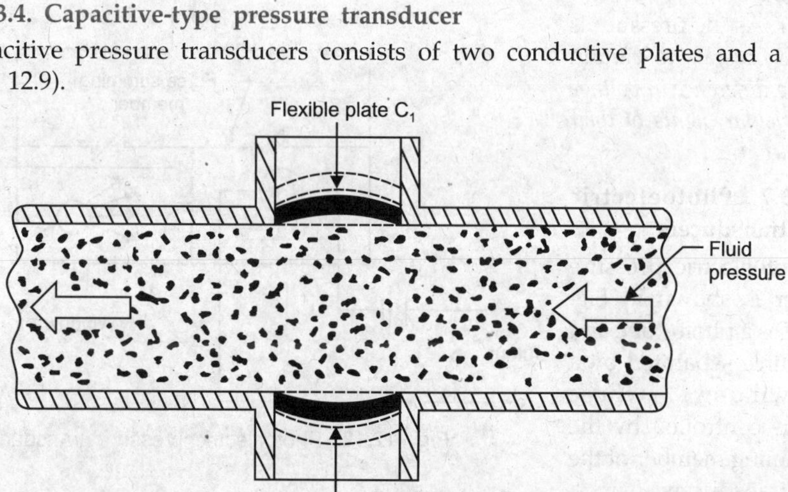


Fig. 12.9. Capacitive-type pressure transducer.

As the pressure increases, the plates move farther apart, *changing the capacitance.* The fluid whose pressure is being measured serves at the dielectric.

**12.1.3.5. Carbon pile pressure transducers**

A carbon pile pressure transducer is shown in Fig. 12.10. It translates pressure change to resistance change. *As pressure change reduces the volume of the carbon pile the resistance decreases.*

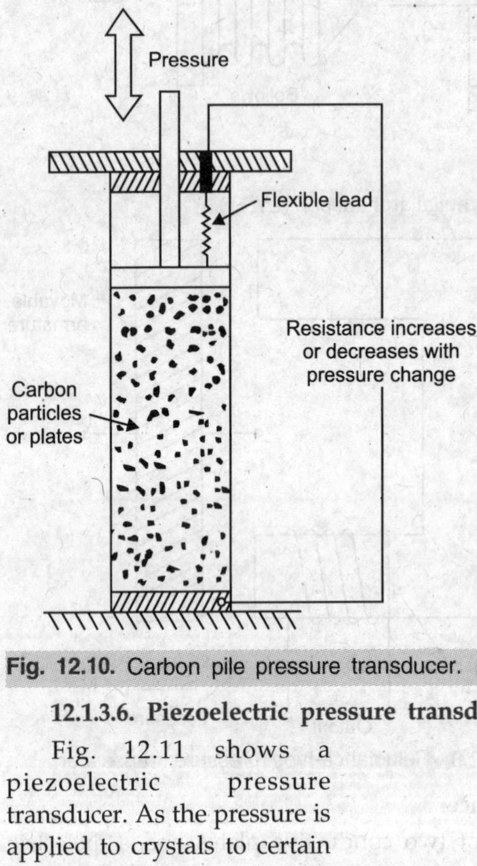


**Fig. 12.10.** Carbon pile pressure transducer.

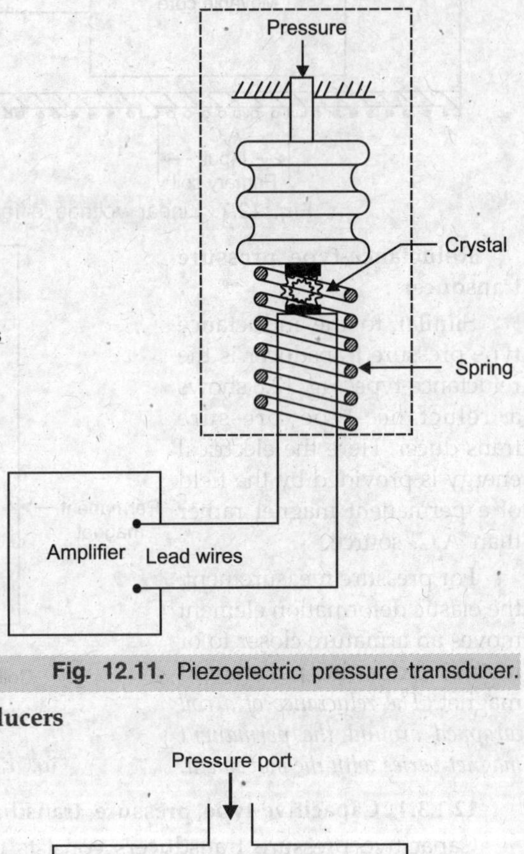


**Fig. 12.11.** Piezoelectric pressure transducer.

**12.1.3.6. Piezoelectric pressure transducers**

Fig. 12.11 shows a piezoelectric pressure transducer. As the pressure is applied to crystals to certain materials a *difference in voltage across particular points of their structure occurs.*

**12.1.3.7. Photoelectric pressure transducers**

A photoelectric pressure transducer is shown in Fig. 12.12: It uses a photo-tube and a light source separated by a small window, whose aperture is controlled by the force summing member of the pressure transducers.

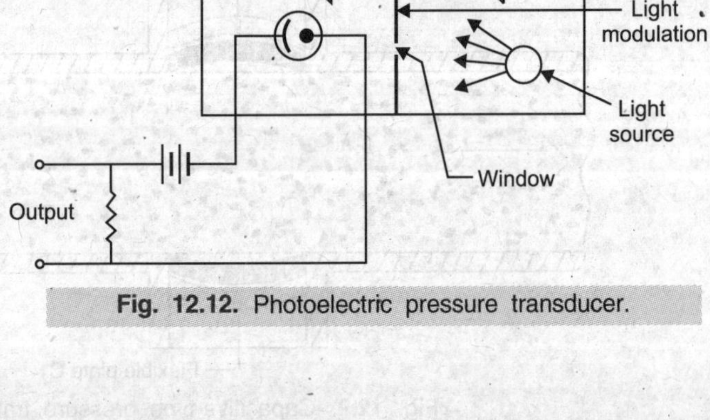


**Fig. 12.12.** Photoelectric pressure transducer.

### 12.1.3.8. Electromagnetic pressure transducers

For many years a technique has been used which consists of coil in plate or grid circuit of an oscillating tube (one that is generating alternating current) to deliver current to some other circuit. The energy may be transferred into some other winding having a transformer action. The coil has to deliver more power to the secondary winding, the additional power must come from the winding in the plate. The increase in plate current causes relays to operate at some predetermined values of current. It can also be used to develop voltages across known resistors. It can cause electromagnets to establish forces used for measuring the force applied which is obtained by pressure bellows.

## 12.1.4. Measurement of High Pressures

Conventional pressure-measuring devices, such as strain-gauge pressure cells and Bourdon-tube gauges may be used at pressure as high as 70000 MPa. *Bourdon tubes* for such pressures are nearly round in section and have a high ratio of wall thickness to diameter. They are, therefore, quite *stiff*, and *deflection per turn is small.* For this reason, high-pressure Bourdon tubes are often made of a number of turns.

For the measurement of very high pressure (above 70000 MPa), the *"electrical gauges" based on the principle of change of resistance with change of pressure are used.*

It is known that resistance of fine wires changes with pressure according to the following linear relationship:

$$R = R_1(1 + b\,\Delta p) \qquad ...(12.3)$$

where, $R$ = Resistance at a pressure of 1 atm., ohms,

$b$ = The pressure coefficient of resistance, and

$\Delta p$ = The gauge pressure.

*A pressure transducer based upon this principle is called a* **Bridgman gauge.**

Fig. 12.13 shows a Bridgman gauge. The sensing element does not actually contact the process medium but is separated therefrom by a kerosene-filled bellows. One end of the sensing coil is connected to a central terminal, as shown, while the other is grounded, thereby completing the necessary electrical circuit.

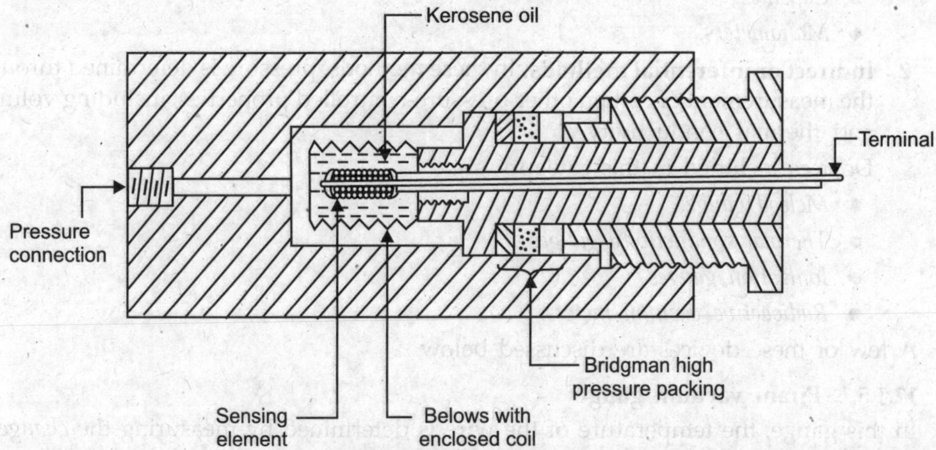


**Fig. 12.13.** Bridgman gauge.

The fine wire used in a typical gauge is made of manganin or gold chrome. Manganin has a pressure coefficient of resistance $25 \times 10^{-12}$ Pa$^{-1}$ and the total resistance of the wire is 100 ohms.

Although pressure coefficient resistance of gold is $10 \times 10^{-12}$ $Pa^{-1}$ (it gives a change in resistance which is $\frac{1}{4}$th of that of manganin for the same change in pressure), it is preferred to manganin in many applications as it has *resistance temperature coefficient* which is one half of that of manganin and, therefore, its use results in *lower errors on account of temperature changes.*

This gauge claims the following **advantages :**

1. Commercially available with scale upto pressures of 15 MPa with accuracies of 0.1 to 0.5% when properly calibrated.
2. Good dynamic response (the resistance wire itself can respond to variations in MHz range but the overall frequency response of the pressure measurement system is limited to *much lower values because of the acoustic response of the transmitting fluid).*

### 12.1.5. Measurement of Low Pressure (Vacuum)

In general, pressures below atmospheric may be called **low pressures** or **vacuums.**

It is impossible to reach an absolute pressure of zero, since a *positive* magnitude of absolute pressure exists at all times, even in a vacuum.

The *micrometer, which is one-millionth of a meter (0.001 mm) of mercury column, is a common unit of "low pressure".*

***Very low pressure*** ...... any pressure *below* 1 mm of mercury (or torr).

***Ultra low pressure*** ...... *less than* a nanometer ($10^{-3}$ μm).

The two basic methods of measurement of low pressure are :

1. **Direct methods.** Here, the displacement deflection caused by the pressure is measured and is correlated to the pressure. The following devices are included in this *category* :
   - *Spiral Bourdon tubes.*
   - *Flat and corrugated diaphragms.*
   - *Capsules.*
   - *Manometers.*
2. **Indirect or inferential methods.** In these methods, pressure is determined through the measurement of certain other pressure-controlled properties, including volume and thermal conductivity.

   Devices included in this category are :
   - *Mcleod gauge.*
   - *Thermal conductivity gauges.*
   - *Ionization gauges.*
   - *Radioactive vacuum meters.*

A few of these devices are discussed below.

#### 12.1.5.1. Pirani vacuum gauge

In this gauge, the temperature of the wire is determined by measuring the *change of resistance.*

The pirani gauge employs a single filament (in the form of four coiled wires of tungsten or platinum connected in parallel) enclosed in a glass tube/chamber, whose pressure is to be measured (See Fig. 12.14).

As the surrounding pressure changes, the filament temperature and hence its resistance also changes. A *compensating cell* is also employed to minimise variation caused by ambient temperature changes (its filament material is same as that of the measuring cell). The resistance change of the filament in the measuring cell is measured by the use of a resistance bridge which is *calibrated in terms of pressure.*

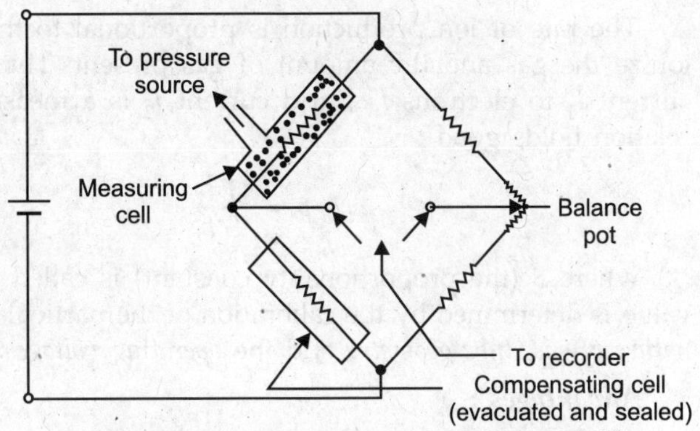


**Fig. 12.14.** Pirani vacuum gauge.

These gauges cover range from about $10^{-5}$ to 1 torr.

***Advantages :***

(*i*) These gauges are rugged, inexpensive and usually more accurate than thermocouple gauges.

(*ii*) The pressure reading range is wider.

(*iii*) Fast response to changes in pressure.

(*iv*) Possibility of process control and remote reading.

***Disadvantages :***

(*i*) Calibration is non-linear and varies from one gas to another.

(*ii*) Require calibration against some pressure standards.

(*iii*) Poor transient response.

(*iv*) Operation requires electrical power.

**12.1.5.2. Ionization gauges**

These gauges are used for measurement of pressures as low as 0.000001 micron (one billionth of a mm of Hg); the maximum pressure which an ionization gauge can measure is about 1 micron.

The ionization gauge (Fig. 12.15) is very similar to the ordinary electronic tube. This gauge therefore, mainly consists of an envelope which is evacuated by the pressure to be measured and contains a heated filament, a positively based grid and a negative based plate. The grid draws electrons from the heated filament, and collision between them and gas molecules causes ionization (Ionization is the process of knocking off an electron from an atom and thus producing a free electron and a positively charged ion) of the molecules. As the plate is negatively charged, the positively charged ions are attracted to the plate of the tube, causing current ($I_1$) flow in the external circuit; the electrons are collected by the grid and a current $I_2$ is produced in the grid circuit.

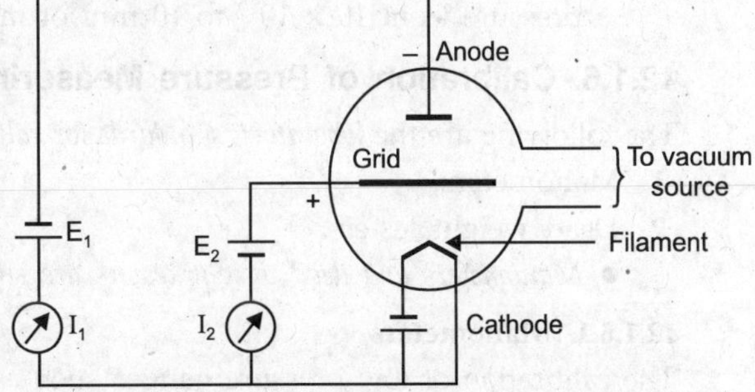


**Fig. 12.15.** Hot-filament ionization gauge.

The rate of ion production is proportional to the number of electrons available to ionize the gas and the amount of gas present. Thus, the ratio of *+ve* ions, *i.e.,* anode current $I_1$ to electrons *i.e.,* grid current $I_2$ is a measure of gas pressure. The following relation holds good :

$$p = \frac{1}{S}\frac{I_1}{I_2} \quad ...(12.4)$$

where $S$ (the proportionality constant) is called the *sensitivity* of the gauge and its value is determined by the calibration of the particular gauge. Sensitivity is a function of *nature of gas, tube geometry* and the *operating voltages.*

***Advantages :***

(*i*) Fast response to changes in pressure.

(*ii*) Possibility of process control and remote indication.

(*iii*) Constant sensitivity for a given gas over a wide range of pressures.

(*iv*) Wide pressure range.

***Disadvantages :***

(*i*) Some gases get decomposed by hot filament.

(*ii*) Contamination of the measured gas by gases forced out of the hot filament.

(*iii*) The filament can burn out quickly if exposed to air when hot or before the pressure is at low value (This disadvantage is eliminated by using a radioactive source of alpha-particles as the ionizing agent).

(*iv*) High cost.

(*v*) Complex electric circuitry.

(*vi*) These gauges must be provided with automatic cut out in order to protect them in case of system leak or break.

(*vii*) These gauges must be carefully installed—in particular absorbed gases must be eliminated.

(*viii*) Calibration varies with gases.

**12.1.5.3. Radioactive vacuum meters**

A radioactive vacuum meter works on the principle that *the current flowing through an ionization chamber depends upon the pressure, or more precisely upon the density of the gas in the chamber.*

- This type of meter permits the measurement and recording of the absolute pressure from $10 \times 10^{-8}$ to 10 mm of mercury.

## 12.1.6. Calibration of Pressure Measuring Equipment

The following are the *laboratory standards for calibrating pressure measuring instruments :*

1. Manometers.
2. Dead weight testers.

- *Manometers and dead weight testers are suitable for static calibration only.*

**12.1.6.1. Manometers**

The calibration of low pressure gauges such as *manometers* and *straight bellows types* is usually done by comparison method using ***precision manometer***. For calibration, simple U-tube manometers with antiparallax devices and moving reservoir type manometers with vernier scales are employed.

- Manometers are used upto 200 kN/m$^2$; *above this pressure dead weight testers are used.*

### 12.1.6.2. Dead weight testers

This tester is a *primary standard for pressure measurement.* It provides good calibration facility over a wide pressure range (0.00007 to 70 N/mm$^2$ gauge in steps as small as 0.01% of range with a calibration uncertainty of 0.01 to 0.05% of the reading).

A dead weight tester is shown in Fig. 12.16. It consists of an accurately machined and finished piston which is inserted into a close fitting cylinder; both of known cross-sectional areas. At the top of the piston is provided a platform to hold standard weights of known accuracy.

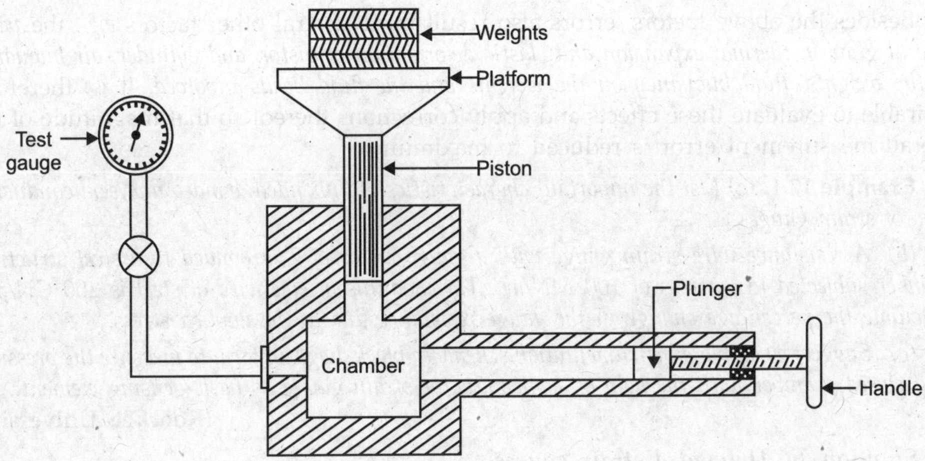


**Fig. 12.16.** Dead weight tester.

***Working :***

- The chamber and the cylinder of the tester are filled with clean oil.
- The gauge to be tested is fixed at its appropriate place and piston is inserted in the cylinder.
- A weight is placed on the piston. Consequently the pressure is exerted on the oil and this pressure is transmitted to the gauge deflecting its pointer. In order to get the desired setting of the gauge pointer, the weights are adjusted.
- Then pressure is varied by forcing the plunger in, by moving the handle. The movement of the handle is continued till enough pressure is built up to lift the piston-weight combination. The piston-weight combination *floats freely under these conditions.* Thus the equilibrium is established with *fluid force balanced against the gravitational force of the weights plus the friction drag.*

*i.e.,* $$p \cdot A = Mg + F_d$$

Hence, pressure, $$p = \frac{Mg + F_d}{A} \quad \text{...(12.5)}$$

where, $M$ = Mass, kg

$g$ = Acceleration due to gravity, m/s$^2$

$F_d$ = Friction drag, $N$, and

$A$ = Effect area of piston-cylinder combination, m$^2$.

The effect of equivalent area ($A$) depends on such factors as *piston cylinder clearance, pressure level, temperature*, and is normally taken as the *mean of the cylinder and piston areas* :

Many kinds of pressure-measuring devices may be calibrated such as :

— *Engine indicators;*

— *Industrial pressure gauges;*

— *Piezoelectrical pressure transducers etc.*

The following *factors affect the accuracy of the dead weight tester* :

- Friction between the piston and the cylinder.
- Uncertainty of value of effective area A.
- Uncertainty of gravitational constant g.

Besides the above factors, errors also result from several other factors *e.g.*, the *value of local gravity, thermal expansion and elastic deformation of piston and cylinder, air buouancy on the weights, fluid buoyancy on the weights and the fluid heads involved.* It is, therefore, desirable to evaluate these effects and apply corrections thereof so that magnitude of the overall measurement error is reduced to maximum.

**Example 12.1.** *(a) List the important characteristics of unbonded, bonded and semiconductor type of strain gauges.*

*(b) A resistance wire strain gauge with a gauge factor of 2 is bonded to a steel structure member subjected to a stress of 100 MN/m$^2$. The modulus of elasticity of steel is 200 GN/m$^2$. Calculate the percentage change in the gauge resistance due to the applied stress.*

*(c) Suggest a complete instrumentation scheme in block diagram form to measure the pressure in a closed chamber with the help of semiconductor strain gauge as basic sensing element.*

**(Roorkee University)**

**Solution.** (*a*) ***Unbonded strain gauges :***

- The range of unbonded strain gauges is ± 0.15% strain with accuracies of better than 0.1% and linearity within 1%.
- The gauges are not cemented but simply screwed at the desired location hence they can be detached and used again.
- Mainly used for measurement of force, pressure, acceleration rather than for the measurement of displacement because of its massive structure.

***Bonded strain gauges :***

- These gauges are bonded directly into the surface of the structure member which is being examined.
- Any changes in strain in the member are directly transmitted to the gauge material.

***Semiconductor strain gauges :***

(*i*) These gauges can measure very small strain, of the order of 0.01 micron, owing to high gauge factor of about ± 130.

(*ii*) Very suitable for measuring *local strains* as they are available in the size of 1.0 mm.

(*iii*) Almost free from hysteresis and creep effects.

(*iv*) Frequency response upto $10^{12}$ Hz.

(*v*) Chemically inert and has good fatigue.

- As for *demerits*, these gauges have :

  — Poor linearity,

— Higher cost, and

— High sensitivity to temperature changes.

(*b*) *Given :* $G_f = 2;\ \sigma = 100\ \text{MN/m}^2;\ E = 200\ \text{GN/m}^2$

$$\text{Strain} = \frac{\text{Stress}}{\text{Modulus of elasticity}} = \frac{100\times10^6}{200\times10^9} = 500\times10^{-6}$$

$$\frac{\Delta R}{R} = \text{Gauge factor} \times \text{strain} = 2 \times 500 \times 10^{-6} = 0.001$$

**Percentage change in resistance** = 0.001 × 100 = **0.1%** **(Ans.)**

(*c*) The four semiconductor strain gauges will be mounted on the four walls of the closed chamber so as to sense the pressure exerted on the walls of the chamber. They will be converted in a bridge arrangement as shown in the Fig. 12.17. Output of the bridge will be amplified and given to the permanent magnet moving coil type instrument calibrated directly in terms of pressure.

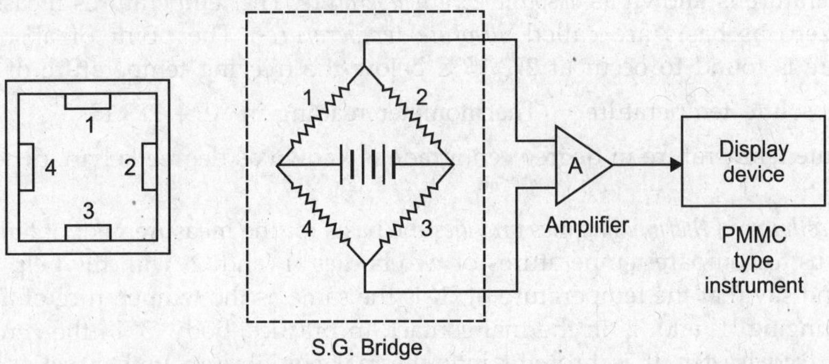


**Fig. 12.17**

**Example 12.2.** *Describe with a suitable diagram an instrumentation scheme to measure gas pressure inside an enclosed vessel using an LVDT and bellows.*

**Solution.** Fig. 12.18(*a*) and (*b*) shows using bellows and LVDT for measurement of pressure or force. The bellows and Bourdon tube act as Primary transducer, *i.e.*, Force summing device. The force of pressure applied to the bellows moves a core inside LVDT. The output voltage of LVDT is a measure of linear motion. The linear motion is measure of pressure. Thus force or pressure is a measure of output voltage of LVDT. In Bourdon tube circular motion is converted into linear motion which moves the core inside LVDT.

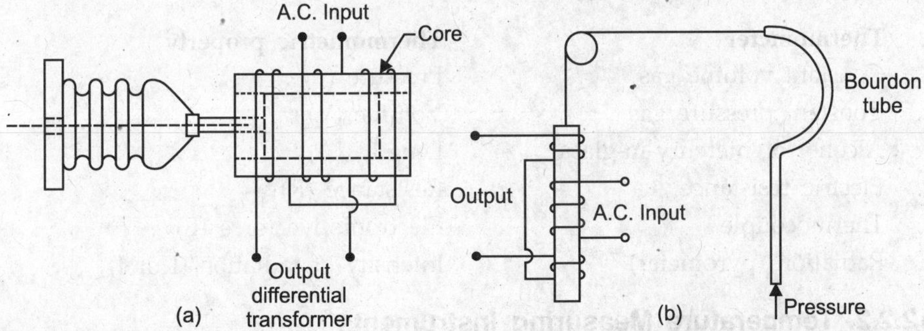


**Fig. 12.18.**

## 12.2 TEMPERATURE MEASUREMENT

### 12.2.1. Introduction

The **temperature** *is a thermal state of a body which distinguishes a hot body from a cold body.* The temperature of a body is proportional to the stored molecular energy *i.e.*, the average molecular kinetic energy of the molecules in a system (A particular molecule does not have a temperature, it has energy; the gas as a system has temperature).

The temperature may also be *defined* as :

- *The measure of the mean K.E. of the molecules of a substance.*
- *The degree of hotness and coldness of a body or an environment measured on a definite scale.*
- *The driving force or potential causing flow of energy as heat.*

*Instruments for measuring ordinary temperatures are known as* ***thermometers*** *and those for measuring high temperatures are known as* ***pyrometers.***

It has been found that a gas will not occupy any volume at a particular temperature. This temperature is known as *absolute zero temperature.* The temperatures measured with absolute zero as basis are called *absolute temperatures.* The point of absolute zero temperature is found to occur at 273.15°C below the freezing temperature of water.

$$\text{Absolute temperature} = \text{Thermometer reading in °C} + 273.15$$

Absolute temperature in degree centigrade is known as degree kelvin, denoted by K (SI unit).

The *zeroth law of thermodynamics provides* the basis for the measurement of temperature. It enables us to compare temperatures of two bodies '1' and '2' with the help of a third body '3' and say that the temperature of '1' is the same as the temperature of '2' without actually bringing '1' and '2' in thermal contact. In practice, body '3' in the zeroth law is called the *thermometer.* It is brought into thermal equilibrium with a set of standard temperature of body '2' and is thus calibrated. Later, when any other body '1' is brought in thermal communication with the thermometer, we say that the body '1' has attained equality of temperature with the thermometer and hence with body '2'. This way, the body '1' has the temperature of body '2' given for example by, say the height of mercury column in thermometer '3'. The height of mercury column in a thermometer, therefore, becomes a *thermometric property.*

These are *six other methods of temperature measurement* which utilize various other properties of materials, these are functions of temperature, as thermometric properties.

Six different kinds of thermometers and the names of the corresponding thermometric properties employed are given below :

| | **Thermometer** | **Thermometric property** |
|---|---|---|
| 1. | Constant volume gas | Pressure (p) |
| 2. | Constant pressure gas | Volume (V) |
| 3. | Alcohol or mercury-in-glass | Length (L) |
| 4. | Electric resistance | Resistance (R) |
| 5. | Thermocouple | Electromotive force (E) |
| 6. | Radiation (pyrometer) | Intensity of radiation (I or J) |

### 12.2.2. Temperature Measuring Instruments

The temperature measuring instruments are based on changes in a broad range of physical properties, among which are the following :

**1. Changes in physical dimensions :**

(*i*) *Liquid-in-glass thermometers.*

(*ii*) *Bimetallic elements.*

- *"Liquid-in-glass thermometer"* continues to appear in both laboratory and household situations, primarily because of its *ease of use and low cost.*
- *"Bimetallic elements"* are used in various low-frequency, low cost applications.

**2. Changes in glass pressure or vapour pressure :**

(*i*) *Constant-volume gas thermometers.*

(*ii*) *Pressure thermometers (gas, vapour and liquid filled).*

**3. Change in electrical properties :**

(*i*) *Resistance thermometers (RTD, PRT).*

(*ii*) *Thermistors.*

(*iii*) *Thermocouples.*

(*iv*) *Semiconductor-junction sensors*

(*a*) Diodes.

(*b*) Integrated circuits.

- *"Electrical sensors"* are widely used particularly when automatic or remote recording is desired.

**4. Changes in emitted thermal radiation :**

(*i*) *Thermal and photon sensors.*

(*ii*) *Total-radiation pyrometers.*

(*iii*) *Optical and two-colour pyrometers.*

(*iv*) *Infrared pyrometers.*

- "Radiant sensors" are used for non-contact temperature sensing, either in high-temperature applications like combustors or for infra red sensing at low temperatures.

**5. Changes in chemical phase :**

(*i*) *Fusible indicators.*

(*ii*) *Liquid crystals.*

(*iii*) *Temperature-reference (fixed-point) cells.*

The temperature measuring instruments may also be *classified* into *two broad categories* as follows :

**1. *Non-electrical methods :***

(*i*) By using change in volume of a liquid when its temperature is changed.

(*ii*) By using change in pressure of a gas when its temperature is changed.

(*iii*) By using changes in the vapour pressure when the temperature is changed.

**2. *Electrical methods :***

(*i*) By thermocouples.

(*ii*) By change in resistance of material with change in temperature.

(*iii*) By comparing the colours of filament and the object whose temperature is to be found out.

(*iv*) By ascertaining the energy received by radiation.

The **thermometers** may also be *classified* as follows :

1. *Expansion thermometers :*
   (*i*) Liquid-in-glass thermometers (*ii*) Bimetallic thermometers.
2. *Pressure thermometers :*
   (*i*) Vapour pressure thermometers (*ii*) Liquid-filled thermometers
   (*iii*) Gas-filled thermometers.
3. *Resistance thermometers*
4. *Thermocouple thermometers*
5. *Radiation pyrometers*
6. *Optical pyrometers.*

Table 12.1 gives the summary of temperature range of different instruments.

Some of these thermometers are discussed in the following Articles.

### 12.2.3. Electrical Resistance Thermometer or Resistance Temperature Detector (RTD)

Electrical methods of temperature measurement are very convenient due to the following *reasons* :

(*i*) They furnish signal that is easily detected, amplified or used for control purposes.

(*ii*) They are usually quite accurate when properly calibrated and compensated.

**Electrical resistance :**

The change of electrical resistance of a wire with temperature changes is another effect of the increased molecular capacity as caused by heat. Resistance thermometers utilise this characteristic and the temperature measurements are conveniently made by the change of resistance of suitable metals of known characteristics. The thermometers of this type are used over a range of temperature from –200 to 650°C. With the increase of temperature the electrical resistance of most of the metals increases in almost direct proportion to the temperature rise. Therefore, *if the electrical resistance of a wire of known and calibrated material is measured, the temperature of the wire can be determined. The resistance element should be of a suitable metal having high temperature coefficient, reproducibility, stability and ability to resist corrosion and oxidation.*

In case of variation of resistance $R$($R = \frac{\rho l}{a}$, where $\rho$ is the resistivity, $l$ length and $a$ is the cross-sectional area of the electrical conductor) with temperature for most of the metals, the following *quadratic* relationship gives good accuracy :

$$R = R_0(1 + \alpha t + \beta t^2) \quad \text{...(12.6)}$$

where, $R_0$ = Resistance at temperature $t = 0$, and

$\alpha, \beta$ = Constants.

For a narrow range, the following *line* relationship may be used with sufficiently good accuracy.

$$R = R_0(1 + \alpha\, t) \quad \text{...(12.7)}$$

where, $\alpha$ = Resistance temperature coefficient, /°C.

**Resistance thermometer construction :**

The thermometer *consists of the following parts :*

(*i*) A resistance element or bulb.

**Table 12.1. Summary of Temperature Range of Different Instruments**

| Type of Instrument | Liquid in glass thermo-ammeter | Bimetallic thermometer | Pressure thermometers | | | Resistance thermometers | | | Thermocouples | | | | Radiation pyrometers | Optical pyrometers |
|---|---|---|---|---|---|---|---|---|---|---|---|---|---|---|
| | *Mercury* | | *Vapour pressure* | *Liquid filled* | *Gas filled* | *Copper* | *Nickel* | *Platinum* | *Iron Constantan* | *Copper Constantan* | *Chromel Alumel* | *Platinum Platinum Rhodium* | – | – |
| *Low temprature limit* | –39°C | –75°C | –250°C | –87°C | –269°C | –140°C | –184°C | –184°C | –196°C | –250°C | –184°C | 0°C | about room temp. | 760°C |
| *High temperature limit* | 600°C | 540°C | 316°C | 650°C | 540°C | 120°C | 316°C | 760°C | 760°C | 350°C | 1260°C | 1540°C | as high as desired. | as high as desired. |
| *Remarks* | Low price, fairly accurate, easily broken. | Easier to read than liquid-in-glass thermometers. | Normally scale non-linear | To be compensated for variation in ambient temperatures | Requires large bulb, linear scale | Very sensitive. Suitable for narrow range Instruments. | | | Widely used in Industry | Constantan is 60% Cu, 40% Ni | Chromel is 20% Cr, 80% Ni Alumel is 2% Al, 98% Ni | Expensive | — | — |

(*ii*) Suitable electrical leads.

(*iii*) An indicating-recording or resistance measuring instrument.

The ***resistance element*** is usually in the form of a coil of very fine platinum, nickel or copper wound non-conductively onto an insulating ceramic former, which is protected externally by a metal sheath.

— A *laboratory type* of resistance thermometer is often wound on a crossed *mica former* (ceramic former in case of industrial type) and enclosed in pyrex tube. The tube may be evacuated or filled with an inert gas to protect the metal wire.

It should be ensured that the resistance wire is *free from mechanical stresses.* A metal which has been strained will suffer a change in the resistance characteristics; the metal is therefore usually *annealed* at a temperature higher than that at which it is to operate.

***Leads*** are taken out of the thermometer for the measurement of the changes in resistance in order to determine the value of temperature.

The change in resistance is usually, measured by a *"Wheatstone bridge"* which may be used either in null (balanced) condition or in the deflection out of balance condition. *The null condition suffices for steady state measurements whereas, transient conditions usually require the use of deflection mode.*

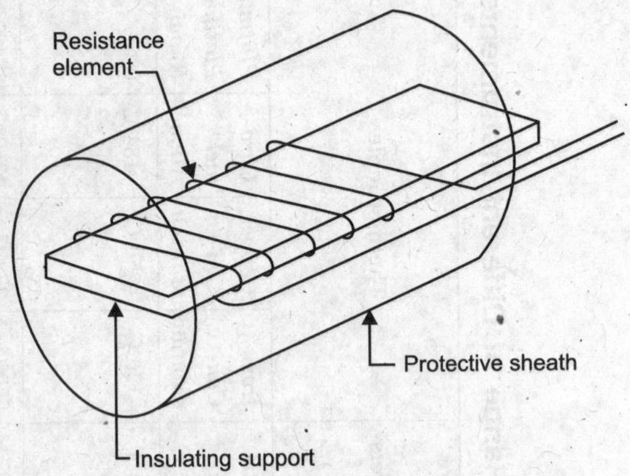


**Fig. 12.19.** Section illustrating the construction of a simple RTD.

Fig. 12.19, shows the section illustrating the construction of RTD.

**Desirable properties of the metal used for fabricating sensing elements :**

The metal to be used for fabricating sensing elements should satisfy the following *properties :*

(*i*) Linearity of resistance.

(*ii*) Resistance to corrosion and oxidation, under the temperature range of interest.

(*iii*) Should be available in conditions providing reproducible and consistent results.

(*iv*) Relatively change in resistance with temperature in order to produce a resistance thermometer with good sensitivity.

(*v*) High sensitivity so that the unit can be fabricated in a compact and convenient size.

(*vi*) No change of phase or state within a reasonable temperature change.

- *Industrial resistance thermometers often referred to as resistance temperature detectors (RTD) are usually made* with elements of *platinum, nickel* and *copper.*

— For precise measurement, platinum is preferred because it is physically stable and has high electrical resistance characteristics.

- The platinum resistance thermometer, owing to its *accuracy, stability* and *sensitivity*, has been used to define International Temperature Scale from the boiling point of oxygen (–182.9°C) to the freezing point of antimony (630.5°C).

**Advantages and disadvantages of resistance thermometers :**

***Advantages :***

(*i*) The resistance thermometer resists corrosion and is physically stable.

(*ii*) The measurement is very accurate.

(*iii*) The distance between the resistance element and the recording element can be made much larger than is possible with pressure thermometers.

(*iv*) It has lot of flexibility with regard to choice of measuring equipment and interchangeability of elements and assembly of components.

(*v*) It requires no reference junction like thermocouples and as such is more effective at room temperature.

(*vi*) The temperature sensitive resistance element can be easily installed and replaced.

(*vii*) Resistance elements can be used to measure differential temperature.

(*viii*) High working signal level, simplicity of lead wire and termination schemes compared with a thermocouple.

(*ix*) The limits of error of a resistance element are claimed to be ± 0.25 per cent of the scale range.

(*x*) The accuracy of the measuring circuit can be easily checked by substituting a standard resistor for the resistance element.

(*xi*) Self-heating effect in the conductor can be reduced to negligible with proper design.

(*xii*) The resistance element response time is of the order of 2 to 10 seconds, depending on the construction.

***Disadvantages :***

(*i*) The resistance thermometers cost more.

(*ii*) They suffer from time lag.

(*iii*) Possibility of current leakage between the resistance element and the ground.

(*iv*) Resistance change due to temperature changes of measuring resistors.

### 12.2.4. Thermistors

A **thermistor** *(contraction of the term "Thermal Resistor") is a thermally sensitive variable resistor made of a ceramic like semiconducting material.*

Thermistors, unlike metals, respond *negatively* to temperature; as the *temperature rises, the thermal resistance decreases.*

Thermistors are often composed of oxides of manganese, nickel, and cobalt in formulations having resistivities of 100 to 450,000 Ω cm; in *cryogenic applications*, doped germanium and carbon-impregnated glass are used. The metallic oxides are milled, mixed in appropriate proportions, are pressed into the desired shape with appropriate binders and finally sintered. The electrical terminals are either embedded before sintering or baked afterwards.

*The electrical characteristics of thermistors are controlled by varying the type of oxide used and physical size and configuration of the thermistor.*

Thermistors may be shaped in the form of beads, disc, washer etc. (Refer to Art. 7.6.2).

Thermistors may be used bare but are usually glass coated or positioned under a thin metal cap. The change in resistance is measured by using circuitry similar to that of metal conductors.

*Advantages :*

(*i*) Fairly less cost.

(*ii*) Availability in very small sizes.

(*iii*) Fast thermal response.

(*iv*) High sensitivity.

(*v*) Easy adaptability to electrical readout devices.

### Comparison between Metal Resistors and Thermistors

| S. No. | Aspects | Metal resistors | Thermistors |
|---|---|---|---|
| 1. | *Resistance change* | Positive (*i.e.*, increase in resistance with temperature increase). | Negative (*i.e.*, reduced resistance with temperature rise). |
| 2. | *Temp.-resistance relationship* | Approximately linear. | $R = R_0 \exp\left[\beta\left(\frac{1}{T} - \frac{1}{T_0}\right)\right]$, where $R$ is the resistance at temperature $T(K)$, $R_0$ is the resistance at reference temperature $T_0(K)$ and $\beta$ is a constant which depends on the thermistor formulation or grade. |
| 3. | *Practical operating range* | –160°C to 600°C (for resistance thermometer) | –100°C to 300°C (for thermistors) |
| 4. | *Stability* | More stable, provide better reproducibility with low hysteresis. | Less stable comparatively. |

- Owing the inherently *high sensitivity* possessed by thermistors :
  - — Very simple electrical circuitry can be used to measure the temperature.
  - — Ordinary ohmmeters maybe used within the limits of accuracy of the meter itself.
  - — More often one of the various forms of resistance bridge is used, either in the null-balanced form or as a deflection bridge.
  - — Simple ballast circuits are also usable.
  - — In some cases, special linearizing circuits are used to obtain an output voltage that varies linearly with temperature.

**Factors to be considered while selecting resistance thermometers :**

The following *factors* need to be considered while making selection of resistance thermometers :

1. The *platinum resistance thermometer* should be selected for measuring temperatures upto *300° to 400°C*; for a temperature range of *0°C to 180°C*, *copper resistance thermometer* is employed.
2. *Adeque length* of the working part of resistance thermometer and also the protruded part of thermometer well should be chosen, since maximum accuracy of measuring is achieved when sensitive element is in the *middle of the stream.*
3. The metal for *preserving casing* should be chosen on the basis of pressure and physical phenomena of the medium. Preserving casings of standard resistance thermometers are made of the following metals :

   (*i*) Brass; (*ii*) Ordinary carbon steel;

   (*iii*) Stainless steel; (*iv*) Alloyed steel.

4. Most of the resistance thermometers cannot operate at places with high vibration or shock-type load. While choosing ratiometer, the following points should be given due considerations :
   - Correspondence of range between ratiometer and resistance thermometer.
   - Value of permissible resistance of leads; ratiometers generally operate with lead resistances of 5 Ω to 15 Ω.
   - Correspondence of scale of an instrument to measured temperatures; an *instrument with a standard scale nearest to the upper limit of measuring should be chosen.*
   - Necessity of a low voltage (4 V) D.C. power supply.

### 12.2.5. Thermocouple Thermometers

For higher range of temperatures *i.e.,* above 650°C, filled thermometers are unsuitable. For higher range of temperatures, *thermocouples* and *pyrometers* are used.

#### 12.2.5.1. Thermoelectric effects

When direct current passes through a pair of thermocouples with junction maintained at different temperatures, the following *five* effects are observed :

1. Seebeck effect.
2. Peltier effect.
3. Thomson effect.
4. Joulean effect.
5. Conduction effect.

#### 12.2.5.2. Thermocouple and thermoelectric pyrometer

A thermocouple, in its simplest form, consists of *two dissimilar metals or alloys which develop e.m.f. when the reference and measuring junctions are at different temperatures.* The reference junction or cold junction is usually maintained at some constant temperature, such as 0°C.

Fig. 12.20, shows a simple circuit of a *thermocouple and temperature measuring device.* In many industrial installations the instruments are equipped with *automatic compensating devices for temperature changes of the reference junction, thus eliminating the necessity of maintaining this junction at constant temperature.*

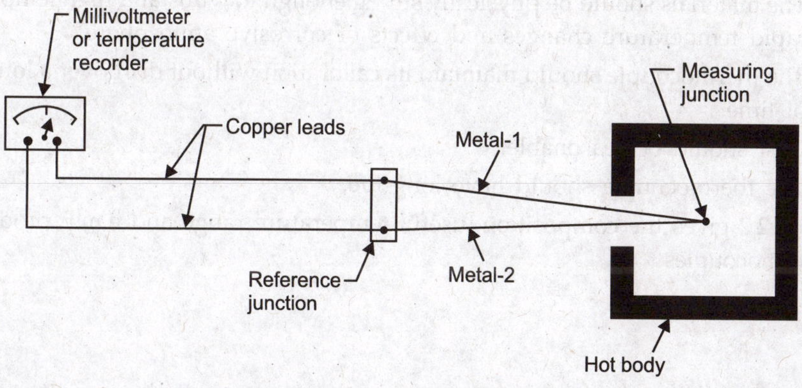


**Fig. 12.20.** Thermocouple.

Fig. 12.21, shows the schematic diagram of *thermoelectric pyrometer*.

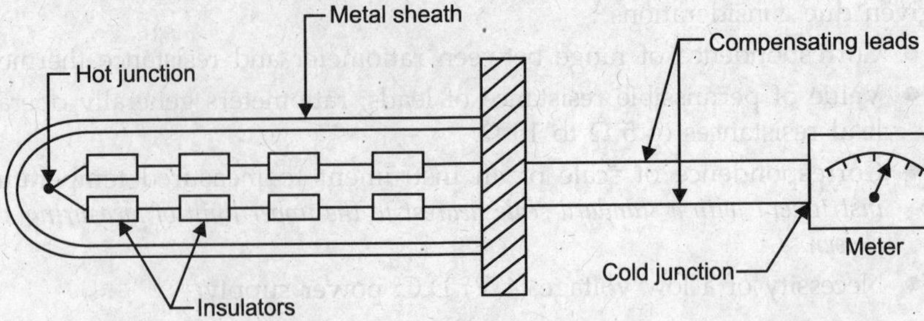


**Fig. 12.21.** Schematic diagram of a thermoelectric pyrometer.

**Compensating leads :**

These are a pair of leads wires made of such material which when connected to a thermocouple, behave such that the effective reference junction or junctions are removed to the other end of the leads. Its material should be such that it has the *same e.m.f. temperature characteristics* as the thermocouple over the range of the ambient temperature and is at the same time *cheap* also. The *resistance* of the compensating leads *should be as low as possible.* Usually they are covered with some protective material suitable for the installation.

**Connectors or extension leads :**

These are generally copper conductors having a *negligible resistance* and are used to connect the compensating leads further to the indicators.

- *In order to prevent the unwanted e.m.f. or unwanted ground currents,* the *thermocouple leads and extension leads must be properly insulated.*

**12.2.5.3. Thermocouple materials**

The thermocouple materials should possess the following *properties* :

1. The temperature-e.m.f. relationship should be reasonably linear and reproducible.
2. Adequate thermo e.m.f. per degree of temperature change to facilitate detection and measurement.
3. The materials should be physically strong enough to withstand high temperatures, rapid temperature changes and effects of corrosive atmosphere.
4. The thermocouple should maintain its calibration without drift over a long period of time.
5. Cost should be reasonable.
6. The thermocouple should have long life.

Table 12.2 gives the composition, useful temperature range and e.m.f. produced for some thermocouples.

**Table 12.2. Composition, useful temperature range and e.m.f. produced for some thermocouples**

| S. No. | Thermocouple | Composition | Temperature (°C) | | Thermoelectric power | | Remarks |
|---|---|---|---|---|---|---|---|
| | | | Useful range | Max. | °C | Millivolt | |
| 1. | *Platinum vs platinum-rhodium* | Pure platinum vs Pt + 10% or 13% RH | 400 to 1450 | 1700 | 0<br>500<br>1000<br>1500 | 0.0<br>4.219<br>9.569<br>15.498 | Used for high temperature measurements. |
| 2. | *Chromel vs alumel* | 90% Ni + 10% Cr vs 95% Ni + 5% (Al + Sn) Ma | –200 to 1200 | 1450 | –200<br>0<br>300<br>600<br>900<br>1200 | –5.75<br>0.0<br>12.21<br>24.90<br>37.36<br>48.85 | High resistance to oxidation. |
| 3. | *Iron vs constantan* | Pure iron vs 45–60% Cu + 55–40% Ni | –200 to 750 | 1000 | –200<br>0<br>300<br>600<br>900 | –8.27<br>0.0<br>16.59<br>33.27<br>52.29 | — |
| 4. | *Copper vs constantan* | Pure copper vs Cu-Ni constantan | –200 to 350 | 600 | –200<br>0<br>200<br>400 | –5.539<br>0.0<br>9.285<br>20.865 | Not suitable in air due to excesssive oxidation. |

**Series combination (Thermopile):**

In a series arrangement, the total output from *n* thermocouples *equals the sum of the individual e.m.fs.*

- The series arrangement provides *considerably more sensitivity* and is frequently used where it is desired to obtain a *substantially large e.m.f. for measurement of small temperature differences between the two junctions.*

**Parallel combination :**

- This arrangement may be used to *obtain an average temperature of a number of points, as in determining the average temperature of gas flowing through a large area passage.*
- The total potential indicated by the potentiometer will be an *average of the junction potentials.*

**12.2.5.4. Advantages and disadvantages of thermocouples**

The following are the *advantages* and *disadvantages* of the thermocouples :

***Advantages :***

1. Thermocouples have a *better response.*
2. They have a *higher range of temperature measurements.*
3. The sensing element of a thermocouple can be *easily installed.*
4. *Cheaper* than resistance thermometers.

5. *Very convenient* for measuring the temperature at *one particular point* in a piece of apparatus.

***Disadvantages :***

1. *Lower accuracy*, and as such they cannot be used for precision work.
2. In order to ensure long life, they need to be *amply protected.* For preventing contamination of thermocouple, when precious metals like platinum or its alloys are being used, the protecting tube has to be made chemically inert and vacuum tight.
3. The *circuitry* for thermocouple is *very complex.*

**Example 12.3.** *Suggest a complete instrumentation scheme to measure high temperature of a furnace using thermocouple. What are the precautions to be taken up while carrying out this measurement ?* **(Punjab University)**

**Solution.** The system comprises of Platinum-Rhodium-Platinum type thermocouples (suitable upto 2200°C) in the form of parallel thermopile for averaging. Reference junction compensation is done by means of an isothermal block.

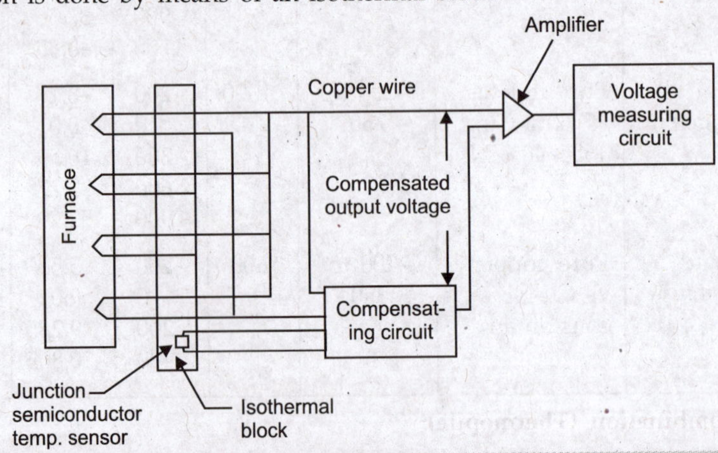


**Fig. 12.22**

For greater accuracy a junction semiconductor temperature sensor may be used as shown in Fig. 12.22, for taking care of the drift caused by ambient temperature variation. The compensated output voltage is amplified and measured with the help of a suitable voltage measuring circuit, directly calibrated in terms of temperatures.

***Precautions :***

1. Reference junction compensation is essential.
2. The thermocouple leads should be as short as possible.
3. The compensating circuit should also provide compensation for any changes in the ambient temperature.

### 12.2.6. Radiation Pyrometers

The principle, that the *radiant energy emitted from a body increases with temperature,* is used in measuring temperature particularly in the *higher ranges.*

Radiation pyrometers are used in the following *situations :*

(*i*) High temperatures.

(*ii*) Where it is not possible to contact the material whose temperature is to be measured.

(iii) Where furnace atmosphere is detrimental to thermocouples and causes erratic measurements and short life.

Following two *"principles"* are used for construction of radiation temperature measuring devices :

(i) ***Total radiation pyrometer.*** In this case the *total radiant energy* from a heated body is measured.

(ii) ***Selective (or partial) radiation pyrometer.*** In this case, we measure the spectral radiant intensity of the radiant energy from the heated body at a *given wavelength.*

Since it is difficult to build a practical pyrometer that is responsive to radiations of all wavelengths or to spectral radiation of one wavelength, therefore, *practical radiation pyrometers response to a wide band of radiation of approximately 0.1 to 8.0 microns width in the visible and infrared band of radiation spectrum;* the actual width of this band, however, depends entirely on the *physical construction of the radiation receiver.*

**Total radiation pyrometer :**

*A device which measures the total intensity of radiation emitted from a body is called total* ***radiation pyrometer.***

The elements of a *total radiation pyrometer* are illustrated in Fig. 12.23. It collects the radiation from an object (hot body) whose temperature is required. A mirror is used to focus this radiation on a thermocouple. This energy which is concentrated on the thermocouple raises its temperature and in turn generates an e.m.f. This e.m.f. is then measured either by the galvanometer or potentiometer method. *Thus rise of temperature is a function of the amount of radiation emitted from the object.*

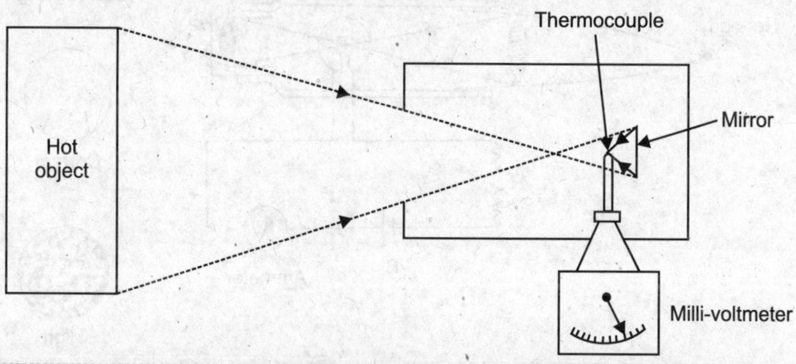


**Fig. 12.23.** A schematic diagram of radiation pyrometer.

**Advantages and disadvantages of radiation pyrometers :**

Following are the *advantages and disadvantages of radiation pyrometers :*

***Advantages :***

(i) Can be used to measure the temperature of an object which may be either stationary or moving, and so adaptable to continuous industrial processing.

(ii) High speed of response.

(iii) Accuracy ± 2% of scale.

(iv) Primarily used to measure temperatures in the range 700–2000°C, where resistance thermometers and thermocouples cannot be used.

(v) Direct contact is not necessary with the object whose temperature is to be measured.

(vi) Suitable for measuring temperatures where the atmospheric or other environmental conditions prevent satisfactory operation of other temperature sensing devices.

(vii) Relatively independent of the distance between the measuring element and the heated body.

***Disadvantages :***

(i) In the situations where the temperature may be high due to operating conditions, *cooling is required* to protect the instrument from overheating.

(ii) The presence of dust and dirt on the mirrors or lens causes the instrument to read too low.

### 12.2.7. Optical Pyrometers

An **optical pyrometer** *works on the principle that matters glow above 480°C and the colour of visible radiation is proportional to the temperature of the glowing matter.* The amount of light radiated from the glowing matter (solid or liquid) is measured and employed to determine the temperature.

Fig. 12.24, shows a *disappearing filament pyrometer.*

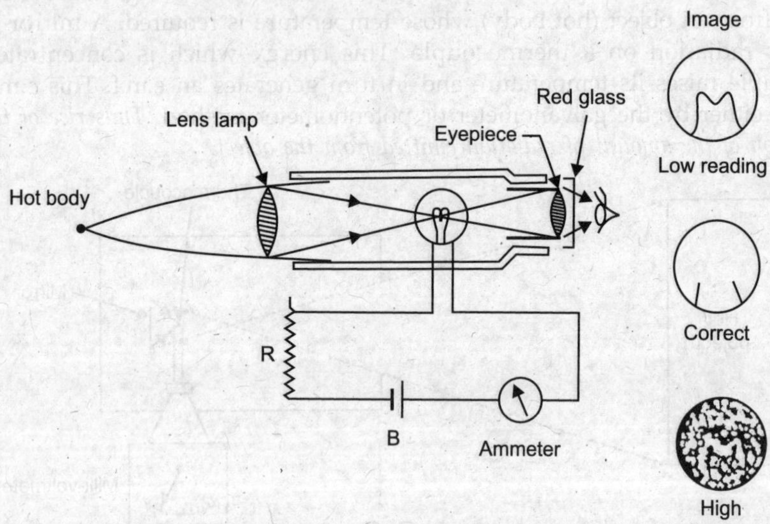


**Fig. 12.24.** Optical pyrometer.

**Operation :**

— The optical pyrometer is sighted at the hot body and focused.

— In the beginning filament will appear dark as compared to the background which is bright (being hot).

— By varying the resistance (*R*) in the filament circuit more and more current is fed into it, till filament becomes equally bright as the background and hence disappears.

— The current flowing in the filament at this stage is measured with the help of an ammeter which is calibrated directly in terms of temperature.

— If the filament current is further increased, the filament appears brighter as compared to the background which then looks dark.

— An optical pyrometer can measure temperature ranging from 700 to 4000°C.

**Uses :**

The optical pyrometer is widely used for *accurate measurement of temperature of :*

— furnaces;

— molten metals.

— other heated materials.

- *This pyrometer has been accepted as the standard means for determining temperatures on the International Temperature Scale from the gold point and upwards.*

***Advantages :***

(*i*) Excellent accuracy within ±5°C for the operating range 700–3000°C.

(*ii*) No direct contact is necessary with the object whose temperature is to be measured. Thus this type of pyrometer can be used in situations where the measuring target is remote and inaccessible such as furnace interiors, molten metals etc.

(*iii*) Measurement is independent of the distance between the target and measuring instrument.

(*iv*) The skill in operating the thermometer can be acquired readily.

***Disadvantages :***

(*i*) The lower measuring temperature is limited to 700°C (The eye is insensitive to wavelength characteristics below this temperature).

(*ii*) Owing to the manual null-balance operation of this pyrometer it is not suitable for continuous reading or automatic control applications.

## 12.3 FLOW MEASUREMENT

### 12.3.1. Introduction

In almost all the flow situations of engineering importance measurements of fluid velocity, flow rate and flow quantity with varying degree of accuracy are a fundamental necessity. Flow of material in a process or system can be measured by a variety of methods depending upon the *material and its condition, the type of flow, the volume, the pressure and temperature range, the accuracy needed and the control required.*

The accurate measurement of flow presents many and varied problems. The flowing medium may be *liquid, a gas, a granular solid.* The flow may be a *laminar or turbulent, steady-state or transient.* In addition, there are several different approaches to the problem of flow measurement.

**Flow characteristics.** When fluids move at very low velocities through uniform conduits, the motions of individual particles are generally along lines paralleling the conduits' walls. The velocity of particles is greatest at the centre and theoretically zero at the wall of conduit with the velocity distribution as shown in Fig. 12.25(*a*), such a flow is called *"laminar"*.

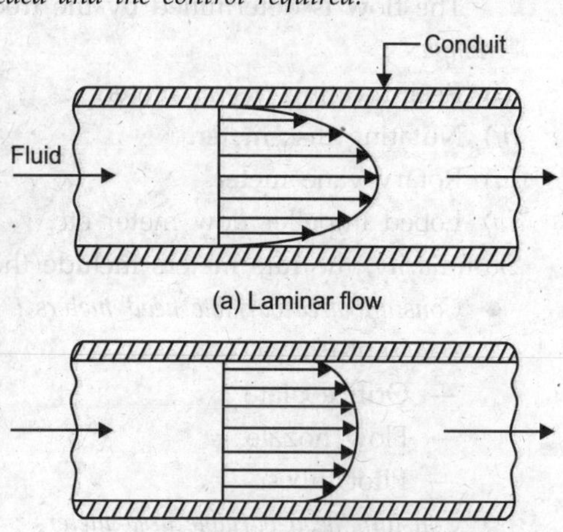


**Fig. 12.25.** Velocity distribution of flow in a pipe or tube.

On increasing the flow, a point is reached where the particles motion becomes more random and complex and flow at *higher rates* is referred to as *"turbulent"* (The approximate velocity at which the change occurs is called the *critical velocity*). The corresponding time-average velocity distribution across a circular tube is shown in Fig. 12.25(*b*).

It has been observed that the critical velocity is a function of several factors that may be put in a dimensionless form called the *Reynolds number, Re,* its value for a circular pipe is given as

$$Re = \frac{\rho V D}{\mu} \quad ...(12.8)$$

where, $\rho$ = Density of fluid, kg/m$^3$,
$V$ = Velocity flow, m/s,
$D$ = Diameter of pipe, m, and
$\mu$ = Dynamic viscosity of fluid, Ns/m$^2$.

When, $Re < 2000$ .... The flow is *laminar* (or viscous)
$Re > 4000$ .... The flow is turbulent

*Re* between 2000 and 4000 (transition zone) .... The flow is unpredictable

### 12.3.2. Flow Measurement Methods/Devices

The flow measurement devices are *classified* as follows :

1. Rate meters
2. Quantity meters.

**1. Rate meters.** They measure either the volumetric flow rate directly or use meters that measure velocity and the volume flow rate can then be calculated with the help of cross-section.

(*i*) ***Inferential type.*** Inferential methods *imply that the flow is not directly measured but is inferred from the measurement of other quantities (e.g., pressure, temperature etc.).*

(*ii*) ***Absolute or positive displacement type.*** These meters basically *capture and release a fixed volume of fluid by some type of pumping action.* They normally count the number of cycles that occur and indicate or register an integrated flow volume. The flow is determined by the frequency of the cycle.

*Examples :*

(*i*) Reciprocating piston meter.
(*ii*) Nutating disc meter.
(*iii*) Rotary vane meter.
(*iv*) Lobed impeller flow meter etc.

Summarily, the rate meters include the following :

- *Constant area-variable head meters :*
  - — Venturi.
  - — Orifice plate.
  - — Flow nozzle.
  - — Pitot tubes.
- *Constant head-variable area meters ;*
  - — Rotameter.
  - — Cylinder and piston arrangement.

- *Variable head and variable area meters :*
  - — Weirs.
  - — Flumes.
- *Variable velocity meters :*
  - — Current meters.
  - — Anemometers.
- *Variable force meters.*
  - — Vane.

The term *obstruction meter* applies to all the above mentioned devices.

Besides the above mentioned rate meters following are some *special meters :*

— Electromagnetic flow meter ;

— Ultrasonic flow meter etc.

**2. Quantity meters.** These meters may be designed for the measurement of either weight or volume. They may be absolute or displacement type. These are generally cited as *positive meters.*

*Examples :*

(*i*) *Weight meters :*

— Weighters and traps.

(*ii*) *Volumetric meters :*

— Tank-rotating impeller piston pumps.

— Nutating piston pump for liquids.

— Bellows and liquid sealed drums for gases.

### 12.3.3. Variable Head Flow for Incompressible Fluids

When the fluid flows through the obstruction meters, such as *venturimeter, orifice meter and flow nozzle,* the fluid accelerates and reduction in pressure takes place. The pressure difference before and after the obstruction is measured by means of a differential pressure sensor and is related to the rate of flow.

#### 12.3.3.1. Venturimeter

A ***venturimeter*** *is an instrument used to measure the rate of discharge in a pipeline and is often fixed permanently at different sections of the pipeline to know the discharge there.*

It is one of the most important practical applications of Bernoulli's theorem. It has been named after the 15th century Italian *engineer Venturi.*

The Venturimeter may be used in any position, horizontal vertical or inclined. In order to measure pressure differentials U-tube manometers are used.

*The size of a venturimeter is expressed in terms of inlet and throat diameters.* For example, a 300 × 150 mm venturimeter fits in a 300 mm diameter pipe and has a throat diameter of 150 mm.

A venturimeter consists of following *three* parts :

(*i*) A short converging part,

(*ii*) Throat, and

(*iii*) Diverging part,

**Expression for rate of flow :**

Fig. 12.26 shows a venturimeter fitted in *horizontal pipe* through which a fluid is flowing.

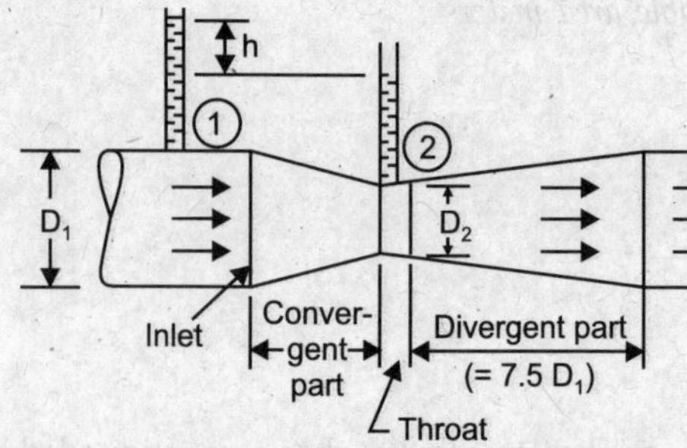


$$\left(\text{Throat ratio } \frac{D_2}{D_1} \text{ varies from } \frac{1}{4} \text{ to } \frac{3}{4}\right)$$

**Fig. 12.26.** Venturimeter.

Let,

$D_1$ = Diameter at inlet or at section 1,

$A_1$ = Area at inlet $\left(=\frac{\pi}{4}D_1^2\right)$,

$p_1$ = Pressure at section 1, $Nm^2$,

$w_1$ = Specific weight of fluid at inlet, $N/m^3$

$V_1$ = Velocity of fluid at section 1 and

$D_2$, $A_2$, $p_2$, $w_2$ and $V_2$ are the corresponding values at section 2.

Applying Bernoulli's equation at sections 1 and 2, we get

$$\frac{p_1}{w_1} + \frac{V_1^2}{2g} + z_1 = \frac{p_2}{w_2} + \frac{V_2^2}{2g} + z_2$$

Here, $z_1 = z_2$ ... since the pipe is *horizontal.*

$$\frac{p_1}{w_1} + \frac{V_1^2}{2g} = \frac{p_2}{w_2} + \frac{V_2^2}{2g}$$

or,
$$\frac{p_1 - p_2}{w} = \frac{V_2^2}{2g} - \frac{V_1^2}{2g} \quad ...(12.9)$$

... For incompressible fluid, $\rho_1 = \rho_2 = \rho$ and $w_1 = w_2 = w$

$\frac{p_1 - p_2}{w}$ = Difference of pressure heads at sections 1 and 2 is equal to h; called *pressure head or piezometer* head.

It is measured by a differential manometer, pressure gauge or any other pressure measuring device.

The continuity equation gives,

$$A_1V_1 = A_2V_2$$

or,
$$V_1 = \frac{A_2V_2}{A_1}$$

Substituting for $V_1$ in equation (12.9), we have

$$\frac{p_1 - p_2}{w} = \frac{V_2^2}{2g} - \frac{A_2^2\,V_2^2}{A_1^2 \times 2g}$$

$$= \frac{V_2^2}{2g}\left(1 - \frac{A_2^2}{A_1^2}\right) = \frac{V_2^2}{2g}\left[1 - (A_2 / A_1)^2\right]$$

or
$$V_2 = \frac{1}{\sqrt{1-(A_2/A_1)^2}}\sqrt{\frac{2g}{w}(p_1 - p_2)} \qquad ...(12.10)$$

For a given venturimeter, $A_1$ and $A_2$ are constants and therefore,

$$V_2 = M\sqrt{\frac{2g}{w}(p_1 - p_2)}$$

where, $M = \frac{1}{\sqrt{1-(A_2/A_1)^2}}$ is a constant called *"velocity approach factor"*

The value of $M$ *is always greater than unity.*

Equation 12.10 has been derived on the basis that there are no losses. However, in actual practice, this is not true and actual velocity is obtained from the theoretical by multiplying the latter by a factor $C_v$, called *"coefficient of velocity"*.

$$(V_2)_{\text{actual}} = C_v M\sqrt{\frac{2g}{w}(p_1 - p_2)} \qquad ...(12.11)$$

Now, Discharge = area × velocity

or
$$Q = A_2 C_v M\sqrt{\frac{2g}{w}(p_1 - p_2)} \qquad ...(12.12)$$

Sometimes, a factor, $C_v$ called *"coefficient of contraction"* is taken into consideration for finding out the actual discharge.

$$\therefore \quad Q_{\text{actual}} = C_c C_v M\, A_2\sqrt{\frac{2g}{w}(p_1 - p_2)} \qquad ...(12.13)$$

or
$$Q_{\text{actual}} = C_d M\, A_2\sqrt{\frac{2g}{w}(p_1 - p_2)} \qquad ...(12.14)$$

where, $C_d = C_c\, C_v$ is called *"coefficient of discharge"*. For venturimeters, the factors $C_d$ and $M$ are applied separately but for nozzles and Orifice plates these factors are combined into a single factor known as *"flow coefficient"* and designated as $K$.

$$\therefore \quad Q_{\text{actual}} = K A_2\sqrt{\frac{2g}{w}(p_1 - p_2)}$$

... For nozzles and orifice plates ...(12.15)

In order to take into account the *changes in flow rate on account of changes in temperature*, a factor $E$ is used. Then the actual discharge is essentially obtained from the following relation :

$$Q_{\text{actual}} = C_d MEA_2\sqrt{\frac{2g}{w}(p_1 - p_2)} \qquad ...(12.16)$$

where, $E$ is the *"thermal expansion correction factor"*

— The value of $C_d$ is the *less than unity* (varies between 0.96 to 0.98). Due to variation of $C_d$ venturimeters are not suitable for *very low velocities.*

**Value of 'h' by differential U-tube manometer :**

**Case I.** *Differential manometer containing a liquid heavier than the liquid flowing through the pipe.*

Let, $S_{hl}$ = Sp. gravity of heavier liquid,

$S_p$ = Sp. gravity of liquid flowing through pipe, and

$y$ = Difference of the heavier liquid column in U-tube

Then, $$h = y\left[\frac{S_{hl}}{S_p} - 1\right] \quad ...(12.17)$$

**Case II.** *Differential manometer containing a liquid lighter than the liquid through the pipe.*

Let, $S_{ll}$ = Sp. gravity of lighter liquid,

$S_p$ = gravity of liquid flowing through pipe, and

$y$ = Difference of lighter liquid column in U-tube.

Then, $$h = y\left[1 - \frac{S_{hl}}{S_p}\right] \quad ...(12.18)$$

***Advantages :***

(*i*) Low head loss (about 10% of differential pressure head)

(*ii*) Due to their extensive use over a long period in the past, their characteristics are well established through practice and consequently their behaviour can be predicted perfectly.

(*iii*) On account of low loss, they have a high coefficient of discharge (0.96 to 0.98) and as such they can be used to measure high flow rates in pipes having diameters of the order of a few meters.

***Disadvantages :***

(*i*) High cost.

(*ii*) Owing to their large size they are unsuitable for applications where there is a limited space.

**Uses :** Venturimeters are used for measurement of flow of :

— Water

— Fluids used in industrial processes;

— Industrial wastes;

— Gases;

— Slurries and dirty liquids;

— Suspended particles.

**12.3.3.2. Orifice plate meter**

An orifice plate is *defined as plate in which an aperture has been cut*. This device is cheaper than venturimeter and is used for measuring the discharge of fluid through a pipe. It also works on the same principle of a venturimeter. It is more frequently used than all other primary elements combined together.

The orifice plates are of the following *four* types :

1. Concentric;
2. Eccentric;

3. Segmental;
4. Quadrant types.

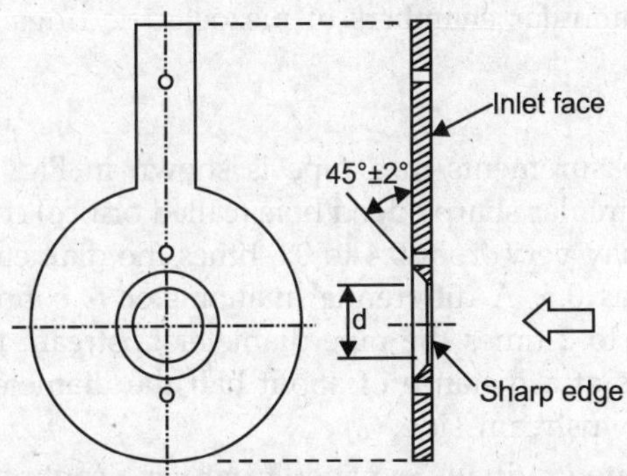


**Fig. 12.27.** Concentration orifice plate with 45° bevelled edges.

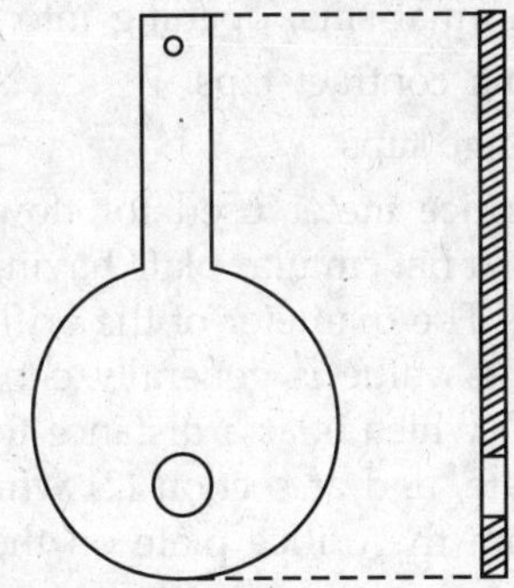

**Fig. 12.28.** Eccentric orifice plate.

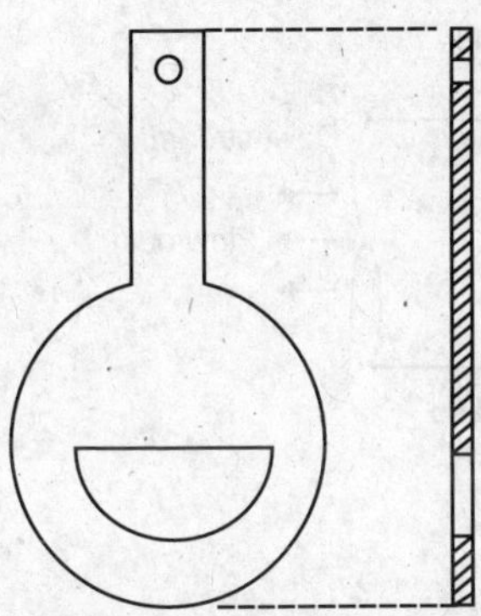

**Fig. 12.29.** Segmental orifice plate.

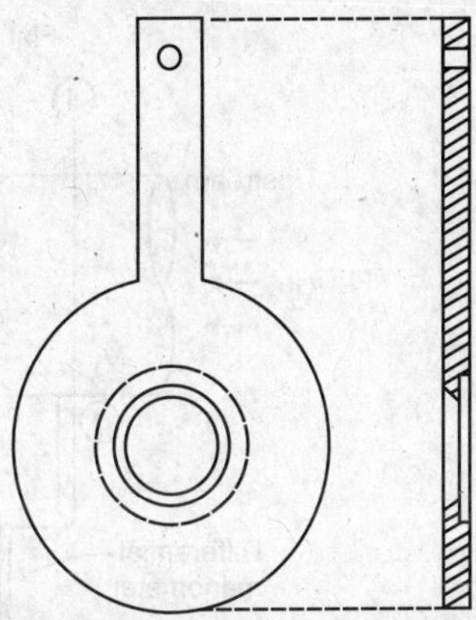

**Fig. 12.30.** Quadrant edge orifice plate.

1. The *"concentric plate"* has a circular hole located *in the centre* (Fig. 12.27).
2. The *"eccentric plate"* has a circular hole located below its centre (Fig. 12.28).
3. The *"segmental plate"* has a hole that is only *partly circular* located *below its centre* (Fig 12.29).
4. The *"quadrant edge orifice plate"* is so constructed that the edge is rounded so as to form a quarter circle (Fig. 12.30).

**Uses :** The type of the orifice plate used depends upon the *characteristics of the fluid that* is to be metered; *the eccentric and segmental types are used for fluids containing solids.*

Orifice plates are made from *steel, stainless steel, monel, phosphor bronze and other such materials, that can withstand the corrosive effect of the fluid.*

— Its thickness is only sufficient to withstand the buckling forces caused by the pressure differential.

— The circular holes are carefully made with 90% square sharp edge.

The *orifice plates can also be* **classified** according to the disposition of pressure taps (upstream and downstream) to measure the differential pressure across it. Following types of *pressure taps* are generally employed.

1. *Corner taps.*
   (*i*) Single tappings (Individual tapes)
   (*ii*) Annual slits, opening into the annular chambers of piezoelectric rings.
2. Vena contract taps.
3. Flange taps

An orifice meter used for flow measurements in a pipe is shown in Fig. 12.31. It consists of a flat circular plate having a circular sharp edged hole (called orifice) concentric with pipe. The diameter of the orifice may very from 0.4 to 0.8 times the diameter of the pipe but its value is generally chosen as 0.5. A differential manometer is connected at section (1) which is at a distance to 1.5 to 2 times the pipe diameter upstream from the orifice plate, and at section (2) which is at a distance of about half the diameter of the orifice from the orifice plate on the downstream side.

The jet coming out of the orifice plate gradually expands from vena contracta to fill the pipe. Fig. 12.31 shows the location of vena contracta point. A part of the K.E. of the jet is converted into eddy currents causing dissipation of energy and loss of head.

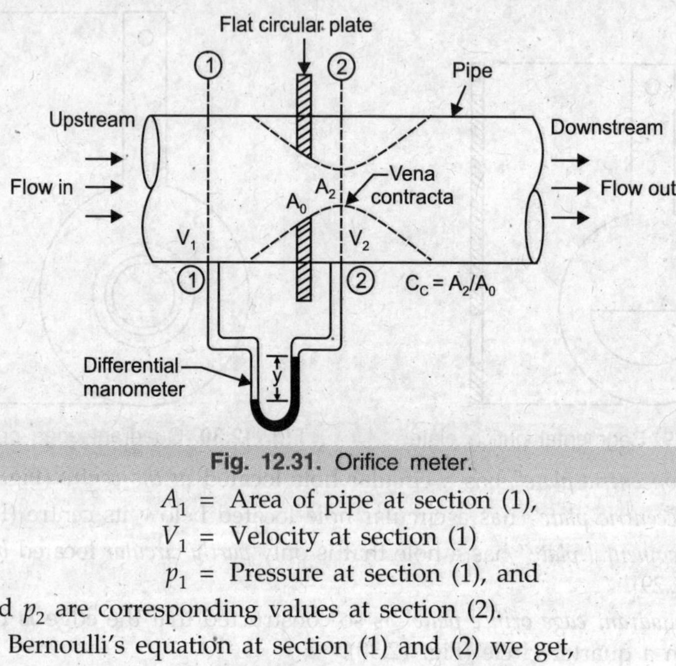


**Fig. 12.31.** Orifice meter.

Let, $A_1$ = Area of pipe at section (1),

$V_1$ = Velocity at section (1)

$p_1$ = Pressure at section (1), and

$A_2$, $V_2$ and $p_2$ are corresponding values at section (2).

Applying Bernoulli's equation at section (1) and (2) we get,

$$\frac{p_1}{w}+\frac{V_1^2}{2g}+z_1 = \frac{p_2}{w}+\frac{V_2^2}{2g}+z_2$$

Here $z_1 = z_2$, since the pipe is horizontal.

$$\therefore \quad \frac{p_1}{w}+\frac{V_1^2}{2g} = \frac{p_2}{w}+\frac{V_2^2}{2g}$$

or

$$V_{2(\text{actual})} = C_v M\sqrt{\frac{2g}{w}(p_1-p_2)} \qquad ...(12.19)$$

$$= \frac{C_v}{\sqrt{1-(A_2/A_1)^2}}\sqrt{\frac{2g}{w}(p_1-p_2)}$$

As the jet comes out of the orifice, it contracts to a minimum area $A_0$ at the vena contracta. Area at the vena contracta, $A_2 = C_v A_o$.

$$\therefore \quad V_{2(\text{actual})} = \frac{C_v}{\sqrt{1-(C_c A_0 / A_1)^2}} \sqrt{\frac{2g}{w}(p_1 - p_2)} \quad \text{...(12.20)}$$

Actual flow rate, $Q_{\text{actual}} = A_2 V_{2(\text{actual})} = C_c A_0 V_{2(\text{actual})}$

$$= C_v C_c \frac{A_0}{\sqrt{1-(C_c A_0 / A_1)^2}} \sqrt{\frac{2g}{w}(p_1 - p_2)} \quad \text{...(12.21)}$$

$$= C_d MEA_0 \sqrt{\frac{2g}{w}(p_1 - p_2)} \quad \text{...(12.22)}$$

$$= KEA_0 \sqrt{\frac{2g}{w}(p_1 - p_2)} \quad \text{...(12.23)}$$

where, $$K = C_d M = \frac{C_d}{\sqrt{1-(C_c A_0 / A_1)^2}}$$

where K is called the *"orifice flow constant"*.

***Advantages :***

(*i*) Low initial cost.

(*ii*) Easy to install.

(*iii*) Simple and less expensive maintenance as compared with a venturimeter.

(*iv*) Requires less space as compared to a venturimeter.

(*v*) The behaviour is predictable and the results are reproducible with good accuracy.

(*v*) Can be used in wide range of pipe sizes (1.25 cm to 150 cm)

***Disadvantages :***

(*i*) Low coefficient of discharge (about 0.6 as compared to almost unity for venturimeters), consequently their sensitivity is much lower than that of venturimeter.

(*ii*) Poor pressure recovery. The overall pressure loss varies form 40 to 90% of the differential pressure.

(*iii*) Susceptible to inaccuracies resulting from erosion, corrosion, and sealing.

(*iv*) Since the flat orifices tend to clog, they are not suitable for measurement of slurries or entrained particles.

(*v*) Owing to the lower physical strength of the orifices, they are likely to be damaged by pressure transients.

**Difference between a venturimeter of an orificemeter :**

• A **venturimeter** is a device which is inserted into pipeline to measure incompressible fluid flow rates. It consists of a convergent section which reduces the diameter to between one half to one-fourth of the pipe diameters. This is followed by a divergent section. The pressure difference between the position just before the venturi and at the throat of the venturi is measured by a differental manometer. *The working of the venturi is based on the Bernoulli's principle, that is when the velocity head increases in an accelerated flow, there is a corresponding reduction in the piezometric head.*

- The **orificemeter** is an opening, usually round, located in the side wall of the tank or reservoir for measuring the flow of a liquid. The main feature of the orifice meter is that *most of the potential energy of the liquid is converted into kinetic energy of the free jet issuing through the orifice.*

*The main points of difference* between a venturimeter and orificemeter are :

1. The venturimeter can be used for *measuring the flow rates of all incompressibly flows* (gases with low pressure variations, as well as liquids), whereas orifice meters are generally used for measuring the *flow rates of liquids.*
2. Venturimeter is installed in *pipeline only,* and the accelerated flow through the apparatus of subsequently decelerated to the original velocity at the outlet of the venturimeter. The flow continues through the pipeline. In the orifice meter the entire potential energy of the fluid converted to kinetic energy, and the jet *discharges freely into the open atmosphere.*
3. In venturimeter, the flow *velocity is measured by noting the pressure difference between the inlet and the throat of the venturimeter,* whereas in the orificemeter the discharge velocity measured by using *pitot tube or by trajectory method.*

### 12.3.3.3. Flow nozzles

In engineering practice, the nozzles are used for the creation of jets and streams for all purposes as well as for fluid metering. When placed in or at the end of a pipeline as metering devices, they are called as "*Flow nozzles*".

In fact, a flow nozzle is nothing else but a *venturimeter without it diverging cones.* It differs from an orifice meter in the sense that it is made curved so that jet contraction is not so large as in an orifice-meter.

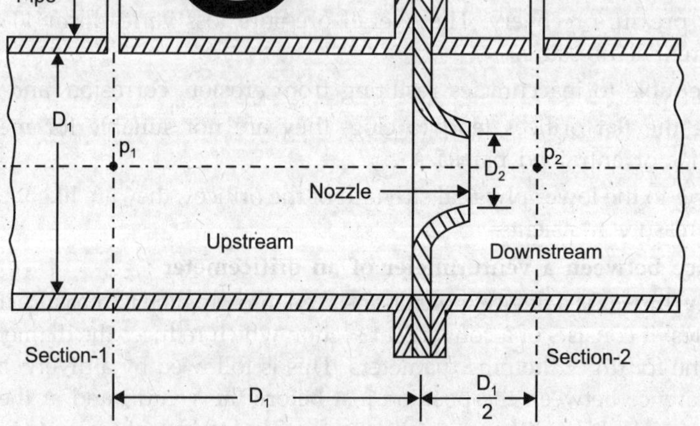


**Fig. 12.32.** Flow measurement through the use of a nozzle.

Fig. 12.32 shows flow measurement by the use of a nozzle. The flow nozzle comprises a smooth, gradual contraction to throat followed by a free, uncontrolled expansion back to the original pipe flow area. In a nozzle the flow is guided smoothly to the section of minimum area. The equation for discharge through a nozzle is given by

$$Q_{actual} = KEA_2 \sqrt{\frac{2g}{w}(p_1 - p_2)} \quad ...(12.25)$$

where,

$$K = \frac{C_d}{\sqrt{1-(A_2/A_1)^2}} = \frac{C_d}{\sqrt{1-(D_2/D_1)^4}} \quad ...(12.26)$$

Since the device contains no provision for an orderly transformation of velocity pressure into static pressure, the nozzle has a pressure loss of from 80 to 90 per cent of the differential pressure obtained.

The *discharge coefficient of a flow nozzle is dependent upon the following factors :*

(*i*) The smoothness of approach to tangency.

(*ii*) The length of cylindrical portion of the nozzle.

(*iii*) The location of the pressure taps. The pipe wall taps located at one pipe diameter upstream and half pipe downstream from the inlet face of the nozzle give best result.

The flow nozzles are usually made of *gunmetal, stainless steel or monel metal.*

**Advantages :**

(*i*) They have higher coefficient of discharge than that of orifice meters.

(*ii*) Cheaper and easier to install.

(*iii*) Physical length is less compared to venturimeter.

(*iv*) Quite useful for measurement of flow of fluids containing solids that settle.

(*v*) Widely used for high pressure/temperature steam flow.

(*vi*) Being more rugged and more resistant to erosion than sharp edged orifice, they can be used for flow measurements at high velocities.

**Disadvantages :**

(*i*) Owing to poor pressure recovery, they are not useful for applications where pressure heads are small or where pressure recovery is a must.

(*ii*) As compared to orificemeter, it is expensive and difficult to install.

(*iii*) Limited moderate pipe sizes, not available above 120 cm.

(*iv*) Higher maintenance.

### 12.3.4. Pitot Tubes

In kinematics and dynamics of fluid flow, flow velocity constitutes an important parameter and as such its measurement is quite essential. *Velocity measurements are made for the following purposes :*

(*i*) To determine the rate of flow (volumetric)

(*ii*) To locate separation points from a knowledge of velocity distribution.

(*iii*) To forecast the weather from wind velocity measurements.

- Instruments like the "*pitot tube*" and "*hot wire anemometers*" measure the "*local velocity*" at point in the channel or ducts through which the fluid is flowing.
- Instruments like the *cup and vane anemometers, the current and turbine meters* measure the "*average velocity*" *of flow.*

**Description of pitot tubes:**

The velocity distribution across a section can be obtained with the help of a *pitot-tube,* which is one of the most accurate devices for velocity measurement.

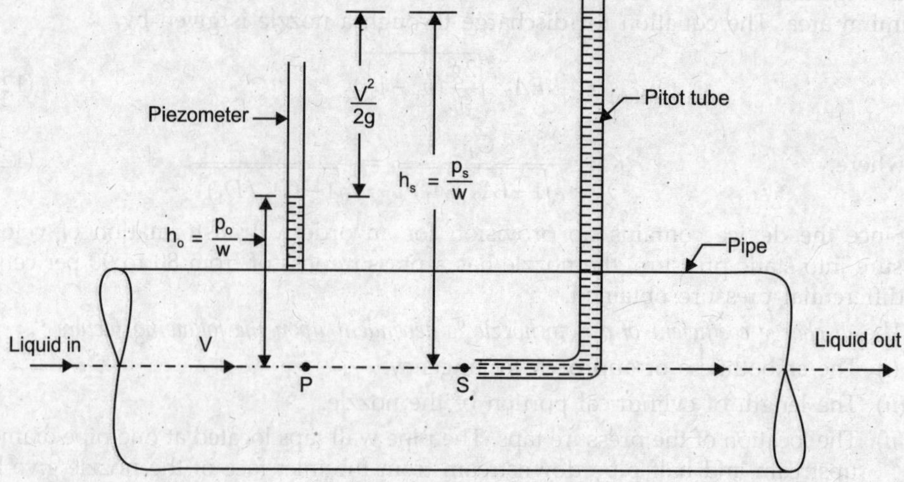


**Fig. 12.33.** Pitot tube.

It consists of a glass tube in the form of a 90° bend of short length open at both its ends. It is placed in the flow with its bent leg directed upstream so that a stagnation point is created immediately in front of the opening (Fig. 12.33). The kinetic energy at this point gets converted into pressure energy causing the liquid to rise in the vertical limb, to a height equal to the stagnation pressure.

Applying Bernaulli's between stagnation point ($S$) and a point ($P$) in the undistributed flow at the same horizontal plane we get,

$$\frac{p_o}{w} + \frac{V^2}{2g} = \frac{p_s}{w}$$

or,
$$h_o + \frac{V^2}{2g} = h_s$$

or,
$$V = \sqrt{2g(h_s - h_o)} \text{ or } \sqrt{2g\Delta h} \quad \text{...(12.27)}$$

where, $p_o$ = Pressure at point '$P$' *i.e.*, static pressure,

$V$ = Velocity at point '$P$' *i.e.*, free flow velocity.

$p_s$ = Stagnation pressure at point '$S$', and

$\Delta h$ = Dynamic pressure

= Difference between stagnation pressure head ($h_s$) and static pressure head ($h_o$)

The height of liquid rise in the pitot tube indicates the stagnation pressure head. The static pressure head may be measured separately with a piezometer (Fig. 12.34).

Both the static pressure as well as stagnation pressure can be measured in a device known as *Pitot-static tube* (Fig. 12.34).

It consists of two concentric pitot-tubes with an annular space in between as shown in the figure. The outer tube has additional two or more holes drilled perpendicular to the direction of flow and thus the liquid level in it gives the static head, while the inner tube works as normal pitot-tube. If a differential manometer is connected to the tubes of a pitot static tube it will measure the dynamic pressure head. If $y$ is manometer difference, then

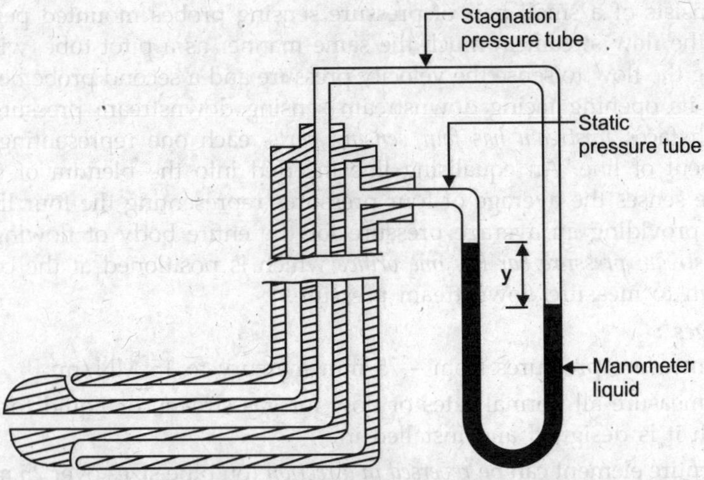


**Fig. 12.34.** Pitot-static tube.

$$\Delta h = y\left(\frac{S_m}{S} - 1\right) \quad ...(12.28)$$

where, $S_m$ = Specific gravity of manometric liquid, and
$S$ = Specific gravity of the liquid flowing through the pipe

When a pitot-tube is placed in the fluid stream the flow along its outer surface gets accelerated and causes the static pressure to decrease. Also, the stem, which is perpendicular to the flow direction, tends to produce an excess pressure head. In order to take these effects into account equation (12.27) is modified to give the actual velocity as :

$$V = C\sqrt{2g\Delta h} \quad ...(12.29)$$

where, C = A corrective coefficient which takes into account the effect of stem and bent leg.

The most commonly used form of Pitot-static tube known as Prandle-Pitot tube is so designed that the effect of stem and bent leg cancel each other *i.e.*, C = 1.

**Advantages :**

(*i*) Negligible loss of head when pitot tube is inserted in the pipe.
(*ii*) Less costly as compared to venturimeter or orificemeter.
(*iii*) Easy to install and remove from the pipeline.

**Disadvantages :**

(*i*) Poor accuracy.
(*ii*) Unsuitable for dirty or sticky liquids.
(*iii*) Requires high flow velocities of about 15 m/s to produce measurable heads.
(*iv*) Sensitive to upstream disturbance, to Reynolds number changes and to dimensions errors.
(*v*) Cannot be used for industrial applications which require an instant readout.

**Annubar tube :**

- An ***Annubar tube*** is a *commercial version of pitot tube and is applicable to liquid, steam and gas flow measurements.* Its simple and unique design is based upon the fundamental Bernoulli's equation for flow balance.

- It consists of a small pair of pressure sensing probes mounted perpendicularly into the flow stream in much the same manner as a pitot tube, with one probe facing the flow to sense the velocity pressure and a second probe behind the first with its opening facing downstream sensing downstream pressure. The probe which faces *upstream has four sensing parts,* each one representing an annubar segment of line. An equalising line inserted into the plenum of the upstream probe senses the average of four pressures representing the four line segments, thus providing an average pressure for the entire body of flowing liquid. The *downstream pressure bar has one orifice* which is positioned at the centre of flow stream to measure downstream pressure.

***Advantages :***

(*i*) Can measure pressures from – 75 mm mercury to 15 $MN/m^2$.

(*ii*) Can measure all normal rates of flow for gas stream, or liquid for the pipe for which it is designed and installed in.

(*iii*) The entire element can be *reversed in direction* (on pipe sizes over 25 mm diameter) to detect reverse flows or act as an off position without a system shut down.

(*iv*) Can measure a wide range of flow rates without deviating from a square root output signal which is compatible with all standard makes of flow instrumentation and control.

(*v*) Relatively less expensive as a flow detector.

(*vi*) Relatively simple to install and light in weight.

(*vii*) Reliable for flow measurements through pipes or circular ducts from 12.5 mm to 225 mm in diameter.

(*viii*) The flow resisitive effect of this meter is almost negligible.

(*ix*) These units are available as complete systems with elements, including, recording and control instrumenation.

### 12.3.5. Weirs and Flumes

These devices are commonly used for *open channel measurements.*

**Weir :** A weir is essentially a dam with a notched opening at the top through which the liquids flows.

The weir make use of *rectangular, V and tapezoidal notches.* The rate of flow is simply measured by measuring the head of water above the lowest point of the weir opening through which the liquid flows. This height is measured by means of a float installed in a box called *stilling well,* which is a part of the total structure. The float is so placed that is not disturbed by the velocity of flow or by turbulence of stream.

- The discharge through a *rectangular notch,*

$$Q = \frac{2}{3} C_d L \sqrt{2g} (H)^{\frac{3}{2}} \quad ...(12.30)$$

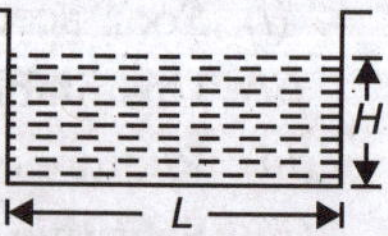

Fig. 12.35. Rectangular notch/wear.

where, $C_d$ = Coefficient of discharge (it varies with the dimensions of the weir and is usually less than 0.7),

$L$ = Length of weir, and

$H$ = Height of water above sill of the notch/wear.

— The rectangular notch is the *most commonly used notch on account of its simplicity, ease of construction and accuracy.*

- The discharge through a V-notch,

$$Q = \frac{8}{15} C_d \sqrt{2g}(H)^{\frac{5}{2}} \tan\frac{\theta}{2} \quad ...(12.31)$$

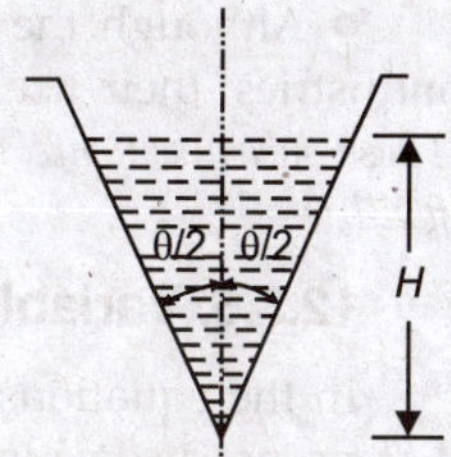


Fig. 12.36. Triangular notch/wear.

where, $\theta$ = angle of notch.

$C_d$ varies with $\theta$ and $H$, kinematic viscosity and surface tension of liquids; typical values are 0.58 to 0.7.

— The V-notch offers the widest range for a single size, the small opening at the lower end can accommodate small flows while the top portion for large flows. However, *because of its shape, it gives the highest loss of head.*

- The discharge through a *trapezoidal notch* (combination of the rectangular and triangular notches),

$$Q = \frac{2}{3} C_d L \sqrt{2g}(H)^{\frac{3}{2}} + \frac{8}{15} C_d \sqrt{2g}\,(H)^{\frac{5}{2}} \tan\frac{\theta}{2} \quad ...(12.32)$$

**Flume :** A *"flume" is an open channel whose sides are arranged sharply to form a throat, followed by a fairly long expansion section.* Stilling wells are provided for measuring the head at both entrance and throat as shown in Fig. 12.38. In this case,

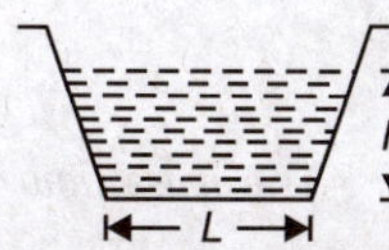


Fig. 12.37. Tapezoidal notch/wear.

$$Q_{\text{actual}} = \frac{CA_2}{\sqrt{1 + (A_2/A_1)^2}} \sqrt{2gh} \quad ...(12.33)$$

C = Venturi flume coefficient (0.95 to 1), and

$h = h_1 - h_2$

When, $h_2 = \frac{2}{3}h$, *maximum value of flow occurs.*

Then, $$Q_{\text{max}} = 1.7 b_2 (H)^{\frac{3}{2}} \quad ...(12.34)$$

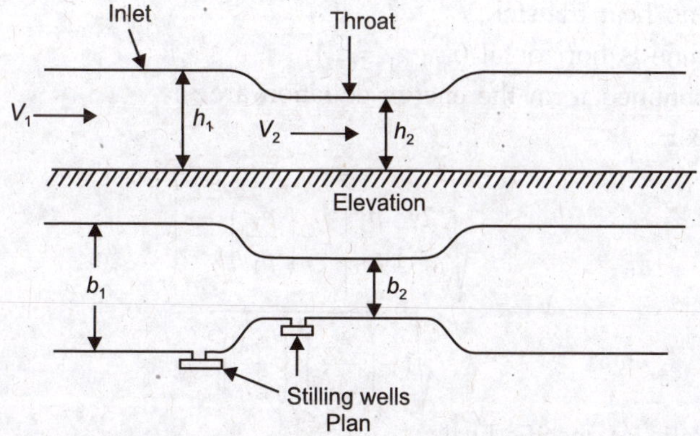


Fig. 12.38. Venturi flume.

The flumes given *better accuracies than weirs* since the loss of head in improved versions of flumes is about $\frac{1}{4}$th that of weirs.

• Although the weirs and flumes have not previously found application in process industries, their use is increasing with the increasing demand for *water pollution control. These measurements are a must when large volumes of water are handled in waste treatment facilities.*

### 12.3.6. Variable Head Meters for Compressible Fluids

In the equations derived so far we have assumed that the density remains constant. Let us now consider the flow of *compressible fluids* in which *density is not constant but changes with pressure.* In order to account for changes in density with pressure, a factor Y, called "**expansion factor**" is introduced. This factor is so chosen that the mass flow rate calculated for incompressible fluids when multiplied by this factor gives the flow rate for compressible fluids.

The volumetric flow rate for a *compressible fluid* for venturies (eqn. 12.16) is given by

$$Q_{actual} = C_d MEA_2 \sqrt{\frac{2g}{w}(p_1 - p_2)}$$

$$= C_d MEA_2 \sqrt{\frac{2}{\rho}(p_1 - p_2)} \qquad ...(12.35)$$

The *mass flow rate* considering the density at the inlet is,

$$m = C_d MEA_2 Y \sqrt{\frac{2}{\rho_1}(p_1 - p_2)} \qquad ...(12.36)$$

The expansion factor $Y$ can be taken as unity for liquids and for atmospheric air in the range $\left(\frac{p_1 - p_2}{\rho}\right)$ less than 0.1 cm of water.

The value of $Y$ can be calculated theoretically for an *adiabatic flow process*. The following *assumptions* are made :

(*i*) The flow is steady and hence the mass entering is equal to mass leaving or $m_1 = m_2$.

(*ii*) There is no heat transfer.

(*iii*) The pipeline is horizontal (*i.e.*, $z_1 = z_2$)

The results obtained form the energy equation are :

Velocity at exit,

$$V_2 = \frac{\sqrt{\frac{2\gamma}{\gamma - 1}\frac{p_1}{\rho_1}\left[1 - \left(\frac{p_2}{p_1}\right)^{\frac{\gamma-1}{\gamma}}\right]}}{\sqrt{1 - \left(\frac{A_2}{A_1}\right)^2\left(\frac{\rho_2}{\rho_1}\right)^2}}$$

where, $\gamma$ = Ratio of specific heats.

The mass flow rate,

$$M_{actual} = C_d A_2 E \, \rho_2 V_2$$

$$= C_d A_2 E \rho_2 \frac{\sqrt{\frac{2\gamma}{\gamma-1}\frac{p_1}{\rho_1}\left[1-\left(\frac{p_2}{p_1}\right)^{\frac{\gamma-1}{\gamma}}\right]}}{\sqrt{1-\left(\frac{A_2}{A_1}\right)^2\left(\frac{\rho_2}{\rho_1}\right)^2}} \quad \text{...(12.37)}$$

**Example 12.4.** *The flow of cooling water in a manufacturing process is measured by a horizontal venturimeter with 200 mm inlet and 100 mm throat. The U-tube mercury manometer connected between inlet and throat of venturi shows a differential pressure of 220 mm of mercury. Calculate the water flow rate if coefficient of discharge is 0.98, specific gravity of mercury 13.6, and density of water 1000 kg/m*$^3$.

**Solution.** Given : $D_1$ = 200 mm = 0.2 m; $D_2$ = 100 mm = 0.1 m

$h$ = 220 mm, $C_d$ = 0.98; Sp. gr. of mercury = 13.6;

$\rho_{water}$ = 1000 kg/m$^3$

**Water flow rate, $Q_{actual}$ :**

Differential pressure head,

$$(p_1 - p_2) = gh\,(\rho_{Hg} - \rho_{water})$$
$$= 9.81 \times 220 \times 10^{-3}\,(13.6 - 1) \times 1000$$
$$\doteq 27.193 \times 10^3 \text{ N/m}^2$$

Velocity approach factor,

$$M = \frac{1}{\sqrt{1-\left(\frac{A_2}{A_1}\right)^2}}$$

(where, $A_1 = \frac{\pi}{4}D_1^2$ and $A_2 = \frac{\pi}{4}D_2^2$)

$$= \frac{1}{\sqrt{1-\left(\frac{D_2}{D_1}\right)^4}} = \frac{1}{\sqrt{1-\left(\frac{100}{200}\right)^4}} = 1.0328$$

The actual discharge through a venturimeter is given by,

$$Q_{actual} = C_d M A_2 = \sqrt{\frac{2g}{w}(p_1 - p_2)} \quad \text{...(12.14)}$$

...(assuming thermal expansion factor $E$ = 1)

$$= 0.98 \times 1.0328 \times \frac{\pi}{4} \times (0.1)^2 \sqrt{\frac{2 \times 9.81}{9.81 \times 1000} \times 27.193 \times 10^3}$$

$$= \mathbf{0.0586\ m^3/s.\ (Ans.)}$$

**Example 12.5.** *The following data relate to an inclined venturimeter :*

*Diameter of the pipeline = 400 mm*

*Inclination of the pipeline with horizontal = 30°*

*Throat diameter = 200 mm*

*The distance between the mouth and throat of the meter = 600 mm*

*Sp. gravity of oil flowing through the pipeline = 0.7*
*Sp. gravity of heavy liquid (U-tube) = 13.6*
*Reading of the differential manometer = 50 mm*
*The coefficient of the meter = 0.98*
*Determine the rate of flow in the pipeline.* **(Punjab University)**

**Solution.** Diameter at inlet, $D_1 = 400 \text{ mm} = 0.4 \text{ m}$

$$\therefore \quad \text{Area of inlet}, A_1 = \frac{\pi}{4} \times 0.4^2 = 0.1257 \text{ m}^2$$

Throat diameter, $D_2 = 200 \text{ mm} = 0.2 \text{ m}$

$$\therefore \text{Area of throat}, A_2 = \frac{\pi}{4} \times (0.2)^2 = 0.0314 \text{ m}^2$$

Reading of the differential manometer (U-tube),

$$y = 50 \text{ mm} = 0.05 \text{ m}$$

Difference of pressure head $h$ is given by

$$h = y\left[\frac{S_{hl}}{S_p} - 1\right]$$

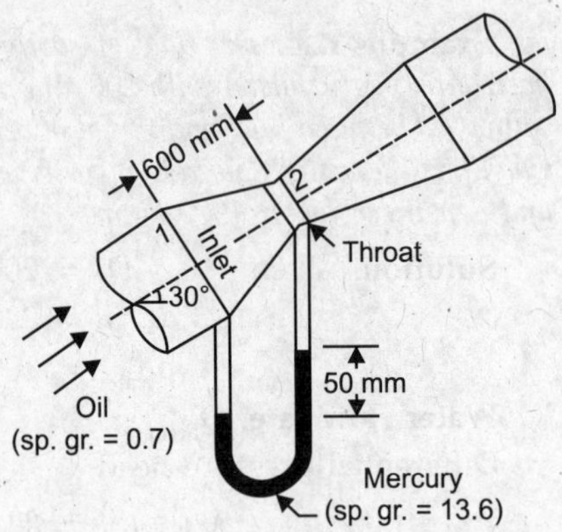

**Fig. 12.38.**

where, $S_{hl}$ = Sp. of gravity of heavy liquid (*i.e.*, mercury) in U-tube = 13.6, and

$S_p$ = Sp. gravity of liquid (*i.e.*, oil) flowing through the pipe = 0.7.

$$\therefore \qquad h = 0.05\left(\frac{13.6}{0.7} - 1\right) = 0.92 \text{ m of oil}$$

Now, applying Bernoulli's equation at sections '1' and '2', we get

$$\frac{p_1}{w} + z_1 + \frac{V_1^2}{2g} = \frac{p_2}{w} + z_2 + \frac{V_2^2}{2g} \qquad ...(i)$$

or, $$\left(\frac{p_1}{w} + z_1\right) - \left(\frac{p_2}{w} + z_2\right) + \frac{V_1^2}{2g} - \frac{V_2^2}{2g} = 0$$

But, $$\left(\frac{p_1}{w} + z_1\right) - \left(\frac{p_2}{w} + z_2\right) = h$$

or, $$\left(\frac{p_1}{w} - \frac{p_2}{w}\right) + (z_1 - z_2) = h$$

It may be noted that *differential gauge reading will include in itself the difference of pressure h and the difference of datum head.*

Thus eqn. (*i*) reduces to :

$$h + \frac{V_1^2}{2g} - \frac{V_2^2}{2g} = 0 \qquad ...(ii)$$

Applying continuity at sections '1' and '2' we get

$$A_1 V_1 = A_2 V_2$$

or, $$V_1 = \frac{A_2 V_2}{A_1} = \frac{\frac{\pi}{4} \times 0.2^2}{\frac{\pi}{4} \times 0.4^2} \times V_2 = \frac{V_2}{4}$$

Substituting the value of $V_1$ and $h$ in equation (*ii*) we get

$$0.92 + \frac{V_2^2}{16 \times 2g} - \frac{V_2^2}{2g} = 0$$

or, $$\frac{V_2^2}{2g}\left(1 - \frac{1}{16}\right) = 0.92 \quad \text{or} \quad V_2^2 \times \frac{15}{16} = 0.92$$

or, $$V_2^2 = \frac{0.92 \times 2 \times 9.81 \times 16}{15} = 19.25 \text{ or } V_2 = 4.38 \text{ m/s}$$

Rate of flow of oil,

$$Q = A_2 V_2 = 0.0314 \times 4.38 = \mathbf{0.1375\ m^3/s\ (Ans.)}$$

**Example 12.6.** *Water flows at the rate of 0.015 $m^3/s$ through a 100 mm diameter orifice used in a 200 mm pipe. What is the difference in pressure head between upstream section, and vena contracta section. Take coefficient of contraction and discharge as 0.60 each.*

**Solution.** *Given* : $Q = 0.015 \text{ m}^3/\text{s}; D_0 = 100 \text{ mm} = 0.1 \text{ m}$

$$D_1 = 200 \text{ mm} = 0.2 \text{ m}; C_c = 0.6; C_d = 0.6$$

$$A_0 = \frac{\pi}{4} D_0^2 = \frac{\pi}{4} \times 0.1^2 = 0.007854 \text{ m}^2$$

$$A_1 = \frac{\pi}{4} D_1^2 = \frac{\pi}{4} \times 0.2^2 = 0.03141 \text{ m}^2$$

**Difference in pressure head :**

The flow rate through an orificemeter is given by

$$Q = KEA_0 \sqrt{\frac{2g}{w}(p_1 - p_2)} \qquad \text{...[Eqn. (12.23)]}$$

Here, $$K = C_d M = \frac{C_d}{\sqrt{1 - \left(\frac{C_0 A_0}{A_1}\right)^2}} = \frac{0.6}{\sqrt{1 - \left(0.6 \times \frac{0.007854}{0.03141}\right)^2}} = 0.6069$$

Assuming $E$ (thermal expansion factor) as 1 and substituting the various values, we get,

$$0.015 = 0.6069 \times 0.007854 \sqrt{\frac{2 \times 9.81}{9810}(p_1 - p_2)}$$

$$= 0.00021317 \sqrt{(p_1 - p_2)}$$

$$\therefore \quad (p_1 - p_2) = \mathbf{4951.5\ N/m^2\ or\ 0.5047\ m\ of\ water\ (Ans.)}$$

**Example 12.7.** *A venturi tube of throat diameter 60 mm has a discharge coefficient of 0.97 and with a flow rate of 1.2$m^3$/s, the pressure differential is 15.5 N/$m^2$. Determine the flow rate when an orifice of 60 mm is used in the same pipe. The discharge coefficient of orifice is 0.6 and pressure differential in same.*

**Solution.** Let the measurements through venturimeter and orifice are denoted by subscripts $v$ and $o$ respectvely.

Then, $$Q_v = C_v M A_2 \sqrt{\frac{2g}{w}(p_1 - p_2)}$$

and, $$Q_o = C_o M A_2 \sqrt{\frac{2g}{w}(p_1 - p_2)}$$

∴ Discharge through the orifice,

$$Q_o = \frac{C_o}{C_v} \times Q_v = \frac{0.6}{0.97} \times 1.2 = \mathbf{0.742\ m^3/s\ (Ans.)}$$

**Example 12.8.** *A sub-marine fitted with a pitot tube moves horizontally in sea. Its axis is 12 m below the surface of water. The pitot tube fixed in front of the sub-marine and along its axis is connected to the two limbs of a U-tube containing mercury, the reading of which is found to be 200 mm. Find the speed of the sub-marine.*

*Take the specific gravity of sea water = 1.025 times fresh water.*

**Solution.** Reading of the manometer, $y$ = 200 mm = 0.2 m of mercury
Sp. gravity of mercury, $S_{hl}$ = 13.6
Sp. gravity of sea water, $S_l$ = 1.025

To find the head, ($h$) using the relation : $h = y\left(\frac{S_{hl}}{S_l} - 1\right)$, we have

$$h = 0.2\left(\frac{13.6}{1.025} - 1\right) = 2.45$$

∴ Velocity of the sub-marine,

$$V = \sqrt{2gh} = \sqrt{2 \times 9.81 \times 2.45}$$
$$= 6.93 \text{ m/s or } \mathbf{24.9\ km/h\ (Ans.)}$$

## 12.3.7. Rotameter and Elbow Meter

### 12.3.7.1. Rotameter

*A **rotameter** is a constant-pressure drop, variable area flow meter.*

**Construction.** Refer to Fig. 12.39. It consists of a tapered metering glass tube, inside of which is located a rotor or active element (float) of the meter. The tube is provided with inlet and outlet connections. The specific gravity of the float or bob material is *higher* than that of the fluid to be metered. On a part of the float spherical slots are cut which cause it (float) to rotate slowly about the axis of the tube and keep it centered. Owing to this spinning accumulation of any sediment on the top or sides of float is checked. However, the stability of the bob may also be ensured by using a guide along which the float would slide.

**Working.** When the rate of flow increases the float rises in the tube and consequently there is an increase in the annular area between the float and the tube. Thus, the float rides higher or lower depending on the rate of flow.

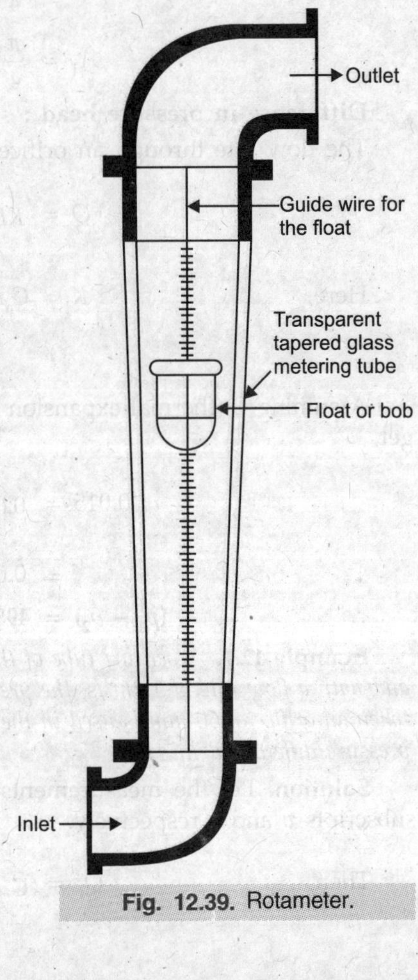


**Fig. 12.39.** Rotameter.

The discharge through a rotameter is given by

$$Q = C_d A_{ann.} \left[ 2g V_{fl} \frac{(\rho_{fl} - \rho_f)}{A_f \rho_f} \right]^{\frac{1}{2}} \quad ...(12.38)$$

where, $Q$ = Volume flow rate,

$C_d$ = Co-efficient of discharge,

$A_{ann.}$ = Annular area between float and tube,

$V_{fl}$ = Volume of float,

$\rho_{fl}$ = Density of float material,

$\rho_f$ = Density of fluid, and

$A_f$ = Maximum cross-sectional area of fluid.

As the flow area $A_{ann.}$ is a function of height of float in the tube, the flow rate scale can be engraved on the tube corresponding to a particular float.

***Advantages :***

1. Simpler in operation.
2. Handling and installation easy.
3. Wide variety of corrosive fluids can be handled.
4. Low cost relatively.
5. Possibility of convenient and visible flow comparisons by mounting several rotameters side by side.
6. Easily equipped with data transmission and recording devices.
7. Rotameters are immune from viscosity effects upto a certain range.
8. They are particularly useful for measurement of low flow rates.
9. They can be used for measuring liquids, gases and vapours.

***Limitations :***

1. Mounted vertically, limited to small pipe sizes and capacities.
2. Less accurate, compared to venturimeter and orificemeter.
3. Glass tube subject to breakage.

**12.3.7.2. Elbow meter**

When liquid flows around a pipe bend there is an increase in pressure with radius, *i.e., the pressure at the outer wall of the bend is more than that at the inner wall. This difference of pressure which exists between the outside and inside of the bend is used for measurement of discharge in a pipeline.*

As shown in Fig. 12.40, the pipe bend is provided with two pressure tappings, one each at the inner and outer walls of the bend. These tappings are connected to the links of U-tube manometer.

As per literature, the following relation between velocity and pressure difference is available :

$$K \frac{V^2}{2g} = \left( \frac{p_o}{w} + z_o \right) - \left( \frac{p_i}{w} + z_i \right)$$

where, $K$ = constant (depends upon the shape and size of the band), *ranges from* 1.3 to 3.2, and

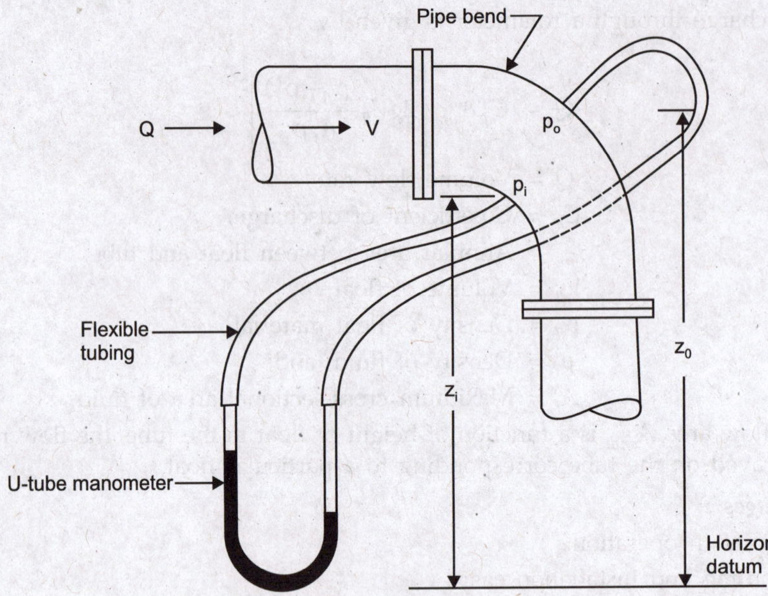


**Fig. 12.40.** Elbow meter.

$V$ = Velocity of flow,

Suffices $o$ and $i$ represent the conditions at the outer and inner walls of the pipe bend.

or, $$V = \frac{1}{\sqrt{K}}\sqrt{2g}\sqrt{\left(\frac{p_o}{w}+z_o\right)-\left(\frac{p_i}{w}+z_i\right)}$$

∴ Discharge, $$Q = AV = C_d A\sqrt{2g}\sqrt{\left(\frac{p_o}{w}+z_o\right)-\left(\frac{p_i}{w}+z_i\right)} \quad ...(12.39)$$

where, $C_d = \frac{1}{\sqrt{K}}$ = Coefficient of discharge, and

$A$ = Cross-sectional area of the pipe.

$C_d$ varies between 0.56 and 0.88.

The following empirical relation has been suggested :

$$C_d = \sqrt{\frac{R}{D}}$$

where, $R$ = Radius of the pipe bend, and

and $D$ = Diameter of the pipe.

- An elbow meter can be conveniently used for the measurement of discharge in pipes which are fitted with elbows and bends.
- Its accuracy, with proper calibration, approaches that of a venturimeter or nozzle.

### 12.3.8. Electromagnetic Flow Meters

These meters are particularly suitable for the flow measurements of *slurries, sludge and any electrically conducting fluid.*

- Flow of an *electrically conducting fluid* may be determined by measuring the e.m.f. developed between a pair of electrodes set in opposite sides of an insulating conduit due to the movement of the liquid through a magnetic field established transversely of the conduit and perpendicular to both the flow and the line joining the electrodes (See Fig. 12.41). *By using an alternating field the effects of electrode polarization may be eliminated. Null measurement of the generated voltage renders the apparatus independent of the resistance of the liquid.*

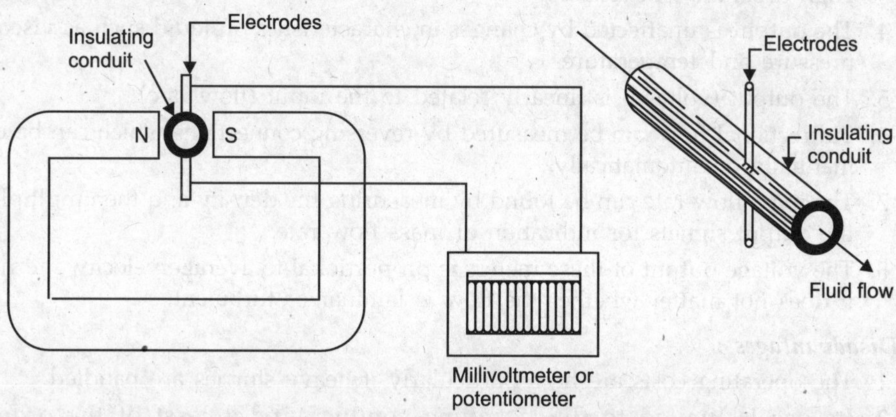


Fig. 12.41. Electromagnetic flow meter.

**Theory :** Refer to Fig. 12.41.

Consider the flow of a conductive liquid through a non-metallic insulating pipe located in the air gap of a permenant magnet (or an electromagnet which may be either A.C. or D.C.) Fluid motion relative to the field generates an induced *e.m.f. i.e.,* the value of which as per Faraday's law is given as :

$$e = Blv \times 10^{-8} \text{ volts} \quad ...(12.40)$$

where, $B$ = Magnetic flux density (in V-s/cm$^2$)

$l$ = Length of the conductor (in cm) and

$v$ = Velocity of the conductor (in cm/s).

The effective length of the conductor corresponds to the inner diameter of the pipe and the velocity of the conductor is proportional to the mean flow velocity. The volume flow rate is give by,

$$Q = \frac{\pi}{4}d^2 v; \quad \text{the eqn. (12.40) then gets modified to}$$

$$e = B \times d \times \frac{4Q}{\pi d^2} \times 10^{-8} \quad \left(\because l = d \text{ and } v = \frac{4Q}{\pi d^2} \text{ from flow equation}\right)$$

or,

$$e = \frac{4QB}{\pi d} \times 10^{-8} \text{ volts} \quad ...(12.41)$$

This shows volume flow rate $Q$ is directly proportional to the induced *e.m.f.* $e$ so long as the flux density remains unaltered.

The *e.m.f.* in the conductor is picked up by means of two electrodes set in the opposite sides of an insulating and non-magnetic pipe. This is to prevent short-circuiting of the induced voltage, and to allow magnetic field to penetrate the liquid. The electrodes are mounted flush with inside faces of the pipe and thus making a direct contact with the flowing liquid.

The *e.m.f.* generated by the instrument, however, is very small (typically of the order of 1 mV) and the resistance of the fluid is often very high. Therefore, the output of the instrument is generally amplified suitably.

***Advantages :***

1. No obstruction to flow.
2. Can handle slurries, greasy materials and liquids containing suspended matter.
3. High accuracy and reliability.
4. The output is unaffected by changes in characteristics of liquid such as viscosity, pressure and temperature.
5. The output (voltage) is linearly related to the input (flow rate).
6. Bidirectional flow can be measured by reversing connections which can be done manually or automatically.
7. The mass flow rate can be found by measuring the density and then multiplying the output signals for indication of mass flow rate.
8. The voltage output of these meters is proportional to average velocity and hence it does not matter whether the flow is laminar or turbulent.

***Disadvantages :***

1. The operating costs are high particularly if heavy slurries are handled.
2. Its use is limited to fluids having conductivity at least of the order of 0.05 μmho/cm.
3. They must be full at all the times because it sees velocity as analogues to volume flow rate.

**Application.** This type of meter in its various forms has been widely used for metering *corrosive acids, cement slurries, sewage, paper pulp, detergents etc. which present difficult handling problems.*

### 12.3.9. Ultrasonic Flow Meters

*The **working principle** of these meters is based on an apparent change in velocity of propagation of sound waves in a fluid with a change in the velocity of the fluid flow. The method uses transmitters and receivers of acoustic energy. "Piezoelectric crystals" are used as both transmitters and receivers.*

Fig. 12.42 illustrates an ultrasonic flow meter for measurement of flow.

***Refer to Fig. 12.42(i)** : Travel time difference method–Single transmitter-receiver system.*

–When flow velocity ($v$) is zero, the transit time of the pulse from the transmitter to the receiver is given by :

$$t_0 = \frac{d}{c} \qquad ...(12.42)$$

where, $d$ = Distance between the transmitter and receiver, and

$c$ = Velocity of sound in fluid.

Now, when the fluid moves with velocity $v$, the transit time becomes :

$$t = \frac{d}{c+v} \qquad ...(12.43)$$

$$= \frac{d(c-v)}{(c+v)(c-v)} = \frac{d(c-v)}{c^2-v^2}$$

...Multiplying numerator and denominator by $(c–v)$

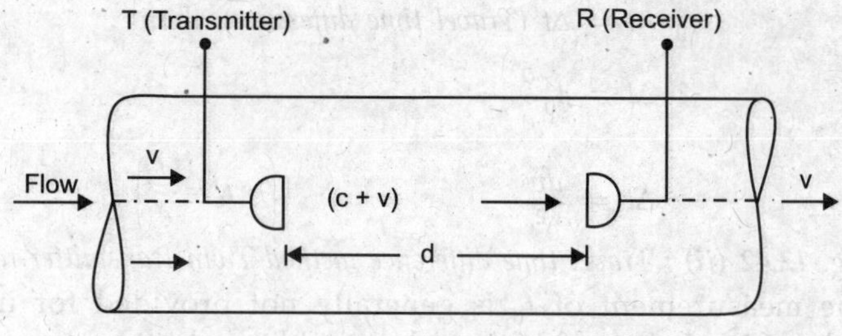


(i) Travel time difference method-single transmitter receiver system

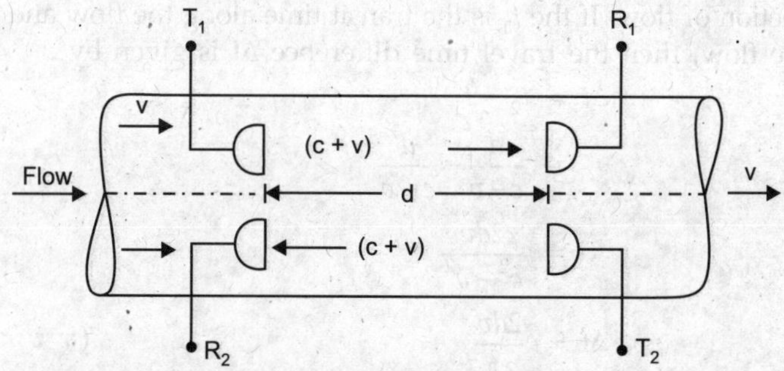


(ii) Travel time difference method-Twin transmitter-receiver system

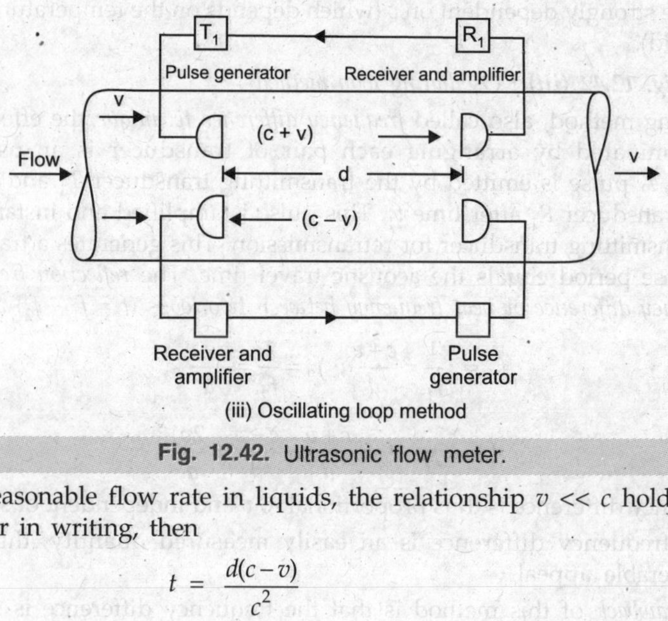


(iii) Oscillating loop method

**Fig. 12.42.** Ultrasonic flow meter.

For any reasonable flow rate in liquids, the relationship $v << c$ holds and there is negligible error in writing, then

$$t = \frac{d(c-v)}{c^2}$$

$$= \frac{d}{c}\left(1-\frac{v}{c}\right)$$

or, $$t = t_0\left(1-\frac{v}{c}\right) \quad ...(12.44)$$

or, $$t_0 - t = t_0\frac{v}{c}$$

Let $t_0 - t = \Delta t$ (*Travel time difference*), then

$$\Delta t = t_0 \frac{v}{c}$$

or,

$$\Delta t = \frac{dv}{c^2} \qquad \left(\because t_0 = \frac{d}{c}\right) \qquad ...(12.45)$$

***Refer to Fig. 12.42 (ii) : Travel time difference method-Twin transmitter-receiver method :***

Because the measurement of $t_0$ is generally not provided for in the present measurement, therefore, it is preferable to have additional set of transmitter-receiver system along with the present system [see Fig. 12.42(*ii*)] to determine the transmit time against the direction of flow. If the $t_1$ is the transit time along the flow and $t_2$ the transmit time against the flow, then the travel time difference $\Delta t$ is given by :

$$\Delta t = t_2 - t_1$$

$$= \frac{d}{c-v} - \frac{d}{c+v}$$

$$= \frac{2dv}{c^2 - v^2}$$

or,

$$\Delta t \simeq \frac{2dv}{c^2} \qquad (\because v << c) \qquad ...(12.46)$$

Thus an output signal proportional to $\Delta t$ is linear in $v$ for constant $c$ but the calibration constant will be strongly dependent on $c$ (which depends on the temperature and pressure of the flow field).

***Refer to Fig. 12.42 (iii) : Oscillating loop method :***

In oscillating method, also called *frequency difference technique* the effect of the sonic velocity is eliminated by arranging each pair of transducer is an oscillating loop [Fig. 12.42(*iii*)]. A pulse is emitted by the transmitting transducer $T_1$ and is received by the receiving transducer $R_1$ after time $t_1$. This pulse is amplified and instantaneously fed back to the transmitting transducer for retransmission. This generates a train of pulses in each path whose period equals the acoustic travel time. The *reflection frequencies* ($f_1$, $f_2$) and the *frequency difference or beat frequency between two loops* ($f = f_1 - f_2$) are given as :

$$f_1 = \frac{1}{t_1} = \frac{c+v}{d}; \quad f_2 = \frac{1}{t_2} = \frac{c-v}{d} \qquad ...(12.47)$$

$$f = f_1 - f_2 = \frac{c+v}{d} - \frac{c-v}{d} = \frac{2v}{d} \qquad ...(12.48)$$

The frequency difference is thus proportional to $v$ and independent of sonic velocity $c$.

— Since frequency difference is an easily measured quantity, this method has considerable appeal.

— The *drawback* of this method is that the frequency difference is small, and the long counting interval needed for a given resolution results in *slow response time.*

***Advantages :*** The following are the *advantages* and *disadvantages* of ultrasonic flow meter :

1. Absence of moving parts.
2. No abstraction to flow.
3. Excellent dynamic response.
4. High accuracy (of the order of ±2% of the full scale values).

5. Linear relationship between output and input.
6. Output is insensitive to variations in viscosity, density and temperature.
7. Suitable for metering both liquids and gases.
8. The output being electrical in nature, the read out can be analog or digital.

***Disadvantages :***

1. Complex construction.
2. Relatively high costs.
3. Use of commercial meters is generally limited to liquids.

**Applications** : This device finds special applications, namely *measurement* of :

— Ocean occurents;

— Water flows in large conduits;

— Flows of various bio-medical and industrial fluids etc.

## 12.3.10. Thermal Flow Meters

This type of flow meter is also useful for the measurement of corrosive liquids. Fig. 12.43 shows a thermal flow meter.

— It consists of a heater placed in the fluid stream, with two temperature sensors such as thermocouples or resistance thermometers.

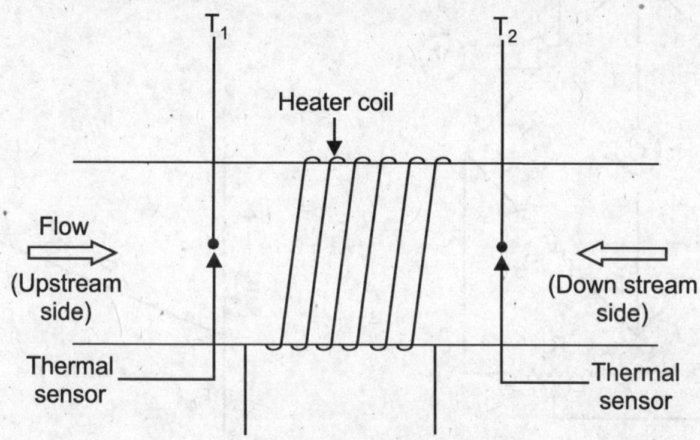


**Fig. 12.43.** Thermal flow meter.

— The temperature sensors are located symmetrically with respect to the heater, one at upstream side and the other at the downstream side, at short distance from the heater.

Under *no* flow conditions, the temperatures $T_1$ and $T_2$ of the sensors will the *same*, and the differential signal measured on the detector will be *zero*. When the *flow increases*, $T_1$ will *fall* and $T_2$ will *rise* until a steady state is reached. The heat transferred to the fluid by the heater is given by :

$$Q = m\, c_p\, (T_2 - T_1) \qquad \text{...(12.49)}$$

where,
$Q$ = Heat transferred,
$m$ = Mass flow rate of the fluid,
$c_p$ = Specific heat of the fluid,
$T_1$ = Temperature of the fluid before heat is transferred to it, and
$T_2$ = Temperature of the fluid after heat is transferred to it by the heater coil

Using the above equation, the value of *m* (*i.e.*, mass flow rate) can be obtained by measuring the temperature difference $(T_2 - T_1)$ and the heat transferred to the fluid $Q$ in terms of constant *power input to the heater*, and knowing the specific heat of $c_p$ of the fluid.

- The dynamic response of this type of flow meter is dependent on the flow conditions, and is usually poor.

### 12.3.11. Wire Anemometers

*These devices are hot wire resistance transducer wire filaments which are used for measurement of flow rates of fluids.*

By suitably locating this transducer (in the form of wire filaments) the flow rates of non-conducting liquids in open channels and closed pipes and of gases in closed pipes can be conveniently measured. The transducer which is in the form of probe is located at the centre of pipe with direction of wire perpendicular to the direction of fluid flow.

The flow sensing element is a short length of 5 μm diameter platinum-tungsten wire welded between the prongs of the probe and heated electrically as a part of Wheatstone bridge. [Fig. 12.44].

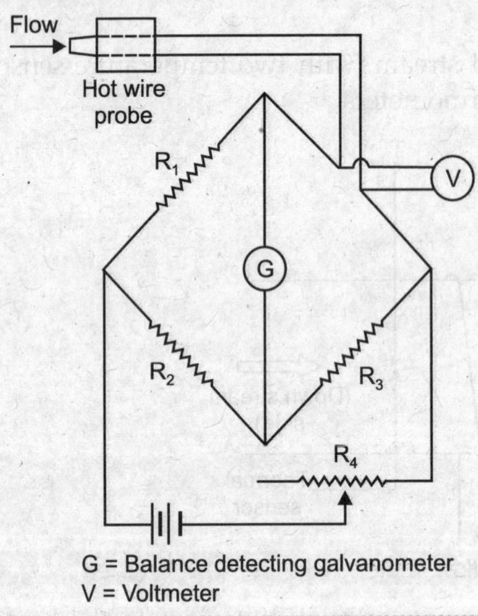


Fig. 12.44. Hot wire bridge circuit.

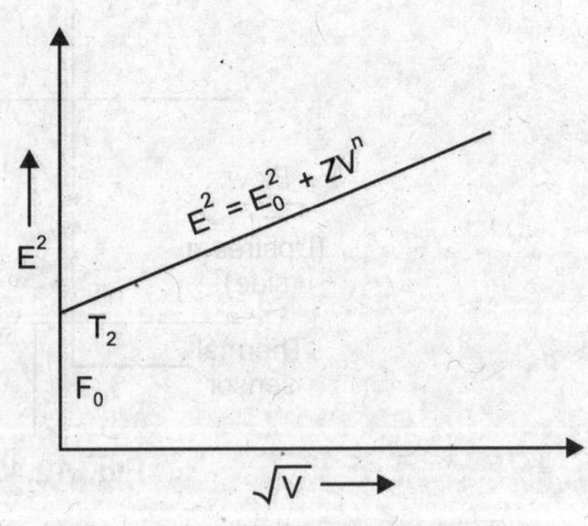


Fig. 12.45. Response of hot wire.

When the probe is introduced into the flowing fluid, it tends to be cooled by the instantaneous velocity and consequently there is a tendency for the electrical resistance to decrease. The rate of cooling of wire depends on the (*i*) dimensions and physical properties of the wire, (*ii*) difference of temperature between the wire and the fluid, (*iii*) physical properties of the fluid, and (*iv*) stream velocity under measurement. For a simple hot wire anemometer the *first three conditions are effectively constant and the instrument response is then a direct measure of the velocity.*

Hot wire anemometers are commonly used in two different modes namely :

1. Constant current mode; 2. Constant temperature mode.

**1. Constant current mode :**

— In this mode the heating current *i.e.*, voltage across the bridge is kept constant.

— Initially the circuit is adjusted such that the galvanometer reads zero when the heated wire lies in stationary air.

— *When the air flows, the hot wire cools, the resistance changes and the galvanometer deflects. The galvanometer deflections are amplified, measured and correlated with air velocity by previous calibrations.*

**2. Constant temperature mode :**

— In this mode operating resistance of the wire and hence its temperature is maintained constant.

— In the event of the tendency of the hot-wire to cool by the flowing fluid, the external bridge voltage is applied to the wire to maintain a sensibly constant temperature. *The bride voltage is varied so as to bring the galvanometer needle to zero; the reading on the voltmeter is recorded and correlated with the air velocity.*

— The response of a hot wire placed normal to the flow velocity $V$ (Fig. 12.45) is described by the relationship.

$$E_2 = E_0^2 + ZV^n \quad \text{...(12.50)}$$

where, $E$ = Instantaneous value of the bridge voltage,

$E_0$, $Z$ = Constants depending on the odd resistance of the wire and on the properties of the fluid medium.

The values of these constants for a given wire are found by calibrating the wire in a uniform flow field against a standard pitot-static probe. Generally with exponent $n$ = 0.5, the response of the hot wire in the form of a plot between $E^2$ and $V^n$ is a straight line.

- *The constant temperature design has the inherent advantage that the wire is protected against burnout.* However, the associated electronic circuitry, specially the D.C. feedback amplifier has an upper limit of the frequency response beyond which the system becomes *unstable* and the *wire temperature oscillates.*
- A modification of the hot wire anemometer is the ***hot film transducer*** which generally enhances its utility. These transducers have great mechanical strength and maybe used for very high temperatures by constructing them with internal cooling water passages.

— They are extremely sensitive to fluctuations in the liquid flow velocity and have been employed for measurements involving frequencies as high as 50 kHz. For such measurements sophisticated electronic instrumentation is required.

***Advantages and limitations.*** The following are the *advantages* and *disadvantages* of hot wire anemometer.

***Advantages :***

1. Suitable for measurement in both gases and liquids (In liquids, the probe has to be modified as the hot film probe).
2. Since the instrument's output is electrical, therefore, the read-out can be analog or digital.
3. Owing to very small probe size there is neither much disturbance nor pressure loss to the flow.
4. The instrument exhibits excellent dynamic characteristics, since its time constant is of the order of $10^{-4}$ to $10^{-6}$ s.
5. The instrument has extended measuring range from very low velocities to supersonic velocities.
6. The instrument shows excellent accuracy of the order of ±0.1% in the measurement of mean velocities and ±2% in the measurement of turbulence levels.

***Disadvantages :***

1. Expensive circuitry.
2. Needs skilled operations.
3. Highly non-linear operating characteristics.
4. The instrument needs frequent calibrations (since the accumulations of dirt/dust on the hot wire changes its sensitivity).
5. The hot wire does not sense whether the flow is in the forward or reverse direction. Its output depends on the cooling of the wire which is same in both the cases.
6. Since the wire is fragile and dust particle striking the hot wire can break the same, therefore, hot wire use and operation has to be done with exceptionally clean fluids.

### 12.3.12. Laser Doppler Anemometer

This type of anemometer measures the instantaneous velocities of gases or liquids flowing in a transparent (usually glass) channel.

*The working principle of this device is based on the Doppler shift in frequency of the light-scattered by an object moving relative to the radiating source.*

Basically the technique consists of focusing laser beams at the point in the fluid where the velocity is to be measured. At this focal point, a photo-detector senses the laser light scattered from the fluid or fluid particles entrained in the fluid. Signal processing of photo-detector output yields the magnitude of Doppler frequency shift which is *directly proportional to the instantaneous velocity of the flow.*

A LASER (Light Amplification by the Stimulated Emission of Radiation) beam is employed, instead of a monochromatic light beam, due to the following *reasons :*

(*i*) Laser beam is coherent.

(*ii*) The laser provides much higher quality of monochromatic (single frequency) light source.

(*iii*) Its wavelength (unlike a monochromatic beam) is affected minimally by changes in the ambient pressure, temperature or humidity.

(*iv*) Its frequency is very stable and as such it can detect the relative Doppler frequency shift ($\Delta f/f$) of the order of $10^{-7}$ for flow velocities of the order 45 km/h.

The following are some of the materials known as active media suitable for production of laser beams :

Ruby; Nd-glass; Nd-Yale; $CO_2$ gas; Neon gas; Ionised argon gas; Helium neon gas; Semiconductor crystal gallium arsenide.

***Advantages and Disadvantages :*** The following are the advantages of laser Doppler anemometer over the usual pitot tube and the hot wire velocity metering techniques :

***Advantages :***

1. Very high accuracies (of the order of ± 0.2 percent).
2. Suitable for measurements in both gas and liquid flows.
3. No transfer functions involved, *i.e.,* the output voltage of the instrument is directly proportional to the instantaneous velocities rather than its inference from pressure difference (pitot tubes) or heat transfer phenomenon (hot wire).
4. It is possible to have measurements in miniature size of volume of the order 0.2 mm side cube.

5. Measurements are of non-contact type (*i.e.*, no physical object is inserted in the flow field and thus flow is undisturbed by the measurement).
6. In this instrument very high frequency response of the order of MHz is possible; consequently, its accuracy in the determination of fluctuating velocities is far superior to that of hot-wire anemometer because of practically no time lag in the measurements.

***Disadvantages :***

1. High cost.
2. Requires a high degree of experience and skill in operation.
3. Unsuitable for clean flows; for such flows, the tracer particles have to be seeded in the flow for scattering the incident light.
4. Need for a transparent channel.

## 12.4 MEASUREMENT OF DISPLACEMENT

### 12.4.1. Measurement of Linear Displacement

*Linear displacement* may be measured by the following *transducers :*

1. Resistive potentiometers.
2. Strain gauges.
3. Variable inductance transducers.
4. Linear variable differential transducers (LVDT).
5. Capacitive transducers.
6. Piezoelectric transducers.
7. Hall effect transducers.
8. Electronic transducers.
9. Ionization transducers.
10. Digital transducers.
11. Acoustic transducers.

• *Mechanical displacement* may be converted into an electrical variable by the simple expedient of adjusting resistance in an electrical circuit. A slide-wire resistor, having a movable contact attached to the part whose displacement is to be measured, may be connected through a 2-conductor circuit to a steady-voltage source in series with an ammeter (or milliammeter) calibrated in terms of the displacement. If the resistor is connected as a voltage divider, the need for a regulated supply is eliminated, and with a 3-conductor circuit the display instrument may be a *ratio-meter or potentiometer.* Such combinations are common and are available for both D.C. and A.C. operations.

• Where deflections are small, of the order of 0.25 mm, measurement may be made by use of a *differential transformer.*

**Measurement of small displacements :**

The linear or differential transformer is the *popular means for measuring small displacements of the order of 0.25 mm.* This device (See Fig. 12.46) is generally produced with a single primary winding and two secondaries, all disposed along a common axis and having in the common magnetic circuit a movable iron core longitudinally displacable with the motion to be measured. The secondaries may be connected additively or differentially and may be included in the circuit of a null type instrument based either by shifting the core of a similar transformer excited from the same source or by the use of a slide wire potentiometer.

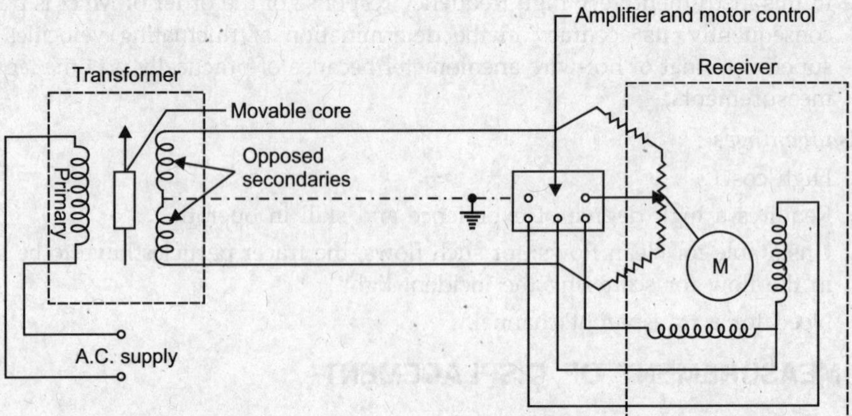


**Fig. 12.46.** Schematic diagram of differential transformer transducer with servo-activated receiver.

*Linear transformer* are regularly supplied for operation at all frequencies upto 30 kHz. The sensitivity, of course, increases with the frequency.

*Linear transformers* may be interconnected in a great variety of arrangements to perform computations or to express desired mathematical functions of measured variables.

### 12.4.2. Measurement of Angular Displacement

The following transducers are used for angular displacement :

1. Resistive potentiometer.
2. Variable inductive transducers.
3. Synchros.
4. Capacitive transducers.
5. Rotary variable differential transformer (RVDT).
6. Variable reluctance transducer.

**Variable reluctance transducer :**

Fig. 12.47, shows a variable reluctance transducer for *measurement of angular displacement*.

$L_1$ and $L_2$ are the two identical coils connected in series opposition. These coils are so placed/positioned with respect to a cam shaped iron core that, when the core is in the *null position, the impedances of both the coils are same.* Hence, voltage across the two coils are the same, though they are 180° out of phase. Any angular displacement from the null position results in a *differential voltage output proportional to the amount of displacement.* The clockwise

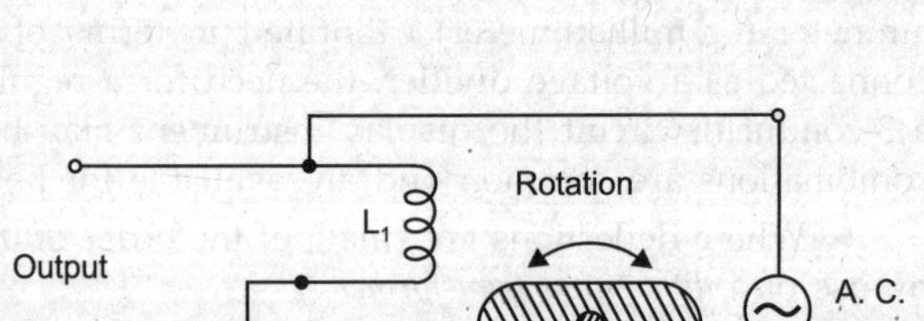


**Fig. 12.47.** Variable reluctance transducer for measurement of angular displacement.

and anticlockwise angular displacements produce different outputs which are *opposite in phase.* Thus the angular displacement may be measured in magnitude and direction.

In these transducers, angular changes from null position to ± 45° may be sensed.

## 12.5 MEASUREMENT OF VELOCITY/SPEED

### 12.5.1. Measurement of Linear Velocity

The following types of transducers are used (in the analogue methods) for measurement of *linear velocity* (first derivative of displacement).

1. Electro-magnetic transducers :
   (*i*) Moving magnet type.
   (*ii*) Moving coil type.
2. Seismic type transducers.

**Electro-magnetic transducers :**

*An electro-magnetic transducer utilizes the voltage produced in a coil on account of change in flux linkages resulting from change in reluctance.* This is the most commonly used transducer for measurement of linear velocities.

**(*i*) Moving magnet type.** A moving type electro-magnetic transducer uses *permanent magnet which provides a constant polarizing field.* Fig. 12.48, shows such a transducer. It consists of rod rigidly coupled to the device whose velocity is being measured. The rod is a permanent magnet which is surrounded by a coil.

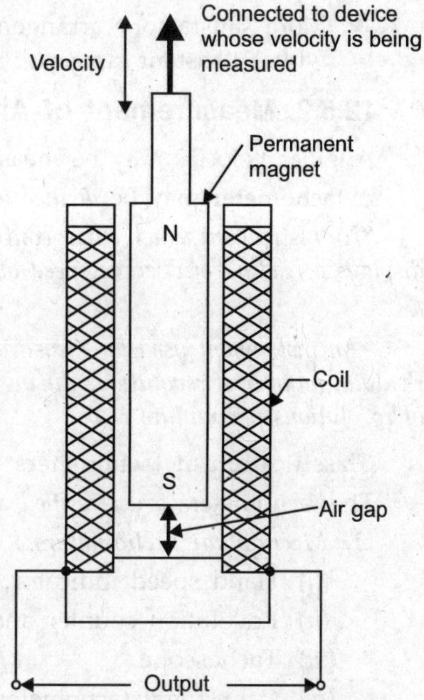


**Fig. 12.48.** Moving magnet type transducer.

***Working.*** When the magnet moves a voltage is induced in the coil which is *directly proportional to the velocity.* The voltage induced in the coil (placed in a magnet field) is given by

$$e = BANV \qquad ...(12.51)$$

*i.e.*,

$$e \propto V \qquad ...(12.52)$$

where, $B$ = Flux density, Wb/m$^2$,
$A$ = Area of coil, m$^2$,
$N$ = Number of turns of coil, and
$V$ = Relative velocity of magnet with respect of coil.

The *direction of motion is determined by the polarity of the output voltage.*

***Advantages :***

1. The output voltage is linearly proportional to velocity.
2. Negligible maintenance (due to absence of mechanical surfaces/contacts).
3. Can be employed as event markers which are robust and inexpensive to manufacture.

***Disadvantages :***

1. Limited frequency response.
2. Unsuitable for measurement of vibrations.
3. Stray magnetic fields adversely affect their performance.

**(ii) Moving coil type.** This type of transducer operates essentially through the *action of a coil moving in a magnetic field; the voltage generated in the coil being proportional to the velocity of the coil.*

The *advantages* of this type of transducer are :

- Reduced stray magnetic fields effects due to the whole device being contained in an antimagnetic case.
- More satisfactory arrangement, as the system forms a closed magnetic circuit with a constant air gap.

### 12.5.2. Measurement of Angular Velocity

Angular velocity may be measured with the help of a *"tachometer"*.

A **tachometer** may be *defined* as follows :

*"An instrument which either continuously indicates the value of rotary speed or continuously displays a reading of average speed over rapidly operated short-intervals of time."*

Or

*"An instrument used for measurement of angular velocity, as of shaft, either by registering the total number of revolutions during the period of contact, or by indicating directly the number of revolutions per minute."*

**Classification of tachometers :**

The tachometers are *classified* as follows :

**I.** ***Mechanical tachometers :***

(*i*) Hand speed indicator.
(*ii*) Revolution counter and timer.
(*iii*) Tachoscope.
(*iv*) Centrifugal tachometers.
(*v*) Vibrating reed tachometers.

**II.** ***Electrical tachometers :***

(*i*) D.C. tachometer generators.
(*ii*) A.C. tachometer generators.
(*iii*) Photoelectric tachometers (or Speed-meters)
(*iv*) Eddy current or drag-cup tachometer.
(*v*) Capacitive tachometer.
(*vi*) Stroboscopic tachometers.

Some of these meters are discussed below :

### 12.5.3. Electrical Tachometers

Electrical tachometers are preferred, as compared to mechanical tachometers, in view of advantages they offer as *electrical transducers.*

**1.** ***D.C. tachometer generator :***

In a D.C. generator the e.m.f. generated depends upon the following two factors :

(*i*) Field excitation (*ii*) Speed.

If for the field system permanent magnet pole pieces are used, then the generated voltage depends only on the speed. Hence the *speed can be computed by measuring the generated e.m.f.*

A D.C. tachometer generator is shown in Fig. 12.49.

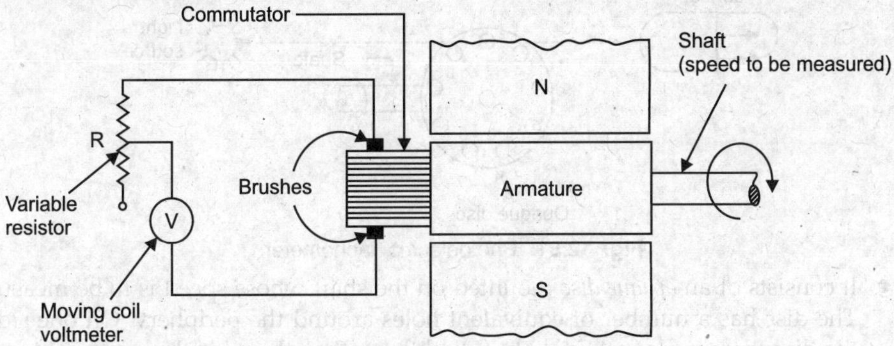


**Fig. 12.49.** D.C. tachometer generator.

- The shaft whose speed is to be measured is coupled to the armature.
- A moving coil voltmeter is connected across the brushes to measure the generated voltage. The variable resistance $R$ is incorporated to limit the current through the voltmeter.

*Since voltage is proportional to speed, the voltmeter may be calibrated in terms of speed. (r.p.m.).*

**2. *A.C. tachometer generator :***

The inherent demerits associated with D.C. tachometer generator, due to the provision of commutor and brushes, are eliminated in A.C. tachometer generator.

An A.C. tachometer generator is shown in Fig. 12.50.

- It consists of, like an alternator, a stationary armature (stator) and a rotating field system (rotor). Owing to the generation of e.m.f. in a stationary coil on a stator, commutation problems no longer exist.
- The alternating e.m.f. induced in the stationary coil is *rectified* and the output D.C. voltage is measured with the help of a moving coil voltmeter ($V$).
- The *ripple content* of the rectified voltage is *smoothened* by the capacitor filter (C).
- As the speed depends on both the amplitude of the voltage and frequency, any one of them can be used as a measure of the speed. In an A.C. tachometer, it is the *induced voltage* that is considered as the required parameter.

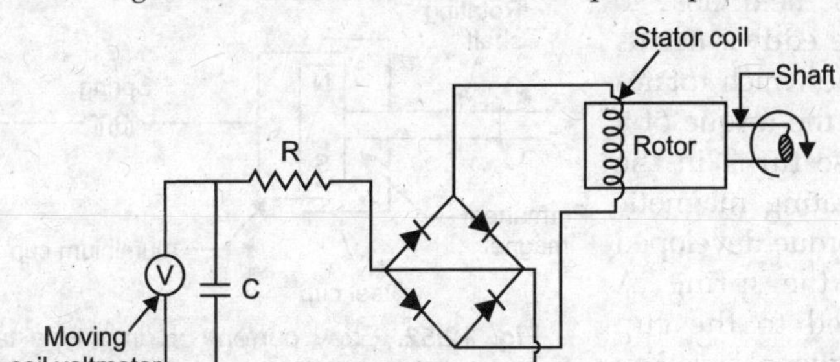


**Fig. 12.50.** A.C. tachometer generator.

**3. *Photoelectric tachometer :***

Fig. 12.51, shows a photoelectric tachometer.

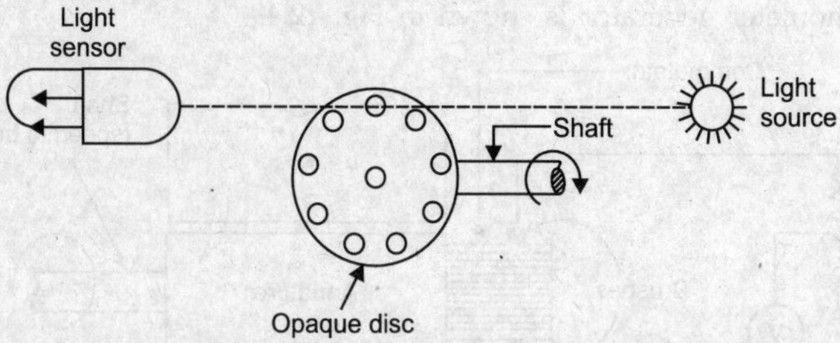


Fig. 12.51. Photoelectric tachometer.

- It consists of an *opaque disc* mounted on the shaft whose speed is to be measured. The disc has a number of equivalent holes around the periphery. On one side of the disc there is a source of light (L) while on the other side there is a light sensor (maybe a photosensitive device or photo-tube) in line with it (light-source).
- On the rotation of the disc, holes and opaque portions of the disc come alternately in between the light source and the light sensor. When a hole comes in between the *two,* light passes through the holes and falls on the light sensor, with the result that an *output pulse is generated.* But when the opaque portion of the disc comes in between, the light from the source is blocked and hence there is no pulse output. Thus *whenever a hole comes in line with the light source and sensor, a pulse is generated.* These pulses are counted/measured through an *electric counter.*

The number of pulses generated depends upon the following *factors* :

(*i*) The number of holes in the disc ;

(*ii*) The shaft speed.

*Since the number of holes are fixed, therefore, the number of pulses generated depends on the speed of the shaft only.* The electronic counter may therefore the calibrated in terms of speed (r.p.m.).

***Advantage.*** It is a *digital instrument.*

***Disadvantage.*** It is required to replace the light source periodically and if the grating period is small then errors might creep in the output.

**4. *Eddy current or drag cup tachometer :***

In this type of tachometer (Fig. 12.52) the test-shaft rotates a permanent magnet and this induces eddy currents in a drag cup or disc held close to the magnet. The eddy currents produce a torque which rotates the cup against the torque of a spring. The disc turns in the direction of rotating magnetic field until the torque developed equals that of the spring. A pointer attached to the cup indicates the rotational speed on a calibrated scale.

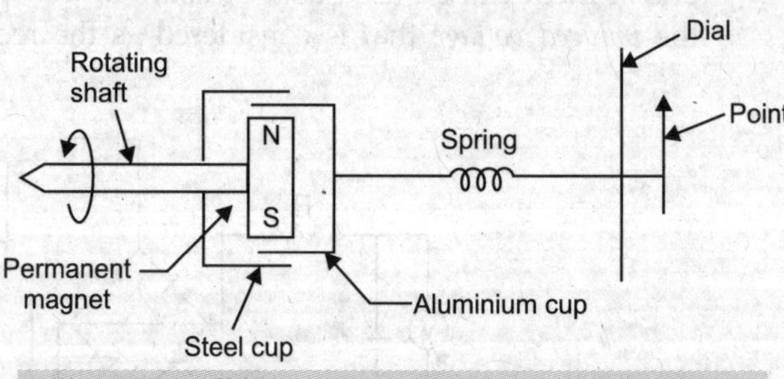


Fig. 12.52. Eddy current or drag type tachometer.

- The *automobile speedometers* operate on this principle.
- These tachometers are used for measuring rotational speeds upto 12000 r.p.m. with an accuracy of ± 3 per cent.

**5. *Capacitive tachometer :***

Refer to Article. 7.8.3.

**6. *Stroboscopic tachometer :***

*Its working principle is to synchronize a flashing light with rotation of shaft, making it appear standstill.* Usually, a reference mark is put on the rotating body and the speed of oscillation of flash produced from stroboscope itself is a measure of r.p.m. of rotating body, *without requiring any contact.*

- These tachometers are most oftenly used in *experimental work.*
- Its *disadvantage* is that it cannot be used where ambient lighting is above a certain value; for it to be most effective, the surrounding light must be subdued.

**Example 12.9.** *Draw an instrumentation scheme to measure the speed of a motor cycle using a magnetic transducer and explain the principle behind its operation.*

**Solution.** The block diagramatic representation of the instrumentation scheme for the measurement of the speed of the motorcycle using a magnetic transducer is given as follows (Fig. 12.53):

The basic principle involved for the above measurement is that the speed of the motor cycle is sensed with the help of an *A.C. tachogenerator* which is coupled with the motor cycle wheel. *The output of the generator is voltage which is direct indication of speed.* The other blocks show the necessary circuit for displaying the output in the desired form.

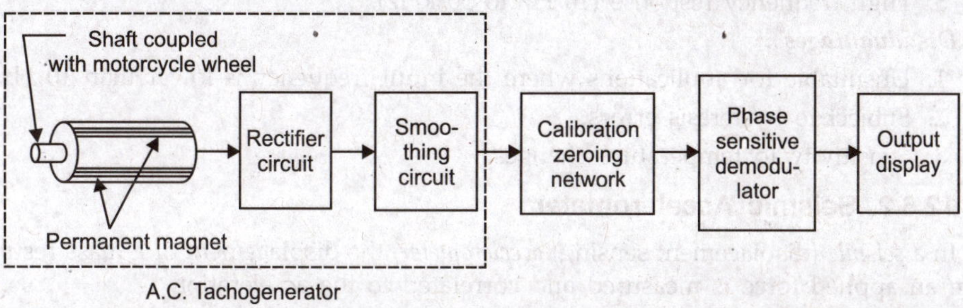


**Fig. 12.53.** Block diagramatic scheme for measurement of motor cycle speed using magnetic transducer.

## 12.6 MEASUREMENT OF ACCELERATION

*The acceleration of a moving body is generally measured by means of sensors called* **accelerometers.**

For measurement of acceleration, generally following types of accelerometers are used :

1. Piezoelectric type
2. Seismic type.

### 12.6.1. Piezoelectric Accelerometer

Refer to Fig. 12.54. A piezoelectric accelerator is probably the *simplest* and *most commonly used transducer for measuring acceleration.*

**Construction.** It consists of a *piezoelectric crystal* sandwiched between two *electrodes* and has *mass* placed on it. The unit is fastened to the base whose acceleration characteristics are to be obtained. The *can* threaded to the base acts as a spring and squeezes the mass against the crystal. *Mass exerts a force on the crystal and a certain voltage output is generated.*

**Working.** When the base is accelerated *downward* inertial reaction force on the base acts upward against the top of the can. This relieves stress on the crystal. According to Newton's second law of motion, *force = mass × acceleration*, since the mass is fixed quantity, the *decrease in force is proportional to the acceleration*. Similarly, an acceleration in the upward direction would increase the force on the crystal in proportion to the acceleration. The resulting change in the output voltage is recorded and correlated to the acceleration imposed on the base.

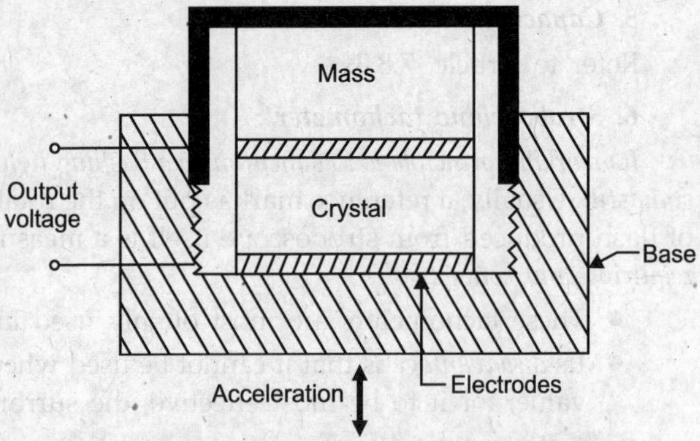


**Fig. 12.54.** Piezoelectric accelerometer.

***Advantages :***

1. Small size and a small weight.
2. High output impedance.
3. Can measure acceleration from a fraction of *g* to thousands of *g*.
4. High sensitivity.
5. High frequency response (10 Hz to 50 kHz).

***Disadvantages :***

1. Unsuitable for applications where the input frequency is lower than 10 Hz.
2. Subject to hysteresis errors.
3. Sensitivity to temperature changes.

## 12.6.2. Seismic Accelerometer

In a *seismic* (displacement sensing) *accelerometer*, the displacement of a mass resulting from an applied force is measured and correlated to the acceleration.

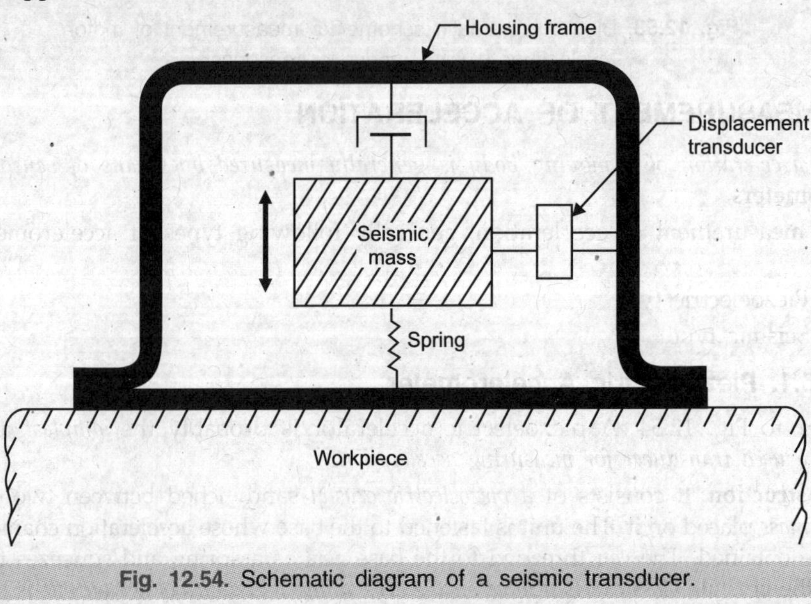


**Fig. 12.54.** Schematic diagram of a seismic transducer.

A schematic diagram of this accelerator is shown in Fig. 12.55. The mass is connected through the parallel spring and damper arrangement to the housing frame. The housing frame is connected to the source of vibrations whose characteristics are to be measured. *The mass has tendency to remain fixed in its spatial position so that the vibrational motion is registered as a relative displacement between mass and housing frame. This displacement is sensed and indicated by an appropriate transducer.*

- This accelerometer may also be used as a *vibration sensor* (vibration refers to the repeated cyclic oscillations of a system).

### 12.6.3. Strain Gauge Acceleration (Vibration) Sensor

Fig. 12.56 shows a strain gauge acceleration/vibration sensor :

— It consists of a Wheatstone bridge in which the semiconductor strain gauges ($R_1$ and $R_2$) one connected in two arms, as shown in the Fig. 12.56. These gauges measure a change in electrical resistance, which is proportional to force due to *acceleration*. Since the acceleration is related to vibrations, the *meter is calibrated to read the voltage ($V_o$) proportional to acceleration.*

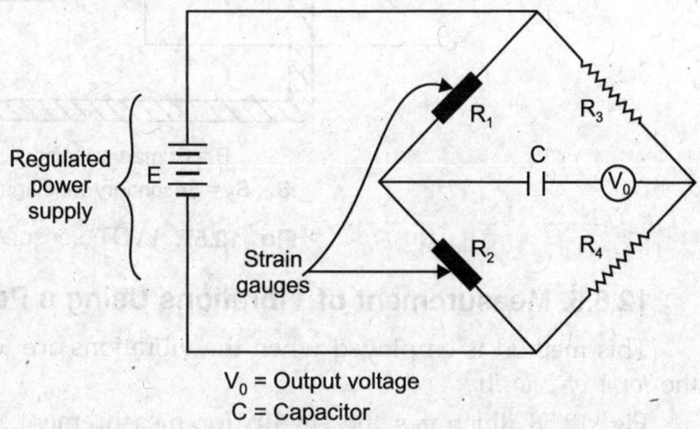


**Fig. 12.56.** Strain gauge acceleration/vibration sensor.

The function of *capacitor* C is to remove the D.C. signal from the output signal.

— The value of $V_0$ (voltage drop across the bridge) is recorded or indicated by the indicator.

- The range of this method is : Vibration amplitude : 5 to 1000 g; Vibration frequency : 0 to 1000 Hz.

### 12.6.4. LVDT Acceleration/Vibration Sensor

This sensor measures vibrations in terms of acceleration.

Fig. 12.57 shows LVDT sensor. It produces an A.C. voltage proportional to displacement of movable ferromagnetic core.

— LVDT's core acts as the mass. When flexible reeds are connected to the core, springs action is produced.

— When vibrations to be measured/monitored are applied to the core, it moves in the upward and downward directions according to the vibration. As a result, A.C. voltage is induced in the secondary windings of LVDT, the magnitude of which is proportional to the amplitude of vibrations. This voltage if required, can be rectified for induction or measurement of vibration amplitude.

- This sensor operates on *higher frequencies* because the mass of LVDT is *small with natural higher frequencies.*
- This system provides *better resolution* and the *operation is frictionless.*

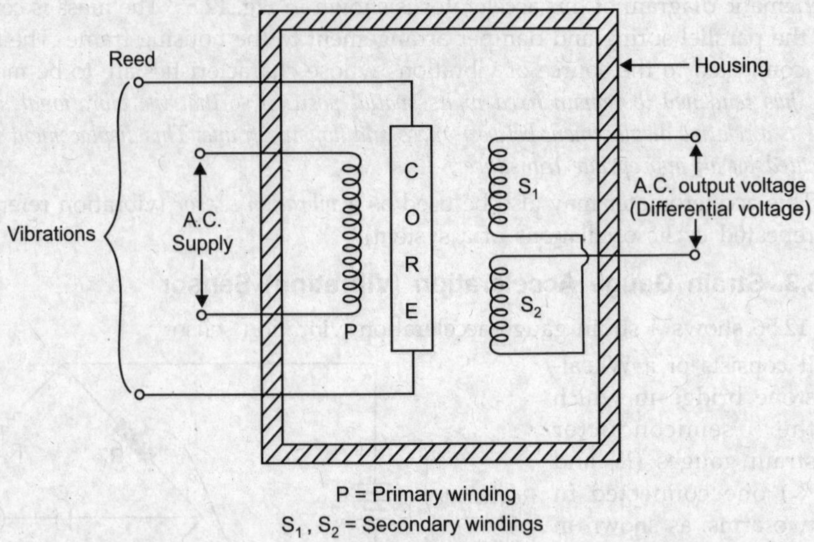


**Fig. 12.57.** LVDT sensor.

### 12.6.5. Measurement of Vibrations Using a Permanent Magnet and a Coil

This method is employed when the vibrations are to be measured or determined in the *form of velocity.*

Fig. 12.58 illustrates the set up for measurement of vibration using a permanent magnet and a coil :

— The system consists of an electrical *coil* suspended by a *spring* in the magnetic field of a *permanent magnet.* The whole assembly is mounted in a *housing,* which is *placed in contact with the vibrating surface.* The coil has low natural frequency by which it moves with housing.

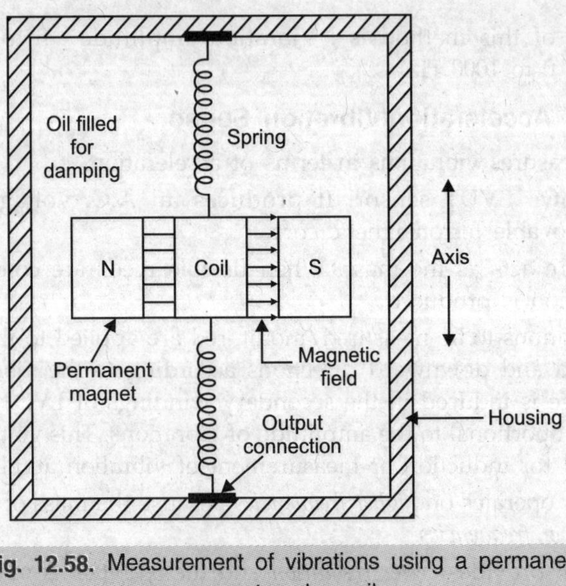


**Fig. 12.58.** Measurement of vibrations using a permanent magnet and a coil.

— When vibration velocity is applied to the coil, a voltage is induced in it. The amplitude of voltage is proportional to vibration velocity, with frequency equal to the vibration frequency.

— The induced voltage is monitored or displayed on a read-out.

## 12.7 MEASUREMENT OF FORCE

**Force** *is defined as any cause that produces, stops or changes the motion of a body or tends to produce these effects.*

An unknown force may be measured by the methods incorporating the following principles.

1. *Balancing the force against a known gravitational force on a standard mass.*
**...Scales and balances**
2. *Translating the force to a fluid pressure and then measuring the resulting pressure.*
**...Hydraulic and pneumatic load cells**
3. *Applying the force to some elastic member and then measuring the resulting deflection.*
**...Proving ring**
4. *Applying the force to a known mass and then measuring the resulting acceleration.*
5. *Balancing the force against a magnetic force developed by interaction of a magnet and a current carrying coil.*

### 12.7.1. Load Cells

Refer to Art 7.14.

**Piezoelectric load cells :**

The principle of working of a piezoelectric load cell is that when *pressure is applied* across certain crystals (quartz etc.) then an *electrical charge proportional to the pressure is developed across certain well-defined faces of the crystal.*

- These are used for measuring forces, particularly of a *transient or rapidly changing variety.*
- Suitable for measuring forces from 5 N to 20 kN.

## 12.8 MEASUREMENT OF TORQUE

### 12.8.1. Introduction

To transmit energy by rotation it is necessary to apply a turning force. In case of a shaft if the force is applied tangentially and in the plane of transverse cross-section the *torque or twisting moment* may be calculated by multiplying the force with the radius of the shaft. If the shaft is subjected to two opposite turning moments it is said to be in *pure torsion* and it will exhibit the tendency of shearing off at every cross-section which is perpendicular to the longitudinal axis.

The measurement of torque is carried out because of the *reasons* : (*i*) It is of considerable interest for its own sake; (*ii*) It is required to obtain load information necessary for stress and deflection analysis; (*iii*) It is often associated with determination of mechanical power.

### 12.8.2. Torque Measurement Methods

Torque may be measured by the following *methods* :

1. Gravity balance method.
2. Mechanical torsion meter.

3. Optical torsion meter.
4. Electrical torsion meter.
5. Strain-gauge torsion meter.

Some of these meters are discussed below.

**12.8.2.1. Electrical torsion meter**

Fig. 12.59, shows such a system using two photoelectric or magnetic transducers, involving two sets of measurements :

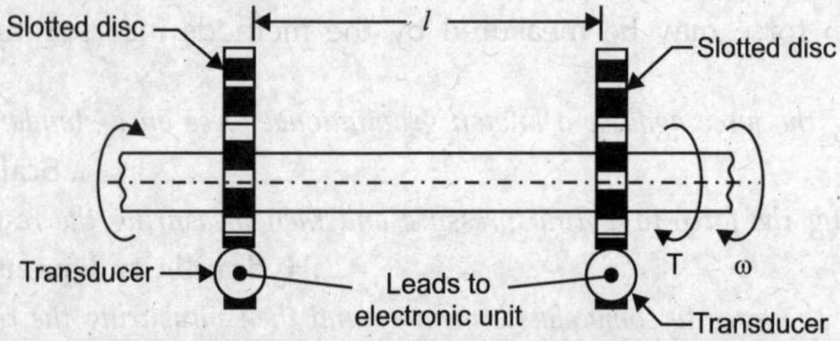


Fig. 12.59. Electrical torsion meter.

(*i*) A count of the pulse from either slotted wheel; this count gives the *frequency or shaft speed.*

(*ii*) A measure of the time between pulses from the two wheels, this signal is proportional to the twist θ and hence *T*, in the shaft.

By combining the above two signals *T* and ω the power being transmitted by the shaft can be calculated.

**12.8.2.2. Strain-gauge torsion meter**

Fig. 12.60, shows a general configuration of a strain gauge bridge circuit widely employed for *torque measurement from a rotating shaft.*

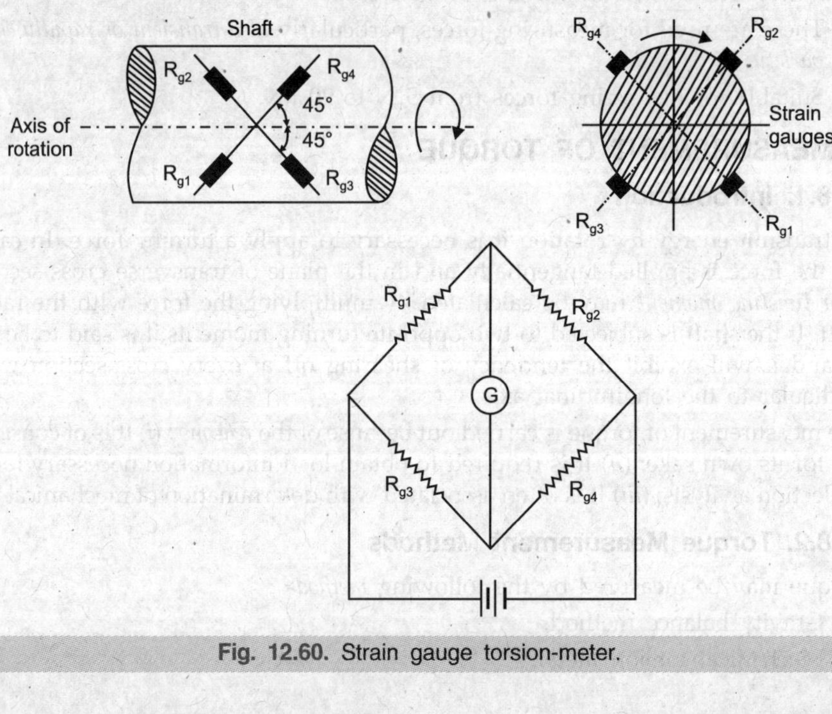


Fig. 12.60. Strain gauge torsion-meter.

***Construction.*** Four bonded-wire strain gauges are mounted on a 45° helix with the axis of rotation; and are placed in *pairs diametrically opposite.* When the gauges are accurately placed and have matched characteristics, the system is temperature compensated and insensitive to bending and thrust or pull effects. Any change in the gauge circuit then results only from torsional deflection.

***Working.*** When the shaft is under torsion, gauges 1 and 4 will *elongate* as a result of the tensile component of a pure shear stress on one diagonal axis, while 2 and 3 will *contract* due to compressive component on the other diagonal axis. These tensile and compressive principal strains can be measured and hence shaft torque calculated.

***Advantages.*** This arrangement has the following *advantages* :

1. Fully temperature compensated.
2. Gives the maximum sensitivity for a given torque.
3. Provides automatic compensation for bending and axial loads.

***Limitation :***

The main limitation/difficulty associated with the use of this arrangement is the connection of the bridge to its power source and display arrangement. *Slip rings* are used for this purpose. These are conducting rings attached to the shaft, but insulated from it, with one of the slip rings connected to each of the bridge terminals. Slip rings are mercury filled and transmit the signal to a stationary member where it is amplified and displayed or recorded.

• *Good results from the strain gauge method are available only when the shaft experiences sufficient strain to produce a measurable output signal.* In case the torque is small the gauge bridge will not give adequate output signal for measurement. In such cases a *flexible coupling* is introduced in the shaft. The coupling incorporates one or more elastic members (elastic member may be a different shaft or a commercial torque meter) to which the gauge may be attached. The elastic members are so designed that they *produce sufficiently large deflection even under light load conditions.* This large deflection produces large strains resulting in large output of the strain gauge bridge.

## 12.9 MEASUREMENT OF SHAFT POWER

**A dynamometer** *is a device used to measure the torque being exerted along a rotating shaft so as to determine the shaft power input or output of power-generating, transmitting and absorbing machinery.*

The dynameters are *classified* as follows :

**1. Absorption dynamometers.** They work on the principle that *the power measured is converted into heat by friction or by other means.* The power absorbed is lost as heat and dissipated to the surroundings, where it serves no useful purpose.

***Examples :***

(*i*) Mechanical brakes :
- Prony brakes.
- Rope brakes.

(*ii*) Hydraulic or fluid friction brakes.

(*iii*) Eddy current dynamometer.

**2. Transmission dynamometers.** They work on the principle that the *power transmitted either to or from the dynamometer is not absorbed or dissipated.* After measurement, the energy is conveyed to the surroundings in a useful mechanical or electrical form.

***Examples :***

(*i*) Torsion and belt dynamometers.

(*ii*) Epicyclic train dynamometers.

(*iii*) Strain gauge dynamometers.

**3. Driving dynamometers.** These instruments *measure power and also supply energy to operate the tested devices.*

***Example :*** *Electric cradled dynamometer* (a typical example).

Some of these dynamometers are discussed below:

## 12.9.1. Eddy Current Dynamometer

*An* **eddy current dynamometer** *(absorption) operates on the principle that when an isolated conductor cuts across magnetic flux, voltage is induced and local currents (called "eddy currents") flow in a short circular path within the conductor. These eddy currents get dissipated in the form of heat.*

Fig. 12.61, shows an eddy current dynamometer. It consists of a *toothed steel rotor* mounted on the shaft of the test-engine, it rotates inside a *smooth bared cast-iron stator;* the clearance between the stator and rotor being very small. The stator carries an *exciting coil*, which is energised by D.C. supply from an external source. The stator is cradled on antifriction bearings and is provided with a brake arm to which is attached a spring balance or a scale pan.

When the dynamo is operated, the rotor turns and causes a constant change in flux density on all stator points. Consequently eddy currents are induced which oppose the rotation of the rotor. The moment of resistance is measured by the brake arm and then torque and shaft power may be calculated.

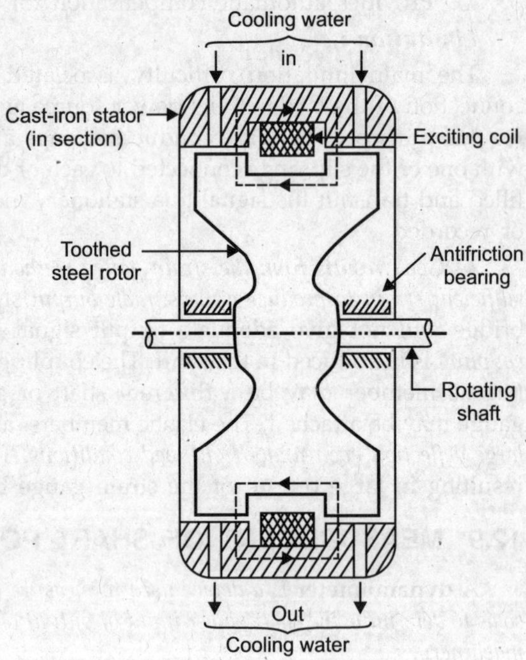


**Fig. 12.61.** Eddy current dynamometer.

The mechanical power supplied to the dynamometer shaft is converted to heat which is carried away partly by air circulation (induced by the rotor teeth) and partly by water circulation (See Fig. 12.61).

- Usual power limit of the dynamometer = 300 H.P.; Speed limit = 6000 r.p.m. (approx.)

***Advantages :***

(*i*) Good control at low rotating speeds.

(*ii*) Comparatively small size for a given capacity.

(*iii*) Suitable for a higher speed range.

## 12.9.2. D.C. Dynamometers

D.C. electric type dynamometer is the most *versatile and accurate dynamometer.*

This is a *cradled dynamometer* and is widely used for power and torque measurements of :

- Internal combustion engines.
- Small steam turbines.
- Pumps and other mechanical equipment.

The basic arrangement of a D.C. dynamometer is shown in Fig. 12.62. The D.C. machine acts as both motor and generator. It works as a *generator* when it is coupled to the machine under test which is a power generating machine. The *power generated by the generator is dissipated in resistance grids* or may be recovered as a useful power.

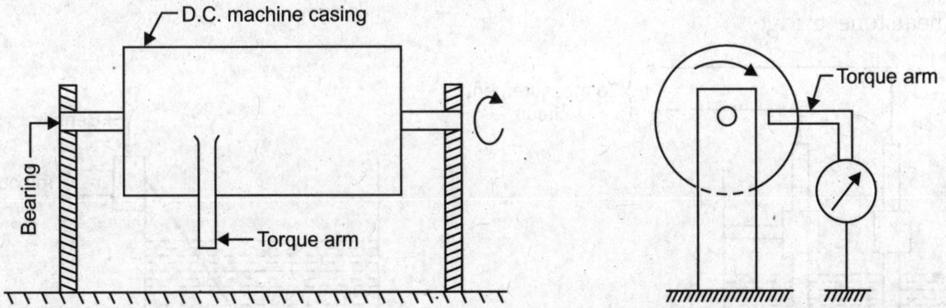


**Fig. 12.62.** D.C. dynamometer.

Thus, the D.C. machine can act both as *absorption dynamometer* and a *transmission dynamometer.* Force and torque is measured through a force/torque arm attached to the casing of the D.C. machine. By controlling the field excitation of the generator the output voltage and hence the power can be controlled.

When the D.C. machine is made to work as a *D.C. motor* it may act as *driving dynamometer.* The D.C. motor may drive a test machine which is a power absorbing device like a pump. The output power of a pump may be smoothly controlled by controlling the speed of the D.C. motor by varying either the armature voltage or field current.

## 12.10 MEASUREMENT OF LIQUID LEVEL

**Liquid level** *refers to the position or height of a liquid surface above datum line.* In order to ascertain the quantity of the liquid held in a container, *level measurements are made.*

Liquid level is an important process variable in industrial plants that use large quantities of liquids such as water, chemicals and solvents. For example, there is need to monitor liquid level in boilers, distillation columns etc.

The liquid level measurements can be grouped into the following *two categories :*

1. *Direct methods :*
   (*i*) Sight glass level gauge.
   (*ii*) Float gauges.
   — Float and tape liquid level gauge.
   — Float and shaft liquid level gauge.
2. *Indirect methods :*
   (*i*) Hydrostatic pressure devices.
   (*ii*) Capacitance type level gauge.

(*iii*) Gamma-rays liquid level gauge.

(*iv*) Ultrasonic liquid level gauge.

Some of these devices are discussed below.

### 12.10.1. Capacitance Type Level Gauge

Fig. 12.63, shows a capacitance type liquid level gauge. It consists of two parallel metal electrodes held firmly at a known fixed distance apart. *The dielectric constant between the electrodes varies with the liquid level and so does the capacitance of the system.* Greater the level the larger will be the capacitance and less the level, the smaller will be the capacitance. The change in capacitance can be measured by some suitable circuitry such as a capacitance Wheatstone bridge.

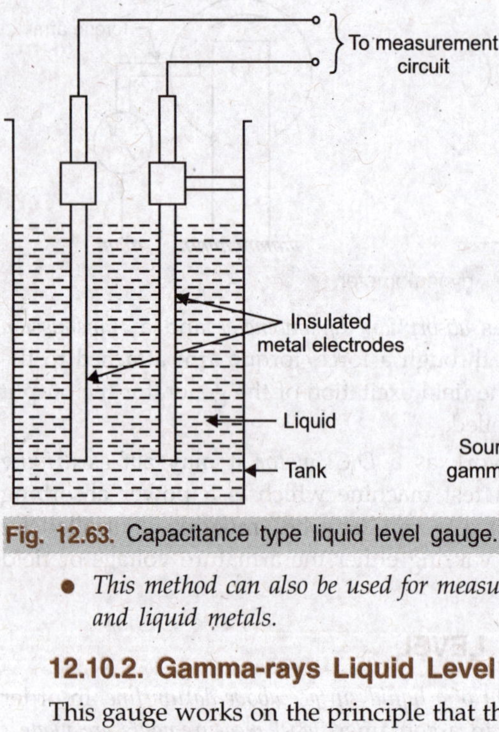


**Fig. 12.63.** Capacitance type liquid level gauge.

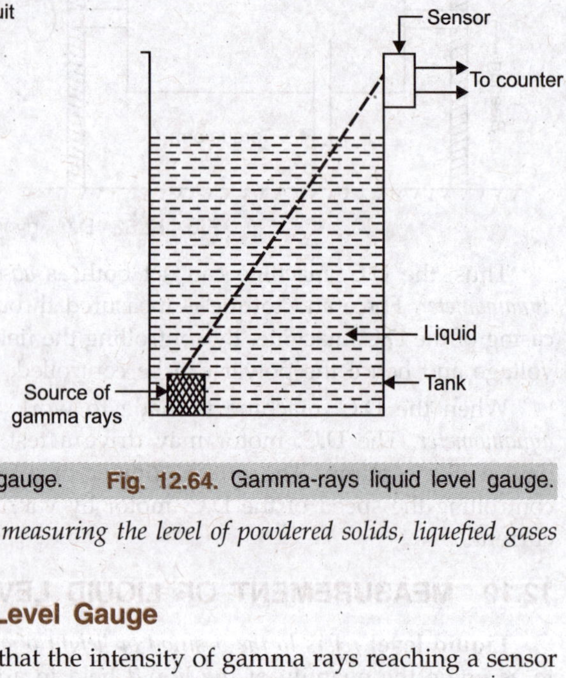


**Fig. 12.64.** Gamma-rays liquid level gauge.

- *This method can also be used for measuring the level of powdered solids, liquefied gases and liquid metals.*

### 12.10.2. Gamma-rays Liquid Level Gauge

This gauge works on the principle that the intensity of gamma rays reaching a sensor of gamma rays (*e.g., Geiger Miller tube*) *varies as the height of liquid column placed between the gamma-rays source and the detector.*

Fig. 12.64, shows the schematic arrangement of the gamma-rays liquid level gauge. As the gamma rays can penetrate the tank walls, the tube senses the rays and greater the intensity of the rays, the greater will be the output of the tube. *The intensity of rays will depend upon the liquid level.* When the tank is empty maximum radiation will reach the Geiger Miller tube. Higher the liquid level, the greater the absorption of gamma rays and consequently lower the detector output. The output is in the form of *pulses* and can be counted by a *counter*, the counter, therefore, may be calibrated in terms of the liquid level.

### 12.10.3. Ultrasonic Liquid Level Gauge

An ultrasonic liquid level gauge works on the *principle of reflection of an acoustic wave from a liquid surface.*

Fig. 12.65, shows the schematic arrangement of an ultrasonic liquid level gauge. The ultrasonic wave sent by a transmitter *'T'*, is reflected at the surface of liquid and is received by a receiver *'R'*. With the change of liquid level the transient time of the wave also changes. The liquid level can be measured either by the *time taken* by the reflected wave to reach the receiver or on the basis of *change of phase* of the wave. Piezoelectric crystals (such as quartz or barium titanate) may be used as transmitter and receiver.

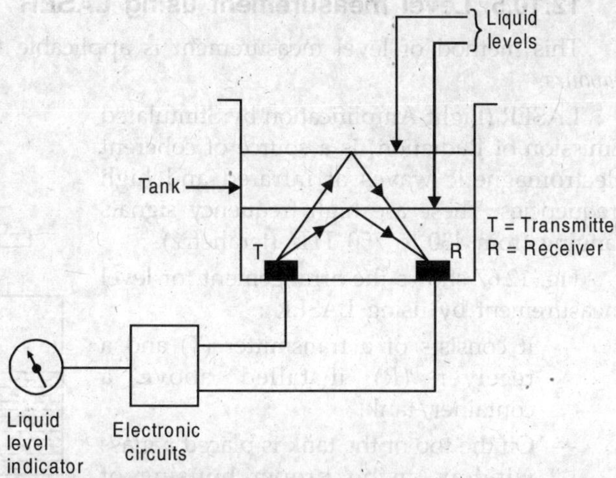


**Fig. 12.65.** Ultrasoznic liquid level gauge.

- Although the gauge works on a simple principle, yet actual instrument involves complex circuitry.

## 12.10.4. Electrical Level Gauges

A simpler way of measuring the level using electrical circuit is the use of *triode tube.* Fig. 12.66, shows the principle of electrical transducers. If the level in tank is below the required level *G*, there is no complete circuit to the grid of the tube, which then accumulates a charge of electrons. This makes it very negative in effect and reduces the tube plate current to a very small amount. Under these circumstances, very little current in the relay coil completes the alarm circuit. The alarm thus indicates the dangerously low level in the boiler tank. A red light signal also serves the same purpose. On the other hand if the level of water rises, it completes the circuit between *G* and other terminal *G′*. The electrons can then be removed from the grid circuit. As the grid is made less negative, the plate current will continue to rise until sufficient current flows in the relay coil to operate it. The triode tube can well be *replaced by transistors.*

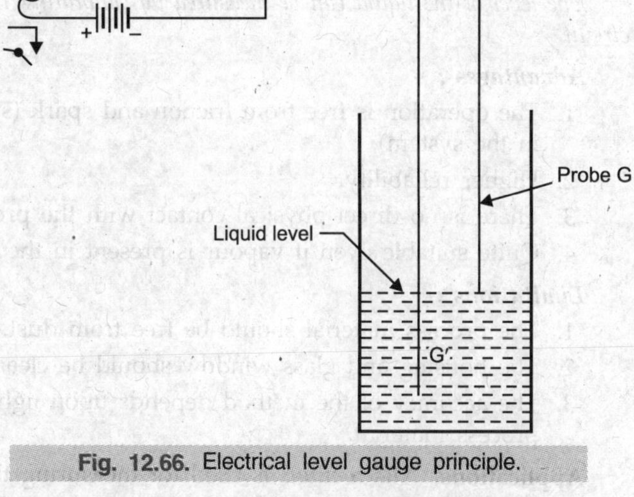


**Fig. 12.66.** Electrical level gauge principle.

### 12.10.5. Level measurement using LASER

This method of level measurement is applicable for *solids, liquids with transparent vapours.*

LASER (Light Amplification by Stimulated Emission of Radiation) is a source of coherent electromagnetic waves at infrared and high frequencies. These are high frequency signals ranging from 430 to 750 THz (terahetrz).

Fig. 12.67 shows the arrangement for level measurement by using LASER :

— It consists of a transmitter (*T*) and a receiver (*R*) installed above a container/tank.

— On the top of the tank is placed a glass window, using proper housing of plastic or rubber.

— A sharp beam of laser is focused towards the liquid level by modulation from the *transmitter* end. When this signal reaches at the surface of the liquid, it gets reflected and is received by the *"receiver"*. In order to measure the liquid level, the time period (*t*) from reflection of the signal to reaching the receiver is calculated from the following the relation :

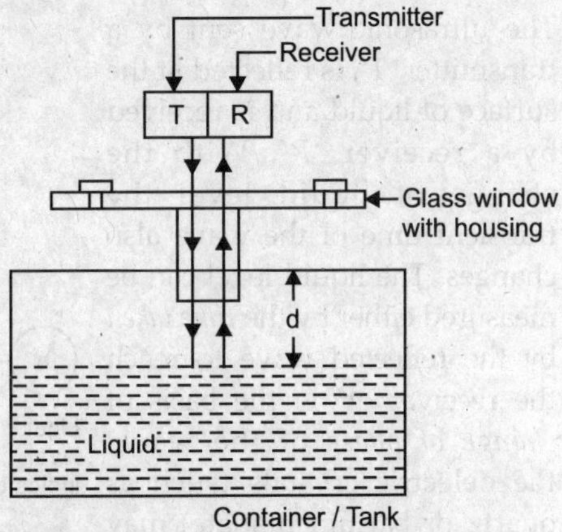


Fig. 12.67. Level measurement by using LASER.

$$t = f(d) \quad \text{...(12.53)}$$

$$\therefore \quad t = \frac{2d}{c} \quad \text{...(12.54)}$$

where, $d$ = Distance from the liquid level, and

$c$ = Velocity of light (= $3 \times 10^8$ m/s).

*The level of the liquid can be measured by calibrating time period in pulses using counter circuit.*

***Advantages :***

1. The operation is free from friction and spark (since no moving part is involved in the system).
2. Higher reliability.
3. There is no direct physical contact with the process material.
4. Quite suitable even if vapour is present in the process material.

***Limitations :***

1. The process material should be free from dust.
2. The housing and glass window should be clean and free from dust.
3. The accuracy of the method depends upon light reflection characteristics of the process material.

**Applications.** This method is used for measuring the *levels of* :

(*i*) Hazardous and toxic liquids and acids;
(*ii*) Liquid aluminium, liquid polyethylene, brown tobacco, etc.;
(*iii*) Wound pulp in paper industries.

### 12.10.6. Microwave Level Detector

In this type of detector electromagnetic radio waves in microwave X-band and K-band of frequency ranges of 8–12.5 GHz and 18–26.5 GHz respectively are used. In order to get better performance solid state devices are employed to generate and detect microwave signals. Microwave signals do not pass through metals but can pass through the vessels having walls of fiber glass, plastic and ceramic.

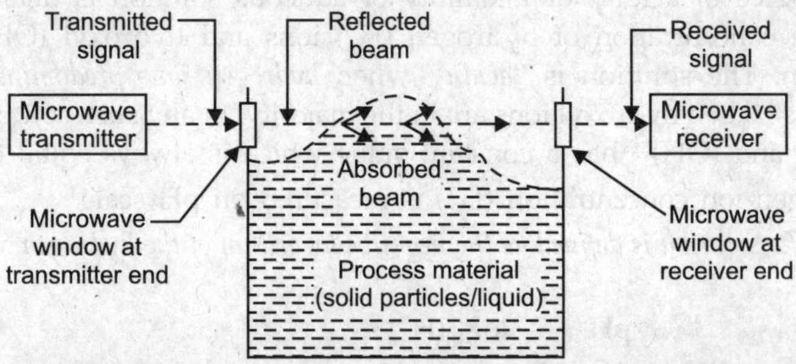


**Fig. 12.68.** Microwave method.

Fig. 12.68 illustrates the arrangement for detecting or measuring the level by microwave method :

— It consists of a microwave transmitter, which transmits the signals either in X-band or K-band through window inside the vessel. The windows are installed at transmitter and receiver ends; these may be made of plastic, teflon, quartz, mica etc.

— When air or vapour is present inside the vessel, strong microwave signals are received at the receiver and detected by the microwave detector. However, with the increase in the level of process material, the transmission of microwaves is retarded and these signals get reflected leading to reduced signal strength detected by the detector. Thus the level of process material can be detected.

- This system is applicable for solids *i.e.*, small size particles, liquids having dielectric upto 10.

***Advantages :***

1. Not affected by dust, foam and mist.
2. Applicable for conductive as well as non-conductive process materials.
3. The operation is free of friction and spark (since no moving part is involved).

***Limitations :***

1. Construction is complex.
2. Suitable for limited dielectric constant of process material.
3. Method is expensive.

**Applications.** It is useful to detect/measure *levels* of :

(*i*) Light powdered materials such as dry saw dust;
(*ii*) Toxic, hazardous chemicals or process materials;
(*iii*) Coal, minerals, and vegetable pulp.

## 12.11 MEASUREMENT OF pH VALUE

— Since *acids* and *bases* are of significant importance in various fields like biology, medicine, food technology, water sewage treatment and the chemical industry, therefore measurement of their concentration and strength are most common.

*The scale on which the acidity or alkalinity is measured is the* **pH scale**.

— The degree of acidity or alkalinity of aqueous solution is determined by the relative concentration of hydrogen ($H^+$) ions and hydroxyl $(OH)^-$ ions in the solution. The solution is *"acidic"* when *hydrogen ions predominate* which it is *"alkaline"* when hydroxyl ions are in the majority. In any such solution the product of ($H^+$) and $(OH)^{-1}$ has a constant value which is always equal to $\mathbf{10^{-14}}$.

— Hydrogen ion concentration ($H^+$) is measured on pH scale.

**pH value** *of a solution is defined as the negative logarithm of the hydrogen ion concentration or,*

$$pH = -\log_{10}(H^+) \qquad \text{...(12.55)}$$

This scale *ranges* from *0 to 14*.

In neutral solution, the concentration of both ($H^+$) and $(OH)^{-1}$ ions are equal *i.e.*, both are $10^{-7}$.

∴ For a neutral solution : $pH = -\log(10^{-7}) = 7$

Thus a neutral solution like pure water has a pH value of 7.

For *"acidic solutions"* the pH value lies *between 0 to 7*. In case of *"alkaline solutions"* the pH value lies in *between 7 to 14*.

- The theory of pH measurement involves the breakdown of certain chemical compounds when combined with water. When these compounds breakdown, tiny electrically charged particles are released. Some of these are positively charged particles–hydrogen ions ($H^+$) and some are negatively charged particles–hydroxyl ions $(OH)^-$. The pH of a solution can be accurately measured by means of an *electronic pH meter*. pH meters usually cover a range of 0 to 14 pH units; however it is possible to make measurements beyond these limits in very concentrated solutions.

**pH cell :**

Fig 12.69 shows a pH cell :

— The pH cell consists of two electrodes (which are immersed into test solution). One of the electrodes called the *"reference electrode"*, is at a constant potential regardless of the pH value of the solution under test. The potential of the other electrode, called the *"measuring electrode"* is determined by the pH value of solution. *Thus the potential difference between the two electrodes depends upon the pH value of the test solution.*

— The *'reference electrode'* is made of glass and consists of an inner assembly containing a solution of calomel (mercury chloride) and mercury. This assembly is surrounded by a larger glass tube, and the space between the two contains an accurate solution of KCl (potassium chloride). A tiny opening in the bottom of the electrode permits KCl to diffuse very slowly into the test solution. In this way the electrical contact is made between this solution and calomel solution of the electrode. Thus the test solution sets the potential level of the reference electrode.

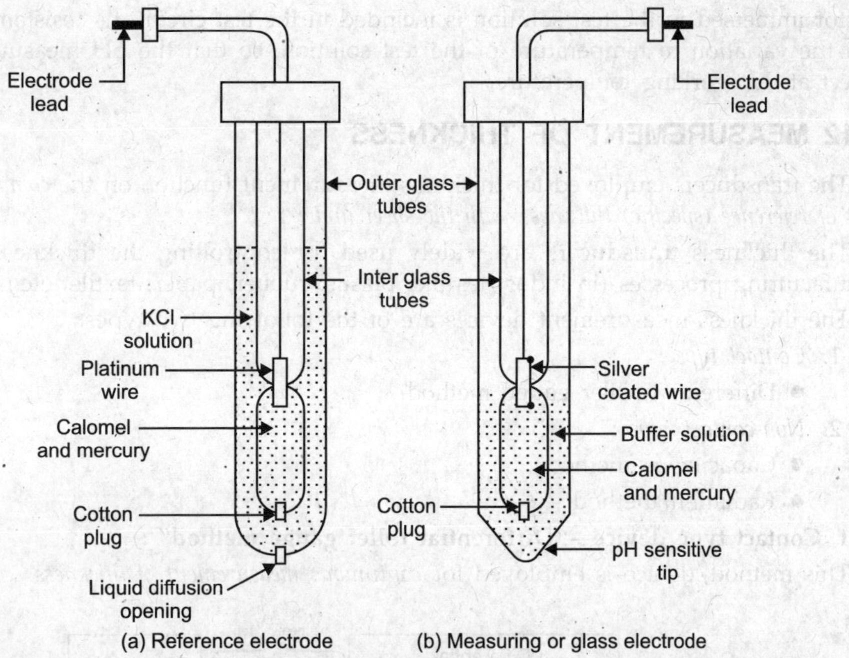


**Fig. 12.69.** pH cell.

In the *measuring electrode* the mercury calomel element is surround by a buffer solution of known and constant pH. There is no opening at the bottom of the outer tube, instead it tapers down to a tip made of thin glass of special composition. *At this tip a potential difference is developed between the buffer solution and the test solution because of difference in the pH value of the two solutions. Since the pH value of the buffer solution is constant the net potential of this electrode is a function of the pH value of the test solution.*

Fig. 12.70 shows the simplified circuit diagram for electronic pH measurement.

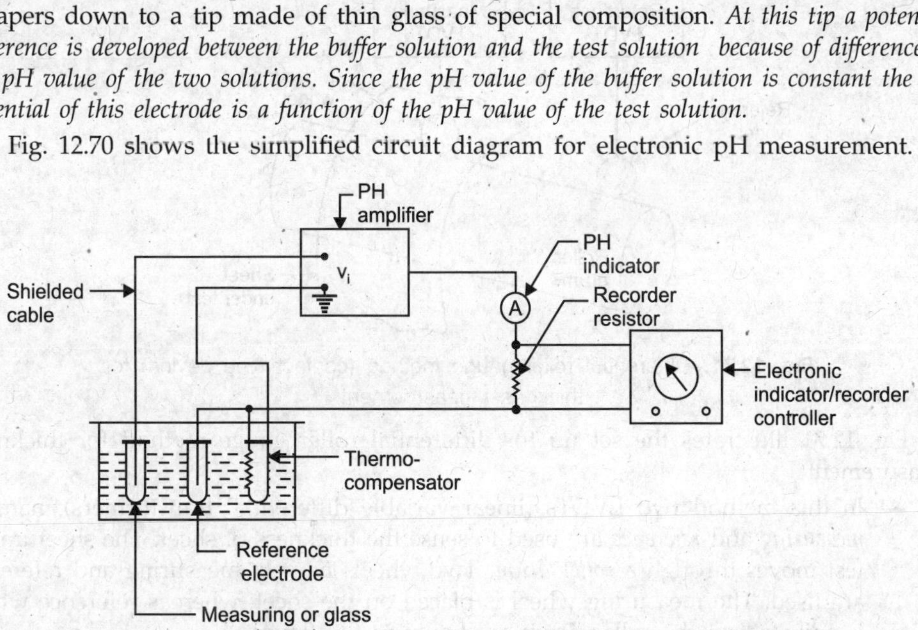


**Fig. 12.70.** Simplified circuit diagram for electronic pH measurement.

The voltage developed between the measuring electrode and the reference electrode is applied as the input voltage, $v_i$ to a null-balance millimeter. A temperature compensating resistor immersed in the test solution is included in the test circuit. Its resistance varies with the variation in temperature of the test solution, so that the pH measurement is correct at the working temperature.

## 12.12 MEASUREMENT OF THICKNESS

The transducers employed for thickness measurement function on the *comparision of sheet of reference (specific) thickness with the sheet under test.*

The thickness transducers are widely used for controlling the thickness during manufacturing processes (in industries like plastic, rubber, paper, textile, etc.)

The thickness measurement devices are of the following two types.

1. *Contact type :*
   - Differential roller gauge method.
2. *Non-contact type :*
   - Capacitance method.
   - Radiation method.

**1. Contact type device—"Differential roller gauge method" :**

This method/device is employed for *continuous measurement of thickness.*

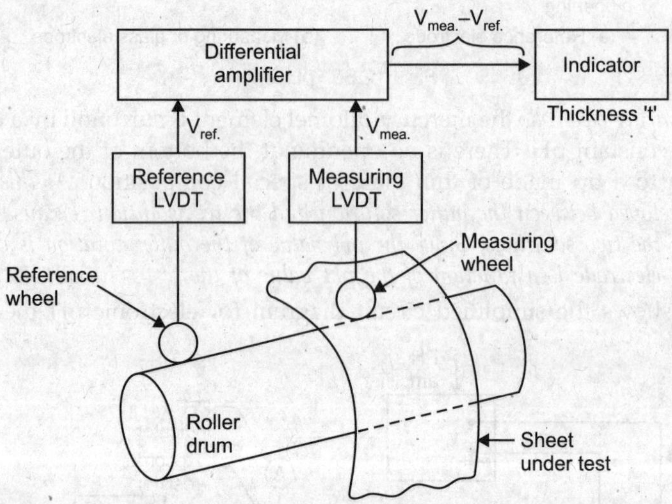


**Fig. 12.71.** Differential roller gauge method (contact type device) for thickness measurement.

Fig. 12.71 illustrates the set up for differential roller gauge method for thickness measurement :

— In this method two LVDTs (linear variable differential transformers), namely *measuring* and *reference* are used to sense the thickness of sheet. The sheet under test moves through a *roller drum*. Two wheels namely measuring and reference are used. The measuring wheel is placed on the sheet; whereas reference wheel is adjusted on the roller drum as shown in the figure.

— The thickness of sheet is determined by the difference of secondary voltages of both the LVDTs due to linear displacements of core or armature positions. The thickness t' of the sheet is determined from the relationship.

$$t = C(V_{mea.} - V_{ref.}) \quad ...(12.56)$$

where, $C$ = The calibration constant,

$V_{mea.}$ = The output voltage of the measuring LVDT,

and $V_{ref.}$ = The output voltage of the reference LVDT.

— The *indicator* indicates the thickness.

- The measurement range of the operating thickness is 0.0025 mm to several cm.

**2. Non-contact type devices :**

(*i*) ***Capacitance method :*** This method is normally employed for measuring thickness of *insulating materials.*

Fig. 12.72 shows the set-up for capacitance method (non-contact type) for thickness measurement.

— In this method two metal electrodes are placed on the two sides of the insulating material sheet under test. This arrangement functions as a *parallel plate capacitor.*

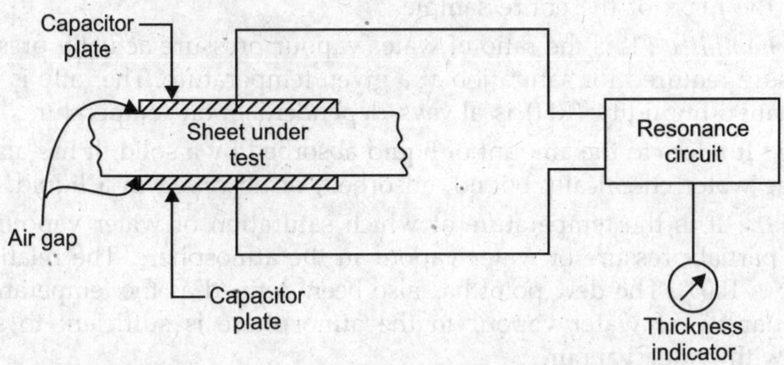


**Fig. 12.72.** Capacitance method (non-contact type) for thickness measurement.

— The capacitance of the capacitor varies according to or in proportion with the thickness of the sheet under test. This *change is capacitance is calibrated in terms of thickness by means of resonance circuit arrangements.* Owing to changes in the value of the capacitance, the resonance ($X_L = X_C$) gets altered which is indicated by the thickness indicator.

(*ii*) ***Radiation method :*** In this method thickness is measured by using nuclear radiations (α, β, γ, X-rays) radiated or generated from radiation source.

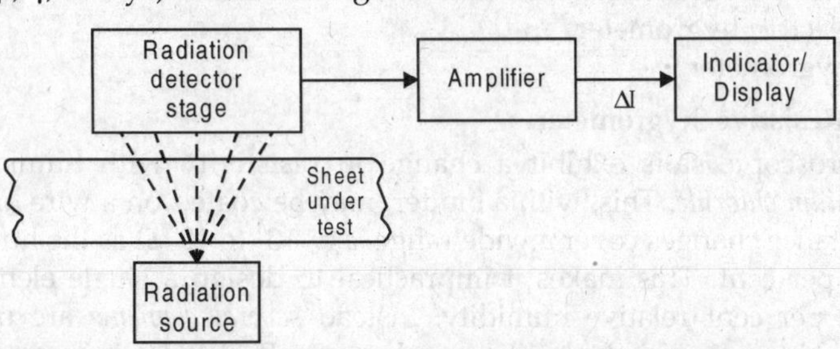


**Fig. 12.73.** Radiation method.

Fig. 12.73 shows nuclear radiation method for thickness measurement :

— The sheet under test is passed through the radiation source and radiation detector stage. The sheet's thickness is determined *in terms of radioactive decay or attenuation of radiations,* the relationship is given by :

$$\Delta I = I_0[1 - e^{(-\mu t)}] \quad ...(12.57)$$

where, $\Delta I$ = The change in ionization current when absorber is inserted,

$I_0$ = The ionization current without the absorber,

$m$ = The absorber coefficient, $cm^2/\mu g$,

$t$ = The thickness in $\mu g/cm^2$.

— The indicator/display is calibrated to read the thickness value.

## 12.13 MEASUREMENT OF HUMIDITY

### 12.13.1. Definitions

***Humidity :*** It is the measure of water vapour present in a gas. It is usually measured as absolute humidity, relative humidity or dew point temperature.

***Absolute humidity :*** It is the mass of water vapour present per unit volume.

***Specific humidity :*** It is the ratio of mass of water vapours present in a sample of a moist gas to the mass of the entire sample.

***Relative humidity :*** It is the ratio of water vapour pressure actually present to water vapour pressure required for saturation at a given temperature. The ratio is expressed in per cent. Relative humidity (RH) is always dependent upon temperature.

***Moisture :*** It refers to the amount of liquid absorbed by a solid. It has also been used to refer to the water chemically bound, adsorbed, or absorbed in a liquid.

***Dew point :*** It is the temperature at which saturation of water vapour pressure is equal to the partial pressure of water vapour in the atmosphere. The relative humidity at dew point is 100%. The dew point has also been defined as the temperature at which the actual quantity of water vapour in the atmosphere is sufficient to saturate this atmosphere with water vapour.

### 12.13.2. Hygrometers

The *"hygrometer" is an instrument which measures humidity directly.* It can be calibrated in terms of absolute humidity. More frequently its output is used to indicate relative humidity.

The various types of hygrometers used are enumerated and described as follows :

1. Resistive hygrometer
2. Capacitance hygrometer
3. Microwave hygrometer
4. Piezoelectric hygrometer
5. Hair hygrometer

#### 12.13.2.1. Resistive hygrometer

Some hygroscopic salts exhibit a change in resistivity with humidity. The most common is *lithium chloride.* This, with a binder, may be coated on a wire or on electrodes. Resulting resistance changes cover a wide range, *e.g.,* $10^4$ to $10^9$ $\Omega$ as the humidity changes from 100 to 0 per cent. This makes it impractical to design a single element to operate from 1 to 100 per cent relative humidity. Instead *several elements* are used, each in a narrow range, with provision for switching elements. Resistance is measured either with a Wheatstone bridge or by a combination of current and voltage measurements.

Most of these must not be exposed to conditions of 100 per cent humidity as the resulting condensation may damage the device. Either they must be operated in a constant temperature environment or temperature corrections must be made. These are accurate to within ±2.5 per cent or ±1.5 percent in some cases. Response times are typically of the order of a few seconds. These are currently the most common electronic hygrometers.

A typical resistance hygrometer is shown in Fig. 12.74. Two metal grids are bonded to a sheet of plastic.

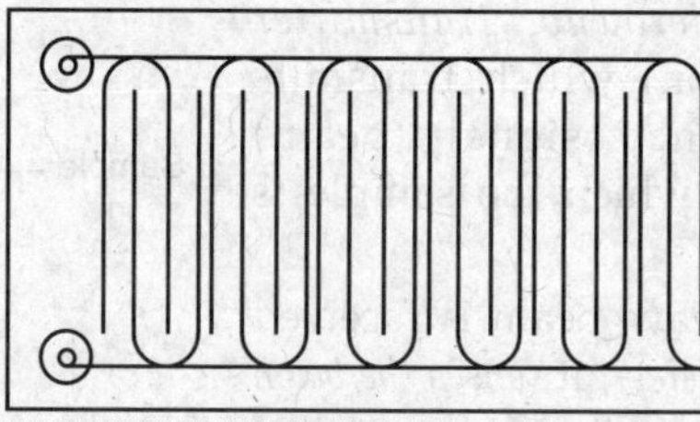

Fig. 12.74. Resistive hygrometer.

The whole is coated with a layer of moisture sensitive chemical such as lithium chloride. The resistance changes when this chemical is exposed to variations in humidity. The higher the relative humidity, the more moisture the lithium chloride will absorb, and the lower will be its resistance.

The resistance of the sensing unit is a measure of the relative humidity. Resistance should be measured by applying A.C. to the Wheatstone bridge. D.C. voltage is not applied because it tends to breakdown the lithium chloride to its lithium and chlorine atoms. The current flow is a measure of the resistance and hence of the relative humidity.

**12.13.2.2. Capacitance hygrometer**

Fig. 12.75 shows a capacitance hygrometer :

— It consists of a conductive aluminium substance over which a layer of porous aluminium oxide is deposited.

— The aluminium oxide is coated with a thin film of gold.

— The aluminium substrate and the gold film act as *electrodes*. These electrodes together with the aluminium oxide as the *delectric*, form a capacitor.

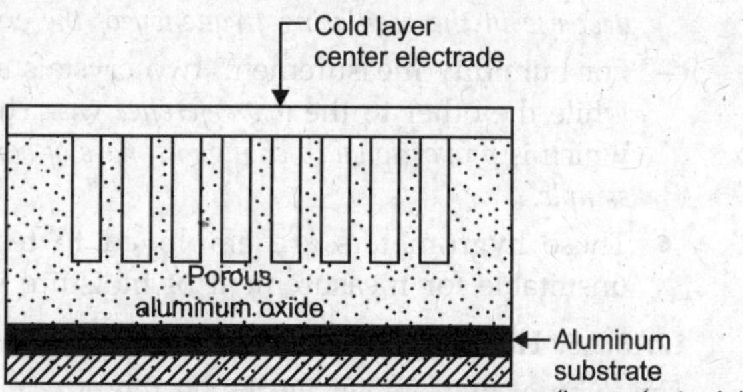


Fig. 12.75. Capacitance hygrometer.

— When this sensor is exposed to air, *the aluminium oxide will absorb the water vapours through the gold layer that changes the dielectric constant of the aluminium oxide. It results in a change in the capacitance which is measured by introducing frequency determining oscillator and other circuits.* The capacitance of the sensor changes with the variation of rate of water content.

— These days, instead of aluminium oxide dielectric, *hygroscopic polymer dielectric is used.*

- It is *small in size* and wide humidity measurement range detects the low dew points. Such a hygrometer is *unsuitable for polar material* (*e.g.*, alcohol) since the capacitor gets short circuited at 15 kHz frequency.

**12.13.2.3. Microwave absorption hydrometer**

In this type of hygrometer microwave frequencies to the tune of 20 to 22 GHz are used; in this frequency range water molecule produces *molecular resonance.*

Fig. 12.76 shows a microwave absorption hygrometer :

— It consists of a *microwave transmitter* (magnetron, klystron) which transmits microwave frequency signal (beam) through the *pipe* in which the sample is flowing.

— The reflected microwave beam is received by the microwave *receiver;* it *senses the beam attenuation and phase shift due to moisture content in the sample.*

— The *display/readout exhibits the mass of moisture per unit volume.*

- This hygrometer is suitable for moisture content measurements in *slurries and pastes.* It is unsuitable for measurement of moisture in gases.

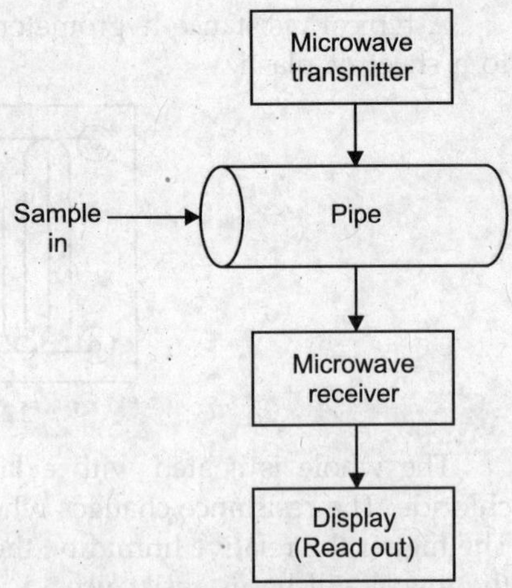


**Fig. 12.76.** Microwave absorption hygrometer.

**12.13.2.4. Piezoelectric hygrometer**

— This type of hygrometer makes use of a piezoelectric crystals like *quartz* crystals, which are coated with hygroscopic material. When this crystal is exposed to the sample, its coating absorbs water content. It results in *increase in the mass and decrease in the oscillating frequency of the crystal.*

— For humidity measurement, two crystals are used. One is exposed to *wet sample* while the other to the *dry reference gas.* Then frequency difference is determined which is *proportional to change in mass of both the crystals and moisture content in the sample.*

- These hygrometers are employed to test *liquid samples.* They are, however, unsuitable for measurement of moisture in liquid phase samples.

**12.13.2.5. Hair hygrometer**

This type of hygrometer works on the *principle that when hair absorbs the moisture, its length changes.*

In the hair hygrometer a strand of hair is attached to an arm through mechanical linkage. The arm is connected to the pointer, which slides over the calibrated scale. A spring is used to maintain proper position of the arm and stand. During moisture measurements when hair absorbs moisture, its length changes. This change in length is applied to the arm which results in sliding of the pointer over the calibrated scale.

- This hygrometer is *small in size, less costly* and can be *installed easily.* However, it is less accurate and requires periodic maintenance.
- It is used in industries where high precision of measurement is not necessary.

## 12.13.3. Psychrometers

The *"psychrometer" is a humidity measuring instrument which utilizes one wet bulb thermometer and one dry bulb thermometer.* The dry bulb thermometer measures the ambient temperature and the wet bulb thermometer measures temperature reduction due to evaporative cooling. A wick, a porous ceramic sleeve, or a similar device saturated with water is in close physical contact with the "bulb" (sensing portion) of the wet bulb thermometer to keep it moist. Relative humidity is determined from the two temperature readings and a reading of the barometric pressure usually with the help of a **Psychrometric Chart**. At any given ambient temperature, the relative humidity decreases as the difference between dry bulb and wet bulb readings increases.

Some of the psychrometers are described below :

**1.** ***Sling psychrometer :*** This device consists of two similar thermometers mounted side by side as shown in Fig. 12.77. The bulb of one of the thermometers is covered by a wet cloth.

When a thin film of water is brought into close contact with air, a portion of it evaporates, and the temperature of the remaining water is lowered due to heat required by the evaporation process. The amount of evaporation and consequent lowering of the temperature depends upon the capability of the air to absorb the moisture. Thus if the relative humidity is lower, air has small moisture and can therefore absorb more. Hence the evaporation is faster and the temperature of the remaining water will be lower.

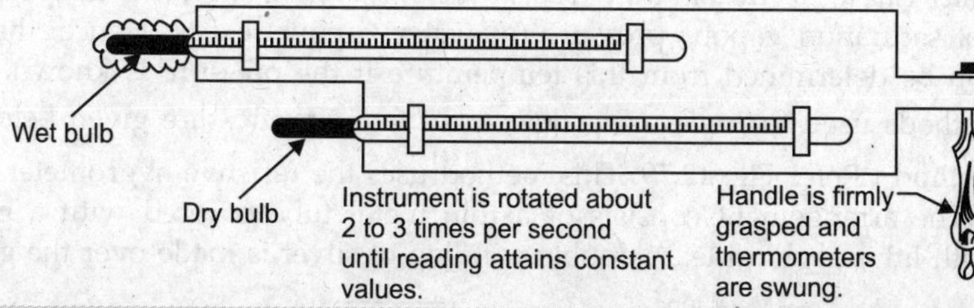


**Fig. 12.77.** Sling psychrometer.

In the sling psychrometer, a handle is provided whereby the thermometer may be whirled, so that the bulbs continuously come in contact with fresh air. If this is not done the stagnant moisture-laden air will engulf the thermometer thereby causing errors.

The dry bulb thermometer is unaffected by the moving air and registers the ambient temperature. The wet bulb thermometer shows a lower temperature due to evaporation. Since the cooling effect is a function of the relative humidity of air, the relative humidity may be determined by comparing the readings of the two thermometers. Relative humidity can directly be read off from psychrometric charts.

**Psychrometric charts** *graphically describe the relationship between the properties of moist air i.e., the dry bulb, the wet bulb and dew point temperatures of the mixture and humidity.*

**2.** ***Commercial type psychrometer :*** A commercial type psychrometer is shown in Fig. 12.78. The filled system thermometer bulbs are placed adjacent to each other and air

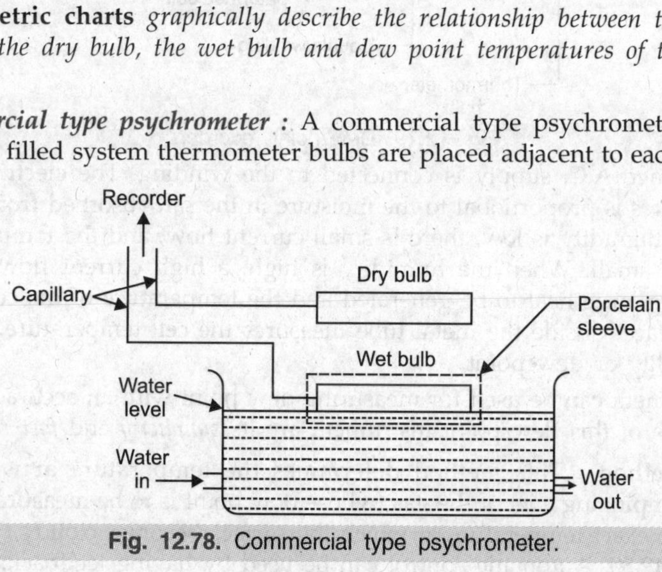


**Fig. 12.78.** Commercial type psychrometer.

is blown across the assembly or they may be placed where there is strong current of air (nearly 5 m/s). One of the bulbs is coated with a knitted or woven cotton which is suspended into a water reservoir whose level is controlled so that the required conditions are satisfied. The temperatures are recorded by suitable recorders.

The two thermometers are connected to a recorder through capillary tubes. In order to avoid errors arising on account of contamination of the wick the bulb may be enclosed in a porcelain sleeve and water is pumped into the annular space at a constant rate. *These instruments are not suited for instantaneous monitoring and control of humidity.*

A variation in the indication and recording arrangement is also possible if the filled in thermometers are repleced by resistance thermometers or thermocouples.

### 12.13.4. Dew Point Measurements

The *'dew point' is a discrete temperature. Specially, it is that temperature at which liquid water and water vapour, or ice and water vapour are in equilibrium.* At this temperature only one value of saturation vapour pressure (of water vapour) exists. Hence, the absolute humidity can be determined from this temperature if the pressure is known.

Two methods used for measuring dew point temperatures are given below :

**First Method :** Refer Fig. 12.79. This method uses the resistive hygrometer described previously. The arrangement consists of a thin metal tube covered with a glass cloth saturated with lithium chloride. A double winding of silver is made over the glass cloth.

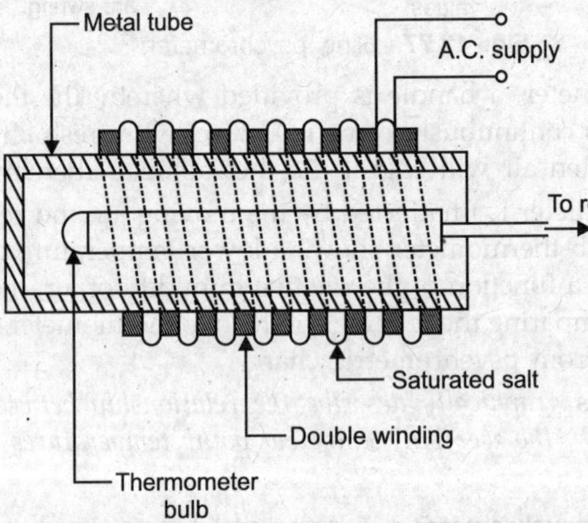


**Fig. 12.79.** Dew point recorder.

A low voltage A.C. supply is connected to the winding. The electric conductivity between the wires is proportional to the moisture in the salt absorbed from surrounding gas. When the humidity is low, there is small current flow, and the temperature rise of the winding is small. When the humidity is high, a high current flows through the winding causing large heat to be generated and the temperature of the cell is high. The thermometer placed inside the metal tube measures the cell temperature, and therefore absolute humidity or dew point.

This instrument can be used for measuring dew point with an accuracy of ±2%. The main advantage of this dew point instrument are its *simplicity* and *fast response*.

**Second Method :** This method determines the temperature at which moisture condenses by impinging a jet of the gas (whose dew point is to be measured) on a mirror and finding the exact temperature at which the surface becomes cloudy. This method is shown in Fig. 12.80. Automatic control can be used by having feedback.

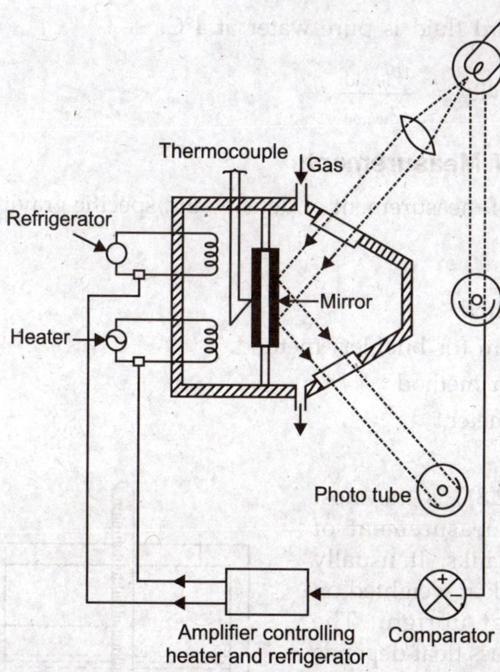


**Fig. 12.80.** Schematic diagram of dew point recorder.

The gas is supplied continuously. This dew point recorder is provided with arrangements for both cooling and heating of the mirror. Light falling on the mirror from a standard light source is reflected on a photo-tube. Another photo-tube receives light straight from the source. The outputs of tubes are then compared by a comparator. If the mirror is foggy due to condensation an error signal is sent and the heater starts functioning and the refrigerator is turned off. If the condensation has not started, the heater is shut off and the refrigerator turned is on. The cycle continues. Thus, the temperature of the mirror tells about the temperature necessary to condense the gas. The temperature is sensed by a thermocouple. The temperature is measured and recorded by a suitable device.

## 12.14 MEASUREMENT OF DENSITY AND SPECIFIC GRAVITY

### 12.14.1. Definitions

***Mass density.*** *It is defined as the mass per unit volume* $\left(\frac{m}{V}\right)$ *at a standard temperature and pressure. It is usually denoted by ρ (rho). Its units are* $kg/m^3$ *i.e.,*

$$\rho = \frac{m}{V} \qquad ...(12.58)$$

***Weight density (or specific weight).*** *It is defined as the weight per unit volume at the standard temperature and pressure.* It is usually denoted by $w$.

$$w = \rho g \qquad ...(12.59)$$

***Specific gravity.*** It is defined as the *ratio of specific weight of the liquid to the specific weight of a standard liquid.* It is dimensionless and has no units. It is denoted by $S$.

For liquids, the standard fluid is pure water at 4°C.

$$\therefore \qquad \text{Specific gravity} = \frac{w_{\text{liquid}}}{w_{\text{water}}}.$$

### 12.14.2. Methods of Measurements

The various methods of measurement of density and specific gravity are enumerated and discussed as follows :

1. Hydrometer.
2. Force balance method.
3. Differential pressure (or bubbler) method.
4. Nuclear absorption method.
5. Acoustic density meter.

#### 12.14.2.1. Hydrometer

An *hydrometer* (Fig. 12.81) is the most popular device for the measurement of liquids in open vessels/tanks. It usually consists of a hollow glass float weighted on one side to make it float upright. The amount of immersion of glass float depends on the density of the liquid. The glass float is usually calibrated in terms of specific gravity (ratio between the density of liquid under measurement to that of the water). Thus the specific gravity of the liquid can be easily determined by noting the amount of immersion from the scale on the glass float.

*Hydrometers are widely used where automatic operation is not required.*

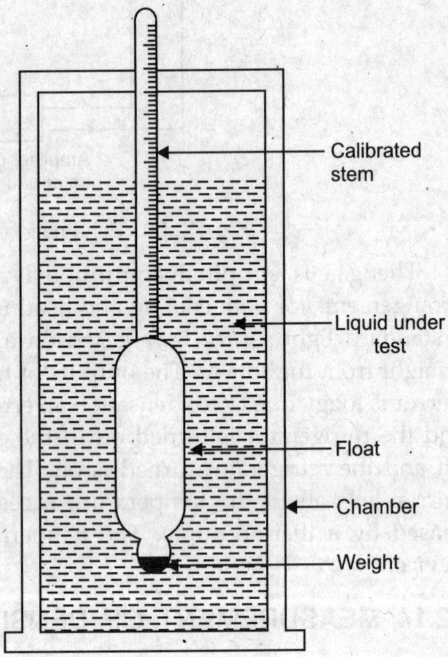


**Fig. 12.81.** Hydrometer.

- ***Photoelectric hydrometer.*** In order to have *continuous monitoring of the density* of a liquid, we use a *"photoelectric hydrometer"*. As opaque stem of the hydrometer rises or falls, it cuts off light rays passing through a slit into a phototube. The output of the phototube is determined by the position of the stem and hence the density of the liquid.

- ***Density measurement using LVDT.*** In the arrangement (Fig. 12.82) an iron cone is provided at the centre of the hydrometer and the body of the hydrometer chamber is surrounded by the windings of the linear voltage differential transformer (LVDT). The sample flows continuously through the hydrometer chamber and the hydrometer will rise or fall in accordance with the changes in specific gravity of the liquid. The output of the differential transformer changes with the position of the iron core. A potentiometer is used to detect any unbalance in the circuit caused by the output from the differential transformer and to convert this signal into a specific gravity/density reading.

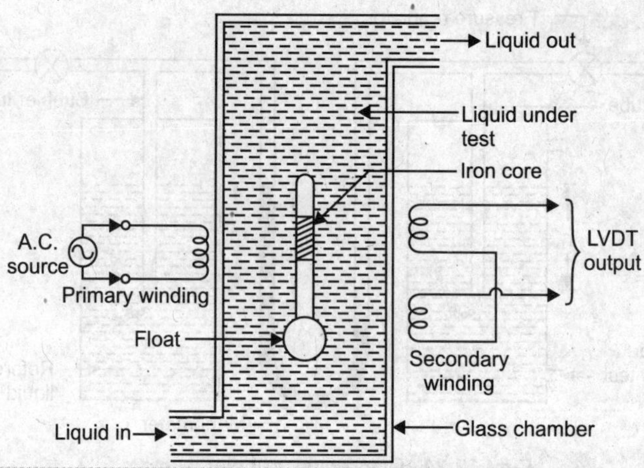


**Fig. 12.82.** Hydrometer-LVDT type.

This method is specially useful for *dynamic measurement of density*. Since the output signal is a voltage, it can be used for display and control of density from a remote location.

**12.14.2.2. Force balance method**

The schematic arrangement of the force balance method is shown in Fig. 12.83. The liquid under test flows through a tube loop of the meter. The tube loop is suspended from a pivoted beam which carries a weight. The system is balanced when the tube is full of sample liquid.

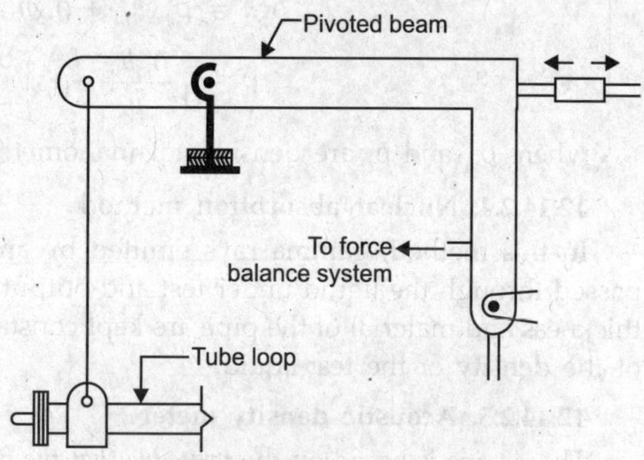


**Fig. 12.83.** Force balance system.

When a liquid of density different from the sample liquid is passed, the beam gets deflected. The beam is connected to a force balance system which keeps the tube loop in balance with the weight. *The force (found in terms of the current-fed to arm electromagnet) is measured on an indicator which is calibrated directly in terms of density.*

**12.14.2.3. Differential pressure (or bubbler) method**

This method utilizes the hydrostatic principle :

$$p = \rho g h \qquad ...(12.60)$$

This means that for a constant depth of liquid from a free surface, the *pressure exerted is directly proportional to density of the liquid* (ρ).

Fig. 12.84 shows a differential bubbler system. It consists of two containers, one containing reference liquid of known density and other connected to the liquid under test in which level/height is maintained constant by an overflow tube. Two bubbler tubes are immersed to equal depths in two containers and compressed air at regulated pressure is supplied. The air pressures to bubbler tubes are so adjusted *that bubbles are released slowly*.

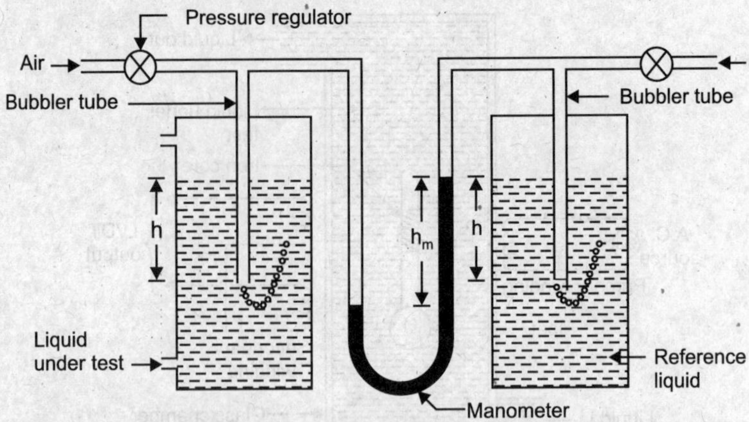


**Fig. 12.84.** Differential bubbler system.

- The two pressures are then proportional to the density of the liquids.
- The reading '$h_m$' of differential manometer is indicative of the specific gravity of the liquid under test.

From the hydrostatic balance for the system, we have

$$\rho g h = \rho_m g h_m + \rho_r g h$$

or,

$$\rho = \left(\frac{\rho_m h_m}{h} + \rho_r\right)$$

where $\rho_m$ and $\rho_r$ are densities of manometric and reference liquids respectively.

**12.14.2.4. Nuclear absorption method**

In this method, gamma rays emitted by an isotope of cesium-137 or cobalt-60 are passed through the liquid under test and output signal detected. If other parameters like thickness and material of the pipe are kept constant, the output signal becomes a measure of the density of the test-liquid.

**12.14.2.5. Acoustic density meter**

The device is based on the *principle that the frequency of oscillation of a tube is dependent on its mass*. When the tube contains a liquid, then its mass is dependent on the density of the liquid. Hence, a change in density of the liquid produces a change in its acoustic resonant frequency.

## 12.15 MEASUREMENT OF VISCOSITY

### 12.15.1. Introduction to Viscosity

**Viscosity** *may be defined as the property of a fluid which determines its resistance to shearing stresses.* It is a measure of the internal fluid friction which causes resistance to flow. It is primarily *due to cohesion and molecular momentum exchange between fluid layers* and as flow occurs, these effects appear as shearing stresses between the moving layers of fluid.

An *ideal fluid has no viscosity.* There is no fluid which can be classified as a perfectly ideal fluid. However, the fluids with very little viscosity are sometimes considered as ideal fluids.

Viscosity of fluids is due to *cohesion and interaction between particles.*

Refer to Fig. 12.85. When two layers of fluid, at a distance '*dy*' apart, move one over the other at different velocities, say $u$ and $u + du$, the viscosity together with relative velocity causes a shear stress acting between the fluid layers. The top layer causes a shear stress on the adjacent lower layer while the lower layer causes a shear stress on the adjacent top layer. This shear stress is proportional to the rate of change of velocity with respect to $y$. It is denoted by $\tau$ (called Tau).

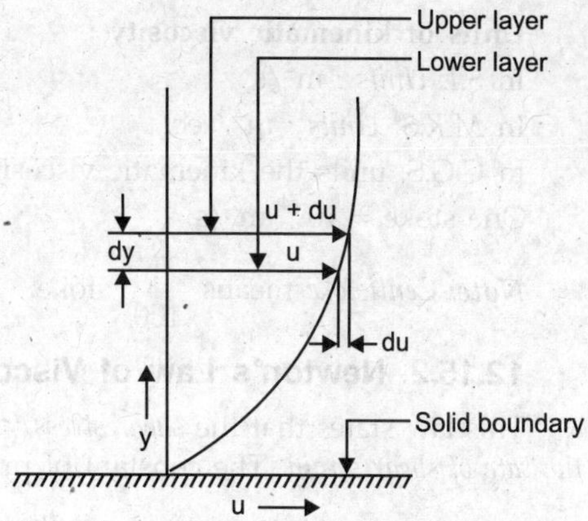


**Fig. 12.85.** Velocity variation near a solid boundary.

Mathematically $\tau \propto \frac{du}{dy}$

or, $$t = \mu \cdot \frac{du}{dy} \qquad ...(12.61)$$

where $\mu$ = Constant of proportionality and is known as *coefficient of dynamic viscosity* or *only viscosity, and*

$\frac{du}{dy}$ = Rate of shear stress or rate of shear deformation or velocity gradient.

From Fig. 12.85, we have

$$\mu = \frac{\pi}{\left[\frac{du}{dy}\right]} \qquad ...(12.62)$$

Thus viscosity may also be defined as the *shear stress required to produce unit rate of shear strain.*

**Units of viscosity :**

In *S.I. Units :* N.s/m$^2$

In *M.K.S. Units :* Kg$_f$ . sec/m$^2$

$$\left[\therefore \mu = \frac{\text{force/area}}{(\text{length/time}) \times \frac{1}{\text{length}}} = \frac{\text{force/length}^2}{\frac{1}{\text{time}}} = \frac{\text{force} \times \text{time}}{(\text{length})^2}\right]$$

The unit of viscosity in C.G.S. is also called *poise* = $\frac{\text{dyne} - \text{sec.}}{\text{cm}^2}$.

One poise = $\frac{1}{10}$ N.s/m$^2$.

***Note:*** The viscosity of water at 20°C is $\frac{1}{100}$ poise or one centipoise.

**Kinematic viscosity :** *Kinematic viscosity* is defined as the *ratio between the dynamic viscosity and density of fluid.* It is denoted by $\nu$ (called nu).

$$\text{Mathematically, } \nu = \frac{\text{Viscosity}}{\text{Density}} = \frac{\mu}{\rho} \qquad ...(12.63)$$

**Units of kinematic viscosity :**

In *S.I. Units* : $m^2/s$

In *M.K.S. Units* : $m^2/sec$.

In C.G.S. units the kinematic viscosity is also known as stoke (= $cm^2/sec$.)

One stoke = $10^{-4}$ $m^2/s$

*Note: Centistoke* means $\frac{1}{100}$ stoke.

## 12.15.2. Newton's Law of Viscosity

This law states that the *shear stress* (τ) *on a fluid element layer is directly proportional to the rate of shear strain.* The constant of proportionality is called the *co-efficient of viscosity.*

$$\text{Mathematically, } \tau = \mu \frac{du}{dy} \qquad ...(12.64)$$

The fluids which follow this law are known as *Newtonian fluids.*

## 12.15.3. Viscosity Measurement Methods

The following methods are used to measure viscosity

1. Rotating cylinder method
2. Falling sphere method
3. Capillary tube method
4. Efflux viscometers.

} Mechanical methods

The devices used for measurement of viscosity are known as ***'Viscometers'.***

5. Electrical method.

### 12.15.3.1. Rotating cylinder method

In this method Newton's law of viscosity is used to measure the viscosity of a fluid. Fig. 12.86, shows a rotating cylinder viscometer. It consists of two concentric cylinders, the annular space between them is fitted with the liquid whose viscosity is to be determined. The outer cylinder is rotated at a constant angular velocity ω with respect to the inner stationary cylinder. The torque transmitted by the enclosed liquid to the stationary cylinder is measured by the torsional strain of the restraining spring attached to the top of the inner cylinder.

According to Newton's law of viscosity

$$\text{Shear stress } (\tau) = \mu \times \text{velocity gradient} \left(\frac{du}{dy}\right)$$

Since the annular space $t = (R_2 - R_1)$ is quite small, (where $R_2$ and $R_1$ are the radii of the outer and inner cylinders respectively), the velocity gradient,

$$\frac{du}{dy} = \frac{V}{t} = \frac{2\pi R_2 N}{60t}$$

(where, $N$ = the rotational speed of the outer cylinder in r.p.m.)

$$\therefore \quad \text{Shear stress } (\tau) = \mu \times \frac{2\pi R_2 N}{60t}$$

$$\text{Viscous drag} = \text{Shear stress} \times \text{area}$$

$$= \mu \times \frac{2\pi R_2 N}{60t} \times 2\pi R_1 h, \text{ where } h = \text{height of liquid.}$$

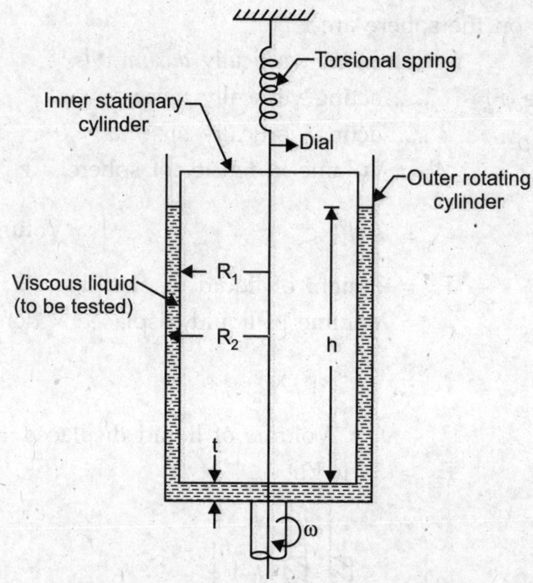


**Fig. 12.86.** Rotating cylinder viscometer.

$$\text{Viscous torque} = \text{Viscous drag} \times \text{radius}$$

$$= \left(\mu \times \frac{2\pi R_2 N}{60t} \times 2\pi R_1 h\right) \times R_1 = \frac{\mu\pi^2 R_1^2 R_2 hN}{15t}$$

Viscous torque must equal to the torque $T$ exerted by the torquemeter.

$$\therefore \quad T = \frac{\mu\pi^2 R_1^2 R_2 \cdot hN}{15t}$$

or,
$$\mu = \frac{15\,Tt}{\pi^2 R_1^2 R_2 hN} \qquad ...(12.65)$$

Thus a rotational type viscometer can be calibrated to directly give μ for given speed of rotation $N$.

### 12.15.3.2. Falling sphere method

Falling sphere method of measuring viscosity of a liquid is based on *Stokes' Law.*

Fig. 12.87, shows a falling sphere viscometer. A small spherical ball is released into the liquid to be tested and it accelerates under the gravitational force until it reaches a maximum/terminal velocity $V$ when the *buoyant and viscous drag forces balance the gravity force.*

Let,
- $d$ = Diameter of the spherical ball,
- $l$ = Distance travelled by the sphere in viscous liquid,
- $t$ = Time taken by the sphere to cover distance $l$,
- $\rho_s$ = Density of sphere,
- $\rho_f$ = Density of fluid/liquid,
- $W$ = Weight of sphere,
- $F_B$ = Buoyant force, acting on the sphere
- $F_D$ = Drag force, and
- $\mu$ = Dynamic viscosity of the liquid under test.

The forces acting on the sphere are :

1. Weight $W$ ...... acting vertically *downwards.*
2. Buoyant force, $F_B$ ...... acting vertically *upwards.*
3. Drag force, $F_D$ ...... acting vertically *upwards*

$$W = \text{Volume} \times \text{density of sphere} \times g$$

$$= \frac{\pi}{6} d^3 \times \rho_s \times g \qquad \left(\because \text{Volume of sphere} = \frac{\pi}{6} d^3\right)$$

$$F_B = \text{Weight of liquid displaced}$$

$$= \text{Volume of liquid displaced} \times \text{density of liquid} \times g$$

$$= \frac{\pi}{6} d^3 \times \rho_f \times g$$

(∴ Volume of liquid displaced = Volume of sphere)

$$F_D = 3\pi\, \mu\, VD \qquad \text{...(Stokes, law)}$$

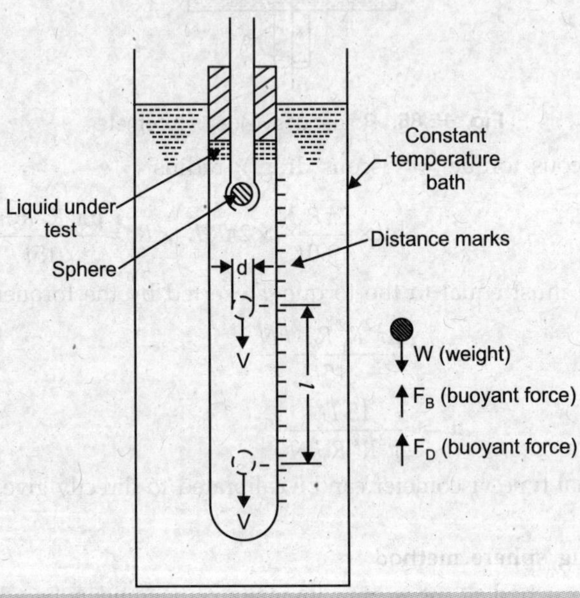


Fig. 12.87. Falling sphere viscometer.

For equilibrium

$$F_D + F_B = W$$

or,

$$F_D = W - F_B$$

$$3\pi\mu\, V_d = \frac{\pi}{6} d^3 \times \rho_s \times g - \frac{\pi}{6} d^3 \times \rho_f \times g \qquad \text{...(substituting the values)}$$

$$= \frac{\pi}{6} d^3 \times g(\rho_s - \rho_f)$$

or,

$$\mu = \frac{\pi}{6} \cdot \frac{d^3 \times g(\rho_s - \rho_f)}{3\pi V d}$$

or,

$$\mu = \frac{gd^2}{18V}(\rho_s - \rho_f) \qquad \text{...(12.66)}$$

From the above equation the value of dynamic viscosity μ can be determined.

*Note:* Stokes' law is essentially valid for Reynolds number below 0.1 where the wall has no effect on the terminal fall viscosity $V$.

### 12.15.3.3. Capillary tube method

*This method makes use of Hagen-Poiseuille equation for laminar flow through circular tubes.*

Fig. 12.88 shows a capillary tube viscometer. It consists of a tank in which the liquid whose viscosity is to be determined is filled. A capillary tube of diameter $D$ and length $L$ is attached horizontally very close to the bottom of tank. The tube is allowed to discharge freely into the atmosphere. The liquid is collected in a measuring tank for a given time. Then the rate of liquid collected in the tank per second is determined. The pressure head is measured at a point far away from the tank.

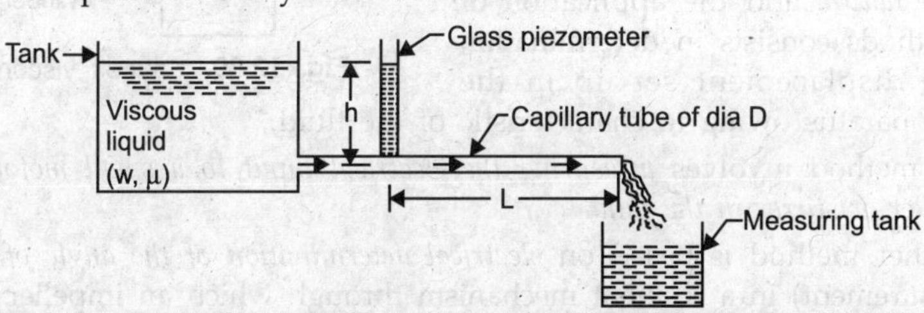


Fig. 12.88. Capillary tube viscometer.

Let, $h$ = Difference of pressure head for length $L$

$\mu$ = Co-efficient of viscosity, and

$w$ = Weight density of the liquid.

According to Hagen-Poiseuilli's equation, we have

$$h = \frac{32\,\mu\,\bar{u}L}{wD^2}$$

But, $$\bar{u} = \frac{Q}{\frac{\pi}{4}D^2}$$ (where, $Q$ = Discharge of liquid through the tube)

$$\therefore \quad h = \frac{32\mu \times \dfrac{Q}{(\pi/4)\times D^2} \times L}{wD^2} = \frac{128\,\mu\,QL}{\pi w D^4}$$

or, $$\mu = \frac{\pi w h D^4}{128\,QL} \quad \text{...(12.67)}$$

### 12.15.3.4. Efflux viscometers

In Efflux viscometers the viscosity is determined by noting the time of efflux, under specific conditions, for a fixed volume of fluid through a specific capillary or aperture (standardized nozzle or orifice).

The operation of *Saybolt*, *Redwood* and *Engler* is based on this principle.

Fig. 12.89 shows a **Saybolt viscometer**. It consists of a tank in the bottom of which is a short capillary tube. The tank is filled with the liquid whose viscosity is to be determined; a constant temperature bath surrounds this tank to ensure that the liquid remains at uniform constant temperature during the test run. The viscosity is determined by measuring the time required for 60 $cm^3$ of liquid at a known temperature to flow out of the reservoir through the tube. The initial level of the liquid in the reservoir is previously adjusted to a standard height. From the time measurement, the kinematic viscosity of the liquid can be determined by the use of empirical formula or the calibration chart.

- These viscometers (laboratory instruments) are widely employed in *petroleum and allied industries.*

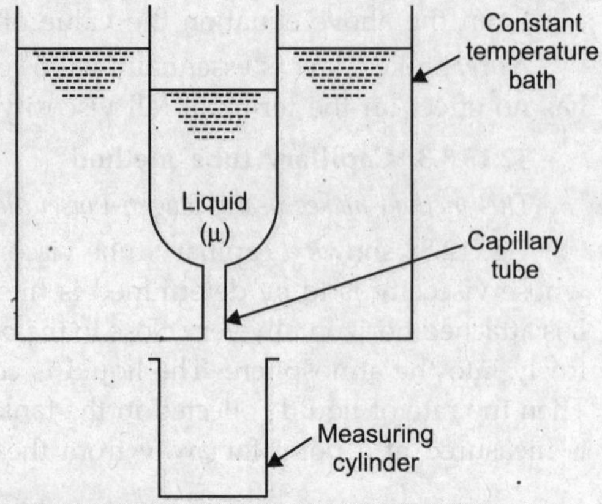


**Fig. 12.89.** Saybolt viscometer.

#### 12.15.3.5. Electrical method

*Viscosity measurement* is essentially *mechanical in nature* and the application of electrical methods consists in determination of stress or displacement set up in the measuring apparatus owing to characteristic of the fluid.

- One method involves *measuring the electrical input to a small motor driving an impeller or stirrer in the fluid.*
- Another method is based on *electrical determination of the angle of lag* (torque measurement) in a resilient mechanism through which an impeller is driven.
- A further method utilizes *mangetostriction to produce longitudinal oscillations* in a steel rod carrying a diaphragm immersed in the liquid. Determination of the electrical loading on the exciting circuit provides a measure of viscosity.

## 12.16 GAS ANALYSERS–MEASUREMENT OF $CO_2$, $O_2$ AND CO

The gas analysis is carried out to determine the concentration of one or more components of a gas mixture. The common types of analysers used for the measurement of $CO_2$ and $O_2$ are described as follows :

### 12.16.1. Measurement of Carbon-dioxide ($CO_2$)

Since gases differ in their ability to conduct the heat, therefore, the *proportion of various constituents of flue gases can be determined by measuring their thermal conductivities.* If the thermal conductivity of air is considered unit of 0°C, the thermal conductivity of $CO_2$ = 0.585; $H_2$ = 7.35; $N_2$ = 1.015; $O_2$ = 1.007 respectively.

Fig. 12.90 shows the commonly used method for measuring $CO_2$ content in flue gases.

— It consists of hot wire thermal conductivity gas analysis cell which comprises of *two chambers,* each containing a wire filament as shown in Fig. 12.90. The gas is passed through one chamber whose conductivity is to be measured and other chamber is filled with a reference gas like air and sealed.

— The platinum wire elements are passed through the cells (as shown in Fig. 12.90) and form a Wheatstone bridge circuit. Under normal working conditions, these elements are heated by the bridge current.

— The temperature of the filament in the measuring chamber rises as the conductivity of the gases passing around it decreases, as the heat of the filament cannot be carried by the gases. Thus the temperature and hence the resistance of the two platinum wire elements changes and unbalances the bridge and provides an *unbalanced voltage which is proportional to the percentage of $CO_2$*. This unbalanced voltage can be used for measuring the percentage of $CO_2$ in flue gases.

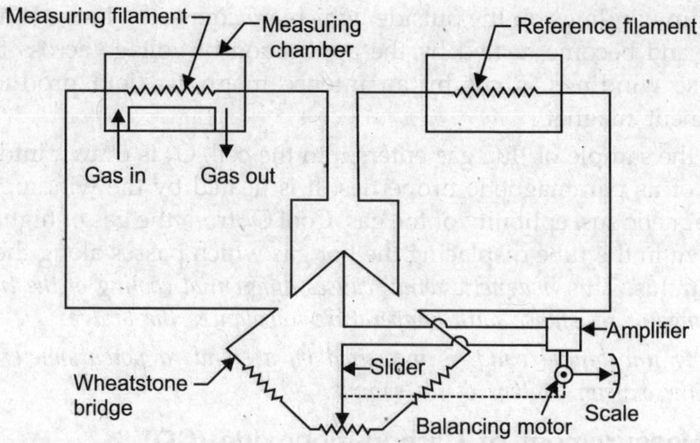


**Fig. 12.90.** Electrical circuit for the measurement of $CO_2$ content in the gases.

— To compensate the error which would arise from variation in temperature of the gas, another resistance is inserted in the circuit (not shown).

This type of instrument gives accurate values *if the proportion of $H_2$ present is small and constant*.

## 12.16.2. Measurement of Oxygen ($O_2$)

Although there are other methods to measure $O_2$, we shall discuss here only "*paramagnetic oxygen analyser*". This analyser works on the principle of *paramagnetic properties of $O_2$*. Oxygen is the only paramagnetic component in the flue gases. It has very high magnetic permeance compared with other gases. The magnetic permeance of paramagnetic cases *decreases with an increase in temperature*. The gas molecules near the heated body in the magnetic field lose part of their magnetic properties and are pushed out of the magnetic field by cooler molecules. The temperature of the cooler molecules entered in the magnetic field where heated body is situated, is increased and in turn they are also pushed out of magnetic field by other molecules. In this way, a convection flow known as *'Magnetic Wind'* is created which cools the heated body. The magnetic wind increases with an increase in $O_2$ content in the flue gases and cooling of the heated body is intensified. *The temperature change of heated body alters its electric resistance which indicates the concentration of $O_2$ in the gas mixture*. Therefore, the $O_2$ content can be measured by measuring an electric resistance of the heated body which can be one of the arms of Wheatstone bridge.

Fig. 12.91 shows the electric circuit used for $O_2$ measurement using the magnetic wind type principle.

— The gas is aspirated into the measuring cell through an annulus with an horizontal by-pass the tube (as shown in Fig. 12.91). This by-pass tube has two identical adjacent

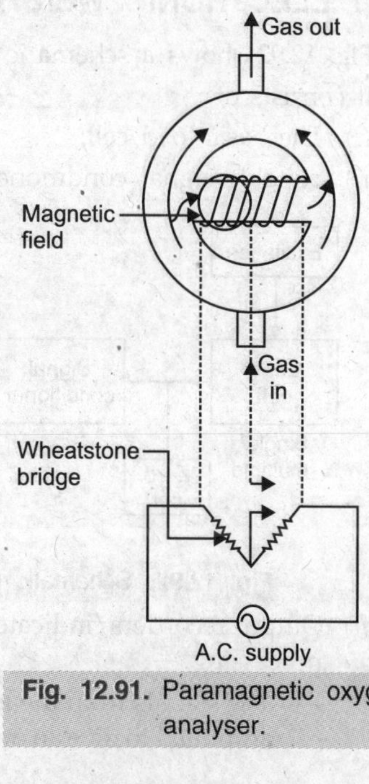


**Fig. 12.91.** Paramagnetic oxygen analyser.

platinum windings on the outside which are connected into a Wheatstone bridge circuit and become heated by the application of voltage across the bridge. One of these windings is cut by an intense magnetic field produced by a large permanent magnet.

— When the sample of flue gas enters into the cell, $O_2$ is drawn into the by-pass by virtue of its paramagnetic properties. It is heated by the winding which reduces the magnetic susceptibility of the gas. Cool $O_2$ from the gas of higher susceptibility is drawn in the tube displacing the hot gas which passes along the tube back into the annulus. This *'magnetic wind' causes differential cooling of the two windings and the change of resistance with temperature unbalances the bridge.*

*The resulting unbalanced e.m.f. is measured by a standard potentiometer recorder and is proportional to the oxygen content of the sample.*

### 12.16.3. Measurement of Carbon-monoxide (CO)

A carbon-monoxide analyser works on the principle of non-dispersive infra-red absorption and is suitable for all biatomic molecules. The principle of the apparatus is based on molecular specific absorption of bands of infra-red radiation. *The different radiation intensities, depending on the gas concentration are first converted in the receiver chamber into flow pulses and then into electrical signals.* The system first shoots beams of infra-red and visible light through the gases in the smoke stack. A molecule of a particular gas hungrily absorbs light energy at particular wavelength. By comparing the original intensities of the beam ($I_0$) with the amount of light that reaches the detector ($I$), the analyser determines the relative amounts of CO in the gases. The resulting signal is amplified and transmitted to the control room. The concentration of the gas is measured in terms of the ratio as log ($I/I_0$). Therefore, any light level change, darking of the window, scattering of particulates or water droplets in the gas stream, effects both $I$ and $I_0$ equally and maintains the ratio unchanged.

## 12.17 ELECTRONIC WEIGHING SYSTEM

Fig. 12.92 shows a schematic block diagram of electronic weighing system :

It consists of :

(*i*) The basic load cell;

(*ii*) Suitable signal conditioners;

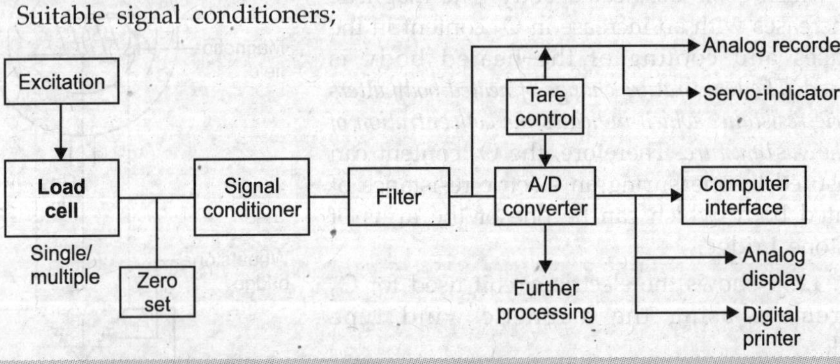


Fig. 12.92. Schematic block diagram of electronic weighing machine.

(*iii*) Output recorders/indicators giving both the analog and digital output for further processing.

— The signals from the load cells (which may be upto 4 or 8 cells) are added and amplified to give an output of 0 to 5 V or 4 to 20 mA for process instrumentation.

This is further converted into digital format with an A/D converter so that the output can be indicated or printed out in a digital format used for processing.

— Taring facility can be provided whenever necessary.

- The electronic system can be designed computer based with suitable software to include all the functions such as totalising, averaging, counting, processing or any other such control operation.

**Applications.** The electronic weighing systems find uses in :

— Platform weighers;
— Truck weighing;
— Crane weighing;
— Tank and hopper weighing;
— Weighing of vehicles in motion;
— Weigh feeters;
— General purpose commodity or product weighing;
— Bulk load sensing (as encountered in chemical plants, steel plants etc.)

***Advantages :*** The main advantages of an electronic weighing system when compared with mechanical systems are :

1. High dependability and ruggedness.
2. The accuracy is quite good even under most adverse and hostile environmental conditions.
3. Small size and compactness.
4. Rapid weighing (useful for in-motion weighing), since response speed is high.
5. Both analog and digital output with print-out facility, remote indication, and parallel display.
6. Computer capability for processing.
7. Availability of tare facility.
8. Excellent flexibility for monitoring multiple loads.

## 12.18 MICROPROCESSOR-BASED INSTRUMENTATION

In microprocessor-based instruments, microprocessor (µp) forms one of the auxiliary functional elements of the instruments. In the word microprocessor or *"micro"* refers to its microminiature size/dimensions, while *"processor"* signifies its vast potential to perform complex computations at fantastically high speeds, together with pre-programmed logic/software which enhances significantly the capabilities and effectiveness of the instruments.

These days, highly versatile microprocessor chips designed and developed on Very Large Scale Integration (VLSI *i.e.,* $10^7$ circuits/$cm^2$) have provided very convenient means of adding Artificial Intelligence (AI) in many of our every day facilities, *examples* being 'automatic washing machines', 'burglar alarms' automatic teller machines (ATMs) etc.

Microprocessor-based instruments are commonly termed as *smart or intelligent instruments.* The micro-miniaturisation has resulted in offering these instruments of the size of a pocket calculator.

Microprocessor, by itself, is an operational computer. In order to shape it in the form of a *digital computer*, it is incorporated with additional circuits for memory and input/output devices.

A computer-based measurement system can process all the inputs and present the data in real time. Fig. 12.93 shows a *typical digital computer-based measurement system.*

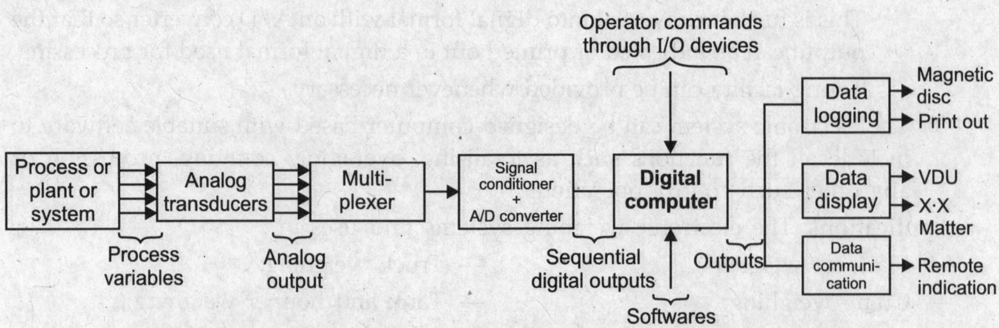


Fig. 12.93. A typical digital computer-based measurement system.

— A 'sequential list of instructions' (termed as a computer program) is fed to a digital computer for suitable processing/manipulation of the data.

— Invariably, Artificial Intelligent (AI) in the form of Artificial Neutral Networks (ANNs) or Fuzzy Logic (FL) may be incorporated. By doing so, the decision-making tasks in various processes are usually accomplished by the computer itself and not by any human operator.

***Advantages and disadvantages of computer-based instrumentation systems :*** The computer-based instrumentation systems entail the following *advantages* and *disadvantages.*

***Advantages :***

1. Lower costs, higher accuracy and more flexibility.
2. Can be battery operated, since they have low consumption of the order of 40 mW.
3. Can be adjusted/programmed with a remote control.
4. Do not require skilled operation due to the incorporation of user-friendly subroutines.
5. Suitably programmed to automatically carry out the mundane tasks of drifts correction, noise reduction/elimination, non-linearity correction, range and span adjustments etc.
6. These instruments have compact, rugged and reliable signal conditional conditioning and display. They can perform in tough, industrial, consumer, military, automobile and environmental conditions.
7. Owing to their compact size, they are mostly of portable type.
8. The measurement, processing and display of the data of the process variables are done in real time *i.e.,* it is one-line type process information in the desired form and units.

***Disadvantages :***

1. These instruments are prone to virus problems and consequently they suddenly become sick and inoperative.
2. They invariably need the processing data in the digital form.
3. They cannot modify the programmes themselves.
4. The commercial computer softwares become absolute very fast and periodically updating the same involves extra expenditure.

## HIGHLIGHTS

### Pressure Measurement

1. *Pressure* is defined as a force per unit area.
2. *Gauges* record the pressure above or below the local atmospheric pressure, since they measure the difference in pressure of the liquid to which they are connected and that of surrounding air.
3. *Manometers* are defined as the devices used for measuring the pressure at a point in a fluid by balancing the column of fluid by the same or another column of liquid. They are treated as *standards* for pressure and differential pressure measurements because they are virtually free from any friction or hysteresis effects.
4. Some important *mechanical gauges* are :
   (*i*) Bourdon tube pressure gauge.
   (*ii*) Diaphragm gauge
   (*iii*) Vacuum gauge.
5. For the measurement of *very high pressures* (above 70000 MPa), the "*electrical gauges*" based on the principle of change of resistance with change of pressure are used.
6. The two basic methods of measurement of low pressure are :
   (*i*) *Direct methods.* The following devices belong to this category :
   — Spiral bourdon gauges;
   — Flat and corrugated diaphragms;
   — Capsules;
   — Manometers.
   (*ii*) *Indirect methods.* Devices included in this category are :
   — Mcleod gauge;
   — Thermal conductivity gauges;
   — Ionization gauges;
   — Radioactive vacuum meters.
7. The following are the laboratory standards for calibrating pressure measuring instruments :
   (*i*) Manometers.
   (*ii*) Dead weight testers.

### Temperature measurement

8. The *temperature* is a thermal state of a body which distinguishes a hot body from a cold body.
9. *Electrical sensors* are widely used particularly when automatic or remote recording is desired.
10. In a *bimetallic thermometer* differential expansion of bimetallic strips is used to indicate the temperature.
11. Gas filled thermometers are generally used for pressures below 35 bar.

12. The sources of static errors in filled systems are :
    (*i*) The ambient temperature effect.
    (*ii*) The heat effect.
    (*iii*) The barometric effect.
    (*iv*) The immersion effect.
13. A *thermistor* is a thermally sensitive variable resistor made of a ceramic-like semiconducting material.
14. A *thermocouple* in its simplest form, consists of two dissimilar metals or alloys which develop e.m.f. when the reference and measuring junctions are at different temperatures.
15. A device which measures the total intensity of radiation emitted from a body is called total *radiation pyrometer.*
16. An *optical pyrometer* works on the principle that matters glow above 480°C and the colour of visible radiation is proportional to the temperature of the glowing matter.

## Flow measurement

17. The flow measurement devices are classified as follows :
    (*i*) Rate meters
    (*ii*) Quantity meters.
18. A *venturimeter* is an instrument used to measure the rate of discharge in a pipeline and is often fixed permanently at different sections of the pipeline to know the discharge there.
19. A *Pitot tube* consists of a glass tube in the form of a 90° bend of short length open at both its ends. In a pitot-static tube both the static pressure as well as stagnation pressure can be measured.
20. A *rotameter* is a constant-pressure drop, variable area flow meter.
21. *Electromagnetic flow meters* are particularly suitable for the flow measurements of slurries, sludge and any electrically conducting fluid.
22. *Hot wire anemometers* are hot wire resistance transducers which are used for measurement of flow rates of fluids.
23. The working principles of *swirl meter* depends upon votex precession.
24. The working principle of *ultrasonic flow meters* is based on an apparent change in the velocity of propagation of sound waves in a fluid with a change in the velocity of the fluid flow.
25. *Lobed impeller flow meter* is primarily used for metering gases.

## Measurement of displacement, velocity, force, torque etc.

26. A *tachometer* is an instrument which either continuously indicates the value of rotary speed or continuously displays a reading of average speed over rapidly operated short intervals of time.
27. The acceleration of a moving body is generally measured by means of sensors called *accelerometers.*
28. An unknown force may measured by : (*i*) Scales and balances; (*ii*) Springs; (*iii*) Proving rings; (*iv*) Load cells.

**29.** *Load cells* are elastic devices that can be used for measurement of force through indirect methods *i.e.*, through the use of secondary transducers.

**30.** Torque may be measured by the following methods :

(*i*) Gravity balance method (*ii*) Mechanical torsion meter
(*iii*) Optical torsion meter (*iv*) Electrical torsion meter.
(*v*) Strain-gauge torsion meter.

**31.** A *dynamometer* is a device used to measure the torque being exerted along a rotating shaft so as to determine the shaft power input or output of power-generating, transmitting and absorbing machinery. The dynamometers are classified as :

(*i*) Absorption dynamometers. (*ii*) Transmission dynamometers.
(*iii*) Driving dynamometers.

## OBJECTIVE TYPE QUESTIONS

### Pressure measurement

**1.** Bellows type of gauges have the advantage that
(*a*) they can be used for low, medium and high pressures
(*b*) they can be used for measurement of dynamic pressures
(*c*) they do not need any temperature compensating devices
(*d*) drift and hysteresis allow their use in applications having ±2% of full span accuracy.

**2.** The disadvantages of strain gauge pressure transducers are:
(*a*) They have a poor frequency response.
(*b*) They do not have a continuous resolution.
(*c*) They cannot be powered by a.c. sources.
(*d*) They have a low output.

**3.** Which of the following types of Bourdon tube shape has a small tip travel and necessitates amplification?
(*a*) C-type (*b*) Spiral
(*c*) Helical shaped (*d*) All of these.

**4.** Manometers measure unknown pressure by
(*a*) measuring liquid levels
(*b*) measuring height of liquid columns
(*c*) balancing the unknown force produced by pressure against a known force
(*d*) noting the deflection of a pointer.

**5.** In a manometer, pressure sensitivity is
(*a*) $1/g\,\rho_m$ (*b*) $g\,\rho_m$
(*c*) $g\,\rho_m\,\rho_f$ (*d*) $1/g\,\rho_m\,\rho_f$
where $g$ = gravitational force and $\rho_m$ and $\rho_f$ are respectively specific gravities of manometric fluid and the transmitting fluid.

**6.** In thermal conductivity gauges, the major source of error is heat lost on account of
(*a*) radiation (*b*) conduction
(*c*) convection (*d*) both conduction and radiation.

7. The desirable property of a manometric fluid is
   (*a*) high viscosity (*b*) high co-efficient of thermal expansion
   (*c*) low vapour pressure (*d*) corrosiveness and sickness.
8. The advantages of manometers are their
   (*a*) higher accuracy, repeatability, wide range of filling liquids and their ability to work as primary standards
   (*b*) fragility and portability
   (*c*) ability to work in hazardous environments
   (*d*) ability to work under all kinds of atmospheric conditions and at all altitudes.
9. Piezometer is used for measuring :
   (*a*) gauge pressure (*b*) vacuum
   (*c*) absolute pressure
   (*d*) pressure difference between two fluids
10. A U-tube differential manometer is used inverted when
    (*a*) pressure difference is small (*b*) pressure difference is large
    (*c*) cannot be used inverted (*d*) any of these.
11. Capacitive type pressure transducers have
    (*a*) good frequency response, suitability for both dynamic and static measurements and continuous resolution
    (*b*) large temperature range
    (*c*) good thermal stability
    (*d*) high input impedance.
12. A typical piezoelectric pressure transducer has a range of
    (*a*) 0–10 kPa (*b*) 0–100 MPa
    (*c*) 0–700 kPa (*d*) 0–20 MPa.
13. In thermocouple gauges, the temperature of heated strip lies between
    (*a*) 0°C–100°C (*b*) 50°C–400°C
    (*c*) 200°C–1000°C (*d*) 10°C–50°C.
14. In order to avoid errors on account of changes in ambient temperature Pirani gauges
    (*a*) are enclosed in temperature controlled chambers
    (*b*) use one active and one dummy element
    (*c*) use bridge measuring circuits
    (*d*) none of the above.
15. Vacuum pressure is
    (*a*) equal to gauge pressure (*b*) equal to atmospheric pressure
    (*c*) lower than atmospheric pressure (*d*) equal to absolute pressure.
16. Velocity pressure is difference of
    (*a*) dynamic pressure and static pressure
    (*b*) stagnation pressure and static pressure
    (*c*) impact pressure and dynamic pressure
    (*d*) impact pressure and stagnation pressure.
17. Bridman gauges are used for measurement of
    (*a*) vacuum (*b*) medium pressure
    (*c*) high pressures (*d*) very high pressures.
18. Bridman gauges use
    (*a*) Monel and stainless steel (*b*) Manganin and gold chrome
    (*c*) Phosphor bronze (*d*) Beryllium copper.

19. A well-type manometer is used in preference to simple U-tube manometer to obtain
(*a*) better accuracy (*b*) better precision
(*c*) a constant zero (*d*) higher sensitivity.

20. Which of the following cannot be used for negative pressures?
(*a*) Piezometer (*b*) Pirani gauge
(*c*) U-tube manometer (*d*) Bourdon gauge.

21. Well type of manometers have
(*a*) poor accuracy (*b*) good accuracy
(*c*) low sensitivity
(*d*) high sensitivity as compared to U-tube manometers having two legs of the same area of cross-section.

22. In barometers the error caused by mercury vapour pressures is of the order of
(*a*) 1 kN/m$^2$ (*b*) 1 N/m$^2$
(*c*) 0.1 N/m$^2$ (*d*) 10 N/m$^2$.

23. The sensitivity of an ionization gauge is defined as
(*a*) $I_i/pI_e$ (*b*) $p/I_iI_e$
(*c*) $I_e/pI_i$ (*d*) none of the above.
where $I_i$ = ion current, $I_e$ = electron current and $p$ = gas pressure.

24. The accuracy of dead weight testers is affected by
(*a*) friction force between the piston and the cylinder
(*b*) uncertainty of the value of effective area of piston
(*c*) uncertainty of the value of gravitational constant
(*d*) all of the above.

25. Signal conditioning is carried out by the capillary tubes which convert gas pressure into a mercury height. This statement pertains to
(*a*) bourdon tube pressure gauge (*b*) pirani gauge
(*c*) Mcleod gauge (*d*) diaphragm pressure transducer.

26. The angular error in a Bourdon tube refers to
(*a*) pointer or hand becomes loose on the spindle
(*b*) gauge tends to give a progressively higher or lower reading
(*c*) gauge reads correctly both at maximum and minimum readings but gives an inaccurate reading at the mid point
(*d*) any of these.

27. A capillary tube when used for damping of pressure signals is called a
(*a*) low pass filter (*b*) high pass filter
(*c*) acoustical filter (*d*) digital filter.

28. McLeod gauge
(*a*) can be used for pressure below 0.1 × 10$^{-3}$ torr.
(*b*) gives continuous output.
(*c*) is sensitive to condensed vapours that may be present in the sample of the gas whose pressure is being measured.
(*d*) cannot be used as standard for vacuum measurements.

29. Inductive pressure transducers
(*a*) have poor dynamic response (*b*) have high sensitivity
(*c*) are sensitive to vibrations (*d*) none of the above.

**30.** Pressure transducers using LVDTs have

(*a*) poor dynamic response (*b*) linear response over a large range
(*c*) low sensitivity (*d*) all of the above.

## Flow measurement

**31.** Which of the following meter is used for measuring flow of clean fluids only?

(*a*) Ultrasonic flow meter (*b*) Turbine flow meter
(*c*) Laser doppler anemometer (*d*) Hot wire anemometer

**32.** Which of the following flow meters is capable of measuring the rate of flow as well as totalized flow?

(*a*) Nutating disc flow meter (*b*) Electromagnetic flow meter
(*c*) Orifice meter (*d*) Lobed impeller flow meter

**33.** Annubar tubes can be used for measurement of flow rates of

(*a*) liquids only (*b*) gases only
(*c*) liquids and gases (*d*) liquids, gases and steam.

**34.** The meter which is suitable for flow totalization is

(*a*) turbine meter (*b*) venturimeter
(*c*) ultrasonic flow meters (*d*) orifice plate.

**35.** The fluid flow between the electrodes of an electromagnetic flow meter generates an e.m.f. which is a function of

(*a*) dynamic pressure (*b*) discharge
(*c*) flow velocity (*d*) all of these.

**36.** A quantitative estimate of density changes can be made by

(*a*) interferometer (*b*) shadograph
(*c*) schlieren system (*d*) any of these.

**37.** The size of a venturimeter is expressed as 300 × 150 mm. It means that

(*a*) the diameter of the upstream pipe is 300 mm and that of downstream pipe 150 mm
(*b*) the diameter of the pipe is 300 mm and that of throat is 150 mm
(*c*) the diameter of pipe is 150 mm and that of throat is 300 mm
(*d*) none of the above.

**38.** Venturimeters have

(*a*) high head loss, low co-efficient of discharge and small size
(*b*) low head loss, low co-efficient of discharge and small size
(*c*) low head loss, low co-efficient of discharge and large size
(*d*) low head loss, high co-efficient of discharge and large size.

**39.** In an ultrasonic flow meter, the error caused by velocity of sound propagation in medium can be eliminated by

(*a*) measuring transit time.
(*b*) measuring phase shift
(*c*) measuring difference of transit times in the direction of flow and against the direction of flow
(*d*) measuring difference of frequency in the direction of flow and against the direction of flow.

**40.** Electromagnetic flow meters

(*a*) produce no head loss, can be used for measurement of flow of corrosive liquids and cannot be used for measurement of bidirectional flow.
(*b*) produce large heads loss, cannot be used for measurement of flow of corrosive liquids and cannot be used for measurement of bidirectional flow

(*c*) produce no head loss, cannot be used for measurement of flow of corrosive liquids and can be used for measurement of bidirectional flow.

(*d*) produce no head loss, can be used for measurement of flow of corrosive liquids and can be used for measurement of bidirectional flow.

**41.** A.C. excitation is used in electromagnetic flow meters because

(*a*) there is polarization, there are thermocouple effects and there is distortion of fluid velocity profile

(*b*) there is no polarization, there are no thermocouple effects and there is no distortion of fluid velocity profile

(*c*) there is polarization, there are no thermocouple effects and there is no distortion of fluid velocity profile

(*d*) there is no polarization, there are thermocouple effects and there is no distortion of fluid velocity profile.

**42.** A pitot-static tube measures

(*a*) static pressure (*b*) dynamic pressure

(*c*) total pressure

(*d*) difference between total and static pressure.

**43.** Which of the following instruments is a rate meter?

(*a*) Venturimeter (*b*) Hot wire anemometer

(*c*) Nutating disk meter (*d*) Current meter.

**44.** A flow meter that measures flow rates which are independent of density is

(*a*) rotameter (*b*) electromagnetic flow meter

(*c*) venturimeter (*d*) orifice meter.

**45.** Rotameter is a

(*a*) drag force flow meter (*b*) variable area flow meter

(*c*) variable head flow meter (*d*) rotating propeller type flow meter.

## Temperature measurement

**46.** Which of the following units are employed where it is not possible to have a physical contact with the system whose temperature is to be measured?

(*a*) Thermocouples (*b*) Resistance thermometers

(*c*) Radiation pyrometers (*d*) Any of these.

**47.** A parallel combination of thermocouples is used for the measurement of

(*a*) small temperature differences between the two junctions

(*b*) large temperature differences between the two junctions

(*c*) average temperature of a number of points

(*d*) all of these.

**48.** Thermopile is

(*a*) a combination of a number of thermocouples connected in series

(*b*) a combination of a number of thermocouples connected in parallel

(*c*) a combination of a number of thermocouples some of which are connected in series and some in parallel

(*d*) none of the above.

**49.** Reference junction compensation in thermocouples can be provided through use of

(*a*) hardware only (*b*) software only

(*c*) both hardware and software (*d*) none of these.

**50.** In pressure thermometers when mercury is used, the bulb and capillary should be made of

(*a*) copper (*b*) alloys of copper
(*c*) stainless steel (*d*) none of these.

**51.** Partial immersion liquid-in-glass thermometers give errors because

(*a*) thermometer is not fully immersed
(*b*) the temperature of the emergent portion is higher than that of the immersed portion
(*c*) the temperature of the emergent portion is lower than that of the immersed portion
(*d*) the temperature of the emergent portion is different than that of the immersed portion.

**52.** Recording is not possible with

(*a*) liquid-in-glass thermometers (*b*) thermocouples
(*c*) filled in system thermometers (*d*) all of these.

**53.** The emf of a thermocouple with junctions at $t_1$ and $t_3$ is the algebraic sum of the emf's of two couples of the same material with junctions at $t_1$ and $t_2$, and $t_2$ and $t_3$ respectively. This statement is known as

(*a*) law of homogeneous circuit
(*b*) law of successive or intermediate temperatures
(*c*) law of intermediate metals
(*d*) none of these.

**54.** The advantages of thermistor are :

(*a*) Resistance of thermistor is high and therefore error due to resistance of leads is negligible.
(*b*) Errors due to self-heating are small.
(*c*) Its sensitivity is about $10^6$ as high as that of platinum resistance thermometers.
(*d*) all of the above.

**55.** Law of intermediate metals in thermocouples allows them to

(*a*) use reference junction compensation
(*b*) use meters for measurement of emf without disturbing the circuit conditions
(*c*) use extension wires of materials other than the one used for making thermocouples
(*d*) both (*b*) and (*c*).

**56.** If the temperature of a radiating body is 1500°C, the wavelength at which maximum radiant energy occurs is

(*a*) 2.8 μm (*b*) 1.63 μm
(*c*) 1 μm (*d*) 20.8 μm.

**57.** Radiation pyrometers are used in the temperature range of :

(*a*) 0–500°C (*b*) 500–1000°C
(*c*) –250 to 500°C (*d*) 1200–2500°C.

**58.** Which metal/non-metal has the highest temperature range?

(*a*) Semiconductors (*b*) Nickel
(*c*) Copper (*d*) Platinum.

**59.** Which arrangement has the manual null-balance range?

(*a*) Optical pyrometer (*b*) Resistance thermometer
(*c*) Liquid-in-glass thermometer (*d*) None of these.

**60.** For the same temperature range, the volume of bulb of a gas-filled thermometer is

(*a*) greater than that of a liquid-filled or a vapour-filled thermometer
(*b*) smaller than that of a liquid-filled or a vapour filled thermometer

(*c*) greater than that of a liquid-filled thermometer but lesser than that of vapour-filled thermometer

(*d*) lesser than that of a liquid-filled thermometer but greater than that of vapour-filled thermometer.

**61.** In filled systems used for temperature measurement, full compensation is used for compensating the effect of

(*a*) changes in ambient temperature, head and barometric pressure

(*b*) changes in ambient temperature where the receiving element is located

(*c*) changes in ambient temperature where both the receiving element and the capillary tube are located

(*d*) all of the above.

**62.** The electrical output from a thermocouple circuit is detected by using

(*a*) Wheatstone bridge (*b*) current sensitive device

(*c*) voltage-balancing circuit (*d*) none of these.

**63.** Which arrangement has the capability to measure the temperature of an object which may be either stationary or moving?

(*a*) Thermocouples. (*b*) Thermistors.

(*c*) Total radiation pyrometer. (*d*) Filled-in-system thermometers.

**64.** Self-compensating capillary tube is used in the filled-in-system thermometer to eliminate or minimise

(*a*) elevation effect (*b*) immersion effect

(*c*) barometric effect (*d*) temperature effect.

**65.** Mercury in used in liquid-filled systems as it gives

(*a*) wide temperature range

(*b*) high sensitivity

(*c*) wide temperature range and approximately linear scale

(*d*) wide temperature range and high sensitivity.

## Measurement of displacement, velocity, force etc.

**66.** Elastic elements used for measurement of force give

(*a*) high sensitivity and slow response if they are stiff

(*b*) low sensitivity and fast response if they are stiff

(*c*) low sensitivity and slow response if they are stiff

(*d*) none of the above.

**67.** Piezoelectric type of load cells can be used for measurement of

(*a*) dynamic forces only.

(*b*) dynamic forces and static forces provided the load cells have a small time constant.

(*c*) dynamic forces and static forces provided that the load cells have a large time constant.

(*d*) none of the above.

**68.** The average speed measurements are given by

(*a*) centrifugal tachometer (*b*) drag cup tachometer

(*c*) revolution counter and a timer (*d*) stroboscope.

**69.** Torque of rotating shafts is measured by mounting a strain gauge bridge on the shaft. The supply to the bridge circuit is given to it and the output from it taken from by using

(*a*) only slip rings (*b*) only rotary transformers

(*c*) slip rings, rotary transformers, telemetry equipment

(*d*) none of the above.

**70.** A sink is

(*a*) a transmission dynamometer (*b*) a driving dynamometer
(*c*) an absorption dynamometer (*d*) any dynamometer.

**71.** The stroboscopic method of speed measurement has the advantage that

(*a*) the method is simple and straight forward
(*b*) the multiples of angular speed can be measured
(*c*) a stationary image can be observed
(*d*) there is no need of any physical contact between the instrument and the rotating shaft.

**72.** The seismic mass transducer for the measurement of acceleration and vibration is a

(*a*) zero order instrument (*b*) first order instrument
(*c*) second order instrument (*d*) none of these.

**73.** Hydraulic load cells are available which have a maximum capacity of

(*a*) 10 kN (*b*) 50 kN
(*c*) 10 MN (*d*) 50 MN.

**74.** Pendulum type of scales used for measurement of force work on the principle of

(*a*) comparing the unknown force with a known force
(*b*) converting the unknown force to power and comparing it with known power
(*c*) converting the unknown force to torque and comparing it with a known torque.
(*d*) all of the above.

**75.** Tachometers are used to measure

(*a*) displacement (*b*) angular velocity
(*c*) vibration (*d*) time.

**76.** The speed of a sealed compressor unit can be measured by a

(*a*) stroboscope (*b*) vibrating reed tachometer
(*c*) capacitive pick-up tachometer (*d*) tachoscope.

## ANSWERS

| | | | | | | |
|---|---|---|---|---|---|---|
| 1. (*a*) | 2. (*d*) | 3. (*a*) | 4. (*c*) | 5. (*a*) | 6. (*a*) | 7. (*c*) |
| 8. (*a*) | 9. (*a*) | 10. (*a*) | 11. (*a*) | 12. (*d*) | 13. (*b*) | 14. (*b*) |
| 15. (*c*) | 16. (*b*) | 17. (*d*) | 18. (*b*) | 19. (*d*) | 20. (*a*) | 21. (*d*) |
| 22. (*b*) | 23. (*c*) | 24. (*d*) | 25. (*c*) | 26. (*c*) | 27. (*c*) | 28. (*c*) |
| 29. (*b*) | 30. (*a*) | 31. (*d*) | 32. (*d*) | 33. (*d*) | 34. (*a*) | 35. (*c*) |
| 36. (*a*) | 37. (*b*) | 38. (*d*) | 39. (*d*) | 40. (*d*) | 41. (*b*) | 42. (*d*) |
| 43. (*a*) | 44. (*b*) | 45. (*b*) | 46. (*c*) | 47. (*c*) | 48. (*a*) | 49. (*c*) |
| 50. (*c*) | 51. (*d*) | 52. (*a*) | 53. (*b*) | 54. (*d*) | 55. (*d*) | 56. (*b*) |
| 57. (*d*) | 58. (*d*) | 59. (*a*) | 60. (*a*) | 61. (*c*) | 62. (*c*) | 63. (*c*) |
| 64. (*d*) | 65. (*c*) | 66. (*b*) | 67. (*c*) | 68. (*c*) | 69. (*c*) | 70. (*c*) |
| 71. (*d*) | 72. (*c*) | 73. (*d*) | 74. (*c*) | 75. (*b*) | 76. (*b*) | |

## THEORETICAL QUESTIONS

### Pressure measurement

**1.** Define the following terms :

(*i*) Pressure (*ii*) Atmospheric pressure

(*iii*) Gauge pressure (*iv*) Absolute pressure

(*v*) Static pressure (*vi*) Total or stagnation pressure.

2. What are the units of pressure?
3. How are pressure measurement instruments classified?
4. What are "manometers"? How are these classified?
5. Why are manometers treated as standards for pressure and differential pressure measurements?
6. How does a differential manometer differ from a simple manometer?
7. List the advantages and limitations of manometers.
8. What are the desirable characteristics of a manometer liquid?
9. Explain briefly the following manometric liquids :
   (*i*) Mercury; (*ii*) Water; (*iii*) Alcohol and kerosene.
10. On what principle do the "Bell gauges" operate? Name the different types of Bell gauges used in practice.
11. Explain briefly any two of the following mechanical gauges :
    (*i*) Bourdon tube pressure gauge (*ii*) Diaphragm gauge
    (*iii*) Vacuum gauge.
12. Explain briefly the following types of pressure elements :
    (*i*) Bourdon tube (*ii*) Diaphragm
    (*iii*) Bellows.
13. State the advantages and disadvantages of "Bourdon tube element."
14. What are the advantages and disadvantages of "Diaphragm elements"?
15. List the advantages and disadvantages of bellow gauges/elements.
16. Explain briefly any three of the following pressure transducers :
    (*i*) Resistance type pressure transducers; (*ii*) Pressure voltage type;
    (*iii*) Capacitance type; (*iv*) Piezoelectric type;
    (*v*) Electromagnetic type.
17. How are very high pressures measured? Explain briefly with a neat and clean sketch the construction and working of a Baridgman gauge used for measuring high pressures.
18. What do you mean by low pressures or vacuums?
19. Draw diagrams to show how LVDTs can be used with bellows, elements and Bourdon tubes for measurement of pressure. Give their advantages and disadvantages.
20. Explain how capacitive type of transducers can be used for pressure measurement.
21. Describe the applications of piezoelectric transducers for measurement of pressure. List their advantages and disadvantages.
22. Describe the construction, working and theory of McLeod gauge for measurement of vacuum. List its advantages and disadvantages.
23. Explain briefly the construction and working of a thermocouple vacuum gauge. State its advantages and disadvantages.
24. Describe briefly the construction and working of a Pirani gauge. Give its advantages and disadvantages.
25. Give the construction and working of a hot-filament ionization gauge. List its advantages and disadvantages.
26. (*a*) Explain how calibration of pressure measuring equipment is carried out.
    (*b*) Describe the construction and working of a "Dead Weight Tester". State the factors which affect the accuracy of dead weight testers.

## Temperature measurement

27. Define the term "temperature".
28. How are temperature measuring instruments classified?
29. Explain briefly the following expansion thermometers :
    (*i*) Liquid-in-glass thermometers.
    (*ii*) Bimetallic thermometers.
30. State the advantages, disadvantages and applications of bimetallic thermometers.
31. Discuss briefly any two of the following pressure thermometers :
    (*i*) Vapour pressure thermometer.
    (*ii*) Liquid-filled thermometer.
    (*iii*) Gas-filled thermometer.
32. State the advantages and limitations of filled-systems.
33. Explain briefly the following sources of static errors in filled systems :
    (*i*) The ambient temperature effect.
    (*ii*) The head effect.
    (*iii*) The barometric effect.
    (*iv*) The immersion effect.
34. Write a short note on "Response speed of filled-systems".
35. Explain briefly the construction and working of a resistance thermometer.
36. What are the advantages and disadvantages of resistance thermometers?
37. What are thermistors? What are their advantages?
38. "Give a comparison between "thermistors" and "metal resistors".
39. List the factors which should be considered while selecting resistance thermometers.
40. What is a "Seebeck effect"?
41. Explain with a neat diagram the construction and working of a thermoelectric pyrometer.
42. What are the desirable properties of thermoelectric materials?
43. State the advantages and disadvantages of thermocouples.
44. Explain briefly the working of a radiation pyrometer.
45. List the advantages and disadvantages of radiation pyrometers.
46. Describe with a neat sketch the working of an optical pyrometer.
47. What are the advantages and disadvantages of an optical pyrometer?
48. State the precautions which should be taken while making temperature measurements.
49. Explain the factors which influence the response of a temperature sensing device.

## Flow measurement

50. List the applications of flow measurements.
51. What are the different factors which influence the choice of method used for the measurement of flow?
52. How are the flow measurements classified?
53. What is the difference between the "Rate meters" and "Quantity meters"?
54. What is the significance of term "inferential" as applied to flow meters?
55. Certain meters are known as variable head meters. Explain clearly what is meant by the designation variable head.
56. Describe briefly the following :
    (*i*) Rotameter; (*ii*) Elbowmeter.
57. Explain with a neat sketch the working of any electromagnetic flow meter.
58. What are hot wire anemometers?

## Measurement of displacement, velocity, force etc.

**59.** How can linear displacement be measured? Explain briefly.

**60.** Explain briefly how small displacement can be measured by a linear or differential transformers.

**61.** Describe briefly with a neat sketch how angular displacement is measured by a variable reluctance transducer.

**62.** Name the transducers which are used for the measurement of linear velocity.

**63.** Explain with a neat diagram the method of measuring linear velocity with the use of moving magnet type transducer.

**64.** How is angular velocity measured?

**65.** What is a tachometer?

**66.** How are tachometers classified?

**67.** Why are electrical tachometers preferred to mechanical tachometers?

**68.** Explain briefly any two of the following electrical tachometers :

(*i*) D.C. tachometer generator (*ii*) A.C. tachometer generator

(*iii*) Photoelectric tachometer.

**69.** Give the description of an eddy current tachometer, with a neat sketch.

**70.** Describe briefly a capacitive pick-up tachometer.

**71.** What are accelerometers?

**72.** Explain briefly the following :

(*i*) Piezoelectric accelerometer (*ii*) Seismic accelerometer.

**73.** How is an unknown force measured?

**74.** What is a load cell?

**75.** Describe briefly the construction and working of strain gauge load cell. State its field of application as well.

**76.** What is the principle of working of a piezoelectric load cell?

**77.** What is 'torque'? How can it be measured?

**78.** Explain briefly any two of the following torque measurement methods :

(*i*) Gravity balance method. (*ii*) Mechanical torsion meter.

(*iii*) Electrical torsion meter.

**79.** Describe briefly how a strain gauge torsion meter is used to measure the torque.

**80.** What is a dynamometer?

**81.** How are dynamometers classified?

**82.** Explain briefly the following :

(*i*) Prony brake dynamometer. (*ii*) Rope brake dynamometer.

**83.** Describe briefly any two of the following dynamometers :

(*i*) Hydraulic dynamometer. (*ii*) Eddy current dynamometer.

(*iii*) Belt transmission dynamometer.

**84.** Explain with a neat diagram the construction and working of a D.C. dynamometer.

CHAPTER

# 13 Microprocessors and Microcontrollers

**13.1 Computers-Brief description:** History and development of computers – Definition of a computer – Characteristics of a computer – Classification of computers – Analog computers – Digital computers – Differences between analog and digital computers – Block diagram of a digital computer – Rating of chips – Computer peripherals – Storage devices – Hardware, software and liveware – Translators – Computer languages – Computer programming process for writing programs – Computing elements of analog computers; **13.2 Microprocessors:** Evolution of microprocessor (μP) – Microprocessor systems – Intel 8085 microprocessor – Microprocessor programming – Digital logic – Types of microprocessors – Processing speed of a processor; **13.3 Organisation of Microcomputers; 13.4 Microcontrollers:** Introduction – Characteristics of microcontrollers – Applications – Classification of microcontrollers – Description of a typical full-featured microcontroller – 8051 microcontroller – Microchip controllers – Comparision of microprocessors and microcontrollers **13.5 Programmable Logic Controllers:** Introduction – Advantages of a PLC – Selection of a PLC – Uses of PLCs – Special features – Architecture/ basic internal structure – Input/output processing and programming. **13.6 Microprocessor-based instrument system – Intelligent instrument**. *Highlights – Objective Type Questions – Theoretical Questions.*

## 13.1 COMPUTER—BRIEF DESCRIPTION

### 13.1.1. History and Development of Computers

- Charles Babbage (an English Mathematician) was responsible for conceiving the concept of the Modern computer, and is called "Father of Computers".
- He designed the early computer called *"Difference Engine"* in the year 1822, with which reliable tables could be produced. In 1833 he improved upon the machine and put forth new idea of *"Analytical Engine"* which could perform the basic arithmetic functions automatically. In this machine punched cards were used as input/output devices for basic input and output.

The concept of use of punched cards was developed further by Horman Hollerith in the year 1889.

- Leonards Torres demonstrated a *digital calculating machine* in Paris in 1920.
- In 1944 Prof. Howard Aiken (Howard University) developed Electromechanical calculators known as Mark-I. This machine could handle about a sequence of 5 arithmetic operations by using memory for previous results.
- On the basis of research done for U.S. army during the World War-II in 1946, the first electronic computer, ENiAC (Electric Numerical Integer and Computer) was designed in 1946. This computer was about 15 metres long and 2 metre high and weighed about 50 tons. It consumed about 200 kW power. This machine did not have any facility for storing program.

- In 1949, the *concept of stored program was adopted.*
- In 1951, was introduced the commercial version of stored program computer UNIVAC-(Universal Automatic Computer)—the first digital computer.

**Generations of computers:**

**First generation** ........ Developed during the years 1951–1959.

- These computers are *"based" on* ***"Vacuum tubes"***.
- Very slow in operation ($10^3$ operations/sec.)
- Big in size and unreliable.
- Short span of life.
- Frequent breakdowns.
- High power consumption and great amount of heat generation.
- Small primitive memories and no auxiliary storage.
- Limited programming capabilities.

*Examples:* UNIVAC-1 and IBM 650.

**Second generation.** Developed during the years 1960–1965.

- These computers are *based on* **"Transistors"**.
- Faster in operation, comparatively ($10^6$ operations/sec.)
- Smaller in size.
- More reliable.
- Consume less power.
- Generate less heat than vacuum tubes.
- Auxiliary memory in the form of magnetic tape was introduced.

*Examples:* UNIVAC 1107, IBM 7090, CDC 1604, Honeywell 800 etc.

**Third generation.** .......... Introduced during 1965–1970, also being used presently.

- These computers are based on ***"Integrated circuits"***, based on silicon technology.
- Much more smaller in size.
- More reliable.
- Faster in operation ($10^9$ operation/sec).
- Less expensive.
- Employ higher capacity internal storage.
- Wide range of peripheral used.
- Make use of new concepts like *multi-programming, multi-processing, high level languages.*

*Examples:* IBM-360/370, Honeywell 6000.

**Fourth generation** ........... Introduced in 70s.

- These computers are based on VLSI (Very large scale integration) chips and microprocessors chips.
- Possess high processing power.
- Low maintenance.
- Faster in operation.
- High reliability.
- Very low power consumption.
- Less expensive.
- Small size.

This generation also includes the following :

— Microcomputers;

— Office automation systems;

— Distributed processing systems.

**Fifth generation** .......... Introduced during late 1990s.

- These computers use optic fibre technology to handle *Artificial Intelligence, Expert Systems, Robotics* etc.
- Possess very high processing speeds.
- More reliable.

### 13.1.2. Definition of a Computer

A **computer** is *a machine that processes data according to set of instructions stored within the machine.*

- It *receives data as input, processes the data, i.e.,* performs arithmetic and logical operations on the same and *produces output in the desired form on output device as per the instructions coded in the program.*
- The processing function of the computer is *directed by the stored program, a set of codes instructions stored in the memory unit, which guides the sequence of steps to be followed during processing.*

### 13.1.3. Characteristics of a Computer

The following are the *characteristics* which make a computer an indispensable unit :

1. Speed
2. Consistency
3. Accuracy
4. Flexibility
5. Reliability
6. Large storage capacity
7. Automatic operation
8. Diligent
9. No emotional ego and psychological problems.

**Limitations of a computer :**

A computer entails the following *limitations* :

1. It does not work on itself, a set of instructions is required for its operation.
2. It cannot take decision on its own, it has to be programmed as per requirements.
3. It is not intelligent, it has to be instructed in detail for the performance of each and every task.
4. It cannot learn by experience, as human beings do.

### 13.1.4. Classification of Computers

The computers may be *classified* as follows:

**1. On the basis of the type of data :**

(*i*) *Analog computers* (These computers process the data in analog form).

(*ii*) *Digital computers* (These computers process the data in digital form).

2. **On the basis of the size and capacity :**
   (*i*) *Microcomputers*
   (*ii*) *Mini computers*
   (*iii*) *Main frame*
   (*iv*) *Super computers.*
3. **On the basis of the type of application :**
   (*i*) *Special purpose computers*
   (*ii*) *General purpose computers.*
4. **On the basis of the number of users :**
   (*i*) *Single user computers*
   (*ii*) *Multi-user computers.*
5. **On the basis of the number of processors :**
   (*i*) *Single processor computers.*
   (*ii*) *Multiprocessor computers.*
6. **On the basis of the type of instructions set :**
   (*i*) *Complex Instruction Set Computers* (CISC).
   (*ii*) *Reduced Instruction Set Computers* (RISC).

### 13.1.5. Analog Computers

- The *principle of operation* of analog computers is *to create a physical analog of mathematical problems.*
- Measure physical variables *continuously.*
- Use signals as input (which may be supplied by devices like barometers, speedometers, thermometers etc.).
- The *result* given by an analog computer is *not very precise, accurate and consistent.*
- These computers find limited applications.

***Example:*** Speedometer of a vehicle (here speed varies continuously).

### 13.1.6. Digital Computers

- The digital computers *accept digits and alphabets as input.*
- Receive data in the form of discrete signals representing ON (high) or OFF (low) voltage.
- The data input can be represented as sets of 0's and 1's representing low and high respectively.
- The digital computers convert data into discrete form before operating on it.
- The *most important characteristic* of a digital computer is that it is general purpose device *capable of being used in a number of different applications.* By changing the stored program, the same machine can be used to implement totally different tasks.

***Example:*** Digital watches.

Digital computers may be further classified based upon : (*i*) Purpose of use (*e.g.*, General purpose, special purpose); (*ii*) Size and capabilities.

On the basis of *size and capabilities,* the digital computers are classified as :

1. Super computers.
2. Mainframe computers.

3. Medium sized computers.
4. Mini computers.
5. Micro computers.

**1. Super computers :**

- These computers are the fastest (speed of calculations upto 1.2 billion instructions per second) and have very high processing speeds.
- They are very large in size and most powerful and costliest.
- Their fields of applications include processing weather data, geological data, genetic engineering etc.
- Word length : 64 bits and more.
- These computers can receive input from more than 1000 individual work stations.

*Example:* PARAM (a super computer developed in India).

**2. Main frame computers :**

- These are large scale general purpose computer systems.
- Possess large storage capacities in several million words.
- Secondary storage directly accessible—of the order of several billion words.
- Can support a large number of terminals (upto 100 or more).
- Faster in operation (100 million instructions/sec. approx).
- Accept all types of high level languages.
- Word length—16 or 32 or 64 bits.

**3. Medium sized computers :**

- Mini versions of mainframe computers.
- They have smaller power than mainframes.
- Processing speeds relatively high with support for about 200 remote systems.

**4. Mini computers :**

- These are general purpose computer systems.
- Reduced storage capacity and performance (as compared to main frame).
- CPU speed—few million instructions/sec.
- Word length—16 or 32 bits.
- Can accept all types of high level languages.
- Can support upto about 20 terminals.

*Note:* In view of fast development in electronics it is difficult to draw a line of demarcation between small main frame computers and large mini-computers.

**5. Micro-computers :**

- These are small sizer computers *utilising microprocessors.* These are popularly known as personal computer (PC).
- CPU is usually contained on one chip.
- Possess low storage capacity (maximum being 256 K words).
- Slow in operation ($10^5$ instructions/sec.).

- Usually provided with *video display unit, floppy drive and printer*. Some micro-computers can support hard disc also.
- Maximum word length is 16 bits; however most of these use 8-bit words.
- Commonly used language—BASIC. However these computers can also accept other high level languages viz., PASCAL, FORTRAN, etc.

***Note:*** **A single chip microcomputer* consists of a single chip on which the central processing unit, input/output and memory units are integrated. This is used for *industrial applications* and also in *product calculators.*

*Its *advantage is the reduction in cost and size, increase in performance and reliability.*

### 13.1.7. Differences between Analog and Digital Computers

The differences between analog and digital computers are given in Table 13.1.

**Table 13.1. Differences between Analog and Digital Computers**

| S. No. | Digital computer | Analog computer |
|---|---|---|
| 1. | It performs calculations by counting and thus counts directly. It is the most *versatile machine.* | It processes work electronically by *analogy*. It does not produce number but produces its results *in the form of graph*. It is *more efficient in continuous calculations.* |
| 2. | It operates on inputs which are on-off type (being digits 0 or 1) and its output is in the form of signals. | It accepts variable electrical signals (analog values) as inputs, and its output is also in the form of analog electrical signals. |
| 3. | It is based on counting operation. | It operates by measuring analog signals. |

These days *digital computers* are being *widely used.*

A *hybrid computer* is combination of both analog and digital computers. It is used for *simulation applications.*

### 13.1.8. Block Diagram of a Digital Computer

Figure 13.1 shows a block diagram of a typical digital computer.

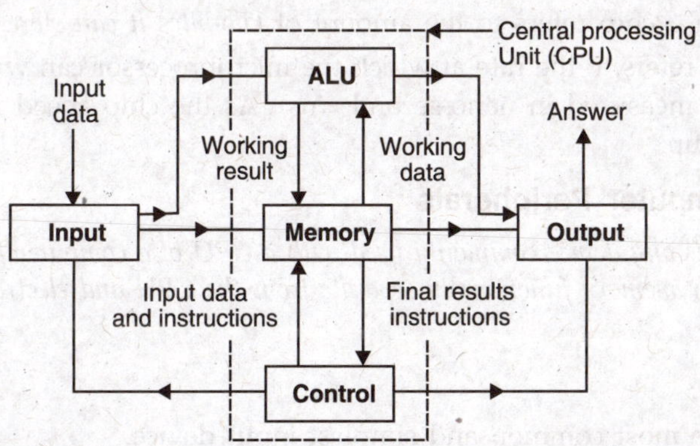


**Fig. 13.1.** Block diagram of a digital computer.

The following are the *five basics elements* of a computer system :

1. **Input :**
   - The data and instructions are first recorded on a machine readable medium, like punched card, and then fed into the computer via a device that codes them in a manner which is suited to conversion into electrical pulses before entering memory.
   - The input supplies data to the computer in digital (binary) form.
2. **Memory :**
   - The memory section within the computer is where data are stored or memorized.
   - Problem to be solved, inputs for the problem, a program of instruction, working data, intermediate results and final results are *types of memory data.*
   - The memory section holds data between high speed computer operation and slower input and output devices.
3. **Arithmetic Logic Unit (ALU) :**
   - ALU performs necessary arithmetical operations on the data and ensures that instructions are obeyed.
   - It also performs *logical operations.*
   - The ALU combined with control unit is called *Central Processing Unit (CPU).*
4. **Control Unit :**
   - It fetches instructions from main memory, interprets them and issues the necessary signals to the components making up the system.
   - It issues commands for all hardware operations necessary in obeying instructions.
5. **Output :**
   - The output is the path for data out of the computer and may include devices for reading out answers.

### 13.1.9. Rating of Chips

Chips are rated in terms of their *capacity* and *speed.*

- *Capacity* of a chip refers to the amount of *kilo-bites it can store.*
- *Chip speed* refers to the rate at which the microprocessor can write to the chip. It is usually measured in nano-seconds (ns). As the chip speed increases, its cost also goes up.

### 13.1.10. Computer Peripherals

A **peripheral** *is any device commonly used with a CPU of a computer for input or output of information or for memory functionally separate from the CPU and electronically detachable.*

**Input devices :**

1. *Keyboard :*
   - It is the most common and simplest input device.
   - It is merely a collection of momentary switches. The outputs of the key switches are fed to electronic circuitry known as *keyboard encodes* which converts them into binary coded values. The values are then fed into the computer which

interprets the key which was pressed. Thus the function of the key changes with the type of work we are doing.

2. ***Mouse :***
   - It is a pointing device and its size is about the size of palm.
   - It is a hand-held device that *controls a pointer on the screen.*
   - It rolls on a small ball. A mouse has one or more buttons on the top. When the user moves the mouse over a flat surface, the screen cursor moves in the direction of the mouse movement.
3. ***Digitizer (or Graphic tablet) :***
   - It is similar to light pen.
   - It consists of a glass plate on which digitizing tablet is moved.
   - It is used for fine drawing works and for image manipulation applications such as Auto-cad.
4. ***Optical Mark Reader (OMR) :***
   - OMR is being used for reading the answer sheet by means of light. It can read upto 150 documents per minute; when on-line with respect to the computer system, can read upto 2000 documents per minute.
   - OMR can also be used for such applications as *order writing payroll, inventory control, insurance, questionnaires* etc.
5. ***Magnetic Ink Character Reader (MICR) :***
   - MICR uses a special ink to print character. These characters can be decoded by special magnetic devices.
   - This system is employed *by banks for processing cheques.*
6. ***Scanner :***
   - It is used for getting existing graphical image (like photographs, mats, etc.) into computer.
   - Once the graphical image is scanned and brought into the computer user can include them into documents or can edit them.
7. ***Light pen :***
   - It consists of a pen like device and photoelectric cell.
   - It is used to draw pictures on the screen.
   - When light pen is in contact with screen, the electron beam activates the photoelectic cell which in turn sends signals into the computer and ultimately a mark is made on the screen where light pen contacted the screen.
8. ***Joy-stick :***
   - It is screen-pointing device.
   - A stick is present with a button at the top. It can be held in the hand and bent in any one of the four directions. As the stick is moved, the action on the screen changes in the appropriate direction.
   - A joy-stick is *widely used for playing computer games.*
9. ***Touch screen :***
   - The touch screen technique involves beam and ultrasonic waves.

- By using touch screen we can issue command to the computer by touching the screen.
- Limited amount of data can be entered via a terminal or a microcomputer that has a touch screen.

**10.** ***Compact Disk Read Only Memory (CDROM) :***

- It is a 120 mm diameter disc with a polycarbonate subtrate, a reflective metalised layer on one side, with a protective lacquer finish.
- Here a laser beam is used to burn a small hole or pit which represent binary '1'. The absence of pit represents '0'. In this way digital information is stored on the disc in large quantities (in Giga Bytes).

**11.** ***Voice Recognition System or Voice Synthesizer :***

- Voice recognition techniques, along with several other techniques, are used to convert the voice signals to appropriate words and device the correct meanings of words. There has been a limited success in this area and these days devices are available commercially to recognize and interpret human voices.

**Output devices :**

**1.** ***Printer :***

- A printer is *device that produces copies of text and graphics on paper.*
- The printers are classified/categorised as follows:

A. *Impact printers :*

(*i*) Solid font

(*ii*) Dot matrix.

B. *Non-impact printers :*

(*i*) Thermal printer (*ii*) Inkjet printer

(*iii*) Laser printer (*iv*) Electrographic printer

(*v*) Electrostatic printer.

**2.** ***Plotters :***

- Plotters are those devices which *reproduce drawings using pens that are attached to movable arms.*
- Platting in different colours is possible.

**3.** ***Monitors or Visual Display Unit (VDU) :***

- A monitor is a television like device, which is used to display information, output and input data.
- It consists of a cathade ray tube (CRT), on which the information is displayed. When the user processes any key on the keyboard, the keyboard encoder generates code of that key which is depressed. This code is then fed to the computer; from there VDU system takes that code and displays it on the screen.

### 13.1.11. Storage Devices

The memory devices in a memory unit (which stores the data, instructions and intermediate results) may be of the following types :

**1. Internal storage device** ..... also known as *main* or *primary* storage device.

The primary storage devices currently in use in computers are :

(*i*) Magnetic core memory device

(*ii*) Thin film memory device

(*iii*) Thin rod memory device

(*iv*) Plated wire memory device.

**2. Auxiliary storage device**

The popular *secondary memory devices* are :

(*i*) Magnetic tape drive

(*ii*) Magnetic disk drive

(*iii*) Magnetic drum

(*iv*) Floppy disk

(*v*) Winchester disk.

**Methods of Input to Backing Stores :**

The following methods are generally used :

(*i*) Key-to-tape

(*ii*) Key-to-cassette/cartridge

(*iii*) Key-to-disk/diskette

**Memory.** The memory is used to store information/data so that it can be retrieved whenever required. There are mainly two types of memories :

1. Primary memory
2. Secondary memory.

**1. Primary memory :**

- It is also known as core memory, main memory, RAM (Random Access Memory).
- It is constructed using purely semiconductor devices, data is stored in the form voltages.
- It is a volatile memory whereas ROM (Read Only Memories) are non-volatile memories.

**2. Secondary memory :**

- Secondary memory, also known as *auxiliary memory*, is used to store large volumes of data.
- Data is stored in the form of magnetic energy and can be stored (in the secondary memory) for large periods.

**Difference between Read Only Memory (ROM) and Random Access Memory (RAM).**

**ROM :**

- As the name implies ROM is a memory unit that performs the read operation only; it *does not have a write capability.* This implies that the binary *information stored in a ROM is made permanent during the hardware production of the unit and cannot be altered by writing different words into it.* Whereas a *RAM is a general-purpose device whose contents can be altered during the computational process.*
- ROM is a type of memory chip that we can read only and we cannot write on it.
- ROM provides permanent storage for program instructions.

- The most important ROM chip in any computer is ROM BIOS (Basic Input/Output System).
- ROM is most oftenly used in microprocessors that always execute the same program such as boot strap loader.

*Disadvantage of ROM :*

(*i*) A ROM is prepared by the manufacturer and cannot be altered once the chip has been made.

(*ii*) It is slow.

The *ROM memory* may be **classified** as follows:

(*i*) *Programmable Read Only Memory (PROM).* Here, the information can be altered, but *not as easily as in the ordinary memory.* Once the operations to be performed have been written into a PROM chip, they are permanent and cannot be changed.

(*ii*) *Erasable Programmable Read Only Memory (EPROM).* This type of ROM can be erased and programmed with the help of special equipment. It has a window at its top, which if exposed to ultraviolet light, allows data to be erased.

(*iii*) *Electrically Erasable Programmable ROM (EEPROM).* In order to erase and reprogramme this type of ROM, it is required to be removed from the socket.

(*iv*) *Flash EPROM.* It is the latest type of ROM. A manufacturer can make changes to the flash EPROM while it remains in the PC, by running a special program.

**RAM :**

- This memory is so named since memory registers can be accessed for information transfer as required.
- RAM chip is made with Metal Oxide Semiconductor (MOS).
- RAM chips may be classified as :
  (*i*) *Dynamic Ram:* It provides volatile storage (*i.e.,* the data stored is lost in the event of a power failure).
  (*ii*) *Static RAM:* These chips are more complicated and take up more space for a given storage capacity than dynamic RMA chips. These chips are also volatile in nature but as long as they are supplied with power, they need not require special regenerator circuits to retain the stored data.
- *Static RAM chips are thus used in specialised applications while Dynamic RAM chips are used in the primary locations.*
- Owing to the volatile nature of these storage elements, a back up *Uninterrupted Power System (UPS) is often installed along with larger computer systems.*

### 13.1.12. Hardware, Software and Liveware

**Hardware :**

The set of *physical components, modules and peripherals comprising a computer system* is called ***Hardware.***

Apart from wires and nut bolts, the major hardware components of computer are :

(*i*) Input-output devices

(*ii*) Control unit

(*iii*) Memory

(*iv*) ALU.

**Software :**

The software is a *set of programs required for data processing activities of the computer*. In other words, the program written in any one of the computer languages, is called *software*.

*System software* includes the following :

(*i*) Operating systems

(*ii*) Language processors (assembles, compilers, interpreters)

(*iii*) Utility program

(*iv*) Subroutine program.

**Liveware :**

All persons concerned with computers, *i.e.*, complier, programmer, etc. are called *liveware*.

### 13.1.13. Translators

A *translator* is a software program which converts statements written in one language into another *e.g.*, converting assembly language to machine code etc. The assembly language program is called *'source program'* and the machine code program is called *'object program'*.

There are three types of *translators* :

1. Assembler  2. Compiler  3. Interpreter.

### 13.1.14. Computer Languages

1. ***Machine language.*** It is a programming language in which the instructions are in a form which allows the computer to perform them immediately, without any further translation. Instructions in machine language are *in the form of a binary code,* also called machine code and are known as machine *instructions.*
2. ***Low level language.*** Low level languages are machine-oriented languages in which each instruction corresponds or resembles a machine instruction. The low level language must be translated into machine language before use.
3. ***High level language.*** The development of high level language was intended to overcome main limitations of level language. The high level languages have an extensive vocabulary of word, symbols and sentences.

Different types of high level languages are :

(*i*) *Commercial languages*. ... The most well commercial language is COBOL (Commercial Business Oriented Language).

(*ii*) *Scientific language*. ... The most well-known languages among this group are :

(*a*) ALGOL (Arithmetic Oriented Language)

(*b*) FORTRAN (Formula Translation)

(*c*) BASIC (Beginner All Purpose Symbolic Instruction Code).

(*iii*) *Special purpose language.*

(*iv*) *Command language.*

(*v*) *Multipurpose language.*

### 13.1.15. Computer Programming Process for Writing Programs

The complete computer programming process followed by programmer for writing comprises the following **steps** :

1. Analysis  2. Flow charting  3. Coding
4. Debugging  5. Documentation  6. Production.

### 13.1.16. Computing Elements of Analog Computers

1. *Attenuators* ...... are used to multiply a variable quantity by a constant.
2. *Summing amplifiers* ...... are used to add or subtract variables as required.
3. *Servo multipliers* ...... are used to multiply two variables.
4. *Function generators* ...... are used to simulate the arbitrary behaviour of variables.
5. *Integrating amplifiers* ...... are used to integrate a variable with respect to time.

## 13.2 MICROPROCESSORS

### 13.2.1. Microprocessor—General Aspects

#### 13.2.1.1. Definition and brief description

A ***microprocessor****, usually abbreviated as* μP, *is a large scale integration (LSI) chip that is capable of performing arithmetic and logic functions as defined by a given programme.* This system by itself does not form an operational computer, and additional circuit for memory and input/output must be supplied and interfaced with the system. The software (it is the programme for controlling the operation of the microprocessor itself) is also to be provided.

Or

*A state machine on a single IC chip with very large scale integration, capable at a desired instant of working as per programme or an instruction of a programme, and which is driven by a clock of frequency of 1 MHz or more, is called a* ***microprocessor***. Such machine is also called a central processing unit (CPU). A CPU forms main part of a computer.

The ***microprocessor*** consists of the following *three segments* (See Fig. 13.2).

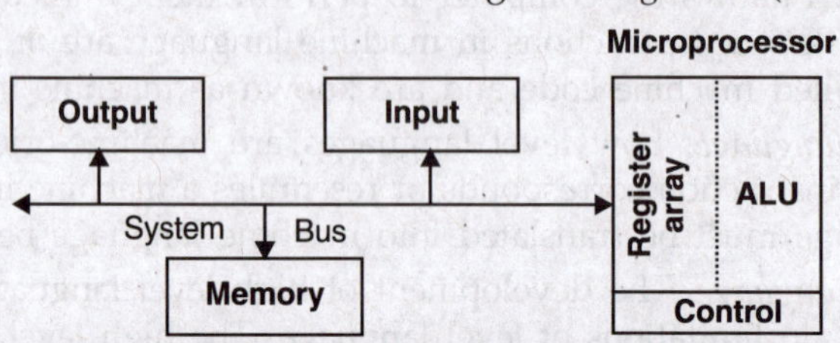


**Fig. 13.2.** Block diagram of a microcomputer.

1. **Arithmetic/Logic Unit (ALU).** In this area of the microprocessor, computing functions are performed on data. The ALU performs arithmetic operations such as addition and subtraction, and logic operations such as AND, OR and exclusive OR. Results are stored *either* in registers or in memory or sent to output devices.
2. **Register unit.** This area of the microprocessor consists of various registers. The registers are used primarily to store data temporarily during the execution of a program. Some of the registers are accessible to the user through instructions.
3. **Control unit.** The control unit provides the necessary timing and control signals to all the operations in the microcomputer. It controls the flow of data between the microprocessor and peripherals including memory.

In short a *microprocessor performs the following functions* :

- Communicates with all peripherals (memory and I/O) using system.
- Controls timing of information flow.
- Performs the computing tasks specified in a programme.

#### 13.2.1.2. Characteristics of microprocessor

In nearly every type of design, with any complexity at all, microprocessors have potential for *drastically reducing component count and shortening design time.* In fact a microprocessor is considered to represent long-awaited next generation of digital building blocks, and that microprocessor will provide the best single approach to the system-level digital integrated circuit.

Some of the characteristics of a microprocessor are listed below :

1. It *handles shorter words than other computers,* usually 4 to as many as 16 bits.
2. It consists of *integrated circuits* from 1 to 30 in number.
3. It contains arithmetic logic unit (ALU), registers, control, random access memory (RAM), data buses and read only memory (ROM) with programmes.

**13.2.1.3. Important features**

The important features of the microprocessors are :

1. *Low cost*
2. *Small size*
3. *Low power consumption*
4. *Versatile* (The versatility of a microprocessor results from its 'stored programme' mode of operation).
5. *Extremely reliable.*

***Note:*** Probably the term 'micro' in the name of the device can be contributed to its *low cost, small size and low power consumption.* The processing capability of a microprocessor should not, however, be underestimated. Currently available 32-bit microprocessors have a processing power similar to that of the mainframe computer of a few years ago. Even the early 8-bit microprocessors are powerful enough to perform several applications.

**13.2.1.4. Functions of Microprocessor**

The functions performed by a microprocessor are list below;

1. Averaging the data;
2. Linearization;
3. Alarm or annuciator settings;
4. Totalization;
5. Setting of set points;
6. Amplifier gain control;
7. Control of I/O devices;
8. To provide control actions to control elements;
9. To format output signals;
10. To rest the system;
11. Interface control, etc.

**13.2.1.5. Uses of microprocessor**

The processing power of the 8-bit microprocessors is more than adequate to satisfy the requirements of most of the *instrumentation applications.* By making an instrument microprocessor-based, it can be made *intelligent by incorporating new features like programmability* which cannot be easily provided in its hard-wired counterpart.

Some important *uses* of microprocessors in instrumentation area are listed below :

1. Frequency meters.
2. Function generators.
3. Frequency synthesizers.
4. Spectrum synthesizers.
5. *Intelligent instruments CRT terminals*
6. Digital millimeters.
7. Oscilloscopes.
8. Counters.
9. **Process control**
   - — Instrumentation
   - — Monitoring and control
   - — Data acquisition
   - — Logging and processing.

10. **Medical Electronics**
    — Patient-monitoring in intensive care unit
    — Pathological analysis
    — Measurement of parameters like blood pressure and temperature.

Under this heading the following instruments/machines are included:

(*i*) Microprocessor based medical instrument.
(*ii*) Microprocessor based ECG machines.
(*iii*) Microprocessor based EEG machines etc.

**Other Applications of microprocessors :**

(*i*) High level language computers.
(*ii*) Replacing hard-wired logic by a microprocessor.
(*iii*) Control of automation and continuous processes.
(*iv*) Computer peripheral controllers.
(*v*) Home entertainment and games.
(*vi*) Inventory control system, pay roll banking etc.

**Some Points about Microprocessors worth noting:**

- A system designed around a microprocessor, like any other digital computer, needs to be programmed. Thus, a sequence of instructions needs to be formulated and input to the microprocessor-based system for effective operation. *A sequence of instructions designed to perform a particular task is known as* "**program**". A set of programs written for a microprocessor-based system is known as *"software"* for that system. In general both *hardware and software design* are of paramount importance in microprocessor-based system design though, in particular cases, one may be more difficult or significant than other.
- **4–6–8–16–32–bits μPs**—In the beginning there was 4-bit μP followed by 8-bit μPs, 16-bit μPs and 32-bit μPs.

— A μP in which most instructions operate upon *n*-bit data is said to be a "*n*-bit μp".

**Example:** The 8085 is an 8-bit μP. It can add two 8-bit numbers and can perform several logical operations on two 8-bit patterns. It can perform many other instructions all on 8-bit data.

Since most instructions operate on data stored in on-chip registers, therefore, an n-bit μP will have registers that can hold *n*-bit data. Again, the 8085 has 8-bit registers to hold data on which its instructions operate. It is not neccessary that an *n*-bit μP will *not* have instructions to perform operations on larger length data. For example, though 8085 is an 8-bit μP, it can perform a few operation on 16-bit data also. However, the majority of its opeations are on 8-bit data.

An *n-bit* μP *can* be used to perform operations on larger data, but such operations will generally be *slower* than n-bit operations. For example, the 8085 *can* be used to perform addition of two 32-bit numbers, however, this has to be done under *program control*. However, a 32-bit μP like the Intel 80386 or the Motorola 68030, can add two 32-bit numbers in just one add operation. For this, and a few other reasons, the 32-bit μPs are *much faster* than the 16 or 8-bit μPs.

**13.2.2. Evolution of Microprocessors (μP)**

The evolution of microprocessors (μP) can be precisely summed up as follows:

- In **1971** the *first microprocessor* was introduced by *"Intel Corporation"*.

— This was "Intel 4004", a processor on a *single chip*. It had the capability of performing simple arithmetic and logical operations, e.g. addition, subtraction, comparison, AND, and OR. It also had control unit which could perform various control functions like fetching an instruction from the memory, decoding it and generating control pulses to execute it. It was a 4-bit microprocessor operating on 4 bits of data at a time.

- In the last few years the microprocessor has changed the calculators to video games and computers. Several microprocessors have been manufactured for all sorts of products, some have succeeded and some have not.
- The first microprocessor to make it into a *home computer* was the *Intel 8080*, a complete 8-bit computer on one chip, introduced in *1974*.
- The first microprocessor to make a real splash in the market was the *Intel 8088*, introduced in *1979* and incorporated into the IBMPC which first appeared around *1982*.
- The PC market moved from the 8088 to 80286 to the 80386 to the 80486 to the Pentium to the Pentium II to the Pentium III, to the Pentium 4. All of these microprocessors are made by Intel and all of them are improvements on the basic design of the 8088. The "Pentium 4" can execute any piece of code that ran on the original 8088 but it does it about 5,000 times faster.

The table 13.2 given below indicates the differences among different processors introduced over the years:

**Table 13.2: Differences among Different Processors**

| Name | Date | Transistors | Data width (Bits) | Microns | Clock speed (MHz) | MIPS |
|---|---|---|---|---|---|---|
| 4004 | 1971 | 2250 | 4 | 10 | 0.1 | 0.06 |
| 8008 | 1972 | 3500 | 8 | 10 | 0.2 | 0.06 |
| 8080 | 1974 | 6,000 | 8 | 6 | 2 | 0.64 |
| 8085 | 1974 | 6,500 | 8 | 3 | 5 | 0.37 |
| 8086 | 1978 | 29,000 | 16 | 3 | 5<br>8<br>10 | 0.33<br>0.66<br>0.75 |
| 8088 | 1979 | 29,000 | 16 | 3 | 5 | 0.33 |
| 80286 | 1982 | 134,000 | 16 | 1.5 | 16 | 1 |
| 80386 | 1985 | 275,000 | 32 | 1.5 | 16 | 5 |
| 80486 | 1989 | 1,200,000 | 32 | 1 | 25 | 20 |
| Pentium | 1993 | 3,100,000 | 32 | 0.8 | 60 | 100 |
| Pentium II | 1997 | 7,500,000 | 32 | 0.035 | 233 | ~300 |
| Pentium III | 1999 | 9,500,000 | 32 | 0.25 | 450 | ~510 |
| Pentium 4 | 2000 | 42,000,000 | 64 | 0.18 | 1500 | ~1,700 |
| Pentium 4 "Prescott" | 2004 | 125,000,000 | 64 | 0.09 | 3.6 Hz | ~7,000 |

*Transistors*: The number of transistors on the chip.

*Data width*: The width of ALU.

*Microns*: The width, in microns, of the smallest wire on the chip.

*Clock speed*: The maximum rate that chip can be clocked at.

*MIPS*: stand for "Millions of instructions per second" and is a rough measure of the performance of a CPU.

## 13.2.3. Microprocessor Systems

Microprocessor systems consist of the following *three* parts :

1. *Central processing unit (CPU) :*
   - — This part uses the microprocessor.
   - — It recognises and carries out program instructions.
2. *Input and output interfaces :*
   - — These interfaces handle communications between the computer and the outside world.
   - — For the interface, the term *port* is used.
3. *Memory :*
   - — Its function is to hold the program instructions and data.

- *"Microprocessors" which have memory and various input/output arrangements on the same chip are called **microcontrollers.***

### 13.2.3.1. The microprocessor

The microprocessor (generally referred to as CPU) is that part of the processor system which carries out the following functions :

(*i*) Processes the data ;

(*ii*) Fetches instructions from memory;

(*iii*) Decodes and executes the instructions.

Following are the various parts of a microprocessor :

**1.** ***Arithmetic and logic unit (ALU):*** This part of the microprocessor *manipulates the data.*

**2.** ***Registers:*** Registers are memory locations within the microprocessor and are employed to store information involved in program execution. The various types of registers used are :

(*i*) *Accumulator register :*

- — It is a temporary holding register for data to be operated on by the arithmetic and logic unit and also, after the operation, the register for holding the results.
- — It deals with all data transfers associated with the execution of arithmetic and logic operations.

(*ii*) *Status register :*

- — The status register (also called flag register or condition code register) contains information concerning the result of the latest process carried out in the ALU.
- — It contains individual bits with each bit having special significance; the bits are called flags.
- — The status of the latest operation is indicated by each flag with each flag being set or reset to indicate a specific status.

(iii) *Program counter register (PC) :*

— This register, also called instruction pointer (IP), contains the address of the memory location that contains the next program instruction.

— This register is *updated* after the execution of each instruction, so that it contains the memory location where the next instruction to be executed is stored.

(iv) *Memory address register (MAR) :*

— This register contains the address of data.

(v) *Instruction register (IR) :*

— This register stores an instruction.

— The control processing unit (CPU), after fetching an instruction from the memory via the data bus, stores it in the instruction register. The microprocessor, after each such fetch, increments the program counter by one with the result that the program counter points to the next instruction waiting to be fetched. The instruction can then be decoded and used to execute an operation. This sequence is known as *fetch-execute cycle.*

(vi) *General purpose registers :*

— These registers serve as temporary storage for data or addresses and *used in operations involving transfers between other registers.*

(vii) *Stack pointer register (SP) :*

— The *stack* is a special area of the memory in which program counter values can be stored when a subroutine part of the program is being used.

— The contents of this register form an address which defines the top of the stack in RAM.

The number and form of the registers depends on the microprocessor concerned.

**13.2.3.2. Buses**

***Buses** are the paths along which digital signals move from one section to another.*

— A bus is just a number of conductors along which electrical signals can be carried. It might be tracks on a printed circuit board or wires in a ribbon cable.

In a microprocessor system there are the following *three forms of bus* :

1. Data bus;
2. Address bus;
3. Control bus.

**1. *Data bus :***

— The data bus carries the data associated with the processing function of the CPU.

— Word lengths used may be 4, 8, 16, 32 or 64.

— Each wire in the bus carries a binary signal, *i.e.,* a0 or a1.

— The more wires the data bus has the longer the word length that can be used.

- The earliest microprocessors were 4-bit (word length : $2^4 = 16$) devices and such 4-bit microprocessors are still used in such devices as toys, washing machines etc. They were followed by 8-bit microprocessors (*e.g.,* Motorola 6800, the Intel 8085A and the Zilog Z80). Now 16-bit, 32-bit and 64-bit microprocessors are available.

2. *Address bus :*

— It carries signals which indicate where data is to be found and so the selection of certain memory locations or input or output ports.

— Each storage location within a memory device has a unique identification, termed its address, so that system is able to select a particular instruction or data item in the memory.

— Each input/output interface also has an address.

— When a particular address is selected by its address being placed on the address bus, *only that location is open to the communications from the CPU.* The CPU is thus *able to communicate with just one location, at a time.*

— A computer with an 8-bit data has typically a 16-bit wide address bus, *i.e.,* 16 wires. This size of address enables $2^{16}$ locations to be addressed. $2^{16}$ is 65 536 locations and is usually written as 64 K, where K is equal to 1024.

3. *Control bus :*

— This bus *carries the signals relating to control actions.*

— It is also used *to carry the system clock signals;* these are to synchronise all the actions of the microprocessor system.

**13.2.3.3. Memory**

In a microprocessor, the memory unit stores binary data and takes the form of one or more integrated circuits (ICs).

— The data may be program instruction codes or numbers being operated on.

— *The size of the memory is determined by the number of wires in the address bus.*

Following are the various forms of memory unit :

1. ROM
2. PROM
3. EPROM
4. EEPROM
5. RAM.

1. *ROM :*

— ROM (*Read Only Memory*) is a memory device in which *data is stored permanently.*

— While manufacturing the integrated circuit ROMs are programmed with the required contents. *No data then can be written into this memory chip in the computer.* The data *can only be read and is used for fixed programs;* memory is not lost when power is removed.

- *Programs stored in ROM are termed as* ***firmware.***

2. *PROM :*

— PROM (Programmable ROM) is used for ROM chips that *can be programmed by the user.*

3. *EPROM :*

— EPROM (Erasable and programmable ROM) is employed for *ROMs that can be programmed and their contents altered.*

— A typical EPROM chip contains a series of small electronic circuits, cells, which can store charge.

— The program is stored by applying voltages to the integrated circuit connecting pins and producing a pattern of charged and uncharged cells. The pattern remains permanently in the chip until *erased by shining ultraviolet light through a quartz window on the top of the device.*

4. *EEPROM :*

— EEPROM (Electrically erasable PROM) is similar to EPROM; erasure, however, is done by the application of a relatively high voltage rather than using ultraviolet light.

5. *RAM :*

— RAM (Random-access memory) is a read/write memory in which data currently being operated on (temporary data) is stored.

— Such a memory can be read or written to.

— When RAM is used for program storage then such programs are referred to as ***software.*** When the system is switched on, software may be loaded into RAM from some other peripheral equipment such as a keyboard or hard disc or floppy disc.

**Difference between a software of a computer and a microprocessor :**

In computer software is loaded into the computer at the beginning of each computation, software in microprocessor is stored within the computer itself in a ROM chip. The modification of the program is achieved by merely replacing ROM IC with another ROM IC containing a different control program. This is a very notable advantage of software implementation in microprocessor.

### 13.2.3.4. Input/Output

*The transfer of data between the microprocessor and the external world is termed as the* ***input/output operation.***

— The pieces of equipment that exchange data with a microprocessor system are called ***peripheral devices.***

— In *input operations* the input device places the data in the data register of the interface chip; this holds the data until it is read by the microprocessor. In *output operations* the microprocessor places the data in the register until it is read by the peripheral.

## 13.2.4. Intel 8085 Microprocessor

### 13.2.4.1. Brief history

— Intel Corporation in early seventies introduced the first microprocessor, *Intel 4004.* This microprocessor was a single chip device which was capable of performing simple arithmetic and logic operations such as addition, subtraction, comparison, AND, and OR. Its control unit could perform various functions such as fetching of an instruction from the memory, decoding it and generating control pulses for executing it. It was a 4-bit microprocessor operating upon 4-bits of data at a time.

— Intel introduced *4040* as modified version of microprocessor 4004.

— Intel, later on, introduced 8-bit microprocessors called 8008 and 8080 which could perform arithmetic and logic operations on 8-bit words.

— These days, modified and better version of 8-bit microprocessor is **Intel 8085** which is most widely used and most popular microprocessor.

Now-a-days 12-bit, 16-bit and 32-bit microprocessors are also available.

**13.2.4.2. Features of 8085 microprocessor**

Features of 8085 microprocessor are given below:

1. It is an 8-bit microprocessor i.e. it can accept, process, or provide 8-bit data simultaneously.
2. It operates on a single +5V power supply connected at $V_{CC}$; power supply ground is connected to $V_{SS}$.
3. It operates on clock cycle with 50% duty cycle.
4. It has a chip clock generator. This internal clock generator requires tuned circuit like LC, RC or crystal. The internal clock generator divides oscillator frequency by 2 and generates clock signal, which can be used for synchronizing external devices.
5. It can opeate with a 3 MHz clock frequency. The 8085A-2 version can operate at the maximum frequency of 5 MHz.
6. It has 16 address lines, hence it can access ($2^{16}$) 64 Kbytes of memory.
7. It provides 8 bit I/O addresses to access ($2^8$) 256 I/O ports.
8. In 8085, the lower 8-bit address bus ($A_0$ – $A_7$) and data bus ($D_0$ – $D_7$) are multiplexed to reduce number of external pins. But due to this, external hardware (latch) is required to separate address lines and data lines.
9. It supports 74 instructions with the following addressing modes:
   (*a*) Immediate (*b*) Register (*c*) Direct (*d*) Indirect (*e*) Implied
10. The Arithmetic Logic Unit (ALU) of 8085 performs:
    (*a*) 8 bit binary addition with or without carry
    (*b*) 16 bit binary addition.
    (*c*) 2 digit BCD addition.
    (*d*) 8-bit binary subtraction with or without borrow
    (*e*) 8-bit logical AND, OR, EX-OR, complement (NOT), and bit shift operations.
11. It has 8-bit accumulator, flag register, instruction register, six 8-bit general purpose registers (B, C, D, E, H and L) and two 16-bit registers (SP and PC). Getting the operand from the general purpose registers is more faster than from memory. Hence skilled programmers always prefer general purpose registers to store program variables than memory.
12. It provides five hardware interrupts: TRAP, RST 7.5, RST 6.5, RST 5.5 and INTR.
13. It has serial I/O control which allows serial communication.
14. It provides control signals ($IO/\overline{M}$, $\overline{RD}$, $\overline{WR}$) to control the bus cycles, and hence external bus controller is not required.
15. The external hardware (another microprocessor or equivalent master) can detect which machine cycle microprocessor is executing using status signals ($IO/\overline{M}$, $S_0$, $S_1$). This feature is very useful when more than one processors are using common system resources (memory and I/O devices).
16. It has a mechanism by which it is possible to increase its interrupt handling capacity.
17. The 8085 has an ability to share system bus with Direct Memory Access controller. This feature allows to transfer large amount of data from I/O device to memory or from memory to I/O device with high speeds.
18. It can be used to implement three chip microcomputer with supporting I/O devices like IC 8155 and IC 8355.

### 7.2.4.3. Description of 8085 Microprocessor

Fig. 13.3, shows the block diagram of Intel 8085 microprocessor.

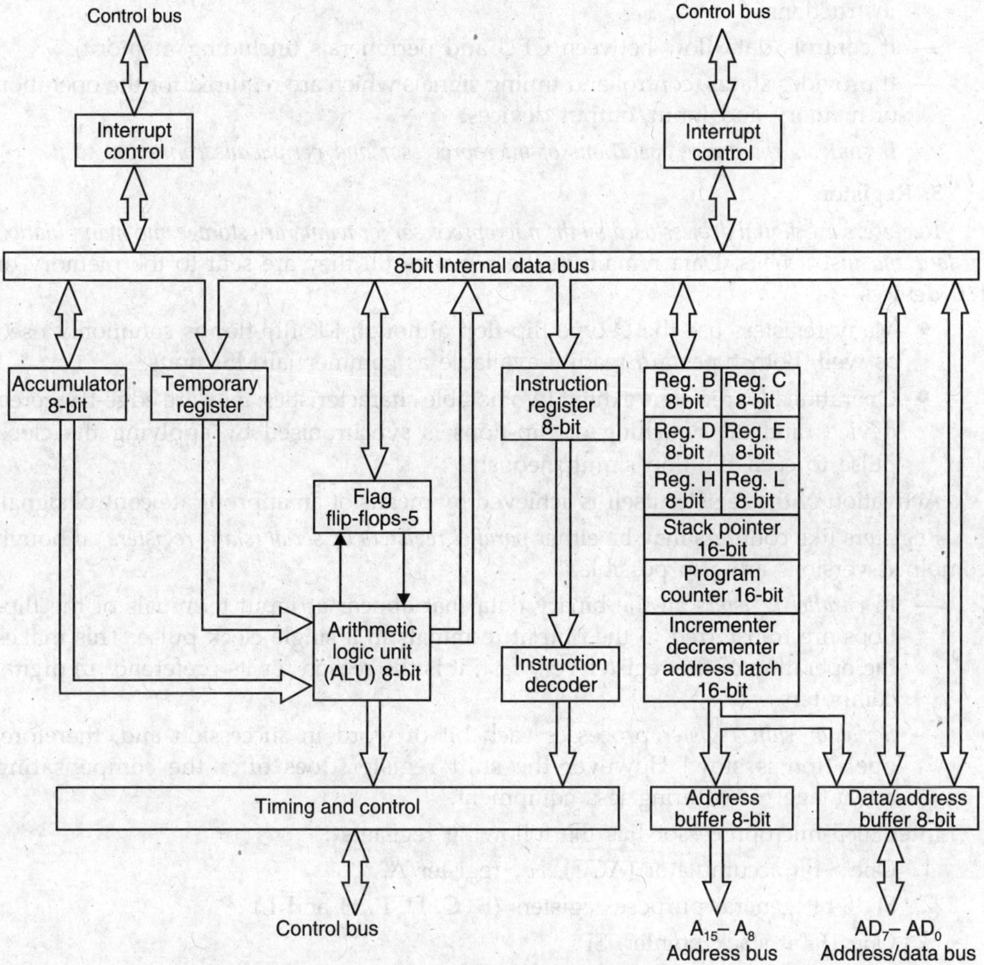


**Fig. 13.3.** Block diagram of Intel 8085 microprocessor.

Some important functional units/parts are enumerated and discussed as follows:

1. Arithmetic and logic unit (AUL).
2. Timing and control unit.
3. Registers
4. Data and address bus
5. Pin configuration

**1. Arithmetic and logic unit (ALU):**

ALU performs the following arithmetic and logical operations :

1. Addition
2. Subtraction
3. Logical AND
4. Logical OR
5. Logical EXCLUSIVE OR
6. Complement (logical NOT)
7. Increment (add 1)
8. Decrement (subtract 1)
9. Left shift, Rotate left, Rotate right.
10. Clear etc.

**2. Timing and control unit:**

— This unit is a section of CPU.

— It *generates timing and control signals* which are necessary for the execution of instructions.

— It controls data flow between CPU and peripherals (including memory).

— It provides status, control and timing signals which are required for the operation of memory and input/output devices.

— *It controls the entire operations of microprocessor and peripherals connected to it.*

**3. Registers**

***Registers** are digital devices used by the microprocessor for temporary storage and manipulation of data and instructions.* Data remain in the registers till they are sent to the memory or I/0 devices.

- Many registers use the D-type flip-flop although J-K flip-flop is commonly used as well. Both types are readily available as commercial MSI units.
- Operationally, registers exhibit two notable characteristics: they are edge-triggered devices and all switching of flip-flops is synchronised by applying the clock pulse to each flip-flop simultaneously.

Activation of the register itself is achieved by means of an appropriate control signal.

Registers like counters may be either *parallel registers* or *serial (shift) registers*, although combined versions are also possible.

— In *parallel registers* all the binary data that appear at input terminals of the flip-flops are transferred to the output terminals in a single clock pulse. This makes the operation of the register very fast; it is the reason for its preference in digital computers.

— *Serial or shift register* processes each bit of word in succession and, therefore, operation is slow. However the shift register does offer the compensating advantage of requiring less equipment.

Intel 8085 microprocessor has the following registers :

1. One 8-bit accumulator (ACC), *i.e.,* register A.
2. Six 8-bit general purpose registers (B, C, D, E, H and L).
3. One 16-bit stack pointer, SP.
4. One 16-bit program counter, PC.
5. Instruction register.
6. Temporary register.

These registers are described below :

**1. *Accumulator (ACC) :***

- It is an 8-bit special purpose register that is a part of the ALU. It is also identified as register A.
- In arithmetic and logical operations the accumulator may store the operand, execute an instruction with the help of other registers, and memory and finally store the result of the operation. In the former case it acts as a source, and in the latter a destination.

**2. *General purpose registers :***

- The 8085 microprocessor contains six 8-bit general purpose registers. These are identified as B, C, D, E, H and L as shown in Fig. 13.3.

- These registers are used in microprocessor for temporary storage of operands or intermediate data in calculations.
- These registers can be used either simply for storage of 8-bit data or in pairs for storage of 16-bit data. When used in pairs, only selected combination can be used for pairing, *i.e.,* B-C, D-E and H-L. When registers are used in pairs the high order byte resides in the first register and low order byte in the second register.

3. ***Stack pointer (SP) :***
   - It is a 16-bit special function register.
   - The *stack* is a sequence of memory locations set aside by a programmer to store/retrieve the contents of accumulator, flags, program counter and general purpose registers during the execution of a program. Any portion of the memory can be used as a stack.
   - In this register, data is stored temporarily on first come and last go basis.
4. ***Program counter (PC) :***
   - It is a 16-bit special-purpose register and is used to hold the memory address of the next instruction to be executed.
   - The contents of the PC are automatically updated by the microprocessor during the execution of an instruction so that at the end of execution it points to the address of the next instruction in the memory.
   - The microprocessor uses the PC for sequencing the execution of instructions.
5. ***Instruction register :***
   - During the execution of a program, microprocessor addresses some memory, which supplies an 8-bit data of instruction code to the data bus which gets stored in the register called the *instruction register.*
   - The instruction register holds the op-code (operation code or instruction code) of the instruction which is being decoded and executed.
6. ***Temporary register :***
   - It is an 8-bit register associated with ALU.
   - It holds data during an arithmetic/logical operation.
   - It is used by the microprocessor and is not accessible to programmer.

***Flags.*** There are five-flops in Intel 8085 microprocessor to serve as status flags; these are : (*i*) Carry Flag (CS), (*ii*) Parity Flag (P), (*iii*) Auxiliary Carry Flag (AC); (*iv*) Zero Flag (Z), and (*v*) Sign Flag (S).

— The flip-flops are set or reset according to the conditions which arise during an arithmetic or logical operation.

— If a flip-flop for a particular flag is set, it indicates 1. When it is reset, it indicates 0.

***Instruction decoder.*** Data from the instruction register is sent to the instruction decoder, where microprocessor decodes it and then translates into specific actions.

**4. Data and address bus**

The data bus of Intel 8085 microprocessor is 8-bit wide and hence, 8 bits of data can be transmitted in parallel from or to the microprocessor. This microprocessor requires a 16-bit wide address bus as the memory addresses are of 16-bits. The 8 *most significant* bits of the address are transmitted by the address bus, A-bus (pins $A_8$ to $A_{15}$). The 8 *least significant* bits of the address are transmitted by address/data bus, AD-bus (pins $AD_0$–$AD_7$).

The address/data bus transmits data and address at different movements. At particular moment it transmits either data or address. Thus *AD-bus operates in time shared mode. This technique is called* ***multiplexing***.

**5. Pin configuration**

Fig. 13.4. shows the schematic diagram of Intel 8085 microprocessor.

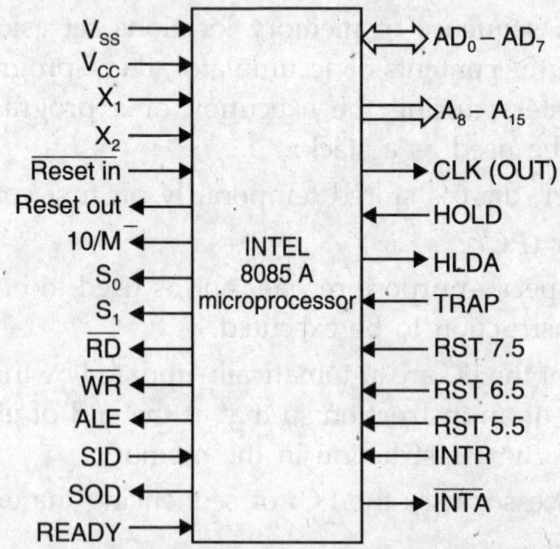


**Fig. 13.4.** Schematic diagram of Intel 8085 microprocessor.

***ADo–AD₇ ( Input/Output):*** They are used for the least significant 8-bits of the memory address of I/O address during the first clock cycle of a machine cycle. Again they are used for data during second and third clock cycles.

***$A_8$–$A_{15}$ (output):*** These are address bus and are used for the most significant bits of the memory address of 8-bits of I/O address.

***IO/$\overline{M}$ (Output):*** It is a status signal which distinguishes whether the address is for memory or I/O. When it goes high the address on the address bus is for an I/O device, whereas, when it goes low the address on the address bus is for a memory location.

***ALE (Output):*** It is an *address latch enable signal.* It goes high during first clock cycle of a machine cycle and enables the lower bits of the address to be latched either into the memory or external latch.

***$S_0$, $S_1$ (Output):*** These are status signals sent by the microprocessor to distinguish the various types of operations given in the table below,

| $S_1$ | $S_0$ | *Operations/Machine cycle* |
|---|---|---|
| 0 | 0 | HALT |
| 0 | 1 | WRITE |
| 1 | 0 | READ |
| 1 | 1 | FETCH |

*$\overline{RD}$ (Output) :*

— It is a signal to control READ operation.

— When it goes low the selected memory or I/O device is read.

*$\overline{WR}$ (Output) :*

— It is a signal to control WRITE operation.

— When it goes low the data on the data bus is written into the selected memory or I/O location.

*READY INPUT :*

— It is used by the microprocessor to sense whether a peripheral is ready to transfer data or not.

— If READY is high the peripheral is ready, if it is low the microprocessor waits till it goes high.

— A slow peripheral may be connected to the microprocessor through READY line.

*HOLD (Input) :*

— It indicates that another device is requesting for the use of the address and data bus.

— The microprocessor after having received a HOLD request relinquishes the use of the buses as soon as the current machine cycle is completed. Internal processing may continue.

— The processor regains the bus after the removal of the HOLD signal.

*HLDA (Output) :*

— This signal indicates that the HOLD request has been received.

— After the removal of a HOLD request the HLDA goes low. The CPU takes over the buses half clock cycle after the HLDA goes low.

*INTR (Input); $\overline{INTA}$ :*

— INTR signal indicates an interrupt request.

— The INTR line is sampled in the last state of the last machine cycle of an instruction. The microprocessor acknowledges the interrupt signals and issues an $\overline{INTA}$ signal.

*$\overline{RESET\ IN}$ (Output) :*

— When this signal is applied the program counter is set to zero and resets the Interrupt Enable and HLDA flip flops. Except the instruction register no other register or flag is affected.

— The CPU remains in the reset condition as long as reset is applied.

*RESET OUT :*

— This is an output signal which shows that CPU is being reset. This can be used as a system RESET.

— The signal is synchronised to the microprocessor clock.

*$X_1$, $X_2$ (Input) :*

— $X_1$ and $X_2$ are the terminals to be connected to an external crystal oscillator which drives an internal circuitry of the microprocessor to produce a suitable clock for the operation of microprocessor.

*CLK (Output) :*

— It is a clock output for user, which can be used for other digital ICs.

— Its frequency is same at which processor operates.

*SID and SOD line :*

— SID is an input line and it is for serial input data. The serial input data at SID can be identified by an instruction called RIM and serial data can be an instruction called SIM.

— In Intel 8085 microprocessor only serial transmission facility is available. The SID and SOD lines pursuit the input and output serial data. The actual transfer of the data is accomplished by software using the RIM and SIM instructions. Both these instructions are single byte and are also used to read or set/reset interrupt masks.

### 13.2.5. Microprocessor Programming

*The process of developing a program is termed as* **Programming**. A microprocessor needs to be programmed before it can perform a useful task.

*A "program" is a sequence of instructions that operate on a data to produce desired results.* The programs, written to perform a particular task, is stored in the semiconductor memory accessible to the microprocessor. The processor, having been instructed the program, fetches one instruction at a time from the memory and executes it.

- **Instructions and Instruction cycle**

An "***instruction***" consists of an *operation code* (abbreviated as *opcode*) and the *address(es) of the data* on which it has to operate. The data on which an instruction operates, is also known as the *operand(s)* for that instruction.

— Most instructions specify directly the address of only one of the operands, the address of the other operand being implicitly known. Some instructions specify the data itself and not the address.

An "***instruction cycle***" consists of a *"fetch cycle"* and *"execute cycle"*. The total time required to execute an instruction is the some of time required to fetch an opcode and the time required to execute it.

— The opcode (the Ist byte of an instruction is its opcode; the instruction may be more than one byte long) fetched from the memory goes to the data register, DR (data/address buffer in Intel 8085 microprocessor) and then to instruction register, IR. From the instruction register it goes to the decode circuitry which *decodes* the instruction. The decoded circuitry is within the microprocessor.

— After the instruction is decoded, *execution begins.* If the operand is in the general purpose registers, execution is immediately performed. The time taken in decoding and execution is one clock cycle. If an instruction contains data or operand address which are still in the memory, the CPU has to perform some read operations to get the desired data. After receiving the data it performs execute operation.

A *read cycle is similar to a fetch cycle.* In case of a read cycle the quantity received from the memory are data or operand address instead of an opcode. In some instructions write operation is performed.

In *write cycle* data are sent from CPU to the memory or an output device.

In some cases an *execute cycle* may involve one or more read or write cycles or both.

— The necessary steps carried out to perform a fetch, a read or write operation constitute a *"Machine cycle"*. An *instruction cycle consists of several machine cycles.*

- **Machine and Assembly Language (ML and AL) Programming:**

  — A *"machine language programe"* is written using binary *opcodes and addresses.*

  — In an *"assembly language program"* the instructions are written using *mueonomic codes and symbolic addresses.*

For a microprocessor it is *easier* to write an assembly language programme than to write a machine language programme. However, a *microprocessor can understand only machine language programs.*

- **High Level Language (HLL) Programming:**
  - — For programming on many microprocessors, higher level languages like *fortran*, C, *Pascal, and colbol* are available.
  - — Though programming in *higher level languages is much easier than in machine or assembly languages,* the former *is possible only when a microcomputer with sufficient memory is available.*
  - — Even when a very small system is being designed, it can be programmed by using a *development system* in high level language. The high level language program can be translated into binary and stored in the memory of the small system.
- **The Programming Process:**
  - — The steps involved in a programming process are clearly depicted in Fig. 13.5.

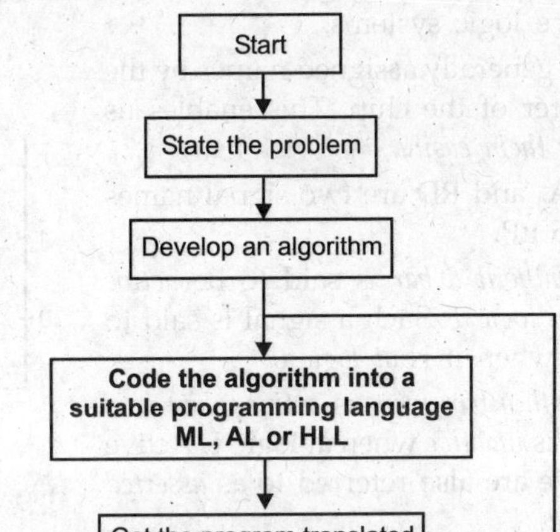


Fig. 13.5. The Programming process.

### 13.2.6. Digital Logic

Since a microprocessor (μP) is a basically a digital device, therefore, all the signals generated by the μP and data processed, are in digital form.

**Analog and digital signals:**

- Fig. 13.6 depicts the difference between analog and digital signals.
  - An *"Analog signal"* is a continuously varying signal. It has *infinitely many levels between a minimum and maximum.*
  - A *"digital signal"* is characterised by just *two levels.* These levels are denoted by *logical 0 and logical 1.*
  - In systems that are *positive logic,* the *higher voltage* denotes *logic 1* and the lower voltage denotes *logic 0.* A logic 1 is also referred to as *true,* and logic 0 as *false.* This terminology gets reversed for negative logic systems.
- Signals are generally assigned names by the manufacturer of the chip. This enables us to *reference them easily.*

  Example: $A_0$ and RD are two signal names in the 8085 μP.

  A signal *without a bar* is said to be *active when it is at logic 1.* Such a signal is said to be *inactive* when it is *at logic 0.*

  A signal *with a bar on it,* is *active when it is at logic 0.* It is *inactive* when at logic 1. Active and inactive are also referred to as *asserted* and *negated.*

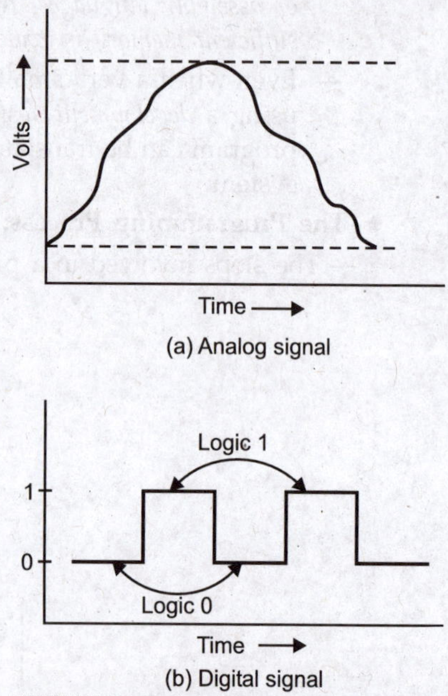


**Fig. 13.6.** Analog and digital signals

- Every digital signal can be in only one of the two stages. This is true in most cases when we are using binary or boolen logic. However some devices generate signals that can be in one of *three possible states.* These states are: *logic 0, logic 1, high impedance.*
  - When in a signal is in high impedanc state, also known as or *hi_Z* a floating state, it is logically *disconnected* from the bus to which it is physically connected.
  - Signals that exhibit these three states, are also known as *tri-state signals.* Such signals are very useful in several applications.

**Digital building blocks:**

When designing with microprocessors, there is often a need to use simple digital devices such as *gates, flip-flops* etc. For details of such devices refer to chapter 6 (Digital electronics).

**Signal levels and device loading:**

- When designing a system, the designer should know what are the *voltage* levels corresponding to these logic levels.
- Beside knowing the *voltage requirements,* the designer needs to be concerned with the *currents requirements* as well.

**Open-collector and Totem-Pole devices:**

— The ouputs of several devices which do not have a *pull-up* resistor in their output stage of the internal circuit are specified to be *open-collector* outputs.

Example: The *7405 hex inverter*.

When connecting a device such as an LED, one must use a pull-up resistor when using an open-collector device.

— In totem-pole devices a pull-up resistor is used.

Example: The 74LS04

When using such devices, one can generally connect many devices to the output of one device. The only limitation is as computed by the current-loading considerations.

### 13.2.7. Types of Microprocessors

The various types of microprocessors are:

1. Vector Processors.
2. Array Processors or SIMD Processors.
3. Scalar and Superscalar Processors.
4. RISC, CISC Processors and EPIC.
5. Digital Signal Processors (DSP)
6. Symbolic Processors
7. I/O Processor
8. Coprocessors
   - Math processors
   - Math coprosessor, Intel 8087
     - Intel 80287, 80C287A
     - Intel 80387 DX
     - Intel 80387 SX
   - Peripheral math processor
   - Graphic processor

**Some Other Microprocessors:**

1. AMD (Advanced Micro Devices) Microprocessors
2. Sun Microprocessors
   - 1 Ultra SPARC I, II, III, 4+, T1
   - Sun's JVM (Java Visual Machine)
   - Sun Pico Java I, II and Sun MicroJava
   - Sun's MAJC processor
3. MIPS Microprocessors
4. Power PC Microprocessors
5. DEC'S Alpha Microprocessors
6. National Semiconductor Microprocessors
7. ARM Microprocessors
8. Cell Processor
9. i.Max and Cold Fire Processors

10. Motorola Microprocessors
    - Motorola MC6800
    - Motorola MC6809
    - Motorola 68000
    - Motorola 68012
    - Motorola 68020
    - Motorola 68030
    - Motorola 68040
    - Motorola 68060
11. MC 6500
12. Transmeta Corporation's Crusoe Chips
13. Via's C3 Processor
14. Hewlett Packard (HP) Processor.

### 13.2.8 Processing Speed of a Processor

- *The processing speed of a microprocessor is usually measured in number of instructions executed per second*. The number of instructions executed per second is also known as ***throughout*** of the microprocessor. It is generally expressed in MIPS (Millions of Instructions per Second).
  - — The MIPS rating of a processor *does not* give correct idea of processor's processing ability because instructions differ from processor to processor for high-level language programs. If the same program runs on a RISC as well as CISC processor, the MIPS rating of the RISC processor will be *higher* because the instructions of RISC processor are *simpler* than those of CISC processor.
  - — The floating-point performance of a processor is given MFLOPS (Millions of floating-Point Instructions per Second). MFLOPS rating gives processor's performance for highly mathematical operations.
  - — SPEC (System Performance Evaluation Committee) ratings are used to specify processor's ratings.
- For on-line transaction processing applications, computer performance is measured in TPS (Transactions per second). Such applications demand rapid, interactive processing for a large number of relatively simple transactions. They are supported by very large databases.
- *LINPACK Ratings* are based on a general purpose FORTRON library of mathematical software for *solving dense linear systems of equations of order 100 or higher*.
- *Dhrystone* is a synthetic testing benchmark and measures *integer* performance in KDhrystones/s. Its drawback is its sensitivity to compilers.
- *Whetstone* is a synthetic testing benchmark and measures both *integer and floating-point* performance in KWhetstones/s.
- iCOMP (intel's Comparative Microprocessor Performance Index) consists of a collection of benchmarks to evaluate an intex of relative performance of Intel microprocessors and is based on *integer, floating point, graphics* and *video performance*.

## 13.3. ORGANISATION OF MICROCOMPUTERS

A "**microcomputer**" *is a small sized and inexpensive computer*. A *'microprocessor' is the heart of a microcomputer*.

- A microcomputer is a device which must be capable of the following *tasks*:

(*i*) *Receiving input* (data and instructions), (*ii*) Performing computations (arithmetic and logical), (*iii*) *Storing data and instructions*, (*iv*) *Displaying the results of any computations*, and (*v*) Controlling all the devices that perform the above mentioned tasks directly or indirectly.

— A device which enables a microcomputer to peform the task (*i*) of the above tasks is known as *"Input device"*:

*Typical input devices* are: Keyboards, mouse, and toggle switches.

— The device which performs the computations task (*ii*) is known as an "Arithmetic Logic Unit (ALU)".

— *"Memories"* are used to perform task (*iii*) *i.e., storing* and *instructions*. Semiconductor memories are the most commonly used ones. In relatively large computers, Cassettes, Floppy disks and hard disks are used together with semiconductor memories.

— The task of *displaying the results* task (*iv*) computed by the μP, is performed by an *"output device"*.

*Some commonly used output devices are*: Character printers, cathode ray tubes (CRT) displays, light emitting diodes (LEDs), and laser printers.

— A microprocessor performs tasks (*ii*) and (*v*) in microcomputers.

Fig. 13.7 shows the block diagram of a typical microcomputer. It exhibits various *input/output devices* and *memories* connected to the microprocessor by *busses*.

— On the *"address bus"* the μP transmits the address of the device or a memory location that it needs to address.

— The *"data bus"* is used by the processor to send and receive data to and from different devices. It may be noted that whereas the *address bus is unidirectional*, the *data bus is bidirectional*.

— The *"control bus"* is used for transmitting and receiving control signals between the processor and various devices.

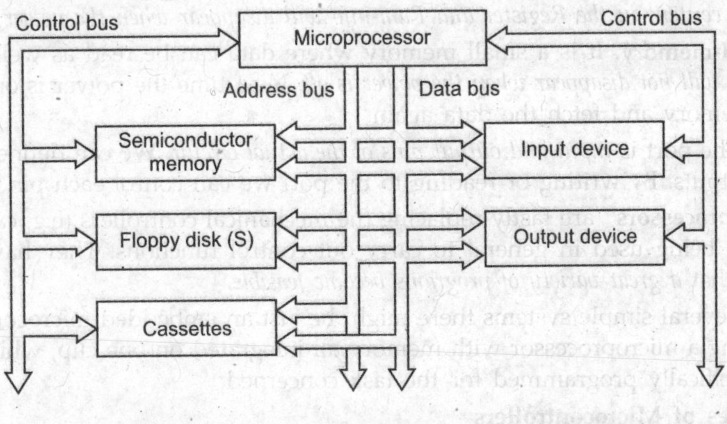


**Fig. 13.7.** Organisation of a microcomputer.

The microcomputer operates as directed by the user:

— The user writes a sequence of instructions known as a *program* and requests the microcomputer to begin executing each instruction starting from the first one.

— The microprocessor fetches the instructions one by one from the memory and executes them.

## 13.4. MICROCONTROLLERS

### 13.4.1. Introduction

*The* **microcontroller** *is the integration of a microprocessor with memory and input/output interfaces, and other peripherals such as timers, on a single chip.* It is basically *a microcomputer on a single IC.*

Fig. 13.8 shows the simplified block diagram of the microcontroller (microprocessor based controller).

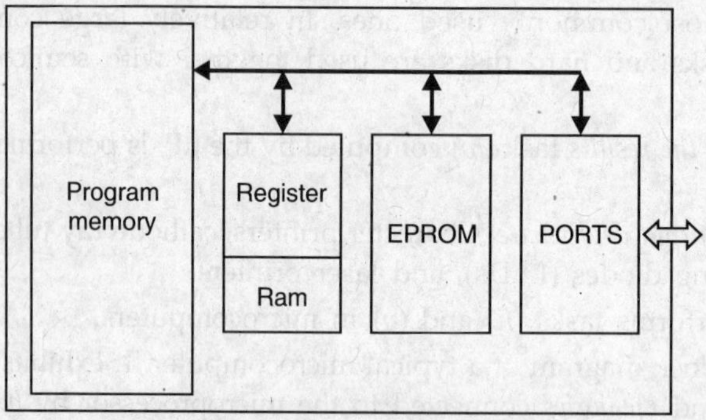

Fig. 13.8. Simplified block diagram of microcontroller.

**Program memory.** It contains the written program. The program is a set of instructions that the microcontroller performs. The software (instructions) is written in a computer and then programmed burned into the "program memory". This memory is a EPROM memory which can be rewritten thousand times.

**Register and Ram box.** It contains all the *internal registers and a small Ram* where data can be stored temporarily. *There are several registers with different functions.*

— The Ram memory is not large, *about 64-128 byte.*

— *The content in the Register and Ram-info will disappear when the power is off.*

**EPROM-memory.** It is a small memory where data can be read as well as written, but the *data will not disappear when the power is off.* Next time the power is on we can go into this memory and fetch the data again.

**Ports.** The port is *input and output pins of the actual circuit.* We can define the pins as input or outputs. By writing or reading to the port we can conrol each pin as we wish.

**"Microprocessors"** are fastly replacing the mechanical controllers (e.g. cam-operated switch) and being used in general to carry out control functions. They have the great ***advantage*** that *a great variety of programs become feasible.*

— In several simple systems there might be just an embedded microcontroller, this being a microprocessor with memory all integrated on one clip, which has been specifically programmed for the task concerned.

**Examples of Microcontrollers**

(*i*) *8-bit microcontrollers (data path 8-bit wide)*:

— The Motorola 68 HC11;

— The Intel 8051;

— The PIC16 C6 X/7X.

(*ii*) *16-bit microcontroller*:

— The Motorola 68 HC 16.

(iii) *32-bit microcontroller:*

— The Motorola 68300.

- Microcontrollers have limited amounts of ROM and RAM. These are *widely used for embedded control systems.*
- A microprocessor system with separate memory and input/output chips is more suitable for processing information in a computer system.

### 13.4.2. Characteristics of Microcontrollers

The *characteristics of microcontrollers* are:

1. Low cost
2. Versatility
3. Ease of programming
4. Small size.

- *Microcontrollers are attractive in mechatronic system design since their small size and broad functionality allow them to be physically embedded in a system to perform all of the necessary control functions.*

### 13.4.3. Applications

Microcontrollers are used in wide array of applications including :

- Entertainment equipment;
- Air planes;
- Home appliances;
- Toys;
- Automobiles;
- Office equipment;
- Trucks.

All these products involve devices that *require some sort of intelligent control,* based on various inputs; ***Examples*** being :

— In a *microwave oven,* the microcontroller monitors the control panel for user input, updates the graphical displays when necessary, and controls the timing and cooking functions.

— In an *automobile,* there are many microcontrollers to control various subsystems, including cruise control, antilock braking, ignition control, keyless entry, environmental control, and air and fuel flow.

— A *toy robot dog* has various sensors to detect inputs from its environment and an on board microcontroller actuates motors to mimic actual dog behaviour based on their input.

— An *office fax machine controls actuators* to feed papers, use photo sensors to scan a page, sends or receives data on a phone line, and provides a user interface complete with menu-driven controls.

All the above mentioned devices are *controlled by 'microcontrollers' and the 'software' running on them.*

- Typically, *'microcontrollers'* have *less than 1 kilobyte to several tens of kilobytes* of program memory, compared with *'microcomputers'* whose ram memory is measured in *megabytes or gigabytes.* Also, microcontroller clock speeds are *slower than those used for microcomputers.*

— A selected microcontroller, for some applications, may not have enough speed or memory to satisfy the needs of the application. The manufacturers of microcontrollers usually provide a wide range of products to accommodate different applications.

— When more memory or Input/Output capability is required, the functionality of the microcontroller can be expanded with *additional components, e.g.,* RAM or EEPROM chips, external A/D and D/A converters, and other microcontrollers.

### 13.4.4. Classification of Microcontrollers

The classification of microcontrollers is made into *four types* on the following bases:

1. Width of data format.
2. Instruction set.
3. Data types handled.
4. Semiconductor technology adopted in their design.

**1. Width of data format:**

The *"width of the microcontroller" means the size of the operand upon which the processor can operate.* Based on this we have 4 bit, 8 bit, 16 bit and 32 bit microcontrollers. Of the above *8 bit is most commonly used; 4 bits are used in domestic appliances.*

— *Wider the width of a data format faster will be processor and expensive are the devices using it.*

**2. Instruction set:**

— On the above basis, microcontrollers may be of two types namely *"Complex Inistruction set computing (CISC)"* or *"Reducing Instruction set computing (RISC)."*

— CISC microcontrollers are *slower* compared to RISC microcontrollers but are *easier to program.*

**3. Data types handles:**

On the above basis, microcontrollers may be *"Floating point type"* and *"Fixed point type"*. These microcontrollers *differ* in the following *two ways:*

(*i*) Their architecture of ALU is different;

(*ii*) A floating point microcontroller can handle data using floating point format, whereas *additional software is necessary for fixed point-controller* to handle floating point data, which results in *increased processing time.*

**4. Semi-conductor technology adopted:**

Following are the *most commonly employed semiconductor technologies:*

(*i*) Transistor-Transistor Logic (TTL).

(*ii*) Complimentary Metal Oxide Semiconductor (CMOS).

(*iii*) Emitter Coupled Logic (ECL).

— TTL technology is *commonly used.*

— CMOS consumers low power because of high input impedance of the basic devices built into it.

— ECL technology is used to *design fast processor.*

### 13.4.5 Description of a Typical Full-featured Microcontroller

Fig. 13.9 shows a block diagram for a *typical full-featured microcontroller,* indicating also the lists of typical external devices that might interface to the microcontroller.

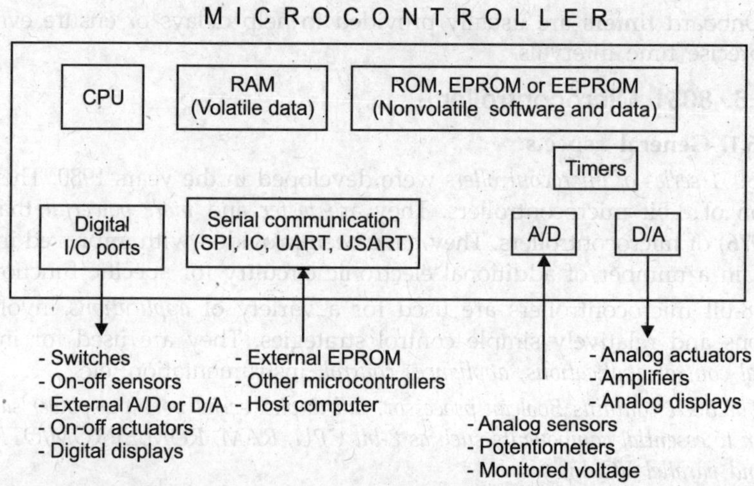


**Fig. 13.9.** Block diagram for a typical full-featured microcontroller.

The components of a microcontroller include the CPU, RAM, ROM, digital I/O ports, a serial communication interface, timers, A/D converters, and D/A converters.

— The **RAM** is used to *store settings and values used by an executing program.*

— The **ROM** is used to *store the program and any permanent data.* A designer can have a program and data permanently stored in ROM by the chip manufacturer, or the **ROM** can be in the form of *EPROM*, or *EEPROM*, which can be reprogrammed by the user.

Software permanently sotred in ROM is referred to as **firmware**.

Microcontroller manufacturers offer programming devices that can download a compiled machine code file from a PC directly to the EPROM of the microcontroller, usually via the PC serial port and special-purpose pins on the microcontroller. These pins can usually be used for other purposes once the device is programmed. Additional EEPROM may also be available and used by the program to store settings and parameters generated or modified during execution. The data in EEPROM is *nonvolatile*, which means the program can access the data when the microcontroller power is turned off and back on again.

— The **digital I/O ports** allow binary data to be transferred to and from the microcontroller using external pins on the IC. These pins can be used to read the state of switches and on-off sensors, to interface to external analog-to digital and digital-to-analog converters, to control digital displays, and to control on-off actuators.

The I/O ports can also be used to transmit signals to and from other microcontrollers to coordinate various functions.

— The microcontroller can also use a *serial port* to transmit to and from external devices, provided these devices support the same serial communication protocol. There are various standards or protocols for serial communication including SPI (Serial peripheral interface), $I^2C$ (Integrated circuit), UART (Universal asynchronous receiver-trasmitter), and USART (Universal synchronous-asynchronous receiver-transmitter).

— The **A/D converter** allows the microcontroller to convert an external analog voltage (*e.g.*, from a sensor) to a digital value that can be processed or stored by the CPU.

The **D/A converter** allows the microcontroller to output an analog voltage to a nondigital device *e.g.*, a motor amplifier.

— Onboard **timers** are usually provided to help delays or ensure events occur at precise time intervals.

### 13.4.6. 8051 Microcontroller

#### 13.4.6.1. General aspects

The *8051 series of microcontrollers* were developed in the years 1980. They are second generation of 8-bit microcontrollers. They are *faster* and *more powerful* than Intel 8048 series (1976) of microcontrollers. They have been provided with improved instruction set and contain a number of additional electronic circuitry for specific functions.

The 8-bit microcontrollers are used for a variety of *applications* involving limited calculations and relatively simple control strategies. They are used for industrial and *commercial control applications, appliances control,* instrumentation, etc.

— *The 8051 contains Boolean processor, full duplex serial port and power saving circuitry in addition to essential components such as 8-bit CPU, RAM, ROM/EPROM/OTPROM, timer/counter and parallel I/O lines.*

Its family-MCS-51 includes 8031, 8051 and 8751 microcontrollers. The summary of MCS-51 microcontrolles is given in the table 13.3 below:

**Table 13.3: Summary of MCS-51**

| Device | Internal memory | | Timer/ Event counters | Interrupts |
|---|---|---|---|---|
| | Program | Data | | |
| 8052AH | 8 K × 8 ROM | 256 × 8 RAM | 3 × 16-Bit | 6 |
| 8051AH | 4 K × 8 ROM | 128 × 8 RAM | 2 × 16-Bit | 5 |
| 8051 | 4 K × 8 ROM | 128 × 8 RAM | 2 × 16-Bit | 5 |
| 8032AH | none | 256 × 8 RAM | 2 × 16-Bit | 6 |
| 8031AH | none | 128 × 8 RAM | 2 × 16-Bit | 5 |
| 8031 | none | 128 × 8 RAM | 2 × 16-Bit | 5 |
| 8751H | 4 K × 8 EPROM | 128 × 8 RAM | 2 × 16-Bit | 5 |
| 8751H-12 | 4 K × 8 EPROM | 128 × 8 RAM | 2 × 16-Bit | 5 |

- The 8051 microcontrollers have *two versions*: COMMERCIAL and EXPRESS. The operating temperature range for commerical versions is 0°C to 70°C and that for the express version is 40°C to +85°C.

Fig. 13.10 shows the block diagram of Intel 8051.

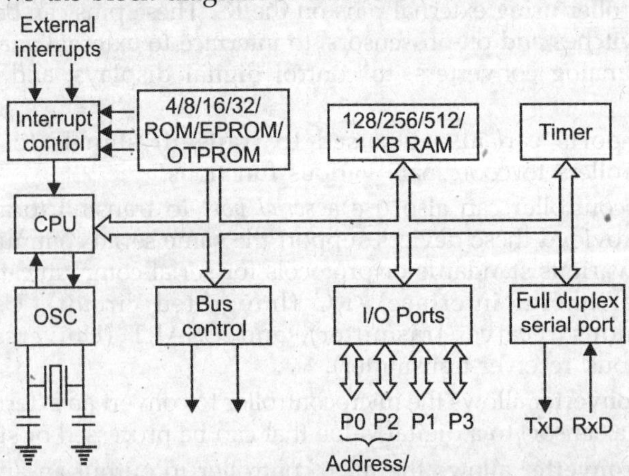


**Fig. 13.10.** Block diagram of 8051 microcontroller.

8051 microcontroller is a *generic part* number that includes entire family of microcontroller ranging from numbers 8031 to 8751 and is available in N-channel intel oxide silicon (NMOS) and complementary metal oxide silicon (CMOS) in variety of packages.

8051, basically, is a 40-pin DIP integrated circuit that includes the following *features*:

1. Internal RAM and ROM
2. I/O ports with programmable pins
3. Timers and counters
4. Serial data communication.

#### 13.4.6.2. Architecture of 8051 micro controller

The *different parts of the 8051 microcontroller* are as follows:

1. Central processing unit (CPU), which includes program counter, ALU, working registers and block circuit.
2. 8-bit CPU with registers (the accumulator) and B
3. 16-bit program counter (PC) and data pointer (DPTR)
4. 8-bit program status word (PSW)
5. 8-bit stack pointer (SP)
6. Internal ROM of 4 KB
7. Internal RAM of 128 bytes
8. Four register banks each containing 8 registers.
   - Sixteen bytes, which may be addressed at bit level
   - Eighty bytes of general-purpose data memory
9. 32-input/output pins arranged as four 8-bit ports $P_0$–$P_3$
10. Two 16-bit timer/counters $T_0$ and $T_1$.
11. Full duplex serial data receiver/transmitter SUBF
12. Control registers—TCON, TMOD, SCON, PCON, IP and IE
13. Two external and three internal interrupt sources
14. Oscillator and clock circuits.

### 13.4.7. Microchip Controllers

- The microchip microcontrollers use a form of architecture termed Harvard architecture (Fig. 13.11). With this architecture, instructions are fetched from program memory using accessible variables.
- Harvard architecture enables faster execution speeds to be achieved for a given clock frequency.

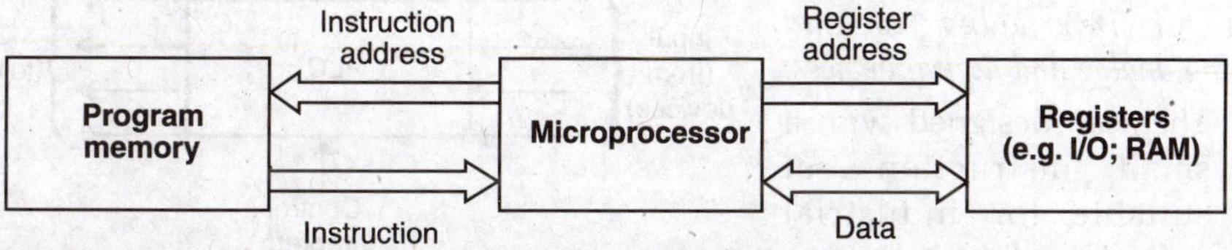


Fig. 13.11. Harvard architecture.

### 13.4.8. Selection of a Microcontroller

While selecting a microcontroller the following *factors* should be considered :

1. Number of input/output pins.

2. Interfaces required.
3. Memory requirements.
4. The number of interrupts required.
5. Processing speed required.

### 13.4.8. Comparison of Microprocessors and Microcontrollers

The comparison of microprocessors and microcontrollers is given the table 13.4 below:

**Table 13.4: Comparison between Microprocessors and Microcontrollers**

| S. No. | Aspects | Microprocessors | Microcontrollers |
|---|---|---|---|
| 1. | *Components/parts* | CPU, interrupt circuits and memory-addressing circuits. | Besides all parts is of microprocessor, they also contain timers, parallel and serial I/O, internal RAM and ROM. |
| 2. | *Access times for memory and I/O devices* | More | Less |
| 3. | *Number of operational codes (for moving data from external memory to CPU)* | Many | One or two |
| 4. | *Nature of deal with rapid movement of codes and data* | For external address to chip. | Within the chip. |
| 5. | *Operation as a digital computer* | By adding external digital parts. | Without adding external digital parts. |
| 6. | *Memory map for data and code.* | Single | Separate |
| 7. | *Hardware used* | More | Less |
| 8. | *Flexibility from design point of view* | More | Less |

## 13.5. PROGRAMMABLE LOGIC CONTROLLERS (PLCs)

### 13.5.1. Introduction

**PLCs** *are specialised industrial devices for interfacing to and controlling analog and digital devices.*

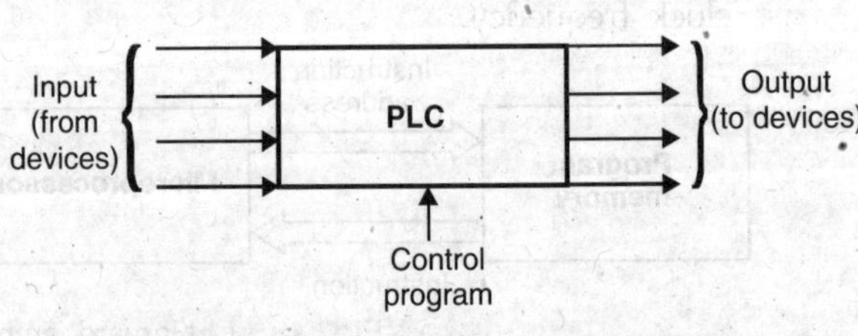


**Fig. 13.12.** Programmable logic controller.

— They are designed with a small instruction set suitable for industrial control applications.

— They are usually programmed with *"ladder logic"*, which is graphical method of laying out the connectivity and logic between system inputs and outputs.

— They are designed with industrial control and industrial environments specifically in mind. Therefore, in addition to being *flexible and easy to program, they are robust and relatively immune to external interference.*

• A programmable logic controller (first conceived in 1968), is shown in Fig. 13.12. *It is a "digital electronic device" that uses a programmable memory to store instructions and to implement functions such as logic sequencing, timing, counting and arithmetic in order to control machines and processes.*

It has been specifically designed to *make programming easy.*

### 13.5.2. Advantages of a PLC:

**The advantages of a PLC are :**

1. *Low cost.*
2. *Easy to install.*
3. Ensures *increased productivity.*
4. *Faster* operational speed.
5. Provides *high reliability and easier maintenance.*
6. Can *withstand* harsh industrial environment/manufacturing environment.
7. Provides *consistency* in manufacturing.
8. Can input/output *both analog and digital signals.*
9. *Reduced cost of scrap and rework* (since the user/designer can change the PLC program without changing any hardware or scraping the present hardware).
10. *Small size.*
11. PLC modules can be *added,* depending upon the input/output requirements.
12. *Easier troubleshooting.*

### 13.5.3. Selection of a PLC

For selection of a PLC, the following *criteria* need to be considered:

1. *Types of inputs/outputs* required, such as:
   — Isolation;
   — Out-board power supply for inputs/outputs;
   — Signal conditioning.
2. *Input/Output capacity* required.
3. *Size of memory* required. This is linked to the number of inputs/outputs and the complexity of program used.
4. *Speed and power* required for CPU—This is linked to the number of types of instructions that can be handled by a PLC.

### 13.5.4. Uses of PLCs

PLCs are widely used and *extend from small-contained units for use with perhaps 20 digital inputs/outputs to modular systems which can be used for large numbers of inputs/outputs, handle digital or analog inputs/outputs, and also carry out PID control modes.*

### 13.5.5. Special features

Although PLCs are similar to computers, yet they have the following *specific features* to their use as controllers:

1. The interfacing for inputs and outputs is *inside* the controller.
2. *Easily programmable.* They have an easily understood programming language.
   — Programming is mainly concerned with *logic* and *switching operation.*
3. *Rugged* and *designed to withstand vibrations, temperatures, humidity and noise.*

### 13.5.6. Architecture basic internal structure

Fig. 13.13 shows the architecture/internal structure of a programmable logic controller (PLC):

A PLC consists of the following *main components* :

1. Central processing unit (CPU);
2. Memory;
3. Input/Output circuitry.

**1. *Central processing unit (CPU) :***

- It controls and processes all the operations within the PLC.

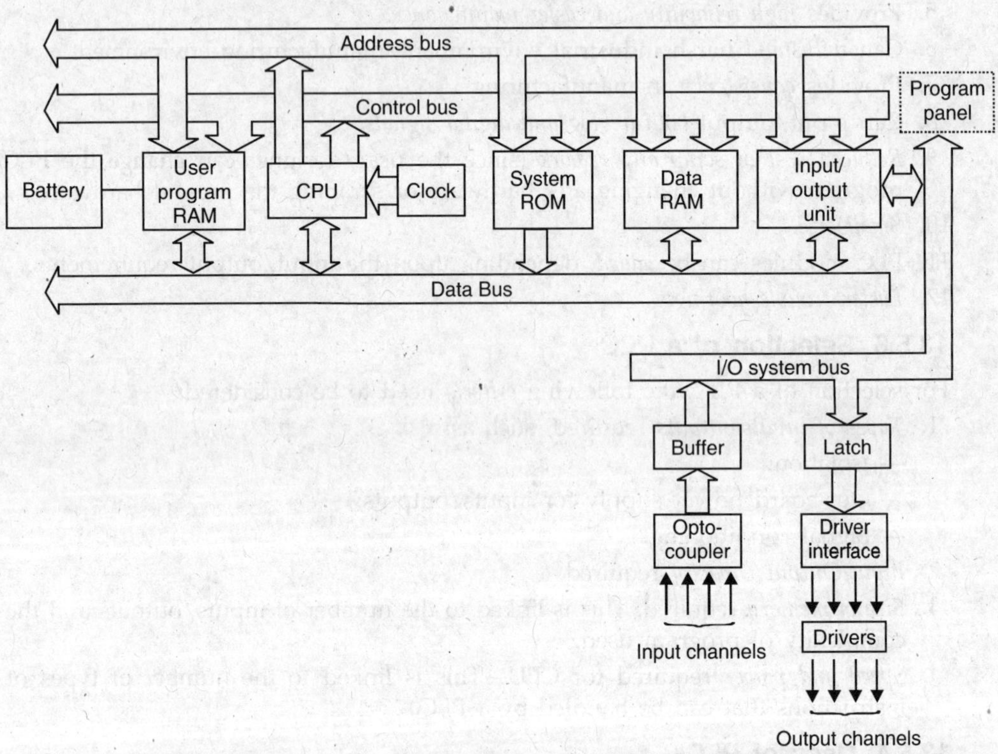


Fig. 13.13. Architecture of a programmable logic controller (PLC).

- It is provided with a *"clock"* with a frequency of typically between 1 and 8 MHz. This frequency determines the operating speed of the PLC and provides the timing and synchronisation for all elements in the system.
- A *"bus system"* carries information and data to and from the CPU, memory and input/output units.

**2.** *Memory:* The various memory elements available in a PLC are:

(*i*) A system ROM to give permanent storage for the operating system and fixed data.

(*ii*) RAM for user's program.

(*iii*) Temporary buffer stores for input/output channels.

- The programs in RAM can be changed by the user. However, to prevent the loss of these programs when the supply is switched off a battery is likely to be used in the PLC to maintain the RAM contents for a period of time.
- Specifications for small PLCs often specify the program memory size in terms of the number of program step (A program step is an instruction for some event to occur) that can be stored. Typically the number of steps that can be handled by a small PLC is of the order of 300 to 1000, which is generally more than adequate for most control situations.

3. ***Input/Output (I/O) circuitry:*** The I/O unit provides the interface between the system and outside world.
   - Programs are entered into the I/O unit from a panel which can vary from small keyboards with liquid crystal displays to those using a visual display unit with keyboard and screen display. The programs, alternatively, can be entered into the system by means of a link to a personal computer which is loaded with an appropriate software package.
   - The I/O channel provides signal conditioning and isolation functions so that sensors and actuators can be generally directly connected to them without the need for other circuitry.
   - The basic form of programming commonly used with PLCs is *ladder programming.* This involves each program task being specified as though a rung of a ladder.

   Following methods can be used for *I/O processing :*

   1. Continuous updating. 2. Mass I/O copying.

***Timers:*** The timers are commonly regarded as relays with coils which, when energised, result in the closing or opening of input contacts after some preset time.

A *timing circuit* is specified by stating the interval to be timed and the conditions or events that are to start and/or stop the timer.

- PLCs are generally provided with only *delay-on timers, i.e.,* a timer which comes on after a time delay.

***Internal relays:***

- These relays are often used when there are programs with multiple input conditions.
- The internal relays are also used for the starting of multiple outputs.

***Counters:*** The use of counters is restored to when there is *a need to count a specified number of contact operations.*

- Counter circuits are supplied as an internal feature of PLCs.

***Shift registers:*** Several internal relays can be grouped together to form a *register* which can provide a storage area for a series sequence of individual bits. Thus a 4-bit and a 8-bit registers would be formed by using four and eight internal registers respectively.

The term *shift register* is used because the bits can be shifted along by one bit when there is a suitable input to the register.

Shift registes have *three inputs*:

- One to load data into the first element of the register (OUT);
- One as the shift command (SFT);
- One for resetting (RST).

### 13.5.7. Input/Output Processing and Programming

**Input/Output processing:**

*"Ladder programming"* is the *basic form of programming commonly used with PLCs. This involves each program task being specified as through a rung of a ladder.* Thus such a rung could specify that the state of switches *A* and *B*, the inputs, be examined and if A and B are both closed then a solenoid, the output, is energised.

A PLC, when carrying out a program, follows the sequence indicated below:

1. Scanning the inputs associated with a rung of the ladder.
2. Solving the logic operation involving these inputs.
3. Setting/reseting the outputs for that rung.
4. Moving on to the next rung and repeating operations 1, 2, 3.
5. Moving on to the next rung and repeating operations 1, 2, 3.
6. Moving on to the next rung and repeating operations 1, 2, 3; continuing until the end of the program.

For *input/output processing the following two methods* are used:

**1. Continuous updating:**

In this method the *CPU scans the input channels* as they occur in the program instructions. Each input point is examined individually and its effect on the program determined.

There will be a *built in delay,* typically about 3 ms, when each point is examined in order to ensure that only *valid* input signals are read by the microprocessor. This delay enables the microprocessor to *avoid counting* an input signal twice, or more frequently, if there is contact bounce at a switch. A number of inputs may have to be scanned, each with a 3 ms delay, before the program has the instruction for a logic operation to be executed and an *output to occur.* The *outputs are latched* so that they *retain* their states until the next updating.

**2. Mass input/output copying:**

In case of continuous updating method since there has to be a 3 ms delay in each point, therefore, the time taken for examining several hundred input/output points can become comparatively *long.*

In order to have a *more rapid execution of program, a specific area of RAM is used as a buffer store between the control logic and input/output unit.* Each input/output has an adress in this memory. At the start of each program cycle the CPU scans all the inputs and copies their status into the input/output addreses in RAM.

As the program is executed logic operations are carried out. The resulting output signals are stored in the reserved input/output section of RAM. *At the end of each program cycle all the inputs are transferred from RAM to the output channels.* The outputs are latched so that they retain their status until the next updating.

**Programming:**

When a PLC programming is carried out, using ladder diagrams, a program is written in a similar manner to drawing a switching circuit.

The ladder diagram consists of two *vertical lines* representing the *power raits. Circuits* are connected as *horizontal lines,* i.e., the rungs of the ladder between these two verticals. While, drawing the circuit line for a rung, *inputs must always precede, outputs* and there *must be at least one output on each line. Each rung must start with an input or a series of inputs and end with an output.* The inputs and outputs are *numbered,* the notation used depending upon the manufacturer of PLC.

- ***Logic function:*** AND, OR, NOR, NAND, EXCLUSIVE – OR(XOR)
- ***Latching:*** Sometimes there are situations where it is necessary to hold a coil energised, even when the input which energises it ceases. The circuit which is used to carry out such an operation is termed as *latch circuit*. It is a *self-manufacturing circuit in that, after being energised, it maintains that state until another input is received; it remembers its last state.*
- ***Sequencing:*** Invariably there are control situations where sequences of outputs are required, with the switch from one *output* to another being conrolled by sensors.

### 13.5.8. Mnemonics

In a *ladder program* each horizontal rung on the ladder represents a line in the program and entire ladder gives the complete program in 'ladder language'. The programmer can enter the program into the PLC using a keyboard with the graphic symbols for the ladder elements, or using a computer screen and a mouse to select symbols, and the program panel or computer than translates these symbols into machine language that can be sotred in the PLC memory.

Alternatively, a *program can also be entered by translating the ladder program into "**mnemonics**"*, each code corresponding to a ladder element, and then enter these into the programming panel or computer. These are then translated into machine language. The mnemonics used by different PLC manufacturers differ.

### 13.5.9. Master and Jump Controls

A whole block of outputs can be simultaneously turned off or on by using the same internal relay contacts in each output rung so that switching it on or off affects every one of the rungs. An *alternative way of programming* to achieve the same effect is to use a ***master relay***.

- *Jumps*. A function which is often provided with PLCs is the *conditional jump* function. Such a function enables programs to be designed so that if a certain condition exists then a *section of program is jumped*.

## 13.6. MICROPROCESSOR-BASED INSTRUMENT SYSTEM – INTELLIGENT INSTRUMENT

*An instrument which operates on microprocessor system or is microprocessor-based itself is called* **Intelligent instrument.**

The *microprocessor-based systems* (in which involvement of human being is completely eliminated or restricted) are employed for wide applications in industries *for monitoring the data as well as controlling the processes.*

Fig. 13.14 shows block diagram of a *microprocessor-based system* :

It consists of *"hardware section"* which includes the following :

— Transducer ;

— Signal conditioning device/element ;

— Analog-to-digital (A/D) converter ;

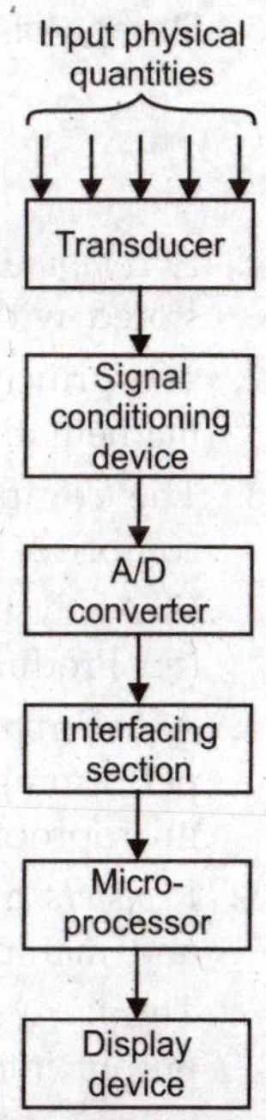


**Fig. 13.14.** Block diagram of microprocessor-based system/instrument.

— Interfacing section ;
— Microprocessor ;
— Display devices.

The "software section" contains the following :

— Programs ;
— Subroutine for controlling various processes ;
— Calibration ;
— Averaging ;
— Totalization etc.

- The overall designing of an intelligent instrument is employed in the following :

— Linear circuits ;
— Digital electronics ;
— Control elements ;
— Controllers ;
— Software.

**Advantages.** The microprocessor-based instruments entail the following *advantages* :

1. Higher flexibility.
2. Increased operational speed.
3. Fast and automatic control actions feasible.
4. Easy to change the system's configuration by altering the software section.
5. Functions like averaging, linearization, totalization, calibration, alarming, etc. become automated.
6. Storage of information or data is available through memory interfacing facility.

## HIGHLIGHTS

1. A *computer* is a machine that processes data according to set of instructions stored within the machine.
2. The principle of operation of *analog computers* is to create a physical analog of mathematical problems. The *digital* computers *accept digits and alphabets as inputs.*
3. The complete programming process followed by programmer for writing comprises the following *steps:*
   (*i*) Analysis; (*ii*) Flow charting; (*iii*) Coding; (*iv*) Debugging; (*v*) Documentation; (*vi*) Production.
4. A "*microprocessor*" is a large scale integration (LSI) chip that is capable of performing arithmetic and logic functions as defined by a given programme. A microprocessor consists of: (*i*) ALU; (*ii*) Register unit; (*iii*) Control unit.
5. *Registers* are digital devices used by the microprocessor for temporary storage and manipulation of data and instructions.
6. The *microcontroller* is the integration of a microprocessor with memory and input/output interfaces, and other peripherals such as liners, on single chip.

## OBJECTIVE TYPE QUESTIONS

**Fill in the blanks or Say 'Yes' or 'No'**

1. Charle's Babbage is called "Father of computers".
2. A ............. is a machine that processes data according to set of instructions stored within the machine.
3. Speedometer is an example of analog computer.
4. ............. are popularly known as personal computer (PC).
5. A hybrid computer is a combination of both analog and digital computers.
6. Chips are rated in terms of their ............. and .............
7. RAM chip is made with Metal Oxide Semiconductor (MOS).
8. ............. are used to multiply a variable quantity by a constant.
9. Servomultipliers are used to multiply two variables.
10. A ............. is a large scale integration (LSI) chip that is capable of performing arithmetic and logic functions as defined by a given programme.
11. Microprocessors which have memory and various input/output arrangements on the same chip are called .............
12. ............. are memory locations within the microprocessor.
13. MAR (Memory address register) contains the address of data.
14. General purpose registers serve as ............. storage for data or addresses and used in operations involving transfers between other registers.
15. ............. are the paths along which digital signals move from one section to another.
16. ............. bus carries the signals relating to control actions.
17. The size of the memory is determined by the number of wires in the address bus.
18. ............. is a device in which data is stored permanently.
19. RAM is a memory that can be read only.
20. Intel 8085 is an 8-bit, NMOS microprocessor.

## ANSWERS

| | | | |
|---|---|---|---|
| 1. Yes | 2. Computer | 3. No | 4. Microcomputers |
| 5. Yes | 6. Capacity, speed | 7. Yes | 8. Attenuators |
| 9. Yes | 10. microprocessor | 11. microcontroller | 12. Registers |
| 13. Yes | 14. temporary | 15. Buses | 16. Control |
| 17. Yes | 18. ROM | 19. No | 20. Yes. |

## THEORETICAL QUESTIONS

1. What is a 'Computer'? Explain.
2. List the characteristics of a computer.
3. What are the limitations of a computer?
4. How are computers classified?
5. How are digital computers classified on the basis of size and capabilities?
6. Explain briefly the following :

   Super computer; Main frame computers; Minicomputer; Microcomputers.
7. What are the differences between analog and digital computers?
8. Draw the block diagram of a computer and explain briefly its various parts.
9. How are chips rated?

10. What do you mean by the term 'Peripheral'? Explain briefly the following devices:
    (*i*) Input devices; (*ii*) Output devices.
11. Explain briefly 'storage devices'.
12. List the steps which are required for computer programming process for writing programs.
13. What is a "Microprocessor"?
14. Draw the block diagram of a microcomputer and explain briefly the three segments (ALU, Register and Control unit) of a microprocessor.
15. What are the characteristics of microprocessor?
16. Mention the important features of the microprocessors.
17. What are the uses of microprocessors?
18. Explain briefly the various parts of a microprocessor system.
19. Explain briefly the following registers:
    Accumulator register; Status register; Program counter register (PC); memory address register; Instruction register; General purpose registers; Stack pointer register.
20. What are 'buses'? Explain briefly the following buses:
    Data bus, Address bus, Control bus.
21. Explain briefly the following forms of memory unit:
    ROM; PROM; EPROM; EEPROM; RAM.
22. Explain briefly Intel 8085 microprocessor with the help of a block diagram.
23. Write a short note on 'Microprocessor programming'.
24. What is a 'Microcontroller'? Explain briefly.
25. How are microcontroller classified? Explain briefly.
26. Give the description of a typical full-featured microcontroller.
27. Describe briefly the 8051 microcontroller.
28. What are 'Microchip controllers'?
29. What are programmable logic controllers (PLCs)?
30. What are the advantages of PLCs?
31. State the special features of PLCs.
32. Explain briefly with a neat diagram, the Architecture of a programmable logic controller (PLC).
33. Explain briefly programming of a PLC.

CHAPTER

# 14 Universities' Questions (Latest) with Solutions

## Section-A: SHORT ANSWER QUESTIONS

**Q. 1.** *What is an instrument? What are the primary requirements of any measuring instrument?*

**Ans.** • ***Instrument*** may be defined as a device for measuring the value or magnitude of a quantity or variable.

• The *'primary requirements'* for any measurement instrument are:

(*i*) Power consumption should be as small as possible.

(*ii*) No alteration in the circuit conditions with the introduction of instrument in the circuit.

**Q. 2.** *List the types of instruments mainly used as ammeters and voltmeters.*

**Ans.** The following types of instruments are mainly used as ammeters and voltmeters: (*i*) PMMC; (*ii*) Moving iron; (*iii*) Electrodynamometer; (*iv*) Hot wire and thermocouple; (*v*) Electrostatic; (*vi*) Induction type; (*vii*) Rectifier.

**Q. 3.** *What are the main advantages of 'Rectifier instruments'?*

**Ans.** Following are the main advantages of rectifier instruments:

(*i*) They can operate well into the *RF* (radio-frequency) range with the proper use of silicon or germanium rectifiers.

(*ii*) Measurements of current and voltage at several hundred megahertz are possible.

(*iii*) This type of meter is more sensitive than any other type of A.C. meter.

**Q. 4.** *What is a 'spectrum analyser'? What information is provided by the spectrum analysis of a signal?*

**Ans.** • ***Spectrum analyser*** is an instrument which graphically presents an energy distribution of the signal as a function of frequency on its CRT.

• The spectrum analysis of a signal provides the information about the following *particulars/parameters:*

(*i*) Bandwidth; (*ii*) Effect of different types of modulation and spurious signal generation; (*iii*) Frequency stability; (*iv*) The component levels; (*v*) Spectral purity; (*vi*) Attenuation.

**Q. 5.** *What are the applications of Linear-variable-differential transformer (LVDT)?*

**Ans.** Following are the *applications* of LVDT:

(*i*) Measurement of material thickness in hot strip or slab steel mills.

(*ii*) In accelerometers.

(*iii*) Jet engine controls in close proximity to exhaust gases.

**Q. 6.** *Differentiate between magnetic tape and digital tape recording.*

**Ans.** The differences between magnetic tape and digital tape recording are given below:

| S.No. | Magnetic recording | Digital recording |
|---|---|---|
| 1. | It is an analog system. | It is a digital system. |
| 2. | More distortion. | Less distortion. |
| 3. | The signal noise ratio is poor. | The signal noise ratio is good. |
| 4. | Vow and flutter are present. | Vow and flutter are totally absent. |
| 5. | Equalisation is required. | Equalisation is not required. |
| 6. | It suffers from accumulation of dirt and dust on its surface. | It is free from such problems due to lacquer coating. |

**Q. 7.** *Nixie tube display is based on what principle?.*

**Ans.** The ***Nixie tube*** works on the *principle* that when a gas breaks down, a glow discharge is produced. There is one common anode and ten cathodes, on which ten (0 to 9) numerals are placed. When a negative voltage is applied to the selected cathode, a gas diode is formed which glows the selected cathode (numeral).

**Q. 8.** *What are the applications of 'strain gauge'?*

**Ans.** The ***strain gauge*** is a measurement transducer for measuring strain and associated stress in experimental stress analysis. Secondly, many other detectors and transducers, notably the load cells, torque meters, diaphragm type pressure gauges, temperature sensors, accelerometers and flow meters, employ a strain gauge as a secondary transducer.

**Q. 9.** *What is 'current transformer (CT)'?*

**Ans.** A ***current transformer*** is an instrument to measure higher currents of transmission lines with the help of a low range ammeter.

**Q. 10.** *List the advantages of Electrodynamic instruments.*

**Ans.** *Advantages:*

1. Can be used on both D.C. as well as A.C. systems.
2. Free from hysteresis and eddy current errors.
3. It is possible to construct ammeters up to 10 A and voltmeters up to 600 V with precision grade accuracy.

**Q. 11.** *Name different types of 'frequency meters'.*

**Ans.** The frequency meters are of following types:

(*i*) Mechanical resonance type; (*ii*) Electrical resonance type, (*iii*) Weston type; (*iv*) Ratiometer type; (*v*) Saturable core type; (*vi*) Digital frequency meters.

**Q. 12.** *Write different methods of measurement of medium resistance.*

**Ans.** The methods of measuring *"medium resistances"* are:

(*i*) Voltmeter ammeter method (*ii*) Ohm-meter method

(*iii*) Bridge method (*iv*) Substitution method.

**Q. 13.** *What is meant by drift?*

**Ans.** By ***drift***, we mean variation in the value of quantity under measurement with time. When it is repeatedly measured, if there is no change in the value, it means zero drift.

**Q. 14.** *Define transformation ratio of a current transformer.*

**Ans.** ***Transformation ratio*** in case of instrument transformers, in general, is defined as the ratio of the primary phasor to the secondary phasor.

Transformation ratio (Actual) for a *CT*,

$$K_{act} = \frac{\text{Primary winding current (Actual)}}{\text{Secondary winding current (Actual)}}$$

Transformation ratio (nominal) for a *CT*,

$$K_{nom} = \frac{\text{Rated primary winding current}}{\text{Rated secondary winding current}}$$

**Q. 15.** *What is ratio correction factor (RCF)?*

**Ans.** The ***ratio correction factor*** is the transformation (actual) ratio divided by nominal ratio.

In other words, $RCF = \frac{K_{act.}}{K_{nom.}}$

- The ratio marked on transformers is their nominal ratio.

**Q. 16.** *How will you differentiate 7 segment display with 14 segment display?*

**Ans.**
- The 7 segment display is used for display of numbers only.
- The 14 segment display is used for display of numbers as well as of alphabets.

**Q. 17.** *What is the role of recorders in measurement system?*

**Ans.** A ***recorder*** records electrical and non-electrical quantities as a function of time. The recorder serves the following *objectives:*

(*i*) It preserves the details of measurement at a particular time.

(*ii*) It provides at a glance the overall picture of the performance of the unit.

(*iii*) It provides immediate reflection on the action taken by the operator.

**Q. 18.** *What is the working principle of 'photoelectric transducer'?*

**Ans.** The ***photoelectric transducer*** operates on the principle that when light strikes special combination of materials then following may result:

(*i*) Electrons may flow.

(*ii*) A voltage may be generated.

(*iii*) A resistance change may take place.

**Q. 19.** *What are the applications of 'strip chart recorders'?*

**Ans.** Some of the important applications of ***strip chart recorders*** are:

(*i*) Temperature recording.

(*ii*) Sound level recording.

(*iii*) Recording amplifier drift.

**Q. 20.** *What VTVM stands for?*

**Ans.** The VTVM stands for vacuum tube volt meter.

**Q. 21.** *What is Cathode Ray Oscilloscope (CRO)?*

**Ans.** ***CRO*** is a device that allows the amplitude of electrical signals, whether they be voltage, current, power etc., to be displayed primarily as a function of time. It is basically an electron beam voltmeter.

It is very useful and versatile laboratory instrument used for display, measurement and analysis of waveforms and other phenomena in electrical and electronic circuits.

**Q. 22.** *Name main components of a cathode ray cube (CRT).*

**Ans.** The main components of a CRT are:

(*i*) Electron gun assembly (*ii*) Deflection plates assembly
(*iii*) Fluorescent screen (*iv*) Glass envelope
(*v*) Base.

**Q. 23.** *List the various applications of a 'wave analyser'.*

**Ans.** The various applications of ***wave analyser*** are:

(*i*) To carry out complete harmonic analysis.
(*ii*) To measure relative amplitudes of single frequency components in a complex waveform.
(*iii*) To measure the signal energy with well defined bandwidth.
(*iv*) To be used to separate and display a large number of harmonics (say 50).
(*v*) The wave analysers are applied industrially in the field of reduction of sound and vibrations generated by rotating electrical machines and apparatus.

**Q. 24.** *What are 'instrument transformers'? How do they differ from power transformers?*

**Ans.**
- ***Instrument transformers*** are devices used with instruments (ammeter, voltmeter etc.) for A.C. measurement. They *increase* the range of the measuring instruments.
- The ***power transformers*** work at 50 Hz and are used for stepping up and down the voltage or current. These are connected between generating station and the receiving station.

**Q. 25.** *Define a transducer and distinguish between a sensor and a transducer.*

**Ans.**
- A ***transducer*** is a device which converts the energy from one form to another.
- A ***sensor*** senses the condition, state or value of the process variable and produces an output which reflects this condition, state or value. A *transducer* transforms the energy of the process variable to an output of some other type of energy which is able to operate some control device. Sometimes a *secondary transducer* may be employed to transform the output of primary senser to still another type of energy.

**Q. 26.** *What is 'LVDT'?*

**Ans.** ***LVDT*** (Linear-variable-differential transformer) is a passive inductive transducer and is commonly employed to measure force (or weight, pressure and acceleration etc., which depend on force) in terms of the amount and direction of displacement of an object.

**Q. 27.** *Enumerate main elements of a digital acquisition system.*

**Ans.** The main elements of a digital acquisition system are:

(*i*) Transducer (*ii*) Signal conditioner
(*iii*) Multiplexer (*iv*) Signal converter

(*v*) Analog to digital converter (*vi*) Digital recorder.

**Q. 28.** *Where are 'Digital data acquisition systems' used?*

**Ans.** ***Digial data acquisition systems*** are used when the physical quantity being monitored has a narrow bandwidth (*i.e.*, when the quantity varies slowly). These systems are also used when high frequency and low cost per channel is required.

These systems are, in general, *more complex* than analog systems, both in terms of instrumentation involved and the volume and complexity of input data they can handle.

**Q. 29.** *What is 'telemetry'?*

**Ans.** ***Telemetry*** is defined as a technology which allows users to collect information from inaccessible and inconvenient locations and transmit it to the accessible places to process, record and display the information in presentable form.

It is used for remote measurement of quantities.

**Q. 30.** *Why is sensitivity of the A.C. section of multimeter less than the D.C. section?*

**Ans.** Due to complex circuitry, the sensitivity of A.C. section of a multimeter is less than the D.C. section.

**Q. 31.** *What are the shortcomings of multimeter?*

**Ans.** It may be noted that A.C. sensitivity of a half-wave A.C. meter is only 45% of the D.C. sensitivity, therefore, the loading effect is greater while measuring A.C. voltages than that while measuring corresponding D.C. voltages.

The rectifying action depends upon the frequency to some extent due to capacitance effect and rectifier instruments show lower readings. Instrument indications may be in error by as much as 0.5% decrease for every 1 kHz rise in frequency.

**Q. 32.** *Why are the rectifier instruments indispensable for measurements in communication circuits?*

**Ans.** The ***rectifier instruments*** are indispensable for measurements in communication circuits because they have very high sensitivity and as such the power required for their operation is small enough to be within the capacities of the circuits.

**Q. 33.** *What are the essentials of indicating instruments?*

**Ans.** ***Indicating instruments*** consist essentially of a pointer moving over a calibrated scale and attached to the moving system pivoted on jewelled bearings. For satisfactory working of indicating instruments the torques required are:

(*i*) Deflecting torque; (*ii*) Controlling torque; (*iii*) Damping torque.

**Q. 34.** *What do you understand by the term 'burden of CT'?*

**Ans.** The product of voltage and current on the secondary side, when the *CT* is supplying the instrument with its maximum rated value of current, is known as ***rated burden*** and is expressed in volt-amperes.

**Q. 35.** *Why the secondary of current transformer is never kept open-circuited?*

**Ans.** The secondary of current transformer is never kept open-circuited because of the following *two reasons:*

(*i*) There will be no secondary m.m.f. and since the primary current (and m.m.f.) is fixed, the core flux will increase enormously. This will cause large eddy current and hysteresis losses and the resulting high temperatures may damage the insulation or even the core.

(ii) A very high voltage will be induced in the multi-turn secondary and this high voltage may be dangerous both to life and to the insulation.

**Q. 36.** *What are prepayment meters?*

**Ans.** ***Prepayment meter*** is a special type of house supply meter, which itself is essentially a standard pattern energy element with the addition of a coin collector box and the prepayment mechanism geared to the recorder train.

**Q. 37.** *List the factors that may lead to inaccuracies in measurement by A.C. bridges.*

**Ans.** The factors that may lead to inaccuracies in measurement by A.C. bridges are:

(i) Stray conductance effect due to imperfect insulation.

(ii) Mutual inductance effect due to magnetic coupling between various components of the bridge.

(iii) Stray capacitance effect due to electrostatic fields between conductors at different potentials.

(iv) Residual-in components, *e.g.*, the existence of small amount of series inductance or shunt capacitance in normally non-reactive resistors.

**Q. 38.** *How can average reading of VTVM be converted to a RMS reading voltmeter?*

**Ans.** The average reading of ***VTVM*** can be converted into a RMS reading voltmeter by calibrating its meter scale in terms of the r.m.s. value of a sinusoidal wave.

**Q. 39.** *What is meant by the deflection factor of a CRO?*

**Ans.** The deflection factor of a CRO is the reciprocal of the deflection sensitivity.

**Q. 40.** *What is a Lissajous pattern?*

**Ans.** ***Lissajous*** is a pattern that results from applying periodic signals to the deflection plates of a CRO.

**Q. 41.** *What is a function generator?*

**Ans.** A ***function generator*** is an instrument that has the capability of producing different types of waveform as its output signal. The most common output waveforms are: Sine waves; Triangular waves; Square waves; Sawtooth waves.

**Q. 42.** *How does a wave analyser functionally differ from a spectrum analyser?*

**Ans.** A *"wave analyser"* is used to measure relative amplitude of signal frequency components in a complex waveform while a *"spectrum analyser"* is used to study energy distribution across the frequency spectrum of a given electrical signal.

**Q. 43.** *What is digital-to-analog (D/A) conversion?*

**Ans.** ***D/A conversion*** basically is the process of taking a value represented in digital code and converting it into a voltage or current which is proportional to the digital value.

**Q. 44.** *What is frequency counter?*

**Ans.** A ***frequency counter*** is a digital instrument that can measure and display the frequency of any periodic waveform.

**Q. 45.** *What is IEEE 488 bus system?*

**Ans.** For a set of instruments to be interfaced with a single computer, a general purpose bus, called the IEEE 488 bus system, has been developed to which a number of instruments can be attached so that communication between the computer and the instruments is possible over a common cable.

**Q. 46.** *What is the difference between 3-digit and $3\frac{1}{2}$-digit digital meters?*

**Ans.** A 3-digit meter has a resolution of $\frac{1}{1000}$th of the range and has a maximum display of 999. In contrast, the $3\frac{1}{2}$-digit digital meter has a resolution of $\frac{1}{2000}$th of the range, and the maximum value displayed is 1999.

**Q. 47.** *Why is dual slope ADC preferred in a digital voltmeter?*

**Ans.** Dual slope ADC is preferred in a digital voltmeter (DVM) because it has low cost and high accuracy.

**Q. 48.** *What are Liquid Crystal Displays (LCDs)?*

**Ans.** ***LCDs*** are passive displays characterised by very low power consumption and good contrast ratio.

**Q. 49.** *List the displays commonly used in digital electronic field.*

**Ans.** Following ***displays*** are commonly used in digital electronic field:

(*i*) Cathode ray tube (CRT)
(*ii*) Light emitting diode (LED)
(*iii*) Liquid crystal display (LCD)
(*iv*) Nixies
(*v*) Electro-luminescent (EL) displays
(*vi*) Incandescent displays
(*vii*) Electrophoretic image displays (EPID)
(*viii*) Liquid vapour display (LVD).

**Q. 50.** *What is a Data Acquisition System (DAS)?*

**Ans.** ***DAS*** may be defined as a system used for data processing, data conversion, data transmission and data storage.

A typical data acquisition consists of individual sensors with necessary signal conditioning, multiplexing, data conversion, data processing, data handling and associated transmission, storage and display system.

**Q. 51.** *What do you understand by oscillographic recorders?*

**Ans.** The oscillographic recorders are instruments which record information on paper or film with ink, powder, electrostatic charge, liquid, electric arc or liqht.

Although galvanometer recorders, pen recorders, X-Y recorders, strip chart recorders and CRT recorders all may be considered oscillographic recorders, yet here the word 'oscillograph' implies either a galvanometer or a CRT recorder–the recorders having *bandwidths more than 20 kHz.*

**Q. 52.** *What is a load cell and what for is it used?*

**Ans.** A ***load cell*** is an electromechanical device that converts weight or force into an electrical signal and is widely used for measurement of static and dynamic forces.

**Q. 53.** *What are resolvers?*

**Ans.** ***Resolver*** is a synchro whose rotor is mechanically driven to translate rotor angle into electrical information corresponding to the sine and cosine of rotor angle. It can interchange rectangular and polar coordinates. It is also called a sine-cosine generator and synchro-resolver.

**Q. 54.** *What is the major difference between the current and voltage telemetering system?*

**Ans.** The major difference between the two systems is that in the current telemetering system the current in the communication circuit is maintained at a measurable value, irrespective of line conditions, whereas the ideal voltage telemetering system draws from the circuit a current so small that the resistive drop in the line is negligible. Thus, in voltage telemetering system voltage at the measuring end is measured by a null or balance method in order to keep the line current zero at the time of measurement.

**Q. 55.** *What is multiplexing in telemetering system?*

**Ans.** Transmission of two or more messages at the same time over a single intercollecting link is called the ***multiplexing*** in *telemetering system, i.e.,* multiplexing is a method of transmitting more than one measurement over a single transmission channel.

**Q. 56.** *Why is it essential to use large moving coil and strong magnetic field in a galvanometer type strip-chart recorder?*

**Ans.** As the friction torque in a recorder is high due to large weight of the moving system and friction between the pen and the chart, so the large deflecting torque is required. Obviously, the controlling torque required will also be large. So in such recorders large moving coil and strong magnetic field are required.

**Q. 57.** *What is meant by instrumentation?*

**Ans.** ***Instrumentation*** refers to the art and science of collection of several instruments and auxiliary equipment and their utilisation for conducting successfully a test or an experiment on a system, process or plant.

**Q. 58.** *What is meant by 'loading effect'?*

**Ans.** The incapability of the system to faithfully measure, record or control the input signal (measurand) in undistorted form is known as ***loading effect***.

**Q. 59.** *What are the special features of vibration galvanometers?*

**Ans.** The special features of vibration galvanometers are large sensitivity, small damping and large restoring constant.

**Q. 60.** *Why is microwave channel most commonly used?*

**Ans.** ***Microwave channel*** is most commonly used because of heavy requirement in industries like power and gas transmission and due to the overcrowding of lower radio-frequency spectrum. The initial cost on transmitter and receiver is high in this system but cost of additional channels is small.

## *Section-B:* QUESTIONS WITH SOLUTIONS

**Q. 1.** (*a*) *What is a 'strain gauge'? Describe briefly semiconductor strain gauges.*

(*b*) *A resistance strain gauge with a gauge factor of 2 is fastened to a steel member, which is subjected to a strain of 1 × 10⁻⁶. If the original resistance value of the gauge is 130 Ω, calculate the change in resistance.*

**Sol.** (*a*) Refer to Art. 7.13.

(*b*) *Given:* Gauge factor, $G_f = 2$; Strain, $e = 1 \times 10^{-6}$; Original resistance, $R = 130\ \Omega$

**Change in resistance, $\Delta R$:**

We know that, $$\frac{\Delta R}{R} = G_f \times e \quad \text{...[Eqn. (7.10)]}$$

Inserting the various values, we get:

$$\frac{\Delta R}{130} = 2 \times 1 \times 10^{-6}$$

$$\therefore \quad \Delta R = 130 \times 2 \times 1 \times 10^{-6} = \mathbf{2.6 \times 10^{-4}\ \Omega\ (Ans.)}$$

**Q. 2.** *A potential transformer of ratio 1000/100 V has primary resistance 94.5 Ω, secondary resistance 0.86 Ω, primary reactance 66.2 Ω, total equivalent reactance 110 Ω, and no-load current 0.02 A at 0.4 power factor. Calculate the phase angle error at no-load.*

**Sol.** *Given:* Turn ratio, $K_t = \frac{1000}{100} = 10$; $R_p = 94.5\ \Omega$; $X_p = 66.2\ \Omega$; $R_s = 0.86\ \Omega$;

$(X_p)_{eq.} = 110\ \Omega$, $I_0 = 0.02\ A$ at 0.4 p.f.

**Phase angle error at no-load $\theta$:**

No-load p.f., $\cos \phi_0 = 0.4$

$$\therefore \quad \sin \phi_0 = \sqrt{1 - \cos^2 \phi_0} = \sqrt{1 - (0.4)^2} = 0.9165$$

Now, $I_w = I_0 \cos \phi_0 = 0.02 \times 0.4 = 0.008$ A

$I_w = I_0 \sin \phi_0 = 0.02 \times 0.9165 = 0.0183$ A

Phase angle, $$\theta = \frac{\frac{I_s}{K_t}\left[(X_p)_{eq} \cos\gamma - (R_p)_{eq} \sin\gamma\right] + I_w X_p - I_m R_p}{K_t V_s} \text{ rad} \quad \text{...[Eqn. (6.31)]}$$

At no load, $I_s = 0$

$$\therefore \quad \theta = \frac{I_w X_p - I_m R_p}{K_t V_s} \text{ rad}$$

or, $$\theta = \frac{0.008 \times 66.2 - 0.0183 \times 94.5}{10 \times 100} = -0.0012 \text{ rad}$$

$$= -0.0012 \times \frac{180}{\pi} \times 60 = \mathbf{4.12'\ (Ans.)}$$

**Q. 3.** (*a*) *Explain the principle of working and operation of a current transformer (CT) and give expressions for the ratio and phase angle errors.*

(*b*) *A 1000/5A, 50 Hz bar type current transformer has secondary burden of 1.5 Ω (non-inductive). Calculate the flux in the core and the ratio error at rated condition of the CT. Assume iron loss in the core to be 1.5 watts. Neglect leakage flux and the magnetizing current.*

**Sol.** (*a*) Refer to Arts. 6.3.1 and 6.3.3.

(*b*) *Given:* Nominal ratio, $K_{nom.} = \frac{1000}{5} = 200$; $N_p = 1$; Secondary impedance = 1.5 Ω (non-inductive); Iron loss in the core = 1.5 W.

**Flux in the core, $\phi_{max}$:**

Let us *assume* turns ratio to be equal to nominal, in the absence of any other data.

Then, turn ratio,

$$K_t = K_{nom.} = 200$$

$$\therefore \quad N_s = K_t N_p = 200 \times 1 = 200$$

Now, secondary induced voltage,

$$E_s = 5 \times 1.5 = 7.5 \text{ V}$$

$$\therefore \quad \phi_{max} = \frac{E_s}{4.44 \times f \times N_s} = \frac{7.5}{4.44 \times 50 \times 200} = \mathbf{0.1689 \times 10^{-3}\ Wb\ (Ans.)}$$

**Ratio error:**

Primary induced voltage,

$$E_p = \frac{E_s}{K_t} = \frac{7.5}{200} \text{ V}$$

Loss component of exciting current,

$$I_w = \frac{\text{Iron loss}}{E_p} = \frac{1.5}{(7.5/200)} = 40 \text{ A}$$

As the burden is purely resistive, therefore, the secondary load angle is zero or $\delta = 0$.

We know that, Actual ratio,

$$K_{act} = K_t + \frac{I_w \sin \delta + I_w \cos \delta}{I_s}$$

$$= K_t + \frac{I_w}{I_s} = 200 + \frac{40}{5} = 208$$

$$\therefore \text{ Ratio error} = \frac{K_{nom} - K_{act}}{K_{act}} \times 100$$

$$= \frac{200 - 208}{208} = \mathbf{-3.846\%\ (Ans.)}$$

**Q. 4.** *(a)* *Discuss working of a Q-meter.*

*(b)* *A coil with a resistance of 10 Ω is connected in the direct measurement mode. Resonance occurs when oscillating frequency is 1 × 10⁶ Hz and resonating capacitor is set at 60 pF. Calculate the % error introduced in the calculated value of Q by 0.02 Ω insertion resistance.*

**Sol.** *(a)* Refer to Art. 2.6.14

*(b)* *Given:* $R = 10\ \Omega, f = 1 \times 10^6$ Hz; $C = 60$ pF; Resistance used across the oscillatory circuit = 0.02 Ω.

**Percentage error:**

The true or effective $Q$ of the coil,

$$Q_{true} = \frac{1}{\omega CR} \qquad \text{...[Eqn. (2.103)]}$$

$$= \frac{1}{(2\pi \times 1 \times 10^6) \times (60 \times 10^{-12}) \times 10} = 265.3$$

The indicated or calculated $Q$ of the coil,

$$Q_{indicated} = \frac{1}{\omega C(R + 0.02)}$$

$$= \frac{1}{(2\pi \times 1 \times 10^6) \times (60 \times 10^{-12})(10 + 0.02)} = 264.7$$

$$\therefore \text{Percentage error} = \frac{Q_{true} - Q_{indicated}}{Q_{true}}$$

$$= \frac{265.3 - 264.7}{265.3} \times 100 = \mathbf{0.226\%\ (Ans.)}$$

**Q. 5.** *(a)* *Differentiate between active and passive transducers.*

*(b)* *A resistance strain gauge has gauge factor of 2 and is fastened to a steel member subjected to a stress of 1050 kg/cm². Modulus of elasticity of steel is 2.1 × 10⁶ kg/cm². Calculate the % change of resistance of the strain gauge.*

**Sol.** *(a)* '*Active transducers*' (also called self-generating type transducers) develop their own voltage or current. The energy required for production of an output signal is obtained from the physical phenomenon being measured.

'*Passive transducers*' (also known as externally-produced transducers) derive the power required for the energy conversion from an external power source. However, they may absorb some energy from the physical phenomenon under study.

*(b)* *Given:* Gauge factor, $G_f = 2$; Stress, $\sigma = 1050$ kg/cm²; Modulus of elasticity, $E = 2.1 \times 10^6$ kg/cm².

**% Change of resistance of the strain gauge:**

$$E = \frac{\text{Stress}}{\text{Strain}} = \frac{\sigma}{e}$$

$$\therefore \qquad e = \frac{\sigma}{E} = \frac{1050}{2.1\times10^6} = 0.0005$$

Also, $$\frac{\Delta R}{R} = G_f \times e \qquad \text{...[Eqn. (7.10)]}$$

$$\therefore \qquad \% \frac{\Delta R}{R} = 2 \times 0.0005 \times 100 = \mathbf{0.1\%\ (Ans.)}$$

**Q. 6.** (a) *What are strain gauge rosettes?*

(b) *A strain gauge having a resistance of 200 Ω and gauge factor of 2.5 is connected in series with a blast resistance of 400 Ω. The total voltage across the combination is 24 V. Determine the change in the output voltage, when a stress of 140 MN/m² is applied. The modulus of elasticity is 200 GN/m².*

**Sol.** (a) ***Rosette gauges*** are used for measuring strain in complex parts. These gauges have three or four separate grids with various angular orientations and they can be cemented to the part with no particular attention being paid to the overall gauge orientation. The resultant strain on each of the grids can then be recorded and the true magnitudes and directions of the significant surface strains in the part can be calculated from these data.

(b) *Given:* $R = 200\ \Omega$; $G_f = 2.5$; $R_s = 400\ \Omega$; $\sigma = 140\ \text{MN/m}^2$; $E = 200\ \text{GN/m}^2$ or $200 \times 10^9\ \text{N/m}^2$.

**Change in the output voltage:**

When no stress is applied, voltage across the strain gauge

$$= 24 \times \frac{200}{(200+400)} = 8\ \text{V}$$

Strain, $$e = \frac{\sigma}{E} = \frac{140\times10^6}{200\times10^9} = 0.0007$$

Now, $$\frac{\Delta R}{R} = G_f \times e \qquad \text{...[Eqn. (7.10)]}$$

$$\therefore \qquad \Delta R = R \times G_f \times e = 200 \times 2.5 \times 0.0007 = 0.35\ \Omega$$

Voltage across the gauge under strain conditions

$$= \frac{(200+0.35)}{(200+0.35)+400} \times 24 = 8.00933\ \text{V}$$

∴ Change in output voltage = 8.00933 – 8 = 0.00933 V or **9.33 mV (Ans.)**

**Q. 7.** (a) *Differentiate between transducers and inverse transducers.*

(b) *A resistance wire strain gauge uses a soft wire of small diameter. The gauge factor is 5.2. Neglecting the piezo-resistive effect, calculate the Poisson's ratio.*

**Sol.** (a) *'Transducers'* are normally associated with a circuit which has non-electrical input and electrical output, *e.g.*, pressure gauge, strain gauge etc.

An *'Inverse transducer'* is a device that converts electrical quantities into non-electrical quantities, *e.g.*, piezo-electric crystal.

(b) *Given:* $G_f = 5.2$

**Poisson's ratio, μ:**

When the piezo-electric effect is neglected, the gauge factor is given by:

$$G_f = 1 + 2\mu \qquad \text{...[Eqn. (7.13)]}$$

or,
$$5.2 = 1 + 2\mu$$

$$\therefore \quad \mu = \frac{5.2 - 1}{2} = \mathbf{2.1\ (Ans.)}$$

**Q. 8.** (*a*) *List the various general configurations of Data Acquisition System.*

(*b*) *Explain with a neat diagram the working of a single-channel data acquisition system.*

**Sol.** (*a*) Refer to Art. 9.2.

(*b*) Refer to Art. 9.3.2.

**Q. 9.** (*a*) *What do you mean by the terms 'Modulation' and 'Demodulation'?*

(*b*) *Explain with a neat diagram the construction and working of a cathode ray oscilloscope.*

**Sol.** (*a*) Refer to Art. 10.2.9.

(*b*) Refer to Art. 2.5.3.

**Q. 10.** (*a*) *What is the difference between 'Accuracy' and 'Precision'?*

(*b*) *A pressure measuring system consists of a piezoelectric transducer, a charge amplifier and an ultraviolet charge recorder; their sensitivities are 6.5 pC/bar, 0.0035 V/pC and 15 mm/V respectively.*

*Determine the deflection on the chart for a pressure change of 18 bar.*

**Sol.** (*a*) **Accuracy :**

The closeness with which an instrument reading approaches the true value of the quantity being measured is called *accuracy.* It is determined as the maximum amount by which the result differs from the true value. Accuracy of an instrument is influenced by factors like static error, dynamic error, reproducibility, dead zone.

**Precision :**

The term 'precise' means clearly or sharply defined. 'Precision' is a measure of reproducibility of measurement. It is usually expressed in terms of the deviation in measurement.

(*b*) With the elements arranged in series, the overall sensitivity of the measuring system ($S$)

= Sensitivity of piezoelectric transducer ($S_1$) × sensitivity of the charge amplifier ($S_2$) × sensitivity of ultraviolet charge recorder ($S_3$).

$$= 6.5\,\frac{\text{pC}}{\text{bar}} \times 0.0035\,\frac{\text{V}}{\text{pC}} \times 15\,\frac{\text{mm}}{\text{V}} = 0.341\,\frac{\text{mm}}{\text{bar}}$$

Also,
$$\text{Sensitivity} = \frac{\text{Change of output signal}}{\text{Change of input signal}}$$

or,
$$0.341\,\frac{\text{mm}}{\text{bar}} = \frac{\text{Output}}{18\ \text{bar}}$$

∴ Deflection on the chart = 0.341 × 18 ≃ **6.14 mm (Ans.)**

**Q. 11.** (*a*) *What are limiting and known errors?*

(*b*) *During a test, the current was measured as 24.4 A flowing in a resistor of 0.13 Ω. Later it was discovered that the ammeter reading was low by 0.95% and marked resistance was high by 0.22%. Determine the true power as a percentage of the power that was originally calculated.*

**Sol.** (*a*) **Limiting errors :**

In most instruments the accuracy is generated to be within a certain percentage of full scale reading. The manufacturer has to specify the deviations from the nominal value of a particular quantity. The *limits of these deviations* from the specified value are defined as 'limiting errors or guarantee errors'.

**Known errors :**

When the error of a quantity of an instrument is known the effect of this error when combined with errors can be computed in a manner similar to the combinations of the limiting errors. However, the difference is that in case of known errors the *signs* of relative errors are given and must be presented in the calculations.

(*b*)
$$\text{True value of current, } I = 24.4\left(1+\frac{0.95}{100}\right) = 24.632 \text{ A}$$

$$\text{True value of resistance, } R = 0.13\left(1-\frac{0.22}{100}\right) = 0.1297\ \Omega$$

$$\text{True value of power} = I^2R = (24.632)^2 \times 0.1297 = 78.694 \text{ W}$$

$$\text{Measured value of power} = (24.4)^2 \times 0.13 = 77.397 \text{ W}$$

$$\therefore \quad \frac{\text{True power}}{\text{Measured power}}\times 100 = \frac{78.694}{77.397}\times 100 = \mathbf{101.67\%\ (Ans.)}$$

**Q. 12.** (*a*) *What is loading effect?*

(*b*) *The coil of a 250 V moving iron voltmeter has a resistance of 450 Ω and inductance of 0.95 H. The current taken by the instrument when placed on 250 V, D.C. supply is 0.04A. Determine the percentage error when the instrument is placed on 250 V, A.C. supply at 100 Hz.*

**Sol.** (*a*) **Loading effect :**

The incapability of the system to faithfully measure the input signal in undistorted form is called the *loading effect*. This results in loading errors. Loading effects occur due to extraction of energy from the measurand.

(*b*)
$$\text{Total ohmic resistance, } R = \frac{250}{0.04} = 6250\ \Omega$$

(Original calibration of the instrument is with direct current)

$$\text{Reactance of the coil, } X_L = 2\pi fL = 2\pi \times 100 \times 0.95 = 596.9\ \Omega$$

$$\text{Coil impedance, } Z = \sqrt{R^2 + X_L^2} = \sqrt{(6250)^2 + (596.9)^2} = 6278.4\ \Omega$$

$$\text{Voltage reading on A.C} = \frac{250\times 6250}{6278.4} = 248.87 \text{ V}$$

$\therefore$ Error $= 248.87 - 250 = -1.13$ V

$$\text{Percentage error} = \frac{1.13}{250} \times 100 = \mathbf{0.452\%\ (Ans.)}$$

**Q. 13.** *(a) What are the advantages of electronic instruments?*

*(b) Describe briefly the Lissajous method of measuring unknown frequency.*

**Sol.** *(a)* Refer to Article 2.3.

*(b)* Refer to Article 2.5.21.6.

**Q. 14.** *(a) What are the advantages of bridge circuits?*

*(b) A circuit consisting of an unknown coil, a resistance and a variable capacitor connected in series is tuned to resonance using a Q-meter. If the frequency is 400 kHz, the resonating capacitor is set at 210 pF, the resistance is 0.9 Ω and Q-meter indicates 105, determine the effective inductance and resistance of the unknown coil.*

**Sol.** *(a)* The *advantages* of bridge circuits are:

*(i)* High measurement accuracy (as the measurement is done by comparing the unknown value with the standard value).

*(ii)* The accuracy is independent of null detector's characteristics (it depends on the values of the components).

*(iii)* The balance equation is independent of the magnitude of the input voltage or its source impedance, the sensitivity and impedance of the null detector or any impedance shunting the detector.

*(iv)* The interchange of the source detector does not affect the balance condition.

*(v)* The bridge circuit can be used in control circuits.

*(b)* *Given:* $f_1 = 400$ kHz; $C = 210$ pF; $R_{sh} = 0.9\ \Omega$; $Q = 105$

**Inductance (L); resistance (R):**

Under resonant conditions, we have:

$$L_{coil} = \frac{1}{\omega^2 C} = \frac{1}{(2\pi \times 400 \times 1000)^2 \times 210 \times 10^{-12}}$$

$$= \mathbf{753.9\ \mu H\ (Ans.)}$$

$$R_{coil} = \frac{\omega L}{Q} - R_{sh} = \frac{(2\pi \times 400 \times 1000) \times (753.9 \times 10^{-6})}{105} - 0.9$$

$$= \mathbf{17.14\ \Omega\ (Ans.)}$$

**Q. 15.** *Explain with a neat diagram the Bolometer method of power measurement.*

**Sol.** Refer to Article 2.6.21.2.

**Q. 16.** *(a) How are signal generators classified?*

*(b) Draw a simple diagram of a pulse generator and explain it briefly.*

**Sol.** *(a)* The signal generators can be *classified* into the following categories:

*(i)* Audio generators
*(ii)* Function generators
*(iii)* Pulse generators
*(iv)* RF generators
*(v)* Frequency synthesizers
*(vi)* Other signal generators.

*(b)* Refer to Article 3.6.1.

**Q. 17.** (*a*) *List the instruments which are used for signal analysis.*

(*b*) *Describe 'Heterodyne Wave Analyser'.*

**Sol.** (*a*) The following instruments are used for signal analysis:

(*i*) Wave analysers (*ii*) Distortion analysers

(*iii*) Spectrum analysers (*iv*) Audio analysers

(*v*) Modulation analysers.

(*b*) Refer to Article 4.2.4.

**Q. 18.** (*a*) *Explain the following characteristics of digital meters:*

*(i) Resolution; (ii) Sensitivity; (iii) Accuracy.*

(*b*) *Describe briefly a "digital multimeter" with the help of a block diagram.*

**Sol.** (*a*) Refer to Article 5.3.7.

(*b*) Refer to Article 5.6.

**Q. 19.** (*a*) *What is burden of an instrument transformer?*

(*b*) *A current transformer rated at 500/5A, 50 Hz with an output of 15 VA has 1 turn primary. At rated load with non-inductive burden, the inphase and quadrature components (referred to the flux) of the exciting mmf are 8 AT and 10 AT respectively. The number of turns in the secondary is 98, and the resistance and leakage reactance of the secondary winding are 0.35 Ω and 0.3 Ω respectively.*

*Determine the following:*

*(i) Current error*

*(ii) Phase angle error.*

**Sol.** (*a*) **Burden of an instrument transformer :**

The load across the secondary winding can be conveniently expressed as the output in volt-ampere at the rated secondary voltage. The *'rated burden'* is the volt-ampere loading which is permissible without errors exceeding the limits for particular class of accuracy.

$$\text{Total secondary burden} = \frac{(\text{Secondary winding } \textit{induced} \text{ voltage})^2}{\left(\begin{array}{c}\text{Impedance of the secondary winding circuit}\\ \textit{including} \text{ impedance of secondary winding}\end{array}\right)}$$

$= (\text{Secondary winding current})^2 \times$ (Impedance of secondary winding circuit *including* secondary winding)

Secondary winding burden due to load

$$= \frac{(\text{Secondary winding } \textit{terminal} \text{ voltage})^2}{(\text{Impedance of load on secondary winding})}$$

$= (\text{Secondary winding current})^2 \times$ (Impedance of *load* in the secondary winding circuit) (*b*) *Given*: $N_p = 1$; $N_s = 98$; Output = 15 VA; Magnetising mmf = 8 AT; Loss mmf = 10 AT; $R_s = 0.35\ \Omega$; $X_s = 0.3\ \Omega$

**(*i*) Ratio (current) error:**

$$K_t = \frac{N_s}{N_p} = \frac{98}{1} = 98; \quad K_{nom.} = \frac{500}{5} = 100$$

$$\text{Magnetising current, } I_m = \frac{\text{Magnetising mmf}}{\text{Primary winding turns}} = \frac{8}{1} = 8\text{ A}$$

$$\text{Loss component, } I_w = \frac{\text{Loss mmf}}{\text{Primary winding turns}} = \frac{10}{1} = 10\text{ A}$$

Resistance of external burden (purely resistive-given),

$$R_e = \frac{VA}{I_s^2} = \frac{15}{5^2} = 0.6\ \Omega$$

Reactance of external burden,

$$X_e = 0\ \Omega$$

Resistance of total secondary circuit burden

$$R_s + R_e = 0.35 + 0.6 = 0.95\ \Omega$$

Reactance of total secondary circuit burden

$$= X_s + X_e = 0.3 + 0 = 0.3\ \Omega$$

$$\text{Secondary phase angle, } \delta = \tan^{-1}\left(\frac{0.3}{0.95}\right) = 17.53°$$

$$\therefore \quad \cos\delta = \cos 17.53° = 0.954 \text{ and } \sin\delta = 0.301$$

$$\text{Actual ratio, } K_{act.} = K_t + \frac{I_m \sin\delta + I_w \cos\delta}{I_s}$$

$$= 98 + \frac{8 \times 0.301 + 10 \times 0.954}{5} = 100.39$$

$$\therefore \quad \textit{Ratio (current) error} = \frac{K_{nom.} - K_{act.}}{K_{act.}} \times 100$$

$$= \frac{100 - 100.39}{100.39} \times 100 = \mathbf{-0.388\%\ (Ans.)}$$

**(*ii*) Phase angle error, θ:**

$$\text{We know that,} \quad \theta = \frac{180}{\pi}\left[\frac{I_m \cos\delta - I_w \sin\delta}{K_t\, I_s}\right] \text{degrees} \quad ...[\text{Eqn. 6.18}]$$

$$= \frac{180}{\pi}\left[\frac{8 \times 0.954 - 10 \times 0.301}{98 \times 5}\right] = \mathbf{0.54°\ (Ans.)}$$

**Q. 20.** (*a*) *Explain with a neat diagram, how pressure is measured with a linear voltage differential transformer (LVDT).*

(*b*) *In a LVDT the output voltage is 1.9 V at a maximum displacement. At a certain load the deviation from linearity is maximum and it is ± 0.0048 V from the straight line through the origin.*

*Determine the linearity at a given load.*

**Sol.** (*a*) Refer to Article 7.7.2.3.

(*b*) *Given*: The output voltage at maximum displacement = 1.9 V

The deviation from a straight line through the origin = ± 0.0048 V

$$\therefore \text{ \% age linearity } = \pm \frac{0.0048}{1.9} \times 100 = \pm \mathbf{0.253\%} \textbf{ (Ans.)}$$

**Q. 21.** (*a*) *What are the applications of Schering bridge?*

(*b*) *The Schering bridge shown in the figure has the following constants:*

*Arm AB : Capacitor of 0.35 μF in parallel with 1.4 kΩ resistance; Arm BC : Resistance of 2.8 kΩ; Arm CD : Unknown capacitor $C_x$ and $R_x$ in series; Arm DA : Capacitor of 0.42 μF; Frequency : 1 kHz*

*Determine: (i) $R_x$ and $C_x$; (ii) Dissipation factor.*

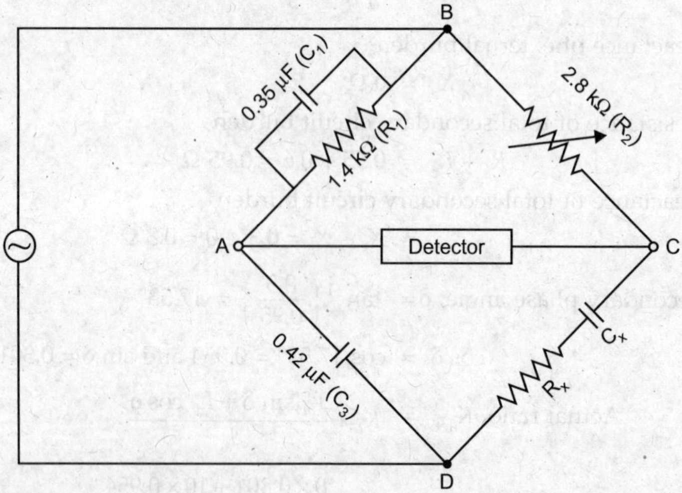


**Sol.** (*a*) Schering bridge (one of the most important A.C. bridges) is used for the measurement of *unknown capacitors, dielectric loss* and *power factor*. It is widely used for testing *small capacitors at low voltages with very high precision*.

(*b*) *Given*: $R_1$ = 1.4 kΩ; $C_1$ = 0.35 μF; $R_2$ = 2.8 kΩ; $C_3$ = 0.42 μF

**(*i*) $R_x$, $C_x$ :**

We know that, $$R_x = \frac{R_2 C_1}{C_3} \quad \text{...[Eqn. (2.58)]}$$

$$= \frac{2.8 \text{ k}\Omega \times 0.35 \text{ μF}}{0.42 \text{ μF}} = \mathbf{2.33 \text{ k}\Omega \text{ (Ans.)}}$$

and, $$C_x = \frac{R_1 C_3}{R_2} \quad \text{...[Eqn. (2.59)]}$$

$$= \frac{1.4 \times 0.42}{2.8} = \mathbf{0.21 \text{ μF (Ans.)}}$$

**(ii) Dissipation factor, D:**

$$D = \omega C_x R_x \quad \text{...[Eqn. (2.61)]}$$

$$= 2\pi f. C_x . R_x = 2\pi \times (1 \times 1000) \times (0.21 \times 10^{-6}) \times (2.33 \times 10^3)$$

$$= \mathbf{3.07\ (Ans.)}$$

**Q. 22.** (a) *Describe with a neat diagram the construction and working of a 'Piezoelectric accelerometer'.*

(b) *A piezoelectric crystal measuring 5 mm × 5 mm × 1.5 mm is used to measure a force. Its voltage sensitivity is 0.052 Vm/N. Calculate the force if voltage developed is 110 V.*

**Sol.** (a) Refer to Article 7.9. 5.

(b) *Given:* $A = 5$ mm × 5 mm = $25 \times 10^{-6}$ m²; $t = 1.5$ mm or $1.5 \times 10^{-3}$ m; Voltage sensitivity, $g = 0.052$ Vm/N; Voltage developed, $E = 110$ V

**Force, F:**

We know that, $E = gtp$ ...[Eqn. (7.6)]

$$110 = 0.052 \times (1.5 \times 10^{-3}) \times p$$

or, Pressure, $$p = \frac{110}{0.052 \times (1.5 \times 10^{-3})} \text{ N/m}^2 = 1.41 \times 10^6 \text{ N/m}^2$$

∴ Force, $F = p \times A = (1.41 \times 10^6) \times (25 \times 10^{-6}) =$ **35.25 N (Ans.)**

**Q. 23.** (a) *Explain with a neat sketch the construction and working of a 'Photoelectric tachometer'.*

(b) *A strain gauge bridge consists of two fixed 120 Ω resistors; one active gauge and one unstrained temperature compensating gauge connected in adjacent arms. The two gauges have an unstrained resistance value of 120 Ω each. The gauge factor is 2.2. Calculate the strain which would be represented by 75000 Ω calibration resistance shunt across one arm of the bridge.*

**Sol.** (a) Refer to Article 7.12.7.

(b) *Given:* $R_g = 120\ \Omega$; $R_{sh} = 75000\ \Omega$; Gauge factor, $G_f = 2.2$

We know that equivalent strain, $e_{eq}$ is given by:

$$e_{eq} = \frac{1}{G_f}\left(\frac{R_g}{R_g + R_{sh}}\right) \quad \text{...[Eqn. (7.39)]}$$

or, $$e_{eq} = \frac{1}{2.2}\left(\frac{120}{120 + 75000}\right) = 7.261 \times 10^{-4}$$ or **726.1 micro strain (Ans.)**

**Q. 24.** (a) *What is 'Wagner earthing device'?*

(b) *A circuit having a capacitance of 150 pF is tuned to a frequency of 1.1 MHz. In this circuit the current falls to 70.7% of its resonant value when the frequency of an e.m.f. of constant magnitude injected in series with the circuit deviates from the resonant frequency by 5.5 kHz. Calculate the Q factor and effective resistance of the circuit.*

**Sol.** (a) **Wagner earth device** connection is one of the most widely used methods for *eliminating some of the effects of stray capacitance in a bridge circuit*. This circuit *eliminates* the troublesome capacitance which exists between the detector terminals and the ground.

(*b*) *Given*: C = 150 pF; $f_0$ = 1.1 MHz

**Values of *Q* and *R* (effective):**

We know that, $Q = \dfrac{\omega_0}{\omega_1 - \omega_2} = \dfrac{f_0}{f_1 - f_2}$ ...[Eqn. (2.127)]

The two frequencies at which current is 70.2% or $\frac{1}{\sqrt{2}}$ time the current at resonant frequency are:

$$f_1 = 1.1 \times 10^6 + 5.5 \times 10^3$$

and, $$f_2 = 1.1 \times 10^6 - 5.5 \times 10^3$$

∴ $$f_1 - f_2 = 11 \times 10^3 \text{ Hz}$$

Hence, $$Q = \frac{1.1 \times 10^6}{11 \times 10^3} = \mathbf{100\ (Ans.)}$$

Now, $$R = \frac{\omega_1 - \omega_2}{\omega_0^2\, C}$$ ...[Eqn. (2.126)]

$$= \frac{f_1 - f_2}{2\pi\, f_0^2\, C} = \frac{11 \times 10^3}{2\pi \times \left(1.1 \times 10^6\right)^2 \times 150 \times 10^{-12}}$$

$$= \mathbf{9.64\ \Omega\ (Ans.)}$$

**Q. 25.** *A bakelite sheet of 4.5 mm thickness is tested at 50 Hz between the electrodes 110 mm diameter. The Schering bridge shown has an air capacitor $C_2$ of 104 pF, a non-reactive resistance $R_4$ of (1000/π) Ω in parallel with a variable capacitor $C_4$ and non-reactive variable resistance $R_3$. Balance is obtained with $C_4$ = 0.52 μF and $R_3$ = 250 Ω.*

*Determine the following:*

*(i) Capacitance; (ii) Power factor; (iii) Relative permittivity of sheet:*

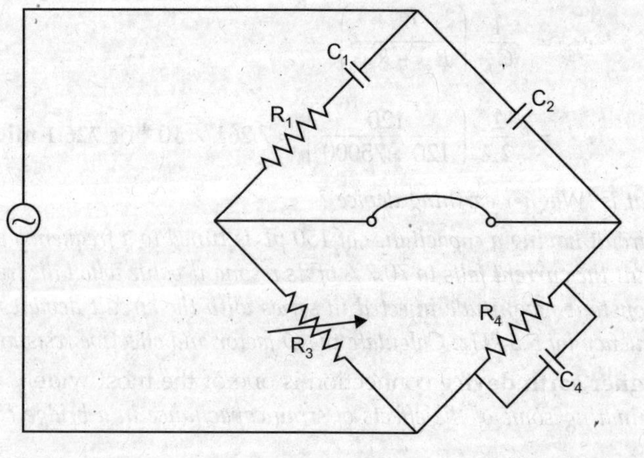


**Sol.** *Given*: $C_2$ = 104 pF = 104 × $10^{-12}$ F; $R_4$ = (1000/π) Ω; $C_4$ = 0.52 μF = 0.52 × $10^{-6}$ F; $R_3$ = 250 Ω

**(i) Capacitance *C*:**

When the bridge is *balanced*, we have:

$$R_1 = \frac{R_3 C_4}{C_2} \qquad \text{...[Eqn. (2.58)]}$$

$$= \frac{250 \times 0.52 \times 10^{-6}}{104 \times 10^{-12}} = 1.25 \text{ M}\Omega$$

and,

$$C_1 = \frac{R_4 C_2}{R_3} \qquad \text{...[Eqn. (2.59)]}$$

$$= \frac{(1000/\pi) \times 104 \times 10^{-12}}{250} = \mathbf{132.4 \ pF \ (Ans.)}$$

**(ii) Power factor:**

$$\text{Power factor} = \omega R_1 C_1 \qquad \text{...[Eqn. (2.61)]}$$

$$= (2\pi \times 50) \times (1.25 \times 10^6) \times (132.4 \times 10^{-12})$$

$$= \mathbf{0.052 \ (Ans.)}$$

**(iii) Relative permittivity, $\in_r$:**

We know that,

$$C = \frac{\in_0 \in_r A}{t}$$

where, $C (= C_1)$ = Capacitance

$\in_0 = 8.854 \times 10^{-12}$

$A$ = Area of the electrodes $= \frac{\pi}{4} \times \left(\frac{110}{1000}\right)^2 = 9.5 \times 10^{-3} \text{ m}^2$, and

$t$ = Thickness of bakelite sheet = 4.5 mm = 0.0045 m

Substituting the values in the above equation, we get:

$$132.4 \times 10^{-12} = \frac{8.854 \times 10^{-12} \times \in_r \times (9.5 \times 10^{-3})}{0.0045}$$

or,

$$\in_r = \frac{132.4 \times 10^{-12} \times 0.0045}{8.854 \times 10^{-12} \times (9.5 \times 10^{-3})} = \mathbf{7.08 \ (Ans.)}$$

**Q. 26.** *(a) What are the applications of a 'vector voltmeter'?*

*(b) In a CRT, the anode to cathode voltage is 2400 V. The parallel deflection plates are 15 mm long and spaced 5 mm. The screen is 450 mm from the centre of the deflecting plates. If mass of electron is $9.109 \times 10^{-31}$ kg and charge on electron $1.602 \times 10^{-19}$ C, find:*

*(i) The beam speed; (ii) The deflection sensitivity of the tube.*

**Sol.** (a) A *'vector voltmeter'* measures the magnitude of a signal at two points in a circuit and simultaneously measures the phase difference between the voltage waveforms at these two points. It is useful in VHF applications and can be used successfully in measurements such as:

(*i*) Complex insertion loss; (*ii*) Amplifier gain and phase shift; (*iii*) Two-port network parameters; (*iv*) Filter transfer function; (*v*) Radio frequency distortion.

(*b*) *Given*: Acceleration voltage, $V_a$ = 2400 V; Length of each *Y*-plate, $l_d$ = 15 mm = 0.015 m; Distance between *Y*-plates, $d$ = 5 mm = 0.005 m; Distance of screen from the centre of the deflecting plates, $L$ = 450 mm = 0.45 m; $m = 9.109 \times 10^{-31}$ kg; $e = 1.602 \times 10^{-9}$ C.

**(*i*) The beam speed, *v*:**

We know that, $$v = \sqrt{\frac{2e\,V_a}{m}} \qquad \text{...[Eqn. (2.1)]}$$

$$= \sqrt{\frac{2 \times 1.602 \times 10^{-19} \times 2400}{9.109 \times 10^{-31}}} = \mathbf{29.05 \times 10^6\ m/s\ (Ans.)}$$

**(*ii*) Deflection sensitivity, *S*:**

$$S = \frac{L\,l_d}{2d\,V_a} \qquad \text{...[Eqn. (2.2)]}$$

$$= \frac{0.45 \times 0.015}{2 \times 0.005 \times 2400}\ \text{m/V} = 2.81 \times 10^{-4}\ \text{m/V}$$

$$= \mathbf{0.281\ mm/V\ (Ans.)}$$

# INDEX